Methods in Enzymology

Volume 152
GUIDE TO MOLECULAR CLONING TECHNIQUES

METHODS IN ENZYMOLOGY

EDITORS-IN-CHIEF

John N. Abelson   Melvin I. Simon

Methods in Enzymology

Volume 152

Guide to Molecular Cloning Techniques

EDITED BY

Shelby L. Berger

NATIONAL CANCER INSTITUTE
NATIONAL INSTITUTES OF HEALTH
BETHESDA, MARYLAND

Alan R. Kimmel

NATIONAL INSTITUTE OF DIABETES AND DIGESTIVE AND KIDNEY DISEASES
NATIONAL INSTITUTES OF HEALTH
BETHESDA, MARYLAND

ACADEMIC PRESS, INC.
Harcourt Brace Jovanovich, Publishers

San Diego   New York   Berkeley   Boston
London   Sydney   Tokyo   Toronto

ACADEMIC PRESS, INC.
1250 Sixth Avenue, San Diego, California 92101

United Kingdom Edition published by
ACADEMIC PRESS INC. (LONDON) LTD.
24–28 Oval Road, London NW1 7DX

LIBRARY OF CONGRESS CATALOG CARD NUMBER: 54-9110

ISBN 0–12–182053–X (alk. paper)
ISBN 0–12–181775–X (comb bound)

PRINTED IN THE UNITED STATES OF AMERICA

89 90   9 8 7 6 5

Table of Contents

Section I. Requirements for a Molecular Biology Laboratory

Section II. General Methods for Isolating and Characterizing Nucleic Acids

Section III. Enzymatic Techniques and Recombinant DNA Technology

Section IV. Restriction Enzymes

Section V. Growth and Maintenance of Bacteria

Section VI. Genomic Cloning

Section VII. Preparation and Characterization of RNA

Section VIII. Preparation of cDNA and the Generation of cDNA Libraries

Contributors to Volume 152

Article numbers are in parentheses following the names of contributors.
Affiliations listed are current.

DEBORAH A. ADAMS (8), *The Agouron Institute, 505 Coast Boulevard South, La Jolla, California 92037*

BARBARA J. B. AMBROSE (56), *Chemical Abstracts Service, Columbus, Ohio 43210*

LYNNE M. ANGERER (68), *Department of Biology, University of Rochester, Rochester, New York 14627*

ROBERT C. ANGERER (68), *Department of Biology, University of Rochester, Rochester, New York 14627*

WAYNE M. BARNES (57), *Department of Biological Chemistry, Washington University School of Medicine, St. Louis, Missouri 63110*

SHELBY L. BERGER (6, 19, 21, 26, 32, 33, 36, 37, 43, 45), *Division of Cancer Biology and Diagnosis, National Cancer Institute, National Institutes of Health, Bethesda, Maryland 20892*

GRANT A. BITTER (70), *Department of Molecular Biology, Amgen, Inc., 1900 Oak Terrace Lane, Thousand Oaks, California 91320*

DAPHNE D. BLUMBERG (1, 2), *Laboratory of Comparative Carcinogenesis, National Cancer Institute, Frederick Cancer Research Facility, Frederick, Maryland 21701*

WILLIAM M. BONNER (7), *Chromosome Structure and Function Section, Laboratory of Molecular Pharmacology, Division of Cancer Treatment, National Cancer Institute, National Institutes of Health, Bethesda, Maryland 20892*

ROY J. BRITTEN (66), *Division of Biology, California Institute of Technology, Pasadena, California 91125*

JOAN E. BROOKS (11), *New England BioLabs, Inc., Beverly, Massachusetts 01915*

FRANK J. CALZONE (66), *Division of Biology, California Institute of Technology, Pasadena, California 91125*

JOHN M. CHIRGWIN (20), *Division of Endocrinology, Department of Medicine, The University of Texas Health Science Center at San Antonio, San Antonio, Texas 78284*

FABIO COBIANCHI (10), *Istituto di Genetica Biochemica ed Evoluzionistica, 27100 Pavia, Italy*

CARL COSTANZI (62), *Department of Neoplastic Diseases, Hahnemann University, Philadelphia, Pennsylvania 19102*

NANCY A. COSTLOW (67), *Department of Molecular Embryology, Memorial Sloan-Kettering Cancer Institute, New York, New York 10021*

KATHLEEN H. COX (68), *Department of Biology, University of California at Los Angeles, Los Angeles, California 90024*

BRYAN R. CULLEN (71), *Department of Molecular Genetics, Hoffmann-La Roche Inc., Roche Research Center, Nutley, New Jersey 07110*

ERIC H. DAVIDSON (66), *Division of Biology, California Institute of Technology, Pasadena, California 91125*

PRESCOTT L. DEININGER (41), *Department of Biochemistry and Molecular Biology, Louisiana State University Medical Center, New Orleans, Louisiana 70112*

ANTHONY G. DILELLA (18, 49), *Department of Cell Biology, Baylor College of Medicine, Houston, Texas 77030*

JAMES DOUGLAS ENGEL (40), *Biochemistry, Molecular Biology, and Cell Biology, Northwestern University, Evanston, Illinois 60201*

WILLIAM H. ESCHENFELDT (36, 37), *Division of Cancer Biology and Diagnosis,*

National Cancer Institute, National Institutes of Health, Bethesda, Maryland 20892

GLEN A. EVANS (65), *Cancer Biology Laboratory, The Salk Institute, La Jolla, California 92138*

GARY FELSENFELD (74), *Laboratory of Molecular Biology, National Institute of Diabetes and Digestive and Kidney Diseases, National Institutes of Health, Bethesda, Maryland 20892*

JOHN C. FIDDES (39), *California Biotechnology Inc., 2450 Bayshore Frontage Road, Mountain View, California 94043*

ANNA-MARIA FRISCHAUF (15, 16, 17), *The Imperial Cancer Research Fund, Lincoln's Inn Fields, London WC2A 3PX, England*

DAVID GILLESPIE (62), *Department of Neoplastic Diseases, Hahnemann University, Philadelphia, Pennsylvania 19102*

EDWARD I. GINNS (28), *Molecular Neurogenetics Unit, Clinical Neuroscience Branch, National Institute of Mental Health, National Institutes of Health, Bethesda, Maryland 20892*

STEPHEN P. GOFF (52), *Department of Biochemistry and Molecular Biophysics, Columbia University, College of Physicians and Surgeons, New York, New York, 10032*

JONATHAN R. GREENE (55), *Department of Biology, Massachusetts Institute of Technology, Cambridge, Massachusetts 02139*

LEONARD GUARENTE (55), *Department of Biology, Massachusetts Institute of Technology, Cambridge, Massachusetts 02139*

UELI GUBLER (34, 35), *Department of Molecular Genetics, Hoffmann-La Roche, Inc., Nutley, New Jersey 07110*

DOUGLAS HANAHAN (39), *Cold Spring Harbor Laboratory, Cold Spring Harbor, New York 11724*

DAVID M. HELFMAN (39, 50), *Cold Spring Harbor Laboratory, Cold Spring Harbor, New York 11724*

LOTHAR HENNIGHAUSEN (73), *Laboratory of Biochemistry and Metabolism, National Institute of Diabetes and Digestive and Kidney Diseases, National Institutes of Health, Bethesda, Maryland 20892*

BERNHARD G. HERRMANN (15), *Laboratory of Molecular Embryology, National Institute for Medical Research, Mill Hill, London NW7 1AA, England*

MICHAEL J. M. HITCHCOCK (28, 29), *Pharmaceutical Research and Development Division, Bristol-Myers Company, Wallingford, Connecticut 06492*

STEPHEN H. HUGHES (50), *Frederick Cancer Research Facility, Frederick, Maryland 21701*

P. DAVID JACKSON (74), *Laboratory of Molecular Biology, National Institute of Diabetes and Digestive and Kidney Diseases, National Institutes of Health, Bethesda, Maryland 20892*

ALLAN JACOBSON (25), *Department of Molecular Genetics and Microbiology, University of Massachusetts Medical School, Worcester, Massachusetts 01605*

ROSEMARY JAGUS (27, 31, 60), *Department of Microbiology, Biochemistry and Molecular Biology, University of Pittsburgh School of Medicine, Pittsburgh, Pennsylvania 15261*

JERRY JENDRISAK (40), *Promega, 2800 South Fish Hatchery Road, Madison, Wisconsin 53711*

FOTIS C. KAFATOS (63, 64), *Department of Cellular and Developmental Biology, The Biological Laboratories, Harvard University, Cambridge, Massachusetts 02138, and Institute of Molecular Biology and Biotechnology and Department of Biology, University of Crete, 71110 Heraclio, Crete, Greece*

P. S. KAYTES (44), *Molecular Biology Research, The Upjohn Company, Kalamazoo, Michigan 49001*

ALAN R. KIMMEL (32, 42, 43, 54, 61), *Laboratory of Cellular and Developmental Biology, National Institute of Diabetes and*

Digestive and Kidney Diseases, National Institutes of Health, Bethesda, Maryland 20892

MARC S. KRUG (26, 33), Division of Cancer Biology and Diagnosis, National Cancer Institute, National Institutes of Health, Bethesda, Maryland 20892

JAMES J. LEE (67), Department of Genetics and Development, College of Physicians and Surgeons, Columbia University, New York, New York 10032

HENRYK LUBON (73), Laboratory of Biochemistry and Metabolism, National Institute of Diabetes and Digestive and Kidney Diseases, National Institutes of Health, Bethesda, Maryland 20892

DENNIS C. LYNCH (24), Department of Medicine, Harvard Medical School, Dana Farber Cancer Institute, Boston, Massachusetts 02115

RAYMOND J. MACDONALD (20), Department of Biochemistry, The University of Texas Health Science Center at Dallas, Dallas, Texas 75235

CAROL J. MARCUS-SEKURA (28, 29), Division of Virology, Office of Biologics Research and Review, Food and Drug Administration, Bethesda, Maryland 20205

BERNARD M. MECHLER (23), Institute of Genetics, Johannes Gutenberg University, D-6500 Mainz, Federal Republic of Germany

JUDY L. MEINKOTH (9, 61), Gene Expression Laboratory, The Salk Institute for Biological Studies, San Diego, California 92138

DOUGLAS A. MELTON (30), Department of Biochemistry and Molecular Biology, Harvard University, Cambridge, Massachusetts 02138

ROBERT C. MIERENDORF (51, 58, 59), Promega, 2800 South Fish Hatchery Road, Madison, Wisconsin 53711

HARVEY MILLER (13), Department of Cell Genetics, Genentech, Inc., 460 Point San Bruno Boulevard, South San Franciso, California 94080

C. GARRETT MIYADA (47), Molecular Biochemistry, Beckman Research Institute of the City of Hope, Duarte, California 91010

JOSEPH R. NEVINS (22), Laboratory of Molecular Cell Biology, Howard Hughes Medical Institute, The Rockefeller University, New York, New York 10021

RICHARD C. OGDEN (8), The Agouron Institute, 505 Coast Boulevard South, La Jolla, California 92037

CHRIS PERCY (51), Promega, 2800 South Fish Hatchery Road, Madison, Wisconsin 53711

DIANA PFEFFER (58, 59), Department of Biological Chemistry, University of California at Los Angeles School of Medicine, Los Angeles, California 90024

REYNALDO C. PLESS (56), BRL/Life Technologies, Inc., Gaithersburg, Maryland 20877

ALAN E. PRZYBYLA (20), Department of Biochemistry, University of Georgia, Athens, Georgia 30602

ROBERT S. PUSKAS (37), Molecular Biology Department, Sigma Chemical Company, St. Louis, Missouri 63178

ELISABETH A. RALEIGH (12), New England BioLabs, Inc., Beverly, Massachusetts 01915

ANURADHA RAY (38), Section of Biochemistry, Molecular and Cell Biology, Cornell University, Ithaca, New York 14853

MARK D. ROSE (53), Department of Molecular Biology, Princeton University, Princeton, New Jersey 08512

MARTIN ROSENBERG (69), Biopharmaceutical Research and Development, Smith Kline and French Laboratories, 709 Swedeland Avenue, Swedeland, Pennsylvania 19479

NADIA ROSENTHAL (72), Howard Hughes Medical Institute and Department of Cardiology, The Children's Hospital, Harvard Medical School, Boston, Massachusetts 02115

THOMAS D. SARGENT (46), *Laboratory of Molecular Genetics, National Institute of Child Health and Human Development, National Institutes of Health, Bethesda, Maryland 20892*

ALLAN R. SHATZMAN (69), *Biological Process Sciences Research and Development, Smith Kline and French Laboratories, 709 Swedeland Avenue, Swedeland, Pennsylvania 19479*

NIKOLAUS A. SPOEREL (63, 64), *Department of Cellular and Developmental Biology, The Biological Laboratories, Harvard University, Cambridge, Massachusetts 02138*

GALVIN H. SWIFT (20), *Department of Biochemistry, The University of Texas Health Science Center at Dallas, Dallas, Texas 75235*

G. VOGELI (44), *Molecular Biology Research, The Upjohn Company, Kalamazoo, Michigan 49001*

GEOFFREY M. WAHL (9, 43, 45, 61, 65), *Gene Expression Laboratory, The Salk Institute, San Diego, California 92138*

DONALD M. WALLACE (4, 5), *Genetics Unit, Glaxo Group Research Limited, Greenford, Middlesex UB6 OHE, England*

R. BRUCE WALLACE (47), *Molecular Biochemistry, Beckman Research Institute of the City of Hope, Duarte, California 91010*

KENNETH F. WERTMAN (14), *Department of Biology, Massachusetts Institute of Technology, Cambridge, Massachusetts 02139*

SAMUEL H. WILSON (10), *Laboratory of Biochemistry, National Cancer Institute, National Institutes of Health, Bethesda, Maryland 20892*

SAVIO L. C. WOO (18, 49), *Howard Hughes Medical Institute, Department of Cell Biology, and Institute of Molecular Genetics, Baylor College of Medicine, Houston, Texas 77030*

WILLIAM I. WOOD (48), *Department of Developmental Biology, Genentech, Inc., 460 Point San Bruno Boulevard, South San Franciso, California 94080*

RAY WU (38), *Section of Biochemistry, Molecular and Cell Biology, Cornell University, Ithaca, New York 14853*

TIYUN WU (38), *Section of Biochemistry, Molecular and Cell Biology, Cornell University, Ithaca, New York 14853*

ARLENE R. WYMAN (14), *Department of Biology, Massachusetts Institute of Technology, Cambridge, Massachusetts 02139*

RICHARD A. YOUNG (40, 51), *Whitehead Institute for Biomedical Research and Department of Biology, Massachusetts Institute of Technology, Cambridge, Massachusetts 02139*

ROBERT A. ZOON (3), *Radiation Safety Branch, National Institutes of Health, Bethesda, Maryland 20892*

Preface

With the birth of recombinant DNA technology, the fields of nucleic acid biochemistry and molecular biology entered an era marked by dramatic and innovative changes in methodology, rapid growth, and an altered perception of how living systems could be studied. As additional possibilities for applying these techniques were realized, it became evident that virtually every area of the life sciences would benefit. For the specialist, this was a very exciting period, but for those who wished to solve specific problems without first becoming full-time molecular biologists, the explosive increase in knowledge was overwhelming. Clearly, the new technology had to be harnessed to serve biologists regardless of scientific background.

This volume represents our contribution. Our aim is to meet the needs of investigators entering molecular biology from other fields and to orient students joining this discipline for the first time. We envisaged a self-contained, concise compendium of state-of-the-art methods that might also appeal to experienced individuals. To impose order on this complex body of information, the book progresses from the basic techniques underlying much of recombinant DNA technology to a series of sections, each addressing a commonly met problem. Topics include genomic cloning, preparation and characterization of mRNA, cDNA cloning, screening libraries, and confirming the identity of selected clones.

The *Guide* contains reliable methods written by leaders in the field. Many articles contrast different approaches for accomplishing the same task in order to highlight strengths and weaknesses in a side-by-side comparison. Others present only the method that was deemed superior. To assist the user, we have been assured that recommended vectors and strains that are not commercially available will be provided by the authors.

Because molecular cloning requires precise attention to detail, the book is heavily edited in the form of cross referencing, Editors' Notes, the Process Guide, and overviews. Pains have been taken to allow the reader to pick and choose among articles without loss of continuity. Cross-referencing helps to clarify the relationships among articles. So, too, do the Editors' Notes. The Process Guide is another integrative device in which fundamental processes in molecular biology are indexed for ready accessibility. Since some methods are used frequently to achieve not quite identical ends, the reader can locate at a glance those that should be considered before choosing a specific technique. Finally, the volume contains overviews to five of the major sections to introduce concepts and strategies and to aid in rapid recognition of relevant material for a particular task.

Within the framework of a one-volume format, choices had to be made. For example, *Drosophila* and *Saccharomyces,* organisms for which specialized methods are abundant, have been short changed. Separate volumes devoted exclusively to these subjects are required for adequate coverage. Mutagenesis has not been emphasized but is covered elsewhere in this series. Some attractive methods were not included because their utility and reproducibility are not firmly established. The *Guide* is primarily intended as an efficient means toward obtaining and characterizing a clone. Within this narrow scope, the contents were chosen based on what would be most important for most investigators most of the time.

SHELBY L. BERGER
ALAN R. KIMMEL

Process Guide

This listing contains the location by chapter number of specific methods for which protocols have been provided. Generally, methods that are mentioned but not presented in detail, or components of a reaction, are not listed here. Thus, Filling in, Nick translation, and Primer extension are in the Process Guide but DNA polymerase I and Klenow fragment require consultation of the Subject Index.

Section I

Requirements for a Molecular Biology Laboratory

[1] Equipping the Laboratory

By Daphne D. Blumberg

So you want to clone a gene or many genes or to make a career out of molecular biology. What is needed to equip a laboratory to accomplish these objectives? A first consideration is the extent to which cloning and molecular biology is to be a major focus of the laboratory and what resources are available. If the object is the one-time cloning and analysis of a single gene, borrowing or making do with a functional although not optimal alternative may be more realistic than purchasing equipment of the highest quality for each step in the process. On the other hand, if cloning and molecular biology are to be the main focus of the laboratory, one will ultimately want state-of-the-art equipment, as procedures are often time-consuming and repetitive, making streamlining of techniques essential in a highly competitive field. Tables I and II give a complete listing of the large and small equipment, accessories, and useful supplies necessary for equipping a "cloning" laboratory. Equipment needs are discussed here to clarify what is inescapably essential, what can be substituted with workable alternatives, and what is nice to have if you can afford it, but you could survive walking three flights of stairs for it. This discussion should provide insight into utilizing scarce resources to the best advantage in order to assemble what is required to get the job done.

The Laboratory Physical Plant, in General, and Considerations on Containment

In the United States all research involving recombinant DNA molecules must be carried out in compliance with the NIII Guidelines. These guidelines can be found in the Federal Register.[1] Two levels (HV-1 and HV-2) of biological and four levels of physical containment (BL-1 to BL-4) are described. BL-2, BL-3, and BL-4 roughly correspond to the National Cancer Institute's three levels of containment for research on oncogenic viruses.[2] The combination of biological and physical containment applicable to a particular cloning project must be determined on an individual basis by consulting the Register.[1] However, in general, most

---

[1] Guidelines for Research Involving Recombinant DNA Molecules, *Fed. Regist.* **49**, Part VI, No. 227, 46266 (1984).

[2] National Cancer Institute, "Safety Standards for Research Involving Oncogenic Viruses," DHEW Publ. No. NIH 75-790. Office of Research Safety, NCI, Washington, D.C., 1974.

TABLE I
LARGE EQUIPMENT AND RELATED ACCESSORIES

Physical plant
  Autoclave (leak-proof autoclaving bags; sterilization tape and Kilit vials)
  Deionization/water purification cartridges and/or glass still
  Crushed-ice machine and ice buckets
  Cold room
  Darkroom with safe lights (GBX-2 filters for X-ray film)
  Glassware washing facility
  Chemical fume hood
  Oven for baking glassware
  Refrigerator
  Freezers: $-20°$, $-70°$, and liquid nitrogen freezer
Cell culture work (bacterial and eukaryotic)
  Laminar flow hood and/or biological safety cabinet
  CO_2 incubator for cell culture
  Microscopes
  If a constant-temperature room is available at $37°$ for bacterial culture: orbital platform
    shaker with clamps for Erlenmeyer flasks culture wheel
  If a $37°$ room is *not* available: closed-environment platform shaker (orbital) with clamps
    for 125-ml, 250-ml, 500-ml, 1-liter, and 2-liter Erlenmeyer flasks
  Incubator for petri plates ($37°$)
  Microwave oven (melting soft agar)
Centrifugation needs
  Ultracentrifuge
  Rotors: swing bucket (i.e., Beckman SW28, large and small buckets), fixed angle (i.e.,
    Beckman Ti50 or 65), vertical (i.e., Beckman VTi50, plus tube heat-sealing unit)
  High-speed centrifuge (Sorvall RC-2B or Beckman J2-21)
  Large capacity rotor (6×250 ml, Sorvall GSA or Beckman JA-14) or (6×500 ml, Sorvall
    GS3 or Beckman JA-10)
  Medium-capacity rotor (8×50 ml, Sorvall HB2 or Beckman JA-20)
  Low-speed centrifuge (IEC-DPR 6000, Beckman J6B or Sorvall RC3 with rotors and
    adapters to accommodate 15-ml tubes up to 1-liter bottles)
  Minifuges (Eppendorf or Beckman or equivalent)
  Refractometer
  Gradient former
  Fraction collector
Darkroom equipment
  Polaroid MP4 camera (or equivalent) and type 57 high-speed print film and Wratten red
    23A filter
  X-ray developer/fixer baths
  Drying racks for X-ray films
  UV transilluminator box (302- or 366-nm wavelength so as not to photo-damage DNA)
  UV-protective plastic glasses
  Cassettes, intensifying screens, film hangers, and X-ray film (XAR and/or SB-5)
Electrophoresis equipment
  Power supplies: high voltage (up to 3000 V), medium voltage (up to 500 V)
  Gel boxes: sequencing gel box, horizontal gel boxes for agarose RNA and DNA gels,
    vertical gel system for regular size polyacrylamide DNA and protein gels, combs,

TABLE I (*continued*)

---

  spacers, glass plates, binder clamps and/or gel sealing tape for assembling some vertical
    gels
Gel dryer
Electrotransfer unit [high current (2–3 A) power supply]
Peristaltic pumps (Markson minipumps) for recirculating gel buffers
Additional equipment
Scintillation counter or Bioscan Quick Count for ^{32}P or ^{125}I
Water baths: shaking (37–68°), standing (10–15°), (37–42°), (68–70°), boiling
Vacuum oven
Vacuum pump
Lyophilizer
Spin vac
Spectrophotometer and cuvettes (glass for visible spectra, quartz micro for UV readings),
    cuvette holders, and cuvette washer
Sonicator (with interchangeable micro and standard probes)
Balances (analytical and top loading)
pH meter and pH indicator papers
Calculator
Personal computer and software packages
Scanning densitometer (for quantitating autoradiographs and scanning gels)

TABLE II

Small Equipment, Accessories, and Useful Supplies

Mechanical pipetting devices
    Electrical or battery-driven device for 1-, 5-, 10-, and 25-ml pipets
    Pipetman and pipet tips (0–20 μl, 0–200 μl, 0–1000 μl)
    Clay Adams micropipet and microcaps: 1–5 ml; 10, 20, 25, 50, and 100 μl
Protective shields for radiation (see this volume [3])
Geiger counter
Vortex mixers
Bunsen burner
Magnetic heater/stirrers
Inoculating loops and needles
Potter's wheel (for spreading bacterial cultures)
Glass spreading rods
Lab timers (minutes/hours)
Stopwatches (seconds)
Hand-held UV (longwave) lamp
Flash unit for preflashing X-ray film
Filtration manifold
Heat lamps (for drying filters)
Slot or spot blotting apparatus
Rocker platform
Heat sealer and plastic bags

(*continued*)

TABLE II (continued)

UV-transparent trays for photographing stained gels
Isothermic containers for liquid N_2 or dry ice–ethanol baths
Equipment for redistillation of phenol and other chemicals
Desiccator
Single-sided razor blades or scalpels (for cutting bands out of gels)
Forceps, scissors, and hemostats
Dialysis clips (and tubing)
Glass wool
Lab coats
Gloves
Lab mat
Plastic wrap such as Saran Wrap
Aluminum foil
Dry ice supply and isothermic storage container
Vacuum tubing
Pyrex baking dishes or refrigerator containers for washing filter blots
Syringes and needles
Hybridization membranes
Whatman 3 MM filter paper and paper towels for Southern and Northern blots
Adhesive marking tape
Alcohol/waterproof lab markers
Glassware and plastic ware: culture tubes and caps (Bellco) 13 × 100 mm, 18 × 150 mm),
    Erlenmeyer flasks [caps (Bellco) or cotton plugs; 125 ml, 250 ml, 500 ml, 1 liter, 2 liter]
Pipets: glass (0.2, 1.0, 5.0, 10.0, and 25 ml), Pasteur pipets, disposable plastic pipets (1, 5,
    10, and 25 ml), pipette wash stands, and autoclave containers
Petri plates: 100 × 15 mm (medium size), 150 × 15 mm (large size for screening libraries)
Centrifuge tubes and bottles
    Polyallomer and ultraclear ultracentrifuge tubes
    Polypropylene bottles (1 liter, 500 ml, 250 ml) and adapters
    Polypropylene tubes (15 and 50 ml)
    Corex glass tubes (15 and 30 ml) and rubber adapters
    Nalgene 15-, 50-, and 250-ml conical tubes and adapters
    Eppendorf tubes (0.5 and 1.5 ml and racks)
Columns: Econo columns, disposable polypropylene columns, tubing

cloning is done under the two least stringent containment levels (BL-1 or BL-2).

A laboratory in compliance with requirements for this level of containment is furnished with a sink for hand washing, windows opening to the outside are fitted with fly screens, and the bench tops are made of material that is impervious to water and resistant to acid, alkali, organic solvents, and moderate heat. Access to an *autoclave* is essential, both for decontaminating materials, as required under the recombinant DNA guidelines, and also for aspects of nucleic acid biochemistry and bacteriology. Leak-proof autoclaving bags are essential for eliminating biohazardous or contaminated material. Additionally, sterilization tape (for 120° and 180°) and *Bacillus stearothermophilus* Kilit ampules are useful for monitoring correct sterilization of glassware and solutions, respectively.

Depending on the level of containment required, a *biological safety cabinet* may be necessary, as well as protection of vacuum lines with high-efficiency particulate air filters (HEPA) and liquid disinfectant traps. These are usually unnecessary at levels below BL-3 and BL-4.

Lab coats and gloves are recommended (BL-1) or required (BL-2 to BL-4) for all personnel. Mouth pipetting is prohibited at all levels of containment. Mechanical pipetting devices must be used. *Automatic devices* (Pipetman, Gilson or Eppendorf) are available for $0-20$ μl, $0-200$ μl, and $0-1000$ μl volumes. Also, a variety of electric or *battery-operated pipetting devices* are available for use with 1- to 25-ml pipets. In addition to meeting the safety requirements, these are of value when doing multiple large-scale phenol extractions where use of the manual rubber pipet bulbs can be extremely fatiguing.

In addition to the equipment described above, mandated by the recombinant DNA guidelines, several other considerations should go into the fundamental laboratory design. Whereas most institutions provide some form of distilled water, it is unlikely to be of sufficient purity for sensitive enzyme reactions. *Deionization units* (e.g., hydro services) attached to the distilled water outlet are often essential. A variety of different cartridge systems exist for this purpose. In some locations, it may also be necessary to redistill the "distilled" water.

An *oven* for baking glassware for use in RNA extractions is also needed. Usually, occasional access to the ovens in glassware washing kitchens will suffice, but if this cannot be arranged, a regular household oven will do the job and may be the least expensive alternative.

A *chemical fume hood* is necessary for a variety of operations ranging from redistilling phenol, running methylmercury or formaldehyde gels, to carrying out sequencing reactions, as well as for routine work with smelly or noxious chemicals. Access is required to a *crushed-ice machine,* as well as a large supply of *ice buckets,* and to working space in a *cold room* and a *darkroom.*

Bacteriological Work

The bacteriological work that is necessary for a cloning laboratory involves primarily growth of bacteria and preparation of bacteriophage, usually λ or M13. A *constant-temperature room* (set to 37°), a *rotating culture wheel* for growing small-volume overnight cultures (in 18 × 150 tubes), and a large *platform shaker* capable of accommodating 12 2-liter Erlenmeyer flasks can satisfy most bacteriological culture needs. Some platform shakers (e.g., New Brunswick) can be purchased with a variety of interchangeable *clamps* for holding 125-ml, 250-ml, 500-ml, 1-liter, 2-liter, 4-liter, and 6-liter flasks, allowing the preparation of smaller scale cultures and phage lysates. If funds are limited, the culture wheel can be omitted and racks of tubes for small cultures can be clamped to the platform shaker.

If a 37° room is not available, an *incubator* (37°) for petri plates and a *closed-environment platform shaker* equipped with the full-size range of clamps are necessary.

Most bacterial work (at BL-1 and BL-2 levels of containment) is performed on bench tops using a *Bunsen burner* for flame sterilization of vessel mouths, pipets, etc. For inoculating cultures or streaking or spreading plates *inoculating loops, needles,* a *potter's wheel* for turning the petri plates while spreading, and a supply of glass rods which can be bent in a hot flame to make *spreading bars* are necessary. For plating phage a useful item is a *microwave oven* for melting stocks of soft agar, although melting agar in an autoclave or boiling water bath, while more cumbersome, will also suffice. A *water bath* set to 45° is used to maintain the melted agar at a temperature which is cool enough to permit survival of cells and hot enough to remain liquid.

Gauze-covered cotton plugs or plastic (Bellco) caps are generally used to cap Erlenmyer flasks or culture tubes (large 18 × 120 or small 13 × 100) for growing bacteria. The remaining necessary supplies are petri dishes. The 100 × 15 mm small-size dishes are used for routine streaking of cultures or titering of phage stocks, but for screening libraries the larger 150 × 15 mm size is desirable, as far more colonies or phage can be plated. Cellulose nitrate filters that fit these large dishes are available and worth the extra expense (Schleicher and Schuell 137-mm circles, New England Nuclear colony/plaque screen). When ordering petri dishes one should note that *bacteriological* petri dishes are one-third to one-half the cost of the same-sized tissue culture dishes.

Centrifugation Needs

Four different types of centrifuges are useful for cloning work: a high-speed centrifuge, a low-speed centrifuge, an ultracentrifuge, and a mini-fuge. A *high-speed refrigerated centrifuge* such as the Sorvall RC5 or Beckman J2-21 equipped with two rotors, a large-capacity (Sorvall GSA, or Beckman JA-14) rotor which holds six 250- or 500-ml bottles, and a smaller (Sorvall HB2 or Beckman JA-20) rotor that holds eight 50-ml tubes is essential. A *low-speed refrigerated centrifuge* such as the IEC-DPR 6000, Sorvall RC3, or Beckman J6B outfitted with rotors and adapters that allow a variety of tubes from 15-ml to 1-liter bottles to be centrifuged is required. Much of the recombinant DNA methodology relies on the use of various forms of CsCl or sucrose gradients, so that frequent access to an *ultracentrifuge,* often for several days at a time, is essential. Having available a large selection of rotors increases flexibility; however, almost all recombinant DNA experiments can be modified to be carried out in only two rotors, a *swinging bucket-type rotor* such as the Beckman SW28 (with large and small buckets) and a *fixed-angle rotor,* such as the Beckman Ti50 or Ti65. If your cloning experiments are limited to one gene and you have available a fixed-angle rotor, you can use this for purifying all plasmid DNA preparations, but if cloning is going to be ongoing, with multiple people preparing plasmid DNAs on a regular basis, it is most economical to purchase a *vertical rotor* (Beckman VTi50 or VTi65, or equivalent). These rotors can reduce centrifugation times from 48 hr or more to just 16 hr. Also, the VTi50 is available with three size adapters which allow one to run both large and small volume gradients with substantial savings in expensive CsCl. Finally, the real workhorse in the cloning laboratory is the *minifuge* (Eppendorf or Beckman or equivalent). This piece of equipment is absolutely essential in cloning, for assembly of reactions in small volumes, and for purification of DNA and RNA with the almost ubiquitous phenol extractions and ethanol precipitations that occur throughout this volume.

A *refractometer,* while not essential, is useful to have somewhere in your building. Most methods detail the number of grams of CsCl per milliliter necessary for preparation of the various gradients, and it is not usually necessary to check the refractive index of the solutions; however, at times a phage or DNA band does not appear where expected and on those occasions a check of the refractive index will usually help locate the missing sample, and allow readjustment of the CsCl concentration so it will reband to its proper location.

Darkroom Work

The darkroom capable of handling two functions plays an essential role in cloning. It must be equipped (1) for exposing and developing X-ray films and (2) for visualizing and photographing nucleic acids in gels or gradients. The first function requires *tanks for developing, washing,* and *fixing X-ray films.* Depending on available resources and level of use, a variety of alternatives can be employed.

For small-scale use, the least expensive method is to prepare X-ray fixer, stop bath, and developer in gallon jugs and pour out what is necessary into trays, manually lifting the X-ray film with tongs from tray to tray. This is obviously time-consuming. If a deep sink is available, one can purchase four 3- to 5-gallon tanks for developer, stop bath, fixer, and water wash which can stand in the sink. Lighttight lids fit the tops of these tanks and they are large enough to accommodate the small 20.3 × 25.4 cm or medium-sized 35 × 43 cm X-ray films. By connecting the sink faucet, with tubing, to the water-wash tank and by positioning the tank so that water can enter and overflow into the sink, one can construct a circulating wash. *Film hangers* are necessary to use this equipment. The only drawback to this arrangement is that all developing and fixing must be done at room temperature. A thermostated running-water bath equipped with 5-gallon developer, stop bath, and fixer tanks, as well as space to hang developed, fixed films in the running-water portion of the bath is the least sophisticated improvement. These units generally have to be connected to the house plumbing system, and function best when particulates are removed with a *dirt/rust filter unit* in the water line. The developer and fixer usually have to be changed once every 4–6 weeks or more often with heavy use. (In many areas costs can be reduced by selling used X-ray fixer to a graphic arts silver recovery service.) A place to hang the wet X-ray films to dry is also necessary. Again, depending on budget, this can be a string with clothes pins, a *wall mount to hold film hangers with a drip tray* below, or a dryer box that blows hot air on the films, drying them in minutes. The most expensive option for developing and fixing X-ray films is the RPX-O-Matic, a fully automated processing machine which develops, fixes, and dries X-ray films in 150 sec.

Darkrooms require *safelights*. For work with X-ray films a Kodak GBX-2 red filter is used. The safelights, themselves, are small, night light-sized 25- or 15-W frosted light bulbs. Two types of *X-ray film* are in general use. For ^{32}P autoradiography, *Kodak XAR double-sided emulsion film* is most sensitive, but for ^{35}S-labeled samples or where high resolution of closely spaced bands is desired *SB-5 film* is preferred. This film has a single-sided emulsion on a blue-tinged support.

For exposing gels or filters to X-ray film, *cassettes* (or film holders) and *intensifying screens* (DuPont Cronex Lightning Plus) for enhancing weak ^{32}P signals are needed. The top-of-the-line film cassettes are lighttight, made of metal, and capable of firmly fixing filter/gel, film, and intensifying screen in place. The ones manufactured by Wolf have a particularly effective closure mechanism which ensures a lighttight environment. These are available for 20.3 × 25.4 cm or 35 × 43 cm size films. The less expensive alternative is a cardboard film holder. Film holders, however, are not always lighttight and must be wrapped with aluminum foil. Also, if highly radioactive gel/filters are to be exposed, they should be further fitted with Plexiglas to protect one film from being affected by radiation from another sample. For most work, space is needed in a −70° *freezer* for exposing X-ray films (see, also, this volume [7]). It is also useful to have a lighttight drawer or box for storing open boxes of X-ray film.

Depending on the nature of your work, you may wish to preflash X-ray film to increase sensitivity and linearity of the film response. This requires a photographic flash unit appropriately fitted with filters and adjusted as described by Laskey[3] and by Bonner (this volume [7]).

The second essential function of the darkroom is for visualizing and photographing nucleic acids in gels or gradients. For this, an *ultraviolet transilluminator* is needed. These are available in short- (254 nm) or longwave (366 nm) models. While shortwave UV can be used to detect smaller amounts of sample, it damages DNA, making it unsuitable for most cloning work where one will want to isolate, transfer, or hybridize the resolved fragments. Most laboratories therefore use either a longwave UV transilluminator or a 302-nm UV transilluminator which minimizes nicking of the sample, while retaining high sensitivity of detection. When using the UV transilluminator, *plastic safety glasses* are required to protect the eyes. DNA and RNA gels are photographed on the transilluminator using a *Polaroid MP4 camera* or equivalent mounted on an adjustable *copy stand* and equipped with a *Wratten red 23A gelatin filter* for UV photography. With the Polaroid camera, the *high-speed type 57 film* is optimal for photographing ethidium bromide-stained gels. The Polaroid cameras, while providing rapid, reasonable quality prints, are quite costly (see, also, this volume [8]). If one is not available, a considerably less expensive, although more time-consuming, option is the use of a 35-mm camera fixed to an inexpensive copy stand. Because of the better optics of the 35-mm cameras, Coomassie-stained or silver-stained protein gels can also be photographed, although for this purpose an X-ray viewing light box is also necessary. For 35-mm photography, technical Pan film 2415 is useful. Using a high-contrast (D19) developer this film gives very high-resolution photographs of DNA or RNA gels. When developing the film with a lower contrast developer (Technidol LC), the film is suitable for use as a continuous tone film for photographing Coomassie- or silver-stained protein gels. Along with the 35-mm camera and accessories, an enlarger for making prints from negatives, as well as trays and forceps for developing, stopping, fixing, and washing the printing paper, and a safe-light equipped with an OC filter are necessary. Most of the photographic papers are fast drying and do not require a paper dryer. They can be hung

[3] R. A. Laskey, this series, Vol. 65, p. 363.

on a clothesline. While more cumbersome for routine cloning work where a fast photograph of a gel is needed before carrying out the next step, the cost of the 35-mm equipment is about 15% of the cost of a Polaroid MP4. Purchasing a Polaroid slide-making kit also allows the preparation of slides from the 35-mm equipment.

The other useful piece of equipment to have in a darkroom is a longwave UV lamp or *hand-held longwave UV monitor* for locating ethidium bromide-stained DNA bands in CsCl gradients. This can take the form of a fluorescent desk lamp fitted with longwave UV bulbs.

Equipment for Carrying out *in Situ* Hybridizations

Hybridization Membranes. A fundamental technique in cloning and molecular biology is the transfer and immobilization of gel-separated DNA or RNA molecules to membrane filters for subsequent analysis by hybridization. Originally, cellulose nitrate was used exclusively and for many applications it is still the matrix of choice. However, because of its fragility and the necessity to transfer under conditions of high ionic strength, a variety of nylon membranes are now commercially available. These are indeed easier to handle but great care and attention to manufacturer's instructions or established protocols must be exercised to avoid serious background problems. The nylon membranes are available in two basic types, charge modified membranes and uncharged membranes. Some can bind DNA or RNA at low ionic strength, making them useful for electrotransfer procedures. Others allow UV cross-linking of DNA or otherwise eliminate the need for baking of filters to immobilize nucleic acids. Some do not require prewetting and are therefore particularly useful for plaque lifting. The properties of individual papers vary sufficiently that careful reading of the manufacturer's specifications is necessary. A third type of transfer membrane is activated paper (diazophenylthioether or diazobenzyloxymethyl paper, Bio-Rad and Schleicher and Schuell) to which DNA or RNA can be covalently bound. The limited stability of this paper plus the requirement for last-minute activation has largely led to replacement by the nylon membranes, which afford many of the same advantages with less effort.

A *vacuum oven* is used for fixing nucleic acids to filter membranes. The house vacuum system is usually adequate; the temperature for baking most filters is 80°.

For some types of experiments it may be preferable to spot RNA or DNA directly on the filter membrane rather than resolve and transfer it from a gel. For this purpose *a spot* or *slot blotting apparatus* is desirable (see this volume [62]) for producing a defined spot that can be excised and counted, or if a slot blot is used, scanned with a densitometer.

The hybridization of filters is usually done in a plastic bag; a large supply of *heat-sealable plastic bags* and a *heat sealer unit* are essential. Prehybridization and hybridization are usually carried out by immersing the sealed plastic bag in a water bath or by taping it to a *rocker platform* in an incubator (see this volume [45, 61]).

Other Useful Equipment. In addition to the usual *vortex mixers, magnetic stirrers, pH meters,* and *regular balances,* access to an *analytical balance* is necessary for weighing out 0.1- to 1-mg quantities of nucleotides and other valuable reagents. It is also useful to have a supply of *pH papers* for pH adjusting 0.5- to 1-ml volumes of nucleotide solutions. It is useful to have a *heater/magnetic stirrer* that has a long rectangular flat support surface (e.g., the Thermoline heater/stirrers) so that water or buffers can be boiled in Pyrex baking dishes for removing radioactivity from filters (see this volume [45, 61]).

Filtration Manifold. For assays that require monitoring precipitable material, a filtration manifold is a useful investment. These are available to fit both 24- and 25-mm diameter filters so it is important to coordinate the filter and manifold purchases. Some manifolds come equipped with a vacuum gauge which allows readily reproducible flow rates to be obtained and allows 10 or more samples to be processed simultaneously (see this volume [6]).

Sonicator. Access to a probe sonicator will be necessary for shearing carrier DNA for hybridizations. As large quantities can be prepared at a time, it may be needed infrequently unless many assays of chloramphenicol acetyltransferase are anticipated. Then, a sonicator equipped with a microprobe will often be required.[4]

Spectrophotometer. For quantitation of nucleic acids, a spectrophotometer is essential. Measurements in the UV range require the use of *quartz cuvettes;* the micro size allows samples containing as little as 0.2 ml to be assayed. Depending on the instrument, *cuvette holders* with adjustable bottoms to raise the *microcuvettes* up to the proper height might be needed. The spectrophotometer can also be used in the visible range for following the density of bacterial cultures. For this purpose, 3-ml *glass cuvettes* and the appropriate holder are used. Plastic disposable cuvettes suitable for measuring the absorbance of bacterial cultures are also available. A cuvette washer attached to the house vacuum allows

---

[4] C. M. Gorman, L. F. Moffat, and B. H. Howard, *Mol. Cell. Biol.* **2**, 1044 (1982).

washing of the cuvettes with acid, water or ethanol, but is not essential; cuvettes can also be washed with a pipe cleaner serving as a brush.

Spin Vac and Lyophilizer. The spin vac is particularly useful for concentrating small samples, for removing ethanol, and for drying DNA or RNA pellets. It is also the best means for drying radioactive precursors. The centrifugal force ensures that the sample remains in the bottom of the Eppendorf tube while it is being concentrated or dried. This is particularly crucial for reactions to be run or pellets to be resuspended in microliter volumes. The spin vac is usually connected to a lyophilizer. The lyophilizer is also a much used piece of equipment because almost all DNA, RNA, and plasmid preparations are ethanol precipitated, and lyophilization is a convenient method for removing these solvents.

Electrophoresis Equipment

Power Supplies. An assortment of power supplies are necessary for the cloning laboratory. For sequencing, a power supply capable of producing a constant voltage of 3000 V and a constant current of 300 mA is desirable. Whereas such a power supply can also be used for horizontal agarose gels or smaller vertical polyacrylamide gels (for resolving small DNA fragments), the maximal voltage requirements for these applications varies from 30 to 150 V with current in the range of 50–100 mA. Thus, a smaller, highly regulated constant-current/constant-voltage power supply delivering 0–500 V and 0–200 mA is convenient. Since one frequently runs multiple gels, several power supplies in the lower voltage output range are useful. If space or budget is a problem, and if the purpose of the electrophoretic fractionation is to monitor digestion or ligation of DNA, for example, the minipower supplies are ideal. They are inexpensive, fit in the palm of a large hand, and deliver up to 100 mA of current at preset voltages of 25, 50, and 100 V. However, they lack an output meter and the ability to control voltage and amperage carefully, attributes essential for polyacrylamide gels and desirable for some applications with agarose gels.

For electrotransfer of DNA from agarose or acrylamide gels to a blotting matrix, a power supply which will deliver up to 2 A of current with a voltage range of 0 to about 250 V is needed. For this application as well, it is useful to be able to both monitor and control the voltage and/or amperage output since it may be necessary to start a transfer at a lower voltage/amperage in order to allow time for small fragments to bind to the matrix, and to finish at a high voltage/amperage in order to obtain efficient transfer of large fragments. Some electrotransfer apparatuses come complete with a built-in power supply.

Gel Boxes. The two basic types of gel boxes used in cloning labs are the horizontal and vertical systems. The horizontal gel systems are used primarily for agarose gel separation of large ($\geq$0.5 kb) DNA fragments and RNA species. These systems are commercially available in a variety of bed sizes and lengths and usually come equipped with a UV-transparent tray for viewing the gel on the UV transilluminator following electrophoresis. Also necessary are a choice of combs of different thicknesses (0.75–3 mm) which form the sample wells. The horizontal systems range in size from the "mini" gel boxes, with bed sizes of about 50 × 75 mm and accommodating five to eight samples (depending on well size) to the extra-long 20 × 25 cm gel bed models which can be used with two rows (10–12 cm apart) of 20-well combs and which accommodate a total of 40 samples. Although preparative combs, which form long single-slot sample wells, are commercially available, more flexibility in sample size can be achieved by taping several "teeth" in the regular comb with Permacel gel sealing tape to make a longer slot on a temporary basis. Less elaborate but equally functional boxes can usually be made by a skilled plastics company for about one-half the price of the commercial variety and with the additional advantage that the boxes can be made to your own specifications.

Vertical gel apparatuses are used for polyacrylamide gels, including sequencing gels and smaller "standard" sized gels for resolving small DNA fragments (less than 0.5 kb) and proteins. Sequencing gels come in two general sizes: the 85 cm long or the 33 × 40 cm long gel systems. These can be purchased complete with glass plates of appropriate thickness (which do not crack upon heating during electrophoresis at the high voltages necessary for sequencing), spacers, combs (shark's tooth or regular square-well varieties), gel boxes, and leads. The glass plates for sequencing gels must be scrupulously cleaned for optimal results, and thus a wash tank large enough to hold them is useful.

A regular vertical gel electrophoresis system which accommodates smaller (usually 15 × 17 cm) gels may be desirable. A variety of different models are available; however, those with the upper buffer chamber mounted in an adjustable fashion on two poles so that gels of varying length can be run are recommended for maximal flexibility.

Three other necessary items are an *electrotransfer unit* for transferring proteins to filter supports (Western blots) or for transferring DNA from polyacrylamide gels to membrane filters; a *gel dryer* for dehydrating gels and fixing them to solid supports; and a *mini-peristaltic pump*. The gel dryer comes in two sizes, a compact model which will dry gels in the size range of 18 × 30 cm and a large model for gels up to 35 × 44 cm. Since any interruption in the vacuum will lead to cracking of the gel, gel dryers are generally connected through a trap immersed in dry ice to a separate *vacuum pump,* rather than to a lyophilizer or to the house vacuum. Finally, for recirculating gel buffers during electrophoresis, the inexpensive Markson *miniperistaltic pumps* are ideal, delivering 0.2 to 24 ml/min, depending on the pump and the size tubing used.

Data Analysis Equipment. If sequencing is planned to be a major portion of the work, a *personal computer* is ultimately a necessity. Every few years the journal *Nucleic Acids Research* devotes an entire issue to a review of the available *computer programs* for analysis of the various types of data obtained from molecular cloning experiments.[5]

It is also useful to have access to a *scanning densitometer,* preferably one that can scan stained gels, as well as autoradiographs.

Additional Equipment

Water Baths. Most reactions such as restriction endonuclease digestion of DNA, cDNA synthesis, and reactions involving removal of or introduction of 5'-phosphate groups into DNA or RNA are carried out in water baths at 37° or 42° and then transferred to water baths at 68–70° to inactivate an enzyme or to melt cohesive ends of DNA. Ligations or nick translations, in contrast, are usually carried out at 12–15°. Additionally, a bath in which DNA can be denatured by boiling is also needed. Therefore, three or four relatively portable, small water baths fitted with lids and capable of maintaining temperatures up to 100° are useful. Low temperatures (12–15°) can be maintained inexpensively by placing one of the baths in a cold room.

A shaking water bath for hybridizations and washing filters is also helpful. The bath must encompass temperatures from 42° to 68° and must be large enough to accommodate a Pyrex baking dish or refrigerator carton containing filter blots of gels.

Columns. Many reactions that are performed with nucleic acids require desalting or removal of unincorporated nucleotides from the end product. In addition, it may be necessary to purify a DNA or RNA fraction by step elution from DEAE-cellulose or to purify mRNA in columns of oligo(dT)-cellulose. For these purposes a supply of columns of different length and diameter is necessary. A variety of inexpensive, disposable columns (Econo columns) are sold by Bio-Rad. These are autoclavable (except for the yellow soft plastic end cap) and are equipped with porous polyethylene bed support disks. For small-scale work, especially for

---

[5] *Nucleic Acids Research* **12,** No. 1 (1984).

oligo(dT)-cellulose chromatography, a disposable, autoclavable poly-propylene column, which holds a bed volume of 2 ml and has a 10-ml reservoir, is also sold by Bio-Rad in bags of 50. The columns can be pur-chased with a variety of fittings and accessories. These are a standard size and are compatible with the fittings on bardic extension tubing (American Hospital Supply), which is ideal for connecting the columns to reservoirs.

A rapid alternative to desalting or removing nucleotides from nucleic acids, chromatographically, is the *spin column*. These columns eliminate the need to collect and assay multiple fractions and make possible the simultaneous desalting of eight or more samples. Minispin columns are commercially available from Cooper Biomedical (formerly Worthington), but are also extremely easy and quick to fabricate, as follows.

1. Plug the bottom of a 1-ml syringe with autoclaved, siliconized glass wool (see below for siliconization method).

2. Pour into the syringe preswollen Sephadex G-25 coarse equili-brated in either water or buffer (10 mM Tris–HCl at pH 7.5 to 8, 1 mM EDTA, 0.1 M NaCl). The fines should have been thoroughly decanted from the swollen Sephadex resin.

3. Place the 1-ml syringe into a centrifuge tube (Falcon 2059 17 × 100 mm, 14-ml tubes) so that the syringe hangs from the "finger grips" with the end a few centimeters from the bottom of the tube and then load the entire assembly into the appropriate adapter in a low-speed centrifuge.

4. Centrifuge at 1600 g for 4 min. The Sephadex will pack down. Add more Sephadex and recentrifuge until the packed volume of the column is about 1 ml, but not less than 0.9 ml.

5. Add 0.1 ml of water or buffer and centrifuge under the identical conditions that were used for packing the column, this time including 1.5 ml-Eppendorf tube with cap cut off in the bottom of the larger centrifuge tube, positioned so that eluate will be collected in the Eppendorf tube. The volume of liquid you collect in the Eppendorf tube after centrifuga-tion should equal 0.1 ml and is equivalent to the void volume of the column.

6. Repeat step 5.

7. Load the DNA sample on the column in the same volume (0.1 ml) as in the previous two steps.

8. Recentrifuge under the identical conditions that were used for packing the column, collecting the sample into a fresh decapped Eppen-dorf tube. It is particularly important not to vary the centrifugation condi-tions, as any change will lead to either incomplete recovery of the sample or reduced purity.

Equipment for Counting Radioactive Samples. Molecular cloning involves ^{32}P-labeled DNA or RNA. Most workers find that a *Geiger counter* allows them to detect rapidly the location of radioactive materials, regardless of whether the purpose is to monitor the distribution of material after a "spin column" or to decontaminate after a spill. It is occasionally useful to be able to assay radioactivity in a *scintillation counter,* but this is not a piece of equipment that you must own. However, a *Bioscan* instrument modified for filters (Washington, DC) is an inexpensive alternative that makes possible quantitative work (see this volume [6]).

Useful Glassware and Plasticware

There are a number of items which are used in large quantities in cloning and molecular biology laboratories. Some of these are summarized below.

Eppendorf Tubes. The two sizes predominantly used are the 0.5-ml and 1.5-ml tubes. In order to centrifuge the 0.5-ml tubes in the Eppendorf centrifuges, a decapped 1.5-ml tube is used as an adapter. Eppendorf tubes are usually autoclaved before use. Various racks that hold Eppendorf tubes are available, but one of the most useful is the serum vial support made two-tiered by removal of the third tier. This holds 72 Eppendorf tubes and withstands temperatures from −70 to 100°. Often on boiling, the caps on the tubes pop open and the contents are forced out of the tubes. To prevent this, racks with "lids" to hold the caps in place are available from some specialty suppliers, but a very inexpensive and equally effective solution is to punch holes the size of the Eppendorf tubes using a cork borer in a thin piece of styrofoam (from the styrofoam boxes that come with wet or dry ice shipments of enzymes). The tubes, held firmly, float on top of the boiling water bath and even if a cap pops, the tube is secure enough that the contents are retained. Also, because the whole rack floats, one does not have to adjust the level of water in the bath.

Pipet Tips. Two useful sizes of pipets are the yellow ones for the 0–20 μl and the 0–200 μl Pipetman and the blue ones for the 0–1 ml Pipetman. These are always autoclaved and can be purchased in packages that are entirely autoclavable.

Syringes and Needles. Syringes and needles are used mostly for removing DNA and phage bands from the sides of centrifuge tubes or for crushing gel bands to extract DNA; the 1, 3, 5, or 10 cm^3 syringes are most useful. For both of these purposes, 23- to 25-gauge, 1-in. to 1.5-in. needles are used. Also, minicolumns and spin columns are often run in 1-ml syringes. For this purpose an 18-gauge needle can often be used to connect small-bore tubing to the column outlet. Hamilton syringes in the 0- to 50-μl range are useful for loading samples on very thin polyacrylamide gels. Drummond makes a pipet specifically for this purpose.

If a sonicator is not available, needles and syringes can be used to shear carrier DNA for hybridizations. Use a 10- or 50-ml syringe equipped with an 18-gauge needle and work down to a 23- or even a 25-gauge needle to perform the task.

Centrifuge Tubes. For high-speed centrifugations, the most useful are 1-liter, 500 ml, or 250-ml polypropylene bottles for large-scale preparations, and 15- and 50-ml hard-walled polypropylene tubes and 15- and 30-ml Corex glass tubes for medium-scale work. For the low-speed centrifuges, Nalgene 15- and 50-ml tubes and 250-ml conical bottles are most useful.

Three types of material for tubes and bottles in general use in cloning laboratories are Corex glass, polypropylene, and polycarbonate. The glass Corex tubes can be centrifuged at 10,000 g (maximum) and must be used with rubber adapters. Since RNA and DNA stick to glass nonspecifically, these tubes should also be siliconized (see below) if they are to be used with small amounts of nucleic acids. The polypropylene and polycarbonate bottles and tubes can be centrifuged to full speed. The 50-ml tubes generally do not require adapters; 250- and 500-ml bottles do. Visualizing the sample is a problem with polypropylene, but not polycarbonate, whereas nonspecific retention of nucleic acids is less of a problem with polypropylene than with polycarbonate.

For ultracentrifugation, polyallomer tubes are generally used with nucleic acids since they are autoclavable, do not nonspecifically bind significant amounts of nucleic acids, and are not damaged by most reagents used in molecular biology. Transparent ultracentrifuge tubes are also available (Beckman Ultraclear). These allow good visualization of pellets and phage bands, but they *cannot* be used with ethanol or solutions above pH 8.0, are not autoclavable, and nonspecifically bind DNA and RNA, making them largely unsuitable for work with nucleic acids but ideal for purifying phage.

Pipets. Most work with nucleic acids is done with disposable (Falcon or Costar) pipets available in sizes of 1, 5, 10, and 25 ml. These pipets do not bind nucleic acids nonspecifically; therefore, they do not need to be siliconized. These pipets, however, cannot be used in the chloroform extraction, an operation which will dissolve the plastic pipets, or in the case of some glass disposable pipets, solubilize the ink used in lettering the pipets.

For bacterial work, a plentiful supply of 0.2-, 1-, 5-, and 10-ml glass pipets is needed, as well as autoclave cans and pipet wash stands for disinfecting used pipets. These should be filled with a diluted solution of Clorox or 7× detergent.

For dispensing scintillation fluid, a *repipet* which can be put on the top of a 1-gallon bottle is useful. For loading agarose gels a *Clay Adams micropipettor* and the 1–5, 10, 20, 50, and 100 μl *graduated capillaries* or Pipetman (Rainin) are used.

Flasks and Culture Tubes for Bacteriological Work. Two types of culture tubes are generally used: 13 × 100 glass tubes for growing very small bacterial cultures and for plating phage, and 18 × 150 glass culture tubes for performing serial dilutions of phage or bacteria and for growing 5-ml overnight cultures. Screw-capped tubes are used for storing phage stocks or preparing bacterial slants for intermediate-term storage of stocks. Nunc vials are useful for freezing glycerol stocks of bacterial strains for long-term storage. Also, very small 1- to 2-ml glass vials with screw caps can be used as stab bottles for shipping bacterial or phage cultures. For larger scale preparations a supply of 125-ml, 250-ml, 500-ml, 1-liter, or 2-liter Erlenmeyer flasks will be necessary. Usually, 125-ml flasks for overnight cultures and 2-liter flasks for preparative cultures are all that are absolutely necessary.

Procedure for Siliconizing Glassware. When recovery of nucleic acids is a problem, siliconization of glassware is important. This should be carried out in a hood and the operator should wear gloves and a lab coat.

1. Prepare a solution of 5% (v/v) dimethyldichlorosilane in chloroform.
2. Soak glassware in the solution for 5–10 min; then remove and let stand 3–4 hr or overnight at room temperature.
3. Rinse glassware thoroughly in distilled water.
4. Bake at 100° for 2–3 hr.

[2] Creating a Ribonuclease-Free Environment

By DAPHNE D. BLUMBERG

The success of any RNA preparation is critically dependent on the elimination of ribonuclease contamination. Ribonucleases, especially of the pancreatic type, are extremely stable enzymes.[1] They are remarkably

[1] J. N. Davidson, "Biochemistry of the Nucleic Acids." Wiley, New York, 1960.

METHODS IN ENZYMOLOGY, VOL. 152

resistant to heating, and even retain considerable activity after being boiled.[2] Additionally, they are active over a wide pH range and have virtually no cofactor requirements.[1] The pancreatic ribonucleases are small enzymes with a molecular weight of about 15,000 consisting of a single polypeptide chain with four disulfide bridges.[3] As a result, they renature readily, even after treatment with most denaturing agents.[4] While the level of ribonuclease present in different cell types and tissues varies, as does the degree to which it is released from cellular organelles upon disruption of cells, ribonuclease is present in virtually all cells. Moreover, because of its stability, it is found in most glassware and solutions. Human beings are, in fact, a major source of ribonuclease contamination. Anything one touches without gloves will undoubtedly be contaminated by "finger ribonucleases." Thus, when isolating and working with RNA, two sources of RNase contamination must be controlled; solutions and glassware must be rendered nuclease free, and RNases from the cells or tissues from which RNA is to be extracted must be inhibited or removed.

Procedures for Creating a Ribonuclease-Free Laboratory Environment

A number of precautions should be taken to ensure that the glassware, solutions, and reagents used in working with RNA are free from RNase contamination.

Glassware that is to be used for RNA preparations should be purchased new and kept separate from the rest of the laboratory supply. After a regular wash it should be soaked in and rinsed with a 0.1% solution of diethyl pyrocarbonate (DEPC), autoclaved to destroy the remaining DEPC, and then baked in an oven at 250° for a minimum of 3 hr. Wherever possible, sterile disposable plasticware should be used. In particular, individually wrapped, sterile plastic pipets, Nalgene 15-, 50-, or 250-ml sterile conical centrifuge tubes and bottles, and individually wrapped Falcon Nalgene tubes have generally proved to be "ribonuclease free." These should be used wherever possible instead of glassware. Eppendorf tubes, pipet tips, and Clay Adams micropipets should be autoclaved before use with RNA.

All solutions to be used with RNA should first be treated with DEPC to inactivate ribonuclease. DEPC is a nonspecific inhibitor of ribonuclease; however, it also reacts with the adenine residues of RNA, rendering

[2] W. Jones, *Am. J. Physiol.* **52,** 203 (1920).
[3] P. Blackburn and S. Moore, *in* "The Enzymes" (P. D. Boyer, ed.), 3rd ed., p. 317. Academic Press, New York, 1982.
[4] M. Sela, C. B. Anfinsen, and W. F. Harrington, *Biochim. Biophys. Acta* **26,** 502 (1957).

it inactive in *in vitro* translation reactions.[5] It is therefore important to remove DEPC from the treated solutions before they are used with RNA. To treat solutions with DEPC, the solution should be adjusted to 0.1% in DEPC and shaken thoroughly or stirred vigorously for 10 min to disperse the DEPC through the entire solution. The solution should then be autoclaved or heated at 70° for 1 hr, or at 60° overnight when gentler conditions are indicated. The decomposition products, CO_2 and ethanol, are volatile and easily released from hot solutions by stirring rapidly using a magnetic stirrer and a sterile Teflon-coated spin bar. It is wise to check the pH of DEPC-treated solutions. Because DEPC also reacts with Tris,[6] Tris buffers should not be directly DEPC treated, but instead should be prepared with DEPC-treated, autoclaved water and then reautoclaved.

Phenol that is to be used with RNA preparations should be redistilled, saturated with DEPC-treated water or buffer, and adjusted to 0.1% with 8-hydroxyquinoline. This compound is an antioxidant and a partial inhibitor of ribonuclease.[7] The phenol, thus prepared, is usually stored frozen until use. Formamide must be extensively deionized. This is done by stirring in the presence of 5 g/100 ml of AG501-X8(D) (20–50 mesh) resin (Bio-Rad) to remove salt impurities, as well as hydrolysis products. After removal of the resin, either by decanting or passing through a Whatman 1 filter paper, the formamide is then sterilized by filtration.

For use with RNA, dry chemicals should be purchased new, reserved only for RNA work, and should only be weighed out with baked spatulas. Additionally, solid CsCl is baked at 200° for 6–12 hr. Gloves should be worn at all times when working with RNA or when touching any of the glassware, plasticware, or reagents to be used with RNA.

On occasion it will be necessary to make up solutions of ribonuclease. Preferably, lyophilized ribonuclease should be banished from the laboratory. If this is not possible, however, great care should be taken to keep it entirely in disposable plasticware, not to spill it on balances, and to weigh it out using disposable spatulas.

While such extensive precautions may seem at times cumbersome, they can often make the critical difference in isolating high molecular weight RNA species which can be translated *in vitro* into large proteins, act as templates for synthesis of full-length cDNAs, or give discrete bands on Northern gels.

[5] L. Ehrenberg, I. Fedorcak, and F. Solymosy, *Prog. Nucleic Acid Res. Mol. Biol.* **16,** 185 (1974).
[6] S. L. Berger, *Anal. Biochem.* **67,** 428 (1975).
[7] K. S. Kirby, *Biochem. J.* **64,** 405 (1956).

Inhibiting Endogenous RNases during Extraction of RNA from Cells and Tissues

The degree to which endogenous ribonucleases present problems in isolating intact RNA depends on the cell type or tissue. Some cell types fortunately have low levels of endogenous ribonuclease or display a subcellular compartmentalization that maintains the separation of RNases and RNA more efficiently during cell disruption than others. At the other end of the spectrum, spleen and pancreas are tissues with high levels of endogenous ribonucleases that readily mix with cellular RNA upon cell lysis and require a more rigorous regimen of ribonuclease inactivation for successful RNA isolation.

Nonspecific Inhibitors of Ribonuclease. A variety of different types of nonspecific inhibitors have been used to reduce ribonuclease activity with variable success. These include polyvinyl sulfate, heparin,[8–10] and 2% solutions of the clays, bentonite or macaloid.[10–12] None is entirely satisfactory. The preparation of bentonite (USP, City Chemical Corp., New York, NY) has been described.[13,14] Briefly, the crude powder is suspended at a concentration of 50 mg/ml in distilled water. That fraction which sediments between 6000 g for 10 min and 10,000 g for 15 min is recovered and resuspended at 10 mg/ml in 0.1 M sodium EDTA (pH 7.5), and stirred for 48 hr at 25° to remove metal ions. The suspension is again subjected to differential centrifugation and the 10,000 g pellet is repeatedly washed with distilled water to remove the EDTA. This material is finally resuspended at a concentration of 5% in 0.01 M sodium acetate (pH 6.0). This suspension of uniformly sized particles is added to the RNA extraction medium and subsequently removed, together with bound RNase, by centrifugation.[14] Macaloid (NL Chemicals, Inc., Hightstown, NJ) is prepared[15] by suspension of the powder in hot 0.05 M Tris–HCl (pH 7.6), 0.01 M magnesium acetate. The suspension is homogenized for 5 min in a high-shear mixer, cooled, and centrifuged. The gelatinous layer of the pellet is recovered, leaving behind the coarser particles. The gel is suspended to a final concentration of 10 mg/ml in 0.05 M Tris–HCl (pH

[8] J. S. Roth, *Arch. Biochem. Biophys.* **44**, 265 (1953).
[9] J. Fellig and C. E. Wiley, *Arch. Biochem. Biophys.* **85**, 313 (1959).
[10] G. H. Jones, *Biochem. Biophys. Res. Commun.* **69**, 469 (1976).
[11] J. C. Gray, *Arch. Biochem. Biophys.* **163**, 343 (1974).
[12] S. L. Berger and H. L. Cooper, *Proc. Natl. Acad. Sci. U.S.A.* **72**, 3873 (1975).
[13] R. L. Watts and A. P. Mathias, *Biochim. Biophys. Acta* **145**, 828 (1967).
[14] H. Frankel-Conrat, B. Singer, and A. Tsugita, *Virology* **14**, 54 (1961).
[15] R. Poulson, *in* "The Ribonucleic Acids" (P. R. Stewart and D. S. Letham, eds.), p. 333. Springer-Verlag, Berlin and New York, 1977.

7.5). Like bentonite, it is added as a suspension to the RNA extraction and then removed, together with the absorbed RNase, by centrifugation.

Deproteinizing agents such as phenol and chloroform,[16] denaturing agents such as sodium dodecyl sulfate (SDS), sarcosyl, and urea,[15] and chaotropic agents such as guanidine hydrochloride[17] or guanidine thiocyanate, used in combination with the nonspecific inhibitors discussed above and the specific inhibitors described below, are the most widely used and effective means of isolating full-length active RNA molecules (see [4] and Section VII in this volume). Digestion of ribonuclease and other proteins by either pronase or proteinase K has also proved useful.[18–21] The activity of proteinase K can be enhanced by the addition of denaturing agents such as 1–2% SDS.[19]

Specific Inhibitors of Ribonuclease. Two very effective specific inhibitors of ribonuclease are vanadyl–ribonucleoside complexes[22] discussed in this volume [21] and RNasin.

RNasin is a ~40,000 molecular weight protein, isolated from human placenta (available from Promega Biotec) or rat liver, which is a potent inhibitor of RNase.[23–25] Its activity is critically dependent upon inclusion of a minimum of 1 mM dithiothreitol, and care must be taken not to expose the RNasin–RNase complexes to denaturing agents such as 7 M urea as the inhibitor will be inactivated and an active RNase will be released. Both inhibitors can be included in certain enzymatic reactions. However, RNasin is compatible with a wider variety of reactions conducted *in vitro* such as nuclear run-on transcription and translation in cell-free systems. The inhibitors are widely discussed throughout this volume.

[16] G. Brawerman, J. Mendecki, and S. Y. Lee, *Biochemistry* **11,** 637 (1972).
[17] J. M. Chirgwin, A. E. Przybyla, R. J. MacDonald, and W. J. Rutter, *Biochemistry* **18,** 5294 (1979).
[18] S. L. Mendelsohn and D. A. Young, *Biochim. Biophys. Acta* **519,** 461 (1978).
[19] H. Hilz, U. Wiegers, and P. Adamietz, *Eur. J. Biochem.* **56,** 103 (1975).
[20] B. Mach, C. Faust, and P. Vassalli, *Proc. Natl. Acad. Sci. U.S.A.* **70,** 451 (1973).
[21] C. H. Faust, H. Diggelmann, and B. Mach, *Biochemistry* **12,** 925 (1973).
[22] S. L. Berger and C. S. Birkenmeier, *Biochemistry* **18,** 5143 (1979).
[23] J. S. Roth, *J. Biol. Chem.* **231,** 1085 (1958).
[24] A. A. M. Gribnau, J. G. G. Schoenmakers, and H. Bloemendal, *Arch. Biochem. Biophys.* **130,** 48 (1969).
[25] G. Martynoff, E. Pays, and G. Vassart, *Biochem. Biophys. Res. Commun.* **93,** 645 (1980).

[3] Safety with ^{32}P- and ^{35}S-Labeled Compounds

By ROBERT A. ZOON

Radioactive materials are commonly used as tracers in many biochemical procedures. Because of the potential for radiation exposure of the bench scientist using these tracers, this chapter has been written about the principles, techniques, and materials used in working safely with the two most commonly used radionuclides in molecular cloning, ^{32}P and ^{35}S. Both are "pure β" emitters, i.e., there are no photons emitted by the nucleus during decay. ^{35}S has a maximum β energy of 167 keV and a half-life of 87.4 days. ^{32}P, with a maximum β energy of 1710 keV (or 1.7 MeV), is considerably more hazardous. Its half-life is 14.3 days.[1]

^{32}P represents, in unshielded form, a significant *external* radiation exposure hazard. Using hundreds of microcuries to tens of millicuries, recommended elsewhere in this volume, one can receive maximum permissible occupational doses (United States or international standards) in a matter of seconds to minutes. To place the following data in perspective, note that in the United States a worker is permitted a maximum skin dose equivalent of 30,000 mRem/annum, distributed at a rate of no more than 7500 mRem/quarter, or a dose equivalent to the lens of the eye, often considered the organ of concern in exposure to ^{32}P, of 5000 mRem/annum or 1250 mRem/quarter. At the surface of a glass "Combi-v-vial" (in which many compounds are shipped) the dose equivalent rate can be as high as 470 mRem/hr-mCi. The exterior surface of a plastic syringe containing a solution of ^{32}P has been measured at over 170,000 mRem/hr-mCi. In a situation without the inherent shielding from the container, such as ^{32}P in direct contact with 1 cm^2 of skin, a dose equivalent of 2×10^6 mRem/hr-mCi is delivered.[2]

^{35}S presents no such external hazard. Beta particles from ^{35}S barely penetrate the dead layer of skin on the hands; they are easily stopped by the walls of typical containers. Thus, intake or *internal* exposure from inhalation and ingestion is the primary hazard.

Basic Precautions. In working with these radionuclides, certain basic techniques must be employed. Disposable rubber gloves should always be worn to prevent contamination of hands by samples or stock solutions. A laboratory coat protects personal clothing and skin. To contain spills and

[1] D. C. Kocher, "Radioactive Decay Tables," TIC 11026. U.S. Dept. of Energy, Washington, D.C., 1981.
[2] "Phosphorus-32 Handling and Hazards." New England Nuclear, Boston, MA (undated).

contamination from microdroplets of materials dispersed during handling (pipetting, vortexing, etc.) the work surfaces should always be covered with plastic-backed absorbent paper. Gloves and paper must be changed frequently and discarded in accordance with the radioactive waste disposal requirements of your institution.

All containers of radioactivity, equipment used with the radioactive materials and therefore, possibly contaminated, and any laboratory areas in which the radioactive materials are manipulated or stored should be labeled to warn other individuals to exercise due caution.

Monitor the work area, equipment, and yourself for contamination frequently, but in no case less than when leaving the laboratory for the day. If contamination is found, decontamination is necessary. A mild detergent solution and warm water can be used for decontaminating equipment, work surfaces, and skin. Commercially produced decontamination solutions offer limited additional capability. It is not considered acceptable to eat, drink, or smoke while working with radioactive materials because ingestion of radionuclides from contaminated hands is possible.

In using ^{32}P, employ the factors of time, distance, and shielding to minimize your radiation exposure. Limit the time spent near radioactive material and maximize the distance from them. Forceps or hemostats for handling samples are advisable. Finally, use shielding where possible to reduce radiation exposure.

Contamination Monitoring Techniques. There are two basic methods for monitoring for contamination. In the first method, filter paper is used to wipe surfaces suspected of contamination and any radioactivity on the paper is determined using liquid scintillation counting. Most liquid scintillation counters have preset channels for ^{14}C/^{35}S and ^{32}P and will automatically count as many as 300 samples. The wipe sample must be placed in liquid scintillation cocktail and shaken to allow the solvent to penetrate the matrix and incorporate the radioactivity. As a guide, any surface whose wipe sample indicates more than 100 dpm/100 cm^2 should be decontaminated and remonitored. This test measures *removable* contamination only. *Fixed* contamination may also be present resulting, in the case of ^{32}P, in an external radiation hazard.

The second technique makes use of a Geiger–Mueller survey meter. Strongly recommended is a "pancake" probe. The pancake detector is considerably more sensitive than an ordinary G–M probe due to its 5-cm diameter window. The Geiger counter can detect ^{35}S, albeit poorly, but is exquisitely sensitive to the energetic β particles emitted by ^{32}P. Use this device with the following caveats: (1) the probe has a very thin entrance window for radiation which is easily damaged by contact with sharp objects; (2) in the absence of contamination, ambient background radiation must be ascertained in order to distinguish genuine contamination. Background levels vary with meter, probe, and geographic location.

Personnel Monitoring. It is prudent, and often legally required, to monitor the external and potential internal exposure for each individual working with radionuclides. External monitoring is accomplished by the wearing of personnel dosimetry badges. These badges contain film and/or thermoluminescent dosimeters and record the radiation exposure to which they are subjected. Fresh badges are usually issued monthly. For persons using or working near sources of ^{32}P, a monitor for the body is required at a minimum. Since the exposure potential is so high and the allowable exposure limits vary for extremities, badges for the hands in the form of watch-style bands or rings are required. Badges must be worn regularly and must be stored in radiation-free areas in order to measure exposure of the individual accurately. Obviously if the badge itself becomes contaminated it should be replaced immediately.

Monitoring for possible internal contamination is done by counting urine specimens. Both ^{35}S and ^{32}P are excreted in urine if ingested or inhaled. Monitoring of the urine should be done at least quarterly for regular users by counting 1 ml of urine for 20 min in a liquid scintillation counter. The minimum detectable activity is a few picocuries per milliliter using these parameters. The specimen should be collected within 48 hr of a typical use of material. If radioactivity is detected, more frequent collection and monitoring may be necessary. Clearly, if positive specimens persist, the causes must be identified and eliminated.

When using ^{35}S, it is prudent to employ a chemical fume hood with a face velocity of at least 100 linear feet per minute to reduce the risk of inhalation of gaseous sulfur by-products.

^{32}P *Shielding.* Due to the high energy of the β particles emitted from ^{32}P, together with *bremsstrahlung,* an effect in which the β particles collide with matter, causing release of X-ray photons, ^{32}P should always be handled in shielded form. The penetration ability of the most energetic β particles of ^{32}P is summarized in Table I for different materials.[3] Recall that β particles have a finite maximum range in matter, whereas photons are diminished exponentially with added shielding.

In spite of the data in Table I, lead, particularly *thin* lead, is generally not recommended for shielding of ^{32}P. Bremsstrahlung, produced in significant quantity when high atomic number material such as lead is bombarded by energetic β particles, also presents a radiation hazard. Note

---

[3] "Radiological Health Handbook," p. 122. U.S. Dept. of Health, Education and Welfare, Washington, D.C., 1970.

TABLE I
MAXIMUM RANGE OF ^{32}P β PARTICLES IN
VARIOUS MATERIALS

Material	Maximum range
Air	6 m
Water	0.84 cm
Plexiglas	0.64 cm
Glass	0.38 cm
Lead	0.045 cm

that whereas thick lead (>0.5 cm) and thin lead both shield the β particles, only thick lead can protect the investigator from the bremsstrahlung.

The recommended shielding material for ^{32}P is 1.0-cm-thick Plexiglas. It is inexpensive, lightweight and transparent. Common Plexiglas shields include the bench shield (Fig. 1), vial and bottle shields, centrifuge and

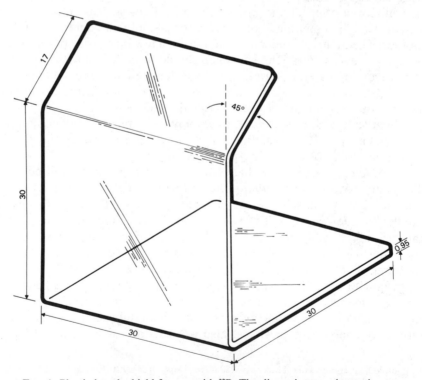

FIG. 1. Plastic bench shield for use with ^{32}P. The dimensions are in centimeters.

microcentrifuge tube shields, tissue culture flask shields, and syringe shields. Since there are few commercial sources supplying these objects, most institutions design and manufacture them in their own shops. Furthermore, adequate shielding should protect all persons working in the laboratory, including those behind and/or to the sides of the source. Sometimes a composite shield of low-density material such as glass or plastic surrounded by a layer of lead to contain bremsstrahlung is the best choice.

Receipt, Storage, and Disposal. Upon receipt of any radioactive material, it is essential to verify that containment has not been compromised during shipment and that contamination during packaging has not occurred. In the United States this is a regulatory requirement satisfied in most institutions by the radiation safety organization. In the absence of such a centralized service, you should test the shipment for contamination by wipe testing the exterior surface of the shipping container and the inner vial. Handle the material as if it were contaminated at all times. If contamination is found, report it to the appropriate authorities and to the supplier of the radioactive shipment.

Proper storage of radioactive materials includes a secure location with shielding where necessary. Storage containers must be properly labeled for contents and to indicate the presence of hazardous radioactivity. Exercise the concept of multiple containment for any radioactive material, i.e., surround the source or sample container with an unbreakable outer container such as a plastic bag or larger sealed plastic bottle. This procedure provides greater assurance of containment in the event of an accident.

Waste disposal is an institution-specific issue, but there are generalizations that can be made. Always dispose of known contaminated or potentially contaminated items as radioactive waste. Some institutions require the separation of combustible and noncombustible radioactive waste, depending on whether waste is incinerated. Liquids and solids should always be separated into distinct containers, since these generally differ significantly in radioactivity content as well as in the ultimate method of disposal. If your institution permits direct disposal of liquids into sink drains, records of the date and amounts of liquid radioactivity discharged must be kept.

The common radioactive materials used in molecular cloning present a significant potential radiation exposure hazard in the quantities normally used. The observation of appropriate precautions in working with the materials, including shielding and proper facilities, protective clothing, proper monitoring, and decontaminating when necessary, will allow use of the tracers with minimum radiation exposure.

Section II

General Methods for Isolating and Characterizing Nucleic Acids

[4] Large- and Small-Scale Phenol Extractions

By DONALD M. WALLACE

Of basic importance to molecular cloning is the ability to extract and purify nucleic acids from a variety of sources. Probably the most common method is phenol extraction. With appropriate modification of the conditions, the technique can be applied to such diverse operations as the isolation of RNA or DNA from complex mixtures of biological origin, e.g., cell extracts, or to the retrieval of DNA from relatively simple mixtures such as those encountered during molecular genetic manipulations *in vitro*. These operations require either large-scale (usually tens to hundreds of milliliters) or small-scale (microliters) phenol extractions, respectively.

A brief review of the literature shows that a plethora of phenol extraction methods have evolved, with little or no explanation of how or why they came about. Rather than describing them, the aim here is to discuss the basic factors which influence the procedure so that rational choices of methodology can be made. In addition, two detailed methods are presented: a large-scale phenol extraction method for RNA from complex mixtures, and a small-scale technique for DNA recovery from simple mixtures. Readers are referred to the relevant chapters of this volume for more detail on DNA and RNA isolation methods (see this volume [13,15,18,20–22] and elsewhere in this series[1]).

The fundamental aim of a phenol extraction is the deproteinization of an aqueous solution containing the desired nucleic acids. In simple terms the phenol reagent is mixed with the sample under conditions which favor the dissociation of proteins from nucleic acids. Centrifugation of the mixture yields two phases: a lower organic phenol phase carrying the protein (much of which actually segregates to the white flocculent interphase) and the less dense aqueous phase containing the intact nucleic acids. Ultimately, recognition of the optimal conditions for a particular extraction procedure is dependent on the nature of the nucleic acids to be isolated and on how they will be subsequently used.

Quality of Phenol

Many commercial suppliers provide liquified phenol which can be used directly, without further purification; it is clear and colorless. For

[1] This series, Vol. 12, Parts A and B.

the small-scale retrieval of DNA from genetic manipulation reactions a small quantity of commercially supplied, pure liquified phenol (10–50 ml) equilibrated with TE buffer (see example 2 for definition), with or without the antioxidant 0.1% 8-hydroxyquinoline (w/v), is a useful stock reagent. Wrapped in foil and stored at 4°, this reagent can be kept for up to 1 month. If no 8-hydroxyquinoline is added, the eventual discoloring of the phenol indicates that it is extensively oxidized and should be discarded.

If the commercially supplied phenol is either crystalline or has a yellow or pink coloration, it is necessary to purify it further by redistillation. The coloration is indicative of the presence of phenolic oxidation products (quinones, diacids, and others) which form during storage. These contaminants must be removed because they cause cleavage of phosphodiester bonds and cross-linking of DNA strands.

Phenol Redistillation. Care must always be taken when handling phenol, especially during redistillation. Phenol burns can be very serious and often require emergency treatment. Redistillation should be carried out in a chemical fume hood with the protective screen in place whenever possible. Operators should wear protective clothing such as rubber gloves, safety glasses, and a plastic/rubber apron. As the redistilled material is stable for long periods, sizable quantities should be prepared in advance and stored frozen.

A round-bottom distillation flask (2 liter size or larger) containing some boiling chips is filled no more than three-quarters full with liquified phenol. If crystalline phenol is used, it should be melted at 65° in a water bath after addition of about 10% water (usually 300 ml water per 3 kg phenol). Do not use phenol containing hypophosphorous acid (H_3PO_2) as an antioxidant; the decomposition products formed on heating are explosive. The mixture is brought to boiling using a heating mantle. The use of electrical heating tape and/or hot water (55–60°) circulated around a water-jacketed condenser is recommended to prevent gaseous phenol from condensing to the solid state in the system during distillation; blocked passages are subject to pressure explosion. At the beginning of the distillation (about 120°) a cloudy liquid which should be discarded emanates from the condenser. Pure phenol emerges from the condenser as a clear liquid in a single phase (about 180°). Collect the phenol in suitable vessels (e.g., screw-capped 100-ml glass bottles). 8-Hydroxyquinoline can be added to the bottles prior to the collection of the redistilled phenol (100 mg 8-hydroxyquinoline per 100 ml phenol). Once the desired volume of phenol is collected, wrap the bottle with foil and store at −20°.

Never allow the volume of phenol in the distillation flask to drop below 100 ml, as the concentrated residue left behind the distillation is explosive. Upon completion of the distillation, permit the apparatus to cool, wash out the debris with hot water, and then rinse the flask with ethanol.

When frozen redistilled phenol is to be used, add a saturating amount of the relevant extraction buffer (30 ml buffer per 100 ml phenol) to the frozen phenol and thaw in a 65° water bath. It is worth noting that the rate of phenol oxidation increases as the pH increases. Therefore, if the equilibrating buffer pH is greater than 8, such phenol reagents should be used fresh and not stored. Other equilibrated phenols may be stored for up to 1 month at 4°. Phenol for RNA extractions should be freshly thawed and the excess should be discarded.

There are several advantages to be gained by having 8-hydroxyquinoline present in the phenol: (1) it retards the rate of phenol oxidation; (2) its bright yellow color in the organic phase of an extraction makes it easy to discern the location of the phases; (3) it is a partial inhibitor of ribonuclease; (4) when extracting RNA in the presence of vanadyl–ribonucleoside complexes (VRC), used as a ribonuclease inhibitor, the extraction can easily be followed by the color change of the 8-hydroxyquinoline from yellow to black when it reacts with the VRC. When the phenol remains yellow, the VRC have been removed from both the aqueous and phenol phases.

Choice of Conditions for Phenol Extractions

Chloroform. Used in conjunction with phenol, chloroform improves the efficiency of nucleic acid extractions. This is principally due to its ability to denature protein, thereby assisting the dissociation of nucleic acid from protein. The high density of chloroform also enhances the separation of the phases, facilitating the removal of the aqueous phase with little cross-contamination with organic material. Removal of lipid from the sample is another benefit of using chloroform. Usually, isoamyl alcohol is added to the chloroform to prevent foaming.

In most situations in which the retrieval of DNA from an *in vitro* reaction is required, a single extraction with phenol (TE-saturated)–chloroform–isoamyl alcohol (24 : 24 : 1, v/v/v) equal in volume to the volume of the aqueous sample is sufficient. The DNA in the aqueous phase can subsequently be recovered by ethanol precipitation (see this volume [5]).

In the case of RNA extraction from complex mixtures, the use of phenol–chloroform has been shown to increase the yield of poly(A)+ mRNA over methods which use phenol alone.[2] Usually, optimal yields

---

[2] R. P. Perry, J. La Torre, D. E. Kelley, and J. R. Greenberg, *Biochim. Biophys. Acta* **262**, 220 (1972).

are obtained by first extracting with an equal volume of buffer-saturated phenol followed by reextraction of the aqueous phase and interphase with an equal volume of phenol mixed 1 : 1 (v/v) with chloroform–isoamyl alcohol (96 : 4, v/v). It is thought that this combination of extractions reduces losses of RNA to the interphase, an undesirable result caused by formation of insoluble RNA–protein aggregates.[3]

Constitution of the Extraction Buffer. The most important consideration in choice of pH is whether DNA or RNA is to be isolated. At pH 5–6 DNA is selectively retained in the organic phase and interphase, leaving RNA in the aqueous phase. An added advantage of using low pH during RNA isolation from biological extracts is the reduced level of activity of many nucleases under these conditions. On the other hand, if DNA is to be extracted, a pH of 8 or higher must be used. At this pH both DNA and RNA will partition to the aqueous phase. A consequence of this pH-dependent partitioning of DNA is the necessity to increase the pH of any *in vitro* manipulations carried out at low pH prior to phenol extraction (e.g., S_1 nuclease digestion at pH 4.5). Basic conditions for extracting DNA are also essential to prevent depurination. It should be noted that if phenol alone is used for the isolation of RNA from biological extracts (as opposed to phenol–chloroform), at pH values below 7.6 poly(A)$^+$ mRNA is lost to the organic phase,[4] and poly(A) tracts may also be cleaved from the poly(A)$^+$ mRNA.[2] However, if such extractions are carried out at pH 9,[4] or at pH 5–9 in combination with chloroform[2] or sodium dodecyl sulfate the poly(A)$^+$ mRNA partitions to the aqueous phase more efficiently.

In the past, the isolation of RNA (especially for mRNA isolation) was commonly carried out in the presence of high salt concentrations (0.2–0.5 *M* NaCl) to reduce the effect of the endogenous ribonucleases. However, with the advent of more effective ribonuclease inhibitors, and in circumstances where ribonuclease is not a problem, optimal RNA yields are obtained at salt concentrations lower than 0.2 *M*. This lower ionic strength reduces nonspecific RNA aggregation, the maximum solubility of RNA occurring at 0.15 *M* Na$^+$. As sodium dodecyl sulfate is commonly used in such RNA extractions the use of potassium salts, which would cause the precipitation of the detergent, is avoided.

For DNA isolation from simple solutions, salt concentration does not significantly influence the efficiency of the extraction. However, if the salt in the aqueous phase is below 50 m*M*, oligomers, in particular, may be lost to the phenol phase. Such losses can be prevented by saturating phenol with TE buffer to which 50 m*M* NaCl or KCl has been added.

[3] R. D. Palmiter, *Biochemistry* **13**, 3606 (1974).
[4] G. Brawerman, J. Mendecki, and S. Y. Lee, *Biochemistry* **11**, 637 (1972).

Detergents are employed to promote the dissociation of proteins from nucleic acids when extracting from biological material, the most commonly used being the ionic detergent sodium dodecyl sulfate (SDS). The most effective release of RNA with the least degradation is obtained by briefly exposing the biological extract to a relatively high concentration of SDS (0.1–2.0%). In order to prevent SDS from precipitating, the exposure should be performed at 10° or higher. An extra benefit of SDS is its mild inhibition of endogenous nucleases. Another frequently used detergent is sodium deoxycholate, which promotes the dissolution of lipid membrane structures (e.g., rough endoplasmic reticulum) as well as the dissociation of proteins from nucleic acids.

It should be noted that Mg^{2+} can mediate aggregation of nucleic acids to each other and to proteins. The presence of ethylenediaminetetraacetic acid (EDTA), the Mg^{2+} chelator, in the extraction buffer inhibits this interaction. Excess EDTA also prevents magnesium hydroxide from precipitating and trapping DNA when alkaline conditions are required. If necessary, dissociation of protein can be further improved by incubation of the cell extract with protease K (50–200 μg/ml, final concentration; 37° for 30 min) in the presence of SDS. (Pronase can be used similarly.) This treatment also results in the reduction of the amount of protein interphase.

For efficient recovery of DNA from manipulations *in vitro* the addition of extra protein-denaturing agents such as those mentioned above is not usually required.

In many cases the ability to obtain undegraded nucleic acids using phenol extractions depends on methods for inhibiting nucleases. For DNA extraction from simple solutions, the presence of EDTA is sufficient to inhibit any contaminant nucleases. In contrast, RNA extraction from biological sources (especially for mRNA isolation) requires the use of powerful ribonuclease inhibitors in most circumstances. Of the many inhibitors suitable for use with the phenol extraction technique, vanadyl–ribonucleoside complexes (VRC)[5,6] (this volume [21]) and RNasin (placental ribonuclease inhibitor)[7] have proved effective enough to permit the isolation of intact mRNA from ribonuclease-rich sources. VRC, added to the biological material prior to cell lysis, should be present at 10 mM in the aqueous sample just prior to phenol extraction. VRC does not normally copurify with the RNA, thus yielding a product which, after ethanol precipitation, is directly translatable in cell-free systems. RNasin is used at 250–1000 units/ml and, being a protein, is extracted into the organic

[5] S. L. Berger and C. S. Birkenmeier, *Biochemistry* **18,** 5143 (1979).
[6] S. L. Berger, D. M. Wallace, G. P. Siegal, M. J. M. Hitchcock, C. S. Birkenmeier, and S. S. Reber, this series, Vol. 79, p. 59.
[7] P. Blackburn, G. Wilson, and S. Moore, *J. Biol. Chem.* **252,** 5904 (1977).

phase and interphase. Heparin, another common ribonuclease inhibitor (see this volume [24]), cannot be removed by extracting with phenol.

Physical Considerations. Extractions of nucleic acids from biological sources are best carried out as quickly as possible in ice with refrigerated buffers to minimize nuclease activity. Only when SDS is present would a higher temperature (10°) be required to prevent precipitation of the detergent. High-temperature phenol extractions (50–60°) have been used for mRNA isolation, although the success of this technique depends on the biological source having relatively low endogenous ribonuclease levels since ribonucleases are not completely inhibited under these conditions.

Good recovery of DNA from *in vitro* manipulations can be achieved by extracting quickly at room temperature.

The timing and vigor of mixing the phenol reagent with the sample are dependent on the volume being extracted and the nature of the nucleic acid to be isolated. Extraction of the DNA products of an *in vitro* reaction (microliter volumes) is carried out by vortexing with an equal volume of phenol or phenol–chloroform–isoamyl alcohol for 1 min in a polypropylene microcentrifuge tube (e.g., Eppendorf tube). Separation of the phases is attained by centrifugation in a bench-top microcentrifuge for 1 min (e.g., Eppendorf microcentrifuge). By careful use of a micropipettor (e.g., Gilson Pipetman) the aqueous phase can be removed without disturbing the organic layer. If the DNA to be extracted is larger than about 10,000 base pairs, a milder mixing method is adopted to reduce damage by mechanical shearing: either inversion by hand or by use of a rotating drum for 10–20 min at 4° is suitable. Also, removal of the aqueous phase is carefully carried out using a wide-mouthed pipet.

In the case of large-scale RNA extractions the phenolized mixture can be shaken vigorously by hand in a glass-stoppered bottle for 5 min, stopping periodically to release any gaseous pressure buildup in the bottle (especially when chloroform is present). The centrifugation is best carried out using 15-ml or 30-ml glass Corex centrifuge tubes, or 250-ml Maxiforce polypropylene centrifuge bottles for larger volumes, at 4000 rpm for 5 min in a fixed-angle rotor at 0–4°. The aqueous phase is withdrawn using a sterile pipet or Pasteur pipet of convenient size, taking care not to disturb the lower organic layer and interphase. If there is an obvious white interphase present at the "final" extraction, the aqueous phase should be reextracted with the phenol–chloroform–isoamyl alcohol mixture described above.

Of fundamental importance for successful phenol extraction of intact RNA from biological sources is the performance of all steps at one session and as quickly as possible. The volume of such RNA extractions is dependent principally on the volume of the cell or tissue extract but also on the type of extract (e.g., polysome pellet, whole tissue extract, cytoplasmic fraction). In general terms, extraction buffer is added to give 10–20 times the volume of the biological material. If large amounts of protein are present, even larger volumes may be required.

If carry-over of trace phenol in the aqueous phase occurs at an undesirable level, this can be removed by subjecting it to one or two extractions with water-saturated diethyl ether. The ether is saturated by shaking well with an equal volume of sterile water in a tight-capped bottle (the upper layer is ether). Mix one volume of water-saturated ether with the final aqueous phase by hand for about 1 min (vortex if microliter volumes) and then permit the phases to separate under gravity for 5 min in ice. Discard the upper phase by pipetting off carefully. Alternatively, as long as the sample is not high-molecular-weight DNA prone to shearing, the phase-separated sample can be placed in dry ice, which freezes the aqueous lower layer while leaving the ether layer liquid, thus facilitating its removal.

Example 1: Large-Scale Phenol Extraction of RNA (for mRNA Isolation)

This protocol has been used successfully for mammalian cells. Other systems may require alterations.

For a postnuclear supernatant of 10^9 mammalian cells in 8 ml cold lysis buffer (an example lysis buffer is 1.25% sucrose, 0.3% NP-40 or Triton N-101, 10 mM NaCl, 3 mM MgCl$_2$, 20 mM Tris–HCl at pH 7.4, and 10 mM VRC) in a 250-ml glass-stoppered bottle:

1. Add 2 ml 10% SDS, 30 ml cold ACE extraction buffer (10 mM sodium acetate, pH 5.1, 50 mM sodium chloride, 3 mM Na$_2$EDTA), 40 ml warmed ACE-saturated phenol.

2. Mix vigorously by hand for 5 min at room temperature and transfer mixture to 30-ml Corex tubes.

3. Centrifuge at 8000 rpm for 10 min (e.g., SS-34 rotor, Sorvall RC-5B) at 4°.

4. Carefully pipet upper aqueous phase to clean cold 250-ml glass-stoppered bottle (use Pasteur pipet when near interphase).

5. Add to the aqueous phase 20 ml chloroform–isoamyl alcohol (94:6,v/v) and 20 ml ACE-saturated phenol.

6. Repeat extraction procedure.

7. If necessary reextract aqueous phase(s) until no white flocculent interphase is present.

RNA in the final aqueous phase is precipitated overnight with ethanol (see this volume [5]).

Notes. This protocol can be modified for whole-cell RNA extractions as follows. To a washed cell pellet containing 0.5 to 2×10^8 cells add 10 ml ACE extraction buffer containing ribonuclease inhibitors [either 10 mM VRC or 0.6% bentonite (see this volume [2]) or both] and 0.5 ml 10% SDS. Invert several times to mix the components. Add 10 ml ACE-saturated phenol at 50° and extract for 5 min at 50°. Centrifuge to separate phases as in step 3 above. Complete steps 4–7 scaled down appropriately for 10 ml of aqueous phase.

All manipulations are carried out in ice unless noted otherwise. All buffers are rendered sterile and ribonuclease free by autoclaving and/or diethyl pyrocarbonate treatment (this volume [2]), and stored at 4°. Solutions of 10% SDS and chloroform–isoamyl alcohol are stored at room temperature. Phenol (0.1% 8-hydroxyquinoline), stored at −20°, is thawed by addition of 30 ml ACE extraction buffer per 100 ml phenol at 65° and maintained at this temperature. Glassware is sterilized by heating in a 160° oven overnight. Gloves are worn at all times to protect RNA from ribonuclease present on hands. The volumes of the reagents may have to be varied according to the protein content of the material to be extracted.

Typical yields: 7.5 mg cytoplasmic RNA per 10^9 fibroblasts and 2.8 mg cytoplasmic RNA per 10^9 leukocytes.

Example 2: Small-Scale Phenol Extraction of DNA (for Recovery of
    DNA from *in Vitro* Manipulation)

A single extraction is normally satisfactory in most situations (e.g., restriction enzyme digests, phosphorylation, ligation, polymerization).

For a 50-μl sample in a 1.5-ml microcentrifuge tube:
1. Add 50 μl TE-saturated phenol–chloroform–isoamyl alcohol (24 : 24 : 1, v/v/v) (TE is 10 mM Tris–HCl at pH 8, 1 mM Na$_2$EDTA; TE-saturated phenol is 3 ml TE per 10 ml phenol.)
2. Vortex for 1 min and centrifuge in a microcentrifuge for 1 min.
3. Carefully transfer the aqueous phase to a clean tube using a micropipettor (e.g., Gilson Pipetman).

DNA is then precipitated by ethanol (see this volume [5]).

Notes. TE-saturated phenol–chloroform–isoamyl alcohol can be stored at room temperature or at −20° for about 1 month. However, since basic conditions are essential for preventing acid depurination of DNA, it is important to monitor the pH of the aqueous phase recovered from stored phenol. Some investigators always reequilibrate phenol solutions with concentrated buffers at pH 8 or 9 before extracting DNA. All manipulations are carried out at room temperature.

Typical yield: where DNA is present at 1 μg or greater the recovery is over 90%. Indeed, the most significant losses occur during the subsequent ethanol precipitation (see this volume [5]).

Special Cases. S_1 nuclease products require that the pH be altered prior to phenol extraction by adding Tris base to a final concentration of 50 mM. Dephosphorylation reactions require more than one extraction to remove bacterial alkaline phosphatase completely (three extractions are recommended). Many investigators treat samples with proteinase K at a final concentration of 50–200 μg/ml for 10 min at 37° before phenol extracting. As noted above, extraction of oligodeoxyribonucleotides may require salt to drive the oligomer into the aqueous phase.

Small-scale RNA extractions are performed similarly with a 1 : 1 mixture of ACE-saturated phenol and chloroform-isoamyl alcohol as noted above for large-scale RNA extractions.

[5] Precipitation of Nucleic Acids

By DONALD M. WALLACE

During the course of a cloning project many occasions occur when it is necessary to concentrate nucleic acid samples or change the solvent in which a nucleic acid is dissolved. Fulfillment of these requirements is met by nucleic acid precipitation techniques.

Ethanol Precipitation

Ethanol precipitation is probably the most versatile method of concentrating nucleic acids and is commonly used to recover DNA or RNA after extraction from biological sources or to retrieve phenol-extracted DNA products of *in vitro* enzymatic manipulations (see this volume [4]).

In brief terms, a precipitate is formed by leaving a mixture of the sample, salt, and ethanol at low temperature (−20° or lower). The precipitated salt of the nucleic acid is then sedimented by centrifugation, the ethanol supernatant removed, and the nucleic acid pellet resuspended in a buffer appropriate to its subsequent use. The choice of salt used for the precipitation is determined both by the nature of the sample and by the intended use of the nucleic acid.

Samples containing phosphate or 10 mM EDTA should not be subjected to ethanol precipitation as these materials will also precipitate along with the nucleic acid; preliminary dialysis of such samples is essential before ethanol precipitation.

[*Editors' Note.* When nucleic acids are precipitated solely to change solvents, one might consider drop dialysis.[1] Drop dialysis represents a practical alternative to conventional dialysis in bags for volumes of 50 μl or less. Float a Millipore (Bedford, MA, catalog no. 01300, VSWP) 0.025μm porosity filter disk, 13 mm diameter, shiny side up, on the surface of 5 ml of the diffusate (the solution one is dialyzing against). A 35-mm petri dish is a convenient vessel. Carefully pipet the retentate (sample) onto the center of the floating disk; it should form a dome-shaped drop. After 2–3 hr, a sample containing approximately 1 μg mRNA should be recovered from the disk with >90% yield and should be found virtually free of labeled triphosphates, salt, or traces of phenol. Since a steady hand is required, a few practice tries with a solution similar to the sample are recommended.]

Basic Method of DNA Precipitation with Ethanol

For recovery of DNA from a typical reaction (e.g., for 1 μg DNA in 20 μl):

1. To 20 μl aqueous DNA sample in a plastic microcentrifuge tube add 2 μl 3 M sodium acetate, pH 5.5 (i.e., 0.1 volume giving 0.3 M Na$^+$), and 40 μl ethanol (i.e., 2 volumes).

2. Mix well by vortexing and immerse the tube into a $-70°$ bath composed of methanol plus dry ice for 15 min. Powdered dry ice can be substituted for the bath. The mixture will freeze or form a slurry in this time.

3. Centrifuge the DNA precipitate in a bench-top microcentrifuge at maximum speed for 10 min (cold room). A whitish pellet of DNA should appear at the bottom of the tube. In general, pellets of 10 μg are visible whereas pellets of 2 μg will be invisible.

4. Remove the ethanol supernatant using a micropipettor (Rainin Pipetman or the equivalent), taking care not to disturb the pellet or the area of the tube where the pellet should be located.

5. Add 100 μl 70% ethanol ($-20°$) to the sample and vortex. This step removes any solute trapped in the precipitate.

6. Resediment the precipitate by centrifugation for 2 min, and remove the supernatant as before.

[1] R. Marusyk, *Anal. Biochem.* **105**, 403 (1980).

7. Dry the pelleted DNA for 1–2 min in a lyophilizer (Speed Vac Concentrator, Savant Instruments Inc.) or vacuum desiccator taking care to release the vacuum gently so as not to dislodge the dried sample.

8. Resuspend the DNA in TE (pH 8) buffer (10 mM Tris–HCl at pH 8; 1 mM Na$_2$EDTA).

Notes and Alternatives. A common alternative to using sodium acetate at 0.3 M is to make the sample 2.5 M in ammonium acetate prior to the addition of ethanol (add 0.5 volume 7.5 M ammonium acetate followed by 2.5 to 3 volumes of ethanol). Cool for 10 min in dry ice or at −70°; the sample will freeze slowly or not at all if left for hours. Besides adequately substituting for the sodium salt, this method also prevents the precipitation of deoxyribonucleoside triphosphates (dNTPs)[1a] and therefore is useful for removing unreacted triphosphates from the products of reverse transcriptase, DNA polymerase, or terminal transferase-catalyzed reactions. Typically, two serial precipitations result in the removal of about 99% of the dNTPs, and greater than 90% DNA recovery. However, if the DNA is to be phosphorylated or tailed, ammonium acetate should be avoided because ammonium ions inhibit the enzymes required for these processes.

Another alternative makes use of Li$^+$ as the cation. To the solution containing DNA or RNA, add 0.1 volume 8 M LiCl (filter sterilized) and 2 to 3 volumes of ethanol. Since LiCl is highly soluble in ethanol-containing solutions at −70°, salt is not coprecipitated with the nucleic acid.

Should the original DNA sample also contain sodium dodecyl sulfate (SDS), the detergent is most effectively removed by making the sample 0.2 M with sodium chloride before ethanol addition (usually by adding 0.04 volume 5 M NaCl followed by 2 volumes ethanol). SDS remains soluble under these conditions.

For small volumes of 1 ml or less containing at least 10 μg/ml DNA in the original sample, the 15-min precipitation at −70° gives at least 80% recovery. If larger volumes are to be handled, the precipitations can be carried out in 15–30 ml glass Corex tubes and the period at −70° increased to permit temperature equilibration (at least 30 min for a 30 ml sample). No matter what the volume being processed, overnight precipitation at −20° is effective even at very low DNA concentrations (10 ng/ml DNA in the original sample). The optimal centrifugation method for precipitate recovery is dictated by the volume of the sample and the quantity of the DNA present. For precipitates of at least 1 μg in 1 ml or less the bench-top microcentrifuge method is adequate. When nanogram quanti-

[1a] H. Okayama and P. Berg, *Mol. Cell. Biol.* **2,** 161 (1982).

ties are to be recovered from 1-ml volumes or less in the absence of carrier, more rigorous ultracentrifugation must be applied. One such technique is that of Shapiro,[2] by which nucleic acid present at 10–1000 ng/ml is recoverable at 70–100% efficiency. The precipitated DNA samples in completely filled microcentrifuge tubes (to prevent cracking) are placed in the buckets of a swing-out rotor (e.g., 0.4-ml tubes in an SW41 bucket; 1.4-ml tubes in an SW27 bucket) which are half-filled with ice-cold 20% ethanol (to prevent freezing). The samples, floating inside the ultracentrifuge buckets, are subjected to centrifugation of 41,000 rpm for 30 min (SW41), or 27,000 rpm for 60 min (SW27) at −2°. The pelleted DNA present at the bottom of the microcentrifuge tube is treated as already stated, except that the centrifugation after the 70% ethanol wash should be carried out in a swing-out rotor in a refrigerated machine (e.g., HB-4 rotor in a Sorvall RC-5B centrifuge at 11,500 rpm for 15 min at −10°). It should be noted that in most situations nanogram DNA quantities can be recovered using the basic protocol as long as a coprecipitant is also used, for example 50 μg/ml tRNA. However, in circumstances when carrier tRNA cannot be used, for instance when the DNA is to be phosphorylated using polynucleotide kinase, ultracentrifugation may be the only method of recovering DNA as a precipitate in ethanol. When dealing with picogram DNA amounts, the use of coprecipitants (tRNA or purified glycogen) is essential for effective DNA recovery.

If larger volumes of material are being handled, high-speed centrifugation (10,000 rpm for 30 min in an HB-4, SS-34, or GSA rotor in a Sorvall RC-5B centrifuge at −10 to 0°) is adequate for 10 μg/ml quantities of nucleic acid, whereas ultracentrifugation (40,000 rpm in an SW41 rotor at 0°) should be used for lower concentrations.

The methods mentioned are suitable for DNA larger than 200 base pairs; for improved recovery of smaller DNAs the sample should be made 10 mM in magnesium chloride before ethanol precipitation.

DNA is best stored at 4° in TE (pH 8) buffer. The EDTA chelates heavy metal ions which are commonly required for DNase activity, while the use of pH 8 minimizes deamidation. For very long term storage (5 years or more) DNA can either be frozen at −70° and not subjected to any freeze–thaw cycles, or it can be dissolved in buoyant CsCl at 4°.

RNA Precipitation with Ethanol

The method for precipitation of RNA is virtually identical to that for DNA. For a 10-ml aqueous RNA-containing sample add 1 ml (0.1 volume)

[2] D. J. Shapiro, *Anal. Biochem.* **110,** 229 (1981).

3 M sodium, potassium, or ammonium acetate (pH 5.5) or 0.1 volume 8 M LiCl followed by 25 ml (2.5 volumes) ethanol. Precipitate and treat as described for DNA precipitation by ethanol. Resuspend the dried RNA pellet in a suitable volume of sterile water or appropriate buffer, and store at −70°, or in liquid nitrogen vapor.

Notes. For milliliter volumes, precipitation is normally carried out overnight at −20° followed by centrifugation in 30-ml glass Corex tubes at 10,000 rpm for 30 min (HB-4 or SS-34 rotor in a Sorvall RC-5B centrifuge at −10 to 0°).

The choice of salt used is dictated by the intended use of the RNA. For instance, if oligo(dT)-cellulose chromatography in the presence of SDS is to be performed, the sodium salt is chosen. If cell-free translation of mRNA is to be carried out, potassium acetate is used. Ammonium acetate can substitute for the potassium salt in this situation. Note that chloride should not be used when translation in cell-free systems is intended, since Cl^- interferes with initiation of protein synthesis. When the RNA to be precipitated also contains SDS and efficient cell-free translation is intended, at least two serial precipitations are required: the first with Na^+ to precipitate the RNA, leaving the SDS in solution, and the second with K^+ to replace the sodium ion in readiness for translation. Li^+ should be avoided when the precipitated RNA is to be reverse transcribed.

The duration and temperature of precipitation, the centrifugation methods of precipitate recovery, and the concentration limits are essentially the same as those stated for DNA ethanol precipitations. However, because the original RNA level is not always known, for example when isolating RNA from a biological source, overnight precipitation at −20° is recommended.

Storage of RNA in aqueous solution at −70° is only recommended if the sample is not going to be subjected to repeated freezing and thawing. Storage of the sample in aliquots at suitable volumes can partially circumvent this problem. Also, the addition of a drop of 0.2 M vanadyl–ribonucleoside complexes (VRC) to the RNA prior to freezing will prevent RNase-catalyzed RNA degradation. VRC does not interfere with the subsequent use of the RNA unless cell-free translation is intended, in which case the VRC can be removed by phenol extraction (translation of mRNA by microinjection into *Xenopus* oocytes is unaffected by a moderate concentration, e.g., 5 mM of VRC).[3] Poly(A)+ mRNA can be stored for years by this method. Alternatively, RNA can be stored securely as an ethanol precipitate at −20°.

[3] R. S. Puskas, N. R. Manley, D. M. Wallace, and S. L. Berger, *Biochemistry* **21**, 4602 (1982).

Note that all buffers, glassware, and plasticware must be rendered free of RNase before working with RNA (see this volume [2]).

Isopropanol Precipitation of DNA

Isopropanol-induced precipitation of DNA minimizes the total volume of the precipitating sample. The precipitation is carried out by adding an equal volume of isopropanol (2-propanol) to the sample, and treating the mixture in the same way as for ethanol precipitations. The drawbacks of this method are that 2-propanol cannot be easily lyophilized due to its relatively low volatility, and also that salts present in the original sample tend to coprecipitate with the DNA. Because of these problems it is common practice to carry out an ethanol precipitation of the sample immediately following a 2-propanol precipitation.

Spermine Precipitation of DNA

The precipitation of DNA by the polyvalent cation spermine[4] is useful for the recovery of DNA from dilute solutions (0.1–100 μg/ml). Due to the selective nature of the precipitation, the procedure yields DNA of relatively high purity. Indeed, DNA can be precipitated from solutions containing proteins; for example, restriction enzyme digests containing bovine serum albumin (BSA) retain 4–5% of the initial BSA in the DNA pellet, and bacterial cleared lysates retain 6–8% of the protein as a contaminant in the recovered DNA. Nucleoside triphosphates are not precipitated by spermine, thus providing a method of recovering dNTP-free DNA; two sequential precipitations are recommended.

For spermine precipitation of a 50-μl sample containing moderate salt typical of genetic engineering reactions (e.g., 0.1 M KCl, 10 mM MgCl$_2$, 1 mM dithiothreitol, 10 mM Tris–HCl at pH 8):

1. To the sample in a 1.4-ml microcentrifuge tube, add 5 μl 0.1 M spermine tetrahydrochloride (0.1 volume, resulting in 10 mM final spermine concentration).

2. Mix well by vortexing and place in ice for 15 min to form the precipitate.

3. Centrifuge the precipitate in a bench-top microcentrifuge for 10 min. For nanogram quantities of DNA, centrifugation at 10,000 rpm for 15 min in a high-speed instrument is required.

4. Remove the supernatant using a micropipettor, taking care not to disturb the pellet.

---

[4] B. C. Hoopes and W. R. McClure, *Nucleic Acids Res.* **9**, 5493 (1981).

5. At this stage a 70% ethanol wash ($-20°$) can be performed to remove any trapped solute (see ethanol precipitation for method).

For removal of the spermine add 0.2 ml cation-exchange buffer in ethanol (1 part 0.3 M sodium acetate or potassium acetate, 10 mM magnesium acetate to 3 parts ethanol; v/v) and vortex to disperse the pellet thoroughly. After leaving in ice for 1 hr with periodical mixing, the sample is subjected to centrifugation as before and the supernatant removed. The pelleted DNA is then treated in the same manner as that stated for ethanol precipitations.

Notes. Because the efficiency of spermine precipitation is sensitive to the ionic strength of the sample, the amount of spermine to be added is determined by the salt content of the sample. At 0.1 mM spermine, DNA will precipitate in the absence of salt. However, if the DNA solution also contains 10 mM EDTA, 6 mM MgCl$_2$, and 10 mM KCl, for example, the final spermine concentration must be increased 10-fold. Precipitation of DNA dissolved in 0.1 M KCl or in 0.1 M KCl, 10 mM MgCl$_2$ requires a final spermine concentration of 2 or 10 mM, respectively.

An alternative method of spermine removal for large amounts of DNA is to dissolve the pellet in a small volume of buffer containing 0.5 M salt and to dialyze against the same buffer (two changes) for 24 hr. The removal of salt from the concentrated DNA sample can then be achieved by conventional ethanol precipitation, or by dialysis against the desired buffer (e.g., TE, pH 8).

Besides precipitating DNA from dilute solutions and from complex mixtures, another application is the differential precipitation of relatively high-molecular-weight DNAs. Under moderate salt conditions the spermine method preferentially precipitates DNA of 0.2 kilobase (kb) and larger; material shorter than 60 base pairs (bp) does not precipitate. A direct application of this phenomenon pointed out by the originators[4] is the purification of linker-bearing high-molecular-weight DNA as a precipitate, by selective removal of the oligodeoxyribonucleotide linker debris generated after restriction endonuclease digestion.

Butanol Concentration of DNA

DNA can be recovered from dilute solutions by extracting several times with butanol to concentrate the sample prior to ethanol precipitation. An equal volume of 2-butanol is mixed vigorously with the sample followed by phase separation on a bench-top microcentrifuge for 1 min (for microliter volumes). The upper organic layer is carefully removed with the aid of a micropipettor. The volume of the aqueous, DNA-containing lower phase will be reduced as water partitions into the butanol phase, thus increasing the DNA concentration. Butanol extraction is carried out until the desired final volume of sample is attained. Because this procedure also concentrates the salts present in the sample, a final ethanol precipitation is carried out and the pelleted DNA resuspended in the desired buffer.

Precipitation of RNA with High Salt Concentrations

Large RNAs (rRNA, hnRNA, mRNA) are insoluble in solutions containing high salt concentrations whereas small RNAs and DNA in general remain soluble.

Differential RNA Precipitation with Lithium Chloride. After preliminary extraction of RNA from a biological source by phenol extraction or a guanidine-based technique the large RNA can be separated from contaminant low-molecular-weight DNA and RNA (tRNA and 5 S rRNA) by subjecting the aqueous sample to LiCl.

1. To the sample add 1 volume of 8 M LiCl.
2. Mix vigorously and incubate at $-20°$ for at least 2 hr.
3. Sediment the precipitated RNA at 10,000 rpm for 10 min at 0° (e.g., GSA or SS-34 rotor in a Sorvall RC-5B at 0°). Remove the supernatant fluid; it contains small RNAs.
4. Resuspend the pellet in water at about 1 mg/ml and repeat the procedure. Finally, resuspend the purified RNA in water and subject it to two conventional ethanol precipitations. (To recover the small RNAs, add 2–3 volumes of ethanol to the LiCl supernatant fluid and process according to conventional ethanol precipitation procedures.)

Should high-molecular-weight DNA contaminate an RNA sample, treat with RNase-free DNase (25 μg/ml DNase in 10 mM MgCl$_2$, 50 mM Tris–HCl at pH 7, at 37° for 30 min) followed by phenol extraction and ethanol precipitation. A simple way for doing this is to add VRC, the RNase inhibitor, to the DNase. The VRC at 10 mM protects the RNA from RNase without affecting the activity of the DNase.[3]

[6] Quantifying ^{32}P-Labeled and Unlabeled Nucleic Acids

By Shelby L. Berger

Recombinant DNA technology depends on detection methods for nucleic acids compatible with amounts ranging from picograms to grams and from tenths of a microliter to liters. In practical terms there are three basic techniques: (1) absorbance methods suitable for a minimum concentration of micrograms per milliliter, (2) fluorescence methods capable of detecting nanograms of DNA and micrograms of RNA, and (3) methods based on the detection of [32]P. Because of the overwhelming importance in molecular biology of the third group, this chapter will stress exquisitely sensitive methods for measuring radioactivity in very small volumes.[1,2] An illustration in which an enzyme-catalyzed reaction performed in 20 μl is monitored by consuming less than 2% of the total volume will be presented.

Absorbance Methods

Both DNA and RNA absorb in the ultraviolet between 250 and 270 nm owing to the spectral characteristics of the four canonical bases. Near neutrality, the molar extinction coefficients of the bases and the wavelength at which maximum absorption occurs are as follows for a 1-cm path length: 1.54×10^4 (259 nm) adenine, 9.1×10^3 (271 nm) cytosine, 1.37×10^4 (253 nm) guanine, 1.0×10^4 (262 nm) uracil, and 7.4×10^3 (260 nm) thymine. (Molar extinction coefficients of the four bases at 260 nm can be found in this volume [47].) These values are useful for determining the concentrations of solutions of nucleoside triphosphates. When the absorption spectrum of either DNA or RNA is compared with the sum of the absorption spectra of the constituent purines and pyrimidines, however, there is a discrepancy: in typical DNA preparations, the intensity of absorption may be 40% less than the intensity of a mixture of the corresponding nucleotides at the same wavelength. This "hypochromic effect" is caused by the interaction between absorbing units when placed in an orderly array.[3] Thus, at 260 nm, an absorbance of 1 measured in a cuvette

---

[1] S. L. Berger, *Anal. Biochem.* **136,** 515 (1984).
[2] S. L. Berger and M. S. Krug, *BioTechniques* **3,** 38 (1985).
[3] I. Tinoco, Jr., *J. Am. Chem. Soc.* **82,** 4785 (1960).

METHODS IN ENZYMOLOGY, VOL. 152

with a 1-cm path length is indicative of double-stranded DNA at a concentration of approximately 50 μg/ml or of RNA or single-stranded DNA at approximately 40 μg/ml. The absorbance of oligonucleotides is usually the sum of the absorbance of the constituent mononucleotides (see this volume [47]). A mixture of random oligomers at 20 μg/ml has an absorbance at 260 nm of 1.

The ratio of the absorbance at 260 to 280 nm is a useful indication of purity. Values for DNA solutions of 1.8 to 1.9 and for RNA solutions of 1.9 to 2.0 are acceptable. The presence of protein, which absorbs at 280 nm, decreases the ratio as does phenol, another likely contaminant.

Fluorescence Methods

DNA and RNA are not themselves fluorescent. When ethidium bromide binds to nucleic acids in solution, the dye, upon excitation in the ultraviolet, fluoresces intensely in the visible range. The orange fluorescence can be used for quantifying both double-stranded and single-stranded molecules, but with markedly different sensitivity. Usually, this goal is achieved by subjecting the sample DNA, together with a series of marker DNAs at different concentrations, to electrophoresis in a gel. The gel is stained with ethidium bromide at 1 μg/ml for 30 min and viewed with a UV transilluminator. (See this volume [8] for a more detailed discussion.) Since bands of equal intensity contain the same mass of material regardless of the size of the DNA, the amount of sample DNA can be estimated by visual comparison with the known standards. Although the eye is adept at matching near-equal intensities, it cannot estimate intensity differences well nor compare a diffuse signal with a sharp one. Therefore, the most precise comparisons are between markers and samples that have size and intensity in common. This approach is capable of detecting ~5 ng of double-stranded DNA and ~0.5 μg of RNA depending on the width and thickness of the band. (See this volume [8] for a more detailed discussion.) Quantitative evaluation of RNA requires marker RNAs or single-stranded DNAs at known concentrations. Methods for staining denaturing RNA gels can be found in this volume [8].

Solution methods for measuring fluorescence have not enjoyed wide popularity. Although potentially quantitative, the signal obtained from ethidium bromide in a solution of DNA depends on the degree of supercoiling of the DNA and the degree of quenching caused by inadvertant contamination. Because neither of these can be assessed easily, most investigators continue to rely on the semiquantitative methods described above. (See this volume [55] for a description of the technique.)

^{32}P-Based Methods

Many of the reactions used to clone and characterize nucleic acids must be performed in small volumes in order to obtain adequate concentrations of enzymes and substrates. Frequently, these reactions involve the incorporation of ^{32}P-labeled substrates into ^{32}P-labeled DNA or RNA. Such reactions can be fully monitored by withdrawing submicroliter quantities for analysis; the majority is retained for further manipulation.

The technique which follows makes use of a filter assay to separate ^{32}P-labeled precursors, such as nucleoside triphosphates or pCp, from labeled polynucleotides. It is based on the observation that precipitation with trichloroacetic acid (TCA) need not be performed in solution. It can be performed equally well by placing a minute drop ($\sim$0.1 to 0.2 μl) of material consisting of a mixture of labeled substrates and products on a glass fiber filter, drying the filter, and washing it with TCA. The acid-soluble substrates are eluted leaving the labeled macromolecules on the filter for subsequent assay in a scintillation counter. (Filtration equipment is described in this volume [1].)

For oligonucleotides too small to be acid precipitable, an analogous technique can be used by substituting DE-81 filters for the glass fiber variety and 0.5 M phosphate at pH 7 for TCA. Oligonucleotides adhere to the filter while nucleotides are removed by washing with the buffer.

The ability to evaluate the progress of a reaction by withdrawing submicroliter volumes depends on methods for assaying radioactivity on dry filters. Cerenkov radiation, which occurs within the glass or plastic wall of a scintillation vial when a dry filter—no liquid whatsoever—is placed flat on the bottom of the vial, is a measure of radioactivity. By determining Cerenkov radiation with the ^{3}H channel of a scintillation counter, one can assay ^{32}P with approximately 25% counting efficiency. The existence of Cerenkov radiation, then, makes possible measurements of radioactivity on filters in a nondestructive manner before they are processed; the counting process has no effect on subsequent procedures performed with the same filter.

It should be clear that the amount of radioactivity spotted on a filter and dried is proportional to the volume applied to that filter. The proportion is valid at any time during a reaction, since it is independent of whether substrates have been converted to products. In contrast, the radioactivity on filters washed with TCA or phosphate buffer reflects only the products of the reaction. With two or more filters, one obtained at zero time and the others after incubation for predetermined intervals, the reaction can be characterized. Given the amount of radioactive substrate initially included in the incubation mixture, the total volume in which the reaction is being conducted, and the efficiency of counting dry filters, one can calculate the precise volume spotted from the radioactivity on each unwashed filter. Given the radioactivity remaining after processing the identical filters, the amount of product per unit volume can be ascertained. The ratio of radioactivity on a washed filter to that of the same filter before it was washed is a measure of the percentage incorporation of precursor into product. Since the precise volume spotted is determined from radioactivity measurements and not from the designated volume of a micropipet, the smallest possible aliquot should be withdrawn, knowing that subsequent withdrawals will result in different but nevertheless precisely measurable volumes.

Cerenkov Radiation

Although a detailed description of Cerenkov radiation is beyond the scope of this chapter,[2,4] it may be useful to summarize the salient features. Cerenkov radiation occurs when a charged particle in a dielectric medium travels faster than the phase velocity of light in that medium. (It is understood that the velocity of the particle cannot exceed the velocity of light *in vacuo*.) Then, a series of pulses of light are emitted that are analogous to the V-shaped pattern of sonic booms produced by aircraft traveling at supersonic speeds. These optical "booms" occur in the visible or near-visible regions of the spectrum with a fixed geometry relative to the moving particle. Like the sonic boom, there is a threshold velocity of the particle below which Cerenkov radiation cannot take place. Thus, energetic ^{32}P particles give rise to Cerenkov radiation in the wall of a scintillation vial or in solution, each at a different efficiency. Both modes can be used to advantage.

Other Methods for Measuring Radioactivity

Recently an inexpensive method for counting ^{32}P dried on filters has become available. The device, Model QC 2000 manufactured by Bioscan (Washington, DC), is a calibrated Geiger tube with a sample holder that maintains a fixed geometry between the filter and the window of the tube. The apparatus responds to β particles entering the window and measures ^{32}P with a counting efficiency of ~30%. It is about 10% the price of a scintillation counter and takes up about a square foot of bench space.

[4] J. V. Jelley, "Cerenkov Radiation and its Applications." Pergamon, Oxford, 1958.

Note that the instrument measures charged particles and *not* Cerenkov radiation.

Procedure for Monitoring Reactions

The analysis of reactions with submicroliter quantities requires that three conditions be met. One must be able (1) to determine the amount of radioactivity on dry filters; (2) to process the dried samples, that is, wash with TCA or phosphate to remove unincorporated labeled substrates; and (3) to count the identical filters again. The procedure is as follows.

1. Assemble the components of the reaction in an Eppendorf tube placed in an ice bucket.
2. With a Pipetman (Rainin) set at 0.2 μl, withdraw the smallest possible aliquot and spot it on the appropriate filter. (Use GF/C glass fiber filters for macromolecules or DE-81 filters for oligonucleotides.)
3. Incubate at the prescribed temperature. As the reaction proceeds, or at the anticipated time of completion, withdraw and spot additional aliquots on separate filters.
4. Dry all filters under a heat lamp for 2 min. Measure the radioactivity on the dry filters using the ^{3}H channel of a scintillation counter or the Bioscan counter.
5. Wash the filters with 10 ml of solution. (Use 5% TCA containing 20 mM sodium pyrophosphate for glass fiber filters or 0.5 M sodium phosphate at pH 7 for DE-81 filters.) Wash GF/C filters with 1 ml ethanol to aid in drying.
6. Dry the filters under heat lamps for 5 min and recount them using either Cerenkov radiation (^{3}H channel), the Bioscan counter, or scintillation fluid (^{32}P channel). The last has a counting efficiency of almost 100% and should be used for samples with low count rates.
7. Calculate the results.

An example taken from a nick translation reaction performed in a total volume of 20 μl is shown below. To exhibit the versatility of the method, the calculation will be carried out without referring to the amount of radioactive substrate initially added. Assume it is unknown.

The data are as follows. The zero time filter, unwashed and washed, contained 596,000 and 6,670 cpm, respectively; the 60-min filter, unwashed and washed, contained 490,000 and 330,000 cpm, respectively. All measurements were performed with dry filters at a counting efficiency of 25%. By inspection of the filter obtained after 60 min of incubation, it can be seen that 68% of the precursor was incorporated into DNA (the background at zero time was negligible).

Since the total radioactivity in the mixture is unknown, the radioactivity measurements obtained from the unwashed filters cannot immediately be converted to volumes. In order to determine the amount of radioactive substrate included in the reaction, the Cerenkov radiation that arises in solution will be utilized. Toward this end, after completing the nick translation, the original 20-μl volume was diluted with 130 μl of buffer to make a final volume of 150 μl. (The amount withdrawn during the incubation was trivial and was ignored.)

8. To each of three 1.5-ml Eppendorf tubes, add 1 μl of the complete diluted reaction mix. Centrifuge briefly to drive the drop to the bottom.

9. Suspend each Eppendorf tube on the rim of a 20-ml scintillation vial and count in the ^{3}H channel. Since the Cerenkov radiation arises in water with improved geometry, rather than in glass, the counting efficiency is now 56%.[5] Note that it is necessary to dilute the reaction so as not to exceed the linear range of a scintillation counter.

10. Recover the three aliquots and return them to the original tube. They are not consumed.

From the average radioactivity of the liquid samples, namely, 1.07×10^6 cpm/μl at 56% efficiency[5] (which is 4.75×10^5 cpm/μl at 25% efficiency[5]), the total radioactivity in the nick translation reaction can be determined. Thus, 4.75×10^5 cpm/μl $\times$ 150 μl or 7.12×10^7 cpm (25% efficiency) was initially added. This value is equivalent to 130 μCi of substrate. It follows that the volumes applied to filters were 0.167 μl (5.96×10^5 cpm $\times$ 20 μl/7.12×10^7 cpm) and 0.138 μl (4.90×10^5 cpm $\times$ 20 μl/7.12×10^7 cpm), respectively, at 0 and 60 min of incubation. The reaction generated ^{32}P-labeled DNA containing 1.90×10^8 dpm, correcting for the zero time value. The complete analysis of the reaction consumed 0.305 μl or 1.5% of the total sample. Clearly, small reaction volumes should *never* deter the investigator from analyzing any reaction in which ^{32}P-labeled substrates can be separated from ^{32}P-labeled products by a filter assay.

---

[5] The efficiency of measuring Cerenkov radiation on dry filters or in solution must be determined for each instrument.

[7] Autoradiograms: ^{35}S and ^{32}P

By WILLIAM M. BONNER

Autoradiography in the strict sense refers to the exposure of film by radioactive particles, whereas fluorography refers to the exposure of film by light generated from the interaction of radioactive particles with added compounds called fluors. Whether fluors are mixed with or are external to the sample depends on the energy and range of the emissions.

Data and comments concerning the relative and absolute sensitivities of the different methods to be described are presented in Table I. The absolute sensitivities are based on the minimum film darkening (0.02 A_{540}) detectable in 24 hr.[1-3] Note that it may take 10–100 times these amounts to get a good exposure overnight. The procedures given below for ^{35}S also apply to ^{14}C and ^{32}P. For ^{3}H, see Bonner and Laskey.[1-3]

^{35}S Autoradiography

Because of the short range of ^{35}S β rays (0.25 mm in water or plastic), careful consideration of several factors is necessary for efficient autoradiography. *Gels* should be prepared as thin as possible and should be fixed or soaked to remove nonvolatile solutes such as urea so that the gel can be dried as thin as possible. Soaking DNA sequencing gels (0.35 mm thick acrylamide) in 5% methanol–5% acetic acid for 15 min before drying increases the autoradiographic efficiency 3-fold.[4] After removing any plastic cover used during drying (see below), the gel is pressed firmly against a piece of Kodak X-Omat AR film and exposed at room temperature. Wet gels, covered with thin plastic wrap, can also be exposed this way but with a very low efficiency. *Blots* are dried in air or vacuum and the surface containing the radioactivity is placed against to the film.

^{35}S Fluorography

Since fluors must be impregnated into the gel matrix for ^{35}S fluorography, the main consideration besides cost is whether the procedures will lead to a significant loss of resolution or material due to diffusion. Impreg-

---

[1] W. M. Bonner and R. A. Laskey, *Eur. J. Biochem.* **46,** 83 (1974).
[2] R. A. Laskey, this series, Vol. 65, p. 363.
[3] W. M. Bonner, this series, Vol. 104, p. 460.
[4] D. L. Ornstein and M. A. Kashdan, *BioTechniques* **3,** 476 (1986).

TABLE I
SENSITIVITIES OF DIFFERENT METHODS OF ^{35}S AND ^{32}P FILM DETECTION[a]

Technique and sample	Temperature (°C)	Sensitivity (dpm/mm²)	Comments
^{35}S Autoradiography			
Gels	+20	20–60	Depends on gel thickness, solute concentration. Wet gels give much lower sensitivity
Blots	+20	20	Radioactivity on the surface
^{35}S Fluorography			
Gels/blots	−70	4	Small molecules could be lost during impregnation
^{32}P Autoradiography			
Gels/blots	+20	5–10	Wet gels give less sensitivity and resolution. Two sheets of lead enclosing film–sample double sensitivity
^{32}P Fluorography			
Gels/blots	−70	1	Screen–film–sample
Gels/blots	−70	0.5	Screen–flashed film–sample–lead (or screen)
Gels/blots	+20	1	Screen–flashed film–sample–lead (or screen). Wet gels give less sensitivity and resolution

[a] These sensitivities are based on the minimum darkening (0.02 A_{540}) detectable in 24 hr. A good exposure overnight may require 10–100 times these amounts. Film is Kodak X-Omat AR film developed by machine. Screen is Cronex Lightning Plus. Lead is a sheet of lead foil. Dry gels were used to compare sensitivities.

nation procedures depend on the type of gel as well as the type of fluorography solution.

Acrylamide gels may be stained, fixed with 10% acetic acid/0–30% alcohol, or prepared directly for fluorography. They may be treated with the original 2,5-diphenyloxazole–dimethyl sulfoxide (PPO–DMSO) solution,[1–3] commercial water-insoluble products, or the more recently developed water-soluble products which process the gel in 15–30 min in one step. Some examples of the latter group are Amplify (Amersham), ENLIGHTNING (New England Nuclear), Fluoro-Hance (Research Products International Corp.), and Autofluor (National Diagnostics). All these methods yield roughly comparable results with fixed gels, but ease of use and cost vary widely. Gels are dried as described below.

In one protocol, DNA sequencing gels are fixed in 5% methanol–5% acetic acid for 15 min, soaked for 30 min in EN³HANCE autoradiography enhancer (New England Nuclear), and washed in cold water for 15 min to precipitate the fluor *in situ* before drying. Enhancement is 4-fold for large oligonucleotides, but minimal for small ones, perhaps due to losses during soaking.

Agarose gels are soaked first in absolute ethanol for at least 30 min to remove water and solutes, then in a 3% (w/v) solution of PPO in ethanol for 3 hr.[5] The gels are then soaked 1 hr in water to precipitate the 2,5-diphenyloxazole in the gel before drying. The commercial water-soluble reagents listed above can also be used for agarose gels.

Blots can be treated by several methods. (1) The dried filters are dipped in a 20% (w/v) solution of PPO in toluene and redried.[6] (2) The dried filters are dipped in slightly warmed (40°) 0.4% PPO in 2-methylnaphthalene, drained, and cooled. The 2-methylnaphthalene solidifies at 34°.[7] (3) The dried filters are sprayed with the commercial spray reagent EN³HANCE and redried.

After preparation and drying (see below), the fluor-impregnated sample is exposed to the flashed face of a piece of Kodak X-Omat AR film and exposed at −70°.

^{32}P Autoradiography and Fluorography

The β ray of ^{32}P has a range of about 6 mm in water or plastic, enabling external fluors in the form of intensifying screens to be used. The major consideration in choosing a ^{32}P detection procedure is whether the sample can be frozen.

Freezable samples include dried acrylamide or agarose gels, and wet or dried blotting membranes. Acrylamide or agarose gels can be dried immediately after electrophoresis (see below). The sample, which can be wrapped in a plastic folder, is placed next to a piece of flashed Kodak X-Omat AR film, with a $CaWO_4$ intensifying screen (Cronex Lightning Plus, DuPont) on the other side of the film (flashed face of film toward screen).[8,9] A second $CaWO_4$ screen or a sheet of lead foil about 0.15 mm thick can be placed behind the sample for approximately 50% increase in speed. This second screen or lead sheet acts by increasing the backscatter of ^{32}P β rays from the sample rather than by any fluorographic process. Expose at −70°.

Thus the most sensitive arrangement is screen–flashed film–sample–

[5] R. A. Laskey, A. D. Mills, and N. R. Morris, *Cell (Cambridge, Mass.)* **10,** 237 (1977).
[6] E. M. Southern, *J. Mol. Biol.* **98,** 503 (1975).
[7] W. M. Bonner and J. D. Stedman, *Anal. Biochem.* **89,** 247 (1978).
[8] R. A. Laskey and A. D. Mills, *FEBS Lett.* **82,** 314 (1977).
[9] R. Swanstrom and P. R. Shank, *Anal. Biochem.* **86,** 184 (1978).

screen (or lead) (Table I). The flashed film and second screen or sheet of lead behind the sample each result in a 50% increase in sensitivity for the fainter bands, even during hour long or overnight exposures. Increases in sensitivity may be considerably larger on weeklong exposures, but flashed film may become too fogged on long exposures with some screens.[2] Another arrangement sometimes seen, screen–film–screen–sample, has lower resolution and about half the sensitivity as the arrangement described above.

Nonfreezable samples such as wet gels are exposed as described for dried gels except at temperatures above 4°.[8] Flashed film and screens are more important for adequate sensitivity above 4°. However, resolution and sensitivity will be less because the sample is thicker. If the wet gel contains sufficient radioactivity to be detected by autoradiography alone (no screens or flashed film), then a convenient procedure is to seal the gel in a plastic bag and lay it on an envelope of "ready pack" film (individually wrapped and sealed). No darkroom is needed until the film is developed.

Drying Gels

To minimize tearing and creasing of the gel during drying, lay it on a thin flexible plastic sheet such as Mylar, let it relax, then lay the paper backing (Whatman 1 or equivalent) on top of the gel, and roll out any air bubbles. Alternatively, ^{32}P-labeled DNA sequencing gels can be left on one glass plate and a sheet of backing paper pressed onto the gel. The backing paper is peeled off bringing the gel with it. The opposite face of the gel is covered with a thin plastic wrap. Then in either case, several sheets of heavier paper (Whatman 3 MM or equivalent) are added next to the backing paper as a support pad and the sandwich is put into the gel dryer (paper pad closest to vacuum source). Acrylamide gels are dried under vacuum at 60–80° (about 1 hr for a 0.8-mm-thick gel). Agarose gels are dried under vacuum with gentle or no heat to prevent melting. Gels containing fluors should not be dried or heated for excessively long times because these compounds are somewhat volatile.

The main problem usually encountered during drying is fracturing of acrylamide gels. This can be avoided if the gels are prepared so that the product of the final concentrations of acrylamide and bisacrylamide (expressed as percentages) equals 1.3[10]; this ensures adequate gel elasticity. In addition, the vacuum should not be released before the gel is com-

---

[10] D. P. Blattner, F. Garner, K. Van Slyke, and A. Bradley, *J. Chromatogr.* **64,** 147 (1972).

pletely dried. Acrylamide gels may be dried without a gel dryer if they are bonded during polymerization to one of the glass plates[11] or to a commercial product such as Gel Bond film (FMC Corporation, Rockland, ME).

For best results, commercial gel dryers should be connected to a corrosion-resistant mechanical vacuum pump with an acid–base vapor trap and a cold trap rather than to a house or aspirator vacuum. Commercial lyophilizers will work but may be damaged in such service.

Cassettes, Screens, Films, and Indexing Marks

Intimate juxtaposition of the sample, film, and screens is important for the best sensitivity and resolution. For ^{35}S autoradiography in particular, the sample should be dried as thin as possible, any plastic covering material used in drying the gel should be removed, and the dried gel must be in direct uniform contact with the film emulsion. For ^{32}P, the wet or dry gels or filters can be wrapped in plastic, taking care to remove any trapped air pockets that might interfere with good uniform contact between the film, sample, and screens. Sometimes, old screens may become wavy, leaving areas of fuzzy bands due to poor contact.

For indexing, a ^{14}C-labeled ink (1 μCi/ml) can be used on top of the plastic (next to the film surface) or a ^{32}P-labeled ink can be prepared for use inside the plastic (use Table I as a guide to the amounts of radioactivity to use). A convenient alternative to radioactive ink is a nonradioactive phosphorescent marker (UltEmit, New England Nuclear). The markings, when activated by ambient light, glow for approximately 2 hr, thus transferring their images to the film.

A commonly used and widely available film–screen combination is Kodak X-Omat AR film (blue-sensitive screen type) and Cronex Lightning Plus screen (CaWO$_4$, blue emitting). Only one screen need be used per cassette, as a thin sheet of lead (about 0.15 mm or 0.006 in. thick) can substitute for the screen behind the sample.

Other film–screen pairs can be used and may be just as sensitive as the above, but for maximum efficiency the film sensitivity must match the screen emission spectrum. One example of a green-emitting screen is Lanex (gadolinium lanthanium oxysulfide; Eastman Kodak Co.) which is paired with a green-sensitive film, Kodak Ortho H. When choosing a screen, check with the manufacturer for the most sensitive compatible film. References 8 and 9 compare several screen-film combinations.

[11] H. Garoff and W. Ansorge, *Anal. Biochem.* **115**, 450 (1981).

Film Development

Film can be stored at room temperature or at 4°, but it should be protected from radiation. Kodak X-Omat AR film can be obtained in individually wrapped "ready packs," which have the advantage that other sheets of film are not inadvertently exposed while removing one sheet. Most X-ray departments have automatic processing equipment for X-Omat films, and it may be possible to make an arrangement with one nearby for developing films. The film can be removed immediately from a cassette at −70°, but cold cassettes will quickly collect moisture from the air so it may be necessary to let them warm up before inserting a second film.

Manual processing can be carried out with four trays in a dark room with a red safelight for blue-sensitive films like Kodak X-Omat AR, or a dark red one for green-sensitive films. For Kodak X-Omat film at 20°, the procedure is to develop for 5 min, rinse for 30 sec, fix for 2–4 min, wash for 5 min in clear running water, and dry in a dust-free area. The chemicals are inexpensive and easily obtained. If a small area can be set aside for a darkroom, then a tank with three chambers can be set up with a water line temperature mixing valve and rust filter for a permanent manual developing system. A film dryer can be set up either in or just outside the darkroom.

Preflashed Film

The sensitivity of film to light is nonlinear. Flashing the film sensitizes it to low intensities of light, making its response more linear.[12]

A film flasher can be made from an inexpensive battery-operated camera flash unit. The unit needs to have a flash duration less than 1 msec, a button so that it can be triggered without a camera, and a flash face that can easily be covered with filters. The light output is greatly decreased by covering the flash face with an orange Kodak Wratten 21 or 22 filter and layers of diffusing neutral density filters such as heavy white paper or partially exposed film, and by blocking off some of the flash face with opaque black paper.

Flashing is done in total darkness. The film is laid on the floor on a yellow backing sheet and the flash unit held approximately eye level. One flash should give a fogging density in the developed film of approximately $0.15\ A_{540}$ (measured in a spectrophotometer by cutting pieces of film to fit a cuvette holder). Fine adjustments can then be made by changing the

---

[12] R. A. Laskey and A. D. Mills, *Eur. J. Biochem.* **56,** 335 (1975).

height of the flash unit or by adding or deleting filters. The flashed face of the film should face the fluor, in the sample for ^{35}S, in the screen for ^{32}P.

Quantitation of Autoradiographs

Film response to autoradiography (no fluors or screens) is linear up to absorbances of about 1 unit. With ^{35}S autoradiography, it is important that the gel matrix be uniform, because different regions of a gradient gel could quench ^{35}S to different extents. Even though the nonlinear film response in fluorography can be corrected by preflashing the film, linearity should be checked with standards prepared with the same isotope.

[8] Electrophoresis in Agarose and Acrylamide Gels

By RICHARD C. OGDEN and DEBORAH A. ADAMS

The scope of this chapter is to present a range of methods by which DNA and RNA molecules can be fractionated and analyzed by means of gel electrophoresis. This chapter will emphasize those techniques which can be simply and routinely applied in the course of molecular cloning and analysis and, wherever appropriate, reference will be made to more exhaustive practical or theoretical considerations of the techniques.

Gel electrophoresis through agarose or polyacrylamide is a very powerful method for rapidly resolving mixtures of nucleic acid molecules which has found wide application in recombinant DNA research. The resolution afforded far exceeds that generally obtained by other sizing techniques. The fractionated nucleic acids can be directly "viewed" *in situ* in the gel and can be readily recovered by a variety of methods tailored to subsequent steps in an experimental protocol. Because it is such an indispensible technique, a great deal of effort has gone into improving its efficacy for particular applications, with the result that many of the original methods have been simplified, scaled down, and improved. This review will cover the various choices to be made when confronted with experiments requiring gel electrophoresis in the order they would typically arise, starting with the type of gel system, choice of running buffer, and equipment, and continuing through gel running, visualization of the separated molecules, extraction of the material from the gel, and workup of preparative samples. Certain specific applications of electrophoresis, for example gels for sequencing purposes (this volume [56, 57])

METHODS IN ENZYMOLOGY, VOL. 152

and transfer of nucleic acids from gels to membranes (this volume [45, 61]), will be covered in separate chapters dealing with these specific techniques.

Gel Electrophoresis of DNA

Gel electrophoresis, as applied to the fractionation of DNA, is a very versatile technique and methods exist for the fractionation of single- or double-stranded molecules ranging in size from a few bases to chromosome-sized duplexes. With an emphasis on the practical application of the technique to molecular cloning, this section will cover the most common usages. DNA may be fractionated in agarose or polyacrylamide gels. Low percentage acrylamide gels are occasionally strengthened by agarose to yield a composite gel which resolves solely according to the acrylamide percentage.

At pH near neutrality, DNA is negatively charged and migrates from cathode to anode with a mobility dependent primarily on fragment size. Normally, smaller linear DNA fragments migrate faster than larger ones. Nondenaturing polyacrylamide gels can be used for separation of double-stranded DNA fragments between 6 bp (20% acrylamide) and 1000 bp (3% acrylamide). Nondenaturing agarose gels can be used for fragments between 70 bp (3% agarose) and 800,000 bp (0.1% agarose). Pulse-field electrophoresis for separation of larger DNA duplexes will be considered later. As little as 1 ng per band of nonradioactive double-stranded sample can be detected (in agarose gels) and both systems are readily amenable to cloning-scale (microgram amounts) preparative work.

Single-stranded DNA can be fractionated by agarose or polyacrylamide gel electrophoresis by inclusion of a denaturing reagent in the gel. There are several denaturing systems available but because of incompatibility between the gel material and denaturant, certain combinations are precluded. The choice of denaturing gel system depends not only on the size of fragments to be resolved but also on the experimental protocol involving the desired fragment after purification. Fractionation of single-stranded DNA can also be achieved by a reversible pretreatment of the sample with glyoxal prior to running the gel at neutral pH. The options available for particular cases will be discussed in the relevant sections below. A more detailed review of gel electrophoresis of DNA with an emphasis on practical approaches has been published.[1]

---

[1] P. G. Sealey and E. M. Southern, in "Gel Electrophoresis of Nucleic Acids—A Practical Approach" (D. Rickwood and B. D. Hames, eds.), p. 39. IRL Press, Washington, D.C., 1982.

Gel Electrophoresis of RNA

Most of the strengths of gel electrophoresis as a technique for fractionating DNA apply equally to RNA. As much of the RNA encountered in the course of molecular cloning is single-stranded, any of the applications of the technique to RNA involve the use of denaturing gels or two-dimensional gels in which the pH or denaturant concentration is varied between the two dimensions such that the structure of the molecules changes. A detailed review of two-dimensional gel electrophoresis techniques has been published.[2]

Agarose and polyacrylamide slab gels are most frequently used for RNA work and the types of gel selected are very similar to those for DNA of comparable size and complexity. It is important to stress that gel electrophoresis of single-stranded RNA, in particular mRNA, should be carried out with equipment and reagents that have been purged of nucleases. For this reason, many laboratories consider it prudent to keep certain electrophoresis equipment for RNA use only.

Choice of Gel System

For the most part, the size range of the sample will serve to determine the choice of gel for both analytical and preparative work. As a generalization, agarose is used for larger molecules, and polyacrylamide for shorter. Low percentage polyacrylamide gels, strengthened with agarose,[3] are used to extend the useful range of polyacrylamide gels.

There is an intermediate size range (70–1000 bp for double-stranded DNA) for which the choice of polyacrylamide or agarose is available. For analytical work, the prime considerations will be the rapidity and ease of the technique, and for this reason the horizontal agarose minigel is widely preferred. For preparative work, polyacrylamide offers many advantages, principally ease of elution and lack of coeluted inhibitors of subsequent enzymatic reactions, but the introduction of inhibitor-free low-melting temperature agarose (see later) has circumvented many earlier problems associated with preparative agarose gel electrophoresis.

The choice of denaturing gel system is determined principally by the aims of the experiment. Denaturants such as formamide, formaldehyde, and methylmercuric hydroxide are toxic and common sense dictates that

[2] R. DeWachter and W. Fiers, in "Gel Electrophoresis of Nucleic Acids—A Practical Approach" (D. Rickwood and B. D. Hames, eds.), p. 77. IRL Press, Washington, D.C., 1982.
[3] A. C. Peacock and C. W. Dingman, *Biochemistry* **7,** 668 (1968).

their use is restricted to essential applications. For many purposes, other less hazardous denaturants such as alkali or urea, or pretreatment with glyoxal can be used without compromising the purpose of the experiment.

Polyacrylamide Gel Electrophoresis of DNA and RNA

Polyacrylamide gels result from the polymerization of acrylamide monomers into linear chains and the linking of these chains with N,N'-methylenebisacrylamide (bis). The concentration of acrylamide and the ratio of acrylamide to bis determine the pore size of the resultant three-dimensional network and hence its sieving effect on nucleic acids of different sizes. The effective range of separation for different concentrations of acrylamide is shown in Table I together with the approximate size of duplex DNA which would comigrate with the marker dyes bromphenol blue and xylene cyanole FF.[4] Two denaturants are commonly used in conjunction with polyacrylamide for the fractionation of single-stranded nucleic acids, urea (final concentration 7 M) or formamide (98%). The preparation and running of denaturing polyacrylamide gels have been discussed previously in this series[5] but a brief outline will be presented here for the sake of completeness. The choice of denaturant is dictated largely by the size of the molecule. Urea is commonly used for sizing and resolving smaller chains (up to ~200 nucleotides). Larger chains and those containing substantial self-complementarity will retain secondary structure in urea, and 98% formamide will be necessary for accurate sizing and good resolution. Thin, urea-containing gels (discussed in the chapters on DNA and RNA sequencing [56,57]) can be run at elevated temperatures, thereby increasing their useful range.

Typical applications in which polyacrylamide gels would be the method of choice include purification of synthetic oligonucleotides, preparative isolation or analysis of DNA (<1 kb), especially if labeled, and resolution of small RNA molecules (e.g., tRNA) by two-dimensional techniques.

Equipment

Polyacrylamide gels of all kinds are invariably poured between two glass plates separated by spacers and run in the vertical position. A typical preparative gel will use 1- to 2-mm spacers and a 0.5- to 1-cm comb. The capacity of such a well is approximately 1 μg per band. As a general rule however, thinner gels (0.5 mm) are preferred since efficient elution by

---

[4] T. Maniatis, A. Jeffrey, and H. V. Van de Sande, *Biochemistry* **14**, 3787 (1975).
[5] T. Maniatis and A. Efstratiadis, this series, Vol. 65, p. 299.

TABLE I

RANGE OF SEPARATION OF ACRYLAMIDE GELS

Acrylamide (% w/v) (acrylamide : bis, 29 : 1)	Optimal range of separation (bp)	Comigration sizes (bp)	
		Bromphenol blue	Xylene cyanole
3.5	100–1000	100	460
5.0	80–500	65	260
8.0	60–400	45	160
12.0	40–200	20	70
20	6–100	12	45

diffusion is possible without crushing, leading to less contamination with acrylamide. Any commercially available apparatus for vertical slab gels can be used and examples of cheaper, homemade apparati abound.[6–8] Many research establishments have access to a machine shop capable of working with Plexiglas to make spacers, combs, and gel stands.

Running Buffers

The best and most common choice of running buffer for nondenaturing or urea-containing denaturing polyacrylamide gels is Tris–borate–EDTA (TBE), pH 8.3.[3] This is commonly stored as a 10× stock solution. Denaturing gels containing formamide are most conveniently run in phosphate buffer.

Nondenaturing Running Buffer

Concentrated stock solution (10× TBE):

Tris base	108 g
Boric acid	55 g
Disodium EDTA · 2H$_2$O	9.3 g
Water to	1 liter

(The pH should be 8.3)

Working solution (1×):

0.089 M Tris–borate, pH 8.3

0.025 M Disodium EDTA

[6] T. Maniatis, E. E. Fritsch, and J. Sambrook, "Molecular Cloning—A Laboratory Manual," p. 175. Cold Spring Harbor Lab., Cold Spring Harbor, New York, 1982.

[7] J. R. Dillon, A. Nasim, and E. R. Nestmann, eds., "Recombinant DNA Methodology," p. 13. Wiley, New York, 1985.

[8] R. DeWachter and W. Fiers, this series, Vol. 21, p. 167.

Denaturing Running Buffer

Urea gels: 1× TBE, pH 8.3 (see above)
Formamide gels (1×):

0.016 M Disodium hydrogen phosphate (Na_2HPO_4)	6.8 g
0.004 M Sodium dihydrogen phosphate (NaH_2PO_4)	1.6 g
Water to	3 liters
pH 7.5	

Preparation of Polyacrylamide Gels

Caution: Acrylamide monomer is highly toxic and readily absorbed through the skin. A mask and gloves should be worn for handling the solid and gloves for handling the solutions. The polymerized material is considered nontoxic.

Concentrated Stock Solutions

30% Acrylamide (29.1 acrylamide : bis)	100 ml
Acrylamide	29 g
N,N'-Methylenebisacrylamide	1 g
Water to	100 ml
Filter	
Ammonium persulfate, 10% freshly prepared	

1. Clean glass plates are rinsed with ethanol and set aside to dry. A variety of plates are commonly used, depending on the design of the electrophoresis apparatus. A general protocol for assembling the plates involves placing the side spacers (and an optional bottom spacer) on the horizontal outer plate, lowering the second plate into position, and clamping and/or taping the sides and bottom of the plates with electrical tape to make a good seal.

2a. *Nondenaturing gels.* The stock acrylamide solution is mixed with water and 10× TBE to yield the appropriate percentage, volume, and buffer concentration (1×), filtered, and gently deaerated by swirling (dissolved oxygen will retard polymerization).

2b. *Denaturing gels containing 7 M urea.* Ultrapure-grade urea (42 g per 100 ml final volume) is added to appropriate volumes of stock acrylamide solution and 10× TBE, and dissolved by stirring, addition of water, and gentle warming (37–42°). After the urea has dissolved, the volume is adjusted with water and the mixture cooled to room temperature. The mixture may be filtered and deaerated if desired.

3. For either of the above types of gel add 50 μl (per 100 ml acrylamide solution) of N,N,N',N'-tetramethylethylenediamine (TEMED) and 1 ml (per 100 ml) 10% ammonium persulfate, the polymerization initiator and catalyst respectively, and mix well.

4. Pour the gel by tilting the plates backward. Fill almost to the top and insert the comb immediately, leaving sufficient room between the top of the glass plate and the top of the teeth to remove the comb after polymerization. Clamping the comb in place results in wells free of thin polyacrylamide films on the walls and allowing the gel to polymerize in a near-horizontal position assures uniform thickness (important for thinner gels) and decreases the chance of leakage. The gel can be "topped up" if necessary to compensate for shrinkage during setting and *slow* leakages. Polymerization times vary with acrylamide percentage; 20% gels will set in approximately 30 min while lower percentage gels take about an hour at room temperature. A useful guide is to watch the unused gel solution, remembering that it will always remain liquid on the surface where air inhibits polymerization.

5. After removing the comb, the wells are rinsed with water, the bottom spacer or tape removed, and the gel attached to the electrophoresis apparatus. The reservoirs are filled with $1\times$ buffer, air bubbles below the bottom of the gel removed with a bent needle and syringe, and the wells flushed with electrophoresis buffer using a Pasteur pipet. It is important to rinse the wells immediately prior to loading, especially for urea-containing gels. It is customary to prerun polyacrylamide gels for a short time (15 min) to equilibrate gel buffer and running buffer and, in the case of thin denaturing gels, to reach running temperature.

The procedure for polyacrylamide gels containing formamide is somewhat different. The following protocol is for 75 ml final gel solution.

1. 100 ml formamide (99%) is deionized within 1–2 days of use with 5 g mixed-bed resin (Bio-Rad AG501-X8, 20–50 mesh) by stirring for 1 hr. After filtration, deionized formamide is stored at $-20°$.

2. Acrylamide and bis (amounts in Table II) are dissolved in a final volume of 74 ml deionized formamide. Disodium hydrogen phosphate ($Na_2HPO_4 \cdot 7H_2O$, 0.32 g), sodium dihydrogen phosphate ($NaH_2PO_4 \cdot$

TABLE II

ACRYLAMIDE GEL CONCENTRATIONS FOR USE
WITH FORMAMIDE

Gel (%)	Acrylamide (g)	N,N'-Methylenebis-acrylamide (g)
4	2.55	0.45
5	3.29	0.56
6	3.82	0.68
7	4.48	0.79
10	6.38	1.13
15	9.55	1.70

H_2O, 0.04 g), and ammonium persulfate (0.10 g) are dissolved in 1 ml water, added to the formamide solution, and filtered. Polymerization is initiated as usual with TEMED and the gel left overnight before use. Formamide gels are not prerun.

Sample Preparation

Generally speaking, DNA samples may be loaded onto nondenaturing polyacrylamide gels simply by adding a concentrated Ficoll solution (to increase sample density) containing marker dyes to assist in sizing. Restriction enzyme digests are most conveniently terminated by adding 0.1 volume 0.5 M EDTA and can then be loaded in Ficoll solution without further workup. Ficoll is preferred over other compounds (e.g., sucrose or glycerol) because it minimizes trailing at the sample edges which could lead to contamination in close-running preparative samples.

The preferred 10× loading solution for nondenaturing polyacrylamide gels is

30% Ficoll
0.25% Bromphenol blue
0.25% Xylene cyanole FF
0.20 M EDTA, pH 8
in 10× TBE (see above; nondenaturing running buffer)

For DNA preparations in low-ionic-strength buffers (e.g., restriction enzyme buffers) and at concentrations less than 1 $\mu g/\mu l$, 0.1 volume of the 10× loading solution is added prior to loading. For some DNA samples, in particular λ DNA, sticky single-stranded ends should be dissociated immediately prior to loading. This is best done by heating 5 min at 65°, followed by rapid cooling at 0°.

In certain cases, most commonly when the sample is too dilute or too high in salt or contains large amounts of protein or other impurities, it is advisable to improve the sample by a combination of phenol extraction and ethanol precipitation (see this volume [4,5]). A convenient way to concentrate a volume of dilute DNA too large to load directly is as follows.

1. Add 2–3 volumes of 2-butanol (*sec*-butyl alcohol) to the sample, mix, and keep the *lower* aqueous phase. Repeat as necessary until the lower phase is at an appropriate volume (typically 100–200 μl).

2. Extract the aqueous phase with 1 volume of chloroform. Keep the *upper* aqueous phase, which contains the DNA. Salt is also concentrated by this procedure.

3. Ethanol precipitate according to the standard protocol.

RNA is invariably phenol extracted and ethanol precipitated at some stage in a protocol prior to loading on a gel. It can then be treated as above. Nucleic acid samples for denaturing polyacrylamide gels are treated as follows. It is advisable, prior to loading, to remove salt from the samples (salt stabilizes duplex DNA). This is done by ethanol precipitation followed by rinsing in 70% ethanol–water.

1. Dissolve the pellet in an appropriate volume of deionized formamide.
2. Add 0.1 volume of 10× marker dyes (0.25% bromphenol blue, 0.25% xylene cyanole FF).
3. Heat 2 min at 100° and load.

Gel Loading

The techniques of loading gels are largely a matter of personal prejudice and whichever method allows a slow, steady delivery of sample should be used. Gels are always loaded with both reservoirs filled with buffer and the power turned off. Thin gels (less than 0.5 mm) are loaded with hand-drawn capillaries (see chapters on DNA sequencing [56–59]). Thicker gels may be loaded with glass capillaries attached to a mouth tube or hand dispenser or with an automatic pipettor according to convenience. The wells should be flushed just before loading and ethanol traces should be removed from the sample to ensure that it sinks.

Size Markers for DNA Gels

Dyes such as bromphenol blue and xylene cyanole FF are routinely used in gel electrophoresis, not so much as size markers but as indicators of how far a gel should be run to obtain the required result. For a more accurate size estimation it is most convenient to run a suitable restriction digest of commonly available DNA alongside the experimental samples. Accurate restriction maps based upon complete sequence data are available for plasmid pBR322 and λcIts857 for example, and a judicious choice of enzyme will generate a simple restriction pattern with molecules spanning the desired size range. Manufacturers' catalogs frequently tabulate this data for easy reference. For example, λ DNA cut with *Hin*dIII and φx174 replicative form DNA cut with *Hae*III make a convenient set of markers from 70 bp to 20 kbp. The products of restriction enzyme digests can readily be isotopically labeled (see this volume [10]) to provide markers for use with labeled samples. It is frequently difficult to obtain multiple size markers in the lower size ranges by restriction digestion but the availability of synthetic DNA, ideal for use as size markers on denaturing polyacrylamide gels, has overcome this problem. It is convenient to keep frozen samples of various markers in loading solution (100 ng/μl) ready for immediate use.

Size Markers for RNA Gels

Because of the structural complexity of single-stranded RNA and the substantial changes in relative mobility known to occur as a function of temperature, ionic strength etc., molecular weight determinations of non-denatured RNA by comparative methods are extremely unreliable. Consequently, when RNA is run with size markers, it is always done so under denaturing conditions. A number of small RNA markers (e.g., tRNAs, 5 S rRNA, 5.8 S rRNA) are appropriate and several well-characterized mRNAs (α- and β-globin) can be used for larger markers on polyacrylamide. Glyoxylation of nucleic acids prior to electrophoresis renders DNA and RNA electrophoretically equivalent for gel electrophoresis and allows glyoxylated DNA restriction fragments of known size to be used as calibration markers for glyoxylated RNA samples.

Electrophoresis Conditions

Polyacrylamide gels are run at room temperature without cooling for most purposes. Generally speaking, large duplex DNA fragments are best resolved at low-voltage gradients (8 V/cm or less) whereas small fragments are run faster to limit diffusion. Denaturing gels are commonly run warm to limit duplex formation. A balance between the considerations of fragment length, resolution required, and available time is often determined empirically by each experimenter in each situation.

Detection of Nucleic Acids in Polyacrylamide Gels

After the electrophoresis run is completed, running buffer is removed from the reservoir and the gel plates detached from the apparatus. After removing any remaining tape and with the gel and plates lying flat, the upper plate is gently pried off from one corner (a spatula is most useful) leaving the gel on the lower plate. The spacers are removed. Several options arise at this point depending on the size, thickness, and percentage of the gel and how the nucleic acid is to be detected. Nonradioactive DNA and RNA are most frequently visualized after staining with the intercalating dye ethidium bromide.[9] As little as 10 ng duplex DNA per band is visible in a polyacrylamide gel. The following protocol is the

[9] C. Aaij and P. Borst, *Biochim. Biophys. Acta* **269,** 192 (1972).

method of choice for staining polyacrylamide gels. *Caution:* Ethidium bromide is mutagenic and gloves should be worn when handling even dilute solutions. It is most conveniently stored as a 5 mg/ml solution in water in a dark bottle.

1. Immerse the gel and supporting glass plate in a solution of $1\times$ TBE containing 0.5 μg/ml ethidium bromide. Staining is carried out for at least 30 min. The solution should just cover the gel.

2. Wearing gloves, remove the gel with care using either the glass plate or a used piece of X-ray film as a support and transfer it onto a piece of plastic wrap (Saran Wrap) on a viewing and photographing surface. In some instances, particularly when it is desirable to reduce background fluorescence, the gel is destained by immersion in water for 15 min prior to viewing.

Ethidium bromide can also be used to detect single-stranded DNA and RNA but the nucleic acid–dye complex is weak and the fluorescent yield low resulting in 5- to 10-fold lower, but still adequate, sensitivity for most purposes. Several other dyes have been used to stain nucleic acids in gels.[10,11] Silver staining[12] has been reported to be more sensitive than ethidium bromide staining and may be preferable when high sensitivity is needed and for denaturing gels.

Methylene blue is an alternative dye for use with preparative samples and avoids the hazards of UV detection. The gel is stained in 0.02% methylene blue, 10 mM Tris–acetate (pH 8.3) for 1–2 hr at 4°. Avoiding direct sunlight, the excess stain is washed out with several changes of water (5–8 hr) and the bands become visible. The limit of detection is about 250 ng/1 cm band.

A common method for detection of synthetic oligonucleotides in denaturing polyacrylamide gels is that of UV shadowing. The gel is removed from the glass plates and placed on a fluorescent chromatography plate (silica gel 60-F254, E. Merck) covered in plastic wrap. A short-wave hand-held UV monitor is used to locate the DNA which appears as a black shadow on a green background and can readily be excised. More than 1 μg DNA per band can be detected by this method. Exposure to the light should be as short as possible. Because it is a shadowing technique,

[10] O. Gaal, G. A. Medgyesi, and L. Vereczkey, "Electrophoresis in the Separation of Biological Molecules." Wiley, New York, 1980.

[11] A. T. Andrews, "Electrophoresis: Theory, Techniques, and Biochemical and Clinical Applications." Oxford Univ. Press (Clarendon), London and New York, 1981.

[12] C. R. Merril, R. C. Switzer, and M. L. Van Keuren, *Proc. Natl. Acad. Sci. U.S.A.* **76**, 4335 (1979).

when excising the DNA the light should be vertically above the band to avoid cutting errors.

Viewing and Photography

Caution: Safety glasses and preferably a face mask should be worn around UV light sources.

The ethidium bromide–nucleic acid complex absorbs UV irradiation at 260 nm (the nucleic acid absorption maximum) or 300 nm (the bound ethidium maximum). The fluorescence of ethidium stacking in duplex DNA is 10 times greater than that of free ethidium bromide and the emission maximum is at 590 nm (red orange). The wavelengths of commercially available UV sources (254, 302, and 366 nm) have been compared for efficiency and deleterious nicking and dimerization of the DNA.[13] The two sources at 254 and 302 nm give the greatest fluorescence, but damage to DNA at 254 nm is extensive and a 302-nm source is to be preferred, especially for preparative work. Unfortunately, these lamps must be used with a filter to eliminate a red light emission which cuts down DNA sensitivity for detection of faint bands. In addition, nicking DNA by illumination at 254 nm may be advantageous if the DNA is large and ultimately will be transferred to membranes for hybridization. Ethidium complexes may be viewed by illumination either from the sides (incident light) with the gel on a black surface or from below (transmitted light). The transillumination system is the method of choice, being both more sensitive and requiring much shorter exposure times. Commercially available UV transilluminators are supplied with a UV-pass, visible blocking filter (an excitation filter). A Polaroid MP4 camera with a Kodak Wratten 22A or 23A red filter (to remove the red emission from the UV lamp) is positioned above the gel. Type 52 (positive) or 55 (positive and negative) film is used with a 545 film holder. Exposure time is typically about 5 sec. Faint bands are often detected by developing the negative.

Autoradiography of Gels

This subject has been covered in more detail elsewhere in this volume [7] and will only be briefly summarized here. Thick polyacrylamide gels (>1 mm) are generally removed from both glass plates, wrapped in plastic wrap, and exposed to X-ray film. Alignment of gel and film is aided by labels prepared with "radioactive ink" (commonly, old ^{32}P added to ink).

---

[13] C. F. Brink and L. Simpson, *Anal. Biochem.* **82,** 455 (1977).

Thin gels or large and unwieldy thick gels are either left on the glass plate, covered with plastic wrap, and exposed to film (the glass plate should be above the gel and film in the cassette to aid good contact) or, for thin gels, peeled off onto a suitable backing material (old X-ray film is best if gel slices are to be excised) and exposed to film.

Sizing and Quantitation of Nucleic Acids in Polyacrylamide Gels

As described previously, and because to a first approximation mobility is independent of sequence or composition, DNA and RNA fragment sizes are estimated from their mobility relative to standards of known size. For most purposes, it is sufficient to obtain an estimate of the fragment size as follows. (1) Photograph the gel containing markers and samples along with a UV-transparent ruler. (2) Construct a standard graph for the markers by plotting mobility (measured in centimeters from the well base) versus the log of molecular weight or size. This is most conveniently plotted on semilog paper for polyacrylamide gels; the size range of fragments is such that this will yield a straight line. (3) The size of the sample fragments can be read directly from the graph using the measured mobilities.

Accurate estimation of sizes is helped, particularly for duplex DNA, by running the gel at low-voltage gradients. A more detailed discussion of sizing of nucleic acids utilizing glyoxal to denature prior to electrophoresis has been presented in this series.[14]

Accurate quantitation of DNA *in situ* in a gel is, for most purposes, not necessary in the context of molecular cloning. Many experimenters are content to estimate by direct visual comparison with known quantities and obtain a feel for what 100 ng of duplex DNA looks like, for example, in an ethidium-stained gel. A sophisticated method involving microdensitometer tracing and comparison to known standards has been previously described.[15] The gel is stained with ethidium bromide and photographed under UV illumination. Microdensitometer tracing of the photographic negative yields pen deflections which are related to the amount of DNA in the band. The absolute quantities are determined by constructing a standard curve for the film, relating pen deflection, exposure, and optical density and using DNA standards of known size. It is more customary and useful for cloning purposes to quantitate nucleic acids after elution from a gel by standard spectrophotometric methods (see this volume [6]).

[14] G. C. Carmichael and G. K. McMaster, this series, Vol. 65, p. 380.
[15] A. Prunell, this series, Vol. 65, p. 353.

Elution from Polyacrylamide Gels

A detailed overview of the recovery of DNA from gels of all kinds has been previously presented in this series.[16] This section will provide a brief summary. The two primary concerns in selecting gel elution techniques are the percentage recovery of the nucleic acid and the nature of contamination which may adversely affect subsequent steps in the protocol. For polyacrylamide gels the most common and reliable elution technique, applicable to both RNA and DNA, is that of crush elution in high-ionic-strength buffer.[17,18] Recovery of nucleic acids (to ~1 kb in length) is almost quantitative especially when thin gels (0.5 mm) are used. The major impurity remaining through cleanup procedures is linear acrylamide, which does not inhibit most enzymes (restriction enzymes, ligase, polymerase) with the exception of reverse transcriptase. If mRNA is fractionated electrophoretically on polyacrylamide gels it will be necessary to repurify by oligo(dT)-cellulose chromatography. A more appropriate protocol for mRNA is presented later. For RNA work, solutions must be sterile and all glassware acid washed. It is also advisable to include 0.1% sodium dodecyl sulfate (SDS) and 1 mM EDTA to impair ribonucleases. A general protocol is as follows.

1. Excise the desired bands from the gel with a razor blade and forceps. For nonradioactive samples, keep exposure to UV light at a minimum.

2. For denaturing gels only, it is advisable to soak the gel slice in water for 15 min to remove urea, formamide, and phosphate buffer. It is particularly important to remove phosphate to prevent a large coprecipitate with gel impurities which can trap nucleic acids. Small oligonucleotides (<30) will begin to elute in water.

3. The gel slice is transferred to a 1.5-ml polypropylene tube, siliconized scintillation vial, or Corex tube depending on its size and is crushed by one of several methods. (This is not necessary for thin gels—one of their advantages—or for small nucleic acids, e.g., synthetic oligonucleotides). An acid-washed spatula can be used to pulverize the gel in the scintillation vial or Corex tube. An efficient way to crush gels is to extrude them from a syringe barrel through a shortened wide-gauge needle (broken with pliers) into a polypropylene tube by low-speed centrifugation (5 min) in a swinging bucket rotor.

[16] H. O. Smith, this series, Vol. 65, p. 371.
[17] W. Gilbert and A. Maxam, *Proc. Natl. Acad. Sci. U.S.A.* **70**, 3581 (1973).
[18] R. Yang, J. Lis, and R. Wu, this series, Vol. 68, p. 176.

4. Add gel elution buffer. The volume depends on the size of gel slice but as a guide 0.5–1 ml is used for a slice 1 cm × 1.5 mm. Gel elution buffer: 0.5 M ammonium acetate, 10 mM magnesium acetate, 1 mM EDTA, 0.1% SDS. Incubate overnight at 37°. Agitation and higher temperature (60°) may be used for larger molecules and larger gel volumes.

5. Fragments of polyacrylamide are removed by centrifugation and filtration. Quik-Sep columns (Isolab) work well for large volumes as they fit 15-ml Corex tubes. A Pasteur pipet plugged with siliconized glass wool can be used for smaller volumes.

6. Ethidium bromide can be removed by extraction with 2-propanol or isobutanol saturated with elution buffer and volumes can be reduced, prior to precipitation, with 2-butanol as described previously. Alternatively, DNA is adjusted to 0.8 M LiCl, 50 mM Tris (pH 9.0) and extracted with an equal volume of phenol–chloroform.

7. Ethanol precipitate and wash with 70% ethanol–water according to standard procedures. This step is not necessary, for example, if the nucleic acid is radioisotopically labeled and is to be used directly for hybridization.

Repeated elution will serve to increase the yield if necessary.

Nucleic acids may also be recovered in high yield from polyacrylamide gels by electroelution. The technique is described in the following section, dealing with agarose gels, where it is more commonly used.

A number of commercially available resins have been introduced in recent years for post-elution purification of nucleic acids. As with DEAE-cellulose, nucleic acids bind to minicolumns of the material in low salt, allowing impurities to pass through by washing and can be eluted in higher salt. Two such materials are NACS-52 (Bethesda Research Laboratories) and Elutip-d (Schleicher and Schuell). Comprehensive applications manuals are provided by the suppliers. A technical description of the properties of a similar resin (RPC-5) has been presented previously in this series.[19,20]

Agarose Gel Electrophoresis

The resolving power of agarose gels is a function of the concentration of dissolved agarose. The migration rate of nucleic acids through agarose gels is additionally dependent upon the molecular size (for linear frag-

[19] R. D. Wells et al., this series, Vol. 65, p. 327.
[20] J. A. Thompson et al., this series, Vol. 100, p. 368.

ments), conformation, and voltage gradient. Under certain conditions, the resolution on the basis of conformation (circular, nicked circle, or linear), is a particularly useful property. The effective range of separation for agarose gels of various percentages is shown in Table III.

Denaturants commonly added to agarose gels for analysis of single-stranded nucleic acids are sodium hydroxide (DNA only), formaldehyde, and methylmercuric hydroxide. The latter is extremely toxic and results in large volumes of hazardous buffer. This disadvantage should be weighed against its undisputed ability to denature nucleic acids completely. In many instances, glyoxylation of DNA or RNA prior to electrophoresis can be effectively substituted. Typical applications of agarose gels are analysis of restriction enzyme digests of cloned DNA, preparation of cut vectors and fragments for cloning, sizing large DNA and RNA, S_1 mapping, and analysis of cDNA cloning intermediates.

Equipment

The horizontal position has emerged in recent years as the preferred position for agarose gel electrophoresis. It has the advantages of simplicity in pouring, loading, and handling, versatility in size using the same apparatus, and support from below, important for low percentage gels. Tanks are available in all sizes commercially including those for minigels and can be cheaply made according to a standard design (ref. 1, p. 42; ref. 6, pp. 154–155, 163). The essence of the design is that the gel is poured on a glass plate with the comb in place. When set, it is transferred on the plate to a platform and is submerged in running buffer. It is common to use a microscope slide if only three or four lanes are required. The gel is poured relying on surface tension and the comb held in place with a clamp.

TABLE III
RANGE OF SEPARATION FOR AGAROSE GELS

Agarose (%)	Optimal range of separation linear DNA (kb)
0.3	60–5.0
0.6	20–1.0
0.7	10–0.8
0.9	7–0.5
1.2	6–0.4
1.5	4–0.2
2.0	3–0.1

Nondenaturing Agarose Gels

Agarose powder comes in many grades. For general analytical electrophoresis purposes the best agarose is type II low endo-osmotic agarose. This type, however, contains contaminants which coelute with DNA and inhibit most commonly used enzymes, which means that DNA must be extensively purified following elution from this kind of gel. An extremely attractive alternative for preparative cloning work involves the use of high-quality, low melting temperature agarose. This agarose melts at 65 and sets at 30°, which allows DNA to remain double-stranded and also allows many enzymes to be used in the liquid agar. T4 DNA ligase reportedly functions efficiently in the solidified gel at 15°. A collection of methods has been recently compiled by Struhl[21] involving low melting temperature agarose gel electrophoresis and subsequent "in-gel" manipulation of fragments including transformation. The source of agarose for this application appears critical and Sea Plaque low melting temperature agarose (Marine Colloids) is recommended. Horizontal agarose gels are generally about 3 mm thick (thinner gels should be run in the vertical position), and are prepared as follows.

1. Powdered agarose is added to the desired electrophoresis buffer (1×) to give the correct percentage. It is common to make 100 ml of gel solution. A typical minigel apparatus will require 25 ml. The agarose is dissolved completely by heating, most conveniently in a microwave oven. Care should be taken to ensure that the solution is homogeneous.

2. The solution is allowed to cool to approximately 50–60° before pouring. Agarose above this temperature can deform some gel trays and combs. Cooling the agarose is also important where surface tension is being used to contain the liquid gel. At this stage, there is the option of adding ethidium bromide (0.5 μg/ml, final) to the gel.[22] This allows the progress of the gel run to be monitored during the run by incident or transmitted longwave UV light (many horizontal gel apparatuses are made of UV-transparent Plexiglas). Incorporation of ethidium will, however, affect the conformation of certain DNA molecules, thereby changing their mobility. To minimize the volume of dilute ethidium bromide and to minimize damage to the DNA ethidium complex by light, it is recommended that gels be run without inclusion of ethidium bromide.

3. Pour the gel into the mold. The precise nature of this operation will depend on the design of the apparatus. In some, tape is used to complete

[21] K. Struhl, *BioTechniques* **3**, 452 (1985).
[22] R. A. Sharp, B. Sugder, and J. Sambrook, *Biochemistry* **12**, 3055 (1973).

the mold in the apparatus itself (e.g., Bethesda Research Laboratory); in others, the gel is poured in a mold separate from the apparatus (e.g., International Biotechnologies Inc.). The comb is generally in place when the gel is poured. It is important that the bottom of the comb does not touch the plate underneath the gel.

4. After setting (~30 min; the gel should look uniformly opalescent), any tape is removed, the gel submerged in buffer, and the comb gently removed. The apparatus is topped up with buffer until the gel surface is just submerged. Ethidium-containing gels may be run in ethidium-containing buffer. Agarose gels are not prerun.

Nondenaturing Running Buffers

Several running buffers are in common use for nondenaturing agarose gel electrophoresis. Tris–borate (see section on polyacrylamide gels) is often used, at 0.5× concentration, but interacts with agarose and leads to low recovery of nucleic acids in preparative gels. Tris–acetate buffer[23] is commonly used and is ideal for preparative work. The buffering capacity is low and so recirculation of buffer on long runs is advisable. Tris–phosphate has good buffering capacity, obviating the need for circulation, but should be avoided if the fractionated sample is to be eluted and ethanol precipitated because of coprecipitation of phosphate. The buffers are stored as concentrated stocks and diluted as needed.

50× Tris–acetate (TAE):

Tris base	242 g
Glacial acetic acid	57.1 ml
0.5 M EDTA (pH 8)	100 ml
Water to	1 liter

TAE working solution:

Tris–acetate	0.04 M
EDTA	0.001 M

10× Tris–phosphate:

Tris base	108 g
85% Phosphoric acid	15.5 ml
0.5 M EDTA (pH 8)	40 ml
Water to	1 liter

Tris–phosphate working solution:

Tris–phosphate	0.08 M
EDTA	0.002 M

[23] V. E. Loening, *Biochem. J.* **102,** 251 (1967).

Sample Preparation and Loading

As with polyacrylamide gel electrophoresis, samples for nondenaturing agarose gels are conveniently loaded by addition of a concentrated Ficoll, dye, EDTA solution (see previous section). The capacity of a standard well (~2 mm deep × 5 mm wide) is about 200 ng per band, and as little as 1–5 ng of double-stranded DNA may be detected by ethidium bromide staining in agarose. Typically, restriction analysis of a recombinant plasmid or phage will involve running 0.5–1 μg DNA per lane. For restriction analysis of genomic DNA where many bands will run such as to produce a smear, 5–10 μg of DNA per lane can be loaded.

Alkaline Agarose Gels[24]

Denaturing agarose gels containing sodium hydroxide are most useful for single-stranded DNA analysis and are frequently used analytically for checking first- and second-strand synthesis in cDNA cloning and the size range of nick-translated DNA (see elsewhere in this volume). Agarose is hydrolyzed at high temperature by alkali and the gels are best made by preparing the gel as in the previous section, but in a neutral buffer (50 mM NaCl, 1 mM EDTA). When set, the gel is soaked by submersion in alkaline running buffer (30 mM NaOH, 1 mM EDTA) for at least 30 min before running.

DNA samples should be phenol extracted and ethanol precipitated and the dried pellet dissolved in

Loading buffer:

50 mM	Sodium hydroxide
1 mM	EDTA
3%	Ficoll
0.025%	Bromocresol green
0.025%	Xylene cyanole FF

The gel, which has been equilibrating in alkaline running buffer for at least 30 min, is drained of excess buffer (leave ~1 mm covering the gel) prior to loading.

Glyoxal Denaturation of Nucleic Acids[14]

Glyoxal denaturation of RNA and of DNA prior to electrophoresis allows molecules to remain denatured throughout electrophoresis, resulting in accurate separation by molecular weight and quantitative transfer to filters for Northern blot analysis. Ease of handling and lack of toxicity

---

[24] M. W. McDonell, M. N. Simon, and F. W. Studier, *J. Mol. Biol.* **110,** 119 (1977).

recommend it over the other commonly used method of formaldehyde denaturation. As these two methods cover the same needs, we recommend the use of glyoxal–DMSO denaturation for electrophoresis. However, greater sensitivity has been obtained in Northern blot experiments (this volume [61]) using formaldehyde gels.

Glyoxal (6 M, 40% solution) should be deionized with a mixed-bed ion-exchange resin such as Bio-Rad AG501-X8 until neutral, then distributed in useful aliquots into tightly capped tubes, and maintained at $-20°$. Glyoxal should not be reused after opening.

To denature, mix

1 M Glyoxal
50% (v/v) Dimethyl sulfoxide
10 mM Sodium phosphate buffer, pH 7.0
DNA or RNA (up to 10 μg/8 μl reaction volume)

Incubate at 50° for 1 hr. Cool in ice.

Add 2 μl loading buffer/8 μl reaction.

Loading buffer:

10 mM Sodium phosphate (pH 7.0)
50 % (v/v) Glycerol
0.4 %   Bromphenol blue

Gel: 0.75–1.5% agarose in 10 mM sodium phosphate (pH 7.0). As markers, use similarly glyoxylated RNAs or DNAs. Run the gel at 3–4 V/cm for 10 hr, with constant recirculation of buffer to maintain pH at 7.0. Glyoxal readily dissociates from DNA and RNA at pH 8.[25]

Nucleic acids can be visualized following electrophoresis by staining with ethidium bromide (0.5 μg/ml in running buffer) and photographed as described above. If the RNA is to be transferred to a filter for hybridization analysis it is recommended that the gel not be stained, since any soaking of the gel prior to transfer has been shown to reduce transfer.[25]

Staining with acridine orange is a more sensitive method to visualize glyoxylated DNA and RNA.[14] Agarose gels are stained for 30 min and polyacrylamide–agarose composite gels for 15 min in 30 μg acridine orange/ml in 10 mM sodium phosphate (pH 7.0). Destaining is accomplished by running hot tap water over the gel for 5–10 min. Stained molecules are then visualized by UV illumination (254 nm). After destaining, double-stranded nucleic acids appear green and single-stranded nucleic acids (i.e., most RNAs and denatured DNA) appear orange. Color photographs using Polaroid 108 color film and a yellow filter can be taken, or black and white photographs using Polaroid 105 positive/negative or 107C positive film and a red filter.

[25] P. S. Thomas, *Proc. Natl. Acad. Sci. U.S.A.* **77**, 5201 (1980).

Methylmercuric Hydroxide Gels[26]

Methylmercuric hydroxide gels are used for accurate sizing of DNA and RNA and for electrophoresis of RNA prior to transfer to membranes for hybridization. These gels are particularly useful when fractionated mRNA is to be eluted and used for *in vitro* translation studies. The risks involved in working with methylmercuric hydroxide should be fully understood and precautions against its potentially lethal effects must be taken. This information has been published previously.[27] *It is poisonous and somewhat volatile and all manipulations must be carried out in a fume hood.* It is strongly recommended that alternative methods (e.g., glyoxylation prior to electrophoresis for molecular weight determination or transfer to membranes for hybridizations) be tried first. Agarose gels containing methylmercuric hydroxide are made as follows.

1. The desired amount of agarose is dissolved in running buffer (see below for methylmercuric hydroxide gel running buffer) by heating.

2. After cooling to 50–60°, add methylmercuric hydroxide to 5 mM final concentration and pour the gel. It is not necessary to include methylmercuric hydroxide in the running buffer.

Denaturation with methylmercuric hydroxide is reversible and nucleic acids can be eluted after soaking the gel in 0.5 M ammonium acetate, 10 mM dithiothreitol, or 2-mercaptoethanol for 1 hr.

The buffer should be recirculated during the run.

Running buffer (1×):
  0.05 M Boric acid
  0.005 M Sodium borate
  0.010 M Sodium sulfate
  pH 8.2

Equal volumes of RNA (10 μg can be loaded per 0.5 cm lane) and 2× loading buffer are mixed.

Loading buffer (2×):

1 M Methyl mercuric hydroxide	25 μl
4× Running buffer	500 μl
30% Ficoll	200 μl
Water to	1 ml
Bromphenol blue	0.1%
Xylene cyanole FF	0.1%

Use of ethidium bromide in RNA gels should be avoided if the RNA is to be transferred to membranes for hybridization. Waste solutions and gels containing methylmercuric hydroxide should be treated before dis-

[26] J. M. Bailey and N. Davidson, *Anal. Biochem.* **70**, 75 (1976).
[27] J. E. Cummins and B. E. Nesbitt, *Nature (London)* **273**, 96 (1978).

posal with excess of a sulfhydryl reagent such as dithiothreitol or 2-mercaptoethanol. For the gel, treatment with 10 mM dithiothreitol for 1 hr is sufficient.

Formaldehyde Gels[28]

Caution: Concentrated formaldehyde solutions should be stored and used in a fume hood.

Formaldehyde–agarose gels are used for fractionation of RNA prior to transfer to hybridization membranes. They are prepared as follows. Formaldehyde (37% in water, 12.3 M) and 5× running buffer (see below) are added to the desired percentage of molten agarose in water (50–60°) to give final concentrations of 1× buffer and 2.2 M formaldehyde.

Running Buffer (1×):
   0.04 M Morpholinopropanesulfonic acid (MOPS), pH 7.0
   0.01 M Sodium acetate
   0.001 M EDTA

RNA samples (up to 20 μg in 5 μl) are incubated at 55° for 15 min or 65° for 5 min in:

5× gel running buffer	2 μl
Freshly deionized formamide	10 μl
Concentrated formaldehyde	3.5 μl

2 μl of the sterile loading buffer is added.
Loading buffer:
   30 % Ficoll
   1 mM EDTA
   0.25 % Bromphenol blue
   0.25 % Xylene cyanole FF

Size Markers

As discussed previously, coelectrophoresis of fragments of known size provides the best method for estimating the size or molecular weight of a sample. The essential condition is to choose a gel system for which the log molecular weight is linearly related to mobility. The useful size range of agarose gels requires the availability of larger known size markers, and frequently restriction digests of λ DNA provide the necessary fragments to construct a standard straight line. A mixed *Eco*RI/*Hin*dIII digest or *Bgl*II digest is frequently used for size markers on agarose for larger fragment separations.

Agarose gels are frequently used to determine the molecular conformation of plasmid DNA: circular, nicked, linear, multimeric, etc. The relative running order and mobilities of the conformers depend on the

[28] H. Lehrach, D. Diamond, J. M. Wozney, and H. Boedtker, *Biochemistry* **16,** 4743 (1977).

buffer, current, and agarose concentration and known circular or linear markers in the appropriate size range should be used. Such gels are frequently run in the presence of ethidium bromide, which (usually at 0.5 μg/ml) increases the mobility of form I (closed circular) DNA and retards that of nicked (form II) and linear (form III) molecules.

For denaturing gels, the most convenient size markers are fully denatured DNA restriction fragments or RNAs of defined size (e.g., several ribosomal RNAs). As a general rule, markers are treated identically to samples and run in whatever type of gel system is chosen. For formaldehyde gels, labeled markers or markers that will be labeled after hybridization are best. Formaldehyde-denatured nucleic acids are not well visualized by ethidium staining. Formaldehyde can be removed from isolated marker lanes by washing in water (4 × 30 min) and 0.1 M ammonium acetate (2 × 30 min) and staining for 1 hr in 0.1 M ammonium acetate, 0.1 M 2-mercaptoethanol with ethidium bromide (0.5 μg/ml). As mentioned earlier, glyoxylated DNA or RNA samples can be sized by using glyoxylated DNA restriction fragments which closely span the desired size range. This is the most convenient and least hazardous method for size determination.

Electrophoresis Conditions

Agarose gels for most purposes are run at room temperature. Exceptions are low percentage agarose gels (<0.5%) which are easier to handle in the cold and low melting temperature agarose gels which may melt if run too fast at room temperature. The best results, in terms of resolution and accurate sizing, are obtained by running agarose gels very slowly (<5 V/cm). It is important to circulate or change buffer in such runs, especially for denaturing gels or glyoxylated samples, in order to avoid pH changes which could begin to reverse glyoxylation (pH >8).

Detection, Viewing, and Photography of Nucleic Acids in Agarose Gels

The essentials of these techniques are identical to those used for polyacrylamide. Ethidium bromide is the dye of choice for both single- and double-stranded nucleic acids, the limits of detection in agarose being somewhat lower than in polyacrylamide (~1–5 ng per band for double-stranded DNA). The sensitivity for single-stranded nucleic acids is 5- to 10-fold less. As discussed above, in particular for looking at intact plasmids, ethidium bromide is frequently incorporated into gel and running buffer during electrophoresis. Otherwise gels are stained after electrophoresis in electrophoresis buffer containing 0.5 μg/ml ethidium bromide for 30 min.

Determination of Fragment Size

As with polyacrylamide gel electrophoresis, it is important to obtain a good linear relationship in the size range of interest between relative mobility and log size. The size of fragments on agarose can be large and errors in measurement potentially great. Errors are minimized by taking the following precautions: (1) Gels are run as slowly as is experimentally realistic. (2) Size markers are selected which closely bracket the unknown sample and thereby provide an accurate standard line. (3) In extreme cases, it is possible to scan photographs with a densitometer and obtain accurate relative mobilities by measuring peak separations. This method is seldom used in the course of routine molecular cloning.

Elution from Agarose Gels

A myriad of techniques exists for elution of nucleic acids from agarose gels—a direct consequence of no entirely satisfactory protocol being available. Reviews of the topic have been presented[16] and will be summarized and updated here. In addition to electroelution and diffusion which are also appropriate for polyacrylamide gels, nucleic acids can be eluted from agarose by gel dissolution and by physical extrusion. It is worth stressing at this point that for many purposes, use of high-quality low melting temperature agarose (discussed previously) eliminates the need for elution and the relatively high cost can be easily offset by frugal use and the time saved. The following methods are most commonly used for DNA.

Electroelution Techniques

Into Dialysis Bags[24]

1. The band of interest is excised with a razor blade and dropped into a dialysis bag tied off at one end and filled with elution buffer. A suitable elution buffer with low conductivity is 5 mM Tris, 2.5 mM acetic acid (pH 8).

2. After removing most of the buffer (the gel slice should be still surrounded) the bag is closed and immersed in elution buffer in an electrophoresis tank (a minigel apparatus works well for small slices). Current is passed through the bag (the gel should be aligned such that its smallest dimension is parallel to the electric field) for 2–3 hr at ~100 V.

3. The current is reversed for 5 min to remove the DNA from the inner walls of the dialysis bag.

4. Using a Pasteur pipet, remove the buffer from the bag. Elution of DNA from the gel slice can be monitored by restaining. The DNA can be

concentrated and further purified by absorption and elution from a mini-column as described above.

Into Troughs

1. The ethidium bromide-stained DNA is localized using a longwave UV light. It is most convenient for this technique to have run the gel in a UV-transparent Plexiglas tray.

2. Cut a small trough directly in front of and 1–2 mm wider than the desired band, remove the buffer covering the gel, and continue electrophoresis. The DNA moves from the gel into the buffer, which is periodically removed and replaced with fresh until all the desired band has been recovered. DNA can be phenol extracted and ethanol precipitated (after removal of any agarose by filtration) and is usually free of inhibitors.

3. A time-saving modification[29] involves making a single incision directly below the band, inserting a piece of Whatman 3 MM paper and dialysis membrane (presoaked 5 min in buffer) into the incision with the paper adjacent to the band, and continuing electrophoresis. Constant monitoring is not required as the DNA sticks to the paper and membrane. The DNA is eluted from the paper with 3×100 μl 0.2 M NaCl, 50 mM Tris (pH 7.6), 1 mM EDTA. This is most conveniently achieved by placing the paper in a plastic pipet tip or punctured polypropylene tube, adding the buffer in 100-μl batches, and recovering the eluate by centrifugation into a polypropylene tube. The DNA is phenol–chloroform extracted and ethanol precipitated. Use of DEAE paper in place of 3 MM and a dialysis membrane is reported to eliminate the need for subsequent cleanup on DEAE, NACS-52, Elutip-d, or similar resins.[30]

Physical Extrusion Techniques

Commonly known as the freeze–squeeze procedure,[31] physical extrusion is ideal for large DNA recovery from low percentage (<0.6%) gels. Yields are normally >50%. The following technique works well.

The gel slice is placed between sheets of Parafilm or plastic wrap and frozen (30 sec on dry ice). As it defrosts, it is squashed between thumb and forefinger or against the lab bench, and the drop of liquid collecting on the plastic contains much of the DNA. The liquid is centrifuged to remove agarose particles and cleaned up as described elsewhere in this chapter. A number of modifications (reviewed in Smith[16]) may suit personal preferences.

[29] S. C. Girvitz, S. Bacchetti, A. J. Rainbow, and F. L. Graham, *Anal. Biochem.* **106,** 492.
[30] G. Dretzen, M. Bellard, P. Sassone-Corsi, and P. Chambon, *Anal. Biochem.* **112,** 295.
[31] R. W. J. Thuring, P. M. Sanders, and P. Borst, *Anal. Biochem.* **66,** 213 (1975).

Isolation of DNA from Low Melting Temperature Agarose

Not all protocols will allow "in-gel" manipulation of DNA in low melting temperature agarose. The following protocol is suitable for recovery of DNA from such gels.[32] The gel is run in 1× TAE.

1. To the excised gel slice add an equal volume of 0.1 M Tris–acetate (pH 7.5), 5 mM EDTA, 0.5 M NaCl and heat to 65° until the agarose has melted.

2. Extract with an equal volume of phenol. A white emulsion forms which separates on centrifugation. Powdered agarose appears at the interface.

3. The aqueous layer is reextracted with phenol–chloroform and chloroform, and the DNA recovered by ethanol precipitation.

RNA, most frequently mRNA, is often transferred from agarose gels to filters for hybridization. These techniques are covered in this volume [61]. The problem associated with mRNA isolation for, for example, translation studies is that standard procedures give poor recoveries and degraded mRNA. A method enabling "in-gel" translation of small (nanogram) quantities of fractionated poly(A)$^+$ mRNA suitable for rapid screening has been described.[33]

RNA may be recovered from low melting temperature denaturing gels (containing methylmercuric hydroxide) as follows.

1. The gel is soaked in 0.1 M dithiothreitol for 30 min and the excised slice dissolved in 4 volumes of 0.5 M ammonium acetate by heating of 65°.

2. Extract with an equal volume of phenol. Reextract with phenol–chloroform and chloroform to remove all agarose from the aqueous layer.

3. The extracted RNA is ethanol precipitated and is generally suitable for use with reverse transcriptase.

Removal of Gel Contaminants

As discussed previously, nucleic acids eluted from conventional agarose gels contain many inhibitors. A combination of phenol extraction and chromatography on DEAE-cellulose or a commercially available reverse-phase resin (discussed for polyacrylamide gels) is usually sufficient to remove contaminants. A typical procedure for DEAE-cellulose is as follows.

1. A small DE-52 (Whatman) column is equilibrated in 0.15 M NaCl, 50 mM Tris (pH 8), 1 mM EDTA, and DNA is loaded in the same or lower buffer concentration (the capacity is >200 μg/ml).

---

[32] L. Weislander, *Anal. Biochem.* **98,** 305 (1979).
[33] T. L. Brandt and P. B. Hackett, *Anal. Biochem.* **135,** 401 (1983).

2. The column is washed with 10 column volumes of loading buffer and the DNA eluted with 1 M NaCl, 50 mM Tris, 1 mM EDTA (1–2 column volumes).

Gel Drying

Agarose and polyacrylamide gels can be dried by heating *in vacuo* using commercially available dryers. This is often performed prior to autoradiography (e.g., ^{35}S-labeled dideoxy sequencing gels). For gels containing denaturants (e.g., urea) it is important to remove them before drying. This is most commonly achieved for urea by soaking for at least 15 min in 10% aqueous acetic acid. Agarose gels containing denaturants should also be washed prior to drying. *Never* dry methylmercuric hydroxide gels without ensuring complete removal of the toxic and volatile denaturant with 2-mercaptoethanol. Gels are placed on two sheets of Whatman 3 MM paper and covered with plastic wrap. For thin gels, the paper is evenly pressed onto the gel (after excess liquid from the 10% acetic acid fixing has been removed with Kimwipes) and the paper with gel attached is gently peeled off the glass plate. The gel is placed on the dryer and covered with a porous plastic sheet and a rubber flap.

Drying times vary with gel thickness and temperature. Sequencing gels typically take 30 min at 80°. Agarose gels should not be heated above their melting point! It is important not to break the vacuum before the gel is completely dry; this can be easily detected by turning off the heat and monitoring whether the gel feels cool (still drying). For autoradiography of ^{35}S, the plastic wrap must be removed from the dried gel.

Conclusion

The powerful techniques described in this chapter are constantly undergoing improvement and finding wider application. Two exciting areas of innovation which will find increasing application in molecular biology are the use of pulse-field gel elecrophoresis for resolution of very large duplex DNA molecules (e.g., chromosomes) (Smith and Cantor,[34] and references therein) and the use of gels to detect, map, and resolve point mutations within otherwise identical DNA molecules (Myers *et al.*,[35] and references therein). The access to hitherto intractable scientific problems afforded by these new applications of gel electrophoresis attest to its continuing central role in molecular cloning.

[34] C. L. Smith and C. R. Cantor, *Nature (London)* **319**, 701 (1986).
[35] R. M. Myers, S. G. Fischer, L. S. Lerman, and T. Maniatis, *Nucleic Acids Res.* **13**, 3131 (1985).

Section III

Enzymatic Techniques and Recombinant DNA Technology

[9] Nick Translation

By JUDY MEINKOTH and GEOFFREY M. WAHL

Nick translation, or more precisely nick translocation, is a specific procedure for incorporating radioactive nucleotides into double-stranded DNA. The method takes advantage of the ability of *Escherichia coli* DNA polymerase I to combine the sequential addition of nucleotide residues to the 3'-hydroxyl terminus of a nick [generated by pancreatic deoxyribonuclease (DNase) I] with the elimination of nucleotides from the adjacent 5'-phosphoryl terminus. Linear, supercoiled, nicked, or gapped circular double-stranded molecules can be labeled to specific activities $>10^8$ cpm/μg with deoxynucleotide 5'-[α-^{32}P]triphosphates by this technique.[1,2] Since the nicks are introduced at random sites in the duplex, the method generates a population of radioactive fragments which partially overlap each other. At saturating levels of nucleotide triphosphates the size of the fragments is determined by the DNase concentration. Fragments approximately 500–1500 nucleotides long produce optimal signal-to-noise ratios when hybridized to immobilized DNA or RNA (see this volume [45, 61]), presumably due to their ability to hybridize with each other in overlapping complementary regions to form "networks" or "hyperpolymers."[2] While experiments consistent with hyperpolymer formation of nick-translated probes have been reported,[3,4] the reproducibility and extent of hyperpolymer formation seem to be difficult to obtain, probably because of the critical dependence on probe size.[2,4] Longer probes have been correlated with higher backgrounds (G. Wahl, unpublished observations).

Nick Translation Reaction

1. Dilute stock DNase in 50 mM Tris–HCl at pH 7.4, 10 mM MgCl$_2$, 1 mg/ml DNase-free bovine serum albumin just before use. A DNase concentration of about 40–80 pg/μl is usually adequate.

2. Mix 2 μl 10× NT (nick translation) buffer [10× − 50 mM MgCl$_2$, 200–250 μM each unlabeled deoxyribonucleoside triphosphate (dNTP), omitting those corresponding to the labeled nucleotide(s), 50 mM Tris–

[1] P. W. J. Rigby, M. Dieckmann, C. Rhodes, and P. Berg, *J. Mol. Biol.* **113**, 237 (1977).
[2] J. Meinkoth and G. Wahl, *Anal. Biochem.* **138**, 267 (1984).
[3] G. M. Wahl, M. Stern, and G. R. Stark, *Proc. Natl. Acad. Sci. U.S.A.* **76**, 3683 (1979).
[4] R. H. Singer, J. B. Lawrence, and C. Villnave, *BioTechniques* **4**, 230 (1986).

HCl at pH 7.4] with 100–400 ng DNA, and 50–300 μCi of deoxyribonucleoside 5'-[α-^{32}P]triphosphate (3000 Ci/mmol, 10 μCi/μl). Add 1 μl diluted DNase and water to 19 μl. Usually three of the four triphosphates (e.g., dATP, dGTP, and TTP) are unlabeled whereas the fourth (dCTP in this example) is supplied as a radioisotope. Any combination of labeled and unlabeled deoxyribonucleoside triphosphates will suffice, however, provided the concentration of each dNTP is greater than 1–3 μM, a value which is probably not saturating for the enzyme. Incubate 2 min at room temperature.

3. Add 1 μl *E. coli* DNA polymerase I (Pol I, 0.8 units/μl) and incubate at 15° for 30 min to 2.5 hr. The progress of the reaction may be monitored by measuring incorporation into trichloroacetic acid-insoluble material using the techniques detailed in this volume [6]. Purify the nick-translated DNA as detailed in steps 4–6.

4. Add 0.1 volume of 2% SDS, 250 mM EDTA, or an equal volume of phenol-saturated H$_2$O to stop the reaction. Phenol-saturated H$_2$O is made by vortexing equal volumes of redistilled phenol (often containing 0.1% hydroxyquinoline as an antioxidant) and H$_2$O for 30 sec. After centrifugation for 30 sec in an Eppendorf Microfuge, or the equivalent, the upper phase consisting of phenol-saturated H$_2$O is recovered.

5. Remove unincorporated nucleotides by passing the reaction through a P-60 (100–200 mesh, Bio-Rad) column poured in a Pasteur pipet or through a spun Sephadex G-50 fine column. Such a column is made in a 1-ml syringe plugged with glass wool by adding aliquots of swelled Sephadex G-50 fine resin (Pharmacia) and centrifuging at ~1000 g for 5 min until a packed volume of ~0.9–1 ml has been attained. Centrifugation is accomplished by placing the column inside a 15-ml conical disposable tube. The lip of the syringe conveniently hangs on the edge of the tube; thus, the column does not come in contact with the effluent. Add 100 μl H$_2$O to the column. Centrifuge at 1000 g for 5 min, transfer the column to a fresh disposable tube, and add the sample brought to 100 μl with H$_2$O. Centrifuge at 1000 g for 5 min. The eluate in the conical tube is virtually free of unincorporated deoxyribonucleoside triphosphates and may be used without further purification. Alternatively, transfer the DNA to a microfuge tube and precipitate it with 5–25 μg tRNA, 0.5 volume of 7.5 M ammonium acetate, and 2.5 volumes of ethanol by incubating 10 min in dry ice. Recover the DNA by centrifuging in a microfuge (12,000 g) for 15 min. The pellet contains the sample. Determine the radioactivity incorporated by any of the methods described in this volume [6]. Dissolve the material in ~100 μl H$_2$O, denature by boiling for 5 min or by heating at 37° in 0.1–0.3 M NaOH for at least 5 min, quench in ice for 2 min, and use for hybridization.

Optimizing Nick-Translated Probe Size

1. Store DNase I (Worthington DPFF) at a concentration of 1 mg/ml in 50% glycerol, 50 mM Tris–HCl (pH 7.4), 10 mM MgCl$_2$, 1 mM dithiothreitol at −70° in small aliquots.

2. Titrate new stocks of DNase to determine the concentration which produces probes 500–1500 bp long as follows. Set up a nick translation reaction as above using a fixed concentration of DNase and perform a time course experiment with the concentrations of dNTPs chosen to mimic experimental conditions. Determine when optimal incorporation occurs. Repeat at different DNase concentrations.

Sizing Nick-Translated Probes

1. Remove approximately 2×10^5 cpm of probe from each reaction.
2. Add NaOH to 0.1 M. Incubate at 37° for 10–15 min.
3. Add 0.2 volume 50% glycerol, 0.5% bromphenol blue.
4. Load samples onto a 2% agarose gel with appropriate denatured size markers (in range of 100–2000 bp). Run at approximately 50 V until the dye has migrated 8–10 cm. Stain, photograph, and then expose gel to film. (See this volume [8].) A good probe migrates as a smear with most of the counts in the range 250–1500 bp. Probes with material which smears from the gel origin have been correlated with high backgrounds.

Notes

1. Nick translation kits are now available from Bethesda Research Laboratories, New England Nuclear, and Amersham. In general they provide instructions, a stock mixture of Pol I and DNase I, and a series of buffers lacking one or more unlabeled deoxyribonucleoside triphosphate(s). The radioactive triphosphates are the sole source of the missing nucleotides.

2. Failure to nick translate a fragment is almost invariably a function of the quality of the DNA. Once the enzymes have been eliminated as the cause by successfully nick translating a known control DNA and once the quality of the radioisotope has been established (by successfully using it to label control DNA), one should focus attention on the sample DNA. The problem can sometimes be alleviated by phenol extracting the DNA (this volume [4]) or by purifying it on a column such as Elutip (Schleicher and Schuell) or NACS (Bethesda Research Laboratories). If neither procedure solves the problem, the concentration of DNase and the incuba-

tion time before Pol I is added may have to be increased. If one is using a kit with premixed enzymes, better results can sometimes be obtained by simply nick translating in a larger volume. Buffer and enzymes are scaled up accordingly, but sample DNA and sometimes radioisotopes are not increased.

[10] Enzymes for Modifying and Labeling DNA and RNA

By FABIO COBIANCHI and SAMUEL H. WILSON

In this chapter, we outline practical applications for 12 of the nucleic acid enzymes now in routine use in the molecular biology laboratory. Properties of these enzymes have been reviewed elsewhere,[1-4] and various methods for their use also have appeared. Here we describe a set of simple procedures we have used along with methods for isolating and monitoring reaction products of interest.

Klenow Fragment of *Escherichia coli* DNA Polymerase I

The Klenow fragment enzyme is a single polypeptide chain obtained by proteolytic digestion of DNA polymerase I or by molecular cloning of the appropriate portion of the gene. The enzyme carries the polymerase and $3' \rightarrow 5'$ exonuclease activities of intact DNA polymerase I, but lacks the $5' \rightarrow 3'$ exonuclease activity of the intact enzyme.

The Klenow fragment is used primarily for filling in and labeling recessed $3'$ ends of double-stranded DNA produced by digestion with restriction enzymes and for repairing the ends of DNA duplexes. Nucleases such as S_1 and Bal 31 and some restriction enzymes often create staggered ends consisting of one or more unpaired bases. Since these unpaired bases prevent subsequent blunt-end ligation (as in the ligation of linkers), Klenow fragment is used to fill in the missing nucleotides. The Klenow fragment is the enzyme of choice in DNA sequencing protocols by the chain termination method (see this volume [58]), although reverse transcriptase is also useful because of its superior ability to read through

[1] This series, Vols. 65, 68, 100.
[2] T. Maniatis, E. F. Fritsch, and J. Sambrook, "Molecular Cloning: A Laboratory Manual." Cold Spring Harbor Lab., Cold Spring Harbor, New York, 1982.
[3] J. G. Chirikjian and T. S. Papas, eds., "Gene Amplification and Analysis" Vol. 2. Elsevier/North-Holland, New York, 1981.
[4] P. D. Boyer, ed., "The Enzymes," 3rd ed., Vol. 15. Academic Press, New York, 1982.

secondary structures. A recent application of the enzyme has been the production of single-stranded probes by primer extension in the single-stranded phage M13 system.[5,6] The probes generated by these methods can be used for *in situ* hybridization and S_1 nuclease analysis.

Filling-in Reaction

Mix, in order, to a total volume of 20 μl using water to adjust the volume,

DNA with 3'-recessed ends	0.1 to 1 μg
10× fill-in buffer [500 mM Tris–HCl at pH 7.5, 100 mM MgCl$_2$, 10 mM dithiothreitol (DTT), 500 μg/ml bovine serum albumin]	2 μl
2 mM each of three deoxyribonucleoside triphosphates (dNTPs)	3 μl
[α-^{32}P]dNTP at 400 Ci/mmol (1–2 μCi total) (Choose the nucleotide not represented among nonradioactive dNTPs)	1–2 μl
Klenow fragment	1 unit

Incubate at 23° for 15 min and add

2 mM dNTP, unlabeled dNTP corresponding to labeled nucleotide above	2 μl

Incubate 10 min at 23°.

Stop the reaction by adding 1 μl 0.5 M EDTA (pH 8.0), add 100 μl of TE at pH 8.0 (10 mM Tris–HCl at pH 8.0, 1 mM EDTA), extract the solution with phenol–chloroform 2 times, and precipitate with ethanol and ammonium acetate as detailed in this volume [5]. Wash the pellet with 70% ethanol. Alternatively, unincorporated triphosphates may be removed using a spun Sephadex G-50 column (this volume [9]) or by drop dialysis (this volume [5]).

Monitoring the Reaction. Incorporation of radioactivity into trichloroacetic acid (TCA)-insoluble material is measured by spotting a tiny aliquot onto GF/C glass filters as detailed in this volume [6].

Production of Single-Stranded Probes

Procedure 1. In this first protocol, the M13 phage recombinant DNA (plus strand) is primed with a commercially available primer upstream of the insert and the primer extended away from the insert in the presence of

[5] J. Meinkoth and G. Wahl, *Anal. Biochem.* **138**, 267 (1984).
[6] A. Seiler-Tuyns, J. D. Eldridge, and B. M. Paterson, *Proc. Natl. Acad. Sci. U.S.A.* **81**, 2980 (1984).

radioactive nucleotides. The resultant radioactive probe consists of a partially duplex molecule in which the specificity is dictated by the insert which is in single-stranded configuration.

In a siliconized Eppendorf tube, mix

500 ng single-stranded M13 recombinant DNA	5 μl
8 ng M13 primer (Bethesda Research Laboratories catalog no. 8238 SA)	4 μl
TM buffer (100 mM Tris–HCl at pH 7.6, 600 mM NaCl, 66 mM MgCl$_2$)	1 μl

Heat at 90° for 5 min, incubate at 23° for 45 min, and add

solution containing 660 μM each dATP, dGTP, TTP and 50 μCi [α-^{32}P]dCTP at 3000 Ci/mmol	3 μl
40 mM dithiothreitol (DTT)	1 μl
1 to 2 units Klenow fragment	1–2 μl

Incubate at 23° for 30 min.

Stop the reaction by adding 1 μl 0.4 M EDTA, 2% sodium dodecyl sulfate (SDS). Remove the unincorporated label by ethanol precipitation as described above. The probe is ready for immediate use and is not to be denatured before hybridization.

Procedure 2. In the second procedure, uniformly labeled single-stranded DNA probes are prepared in which only the cloned insert is specifically labeled. In this protocol, the M13 phage DNA-containing insert is primed with one of the standard sequencing primers (downstream of the insert), and the primer is extended to traverse the insert in the presence of radioactive nucleotides. The partially duplex product is digested with a restriction enzyme that will cut at the end of the insert distal to the primer. The strands are dissociated, and the single-stranded labeled probe is resolved and recovered by acrylamide gel electrophoresis in the presence of urea.

In a siliconized Eppendorf tube, mix

500 ng single-stranded M13 recombinant DNA	5 μl
3 ng M13 primer (New England BioLabs catalog no. 1211)	1 μl
TM buffer (see above)	2 μl
Water	2 μl

Incubate at 60° for 1 hr and add

solution containing 1 mM each dCTP, dGTP, TTP	10 μl
50 μCi [α-^{32}P]dATP at 600 Ci/mmol	5 μl
50 mM DTT	1 μl
Water	3 μl
5 units Klenow fragment	1 μl

Incubate 30 min at 23° and add

chase solution (2 mM each dATP, dCTP, dGTP,   1 μl
    TTP)
Incubate at 23° for 20 min.

Inactivate the Klenow fragment by incubating at 70° for 5 min. Add 4 μl of the appropriate 10× restriction buffer, 1 μl restriction enzyme, and water to 40 μl. Incubate at 37° for 1 hr. Purify the single-stranded labeled DNA probe by gel electrophoresis under denaturing conditions as described in this volume [8].

Procedure 3. This technique is aimed at synthesizing highly radioactive oligomers for use in hybridization experiments. It is basically a variation of procedures 1 and 2 using an oligomer as template and a smaller oligomer complementary to the 3′ end of the template as the primer.[7] Since the primer and template are often identical in size after the reaction, a method has been developed to facilitate electrophoretic separation of the two fragments in gels. The blocking group at the 5′ end of the primer, a dimethoxytrityl group, is not removed following organic synthesis of the primer. Since this group is very large, it decreases mobility of the labeled primer. The extended primer may be excised from a denaturing polyacrylamide gel without fear of contamination by unlabeled template. The blocking group does not interfere with subsequent hybridization.[7] As an alternative, an internal primer that does not form a flush 3′ end with the template may be used.

Mix in a total of 10 μl, using water to adjust the volume,
    extension buffer (600 mM Tris–HCl at pH 7.5, 600   1 μl
        mM NaCl, 60 mM MgCl$_2$, 20 mM DTT)
    primer oligomer with blocked 5′ end (4-fold molar   100 ng
        excess over template)
    template oligomer   60 ng
Heat at 55° for 10 min; cool slowly to room temperature.
Transfer the entire mixture to a fresh tube containing
    300 μCi [α-^{32}P]dNTP at 3000 Ci/mmol (all four   dry
        dNTPs may be labeled; unlabeled dNTPs each at
        10 μM)
Place in ice 5 min and add
    Klenow fragment (Boehringer-Mannheim)   2.5 units
Incubate 2–3 hr in ice.

Purify labeled strand in a 20% polyacrylamide–urea gel using the methods in this volume [8].

Monitoring Reactions in Procedures 1 through 3. Use methods described under Filling-in Reaction.

---

[7] A. B. Studencki and R. B. Wallace, *DNA* **3**, 7 (1984).

Notes. The filling-in reaction described is intended as a "repairing reaction" for generating blunt ends in DNA molecules. In order to monitor the activity of the enzyme the labeled deoxyribonucleoside triphosphate is added at low specific activity, 400–600 Ci/mmol, as a tracer. Since the probe prepared by procedure 2 is to be used in S_1 analysis experiments, the specific activity of the labeled triphosphate is chosen low in order to minimize the formation of breakdown products. The specific activity of probes obtained with procedures 1 and 2 should be in the order of 10^9 and 10^8 cpm/μg of DNA, respectively. Procedure 3 provides the highest specific activities possible, $>10^9$ cpm/μg.

Second-Strand DNA Synthesis with Random Oligodeoxynucleotides as Primers

It is well known that DNA polymerases can use oligonucleotides as primers for second-strand synthesis from a single-stranded DNA template.[8,9] A technique was developed by Feinberg and Vogelstein[10] in which the Klenow fragment is used to copy single-stranded DNA templates primed with random oligodeoxynucleotides prepared from calf thymus DNA or random hexamers obtained commercially.

Using this technique a stable high specific activity DNA can be obtained using very small amounts of DNA template. Probes obtained by this procedure are suitable for filter hybridization techniques such as Southern and Northern analysis, colony screening, plaque hybridization, and *in situ* hybridization.

In an Eppendorf centrifuge tube in ice mix

10 mg/ml bovine serum albumin (nucleic acid grade)	1 μl
LS solution (440 mM HEPES–NaOH at pH 7.6, 44 μM each dATP, dGTP, TTP, 110 mM Tris–HCl at pH 7.6, 11 mM MgCl$_2$, 22 mM 2-mercaptoethanol, oligodeoxyribonucleotide hexamers at 300 μg/ml, commercially available, or prepared by DNase I digestion of calf thymus DNA)	11.5 μl
Water	12.5 μl

The purified DNA restriction fragment (1–5 μl, 20–100 ng in water) is sealed in a capillary tube and heated at 100° for 2 min, immediately cooled to 0°, and added to the reaction mixture. To the primers–template mixture in ice add

50 μCi [α-^{32}P]dCTP at 3000 Ci/mmol	5 μl
(2.5 units) Klenow fragment	1 μl
Incubate at 23° for 3 hr.	

[8] M. Goulian, *Cold Spring Harbor Symp. Quant. Biol.* **33,** 11 (1984).
[9] J. M. Taylor, R. Illmensee, and J. Sumers, *Biochim. Biophys. Acta* **442,** 324 (1976).
[10] A. P. Feinberg and B. Vogelstein, *Anal. Biochem.* **132,** 6 (1983).

Stop the reaction by adding 1 μl 0.5 M EDTA and process the sample as described for the Filling-in Reaction.

Monitoring the Reaction. Incorporation of radioactivity into acid-insoluble material is monitored by TCA precipitation on filters as described in this volume [6]. The reaction product is analyzed by polyacryamide gel electrophoresis under denaturing conditions (this volume [8]). From a 1.2-kb restriction fragment, the average size of single-stranded probe was 0.8 kb. Over 70% of the precursor triphosphate is incorporated into DNA and a specific activity of 10^9 cpm/μg of DNA can be obtained.

Removing 5′-Terminal Phosphates from DNA

Removal of 5′-phosphates from DNA or RNA is performed prior to labeling 5′ ends with ^{32}P or to prevent undesired intramolecular ligation of either vector DNA or RNA. Bacterial alkaline phosphatase (BAPF) and calf intestine alkaline phosphatase (CIP) are commercially available and are used for this purpose. The amount of enzyme necessary may vary depending upon the substrate. Therefore, it is best to determine the amount of enzyme required empirically. Generally 1 unit of enzyme is needed for the complete removal of terminal phosphates from 100 pmol of 5′ ends of DNA. Both enzymes also cleave 3′-phosphates.

Bacterial Alkaline Phosphatase Reaction

Mix in a total of 50 μl, using water to adjust the volume
DNA with 5′-terminal phosphates   10 pmol of ends
500 mM Tris–HCl at pH 8.0   5 μl
Bacterial alkaline phosphatase   0.1 unit
Incubate at 37° (5′-protruding or flush ends) or at 60° (recessed 5′ ends) for 30 min.

Calf Intestine Alkaline Phosphatase Reaction

Mix in a total of 50 μl, using water to adjust the volume
DNA with 5′-terminal phosphates,   10 pmol of ends
10× CIP buffer (500 mM Tris–HCl at pH   5 μl
  9.0, 10 mM MgCl$_2$, 1 mM ZnCl$_2$, 10 mM
  spermidine)
Calf intestine phosphatase (CIP)   0.1 unit
Incubate at 37° (5′-protruding ends) for 30 min, add another 0.1 unit of CIP, and continue incubation for an additional 30 min. For blunt ends or recessed 5′ ends use 56°.

Stop the reaction by adding 1 μl 0.5 M EDTA at pH 8.0. Add 100 μl TE at pH 8.0. Extract twice with an equal volume of phenol–chloroform (1 : 1) saturated with TE at pH 8.0, and twice with chloroform. Extract with abundant ether; add sodium acetate at pH 5.3 to a final concentration of 0.3 M and precipitate the DNA with 2.5 volumes of ethanol. Centrifuge and resuspend the pellet in a minimal volume of TE at pH 8.0. The DNA can be further purified by chromatography in Sephadex G-50 and is now ready for subsequent ligation or end labeling.

Monitoring the Reaction. If some of the DNA has already been end labeled and is available as tracer, the disappearance of radioactivity from trichloroacetic acid-precipitable material, hence the extent of the reaction, can be measured by means of a filter assay (this volume [6]). Alternatively the extent of dephosphorylation of vector DNA can be followed by self-ligation (of the vector DNA) and subsequent bacterial transformation. The efficiency of transformation with dephosphorylated DNA product above should be less than the efficiency observed with vector not dephosphorylated. See elsewhere in this chapter for ligation methods.

T4 Polynucleotide Kinase: Uses for 5'-End Labeling with ^{32}P

Bacteriophage T4-induced polynucleotide kinase catalyzes transfer of the γ-phosphate of ATP to the 5'-hydroxyl terminus of either DNA or RNA.[11] Because of the requirement for a 5'-hydroxyl terminus, the substrate often must be dephosphorylated with alkaline phosphatase prior to the [^{32}P]phosphotransfer reaction. Alternatively, a 5'-phosphoryl terminus can be labeled by phosphate exchange, taking advantage of the reversibility of the T4 polynucleotide kinase-catalyzed reaction.[12]

Labeling 5'-Hydroxyl Termini (Forward Reaction)

For protruding 5' ends, mix in an Eppendorf tube in ice, adjusting the volume to 50 μl with water

Dephosphorylated DNA	1–50 pmol of 5' ends
10× kinase buffer I (500 mM Tris–HCl at pH 7.6, 100 mM MgCl$_2$, 50 mM DTT, 1 mM spermidine hydrochloride, 1 mM EDTA)	5 μl
150 μCi [γ-^{32}P]ATP at 3000 Ci/mmol	15 μl
T4 polynucleotide kinase	20 units

Incubate at 37° for 30 min.

For flush or recessed 5' ends, mix in an Eppendorf tube in ice, adjusting the volume to 40 μl with water

Dephosphorylated DNA	1–50 pmol of 5' ends
10× TSE buffer (200 mM Tris–HCl at pH 9.5, 10 mM spermidine hydrochloride, 1 mM EDTA)	4 μl

---

[11] C. C. Richardson, *Proc. Natl. Acad. Sci. U.S.A.* **54,** 158 (1965).

[12] J. H. Van de Sande, K. Kleppe, and H. G. Khorana, *Biochemistry* **12,** 5050 (1973).

Incubate at 90° for 2 min, chill by placing in ice water, and add

10× kinase buffer II (500 mM Tris–HCl	5 μl
at pH 9.5, 100 mM MgCl$_2$, 50 mM DTT,	
50% glycerol (v/v)	
150 μCi [γ-^{32}P]ATP at 3000 Ci/mmol	dry
T4 polynucleotide kinase	20 units
Water to 50 μl	

Incubate at 37° for 15 min.

Exchange Reaction

Mix in a total of 50 μl, using water to adjust the volume

DNA 5′-PO$_4$	1–50 pmol of 5′ ends
10× kinase buffer III (500 mM imida-	5 μl
zole–HCl at pH 6.6, 100 mM MgCl$_2$,	
50 mM DTT, 1 mM spermidine hy-	
drochloride, 1 mM EDTA)	
5 mM ADP	3 μl
100 μCi [γ-^{32}P]ATP at 3000 Ci/mmol	10 μl
T4 polynucleotide kinase	20 units

Incubate at 37° for 30 min.

Coupled Phosphatase–Kinase Reaction

This method is particularly useful for end labeling of restriction fragments since the restriction digestion, dephosphorylation, and terminal labeling can be performed sequentially in one reaction tube. In this procedure, the dephosphorylation of the substrate is carried out as usual with calf intestine alkaline phosphatase; subsequently, the labeling reaction is performed in the presence of inorganic phosphate under conditions that will inhibit alkaline phosphatase but not the phosphorylation reaction.[13]

In an Eppendorf tube in ice mix

50 pmol 5′ ends of restricted DNA in pre-	
vious restriction buffer	10 μl
Calf intestine alkaline phosphatase	0.1 unit

Incubate at 45° for 45 min. Add

10× kinase buffer IV (500 mM glycine–NaOH	5 μl
at pH 9.2, 100 mM DTT, 50 mM MgCl$_2$)	
Sodium phosphate at pH 9.2	see note below
[γ-^{32}P]ATP at 1000 Ci/mmol	see note below
T4 polynucleotide kinase	4 units
Water to a final volume of	50 μl

Incubate at 37° for 15 min.

---

13 G. Chaconas and J. H. Van de Sande, this series, Vol. 65, p. 75.

Stop the reaction with 1 μl of 0.5 M EDTA and purify the DNA by phenol extraction and ethanol precipitation as described for Klenow fragment filling-in reaction.

Notes. For restriction fragments with flush or 3' protruding ends, the ATP concentration in the reaction should be 10–30 μM with an ATP to 5' ends ratio of at least 10:1 and a sodium phosphate concentration of 1 mM. For restriction fragments with 5'-protruding ends, an ATP concentration in the reaction of 10 μM is recommended with an ATP to 5' ends ratio of 10:1 and sodium phosphate concentration of 5 mM.

Monitoring the Reactions. Incorporation of ^{32}P can be monitored by TCA precipitation as described in this volume [6]. For oligonucleotides of 10 residues or less, use DE-81 filters and wash with concentrated phosphate as described in this volume [6]. Under conditions described above, the forward reaction yields 2×10^6 cpm/pmol end, the exchange reaction yields 6×10^5 cpm/pmol end, and the coupled reaction yields 10^5 cpm/pmol end.

The 3'-Phosphatase Activity of T4 Polynucleotide Kinase

T4 polynucleotide kinase possesses a 3'-phosphatase activity,[14] in addition to the activity that transfers the γ-phosphate of ATP to the 5'-OH terminus of DNA. The kinase and phosphatase activities of the enzyme may be used independently by choosing appropriate reaction conditions. By incubating the substrate at low pH in the absence of ATP or ADP, 3'-phosphates will be removed from DNA or RNA molecules without altering the structure of the 5' terminus. Note that a commercially available cloned polynucleotide kinase lacks the 3'-phosphatase activity of the native enzyme and cannot be used for this reaction.

Reaction Conditions[15]

Mix in a total volume of 50 μl, using water to adjust the volume
    DNA with 3'-phosphate ends   1–50 pmol of ends
    10× reaction buffer (1 M Tris–HCl at pH   5 μl
        6.5, 1 M magnesium acetate, 50 mM 2-
        mercaptoethanol)
    T4 polynucleotide kinase   6 units
    Incubate at 37° for 12 hr.

Stop the reaction by adding 1 μl 0.5 M EDTA and recover the DNA by phenol extraction and ethanol precipitation as described above.

Monitoring the Reaction. If 3'-end-labeled DNA is available as tracer,

[14] V. Cameron and O. C. Uhlenbeck, *Biochemistry* **16**, 5120 (1977).
[15] B. Royer-Pokorna, L. K. Gordon, and W. A. Haseltine, *Nucleic Acids Res.* **9**, 4595 (1981).

the extent of the reaction can be followed using either loss of trichloroacetic acid-precipitable material or loss of DE-81 binding material (this volume [6]).

3′-Terminal Labeling of RNA with T4 RNA Ligase[16]

Bacteriophage T4 RNA ligase catalyzes the ligation of a 5′-phosphoryl donor to a 3′-hydroxyl acceptor. Although the minimal acceptor must be a trinucleoside diphosphate, mononucleoside 3′,5′-bisphosphates (pNps) are effective donors.[17,18] One consequence of this observation is that a convenient method for labeling the 3′ end of RNA molecules *in vitro* is available. By using a [5′-^{32}P]pNp as a donor and RNA as acceptor, the product of the reaction is an RNA molecule extended by one nucleotide and containing a 3′-terminal phosphate and a [^{32}P]phosphate in the last internucleotide linkage.

$$\overset{5'}{\cdots} \text{-pC-pG-pA-}\overset{3'}{\text{OH}} \xrightarrow[\text{ATP}]{[5'-^{32}\text{P}]\text{pCp}} \overset{5'}{\cdots} \text{-pC-pG-pA-[}^{32}\text{P}]\text{pCp}^{3'}$$

In an Eppendorf tube in ice mix

RNA	1 μg
0.15 mM ATP	1 μl
0.75 M HEPES–NaOH at pH 8.3	2 μl
0.3 M MgCl$_2$	1 μl
0.1 M DTT	1 μl
Dimethyl sulfoxide	3 μl
300 μg/ml BSA	1 μl
100 μCi [5′-^{32}P]pCp at 3000 Ci/mmol	10 μl
T4 RNA ligase at 3000 units/ml	4 μl
Water to final volume of	30 μl

Incubate at 4° for 24 hr.

Monitoring the Reaction. The reaction is monitored by measuring incorporation of radioactivity into trichloroacetic acid-precipitable material. The reaction products can be analyzed by polyacrylamide gel electrophoresis under denaturing conditions; the labeled RNA can be purified by drop dialysis (this volume [5]). The yield may vary, depending on the sequence and the secondary structure at the 3′ end of the RNA; between 10^4 and 10^5 cpm/μg of RNA is expected. In general, poly(A)$^+$ RNA and tRNA are better substrates than rRNA.

---

[16] T. E. England, A. G. Bruce, and O. C. Uhlenbeck, this series, Vol. 65, p. 65.

[17] T. E. England, R. I. Gumport, and O. C. Uhlenbeck, *Proc. Natl. Acad. Sci. U.S.A.* **74,** 4839 (1977).

[18] Y. Kikuchi, F. Hishinuma, and K. Sakaguchi, *Proc. Natl. Acad. Sci. U.S.A.* **75,** 1270 (1978).

3'-End Labeling of DNA Using [α-^{32}P]ddATP

Recently, a method[19] for efficient 3'-end labeling of DNA was developed in which blunt or 3'-protruding ends are labeled with [α-^{32}P]ddATP using terminal transferase (TdT). In the past, the method of choice for labeling 3'-protruding ends had been with cordycepin 5'-[α-^{32}P]triphosphate ([α-^{32}P]KTP). Low incorporation of precursor was a problem, however, especially for subsequent sequencing (with the chemical method) from the 3' end. Since the new ddATP method of labeling is more efficient than the KTP method, the 3'-end labeling of DNA is now a valid alternative to 5'-end labeling before DNA sequencing.

In a total volume of 50 μl, using water to adjust the volume mix

DNA fragment	10 pmol of 3' ends
5× TdT buffer (700 mM sodium cacodylate at pH 7.2, 5 mM CoCl$_2$, 0.5 mM DTT)	10 μl
150 μCi [α-^{32}P]ddATP at 3000 Ci/mmol	15 μl
TdT	10 units

Incubate at 37° for 1 hr.

Stop the reaction by adding 1 μl 0.5 M EDTA, and purify the DNA by phenol extraction and ethanol precipitation as described in this chapter or this volume [4, 5].

Monitoring the Reaction. The reaction is monitored by measuring incorporation of [α-^{32}P]ddATP into acid-precipitable material as described in this volume [6] and can be expected to yield 3×10^6 cpm/pmol of 3' ends.

Exonucleases

Nuclease Bal 31

Bal 31 nuclease, purified from the culture medium of *Alteromonas espejiana,* carries a double-stranded DNA exonuclease activity that progressively and bidirectionally removes mononucleotides from both strands of linear DNA.[20] Since degradation occurs only at ends of the duplex DNA molecule, the enzyme is useful for such applications as restriction enzyme fragment mapping and controlled removal of sequences from the ends of molecules to be cloned or expressed. The enzyme is highly sequence dependent, and the rate of shortening of DNA molecules can vary considerably. The incubation time required to achieve

---

[19] S. I. Yousaf, A. R. Carroll, and B. E. Clarke, *Gene* **27,** 309 (1984).

[20] H. B. Gray, Jr., D. A. Ostrander, J. L. Hodnett, R. J. Legersky, and D. L. Robberson, *Nucleic Acids Res.* **2,** 1459 (1975).

the desired resection of a DNA molecule can be estimated according to the following equation[21]:

$$M_t = M_0 - 2M_n V_{max} t/(K_m + S_0)$$

where M_0 is the original MW of the DNA, M_t is the MW after t minutes of incubation, M_n is the average MW of a mononucleotide (330), V_{max} is the maximum reaction velocity, K_m is the Michaelis constant, and S_0 is the molar concentration of duplex termini. For the conditions given below, a value of $K_m = 2 \times 10^{-8} M$ and $V_{max} = 3 \times 10^{-6}$ mol/liter-min should be used.

In an Eppendorf tube in ice mix

Linear DNA	1.5–15 μg
10× Bal31 buffer (200 mM Tris–HCl at pH 8.1, 125 mM MgSO$_4$, 125 mM MgCl$_2$, 6 M NaCl, 10 mM EDTA)	10 μl
Bal 31	1–4 units per μg of DNA
Water to a final volume of	100 μl

Incubate at 23°.

Aliquots of the reaction are removed at various times and the reactions are stopped by adding EGTA to 20 mM final concentration. Measure the extent of digestion by analyzing the products electrophoretically (this volume [8]).

Exonuclease III

Exonuclease III of *Escherichia coli* catalyzes the release of mononucleotides from the 3' end of a DNA strand. In addition, the enzyme is a DNA 3'-phosphatase, an apurine endonuclease that can cleave phosphodiester bonds at apurinic sites, and an RNase H preferentially degrading RNA in a DNA–RNA hybrid. The enzyme will attack a nick, but will not cleave either single-stranded DNA or double-stranded DNA with a 5' recessed end.[22,23] Exo III followed by S$_1$ nuclease can be used to shorten double-stranded DNA in a controlled manner for production of deletions in particular regions of interest.[24] Wu and colleagues[25] have developed a method for the control of exonucleolytic degradation of duplex DNA with Exo III in which the products are molecules with 3' ends shortened to varying lengths. These 3' ends can then be extended or sequenced using the chain termination procedure. In the following protocol, using a rela-

[21] R. J. Legersky, J. L. Hodnett, and H. B. Gray, Jr., *Nucleic Acids Res.* 5, 1445 (1978).
[22] C. Richardson, I. Lehman, and A. Kornberg, *J. Biol. Chem.* 239, 51 (1964).
[23] S. Rogers and B. Weiss, this series, Vol. 65, p. 201.
[24] T. Roberts, R. Kacicj, and M. Ptashne, *Proc. Natl. Acad. Sci. U.S.A.* 76, 760 (1979).
[25] G. Li-He and R. Wu, this series, Vol. 100, p. 60.

TABLE I
DIGESTION OF DNA WITH EXONUCLEASE III[a]

Number of nucleotides removed from each end of DNA	Incubation (min)	Exo III (units/pmol DNA)
100–250	10–15	15–20
250–500	25–50	20–25
500–750	50–75	25–30
750–1000	75–100	30–35
1000–1500	100–150	35–45

[a] To remove approximately 10 nucleotides per minute from each 3′ end of DNA, the concentration of *Escherichia coli* exonuclease III (Exo III) needed is given in the table. A unit of exonuclease III is the amount of enzyme that liberates 1 nmol of mononucleotides from a sonicated DNA substrate in 30 min at 37°. Table reproduced from Li-He and Wu.[25]

tively high ratio of enzyme to DNA ends and a moderate concentration of salt, the digestion is synchronous, removing approximately 10 nucleotides per minute per 3′ end at 23°.[25]

In an Eppendorf tube in ice mix

DNA with 3′ recessed or blunt ends	1 pmol
10× Exo III buffer (660 mM Tris–HCl at pH 8.0, 770 mM NaCl, 50 mM MgCl$_2$, 100 mM DTT)	4 μl
Exonuclease III	(see Table I)
Water to a final volume of	40 μl

Incubate at 23° for the required time (see Table I).
Stop the reaction by addition of 2 μl of 0.5 M EDTA.

λ Exonuclease

Several different catalytic properties of λ exonuclease make this enzyme a useful reagent for the analysis and modification of DNA structure. The enzyme processively degrades duplex DNA from the 5′ end (flush or recessed) 100 times faster than single-stranded DNA, with preference for a 5′-phosphoryl group over a 5′-hydroxyl group. To cleave each molecule at a uniform rate, a molar excess of enzyme is required. The enzyme cannot start digestion of duplex DNA at a nick or a gap. λ exonuclease has been used in a variety of different applications such as determination of DNA structure at termini, restriction mapping, preparation of substrate for terminal transferase, determination of direction of transcription, preparation of substrate for DNA sequencing, and preparation of substrates for other enzymes. See Little[26] for a review.

[26] J. W. Little, *in* "Gene Amplification and Analysis" (J. G. Chirikjian and T. S. Papas, eds.), Vol. 2, p. 135. Elsevier/North-Holland, New York, 1981.

λ *Exonuclease Reaction for Preparation of Template for Terminal Transferase.* Jackson *et al.*[27] demonstrated that the low efficiency of 3' labeling by terminal deoxynucleotidyl transferase in the presence of Co^{2+} can be enhanced by using λ exonuclease to remove 5' unpaired terminal nucleotides that are often generated when DNA fragments are prepared by restriction enzyme digestion.

In an Eppendorf tube in ice mix

DNA fragment	10 pmol of 5' ends
10× exonuclease buffer (670 mM glycine–KOH at pH 9.4, 30 mM MgCl$_2$, 5 mM DTT)	10 μl
λ exonuclease	50 units
Water to a final volume of	100 μl

Incubate at 14° for 4–5 min, and stop the reaction by adding 2 μl of 0.5 M EDTA.

T4 DNA Polymerase

In the absence of deoxyribonucleoside triphosphates, the T4 DNA polymerase 3' exonuclease activity can be used to resect molecules of linear duplex DNA. The product of the reaction is a partially single-stranded template that can be used in resynthesis of double-stranded DNA by adding deoxyribonucleoside triphosphates. The resulting DNA molecule can be full-length, double-stranded, unnicked, and radioactive with high specific activity[28] (see Fig. 1). In addition, in the presence of a saturating concentration of the four canonical triphosphates, the enzyme will create blunt-ended molecules starting with either 5'-protruding or 5'-recessed ends.

In an Eppendorf tube in ice mix

DNA fragment	1–2 μg
10× T4 buffer (330 mM Tris–acetate at pH 7.9 660 mM potassium acetate, 100 mM magnesium acetate)	5 μl
T4 DNA polymerase	5–10 units
Water to a final volume of	50 μl

Incubate at 37° for the appropriate time (20–120 min) (resection reaction). Add

Resynthesis mix (1× T4 buffer with 50 mM each dATP, dGTP, TTP and 5 μM [α-^{32}P]dCTP at 400 Ci/mmol, and 14% (v/v) glycerol)	150 μl

Incubate at 37° for the appropriate time (40–120 min) (resynthesis reaction).

[27] D. A. Jackson, R. H. Symons, and P. Berg, *Proc. Natl. Acad. Sci. U.S.A.* **69**, 2904 (1972).
[28] K. C. Deen, J. A. Landers, and M. Berninger, *Anal. Biochem.* **135**, 456 (1983).

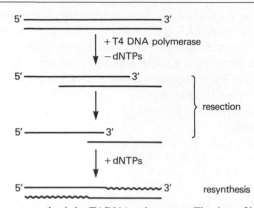

FIG. 1. Replacement synthesis by T4 DNA polymerase. The time of incubation for the 3'-exonuclease reaction depends on the length of the DNA fragment and must be determined experimentally. In order to maximize label incorporation into a DNA fragment, the resection reaction is allowed to continue until just before the middle point of the DNA molecules. This is done by removing aliquots from the resection reaction mixture (4 μl) and incubating with 11 μl of resynthesis mix at 37° for a time that is at least twice as long as the resection reaction.[29] Aliquots (1 μl) of the resynthesis reaction mixture are precipitated with 10% TCA to determine the extent of incorporation. Figure redrawn from Deen et al.[28]

Stop the reaction by adding EDTA to 20 mM final concentration. The reaction products are purified by phenol extraction and ethanol precipitation as described in this volume [4,5]. Under these conditions, specific activity as high as 10^8 cpm/μg can be obtained.

Monitoring the Reaction. Exonuclease reactions are monitored by characterizing the products electrophoretically in agarose or polyacrylamide gels (depending on the expected size of the products) and then, more precisely, by subcloning and nucleotide sequence analysis.

[*Editors' Note.* Conditions for the use of exonuclease VII are found in this volume [66]. The single-strand specific nucleases, S_1 and mung bean nuclease, are discussed in this volume [34] and [35], respectively.]

Ligation

From the many sets of reaction conditions described in the literature, it is apparent that conditions for enzymatic activity of T4 DNA ligase sometimes are suboptimal for use in cloning experiments. It is well known, for example, that a high concentration of ATP in the ligation reaction leads to accumulation of 5'-adenylated intermediates[29] with consequent low yield of recombinant molecules. Lathe et al.[30] developed a method for linker tailing DNA molecules without the need for restriction enzyme digestion of the linkers and thus for the methylation of the tailed

[29] V. Sgaramella and H. G. Khorana, *J. Mol. Biol.* **72**, 427 (1972).
[30] R. Lathe, M. P. Kieny, S. Skory, and J. P. Lecocq, *DNA* **3**, 173 (1984).

molecules. In this study, low ionic strength (30 mM NaCl) and low ATP concentration (0.25 mM) were found to be optimal. Recently, King *et al.*[31] undertook a study to determine optimal T4 DNA ligase reaction conditions (for blunt end ligation) using the efficiency of bacterial transformation as assay.

Blunt End Ligation (Optimal); Vector/Insert Molar Ratio = 3

In a siliconized Eppendorf tube in ice mix

Dephosphorylated vector DNA (~4 kb)	160 ng
DNA fragment (~1 kb)	13 ng
10× ligase buffer I (250 mM Tris–HCl at pH 7.5, 50 mM MgCl$_2$, 25% (w/v) polyethylene glycol 8000, 5 mM DTT, 4 mM ATP) (no effect on the efficiency of transformation was detected when the ATP final concentration was varied from 10 μM to 1 mM)	4 μl
T4 DNA ligase	1 unit
Water to a final volume of	20 μl

Incubate at 23° for 4 hours and stop the reaction by adding 1 μl of 0.5 M EDTA. Dilute 5-fold before adding the mixture to competent cells for transformation.

Sticky End Ligation; Vector/Insert Ratio = 0.5

In a siliconized Eppendorf tube in ice mix:

Vector DNA (~4 kb)	160 ng
DNA fragment (~1 kb)	80 ng
10× ligase buffer II (250 mM Tris–HCl at pH 7.5, 100 mM MgCl$_2$, 100 mM DTT, 4 mM ATP)	2 μl
T4 DNA ligase	0.01 unit
Water to a final volume of	20 μl

Incubate overnight at 8°.

Linker Tailing[31]

In a siliconized Eppendorf tube in ice mix

Linear blunt-ended plasmid DNA	5 μg
10× ligation buffer III (300 mM Tris–HCl at pH 7.5, 300 mM NaCl, 80 mM MgCl, 20 mM DTT, 2 mM EDTA, 70 mM spermidine hydrochloride, 1 mg/ml bovine serum albumin, and 2.5 mM ATP)	2 μl
Unphosphorylated linker	500 ng if 10 or 12-mer
	12.5 μg if 8-mer

[31] P. V. King and R. W. Blakesley, *Focus (Bethesda, Md.)* **8,** No. 1, 1 (1986).

| T4 DNA ligase | 2 units |
| Water to a final volume of | 20 μl |

Incubate overnight at 4°.

Monitoring the Reaction. Ligation reactions are monitored by characterizing the products by agarose gel electrophoresis and by bacterial transformation.

DNase I

Pancreatic deoxyribonuclease I is an endonuclease catalyzing the hydrolysis of both double- and single-stranded DNA to produce 5'-phosphoryloligodeoxynucleotides. In molecular biology, DNase I is used, for example, to introduce random nicks in DNA substrates, as in the preparation of substrate for the nick translation reaction.[32] The enzyme has been used, in the presence of ethidium bromide, to introduce one nick at random per circular DNA molecule, in order to obtain sensitive sites for the bisulfite-mediated mutagenesis protocol.[33] Other applications have been described such as production of random DNA fragments for shotgun cloning and sequencing in the M13 system,[34] studies of chromatin structure,[35] and analysis of DNA–protein complexes.[36]

Reaction Conditions for the Production of Random Fragments.[37] Resuspend the plasmid DNA to be cloned and sequenced in 6× SSC at a concentration of 12 mg/ml. Incubate for 10 min in boiling water (relaxed form production). Cool the solution slowly by holding it for 19 min at 65° and then for 10 min at 23° (renaturation). Precipitate the DNA with ethanol and resuspend in 50 mM Tris–HCl at pH 7.6, to a final concentration of 1 mg/ml.

In an Eppendorf tube in ice mix

Resuspended DNA from previous step	50 μl
10× DNase buffer (500 mM Tris–HCl at pH 7.5, 10 mM MgCl$_2$, 1 mg/ml bovine serum albumin)	10 μl
DNase I (electrophoretically pure)	10 ng
Water to a final volume of	100 μl

Incubate at 23°. Remove aliquots of 25 μl every 5 min and stop the reactions by addition of 1 μl of 0.5 M EDTA.

Monitoring the Reaction. The digestion is checked by electrophoresis of the DNA in a 6% polyacrylamide gel using a portion of each sample (10 μl).

[32] P. W. Rigby, M. Dieckmann, C. Rhodes, and P. Berg, *J. Mol. Biol.* **113,** 237 (1977).
[33] L. Greenfield, L. Simpson, and D. Kaplan, *Biochim. Biophys. Acta* **407,** 365 (1975).
[34] S. Anderson, *Nucleic Acids Res.* **9,** 3015 (1981).
[35] H. Weintraub and M. Groudine, *Science* **193,** 848 (1976).
[36] A. Prunell, R. D. Kornberg, L. Lutter, A. Klug, M. Levitt, and F. H. C. Crick, *Science* **204,** 855 (1979).
[37] J. Messing, this series, Vol. 101, p. 20.

Section IV

Restriction Enzymes

[11] Properties and Uses of Restriction Endonucleases

By JOAN E. BROOKS

This chapter focuses on the properties and uses of restriction endonucleases. The large battery of endonucleases now commercially available will first be described in terms of nomenclature and properties. The next section will cover basic methods for their use in mapping and genetic engineering experiments, and a final section will focus on the more detailed aspects of selecting endonucleases for generating ends compatible with subsequent steps of construction.

Definition of Restriction Endonucleases and Nomenclature

Many bacteria contain systems to guard against invasion of foreign DNA. The cells contain specific endonucleases that make double-strand scissions in invading DNA unless the DNA has been previously been modified, usually by the appropriate DNA methylase (but see this volume [12]). The endonuclease with its accompanying methylase is called a restriction modification system.[1-3]

Three distinct types of restriction modification systems have been characterized on the basis of the subunit composition, cofactor requirements, and type of DNA cleavage.[4] Type I systems are the most complex. The endonucleases contain three different types of subunits and require Mg^{2+}, ATP, and S-adenosylmethionine for DNA cleavage. Their recognition sites are complex, and DNA cleavage occurs at nonspecific sites 400–7000 base pairs from the recognition site. Type III systems are somewhat less complex. The endonucleases contain only two types of subunits. Although Mg^{2+} and ATP are required for DNA cleavage as in type I enzymes, S-adenosylmethionine stimulates enzymatic activity without being an absolute requirement. DNA cleavage again occurs distal to the recognition site, but only by 25–27 base pairs. Type II systems are much simpler. The endonucleases contain only one type of subunit, and Mg^{2+} alone is required for DNA cleavage. DNA cleavage occurs at specific sites within or adjacent to the enzyme's recognition site. It is therefore this class of restriction endonucleases that has proved useful to molecular biologists.

[1] W. Arber, *Prog. Nucleic Acid Res. Mol. Biol.* **14,** 1 (1974).
[2] H. O. Smith, *Science* **205,** 455 (1979).
[3] P. Modrich, *Q. Rev. Biophys.* **12,** 315 (1979).
[4] R. Yuan, *Annu. Rev. Biochem.* **50,** 285 (1981).

A large number of site-specific endonucleases with cofactor requirements and cleavage sites similar to known type II restriction endonucleases have been found in many bacteria. For the vast majority of these, there is no evidence that restriction of foreign DNA occurs *in vivo,* but by analogy with bona fide type II restriction endonucleases these have been termed restriction endonucleases.[5]

In 1973, when it became clear that a large number of these endonucleases existed and would be extensively utilized, Smith and Nathans[6] proposed the nomenclature system that is still followed. A three-letter abbreviation for the parent organism (*Hin* for *Haemophilus influenzae* or *Bam* for *Bacillus amyloliquefaciens*), an additional letter if necessary to identify strain or serotype (*Hind* or *Bam*H), and then a Roman numeral to reflect the order of identification or characterization (*Hind*III or *Bam*HI). A complete list of named and characterized restriction endonucleases has been compiled and is updated yearly by Roberts.[7]

Because of their practical uses for molecular dissection of DNA, the restriction endonucleases have been characterized primarily with respect to their recognition sequences and cleavage specificity rather than their protein properties.[8] The majority recognize sequences 4–6 nucleotides in length, but some have now been found with 7- and 8-base recognition sites.[7] Most, but not all, recognition sites contain a dyad axis of symmetry and in most cases all the bases within the site are uniquely specified. (The symmetrical recognition sequence of these endonucleases have been termed *palindromes*.) Those with degenerate or "relaxed" specificities can recognize multiple bases at some positions. Endonucleases with symmetrical recognition sites generally cleave symmetrically within or adjacent to the recognition site, but those that recognize asymmetric sites tend to cut at a distance from the recognition site.

In addition to recognition site, the nature of the DNA cleavage is of paramount importance to uses of restriction endonucleases because the nature of the ends determines the suitability of the fragments for subsequent procedures. All restriction endonucleases cleave their DNA substrates to form 5′-phosphate and 3′-hydroxyl termini on each strand.[9] The

---

[5] R. J. Roberts, *CRC Crit. Rev. Biochem.* **4**, 123 (1976).

[6] H. O. Smith and D. Nathans, *J. Mol. Biol.* **81**, 419 (1973).

[7] R. J. Roberts, *Nucleic Acids Res.* **13**, r165 (1985).

[8] P. Modrich and R. J. Roberts, *in* "Nucleases" (S. Linn and R. J. Roberts, eds.), p. 109. Cold Spring Harbor Lab., Cold Spring Harbor, New York, 1982.

[9] The one possible exception found to date is *Nci*I which has been reported to liberate 3′-phosphate and 5′-hydroxyl termini after digestion of DNA (A. W. Hu and A. H. Marschel, unpublished observations). However, recent experimentation using highly concentrated T4 DNA ligase shows that the *Nci*I ends will religate, suggesting they do have intact 5′-phosphate groups (E. Rosenvold, unpublished observations). Further experimentation needs to be done.

breaks can be staggered, generating either 5'-phosphate extensions on each strand or 3'-hydroxyl extensions on each strand, or they can be opposed, generating "blunt" ends. (See section below on Generating the Proper Ends for Subsequent Steps.)

Among the more than 400 different endonucleases isolated from bacterial strains many share common specificities. Restriction endonucleases which recognize identical sequences have been called "isoschizomers" by Roberts.[5] Although the recognition sequences of isoschizomers are the same, they may vary with respect to the site of cleavage (XmaI vs SmaI[10]), sensitivity to methylation (MspI vs HpaII[11]), and in cleavage rates at various sites (XhoI vs PaeR7 I[12]).

Restriction Endonuclease Selection and Use

Choice of Enodnuclease

The size of DNA fragments to be generated is often a primary consideration in choosing a restriction endonuclease. Fragments of a few hundred bases in length are useful for fine-structure restriction mapping, for "shotgunning" small DNA fragments into vectors such as M13 for DNA sequence analysis, or for generation of more or less random breaks by partial digestion for site-specific mutagenesis or cloning experiments.[13]

Fragments of 1–10 kilobases (kb) are useful for mapping larger DNA regions and for cloning whole genes complete with introns (from eukaryotes) and control sequences into plasmid or λ phage vectors.[14] Still larger fragments 5–50 kb in length are suitable for cloning into cosmid vectors, cloning whole operons, or genome "walking."[15] Finally, techniques currently available to resolve DNA fragments of 1 megabase (1000 kb) or larger can be used to separate and purify whole or partial chromosomes for mapping and cloning.[16,17]

It is difficult to predict the frequency of occurrence of any restriction endonuclease cleavage site within a new DNA substrate. Within a perfectly random DNA sequence that contained 50% GC, a 4-base recognition sequence would occur on average every 256 bases, a 6-base sequence would appear on average at 4-kb intervals, and an 8-base sequence would

[10] S. A. Endow and R. J. Roberts, *J. Mol. Biol.* **112**, 521 (1977).
[11] C. Waalwijk and R. A. Flavell, *Nucleic Acids Res.* **5**, 3231 (1978).
[12] T. R. Gingeras and J. E. Brooks, *Proc. Natl. Acad. Sci. U.S.A.* **80**, 402 (1983).
[13] J. Messing, this series, Vol. 101, p. 20.
[14] A.-M. Frischauf, this volume [17].
[15] A. G. DiLella and S. L. C. Woo, this volume [18].
[16] D. Schwartz and C. R. Cantor, *Cell (Cambridge, Mass.)* **37**, 67 (1984).
[17] G. F. Carle, M. Frank, and M. V. Olsen, *Science* **232**, 65 (1986).

occur at 65-kb intervals. These probabilities are, of course, substantially different for DNA with a skewed GC content (actual DNA can vary from 22 to 73% GC[18]). The level of nonrandomness increases at higher levels of complexity: some di- and trinucleotides occur at distinctly nonrandom ratios. A well-known example of underrepresentation of a dinucleotide is the CpG dinucleotide in mammalian DNA.[19] In addition, the high level of repetitive DNA in higher eukaryotes further biases restriction site representations.[20,21]

Theoretical predictions of site frequency, based on known base composition and repetitive sequences, are being compiled for a variety of organisms.[22] This knowledge should be useful in endonuclease selection, especially for experiments requiring megabase-sized fragments.

Choice of Restriction Endonucleases

Restriction endonucleases are relatively stable proteins. Their purification to homogeneity is not often necessary and optimal reaction conditions are rarely ascertained. A typical restriction reaction will contain, in addition to the DNA substrate and restriction endonuclease(s), Tris buffer, Mg^{2+}, NaCl, 2-mercaptoethanol, and bovine serum albumin. All restriction endonucleases require Mg^{2+} as a cofactor, and most are active in the pH range of 7.2–7.6. The predominant difference among the endonucleases is their dependence on ionic strength. Most manufacturers now include tables in their catalogs that give activities of their endonucleases in standardized buffers with varying ionic strengths.[23,24] These are particularly valuable for molecular biologists who use dozens of endonucleases and may want to do simultaneous double and tripe digests on the same DNA substrate.

Aside from ionic strength and cation preferences, restriction endonucleases may vary in temperature optima. Most restriction digests are routinely done at 37°, but a few endonucleases (notably *Sma*I) prefer a lower incubation temperature and several, mainly those isolated from thermophiles (such as *Taq*I), require much higher temperatures.

[18] W. M. Normore, H. S. Shapiro, and P. Setlow, *in* "CRC Handbook of Biochemistry and Molecular Biology" (G. D. Fasman, ed.), p. 65. CRC Press, Cleveland, Ohio, 1976.

[19] J. Josse, A. D. Kaiser, and A. Kornberg, *J. Biol. Chem.* **237**, 864 (1981).

[20] K. Tartoff, *Annu. Rev. Genet.* **9**, 355 (1975).

[21] M. Botchan, G. McKenna, and P. A. Sharp, *Cold Spring Harbor Symp. Quant. Biol.* **38**, 383 (1974).

[22] M. Nelson and M. McClelland, *in* "Gene Amplification and Analysis" (J. G. Chirikjian, ed.), Vol. 5 (in press).

[23] R. D. Wells, R. D. Klein, and C. K. Singleton, *in* "The Enzymes" (P. D. Boyer, ed.), 3rd ed., Vol. 14, p. 157. Academic Press, New York, 1981.

[24] R. Fuchs and R. Blakesley, this series, Vol. 100, p. 3.

Units of endonuclease activity are usually measured with λ DNA as a substrate using the (nonstandardized) buffer recommended by the manufacturer for the particular endonuclease. Endonuclease activity also varies greatly with the DNA substrate, and activity for each site can be modified by the neighboring sequences. The classic example of differential cleavage rate is the nearly 10-fold difference in reaction rate observed between two *Eco*RI sites in λ DNA.[25] A greater than 50-fold difference in cleavage rates exists among sites for both *Nar*I and *Nae*I on pBR322 DNA.[26] Furthermore, *Pae*R7 I does not cut one of its canonical sites on adenovirus 2 DNA.[12] Rate variability can be different for different isoschizomers; *Xho*I does not show the same variability as *Pae*R7 I on Ad 2 DNA.[12] And, in most cases, such rate differences have not been systematically investigated.

Therefore, titration of previously untried endonucleases on one's particular DNA substrate under reaction conditions to be used is recommended, especially if the reaction buffer is different from the manufacturer's prescribed buffer. If it is desirable to dilute the restriction endonuclease before use, this should be done in the manufacturer's suggested dilution buffer rather than assay buffer to maintain activity.

The activity of most restriction endonucleases is also adversely affected by the proximity of the recognition site to the end of a DNA molecule.[24] This parameter is especially important when one is trying to cleave at adjacent sites within a linker region of a cloning vector.

A word should be mentioned about star activity.[27] This phenomenon was first observed and has been best studied with *Eco*RI endonuclease.[28] Under "altered" reaction conditions (including low ionic strength, high pH, Mn^{2+} substitution of Mg^{2+}, presence of organic solvents such as glycerol or dimethyl sulfoxide, and high endonuclease concentration), *Eco*RI can be shown to cleave at noncanonical sites (within sequences other than GAATTC[28-32]). These sites are not random, however. They all

[25] M. Thomas and R. W. Davis, *J. Mol. Biol.* **91**, 315 (1975).
[26] G. Wilson, D. Comb, L. Greenough, and I. Schildkraut, unpublished observations.
[27] P. Modrich, *CRC Crit. Rev. Biochem.* **13**, 287 (1982).
[28] B. Polisky, P. Greene, D. E. Garfin, B. J. McCarthy, H. M. Goodman, and H. W. Boyer, *Proc. Natl. Acad. Sci. U.S.A.* **72**, 3310 (1975).
[29] Recognition sequences are written 5' → 3'; one strand only is given. Other nucleic acid nomenclature includes the following conventions. Only bases are indicated (deoxyribonucleotide residues are often not differentiated from their ribonucleotide counterparts). The presence of phosphodiester bonds is understood (internal phosphates are usually not written). Oligonucleotide sequences which terminate in either a 5'- or 3'-phosphate monoester are explicitly stated, e.g., p(T)₆ means a hexamer of thymidylic acid residues with a 5'-terminal phosphate. A vertical arrow (↓) indicates a cleavage site usually generating a terminal 5'-phosphate (see ref. 9 for a possible exception).
[30] M. Hsu and P. Berg, *Biochemistry* **17**, 131 (1978).

have the sequence AATT. Even under star conditions, canonical *Eco*RI sites (GAATTC) are cleaved first and a hierarchy exists as to the order in which secondary sites are cleaved.[33,34] It is extremely difficult to generate complete star digests, so star activity is often more problematic than useful.

Under star conditions many other restriction endonucleases have been reported to have secondary star activity. These include *Hind*III, *Hha*I, *Bsu*I, *Xba*I, *Sal*I, *Pst*I, *Bam*HI, and *Sst*I.[24] However, since no other endonuclease has been purified and characterized as well as *Eco*RI, it is often difficult to determine whether the secondary activity is, in fact, associated with the aforementioned endonucleases. It is possible that some star activities are actually traces of contaminating second endonucleases. To avoid star activity, manufacturers recommend keeping glycerol and endonuclease concentrations low, ionic strength high, and digestion times as short as possible.

A thorough guide for the preparation of DNA substrates and use of restriction endonucleases is given in a previous volume of this series.[24] It also includes a useful troubleshooting appendix which addresses the types of problems most frequently encountered in the use of restriction endonucleases.

Mapping

Most applications of recombinant DNA technology are facilitated by the generation of a physical map of restriction sites within the DNA being studied. A number of strategies can be used to construct these maps, and a combination of several are often necessary to obtain maps that are sufficiently accurate and detailed.

To start a map, one usually digests the DNA of interest with a series of single restriction endonucleases. Products of the endonuclease digestion are resolved by analytical gel electrophoresis on agarose or polyacrylamide gels.[35] Visualization of the DNA is achieved by ethidium bromide staining followed by ultraviolet light monitoring, or by autoradiography of

---

[31] T. I. Tikchonenko, E. E. Karamov, B. A. Zavizion, and B. S. Naroditsky, *Gene* **4,** 195 (1978).

[32] M. Nasri and D. Thomas, *Nucleic Acids Res.* **14,** 811 (1986).

[33] C. P. Woodbury, O. Hagenbuchle, and P. H. von Hippel, *J. Biol. Chem.* **225,** 11534 (1980).

[34] R. C. Gardner, A. J. Howarth, J. Messing, and R. Shepard, *DNA* **1,** 109 (1982).

[35] R. C. Ogden and D. A. Adams, this volume [8].

radioactively labeled DNA substrates. To estimate size, a standardization curve is first generated by plotting the relative mobility of DNA fragments of known molecular weight against the log of their molecular weights. Then the mobility of each DNA fragment in question is fitted to the curve.[36] After resolving the restriction products on a gel, it is easy to determine how many cleavage sites occur on the DNA in question. On a linear DNA molecule, the first cleavage will yield two fragments and then each additional site adds one more fragment. When starting with a circular DNA substrate, however, the first cleavage will linearize and it will remain one fragment; then each additional cut will yield one more fragment.

It is difficult to size circular DNA substrates accurately on a gel. Supercoiled molecules migrate aberrantly fast and differ in mobility from nicked circles. It is usually more accurate to linearize the DNA before attempting to estimate its size.

The first step in ordering the DNA fragments for a map is generating a limit digest with the endonuclease of choice. A characteristic feature of limit digests is that all cleavage products are present in equimolar amounts. Therefore, when uniformly labeled DNA is cleaved the amount of radioactivity in each limit product is directly proportional to its size. Similarly, when cleaved DNA is stained with ethidium bromide and visualized under ultraviolet light, the intensity of the ultraviolet absorption by each fragment is directly proportional to its size.

The most frequent problem encountered in restriction site mapping is identifying partial digestion products and double bands (or "doublets") on a gel. In both cases a band of variant intensity is produced. "Partials" usually appear as bands that are too light (or underlabeled); if one adds up the total DNA length of all the fragments one gets too large a value. Sometimes partials can be eliminated by digesting the DNA substrate for a longer time with more endonuclease.

Doublets occur when two products are similarly sized and therefore unresolved on the gel. The doublet band appears more intense (or radioactive) than its neighbors; fragments add up to too small a value. Sometimes double bands can be separated on a higher resolution gel. Often their presence can be verified by digesting one member away with a second endonuclease.

Another problem that might occur when attempting to generate a restriction map is that of DNA fragments migrating aberrantly on a gel. For example, on polyacrylamide gels, high-molecular-weight fragments with high GC contents migrate considerably faster than fragments with low GC

[36] K. J. Danna, this series, Vol. 65, p. 449.

contents.[37] Anomalous fragment mobilities can occur on both polyacrylamide and agarose gel matrices, and the problem is particularly severe with short DNA fragments. The sizing of small fragments is more accurately done on denaturing gels.[38]

After assessing the fragment profile for a number of restriction endonucleases on the DNA substrate of interest, a whole variety of mapping techniques may be employed. Most commonly the DNA is then digested with combinations of endonucleases whose single profiles have been characterized. Also individual fragments may be isolated and then digested with various other endonucleases.

Another relatively easy and popular mapping method involves labeling a DNA fragment specifically at one terminus and then partially digesting the fragment in a series of reactions with various endonucleases that cut the substrate several times.[39] In a variation of this method, DNA fragments generated by partial endonucleolytic digestions can be isolated and then individually digested to completion to establish fragment order.[34]

Finally, a DNA fragment to be mapped can be digested to different extents with a processive double-stranded exonuclease (most commonly Bal 31). DNA isolated at each time point can then be digested by one or more restriction endonucleases. Fragments disappear in the order in which the restriction site appeared on the DNA.[40]

Using a combination of mapping methods and a good number of endonucleases it is relatively easy to generate a restriction map that is detailed and accurate down to the 100–200 base pair level.

Generating the Proper Ends for Subsequent Steps

Table I contains a list of commercially available restriction endonucleases. They have been grouped on the basis of recognition and cleavage site, and they will be discussed mainly in these groups. Group 1 endonucleases are those that cut defined, palindromic sequences. Group 2 endonucleases are those that cut interrupted palindromes. Group 3 contains endonucleases recognizing "relaxed" specificities. Group 4 endonuclease are those that recognize nonpalindromic sequences.

Within each group, class A endonucleases cleave to produce 5' extensions, class B to produce 3' extensions, and class C, blunt ends.

[37] R. S. Zeiger, R. Salomon, C. W. Dingman, and A. C. Peacock, *Nature (London), New Biol.* **238,** 65 (1972).

[38] T. Maniatis, A. Jeffrey, and H. Van de Sande, *Biochemistry* **14,** 3787 (1975).

[39] H. O. Smith and M. L. Birnstiel, *Nucleic Acids Res.* **3,** 2387 (1976).

[40] R. J. Legerski, J. L. Hodnett, and H. B. Gray, Jr., *Nucleic Acids Res.* **5,** 1445 (1978).

TABLE I
COMMERCIALLY AVAILABLE RESTRICTION ENDONUCLEASES

Endonuclease	Recognition sequence

Group 1. Endonucleases cutting within palindromic sequences

A. 5′ Extensions

Tetrameric recognition site

1. HinPI		G	↓ C	G	C		
2. HpaII, MspI		C	↓ C	G	G		
3. MboI, Sau3AI, NdeI, NdeII			↓ G	A	T	C	
4. TaqI		T	↓ C	G	A		
5. MaeI		C	↓ T	A	G		
6. MaeII		A	↓ C	G	T		

Hexameric recognition site

7. NarI	G	G	↓ C	G	C	C	
8. XmaI, XcyI		C	↓ C	C	G	G	G
9. BspMII		T	↓ C	C	G	G	A
10. XmaIII, EagI		C	↓ G	G	C	C	G
11. BssHII		G	↓ C	G	C	G	C
12. MluI		A	↓ C	G	C	G	T
13. EcoRI		G	↓ A	A	T	T	C
14. NdeI	C	A	↓ T	A	T	G	
15. BamHI, BstI		G	↓ G	A	T	C	C
16. BglII		A	↓ G	A	T	C	T
17. BclI		T	↓ G	A	T	C	A
18. ClaI, BanIII	A	T	↓ C	G	A	T	
19. XhoI, PaeR7I		C	↓ T	C	G	A	G
20. SalI		G	↓ T	C	G	A	C
21. AsuII	T	T	↓ C	G	A	A	
22. HindIII		A	↓ A	C	G	T	T
23. NheI		G	↓ C	T	A	G	C
24. SpeI		A	↓ C	T	A	G	T
25. XbaI		T	↓ C	T	A	G	A
26. AvrII		C	↓ C	T	A	G	G
27. NcoI		C	↓ C	A	T	G	G
28. Asp718		G	↓ G	T	A	C	C

Octameric recognition site

29. NotI	G	C	↓ G	G	C	C	G	C

B. 3′ Extensions

Tetrameric recognition site

1. HhaI, CfoI		G C	G	↓ C	
2. NlaIII	C	A T	G	↓	

Hexameric recognition site

3. SacII, SstII	C	C G C	↓ G	G	
4. ApaI	G G	G C C	↓ C		
5. SphI	G C	A T G	↓ C		

(continued)

TABLE I (*continued*)

Endonuclease	Recognition sequence
6. *Kpn*I	G G T A C ↓C
7. *Pvu*I, *Xor*II, *Rsp*I	C G A T C ↓G
8. *Sac*I, *Sst*I	G A G C T ↓C
9. *Pst*I	C T G C A ↓G
10. *Nsi*I	A T G C A ↓T
11. *Aat*II	G A C G T ↓C

C. Blunts

Tetrameric recognite site

1. *Fnu*DII, *Tha*I	C G ↓C G
2. *Hae*III, *Pal*I	G G ↓C C
3. *Alu*I	A G ↓C T
4. *Rsa*I	G T ↓A C
5. *Dpn*I	G ^{me}A ↓T C

Hexameric recognition site

6. *Nae*I	G C C ↓G G C
7. *Sma*I	C C C ↓G G G
8. *Stu*I, *Aat*I	A G G ↓C C T
9. *Bal*I	T G G ↓C C A
10. *Nru*I	T C G ↓C G A
11. *Fsp*I, *Mst*I	T G C ↓G C A
12. *Pvu*II	C A G ↓C T G
13. *Sna*BI	T A C ↓G T A
14. *Sca*I	A G T ↓A C T
15. *Eco*RV	G A T ↓A T C
16. *Hpa*I	G T T ↓A A C
17. *Ssp*I	A A T ↓A T T
18. *Dra*I	T T T ↓A A A

Group 2. Endonucleases cutting interrupted palindromes

A. 5′ Extensions

1. *Dde*I	C ↓T N A G
2. *Fnu*4HI	G C ↓N G C
3. *Hinf*I	G ↓A N T C
4. *Mae*III	↓G T N A C
5. *Sau*96I	G ↓G N C C
6. *Scr*FI	C C ↓N G G
7. *Bst*EII	G ↓G T N A C C
8. *Mst*II, *Cvn*I, *Sau*I	C C ↓T N A G G
9. *Tth*111I	G A C N ↓N N G T C

B. 3′ Extensions

1. *Dra*III	C A C N N N ↓G T G
2. *Bgl*I	G C C N N N N ↓N G G C
3. *Bst*XI	C C A N N N N N ↓N T G G
4. *Sfi*I	G G C C N N N N ↓N G G C C

TABLE I (*continued*)

Endonuclease	Recognition sequence
C. Blunts	
1. *Nla*IV	G G N ↓ N C C
2. *Xmn*I, *Asp*700	G A A N N ↓ N N T T C
Group 3. Endonucleases recognizing sequences with degeneracies	
A. 5′ Extensions	
Pentameric recognition site	
1. *Eco*RII	↓ C C A/T G G
2. *Bst*NI, *Apy*I	C C ↓ A/T G G
3. *Ava*II	G ↓ G A/T C C
4. *Nci*I	C C ↓ C/G G G
Hexameric recognition site	
5. *Xho*II	Pu ↓ G A T C Py
6. *Aha*II	G Pu ↓ C G Py C
7. *Ava*I, *Nsp*III	C ↓ Py C G Pu G
8. *Ban*II	G ↓ Pu G C Py C
9. *Ban*I	G ↓ G Py Pu C C
10. *Acc*I	G T ↓ A/C G/T A C
11. *Sty*I	C ↓ C A/T A/T G G
Heptameric recognition site	
12. *Eco*O109, *Dra*II	Pu G ↓ G N C C Py
13. *Ppu*MI	Pu G ↓ G A/T C C Py
14. *Rsr*II	C G ↓ G A/T C C G
B. 3′ Extensions	
Hexameric recognition site	
1. *Hae*II	PuG C G C ↓ Py
2. *Hgi*AI	G T/A G C T/A ↓ C
3. *Bsp*1286, *Nsp*II	G G/A/T G C C/A/T ↓ C
C. Blunts	
1. *Hinc*II, *Hind*II	G T Py ↓ Pu A C
Group 4. Endonucleases recognizing nonpalindromes	
A. Cutting within sequence, 3′ extension	
1. *Bsm*I	G A A T G C N↓
	C T T A C ↑ G N
B. Cutting away 5′ Extensions	
1. *Hph*I	G G T G A N8 ↓
	C C A C T N7 ↑
2. *Mbo*II	G A A G A N8 ↓
	C T T C T N7 ↑

(continued)

TABLE I (*continued*)

Endonuclease	Recognition sequence			
3′ Extensions				
1. *Bbv*I	G C A	G C	N8	↓
	C G T	C G	N12	↑
2. *Fok*I	G G A	T G	N9	↓
	C C T	A C	N13	↑
3. *Hga*I	G A C	G C	N5	↓
	C T G	C G	N10	↑
4. *Sfa*NI	G C A	T C	N5	↓
	C G T	A G	N9	↑
5. *Bsp*MI	A C C T	G C	N4	↓
	T G G A	C G	N8	↑
C. Blunts				
1. *Mnl*I	C C	T C	N7	↓
	G G	A G	N7	↑

Ligation

In genetic engineering one of the most useful features of DNA cleaved by restriction endonucleases is the ability to ligate different fragments together at their common restriction sites. Only the sequence of the DNA within the single-stranded region (if any) of the restriction cleavage site has to be complementary for ligation to occur. The surrounding DNA sequence does not affect ligation. Many factors affect the ease of ligation of DNA molecules but an overriding one is whether protruding or "blunt" ends are generated. In addition, the length of the protruding end and stability of the hydrogen-bonded structure formed are important. Any contaminating phosphatase, endonuclease, or exonuclease activity in a restriction endonuclease will also markedly reduce the ability of the ends generated to be ligated.

In Table I, group 1 endonucleases are those that cut defined (nondegenerate) sequences. Each such endonuclease generates ends that are compatible with each other. Within this group, class A and class B endonucleases producing 5′ and 3′ extensions, respectively, generate ends that ligate, in general, with high efficiency. In both groups, as a rule, 4-base extensions ligate better than 2-base extensions, and extensions with GC pairs ligate more readily than those with AT.[41] Group C endonucleases produce blunt ends, with no extensions. Although they are advan-

---

[41] N. P. Higgins and N. R. Cozzarelli, this series, Vol. 68, p. 50.

tageous in the sense that any endonuclease within this class produces ends that are compatible with all other members of the class, blunt ends are much more difficult to ligate than either type of extension. The ligation reaction usually requires 20 to 100 times more T4 DNA ligase and higher DNA concentrations for the ends to rejoin efficiently.[42]

Within class B there is the interesting case of PvuI. This endonuclease cuts at the sequence CGAT ↓ CG, leaving a 2-base 3′ extension. The sites religate at high efficiency unless the DNA has come from a dam⁺ bacterial host. N^6-Methyladenine within the sequence CG^{me}ATCG does not interfere with PvuI cleavage, but it inhibits ligation of the site.[43] This is the only known case where methylation affects ligation but not cleavage. The effect appears to be mediated by an altered melting temperature of the AT base pair.[44]

Group 2 endonucleases are those that cut interrupted palindromes. Ligation of fragments with 3′ or 5′ extensions generated by those endonucleases is usually not possible due to the heterogeneity of sequence within the interrupted part of the site. Blunt cutters, like NlaIV and XmnI which cut symmetrically within the interruption, produce DNA fragments that can be rejoined with regeneration of the recognition site.

Group 3 endonucleases are those recognizing "relaxed" specificities. Only a subset of such fragments will be compatible and able to ligate. The exceptions are XhoII, AhaII, and HaeII, where the degeneracies lie outside the protruding ends, and HincII, which generates symmetrical, blunt, and therefore compatible ends.

Within groups 2 and 3, a word should be said about ScrFI, Fnu4HI, BstNI, and Tth111I. These endonucleases all cleave to generate a single base extension. These are extremely difficult to ligate—much more so even than blunt ends. Such DNA is best treated to remove or fill in the base before subsequent steps.[45]

Finally, group 4 endonucleases are those recognizing nonpalindromic sequences. All contain at least one degeneracy within the cleavage site, and most cleave away from their recognition site. Therefore it is not usually possible to get the generated fragments to religate. If, however, the ends of the fragments are treated to generate flush ends, the recognition sites are left intact and always suitable for recutting. The exception within this group is BsmI, where cleavage occurs within the recognition site on one strand.

[42] V. Sgaramella, Proc. Natl. Acad. Sci. U.S.A. 69, 3389 (1972).
[43] Specific methylation sites are designated by a superscript "me" 5′ to the modified base.
[44] J. J. Sninsky and M. Myers, unpublished observations.
[45] I. Schildkraut, unpublished observations.

Compatible Ends

Often it is desirable to join DNA cut by two different restriction endonucleases together. This is possible with a number of different restriction endonucleases that generate the same type of ends. However, in general, when this is done, a new chimeric site is formed that is no longer cleavable by the original endonucleases used.

The largest group of compatible ends are those generated by any endonuclease whose cleavage leaves a blunt end. Endonucleases generating other families of compatible ends are listed in Table II.

Other Subsequent Steps

Many manipulations other than directly ligating restriction fragments can be carried out on cleaved DNA fragments. The types of manipulations possible depend on the ends that have been generated.

5' Extensions. The 3'-OH recessed ends can readily be filled in to form blunt ends or to carry label by the addition of Klenow enzyme and the appropriate deoxynucleotides.[46] Sometimes it is desirable to cleave a restriction site that produces a 5' extension, fill in with Klenow, and then religate the site. This procedure will often generate a new restriction site. A list of sites that can be generated in this manner is given in Table III.

The 5' ends are also readily labeled by first removing the phosphate group with alkaline phosphatase and then replacing the phosphate moiety with [γ-^{32}P]ATP and T4 polynucleotide kinase.[46] Alternatively, 5' extensions can be readily removed by treatment of the DNA with S_1 or mung bean nucleases.[47,48]

Nucleotides from the recessed 3'-hydroxyl end of a 5'-phosphate extension may be sequentially removed by treatment with exonuclease III.[46]

3' Extensions. There is no counterpart of Klenow treatment that can be used to fill in 5' recessed ends. The 3' extension can be digested away by treatment with T4 DNA polymerase in the absence of deoxynucleotides, or Klenow enzyme in the presence of deoxynucleotides. Alternatively, like 5' ends, the 3' protrusions can be digested away with mung bean or S_1 nuclease. 3' Extensions are also excellent substrates for tailing with terminal transferase.[49] This can often result in the regeneration of the restriction recognition site; when using the *Pst*I or *Sph*I sites on pBR322, G and C tailing respectively, will restore the site on the recombinant

[46] F. Cobianchi and S. H. Wilson, this volume [10].
[47] V. M. Vogt, *Eur. J. Biochem.* **33,** 192 (1973).
[48] D. Komalski, W. D. Kroeker, and M. Laskowski, Sr., *Biochemistry* **15,** 4457 (1976).
[49] W. H. Eschenfeldt and S. L. Berger, this volume [37].

TABLE II

ENDONUCLEASES GENERATING ENDS COMPATIBLE
FOR LIGATION

Endonucleases	Cleavage site
5' CG family	
1. *Msp*I, *Hpa*II	C ↓ **C G** G
2. *Taq*I	T ↓ **C G** A
3. *Hin*PI	G ↓ **C G** C
4. *Mae*II	A ↓ **C G** T
5. *Aha*II	G Pu ↓ **C G** Py C
6. *Cla*I, *Ban*III	A T ↓ **C G** A T
7. *Nar*I	G G ↓ **C G** C C
8. *Asu*II	T T ↓ **C G** A A
5' GATC family	
1. *Mbo*I, *Sau*3AI, *Nde*II	↓ **G A T** C
2. *Xho*II	Pu ↓ **G A T** C Py
3. *Bam*HI, *Bst*I	G ↓ **G A T** C C
4. *Bgl*II	A ↓ **G A T** C T
5. *Bcl*I	T ↓ **G A T** C A
5' CTAG family	
1. *Nhe*I	G ↓ **C T A** G C
2. *Xba*I	T ↓ **C T A** G A
3. *Spe*I	A ↓ **C T A** G T
4. *Avr*II	C ↓ **C T A** G G
5' CCGG family	
1. *Bsp*MII	T ↓ **C C G** G A
2. *Xma*I	C ↓ **C C G** G G
3. *Not*I	G C ↓ **C C G** G G C
5' TCGA family	
1. *Xho*I, *Pae*R7I	C ↓ **T C G** A G
2. *Sal*I	G ↓ **T C G** A C
3' CATG family	
1. *Nla*IV	**C A T G** ↓
2. *Sph*I	G **C A T G** ↓ C
3' TGCA family	
1. *Nsi*I	A **T G C A** ↓ T
2. *Pst*I	C **T G C A** ↓ G

plasmid. To remove nucleotides from the 5' end of a recessed 5' strand, one treats with λ exonuclease.[46]

Blunt Ends. All flush ends are suitable for ligation to one another. In addition, one can easily create a new restriction site at these ends by ligating phosphorylated linker or adaptor molecules to their 3'-OH ends before ligating the fragments together.[50]

[50] R. Wu *et al.,* this volume [38].

TABLE III
5′ EXTENSION FILL-IN SITES GENERATED

Endonucleases	Cleavage site	Fill in	New site generated
1. *Mbo*I, *Sau*3AI, *Nde*II	↓ G A T C	**G A T C G A T** C	*Cla*I
2. *Bam*HI, *Bst*I	G ↓ G A T C C	G **G A T C G A T** C C	*Cla*I
3. *Bcl*I	T ↓ G A T C A	T **G A T C G A T** C A	*Cla*I
4. *Bgl*II	A ↓ G A T C T	A **G A T C G A T** C T	*Cla*I
5. *Xho*II	Pu ↓ G A T C Py	Pu **G A T C G A T** C Py	*Cla*I
6. *Hin*PI	G ↓ C G C	**G C G C G C**	*Bss*HI
7. *Nar*I	G G ↓ C G C C	**G G C G C G C** C	*Bss*HI
8. *Bss*HI	G ↓ C G C G C	G **C G C G C G C** G C	*Bss*HI
9. *Mlu*I	A ↓ C G C G T	A **C G C G C G C** G T	*Bss*HI
10. *Avr*II	C ↓ C T A G G	C **C T A G C T A** G G	*Alu*I
11. *Nhe*I	G ↓ C T A G C	G **C T A G C T A** G C	*Alu*I
12. *Spe*I	A ↓ C T A G T	A **C T A G C T A** G T	*Alu*I
13. *Xba*I	T ↓ C T A G A	T **C T A G C T A** G A	*Alu*I
14. *Taq*I	T ↓ C G A	**T C G C G A**	*Nru*I
15. *Cla*I, *Ban*III	A T ↓ C G A T	**A T C G C G A** T	*Nru*I
16. *Asu*II	T T ↓ C G A A	**T T C G C G A** A	*Nru*I
17. *Bsp*MII	T ↓ G G C C A	T **G G C C G G C** C A	*Nae*I
18. *Xma*III, *Eag*I	C ↓ G G C C G	C **G G C C G G C** C G	*Nae*I
19. *Not*I	G C ↓ G G C C G C	G C **G G C C G G C** C GC	*Nae*I
20. *Xho*I, *PaeR7*I	C ↓ T C G A G	C T **C G A T C G** A G	*Pvu*I
21. *Sal*I	G ↓ T C G A C	G T **C G A T C G** A C	*Pvu*I
22. *Hpa*II, *Msp*I	C ↓ C G G	**C C G C G G**	*Sac*II
23. *Aha*II	G Pu ↓ C G Py C	C Pu **C G C G** Py C	*Fnu*DII
24. *Xma*I	C ↓ C C G G G	C **C C G G C C G** G G	*Xma*III
25. *Mae*II	A ↓ C G T	**A C G C G T**	*Mlu*II
26. *Hind*III	A ↓ A C G T T	A **A G C T A G C** T T	*Nhe*I
27. *Asp*718	G ↓ G T A C C	G **G T A C G T A** C C	*Sna*BI
28. *Nco*I	C ↓ C A T G G	C **C A T G C A T** G G	*Nsi*I
29. *Eco*RI	G ↓ A A T T C	**G AA T T A A T T C**	*Xmn*I
30. *Mae*I	C ↓ T A G	**C T A T A G**	(none)
31. *Nde*I	C A ↓ T A T G	**C A T A T A T G**	(none)

Some new uses are also being made of the more exotic "degenerate" and "nonpalindromic" endonucleases. Neither group is heavily used in typical recombinant DNA work because of the heterogeneity at their cleavage sites. But in the case of the degenerate endonucleases, new cleavage specificities are being elicited by first methylating the substrate with a methylase of overlapping specificity before restriction digestion. For example, pBR322 is cleaved at two positions by *Hinc*II endonuclease (GTPyPuAC).[51] However, if the DNA is first methylated by M. *Taq*I (TCGA), one site is blocked, resulting in cleavage at a single unique site in the ampicillin gene (GTCAAC).[52]

Szybalski has recently developed a method to cleave cloned single-stranded DNA at any site desired using the nonpalindromic endonucleases. An oligonucleotide adaptor is synthesized that has a hairpin containing the recognition site for the endonuclease and a long single-stranded region. The single-stranded region contains the sequence complementary to the site to be cleaved, at a precise distance from the recognition site. The oligonucleotide is hybridized to the single-stranded DNA substrate and then treated with endonuclease, resulting in cleavage. This method has been successfully used with *Fok*I endonuclease.[53,54]

Summary

It is clear that we have still not exhausted all the restriction endonuclease specificities to be found in nature. Recently discovered *Bsm*I is the first endonuclease recognizing a nonpalindromic sequence that cleaves within the site.[55] Certainly other endonucleases belonging to this class will soon be discovered. More endonucleases are now being sought that recognize longer recognition sequences, because large fragments can now be readily separated by pulse-field electrophoresis. New sources of endonucleases are also being found; for example, a group of viruses that grow on *Chlorella* algae produce type II-like site-specific endonucleases.[56]

As the number and variety of known restriction endonucleases increase, the number and variety of applications keep pace. There is still no end in sight.

---

[51] Py indicates pyrimidines; Pu, purines.
[52] M. Nelson, C. Christ, and I. Schildkraut, *Nucleic Acids Res.* **12,** 5165 (1985).
[53] W. Szybalski, *Gene* **40,** 169 (1985).
[54] A. J. Podhajska and W. Szybalski, *Gene* **40,** 175 (1985).
[55] C. Christ and D. Ingalls, unpublished observations.
[56] Y. Xia, D. E. Burbank, L. Uher, D. Rabussay, and J. L. van Etten, *Mol. Cell. Biol.* **6,** 1430 (1986).

[12] Restriction and Modification *in Vivo* by *Escherichia coli* K12

By Elisabeth A. Raleigh

This chapter focuses on how restriction of newly introduced DNA by *Escherichia coli* can interfere with cloning and subcloning work, and in particular on how the pattern of methylation of the DNA influences this. A principal aim is to acquaint the reader with three *E. coli* restriction systems that attack DNA only when it is appropriately methylated.

The second part of this chapter describes biological restriction and modification in general terms. The third part discusses the particular restriction systems found in *E. coli* K12, first briefly the familiar K and P1 restriction systems, and then in detail the methyl-specific McrA, McrB and Mar systems. Some common strains are discussed with special reference to their restriction phenotypes. The fourth part briefly reviews the *E. coli* methylation functions, Dam and Dcm, as they affect sensitivity to digestion of DNA *in vitro*.

How DNA Acquires Sensitivity or Resistance to Restriction

DNA from many organisms contains methylated bases. 6-Methyladenine, 5-methylcytosine, and N^4-methylcytosine have been found in different systems.[1,2] In bacteria, one function of the methylation pattern of DNA is to distinguish foreign DNA from native DNA. Foreign DNA has an inappropriate methylation pattern and is often therefore restricted by one or more sequence-specific restriction functions, while native DNA is protected because its methylation pattern is appropriate. The biological result of restriction is the death of the replicon; the corresponding molecular event, where it has been examined, is double-strand cleavage of the DNA.

In most familiar cases, such as *Eco*K or *Eco*RI, the presence of methylation at a restriction site protects the site from attack. However, a small number of restriction functions act on a sequence *only* when it is methyl-

---

[1] H. O. Smith and S. V. Kelley, *in* "DNA Methylation: Biochemistry and Biological Significance" (A. Razin, H. Cedar, and A. D. Riggs, eds.), p. 39. Springer-Verlag, Berlin and New York, 1984.
[2] A. Janulaitis, S. Klimasauskas, M. Petrusyte, and V. Butkus, *FEBS Lett.* **161**, 131 (1983).

ated.[3-6] The DNA is then protected by the absence of methylation of that sequence. Such methyl-specific restriction functions could be common in nature, since they would not be detected by most screening procedures, due to a lack of appropriately methylated substrate DNA. The discovery that *E. coli* K12 contains three methylcytosine-specific restriction systems, McrA,[4] McrB,[4] and Mar,[6] is of particular importance to those planning to manipulate DNA using *E. coli* as a host.

In practical terms, a strain displays restriction *in vivo* when native DNA (e.g., a plasmid or phage grown in the same strain) transforms or infects at high efficiency, but foreign DNA of the same sequence (the same vector, but grown in a different strain) transforms or infects at much lower efficiency. Some of the input phage or plasmids escape restriction, at low frequency, and replicate in the presence of the associated modification function. The progeny are thus protected from further restriction, being recognized as native. Protection is conferred by alteration of the methylation pattern, either by methylation of the appropriate sequence or by failure to methylate it, depending on the system. Survivors of restriction are found at frequencies ranging from 10^{-5} (for *Eco*RI, e.g.) to 10^{-1} (for McrA, see below).

Restriction can be bypassed by first propagating the foreign DNA in a strain that lacks the restriction function but still possesses the protective function. Depending on the restriction system, this means either that the strain must possess the cognate modification methylase (as in the case of *Eco*K or *Eco*RI), or that it must lack the methylase(s) that confer sensitivity to restriction (as in the case of McrA, McrB, and Mar; see below).

Characteristics of *in Vivo* Restriction Systems in *E. coli* K12

*Eco*K and *Eco*P1 Restriction

Most laboratory strains of *E. coli* K12 contain a restriction system (*Eco*K, specified by the *hsdRMS* genes) that recognizes the sequence 5′ A[me?]ACNNNNNNGTGC 3′[6a] (and its complement, 5′ GC[me]ACNNNNN-NGTT 3′) and cleaves the DNA, unless the indicated A residues are

[3] S. Lacks and J. R. Greenberg, *J. Mol. Biol.* **114**, 153 (1977).

[4] E. A. Raleigh and G. Wilson, *Proc. Natl. Acad. Sci. U.S.A.* **83**, 9070 (1986)

[5] T. L. Sladek, J. A. Nowak, and J. Maniloff, *J. Bacteriol.* **165**, 219 (1986).

[6] J. Heitman, personal communication.

[6a] Recognition sequences are written 5′ → 3′; specific methylation sites are designated by a superscript "me" 5′ to the modified base. For example, N[me]C indicates that the C is modified. A question mark indicates that the methylation is inferred but not proved to occur at the site indicated.

methylated.[7] Only about one in a thousand unmodified λ phage will survive to form plaques on *Eco*K[+] strains. Normally, however, all five of the *Eco*K sites present in the wild-type λ genome will be K modified, since most (but not all) *E. coli* K12 derivatives contain the *Eco*K methylase, with or without the restriction function. HB101 and RR1 are common strains that do not contain the *Eco*K methylase (see Strain Choice, below).

The DNA of organisms other than *E. coli* K12 and its phages will be unmethylated at *Eco*K sites, and hence newly constructed primary clones (from ligation reactions) will be subject to lethal restriction. This will be reflected in poor cloning efficiency; ligation mixtures with *Eco*K-sensitive insert DNA will give smaller numbers of transformants than a ligation mixture using insert DNA prepared from the *E. coli* host but otherwise identical. The overall restriction factor may be relatively small in a ''shotgun'' experiment, since the 7-base *Eco*K site will occur only about once in 8 kb. For example, 60% of all possible 5-kb inserts will be subject to restriction, reducing the size of the library only by a factor of two. Such restriction may, however, result in substantial underrepresentation in libraries of specific fragments containing *Eco*K sites.

This problem can be circumvented by the use of mutant strains lacking the *Eco*K endonuclease activity (Table I). There are two kinds of strains in use: those that have lost both methylase and endonuclease activities (R[−]M[−] phenotype), and those that have lost only the endonuclease (R[−]M[+] phenotype). An advantage of R[−]M[+] strains is that, once a clone has been passed through such a strain, the DNA is methylated and thus protected, so that the clone will no longer be subject to restriction by $R_K{}^+$ derivatives of K12.

Some *E. coli* K12 strains are lysogenic for the P1 prophage, and therefore contain the phage-encoded *Eco*P1 restriction system. Of commonly used strains, only JM103[8] carries this system. However, it is useful to check strains newly constructed by P1 transduction for the ability to plate a heterologous phage (e.g., λ) at normal efficiency in a standard plaque assay.[9] If the plating efficiency is reduced compared with the parent strain, the new strain probably carries a P1 prophage.

[7] T. Bickle, *in* ''Nucleases'' (S. M. Linn and R. J. Roberts, eds.), p. 85. Cold Spring Harbor Lab., Cold Spring Harbor, New York, 1982.

[8] C. Janisch-Perron, J. Vieira, and J. Messing, *Gene* **33,** 103 (1985).

[9] J. H. Miller, ''Experiments in Molecular Biology.'' Cold Spring Harbor Lab., Cold Spring Harbor, New York, 1972.

TABLE I

RESTRICTION PHENOTYPES OF COMMON STRAINS OF *E. coli* K12[a]

Strain	EcoK		Mcr		Comments	Reference[b]
	R	M	A	B		
MM294*	−	+	+	+	Host for plasmid cloning; *endA supE*	1, 2
HB101	−	−	+	−	Host for plasmid cloning. K12-B hybrid; *mar supE recA*	1, 3
RR1	−	−	+	−	*recA+* derivative of HB101	1, 3
C600	+	+	−	+	Host for many λ vectors; *supE*	1, 4
K802	−	+	−	−	Host for λ vectors and libraries; *supE*	1, 3
LE392	−	+	−	+	Host for plasmid cloning and for λ libraries; grandparent of Y1088; *supE supF*	1, 3
JM101	+	+	+	nt	Host for M13 sequencing vectors; *supE*	1, 5
JM103	−	+	nt	nt	Host for M13 sequencing vectors; carries *EcoP1* restriction; *supE*	5
JM105	−	+	nt	nt	Host for M13 sequencing vectors; nonsuppressing (*sup⁰*) host	5
JM107*	−	+	−	+	Host for M13 sequencing vectors; *supE*	1, 5
Y1088	−	+	(−)	+	Lytic host for amplification of λgt11 libraries; *supE supF*	1, 6
Y1084*	+	+	−	+	Parent of Y1090; *supF hsdR⁻ mcrB⁻* derivative available	1, 6
Y1090	+	+	(−)	+	Lytic host for immunological screening of amplified λgt11 libraries; *supF*	1, 6
K12	+	+	+	+	Strain of origin	1, 4
W3110*	+	+	+	+	*sup⁰* host	1, 4
GM2163	−	+	−	−	*dam dcm supE*; used for preparing unmethylated DNA	1, 7

[a] Abbreviations and special designations: nt, not tested; *, *mcrB⁻* derivatives are available from the author; ( ), phenotype inferred from the phenotype of plasmidless ancestor.

[b] References: (1) E. A. Raleigh and G. Wilson, *Proc. Natl. Acad. Sci. U.S.A.* **83,** 9070 (1986). (2) M. Meselson and R. Yuan, *Nature (London)* **217,** 1110 (1968). (3) T. Maniatis, E. F. Fritsch, and J. Sambrook, "Molecular Cloning: A Laboratory Manual." Cold Spring Harbor Lab., Cold Spring Harbor, New York, 1982. (4) B. J. Bachmann, *Bacteriol. Rev.* **36,** 525 (1972). (5) C. Janisch-Perron, J. Vieira, and J. Messing, *Gene* **22,** 103 (1983). (6) R. A. Young and R. W. Davis, *Science* **222,** 778 (1983). (7) M. Marinus, M. Carraway, A. Z. Frey, and J. A. Arraj, *Mol. Gen. Genet.* **192,** 288 (1983).

McrA and McrB Restriction

What the Systems Are. Another potentially serious problem arises because of the newly described methylcytosine-specific restriction systems present in *E. coli* K12. These two restriction systems were discovered as loci that interfered with the cloning of modification methylase genes from various bacteria, and are designated McrA and McrB (for *m*odified *c*ytosine *r*estriction). Although the methylcytosine-restricting function of McrA and McrB has only recently been described, these systems are probably functionally related, if not identical, to the RglA and RglB systems, which are known to restrict DNA containing 5-hydroxymethylcytosine.[10,11] The restriction genes map at 25 min (*mcrA*) and 99 min (*mcrB*) on the *E. coli* map.[12,13] Table I includes information on the Mcr phenotypes of those commonly used strains that have been tested.

Specificity of McrA and McrB. Restriction by both of these systems has been shown to depend on a specific pattern of methylation, although the precise recognition sequences are not yet known. pBR322 transforms Mcr+ strains well when unmethylated, but transformation efficiency declines in a methylation-dependent manner to about 1% of the original value when it is fully methylated by the *Alu*I methylase (M.*Alu*I; recognition sequence AG^{me}CT). This decline is not seen in isogenic McrB− strains. Similarly, methylation by M.*Hpa*II (recognition sequence C^{me}CGG) results in a maximal restriction of 10-fold. This restriction is abolished in isogenic McrA− strains. The two effects are genetically independent. An *mcrB* mutation does not affect restriction of *Hpa*II-methylated DNA, and an *mcrA* mutation does not affect restriction of *Alu*I-methylated DNA. None of six adenine methylases tested resulted in detectable Mcr-dependent restriction (but see section below, Mar Restriction).

Since only M.*Hpa*II has so far been found to confer sensitivity to McrA restriction, it is impossible to say much about McrA specificity other than that it is distinct from McrB and at least some *Hpa*II-methylated sites are contained in or overlap McrA recognition sites.

The McrB function clearly recognizes modified cytosine and can discriminate among different sequences containing methylated cytosine. A comparison of the sequences modified by the 14 methylases that confer sensitivity suggests that the recognition site may include the sequence

[10] H. Revel, *Virology* **31**, 688 (1967).
[11] H. Revel, *in* "Bacteriophage T4" (C. K. Mathews, E. M. Kutter, G. Mosig, and P. Berget, eds.), p. 156. Am. Soc. Microbiol., Washington, D.C., 1983.
[12] B. Bachmann, *Microbiol. Rev.* **47**, 180 (1983).
[13] E. A. Raleigh, R. Trimarchi, and H. Revel, unpublished observations.

$G^{me}C$.[4] Cytosine methylases that do not confer sensitivity to McrB restriction include M.*Hpa*II ($C^{me}CGG$), M.*Bam*HI ($GGAT^{me}CC$), M.*Hph*I ($T^{me}CACC$), and Dcm ($C^{me}C(A/T)GG$). The apparent specificity of McrB is therefore quite broad, and many targets may be expected if a DNA sequence contains a significant number of methylated cytosines.

It is presumed that 5-methylcytosine, not N^4-methylcytosine, is the modified base required for restriction, for two reasons. First, M.*Msp*I is known to methylate the C-5 position[14]; second, the hydroxymethyl group in T-even phage DNA is at the 5 position[15] and confers sensitivity to the related or identical RglA and B functions. However, the position of base methylation has been determined rigorously in very few cases; it has generally been assumed that biological cytosine methylation occurs at the 5 position. Restriction of DNA containing N^4-methylcytosine by McrA or McrB or by some other system has not been ruled out.

Degree of Restriction. Restriction of a plasmid bearing the gene for a susceptible methylase is very strong, about 10^{-5}. This is much greater than the maximal restriction of a plasmid methylated by the same methylase but not carrying the gene (about 10^{-2}). The difference is expected, because the methylated plasmid will lose restriction recognition sites as it replicates (since the newly replicated sites will remain unmethylated); but the plasmid bearing the methylase gene will not only not lose sites, it will create new ones in the host chromosome as the methylase is expressed.

Effect on Cloning Experiments. The magnitude of the effect of the Mcr restriction systems on actual cloning experiments is not known in detail. The specificity of McrB restriction seems to be quite broad (see above), with a recognition site that may be as small as 2 base pairs. Since restriction increases as more methyl groups are introduced (in transformation experiments with pBR322 DNA methylated *in vitro*[4]), it can be assumed that the more highly methylated the DNA, the greater the likely effect. The maximum effect is found to be 10- to 100-fold in the methylation experiments. However, this maximum is fixed by the number of methylase targets that are also Mcr targets, not by the number of potential Mcr targets.

A specific experiment relevant to the magnitude of Mcr restriction in cloning experiments was reported by Hanahan.[16,17] pBR322 DNA was

[14] R. Y. Walder, J. L. Hartley, J. E. Donelson, and J. A. Walder, *J. Biol. Chem.* **258,** 1235 (1983).

[15] G. R. Wyatt and S. S. Cohen, *Biochem. J.* **55,** 774 (1953).

[16] D. Hanahan, D. Lane, L. Lipsich, M. Wigler, and M. Botchan, *Cell (Cambridge, Mass.)* **21,** 127 (1980).

integrated into mouse chromosomal sequences, the mouse cells propagated, and pBR322 sequences were recovered by restriction digest. These mouse-modified pBR322 sequences transformed DH1 (a descendant of MM294) at a frequency of 10^{-2} to 10^{-3} relative to the same amount of *E. coli*-modified pBR322 mixed with mouse DNA in a reconstruction experiment. This is likely to be the result of McrA and McrB restriction, since mammalian cells are known to contain methylcytosine (see Strain Choice below).

Mcr restriction may also account for reports that plant DNA (with up to 25% of the C residues methylated; see Levels of Methylation, below) is particularly difficult to clone. For comparison, methylation of all the *Msp*I sites in pBR322 gives a methylation level of about 2% of the C residues, and results in 100-fold restriction. As in the case of *Eco*K restriction, some sequences may be underrepresented in genomic libraries, even when the overall methylcytosine content is not very high, since those sequences carrying a concentration of methylated cytosines are likely to be preferentially lost from the library.

Effect on Other Experiments. Just as the pattern of methylation acquired *in vivo* determines sensitivity or resistance to restriction *in vivo,* the methylation pattern can be altered *in vitro* to change the apparent specificity of restriction endonuclease digestion *in vitro,*[18] or simply to protect DNA from digestion, as when adding linkers to DNA to be cloned.[19] When cytosine methylases are used in such procedures, the resulting DNA will be sensitive to Mcr restriction *in vivo,* and McrA⁻B⁻ hosts should be used (Table I and above). Two specific examples (mentioned above) of the use of methylation in cloning and in the remodeling of clones are discussed at greater length by Nelson and Schildkraut,[20] and by Wu *et al.*[21]

Mar Restriction

There also appears to be a restriction function in *E. coli* K12 that attacks DNA methylated at adenine residues.[6] Plasmid clones carrying M.*Pst*I (CTGC^meAG) or M.*Hha*II (G^meANTC), but not M.*Eco*RI

[17] D. Hanahan, *in* J. L. Ingraham, B. Magasanik, M. Schaechter, K. B. Low, F. C. Neidhardt, and H. E. Umbarger, eds., *"Escherichia coli* and *Salmonella typhimurium:* Cellular and Molecular Biology." Am. Soc. Microbiol., Washington, D.C., 1986.

[18] M. McClelland and M. Nelson, *Nucleic Acids Res.* **13,** r201 (1985).

[19] T. Maniatis, E. F. Fritsch, and J. Sambrook, "Molecular Cloning: A Laboratory Manual." Cold Spring Harbor Lab., Cold Spring Harbor, New York, 1982.

[20] M. Nelson and I. Schildkraut, this series, Vol. 155, in press, 1987.

[21] This volume [38].

(GA^{me}ATTC) or Dam (G^{me}ATC), damage the DNA of Mar$^+$ hosts and induce synthesis of repair functions (the SOS system). λ DNA methylated by the *Hha*II methylase *in vivo* is restricted about 10-fold by Mar$^+$ cells. The genetic locus in question is also located near 99 min, near the *mcrB* and *hsdRMS* (*Eco*K) loci, but is genetically independent of both. The effects of Mar restriction on foreign DNA are likely to be similar to the effects of Mcr restriction, but with different specificity. The status of the *mar* locus in most laboratory strains is not known at present.

Levels of Methylation in Natural DNA

The importance of Mcr and Mar restriction will depend on the amount of 5-methylcytosine and 6-methyladenine in the DNA to be cloned. The presence of methylation in DNA may also result in discrepancies between restriction maps of clones and of the original DNA (see last section below). Some generalizations can be made here about the levels of methylation likely to be encountered; more detailed information is available elsewhere.[22,23]

There are three common kinds of methylated bases in DNA: 6-methyladenine, N^4-methylcytosine, and 5-methylcytosine. 6-Methyladenine is frequently found in bacteria, sometimes in association with restriction systems.[1,24] It is also found in lower but not higher eukaryotes.[25–27] 6-Methyladenine-requiring restriction has been observed in *Pneumococcus*[3] and, very recently, in *E. coli*.[6] Usually 0.2–2.5% of total bases are 6-methyladenine, when this methylation is present.[1,26]

N^4-Methylcytosine has recently been found in bacteria,[2,28] and is not known to occur in other organisms. It is not known whether N^4-methylcytosine-requiring restriction systems exist in *E. coli* or elsewhere. N^4-Methylcytosine comprises 0.05–0.4% of bases when present.[28]

5-Methylcytosine is widely distributed, occurring in higher plants and animals as well as in bacteria and lower eukaryotes. As described above, two 5-methylcytosine-requiring restriction systems are found in *E. coli*

[22] R. H. Hall, "The Modified Nucleosides in Nucleic Acids," p. 281. Columbia Univ. Press, New York, 1971.

[23] H. S. Shapiro, *in* "Handbook of Biochemistry" (H. A. Sober, ed.), p. H-89. CRC Press, Cleveland, Ohio, 1970.

[24] M. Marinus, *in* "DNA Methylation: Biochemistry and Biological Significance" (A. Razin, H. Cedar, and A. D. Riggs, eds.), p. 81. Springer-Verlag, Berlin and New York, 1984.

[25] M. A. Garovsky, S. Hattman, and G. L. Pleger, *J. Cell Biol.* **56,** 697 (1973).

[26] P. M. M. Rae and R. E. Steele, *BioSystems* **10,** 37 (1978).

[27] M. Ehrlich and R. Y.-H. Wang, *Science* **212,** 1350 (1981).

[28] M. Ehrlich, M. A. Gama-Sosa, L. H. Carriera, L. G. Ljungdahl, K. C. Kuo, and C. G. Gehrke, *Nucleic Acids Res.* **13,** 1399 (1985).

K12. The levels of 5-methylcytosine range from 0.5–2% of bases in bacteria[1], to 0.7–2.8% in mammals, to as much as 7% of bases (25% of cytosines) in higher plants.[27,29] In mammals, cytosine methylation is usually found in ^{me}CpG or ^{me}CpNpG sequences. It is important to note that both McrA and McrB could attack such sequences, depending what bases are found in the immediate vicinity. No 5-methylcytosine has been detected in *Drosophila melanogaster*[30] or *Saccharomyces cerevisiae*.[31] In the latter case, a conflicting result was reported by the same laboratory earlier, using a less reliable method.[32]

Strain Choice

Some of the strains shown in Table I are widely used in cloning experiments of one sort or another. A brief discussion of the uses of these strains follows, with particular reference to their restriction phenotypes. However, the Mar phenotype is known only for HB101, and by inference for RR1. For more detailed discussion of strain choice, see this volume [18].

MM294, **HB101**, and **RR1** are used as hosts for plasmid clones. All have the advantage of efficient transformation to begin with, and all lack the *Eco*K restriction activity. All contain the *supE44* allele and can be used to propagate some phage vectors that contain amber mutations; however, none contains the *supF* suppressor, and cannot be used as lytic hosts for λ phages carrying the S100 or S7 mutations. MM294 retains the *Eco*K modification activity, which allows clones propagated in it subsequently to be transferred into *Eco*K⁺ strains, but both Mcr activities are present. HB101 and RR1 lack the McrB and Mar functions, reducing the probability of loss of clones containing methyl-C or methyl-A, but they also lack the K modification activity, reducing accessibility to other strains for clones propagated in it. HB101 is also RecA⁻.

C600, **K802**, and **LE392** are common hosts for propagating vector phages. Under normal circumstances C600 is not used for primary cloning. K802 and LE392 have been used for preparation of λ libraries. K802 has the advantage of lacking the *Eco*K, McrA, and McrB restriction systems. All carry the *supE* (glutamine-inserting) suppressor. LE392 also carries the *supF* (tyrosine-inserting) suppressor, and will propagate phage with the S7 or S100 mutations. LE392 is also known as ED8654, BHB2600, and BNN45.

[29] S. Hake and V. Walbot, *Chromosoma* **79**, 251 (1980).
[30] S. Urieli-Shoval, Y. Gruenbaum, J. Sedat, and A. Razin, *FEBS Lett.* **146**, 148 (1982).
[31] J. H. Proffitt, J. R. Davie, D. Swinton, and S. Hattman, *Mol. Cell. Biol.* **4**, 985 (1984).
[32] S. Hattman, C. Kenny, L. Berger, and K. Pratt, *J. Bacteriol.* **135**, 1157 (1978).

Y1084, **Y1088**, and **Y1090**. Y1084 is the plasmidless parent of Y1090, not usually used for cloning experiments. Y1088 and Y1090 are lytic hosts for the expression vector λgt11. In this cloning system, clones can be identified because the inserts inactivate the phage-borne *lacZ* gene. This cloning system is particularly useful because efficient and tightly regulated expression of fusion genes (consisting of joined *lacZ* and insert DNA coding regions) can be obtained. Both of these hosts contain a plasmid, pMC9, carrying the *lacI* gene (this is the plasmid missing in Y1084). The plasmid allows repression of the *lac* promoter carried on the phage. Typically, Y1088 is used for initial infection by packaged ligation mixtures, since it is *Eco*K R⁻M⁺. However, it still carries the McrB restriction system, so that sequences that are C-methylated in the original organism are likely to be restricted. Y1090 is normally used as host for secondary passage of clones, since it is *Eco*K R⁺, as well as McrB⁺. Its particular usefulness is that it lacks the *lon* gene product, a protease that frequently degrades abnormal proteins. Fusion proteins are thought to be targets of the Lon protease.

JM101 and **JM107** are primarily hosts for M13 derivatives to be used in Sanger dideoxy sequencing procedures. Clones that have been identified in some other system are subcloned into these derivatives (M13mp8, 9, 18, or 19 are widely used). Consequently, the DNA to be propagated has usually already been passed through *E. coli* K12 and will not be subject to restriction by any of the systems present. If the DNA to be subcloned is prepared in a strain lacking the *Eco*K methylase (for example, HB101 or RR1), it should be remembered that JM101 carries the *Eco*K restriction activity. When subcloning procedures make use of cytosine methylases (as described above and in refs. 19 and 20), it should be remembered that JM101 is McrA⁺ (McrB not tested); JM107 is McrA⁻B⁺. JM103 and JM105 have not been tested for Mcr phenotype.

The remaining strains in Table I (K12, W3110, and GM2163) are not themselves frequently used in cloning experiments. K12, obviously, is the ancestor of all strains. W3110 and C600, above, frequently are ancestral to other strains that are used in cloning work. GM2163 is Dam⁻ Dcm⁻, and is sometimes used to prepare DNA that can be cut, for example, by *Bcl*I and *Eco*RII (see below and Table II). It also has the advantage of lacking *Eco*K, McrA, and McrB.

Effects of *in Vivo* Methylation on *in Vitro* Restriction

Escherichia coli K12 also contains two methylases, Dam (for DNA adenine methylase) and Dcm (for DNA cytosine methylase).[24] No restriction functions are associated with these methylases. However, *in vivo*

TABLE II
ENZYME ACTIVITIES SENSITIVE TO Dam OR
Dcm METHYLATION

Enzyme	Sequence[a]
Dam-sensitive enzymes	
BclI	TGATCA
ClaI	GATCGATC
HphI	GGTGATC
MboI	GATC
MboII	GAAGATC
NruI	GATCGCGA
TaqI	GATCGA
XbaI	TCTAGATC
Dcm-sensitive enzymes	
AvaII	GG(AT)CC(AT)GG
EcoRII	CC(AT)GG
Sau96I	GGNCC(AT)GG
StuI	AGGCCTGG

[a] Italic type indicates restriction enzyme recognition site.

methylation by the Dam and Dcm methylases can mask certain endonuclease recognition sites by methylation protection (Table II). The endonucleases in question either contain or overlap a Dam site, or contain or overlap a Dcm site.

For example, ClaI (ATCGAT) and Dam (GATC) recognition sites share the sequence ATC. Approximately seven-sixteenths of all ClaI sites will be preceded by a G or followed by a C, and thus overlap a Dam site. The precise proportion will depend on the base composition of the DNA in question. Methylation of the A by the Dam methylase will protect these sites from ClaI digestion. This phenomenon may be encountered inadvertently when clones are grown in E. coli K12 strains, most of which contain the Dam methylase. For example, one of two ClaI sites in the lacZ gene overlaps a Dam site, and is not digested unless the DNA is prepared from a Dam⁻ strain.[33,34] The effect of Dam methylation is thus to increase the specificity of ClaI digestion. Its effective recognition site becomes (C/T/A) ATCGAT (T/A/G), instead of N ATCGAT N.

A similar interference may be observed when comparing a restriction map of a clone with a restriction map of the original DNA (e.g., using

[33] A. Kallnins, K. Otto, V. Ruther, and B. Müller-Hill, EMBO J. **2,** 593 (1983).
[34] E. Raleigh, unpublished.

Southern blot analysis of genomic DNA with the clone as probe). Sites mapped on the clone may be blocked by adenine or cytosine methylation in genomic DNA of the original organism.

Some endonucleases can discriminate among the possible base modifications. For example, *Bam*HI, *Bgl*II, *Sau*3AI, *Pvu*I all recognize sequences containing GATC, but are not blocked by Dam methylation.[18] Their cognate methylases apparently modify the C in GATC instead of the A. *Bst*NI (CC (A/T) GG) is not blocked by Dcm methylation. The protective methylation for the *Bst*N1 endonuclease is not known, but could be N^4-methylcytosine.[18,28]

Section V

Growth and Maintenance of Bacteria

[13] Practical Aspects of Preparing Phage and Plasmid DNA: Growth, Maintenance, and Storage of Bacteria and Bacteriophage

By Harvey Miller

The practical aspects of cloning can be divided into four major topics: (I) handling, culturing, and storing bacteria; (II) preparation of competent cells for transformation; (III) growth of phage and the preparation of phage DNA; and (IV) isolation and purification of plasmids. A section of this chapter is devoted to providing detailed instructions for methods in each of these broad categories. Brief explanations of theory are included where necessary. There are also three tables that contain basic information about bacterial strains used in cloning (Table I), buffers, media, and agar (Table II), and antibiotics (Table III).

Handling, Culturing, and Storing of Bacteria

Successful recombinant DNA techniques require the proper handling of a substantial number of bacterial strains. Although all are derivatives of *Escherichia coli* K12 strains, the strains have different characteristics making them suitable for a particular purpose. For example, many strains lack host restriction or modification systems which allow propagation of unmodified DNA. Some carry mutations of the *lac* operon which allows detection of inserts into phage and plasmids of the pUC and M13mp type. Strains containing nonsense suppressors permit the growth of specialized phage cloning vectors. Table I lists many common bacterial strains, their genotypes, and uses.

It is important when working with these bacteria that the cultures be kept pure, that the phenotypes be verified prior to use, and that they be stored properly. It is assumed that sterile glassware and media are available. A thorough discussion of basic microbiological techniques can be found in *Experiments in Molecular Genetics*.[1]

Pure Cultures. Pure cultures can be obtained by propagating bacterial cultures from single, isolated colonies on agar plates. The simplest method of obtaining isolated colonies is by dilution streaking with an inoculating loop. A small inoculum of bacterial culture is picked up with a

---

[1] J. H. Miller, "Experiments in Molecular Genetics." Cold Spring Harbor Lab., Cold Spring Harbor, New York, 1972.

TABLE I

Common Bacterial Strains Used in Cloning

Strain	Genotype	Comments
C600	F^-, *thi-1, thr-1, leuB6, lacY1, tonA21, supE44*	This strain is phage T1 resistant (*tonA*) and an amber suppressor. Restriction/modification mutants are also available
DH1	F^-, *recA1, endA1, gyrA96, thi-1, hsdR17, supE44*	An MM294 derivative that is naladixic acid resistant (*gyrA*), *recA*, and suppressor containing
MC1061	F^-, *araD139*, Δ(*ara,leu*) *7696*, Δ*lacY74*, *galU*$^-$, *galK*$^-$, *hsdR*$^-$ *strA*	
JM103	*thi*$^-$, *strA, supE, endA, sbcB, hsdR*$^-$ Δ(*lac pro*) *F'traD36 proAB lacI*^{q}, *lacZ* Δ*M15*	Used for plating M13 derivatives. A male strain containing a nontransferring plasmid, a suppressor, *lac* repressor overproducer, and *lac* α complementing activity
JM109	*recA1, endoA1, gyrA96, thi*$^-$, *hsdR17, supE44, relA1* Δ(*lac,pro*) *F'traD36 proAB lacI*^{q}, *lacZ* Δ*M15*	Similar to JM103 but *recA*
HB101	F^-, *hsdS20, recA13 ara-14, proA2, lacY1, galK2, straA xyl-5, mtl-1, supE44*	Highly transformable strain; restriction and modification defective and *recA*
x1776	F^-, *tonA53, dapD8, minA1, supE42*, Δ(*gal-uvrB*)*40*, *minB2, rfb-2, gyrA25, thyA142, oms-2, metC65, oms-1*, Δ(*bioH-asd*)*29*, *cycB2, cycA1, hsdR2*	A highly transformable strain, especially with the Hanahan protocol. It was once required for containment reasons, but due to its many nutritional requirements and sensitivity to detergents, is not frequently used
MM294	F^-, *endoA1, thi*$^-$, *pro*$^-$, *hsdR*$^-$, *supE44*	Highly transformable, well-growing strain which gives high yields of plasmid DNA
Y1088	Δ*lacU169 supE supF hsdR*$^-$ *metB trpR tonA21 proC*::Tn*5* (*pMC9*)*pMC9*= *pBR322lacI*^{q}	A permissive host for growing λgt11. Contains the *supF* suppressor to allow lysis when the phage contain the S100 mutation and also represses the *lac* promoter carried by λgt11
N99	F^-, *strA, galK2* λ^-, *IN*(*rrnD-rrnE*)*1*	Strain MM28 of Meselson, a nonsuppressing strain
BNN93	*hsdR*$^-$, *supE, thr, leu, thi, lacY1 tonA21*	Permissive host for λgt10
BNN102	Same as BNN93, but *hflA150*(*chr*::Tn*10*)	Nonpermissive host for λgt10 but permissive if DNA insert interrupts λ *c*I gene
Y1089	*dlacU169, d*(*lon,araD*)*169 strA, hflA150* (*chr*::Tn*10*) (pMC9)	Used for making lysogens at high frequency with recombinant λgt11 clones. The *lon* deletion stabilizes β-galactosidase fusion proteins

TABLE I (*continued*)

Strain	Genotype	Comments
Y1090	Δ*lacU169* Δ(*lon,araD*)*139* *strA supF*(*trpC*::Tn*10*) (pMC9)	Permissive host for plating λgt11 recombinants for plaque screening. The *lon* deletion stabilizes fusion proteins and the *supF* suppresses the S100 mutation
RR1	RecA⁺ derivative of HB101	

TABLE II

BUFFERS, AGAR, AND MEDIA

Substance	Composition
TM buffer	10 mM Tris–HCl at pH 7.4, 10 mM MgSO$_4$
TMG buffer	10 mM Tris–HCl at pH 7.4, 10 mM MgSO$_4$, 0.01% gelatin
TE buffer	10 mM Tris–HCl at pII 7.4, 1 mM EDTA
TEN	10 mM Tris–HCl at pH 8, 100 mM NaCl, 1 mM EDTA
LB broth	10 g Bacto-tryptone (Difco), 5 g yeast extract (Difco), 5 g NaCl, in 1 liter water. Adjust pH to 7.2. Autoclave
LB agar	LB broth with 10 g Bacto-agar (Difco). Autoclave
TB agar	10 g Bacto-tryptone, 10 g Bacto-agar, 5 g NaCl, in 1 liter water. Adjust pH to 7.2. Autoclave
TB top agar	TB agar but with 7 g Bacto-agar. Autoclave
TCMG agar	10 g BBL trypticase (not soy), 8.5 g Bacto-agar, 5 g NaCl, in 1 liter water. Adjust pH to 7.2. Autoclave. Add 10 ml sterile 1 M MgSO$_4$
MacConkey agar	40 g Difco MacConkey base, 10 g sugar in 1 liter water. Autoclave
Stab agar	10 g Bacto-tryptone, 7 g Bacto-agar, in 1 liter water. Heat to dissolve. Aliquot to vials. Autoclave
M9 medium	6 g Na$_2$HPO$_4$, 3 g KH$_2$PO$_4$, 0.5 g NaCl, 1 g NH$_4$Cl, in 1 liter water. Adjust pH to 7.4. Autoclave. Add 2 ml 1 M MgSO$_4$ (autoclaved or filter sterilized)
M9 (glucose) medium	M9 1 liter, 10 ml 20% glucose (autoclaved or filter sterilized)
M9 (with casamino acids) medium	M9 1 liter; 10 ml 20% casamino acids (autoclaved)
YT medium	8 g Bacto-tryptone (casein hydrolyzate), 5 g yeast extract, 5 g NaCl, in 1 liter water. Autoclave
YT agar	YT medium with 10 g Bacto-agar (Difco). Autoclave
YT top agar	YT agar but with 7 g Bacto-agar. Autoclave

TABLE III
COMMON ANTIBIOTICS[a]

Antiobiotic	Useful concentration (μg/ml)	Solvent	Remarks
Tetracycline	15	Ethanol	Light sensitive
Streptomycin	50	Water	
Kanamycin	25	Water	
Chloramphenicol	12.5	Ethanol	
Ampicillin	50	Water	Unstable. Depleted from media by resistant bacteria

[a] Antibiotic solutions should be made in 100-fold concentrated stock solutions and filter sterilized (not necessary for ethanol solutions), and stored in the cold. Tetracycline solutions should be wrapped in foil. When making agar plates, cool agar to 55° before adding antibiotic.

flame-sterilized and cooled inoculating loop. Starting at one edge of the plate, streak the loop back and forth until one-half of the plate is covered. Reflame the loop. Rotate the plate counterclockwise a quarter-turn and streak half of the plate again. Flame the loop, rotate another quarter-turn, and streak again. There should be isolated colonies at least in the last sector streaked. Pick an isolated colony and restreak a new plate to obtain a pure culture master plate. This plate can be stored for several weeks at 4° if the edges are wrapped in Parafilm.

Titering Bacteria. Overnight cultures of most strains used in cloning produce ~4 × 10^9 bacteria/ml ($A_{550} \simeq 10$) depending upon the medium, degree of aeration, the strain, and the temperature. To determine the concentration more precisely, appropriate dilutions of the culture should be plated. An 82-mm plate can accommodate 50–200 well-separated colonies.

To plate bacteria, place a drop containing 0.01–0.5 ml of culture at the center of the plate and spread the solution rapidly over the surface of the agar using a sterile bent glass rod. A rotating wheel aids even, rapid spreading but it is not essential. The bent glass rod must be sterilized between samples. Dip it into 70% ethanol, place briefly in the flame of a Bunsen burner, and allow the ethanol to burn off. Touch the heated rod to the agar surface of the plate to cool it before making contact with the bacteria.

During screening of densely plated plasmid libraries (>1000 colonies/ plate), it is often not possible to obtain the colony responsible for a positive signal without also selecting others. In such cases, it is necessary to replate the clones in the vicinity of interest at a much lower density so that, on rescreening, the signal can be assigned to a specific colony (see this volume [45]). This is done by smearing the plate with a sterile loop over an area of about 1 cm in diameter surrounding the positive signal, dipping the loop into about 1 ml of solution (e.g., LB broth), and suspending any clumps by vortexing. Since the titer is completely unknown, dilutions spanning about four orders of magnitude are usually plated. The remaining culture can be stored at 4° for a few days as a backup source of bacteria.

Storage of Pure Cultures. Cultures can be stored for long periods of time in stab vials, as frozen glycerol cultures, or as lyophilized cultures. For most purposes, stabs and glycerol cultures are the methods of choice. Stabs should be transferred every 2 years whereas frozen glycerol cultures can be kept indefinitely.

Stab Cultures. Use 2- to 3-ml glass vials, with tight-fitting caps preferably with rubber gaskets. Fill the vial two-thirds full with molten stab agar. With the caps loosened, autoclave the media and let cool. The caps should then be tightened. Pick an isolated colony with a sterile inoculating needle or toothpick and stab several times into the vial. Tighten the cap. To ensure an airtight seal, dip the top of the sealed vial in melted paraffin. Store the vials at room temperature in the dark. It is not necessary or desirable to add antibiotics to the stab media since they degrade with time and little opportunity for segregation of plasmids exists during the short period of growth in the vial.

Glycerol Cultures. Bacterial culture can be stored at −70° in solutions of 15–50% glycerol. In one method, 0.15 volumes of sterile (autoclaved) glycerol is added to 0.85 volumes of LB broth culture. The glycerol is mixed with the media in a sterile freezer-proof screw-capped plastic vial. The vial is frozen in ethanol–dry ice or liquid nitrogen and stored at −70°. Alternatively, cell paste scraped from an agar plate or a pellet from an overnight culture can be thoroughly mixed with TM buffer made 50% with glycerol. Again, the cultures are frozen in plastic screw-capped tubes and stored at −70°.

To recover the bacteria, it is not necessary to thaw the tube. The surface of the frozen culture can be scraped with a toothpick or a sterile inoculating loop and directly streaked on an agar plate. Repeated thawing and freezing should be avoided. It is best to make several tubes of each culture and use only one for subculturing.

Testing for Phenotype. Most strains used for recombinant DNA are rather stable. However, after many generations of growth some strains may segregate plasmids, revert mutations, or develop undesirable mutations. In addition, airborne bacterial and phage contaminants are always a problem even when strict sterile technique is employed. After a short period of time working with *E. coli,* one should be able to spot bacteria with different color, odor, and colony shape from *E. coli.* These should be discarded. In addition, if large, rapidly growing plaques appear on pour plates of *E. coli* (usually phage T1), these plates should be autoclaved. T1 phage is stable in air and can decimate culture collections. It is necessary to scrub laboratory benches with strong antiseptic to kill T1.

The majority of phenotypic markers in most bacterial strains are irrelevant to their usefulness in recombinant DNA techniques. For example, many strains contain auxotrophic or sugar fermentation mutations which were present in starting strains or useful in strain construction. These mutations, however, are useful in identifying the proper strain. Auxotrophic markers can be easily tested by growing bacterial strains on minimal media (M9 for example) with and without the required supplement. Sugar fermentation markers can be tested on MacConkey agar plates with the proper sugar added.

The *recA* mutation (recombination deficient) can be tested by the sensitivity of the strain to ultraviolet radiation or chemicals such as mitomycin C or methylmethane sulfonate. The simplest method is to streak half of a plate with the putative *recA* strain and the other half with a *rec+* control. Cover half of each streak on the open plate with aluminum foil and expose the open plate to a bacteriocidal ultraviolet lamp for several seconds. Replace the cover and incubate. The *recA* strain should be killed much more effectively than the wild-type control. *recA* mutations are important for stabilizing clones in which repeated sequences can delete or invert by recombination.

Suppressor Mutations. Bacteria containing suppressor mutations carry altered tRNAs that are capable of inserting amino acids at the positions of nonsense mutations. Since many phages used in recombinant DNA work contain nonsense (usually amber) mutations that restrict growth, it is important to verify that strains contain the appropriate suppressor. For most phages, the mutations were included for containment reasons. However some, such as the S mutations of λ, are useful for propagation of high titer lysates. Suppressor mutations are most easily scored by their ability to plate the appropriate λ amber phages. For example, λ S mutations are only suppressed by *supF*-containing strains whereas the *Pam3* mutation is only suppressed by *supE* or *supD.*

Host Restriction and Modification. K12 strains of *E. coli* contain a restriction and modification system (*hsdR* and *hsdM*) which will restrict nonmodified DNA such as that originating from eukaryotic sources (see also this volume [12]). Therefore, many strains have been developed that lack the restriction system and thus do not degrade non-*E. coli* DNA, but still modify the DNA, so the DNA or phage derived from these strains can be propagated in any K12 strain. To test for restriction, compare the titer of a λ phage grown on a known nonmodifying host (*hsdR⁻ hsdM⁻*) on the nonrestricting host and the strain to be tested. If the titers are roughly equal, the strain is nonrestricting. If the titer is greatly reduced (10-fold lower), the strain is restricting (use as a positive control phage grown on a strain with the host modification system).

hfl Mutant Strains. The cDNA cloning and expression vectors, λgt10 and λgt11,[2] respectively, require the use of *E. coli hfl⁻* mutants. After infecting an *hfl* mutant, λ phage with a normal repressor (*c*I) gene over-produce repressor protein, resulting in greatly reduced burst size and therefore, minute plaques. The lysogeny frequency is greatly enhanced. For λgt10, the ability to plaque on *hfl* bacteria is the basis for selecting phage with DNA inserts, since disruption of the *c*I gene by DNA insertion permits phage growth. For λgt11, *hfl* mutants facilitate lysogenization for phage expressing fusion proteins.

The Hfl⁻ phenotype can be scored by the ability of the strain to plate λ *c*I⁻ or *c*II⁻ phage but not λ wild type.

Testing for λ Lysogens. The presence of a λ prophage in a bacterium (the combination known as a lysogen) can be verified by spotting phage of the same immunity on a lawn of the strain to be tested. The phage should not grow since the presence of λ repressor produced by the prophage blocks development of the infecting phage. The ability of a phage with a different immunity, such as λimm434, to grow indicates that the strain is a true lysogen, not just resistant to phage adsorption. In some cases, when the prophage to be tested carries the *c*I857 thermoinducible repressor, the failure of the lysogen to grow at high temperature (42°) is sufficient evidence that the prophage is present.

Growth on P2 Lysogens. λ phages containing the *red* and *gam* genes of λ are growth inhibited by P2 lysogens. This is called the Spi⁺ phenotype (sensitive to P2 inhibition). Some λ cloning vectors such as the EMBL3 and EMBL4 vectors exploit this phenotype by containing stuffer fragments which contain *red* and *gam*. When the stuffer is replaced with recombinant fragments, the recombinant phage can grow on P2 lysogens. The P2 prophage can be detected by immunity to P2 phage infection or by the exclusion of λ but not λ *red⁻ gam⁻*. (See this volume [14] for a more detailed discussion of λ.)

Preparation of Competent Cells

There are two methods for introducing DNA into bacterial cells, transformation or transfection into competent cells and infection of cells after

---

[2] T. V. Huynh, R. A. Young, and R. W. Davis, *in* "DNA Cloning" (D. M. Glover, ed.), Vol. 1, p. 49. IRL Press, Washington, D.C., 1985.

in vitro packaging of DNA containing λ phage cohesive ends. Prior to the development of efficient phage cDNA cloning vectors such as λgt10, which can take advantage of the highly efficient *in vitro* packaging systems, transformation of competent cells with plasmid vectors was the only method of cDNA cloning. For this reason, high efficiency transformation yielding >10^7 transformants/μg of plasmid DNA, was essential. Variations of the original Mandel and Higa[3] method of preparing competent *E. coli* cells by $CaCl_2$ treatment could not reproducibly reach these levels of competency. However, by utilizing a demanding protocol and certain *E. coli* strains, Hanahan[4] was able to achieve high competency. Since most cDNA cloning now employs phage vectors such as λgt10, the high frequencies of transformation obtainable by the Hanahan procedure are not essential for most general-purpose plasmid construction and mutagenesis. Therefore, this procedure will be omitted from this chapter. The reader can refer to Hanahan[4] for a detailed protocol of this procedure.

The frequency of transformation depends greatly on the strain used. When high frequencies of transformation are required, strains such as HB101 and MM294 are essential. However, almost any *E. coli* strain can be transformed at some frequency with circular plasmid DNA. It is recommended that plasmid constructions be identified in a highly competent strain, and subsequently be introduced into a special purpose strain (such as a *lac* mutant).

Prepared competent cells can be frozen in glycerol and stored at $-70°$ indefinitely. Batches of competent cells prepared in exactly the same manner vary in efficiency of transformation. Therefore, it is recommended that large batches of competent cells be prepared for freezing. When a particularly good batch is prepared, you will then have several hundred aliquots to use, obviating the need to prepare cells each time they are needed.

Several factors are critical in the procedure given below. Particular attention should be paid to (1) the density of the cells at harvest, (2) the necessity for maintaining the cells at $0-4°$ throughout the processing, and (3) the long incubation time at $4°$ prior to freezing.[5]

Procedure

1. Grow a fresh overnight culture of cells in LB broth at $37°$.
2. Dilute the cells 40-fold into 1 liter of fresh medium. Incubate at $37°$ with good aeration until an absorbance at 550 nm of 0.4–0.5 is reached.

[3] M. Mandel and A. Higa, *J. Mol. Biol.* **53,** 154 (1970).
[4] D. Hanahan, *in* "DNA Cloning" (D. M. Glover, ed.), Vol. 1, p. 109. IRL Press, Washington, D.C., 1985.
[5] M. Dagert and S. D. Ehrlich, *Gene* **6,** 23 (1979).

3. Immediately chill the culture by swirling in an ice-water bath.

4. When the cells are chilled, centrifuge the culture at 4° at 5000 rpm for 10 min.

5. Decant the supernatant and place the pellets in ice.

6. Resuspend the pellets in 500 ml ice-cold 100 mM CaCl$_2$. It is easier to resuspend the pellets if they are first vortexed before the CaCl$_2$ is added as some clumping can occur. The cells can be uniformly suspended by sucking up and down with a 25-ml pipet.

7. Once the cells are resuspended, incubate in ice for 30 min with occasional swirling.

8. Pellet the cells once again at 5000 rpm for 10 min at 4°.

9. Resuspend in 40 ml of ice-cold 100 mM CaCl$_2$, 15% glycerol.

10. Distribute aliquots of 0.2 ml of cells into sterile Eppendorf tubes in ice.

11. Keep in ice at 0–4° for 12–24 hr. This step is essential for high competency although the cells are considerably competent at this stage.

12. Freeze the tubes in ethanol–dry ice or liquid nitrogen and place immediately at −70°. The cells retain their competency for months if kept in this manner.

Transformation of Competent Cells

1. Thaw a tube of frozen competent cells at 4° (or in ice).

2. Add DNA. Ligation buffer is fine. If using a volume greater than 50 μl for 200 μl competent cells, bring the DNA solution to 100 mM CaCl$_2$. (See this volume [39] for the amounts of DNA that can be accommodated.)

3. Incubate in ice for 30 min.

4. Heat shock for 2–5 min in a 43° water bath.

5. Add 0.4 ml of LB broth at room temperature to each tube and incubate for 1 hr at 37°.

6. Spread on agar plate containing appropriate antibiotics. Spread dilutions if more than 500 colonies are expected. Incubate the plates inverted overnight at 37°.

Methods for transformation of the appropriate bacteria with double-stranded (RF) M13 DNA are presented in the next section.

[*Editors' Note*. Variations on this transformation procedure abound. Many investigators find that ligation buffer reduces the efficiency of transformation, and therefore they dilute DNA in ligation buffer 5-fold to optimize yields. Usually 1–5 μl of diluted DNA solution per 200 μl of competent cell gives acceptable results. The amount of DNA should be limited to ~1–10 ng since the number of transformants may not increase with increased DNA above 10 ng. The timing of the heat shock depends on the volume of bacteria and the shape of the test tube. Thus, once optimal conditions have been established, it is wise to scale up by performing many small reactions rather than one large reaction. Transformation efficiencies for supercoiled plasmid DNA and for various types of recombinants can be found in this volume [39].

Investigators who have never transformed bacteria and who do not have knowledgeable colleagues nearby may wish to purchase competent cells (Bethesda Research Laboratories, Gaithersburg, MD). When used according to the supplier's instructions, these cells are reliable, albeit prohibitively expensive, especially for large-scale work.]

Growth of Phage and the Preparation of Phage DNA: λ and M13

The biologies of phages λ and M13 differ greatly and, therefore, the methods of preparing their genome DNA is different. The λ genome can exist in a prophage form, integrated into the bacterial genome, or in a free viral form. The viral form must be propagated by infection of sensitive bacteria whereas the prophage form can be more conveniently propagated by simple induction of a lysogenic bacterial host. Since most λ cloning vectors, by design, have removed the integration and excision apparatus from λ, propagation by infection is the predominant method of preparation of λ phage particles, from which the DNA is extracted. Once a high-titer lysate of λ is obtained, the particles are concentrated by polyethylene glycol precipitation[6] and/or CsCl gradient centrifugation and the DNA extracted by phenol extraction.

Infection of sensitive male (containing F factor) bacteria with M13 phage derivatives does not lead to cell death and lysis as does infection by lytically growing λ. Rather, the cells become chronically infected and produce phage particles by budding. Therefore, supernatants from cultures of M13-bearing cells are a relatively pure source of M13 particles. The particles need only be concentrated, again by polyethylene glycol, and phenol extracted to remove protein from the DNA. This procedure yields single-stranded closed circular phage genomes. This form of the DNA is useful for sequencing by the Sanger dideoxy method and for *in vitro* mutagenesis. For cloning DNA fragments into M13 vectors, the closed circular double-stranded replication form (RFI) is necessary. This form can be isolated from M13-bearing cultures in a manner identical to multicopy plasmid isolation (see Plasmid Isolation, below). However, no plasmid amplification is possible or necessary.

---

[6] K. R. Yamamoto, B. M. Alberts, R. Benzinger, L. Lawhorne, and G. Treiber, *Virology* **40**, 734 (1970).

Growth of Phage λ. Before proceeding to grow λ derivatives, it is important to verify the phage genotype, to work with the progeny of a single pure plaque, and to titrate the number of phage particles accurately.

Preparation of Bacterial Lawns

λ phage grows on *E. coli* strain K12 derivatives. The phage tail fibers bind to the *lamB* (maltose-binding protein) receptor, which is induced by maltose in the growth media, but catabolite repressed by glucose. Therefore, plating bacteria are grown in broth containing maltose to maximize adsorption. Magnesium ion is necessary for adsorption and phage integrity and therefore, the maltose-grown culture is resuspended in 0.01 M $MgSO_4$.

Procedure. A fresh colony from an agar plate is used to inoculate a flask of sterile LB broth with 0.2% maltose and grown overnight at 37° to saturation. The culture is centrifuged at 5000 rpm for 10 min. The supernatant is discarded and the pellet is resuspended in 0.01 M $MgSO_4$ at one-quarter the original volume. This lawn can be kept for a month at 4°. One drop (0.2 ml) of this lawn is sufficient for a pour plate.

Titration of Phage

Most phage lysates contain between 10^7 and 10^{11} particles/ml. In order to determine the number of particles, the number of plaque-forming units (pfu) which for λ, is about one-half of the number of particles, is determined. The lysate is serially diluted and samples of the dilutions are mixed with bacterial lawn and molten top agar media which is then poured on an agar plate. Viable phage form an area of clearing (plaque) on the confluent lawn indicating the phage have infected and lysed the bacteria. The number of plaques is counted and total pfu is determined from the dilution factor.

1. Fill two sterile culture tubes with 10 ml of TMG buffer. Pipet 10 μl of lysate into the first tube. Vortex well. This tube is a 10^{-3} dilution.

2. Remove 10 μl from this tube into a second tube using a new pipet. Vortex well. This tube is a 10^{-6} dilution.

3. Pipet 100 μl and 10 μl, respectively, into two small tubes containing one drop (0.2 ml) plating lawn and incubate at room temperature for 5 min.

4. Add 2.5 ml molten top agar (at 55°) and immediately pour on the surface of an agar plate, touching the tip of the tube to the plate to remove the last drop. Gently tip the plate from side to side to cover the surface of the plate evenly.

5. After the agar hardens, invert the plates and incubate overnight. The plate with 100 μl of diluted lysate is labeled a 10^{-7} dilution and that with 10 μl a 10^{-8} dilution.

6. Count plates having between 30 and 300 plaques. Multiply by the dilution to derive the original titer.

If no plaques appear, either the wrong bacterial strain was used (some phage mutants require special suppressor strains) or the lysate was too dilute.

For general-purpose titrations, TB agar plates should be used. For phages that make exceedingly small plaques, use TCMG agar plates. The lower agar concentration, high magnesium concentration, and poorer growth medium of TCMG plates allow more rapid diffusion of the phage. If turbid plaques need to be distinguished from clear plaques, LB agar plates should be used. The rich medium favors lysogeny. In any case, the plates should be poured thick (40 ml for a standard 82-mm petri dish) and dried overnight inverted in a warm room. They can then be stored in plastic bags at 4° for several months.

Induction of Lysogens

This is the most efficient method for preparing phage lysates. Phages such as λgt11 that carry an intact integration–excision system can be prepared this way (λgt10 cannot). In most cases, the phage to be induced contains two mutations that simplify the induction and purification of the phage. The first is the cI857 thermoinducible repressor which allows induction of the lysogen merely by elevating the culture temperature above 37°. The second is the S7 or S100 lysis defective mutation. Phages carrying the latter mutation require the SupF suppressor for bacterial lysis to occur. In nonsuppressing hosts, lysis of the induced bacteria is prevented and phage yield is enhanced. In addition, the phage containing cells can be concentrated by centrifugation. The phage are released by treatment with chloroform. Concentrations of $>10^{12}$ pfu/ml can be obtained without any additional steps. Prior to preparing a culture for induction, the temperature-sensitive phenotype of the lysogen should be verified. Lysogens of cI857-containing phage must be grown at temperatures below 32° and fail to grow above 37° due to phage induction. Therefore, plate in replicates single colonies from a plate grown at 32° to two plates, one grown at 32° and the other at 42°. Prepare an overnight culture in LB at 32° from a colony that failed to grow at 42°.

1. Inoculate 1 liter of LB with 25 ml of the overnight culture of a λ lysogen in a nonsuppressing strain such as N99. A minimum flask size of 4 liters should be used. Incubate at 32° with vigorous shaking.

2. Have ready two shaking water baths at 45 and 37°. Monitor the $A_{550\,nm}$ of the culture. When the absorbance reaches 0.6 (~3 × 10^8/ml), rapidly raise the temperature of the culture to 45°. This is best accomplished by shaking in a sink full of 70° water. Use a thermometer in the flask. Alternatively, the culture can be grown to A_{550} = 1 in 500 ml of LB broth and an equal volume of LB broth preheated to 60° rapidly added. Avoid merely shifting the culture to a 45° bath since the slow shift causes asynchronous induction.

3. Once the temperature has been raised to 45°, place the flask in the 45° bath, and shake vigorously for 15 min.

4. Then shift to the 37° bath. The 45° incubation ensures irreversible induction yet the cell growth and therefore phage yield improve at 37°.

5. After ~2 hr at 37°, remove 1 ml to a small glass test tube and add several drops of chloroform. Vortex vigorously. If the bacterial suspension clears and you can see fibrous cell debris, the induction has been successful. Titer this lysate as described earlier in this section.

6. Chill the culture and centrifuge at 5000 rpm for 20 min. Decant the supernatant and resuspend the pellet in 0.05 volume (50 ml) of TM buffer.

7. Stir with 5 ml chloroform in a 100-ml beaker for at least 1 hr at room temperature or 37° (chloroform lysis at this density requires substantial time).

8. Centrifuge the lysate at 10,000 rpm for 10 min to remove cell debris. The resulting lysate should have a titer of at least 10^12/ml. It is now ready for CsCl density centrifugation. DNase may be added at this stage to remove host cell DNA (10 μg/ml followed by 37° incubation for 30 min).

Growth by Infection of Sensitive Bacteria

Two commonly used methods for preparing lysates by infection are the confluent plate lysate method and the liquid lysis method. The plate lysate method, although yielding generally higher titers, is not suitable for large-scale DNA preparation since only ~2–3 ml of lysate is extracted from each plate. The liquid lysis method is applicable at any scale. The important parameters for both methods are the cell density and the ratio of phage to cell or multiplicity of infection (moi). Since the phage are in contact with cell debris containing phage receptors for extended periods of time, loss of titer by adsorption to cell debris is the biggest problem. This can be minimized by using bacteria not grown in maltose, omitting magnesium from the culture, and including 5 mM CaCl$_2$. Sufficient receptors are present to allow some phage adsorption when maltose is omitted, in contrast to plating lawns where maximum adsorption is desirable. The moi and cell density are important since the phage must go through repeated cycles of growth, and sufficient numbers of growing cells must be present for the last round of infection. The plate lysate method has the advantage that rapidly growing mutants of λ, such as spontaneous clear plaque mutants of λgt10, do not overgrow the starting phage since diffusion of the mutant plaques is limited by the agar.

Procedure—Plate Lysates

1. Prepare a phage stock from a single plaque by punching out a plaque with a sterile Pasteur pipet into 1 ml of TMG buffer. Add a drop of chloroform and vortex briefly. A healthy plaque contains ~10^6 pfu. You can prepare a plate of plaques either by the titration procedure or by streaking an inoculating loop of lysate gently over the surface of a plate poured with a drop of lawn in top agar.

2. In a manner similar to titering phage, put 0.1 ml of the lysate into a drop of bacterial lawn and adsorb for 10 min at room temperature.

3. Add 2.5 ml TB and 2.5 ml TB top agar. Quickly pour on the surface of an 82-mm LB plate and let solidify.

4. Incubate the plate without inverting. After about 4–5 hr, the plaques should have spread entirely over the plate.

5. Scrape the loose top agar off of the plate with a bent glass rod into a centrifuge tube and vortex with several drops of chloroform to kill the remaining bacteria and to complete lysis.

6. Centrifuge the lysate at 10,000 rpm for 10 min and save the supernatant. The phage can be stored over chloroform at 4°. This method can be scaled up for large petri dishes (150 mm). Use four times the amounts as for small plates.

Procedure—Liquid Lysates

1. Inoculate LB medium plus 5 mM CaCl$_2$ with an overnight culture of appropriate bacteria such that the $A_{550} < 0.2$. Use a flask that is at least four times the volume of medium.

2. Grow the bacteria at 37° with vigorous shaking until an absorbance of 0.2 is reached (~10^8/ml).

3. At this point, add phage at an moi of 0.01 (10^9 phage for 1 liter). An adsorption period is not necessary at this high cell to phage ratio. (For small-scale liquid lysates use 1 to 2 ml of bacterial culture and a single plaque.)

4. Vigorously shake the flask in a water bath at 37° for ~3–4 hr, or until visible lysis occurs. If lysis is not seen, remove an aliquot and vortex with chloroform. If the culture now lyses or if previously lysed, add chloroform to 10 ml/liter and shake at 37° for an additional 15 min.

5. Centrifuge the lysate at 10,000 rpm for 15 min to remove cell debris.

6. Titer the lysate. The lysate is now ready for polyethylene glycol precipitation. The titer should be between 10^{10} and 10^{11}/ml.

If no lysis occurs, it may be worthwhile titering to see if sufficient phage are present to merit purification. If not, you can experiment with a higher moi of phage for infection or a lower cell density.

Polyethylene Glycol Precipitation

The phage in the lysate can be concentrated by polyethylene glycol (PEG) 6000 phase separation in high salt.[6] The PEG–phage pellet is recovered by centrifugation and the phage extracted in low-salt buffer.

1. Measure lysate. For each liter of lysate, stir in 40 g NaCl.
2. Gradually stir in 140 g PEG 6000 until dissolved at room temperature.
3. Leave at 4° at least 2 hr, preferably overnight. Shake the flask to resuspend any sediment and centrifuge at 5000 rpm for 20 min.
4. Decant the supernatant and titer to check recovery. Resuspend the pellet in 50 ml TM buffer.
5. Stir in a beaker at 37° with 5 ml chloroform until the pellet is uniformly suspended. Most loss occurs at this stage so the length of time of stirring is important (at least 1 hr).
6. Centrifuge the resuspended pellet at 10,000 rpm for 20 min. Decant and save the supernatant. The supernatant will still be milky gray but further centrifugation will not help.
7. Titer the supernatant to check recovery. If recovery is poor, try reextracting the pellet with chloroform for a longer time. The lysate is now ready for CsCl centrifugation. (CsCl centrifugation of phage is usually omitted when small-scale preparations are analyzed.)

CsCl Gradient Centrifugation

Two basic types of CsCl gradient centrifugations are used for phage purification, isopycnic or equilibrium gradients and step gradients. With isopycnic gradients, the phage suspension is mixed with CsCl to a density equal to that of the phage (1.5 for λ; however, deletion and insertion phage have lesser or greater density, respectively), and centrifuged at high speed. The CsCl forms a density gradient with the phage eventually banding at its characteristic density. With step or block gradients, the gradient is preformed by layering different density solutions of CsCl, overlaying the lysate, and centrifuging the phage through the steps until they eventually stop at the step corresponding to a density equal or greater then their own. The step gradients are more conveniently used for small volumes.

CsCl Solutions. With either method, CsCl solutions of proper density must be prepared. For wild-type λ, a 45.5% (w/w) solution will give a density of 1.5 g/ml. The amount of solid CsCl to add is calculated from the weight or volume of the solution from the formula

$$45.5/54.5 \times \text{weight of solution} = \text{g CsCl}$$

for a density of 1.5 g/ml. After the CsCl dissolves, the density can be checked either with a refractometer, or, more accurately, by weighing 100 μl of the solution. The weight of this aliquot should be 0.15 g. If not, add sufficient CsCl to bring the entire solution to 1.5 g/ml. If too dense, add buffer until the proper density is reached. The refractive index at 20° of this solution should be 1.3810.

Equilibrium Centrifugation

1. Weigh the supernatant from the PEG precipitation and add CsCl as described above to a density of 1.5 g/ml.

2. Fill cellulose nitrate ultracentrifuge tubes for a Beckman 50Ti or equivalent rotor and centrifuge at ~37,000 rpm for 24 hr at 4°.

3. After centrifugation, the phage band should appear as a bluish-white band in the center of the tube. It is easier to visualize the phage band if black velvet is placed behind the tube.

4. Remove the phage band by side puncture of the tube with a 26-gauge needle attached to a 5-ml syringe. Make sure to remove the top of the tube. Place a bit of vacuum grease on the tube under the phage band to prevent leakage. Insert the needle under the band and slowly draw out the band. The band will seem never to disappear since an interface will always be present. Do not take more than 1.5 ml.

5. Pool the collected bands and adjust the density to 1.5 g/liter. Fill another tube with the phage and CsCl solution of the proper density and centrifuge again. Collect the band as before and dialyze the phage against TM buffer. The phage is now ready for DNA extraction.

Step Gradients

1. Prepare three CsCl solutions in TM buffer of densities 1.4 g/cm³ (39/61 × volume), 1.5 g/cm³ (as above), and 1.7 g/cm³ (56.5/43.5 × volume).

2. In a swinging bucket rotor tube (SW27 for example), add 25 ml of phage lysate.

3. Carefully underlay the lysate with 4 ml of the 1.4 g/cm³ CsCl solution. This is best done with a syringe.

4. Next, underlay the CsCl with 4 ml of the 1.5 g/cm³ solution.

5. Finally, underlay with the 1.7 g/cm³ solution.

6. Spin at 27,000 rpm for 2 hr at 4°. The phage band should be located near the interface of the 1.4 and 1.5 g CsCl/cm³ solutions.

7. Collect the band as described above. The phage can be pooled and rebanded by equilibrium centrifugation.

Extracting DNA from Phage

The DNA is extracted from the phage by phenol–chloroform extraction. Use redistilled phenol that has been adjusted to pH 7–8 with 1 M Tris (see below). The DNA concentration of the phage lysates after dialysis can be determined by the absorbance at 260 nm. When $A_{260} = 1$, the solution has a DNA concentration of 50 μg/ml or 10^{12} phage/ml. It is best to dilute the phage solution to 200–300 μg/ml prior to phenol extraction. Avoid ethanol precipitation of large amounts of phage DNA because of the difficulty in resuspension (λ forms large concatamers by hydrogen bonding at the ends).

1. Dilute the dialyzed phage solution to an absorbance at 260 nm of 4–6.

2. Add an equal volume of treated phenol to the phage in a capped glass centrifuge tube. Mix by inversion for 5 min at room temperature.

3. Add chloroform equal to the volume of phenol and mix another 5 min.

4. Centrifuge at 5000 rpm for 5 min to separate the phases.

5. Remove the top layer with a Pasteur pipet whose tip has been bent into a U in a flame. Suck the top layer down from underneath so as not to disturb the proteinaceous interface.

6. Repeat the phenol extraction twice more (steps 2–5) followed by a final extraction of the aqueous phase with an equal volume of chloroform, alone.

7. The final aqueous phase is dialyzed against three changes of TE buffer.

8. Read A_{260} to calculate the DNA concentration and A_{280} to measure contamination. Pure DNA should have an A_{260}/A_{280} ratio of 2. If less than this, protein is still contaminating the DNA and more phenol extractions are required. The DNA can be frozen or stored with a few drops of chloroform at 4°.

Preparation of Phenol

1. Melt a bottle of redistilled phenol in a 65° water bath.

2. Add an equal volume of 1 M Tris–HCl (pH 7). Shake for 1 min. Replace at 65° until phases separate and cloudiness disappears.

3. Check pH of aqueous phase with pH paper. If well below pH 7, remove aqueous phase with a pipet (do not mouth pipet) and add more 1 M Tris–HCl (pH 7).

4. When pH is 7, remove aqueous phase and replace with 10 mM Tris–HCl (pH 7), 1 mM EDTA.

5. Shake, separate phases, remove the aqueous layer, and replace with TE buffer.

6. Add solid 8-hydroxyquinoline to 0.1%. Phenol prepared in this manner and stored at 4° can be kept for several months. When the yellow 8-hydroxyquinoline turns orange, discard.

Transformation of Competent Cells with Replicative Form (RF) DNA from M13

The biology of M13 phage exhibits certain similarities to the replication of plasmid DNA and certain properties common to bacteriophage λ production. Therefore, infection of *E. coli* with M13 DNA combines procedures derived from plasmid transformation with those of λ transfection.

Replicative form (RF) double-stranded M13 DNA is functionally analogous to plasmid DNA. Thus, the initial phase of M13 transformation involves the introduction of M13 RF DNA into calcium-treated competent cells. Since cells infected with M13 DNA will produce mature M13 phage and release them into the medium, the addition of rapidly growing male bacteria provides a fresh source of cells susceptible to infection with concomitant amplification of the number of individual phage particles eventually produced.

1. Remove one colony of an appropriate bacterial strain (e.g., JM103) from a minimal medium plate and transfer to 3 ml of minimum medium. Minimal plates and media must contain 1 mM thiamine HCl to support growth of JM101, 103, etc. Grow at 37° overnight.

2. Add ~1 ml of the overnight culture to 100 ml YT medium and grow for 1.5 hr (A_{550} should be about 0.3). Then add ~250 μl of this culture to 25 ml YT medium and continue to grow at 37°. The remaining cells (~100 ml) are collected by centrifugation (7000 rpm, 5 min) and resuspended in 50 ml of sterile 50 mM CaCl$_2$ at 4°. Keep at 4° for 20 min.

3. Centrifuge the CaCl$_2$-treated cells from step 2 and resuspend in 10 ml sterile 50 mM CaCl$_2$ at 4°. Mix 300 μl of these cells with the DNA to be transformed and keep in ice for 40 min. Then heat shock for 2 min at 42°.

4. Add 10 μl isopropyl-β-D-thiogalactopyranoside (IPTG) (100 mM), 50 μl 2% 5-bromo-4-chloro-3-indolyl-β-D-galactopyranoside (X-Gal), 300 μl of freshly growing cells from step 2, 3.5 ml YT top agar, and plate on 2× YT agar and incubate overnight.

5. Turbid plaques should appear on a lawn of bacteria. "Blue" plaques indicate nonrecombinant phage. "White" plaques indicate that a foreign fragment of DNA has been inserted into the β-galactosidase gene,

resulting in loss of activity. Restriction mapping is needed for confirmation.

Preparation of M13 Phage DNA

As discussed above, M13 phage is simpler to grow and isolate since the phage is secreted into the medium and can be isolated, concentrated, and extracted without the contamination of cellular DNA. The phage lysate is prepared by simply growing a chronically infected culture to saturation, removing all bacterial cells by centrifugation, PEG precipitating the phage, and then deproteinizing with phenol–chloroform.

1. Prepare a pour plate containing the desired M13 plaques. This can be done either by transforming competent cells with double-stranded replicative form M13 phage DNA (RF) and pouring a plate, as described above, or merely by streaking phage on the appropriate lawn. A male (F$^+$) strain of *E. coli* such as JM103 must be used because M13 phage only adsorb to the F pilus of male strains.

2. Pick a plaque with a toothpick or inoculating needle and inoculate a broth culture (either LB or 2× YT). Grow the culture to saturation at 37°.

3. Pellet the bacteria by centrifugation at 15,000 rpm for 15 min. Save supernatant. (This supernatant contains mature single-stranded M13 phage.)

4. Centrifuge again until no more cell pellet is visible. It is important to remove all bacteria from the lysate. You may wish to save some of the pellet to inoculate future preparations.

5. Next add 0.25 volume of 2.5 *M* NaCl, 20% PEG 6000. Mix thoroughly and let sit at room temperature for 15 min.

6. Centrifuge at 15,000 rpm for 15 min. A pellet should be visible at this stage. If not, there were no phage in the lysate. Try growing the phage again.

7. If there is a substantial pellet, remove all supernatant and discard supernatant. Resuspend pellet in 0.2 volume of TE buffer. Vortex vigorously to resuspend pellet.

8. Heat at 55° for 5 min. Vortex vigorously again.

9. Add an equal volume of prepared phenol (see above) and vortex vigorously for 1 min.

10. Add a volume of chloroform equal to the phenol, vortex, and centrifuge at 5000 rpm for 5 min to break the phases.

11. Remove the aqueous phase, being careful not to disturb the protein interface.

12. Reextract with phenol–chloroform and then finally with chloroform.

13. Add 0.1 volume 3 M sodium acetate at pH 4.5 and 2 volumes cold ethanol. Chill in ice for 10 min.

14. Centrifuge at 1000 rpm for 10 min at 4°. Decant the supernatant.

15. Carefully rinse the pellet with cold 70% ethanol and decant the ethanol. Dry the pellets under vacuum.

16. Resuspend in a convenient volume of distilled water. Determine the concentration of DNA. An A_{260} of 1 for single-stranded DNA is equivalent to a concentration of 35 μg/ml. The yield should be 0.5–1.0 mg/liter culture.

Analyzing Insert Orientations in M13

If different M13 DNAs contain identical DNA sequences inserted in opposite orientations, they are capable of hybridizing to one another, forming a DNA structure of molecular weight higher than either of the parental DNAs. This molecule of unique size can be resolved by agarose gel electrophoresis.

1. Mix 10 μl of medium supernatant (from step 3 above) from each of the M13 constructions to be tested with 10 μl 1% sodium dodecyl sulfate (SDS), 10 μl 1× TBE (this volume [8]), 10 μl 20% Ficoll.

2. Incubate at 65° for 30 min and 50° for 30 min.

3. Separate by electrophoresis in a 0.7% agarose gel at ~10 V/cm for ~5 hr.

4. Run controls using only one type of each construct to be tested. If the mobility of the sample is less than that of either of the controls, the sample contains inserts in the opposite orientation.

Isolation of Plasmids: Large and Small Scale

Plasmid DNA can be separated from bacterial chromosomal DNA by several techniques due to its compact size and closed circular superhelical form. The bulk of chromosomal DNA can be pelleted as a DNA–protein complex by centrifugation following gentle cell lysis, leaving the plasmid DNA in solution. Alternatively, chromosomal DNA can be denatured with alkali or heat to the single-stranded form, yet the closed circular nature of plasmid DNA allows rapid renaturation following neutralization or cooling. Complete purification requires deproteination and banding in CsCl–dye bouyant density gradients. If complete purity is not necessary, gel filtration can remove the bulk of contaminating nucleic acids.

Plasmids containing the ColE1 replication origin can be amplified by chloramphenicol treatment.[7] However, since the chloramphenicol stops cell growth at the low density at which it is added and must be added at a particular cell density, it is often more convenient to extract plasmid DNA from saturated cultures. Although there is more debris using this technique, it is efficiently removed by the procedure.

M13 phage RF DNA can be purified using the saturated culture method. The culture is grown using the same procedure as for preparing the phage DNA. No amplification is possible.

Small-Scale Plasmid Preparations ("Minipreps")

The most convenient method for preparing small amounts of DNA suitable for 5–10 restriction analyses is the alkali method of Birnboim and Doly.[8] The cells are lysed by SDS at high pH and then neutralized. The plasmid DNA renatures but the chromosomal DNA remains denatured and precipitates in a protein–DNA–SDS complex. This procedure can be scaled up to any volume. However, if more than a 10-ml culture is prepared, it is necessary to deproteinize with phenol prior to ethanol precipitation.

1. Prepare 5-ml cultures in LB broth containing the appropriate antibiotic. The cultures can be grown in disposable 14-ml plastic centrifuge tubes by picking colonies with a toothpick and dropping the toothpick into the tube. Cap the tube and incubate at 37° with shaking (250 rpm) for 6 hr to overnight.

2. Pellet the cells by centrifugation at 5000 rpm for 5 min. Discard supernatant and toothpick.

3. Add 100 μl of 50 mM glucose, 25 mM Tris–HCl (pH 8), 10 mM EDTA, 2 mg/ml lysozyme (freshly prepared). Resuspend pellet and incubate in ice for 10 min.

4. Add 200 μl 0.2 N NaOH, 1% SDS. Mix gently. Incubate in ice for 10 min. The SDS–NaOH solution must be made up just prior to use and kept at room temperature. Mix 3.5 ml water, 1 ml 1 N NaOH, and 0.5 ml 10% SDS.

5. Add 150 μl 3 M sodium acetate (pH 4.8). This is prepared by bringing a 3 M solution to pH 4.8 with glacial acetic acid. Mix gently. Incubate for 10 min in a freezer. The solution will freeze if left longer. A white precipitate will form.

6. Centrifuge for 15 min at 15,000 rpm at 4°.

7. Pour supernatant into Eppendorf tubes and fill with cold ethanol. Incubate in ice for 10 min.

[7] D. B. Clewell, *J. Bacteriol.* **11,** 667 (1972).
[8] H. C. Birnboim and J. Doly, *Nucleic Acids Res.* **7,** 1513 (1979).

8. Pellet the DNA for 1 min in a microfuge and aspirate off supernatant. Add 0.5 ml cold 70% ethanol and aspirate off.

9. Dry under vacuum. Resuspend in 50 μl distilled water containing 10 μg/ml pancreatic ribonuclease (RNase). RNase is prepared by dissolving deoxyribonuclease-free RNase in water or 10 mM Tris–HCl (pH 7.5) and heating to 90° for 10 min. The RNase solution can be stored frozen, and thawed when needed.

10. The DNA at this stage is suitable for restriction analysis or fragment preparation. Use 5–10 μl/reaction. If difficulty in cutting is observed, the DNA can be phenol extracted and ethanol precipitated.

Small-Scale M13 Plasmid Preparations

1. Inoculate 4 ml of YT medium with 20 μl of an overnight culture of the appropriate bacterial strain (see this volume [12]) and with a pure plaque from a transformation.

2. Grow at 37° for 8 hr.

3. Centrifuge cells and decant medium supernatant. Save both and keep at 4°.

4. M13 DNA can be prepared from the cell pellet as described for small-scale plasmid DNA preparations above.

5. Media supernatant can be saved at 4° as a phage stock. It can also be used for preparing single-stranded DNA (see this chapter), for sequencing, or for generating specific radiolabeled probes (see this volume [10]).

Large-Scale Plasmid Preparation

As discussed above, plasmid preparations yielding 1–2 mg of DNA can be prepared using both amplified or unamplified cultures. The unamplified cultures are prepared by a scale-up of the small-scale plasmid preparation with some modifications. The chloramphenicol-amplified cultures are prepared by a gentle detergent lysis procedure.

Unamplified Culture Procedure

1. Grow a 1-liter overnight culture in LB with the proper antibiotic.

2. Pellet the cells and resuspend in 20 ml of the Tris–glucose–EDTA lysozyme solution given above. Divide into four 30-ml Corex centrifuge tubes.

3. Incubate as above and add 10 ml of the NaOH–SDS solution to each tube.

4. Incubate in ice for 10 min. Add 7.5 ml of the 3 M sodium acetate solution to each tube.

5. Chill in the freezer for 15 min (more time is needed because of the bigger volume).

6. Centrifuge at 15,000 rpm for 15 min. Pour supernatant into four clean Corex tubes and centrifuge another 10 min at 15,000 rpm.

7. The supernatant should be clear at this point. Pour into a 250-ml centrifuge bottle. There should be ~80 ml. Add an equal volume of phenol, mix vigorously ~5min, add an equal volume of chloroform, mix again. Centrifuge at 5000 rpm for 10 min to separate phases.

8. Remove supernatant into a clean bottle and extract with an equal volume of chloroform. Centrifuge 10 min at 5000 rpm and remove top layer to a clean bottle. Add 0.1 volume 3 M sodium acetate (pH 4.8) and 2 volumes ice-cold ethanol.

9. Chill in ice 10 min. Centrifuge at 8000 rpm for 10 min. Pour off supernatant and drain pellet. Without disturbing pellet rinse with ice-cold 70% ethanol. Drain pellet. Dry under vacuum.

10. Resuspend the pellet in 1 ml distilled water. Add RNase to 1 μg/ml and incubate at 37° for 30 min. If a precipitate forms, centrifuge and pour supernatant into a clean tube.

11. The DNA can be used for most purposes at this point. If greater purity is necessary, the DNA can be subjected to gel filtration or CsCl–ethidim bromide banding.

Large-Scale M13 Plasmid Preparations

1. Grow appropriate bacterial strain (see this volume [12]) in minimal medium overnight. Dilute with an equal volume of YT medium. For each culture (final volume ~8 ml) add (a) 28 μl of medium supernatant from step 3 of Preparation of M13 Phage DNA or (b) an isolated plaque from a plate.

2. Grow at 37° for 3–4 hr and use to inoculate 500 ml YT medium; grow an additional 8 hr at 37°C.

3. DNA is prepared as described in this chapter for large-scale plasmid preparations.

4. Medium supernatant, after removal of bacteria by centrifugation, can be saved at 4° for a source of phage.

Purification by Gel Filtration

The plasmid DNA can be separated from the majority of contaminating single-stranded DNA and RNase-resistant RNA by gel filtration through BioGel A-50. The high-molecular-weight plasmid DNA is excluded from the column and elutes in a peak before the small, included contaminants.

1. Pour a 1 × 10 cm column of BioGel A-50.

2. Wash the column with 10 volumes of TE buffer. Clamp off the bottom.

3. Mix DNA sample with bromphenol blue solution such that sample is visibly blue. Gently layer sample (<1 ml) onto column. Unclamp and let sample enter gel. Connect column to a reservoir of TE buffer.

4. Collect 1-ml fractions. The plasmid DNA elutes when the bromphenol blue moves one-third to one-half way down the column.

5. Measure the absorbance at 260 nm of the fractions. The plasmid DNA should elute as a small peak followed by a large peak of contaminating nucleic acid. Alternatively, aliquots of each fraction can be added to 200 μl of a 1 μg/ml solution of ethidium bromide sequentially in a microtiter plate and the fluorescence viewed under ultraviolet light to locate the peaks.

6. Pool the peak fractions of plasmid DNA and recover by ethanol precipitation (see this volume [5]).

Purification by CsCl–Ethidium Bromide Centrifugation

As with phage purification by bouyant density centrifugation, DNA can also be banded in CsCl density gradients. The bouyant density of DNA is ~1.7 (depending on GC content). However, the binding of intercalating dyes, such as ethidium bromide, reduces the bouyant density. Closed circular DNA binds less ethidium bromide than linear or nicked circular DNA due to topological restraints. Therefore, the bouyant density of closed circular plasmid DNA will be less than the other forms of double-stranded DNA in saturating ethidium bromide. This density difference can be exploited to purify plasmid DNA.

1. Measure the volume of the DNA solution. To each 3.7 ml of DNA solution, add 0.3 ml of a 10 mg/ml solution of ethidium bromide. If a precipitate forms, centrifuge the solution and discard the pellet. Add 4 g CsCl and dissolve. This should give a density of 1.59 (see phage purification). Adjust if necessary. The refractive index at 20° should be 1.3893. Put solution (~5 ml) into cellulose nitrate centrifuge tubes suitable for a 50Ti rotor. Fill the tube with mineral oil. Do not fill the tubes with more than 5–6 ml solution as a steeper gradient will be formed with less resolution.

2. Centrifuge the tube at 37,000 rpm for ~36 hr at 4°. Remove the tube carefully and observe in either ordinary or ultraviolet light. There will be two bands visible. The lower one is the supercoiled plasmid DNA and the upper band is linear and nicked circular DNA. RNA forms a pellet at this density of CsCl.

3. Remove lower band with a hypodermic syringe (see discussion of phage purification).

4. Remove the ethidium bromide by shaking with an equal volume of CsCl-saturated 2-propanol. The ethidium bromide will partition into the upper 2-propanol phase and the DNA will remain in the lower. Remove and discard the upper phase. Repeat the extraction until all traces of pink are removed and then once more. If CsCl precipitates due to loss of water from the aqueous phase, add enough water to redissolve the salt.

5. Ethanol precipitate the DNA. First dilute the DNA solution by adding 2 volumes distilled water (so the salt will not precipitate). Add 6 volumes ice-cold ethanol and chill in ice for 10 min.

6. Centrifuge at 10,000 rpm for 10 min at 4°. Discard supernatant. Wash the pellet with 70% ethanol and dry the pellet in a vacuum. Resuspend the pellet in distilled water. The DNA can be stored frozen.

Isolation of Plasmid DNA from Amplified Cultures

Plasmids containing the ColE1 replicon continue to replicate in bacteria when protein synthesis is blocked by chloramphenicol.[7] This fact can be exploited to improve the yield of plasmid DNA to ~2 mg/liter of bacteria. It is important that the chloramphenicol be added to a low-density, rapidly growing culture. The procedure given uses a defined media. The use of broth cultures is reported to work as well.[9]

1. Grow an overnight culture of the strain in M9 medium containing 0.2% glucose, casamino acids, and antibiotics at 37°.

2. Inoculate a 4-liter flask containing 1 liter of prewarmed media with 10 ml of the overnight culture and shake at 37° until an absorbance at 550 nm of 0.5–0.8 is reached.

3. Add chloramphenicol to 170 μg/ml (1.7 ml of a 100 mg/ml solution in ethanol). Incubate overnight at 37° with shaking.

4. Pellet the cells at 5000 rpm for 10 min. Wash the cells in 50 mM Tris–HCl (pH 8), 50 mM NaCl, 5 mM EDTA. Pellet the cells in a plastic centrifuge tube. Freeze the pellet. The pellet may be stored frozen.

5. Thaw the pellet. Resuspend in 10 ml 50 mM Tris–HCl (pH 8), 25% sucrose. Add 2 ml of a 5 mg/ml fresh solution of lysozyme in 0.25 M Tris–HCl (pH 8). Mix gently. Incubate in ice for 5 min.

6. Add 4 ml 0.25 M EDTA (pH 8). Incubate 10 min in ice.

7. Add 16 ml of 50 mM Tris–HCl (pH 8), 0.0625 M EDTA, 0.1% Triton X-100. Mix gently.

[9] T. Maniatis, E. F. Fritsch, and J. Sambrook, "Molecular Cloning: A Laboratory Manual." Cold Spring Harbor Lab., Cold Spring Harbor, New York, 1982.

8. As soon as the culture clears (0–15 min) centrifuge at 19,000 rpm for 25 min.

9. Carefully decant the supernatant into a fresh tube. Pellets from this procedure are extremely loose. Sacrifice some of the supernatant rather than contaminate with pellet.

10. Add RNase to 10 μg/ml and incubate at 37° for 20 min.

11. Phenol–chloroform extract and ethanol precipitate as described for large-scale Plasmid Preparation, Unamplified Culture Procedure. Dissolve the pellet in distilled water. The plasmid DNA can be further purified by BioGel chromatography or CsCl–ethidium bromide banding.

Section VI

Genomic Cloning

Editors' Note: Before cloning genomic DNA, the advantages and disadvantages of the two primary cloning systems must be considered. Bacteriophage λ is more commonly used. Cloning efficiency in λ is relatively high and the manipulation of the >100,000 clones needed to ensure a complete library of genomic sequences is convenient. The average size of insert DNA in bacteriophage λ recombinants is ~15,000 base pairs. In contrast, the insert size accommodated by cosmids is 2- to 3-fold greater than the maximum found in λ libraries. When entire genes are to be studied, this represents an advantage, sufficient to offset the technical difficulties of handling cosmids.

[14] Host Strains That Alleviate Underrepresentation of Specific Sequences: Overview

By ARLENE R. WYMAN and KENNETH F. WERTMAN

The purpose of this chapter is to describe cloning hosts and vectors which address the problem that certain types of sequences may become underrepresented in genomic libraries constructed in bacteriophage λ. What causes a library to deviate in content from the genome is the unequal growth and amplification of the individual plaques. This discrepancy may be caused by the inviability of individual phage clones, where viability is defined simply as the ability of a clone to form a plaque on a lawn of host bacteria. Certain types of sequences whose inviability as λ clones on wild-type hosts has been documented may be recovered using mutant hosts which are relatively permissive for such sequences. In addition there are some sequences which, though viable, are prone to rearrangement (such as deletion) on a wild-type host; these clones may be grown on alternative hosts which maximize stability. Thus, underrepresentation of sequences can be avoided by utilization of the appropriate host.

Background

DNA Metabolism of Bacteriophage λ

Viability of a particular genomic clone depends on viral replication and packaging, which is the outcome of a complex interaction of the *rec* genes of the host, the *red* and *gam* genes of the vector, and the chi element. During the course of normal λ growth, two distinct types of replicating chromosomes coexist within the cell[1] (Fig. 1). The first is called θ replication, which is a typical bidirectional replication mode that forms replication bubbles in circular molecules. The second type is σ, or rolling circle replication, in which one of the chromosome strands is nicked and displaced by the procession of continuous DNA synthesis on the circular template. The displaced strand is copied in a normal fashion by discontinuous DNA synthesis. These types of replication generate different products: θ replication produces monomer circles, while rolling circle replication produces concatenated linear molecules. The substrate

[1] M. E. Furth and S. H. Wickner, *in* "Lambda II" (R. W. Hendrix, J. W. Roberts, F. W. Stahl, and R. A. Weisberg, eds.), p. 145. Cold Spring Harbor Lab., Cold Spring Harbor, New York, 1983.

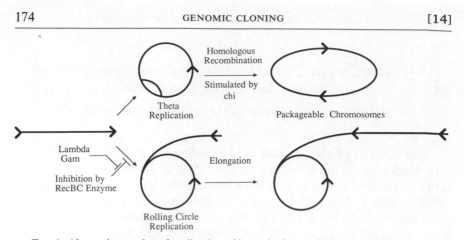

Fig. 1. Alternative modes of replication of bacteriophage λ DNA. RecBC inhibits replication by rolling circles; λ GAM function will promote rolling circle replication by inhibiting RecBC.

for the λ packaging machinery must contain at least two complete cohesive end (*cos*) sequences in the same orientation[2] (this volume [17,18]). Products of σ replication are packaged directly. In contrast, since the monomeric circles generated by θ replication contain only a single *cos* site, they can be packaged only if recombination between monomers takes place. Thus most of the packaged chromosomes are products of rolling circle replication.

Early in the λ lytic cycle in normal infections, replication is primarily by the θ mode. At later times, rolling circle type intermediates are seen in addition to the θ molecules.[3] This derepression of rolling circle replication is probably due to the accumulation of λ gamma protein (the *gam* gene product), which is an inhibitor of the host RecBC enzyme (exonuclease V, EC 3.1.11.5).[4] These observations have led to the suggestion that rolling circle structures (or their precursors) are sensitive to an activity of RecBC enzyme.

Growth of gam⁻ Phages

These properties of λ replication and packaging explain many phenotypes of the λ *gam*⁻ phage. Because *gam*⁻ phages fail to inhibit RecBC enzyme, they are unable to produce concatemers via rolling circle replica-

[2] M. Feiss and A. Becker, *in* "Lambda II" (R. W. Hendrix, J. W. Roberts, F. W. Stahl, and R. A. Weisberg, eds.), p. 305. Cold Spring Harbor Lab., Cold Spring Harbor, New York, 1983.
[3] M. Better and D. Freifelder, *Virology* **126**, 168 (1983).
[4] L. W. Enquist and A. Skalka, *J. Mol. Biol.* **75**, 185 (1973).

tion and are thus dependent on homologous recombination for the formation of packageable chromosomes. This recombination requirement can be fulfilled by host RecA-dependent pathways or the λ Red pathway. It follows that phages which are mutant in both *gam* and *red* genes do not plaque on *recA⁻* strains and make small plaques on recombination proficient strains. The introduction of sequences which stimulate recombination (chi) improve phage growth,[5] indicating that recombination frequency limits the production of phage particles in *red⁻ gam⁻* infections.

This limitation for packageable DNA causes the plaque size of *red⁻ gam⁻* phages to be sensitive to conditions which affect the recombination of monomer circles. (Phages that are *red⁻ gam⁻* exhibit the Spi phenotype, described in this volume [13,17].) Theoretically, a reduction in packaging substrate can be realized by either a reduction in the synthesis of monomer circles, or by an inhibition of their recombination to form multimers. In the context of DNA cloning, the former mechanism is more reasonable, in that cis effects on the progression of replications forks are more readily visualized. Regardless of the mechanism, all detrimental effects of *gam⁻* mutations are relieved by host mutations in the structural genes encoding the RecBC enzyme (*recB*, *recC*, and *recD*), and thus restore rolling circle replication and concatemer formation.

Sequences Prone to Underrepresentation in Genomic Libraries

There are two types of sequences which have been shown to become underrepresented in genomic libraries, those which contain inverted repetitions and those which contain direct repetitions. These shall be discussed separately.

Inverted Repetitions

Palindromes, or inverted repetitions, have been seen to be unstable or completely inviable in bacteriophage λ[6] as well as plasmids[7,8] in wild-type *Escherichia coli* hosts. It is not yet clear whether palindromes are inherently unstable, or if they impose severe problems which provide a strong selection for deleted derivatives.

Efficient propagation of palindromes in bacteriophage λ may be achieved through the use of *E. coli* hosts having mutations in *recB* and/or *recC* and *sbcB* genes.[6] Mutation in the *sbcB* gene was originally identified

[5] S. T. Lam, M. M. Stahl, K. D. McMillen, and F. W. Stahl, *Genetics* **77**, 425 (1974).
[6] D. R. F. Leach and F. W. Stahl, *Nature (London)* **305**, 448 (1983).
[7] J. Collins, *Cold Spring Harbor Symp. Quant. Biol.* **45**, 409 (1980).
[8] D. M. J. Lilley, *Nature (London)* **292**, 380 (1981).

as a suppressor of recombination deficiency of $recB^-$ $recC^-$ strains.[9] Inclusion of the $sbcB$ mutation improves the growth properties of $recB^-$ $recC^-$ cells and is necessary for the propagation of long inverted repetitions of at least 1.6 kb,[6] for unknown reasons. Although $recB^-$ $recC^-$ mutations also increase the stability of palindromes in plasmids,[10,11] these strains are not in general suitable hosts for the maintenance of ColE1-derived vectors (e.g., pBR322) due to instability[12] of the replicon (replicating unit). This problem is overcome by additional mutations in either $recA$[11,12] or $recF$,[10,12] which are derepressed in $recB^-$ $recC^-$ $sbcB^-$ cells. These mutations should as well stabilize direct repetitions by eliminating homologous recombination.

Palindromes are common features of eukaryotic DNA. In human DNA most inverted repetitions consist of elements approximately 300 bp in length (members of the so-called Alu family of repeated sequences[13]). Such palindromes are too short to cause λ clone inviability.[14] Aside from these, however, a significant fraction of human DNA contains longer palindromic sequences and fails to grow as λ clones on a wild-type host, but will grow on a $recB^-$ $recC^-$ $sbcB^-$ host. These clones represent approximately 9% of random 20-kb human genomic fragments.[15] A more extreme example was found in the case of the *Physarum* genome, where over 80% of clones propagated on $recB^-$ $recC^-$ $sbcB^-$ failed to plaque on a wild-type host.[16] It is therefore strongly recommended that genomic libraries in bacteriophage λ vectors be plated, screened, and amplified on a host, such as CES200, which is $recB^-$ $recC^-$ $sbcB^-$.

The use of this host has been successful in several instances where genomic sequences appeared otherwise to resist cloning. These include the human gene for an inhibitor of leukocyte proteases, called secretory leukocyte protease inhibitor, or SLPI,[17] a genomic segment representing a deletion within the low-density-lipoprotein receptor gene from an individual with familial hypercholesterolemia,[18] and a hypervariable locus

[9] S. D. Barbour, H. Nagasishi, A. Templin, and A. J. Clark, *Proc. Natl. Acad. Sci. U.S.A.* **67,** 128 (1970).

[10] R. Boissy and C. R. Astell, *Gene* **35,** 179 (1985).

[11] J. Collins, G. Volckaert, and P. Nevers, *Gene* **19,** 139 (1982).

[12] C. L. Bassett and S. R. Kushner, *J. Bacteriol.* **157,** 661 (1984).

[13] C. M. Houck, F. P. Rinehart, and C. W. Schmid, *J. Mol. Biol.* **132,** 289 (1979).

[14] David R. F. Leach, personal communication (1985).

[15] A. R. Wyman, L. B. Wolfe, and D. Botstein, *Proc. Natl. Acad. Sci. U.S.A.* **82,** 2880 (1985).

[16] W. F. Nader, T. D. Edlind, A. Huettermann, and H. W. Sauer, *Proc. Natl. Acad. Sci. U.S.A.* **82,** 2698 (1985).

[17] G. L. Stetler, M. T. Brewer, and R. C. Thompson, *Nucl. Acids Res.* **14,** 7883 (1986).

[18] M. A. Lehrman, D. W. Russell, J. L. Goldstein, and M. S. Brown, *J. Biol. Chem.* **262,** 3354 (1987).

(D14S1) on human chromosome 14.[19] Host CES201 (*recA⁻ recB⁻ recC⁻ sbcB⁻*) has been used to recover recombinant plasmids consisting of adeno-associated virus DNA, which contains inverted repetitions, cloned in pBR322.[20]

Direct Repetitions

Sequences which are present in multiple copies within the same λ clone may be gradually lost when the clones are grown on a wild-type host. For example, λ clones of the α-globin locus in human DNA readily give rise to deletions,[21] presumably brought about by unequal crossing-over between homologous regions. Similar deletions are found in clones of the immunoglobulin λ light-chain genes,[22] which are tandemly duplicated in the genome. In this case, deletion is prevented by subcloning and growing on a *recA⁻* host. Another case of instability of a tandem array is found in the 5'-upstream region of the human insulin gene.[23,24]

A reasonable approach to the maintenance of tandem duplications in phage clones is to use a host which is *recA⁻*.[25] However, growth of *red⁻* phages (which most cloning vectors are) on a *recA⁻ (recB⁺ recC⁺)* host obligates the presence of an intact *gam* gene in order to allow multimer formation by rolling circle replication. Vectors which are *gam⁺* are available, for example Charon 35.[26] Alternately a *recA⁻ recB⁻ recC⁻* host may be used when cloning in *gam⁻* vectors such as EMBL3.

Comparison of Factors Which Influence Plaque Formation

It has been shown[27] that wild-type cells are not totally restrictive for phages containing palindromic DNA sequences but support a small phage burst. When the replication of a *red⁻ gam⁻* phage, already impaired on a *recBC⁺* host, further falters in the presence of an inverted repetition, insufficient phage are produced to permit the formation of a visible plaque. A remedy for this problem is to increase the production of pack-

[19] A. R. Wyman, unpublished (1986).
[20] L.-S. Chang and T. Shenk, personal communication (1986).
[21] J. Lauer, C.-K. J. Shen, and T. Maniatis, *Cell (Cambridge, Mass.)* **20**, 119 (1980).
[22] R. A. Taub, G. F. Hollis, P. A. Hieter, S. Korsmeyer, T. A. Waldmann, and P. Leder, *Nature (London)* **304**, 172 (1983).
[23] G. I. Bell, M. J. Selby, and W. J. Rutter, *Nature (London)* **295**, 31 (1982).
[24] A. Ullrich, T. J. Dull, A. Gray, J. A. Philips, III, and S. Peter, *Nucleic Acids Res.* **10**, 2225 (1982).
[25] A. J. D. Bellett, H. G. Busse, and R. L. Baldwin, in "The Bacteriophage Lambda" (A. D. Hershey, ed.), p. 501. Cold Spring Harbor Lab., Cold Spring Harbor, New York, 1971.
[26] W. A. M. Loenen and F. R. Blattner, *Gene* **26**, 171 (1983).
[27] K. F. Wertman, A. R. Wyman, and D. Botstein, *Gene* **49**, 253 (1986).

ageable DNA. In principle, this can be accomplished either by enhancing recombination between monomers produced by θ replication, or by ensuring that replication will occur by the rolling circle mode. In practice, these two strategies are not equally successful.

When *red⁻ gam⁻* phages are grown on *recA⁺ recB⁺ recC⁺* hosts, the level of packageable DNA is increased if one or the other of the phage arms contains a chi sequence as, for example, in vectors EMBL3[28] and λ2001.[29] However, conflicting results have been obtained with this approach. In the case of human genomic clones, the use of the chi⁺ vector EMBL3 reduces the number of clones which fail to grow on wild-type hosts from 9 to 0.5%.[30] In contrast, a *Physarum* library was also made in the EMBL3 vector, yet 80% of the clones failed to form plaques on wild-type cells.[16] One is forced to conclude that the nature of the palindromic sequences (length, symmetry) may affect the success of the chi strategy; a severe replication block may depress θ production of monomers to such an extent that chi-stimulated recombination is limited for substrate.

The more satisfactory approach is one that ensures that the rolling circle mode of replication is utilized. Inherent in this strategy is the assumption that replication is not totally blocked by the problematic sequence and that the burst of phage is sufficient for plaque formation. Rolling circle replication is ensured by eliminating host RecBC activity, either by inhibition with phage-encoded Gam⁺ function or by mutation in one or more of the structural genes for the enzyme (*recB*, *recC*, and *recD*). We have found that the inclusion of the *gam* gene in the vector is sufficient to permit propagation of sequences which are otherwise inviable on wild-type cells.[27,30] Our experience, however, is that greater phage production is achieved through host mutations.[30]

The original permissive host, CES200, had null mutations in *recB*, *recC,* as well as *sbcB* mutation, which was necessary for host and phage viability (see above). Subsequently we have described[27] a second class of permissive hosts which are mutant in a third subunit of the RecBC enzyme, the *recD* gene product.[31] Strains which lack *recD* product are deficient in several RecBC enzyme activities, most notably, that which blocks rolling circle replication. We prefer to use *recD⁻* strains for pre-

[28] A.-M. Frischauf, H. Lehrach, A. Poustka, and N. Murray, *J. Mol. Biol.* **170,** 827 (1983).
[29] J. Karn, H. W. D. Matthes, M. J. Gait, and S. Brenner, *Gene* **32,** 217 (1984).
[30] A. R. Wyman, K. F. Wertman, D. Barker, C. Helms, and W. H. Petri, *Gene* **49,** 263 (1986).
[31] S. K. Amundsen, A. F. Taylor, A. M. Chaudhury, and G. R. Smith, *Proc. Natl. Acad. Sci. U.S.A.* **83,** 5558 (1986).

parative procedures because of their superior health and greater phage yield in liquid and plate lysates. We are hesitant to suggest their use for primary library screening, however, because the evidence concerning whether they allow growth and amplification of all clones is somewhat ambiguous. In our experiments with Charon 30, of 105 clones which failed to grow on a wild-type host we never found one which formed plaques on $recB^-$ $recC^-$ $sbcB^-$ cells but not on $recD^-$ cells. On the other hand, a phage bearing a constructed, perfect 3.6-kb palindrome[32] does not plate on a $recD^-$ host.[33] It is worth emphasizing that the use of $recD^-$ hosts is advantageous regardless of the presence of palindromes in the phage being studied, since all red^- gam^- phages grow better if allowed to replicate by rolling circle replication. As a further improvement, we have constructed a $recD^-$ $recA^-$ host which should be helpful in stabilizing direct repetitions.

Conclusion

We recommend that genomic libraries which are to be constructed in bacteriophage λ vectors be propagated, screened, and amplified on a $recB^-$ $recC^-$ $sbcB^-$ host such as CES200, or its $recA^-$ derivative CES201. The use of these strains negates the relevance of chi as well as the gam gene product, since hosts lacking RecBC enzyme are indifferent to their presence. This allows the choice of vector to be based on the cloning logistics (e.g., choice of restriction sites, insert size). The EMBL3 and EMBL4 vectors and λ2001 have the advantage of convenient multiple cloning sites and large insert capacity, and obviate purification of vector arms. However, in the event that one wishes to discriminate between phages containing long inverted repetitions and those which do not,[15] one must use a vector lacking chi and $gam,$ for example Charon 21 or 30. On $recBC^+$ hosts, such vectors containing long inverted repeats cannot be propagated.

Phages selected from screening on CES200 should be tested for viability on a $recD^-$ host (DB1316), which can be used for preparative growth of phages. A $recA^-$ $recD^-$ strain (DB1317) is also available to prevent the deletion of tandem duplications while permitting propagation of inverted repetitions. For studies using vectors with amber mutations (e.g., Charon 4A or 21A), amber-suppressing ($supF^-$) derivatives of these strains

[32] C. E. Shurvinton, M. M. Stahl, and F. W. Stahl, *Proc. Natl. Acad. Sci. U.S.A.* **84,** 1624 (1987).
[33] F. W. Stahl, personal communication (1986).

(DB1318 and DB1319, respectively) are available. All the strains described herein can be acquired from the *E. coli* Genetic Stock Center.[34]

[34] B. J. Bachmann, *E. coli* Genetic Stock Center, Department of Biology, 25 OML, Yale University, P.O. Box 6666, New Haven, Connecticut..

[15] Isolation of Genomic DNA

By BERNHARD G. HERRMANN and ANNA-MARIA FRISCHAUF

Mammalian high molecular weight DNA can be prepared from a variety of sources. The choice is dictated by the availability of materials and the ease of preparation. Tissue culture cells are particularly convenient to work with. Isolation of nuclei prior to extraction of DNA is now done much more rarely than in the past, because, in most cases, mitochondrial DNA does not cause problems. The presence of small amounts of RNA does not inhibit restriction enzymes later on. The preparation has to be nuclease free, of course. Ethanol-precipitable material other than nucleic acids does not usually have a negative effect on further manipulations except that it may cause difficulties in dissolving high-molecular-weight DNA.

If high-molecular-weight DNA is isolated for construction of genomic λ libraries or genomic Southern blots the DNA should have a minimal size of 100–200 kb. For cosmid libraries (this volume [18]) larger DNA is needed. The protocol described below, a modification of that by Blin and Stafford,[1] yields DNA of good quality with a standardized procedure designed to minimize the number of steps and manipulations. This is especially convenient if several samples are prepared in parallel.

The size of the DNA can be ascertained in 0.3% agarose gels run at a low-voltage gradient. It is easy to decide by this criterion whether the size of the DNA is greater than the required lower limit. For a better view of the size distribution of the DNA, pulsed-field gradient gel electrophoresis[2] can be used, but in practice DNA preparations that pass the standard gel test are satisfactory.

[1] N. Blin and D. W. Stafford, *Nucleic Acids Res.* **3**, 2303 (1976).
[2] D. C. Schwartz and C. R. Cantor, *Cell (Cambridge, Mass.)* **37**, 67 (1984).

Methods

DNA from Whole Organs

Liver, spleen, or kidney can be used. Spleen usually gives very good yields of pure DNA. When using liver the animal (mouse) should be starved for 24 hr to deplete glycogen levels before killing. The tissue is cut into small pieces and dropped into liquid nitrogen. Frozen tissue can be stored at $-80°$ for more than a year.

Prepare dry ice, a precooled mortar and pestle, liquid nitrogen, a metal spoon, and a 400-ml beaker containing 20 ml TEN9, DNase-free RNase A (100 μg/ml). The RNase is described in this volume [13]. All tools must be sterilized and the liquid nitrogen container should be rinsed with water before use to avoid contamination with DNA.

Pour liquid nitrogen into the mortar and add the frozen tissue pieces. Use the pestle to break the pieces gently into smaller bits in the presence of liquid nitrogen. This helps to avoid spilling. Then quickly grind the tissue to powder without further addition of liquid nitrogen. Chill one end of a metal spoon by dipping it into liquid nitrogen. Distribute the powder in small portions onto the surface of the buffer in the prepared beaker. Allow the powder to spread on the surface; then swirl the beaker a little to submerge the material. Continue until everything is in solution. Transfer the solution into a 50-ml plastic tube and shake repeatedly to obtain a more homogeneous solution. Leave the tube on a rocking platform (30 rpm) at room temperature for 10 min. Add 1 ml 20% sodium dodecyl sulfate (SDS), invert the tube a few times, and continue rocking for 10 min. Add 1 ml proteinase K solution (10 mg/ml in water). Invert and gently shake the tube, and then incubate at 37 or 55° overnight on a rocking platform. The solution should be reasonably clear after proteinase digestion. Transfer the solution into a flat-sided glass bottle with a tight cap and add 20 ml phenol. The bottle should allow the solution to cover a large surface to facilitate mixing with phenol. Put the bottle on a rocking platform for 1–3 hr at room temperature. Centrifuge the mixture in a 50-ml plastic tube in a table-top centrifuge for 10 min at 3000 rpm to separate the phases. To remove the phenol phase proceed as follows. Take a Pasteur pipet connected to a flask with a piece of tubing and from there to an aspirator. While keeping the clamp on the tubing closed, put the tip of the pipet through the aqueous layer to the bottom of the tube. Wait until the viscous thread on the end of the pipet has detached. Gently apply suction and remove the phenol. Close the clamp again and pull out the pipet. Discard the phenol.

If not all particles have gone to the interphase, 20 ml phenol is added and the procedure is repeated. If the aqueous phase (containing the DNA) is clear, transfer the DNA solution into a Corex tube and centrifuge in a SS-34 rotor (Sorvall) at 9000 rpm for 20 min at 25° (to avoid precipitation of the SDS). Protein and undissolved materials stick to the glass. Carefully pour the homogeneous supernatant into a 50-ml plastic tube, and then dialyze it against TE (dilution factor, 1000) first at room temperature to avoid SDS precipitation, then in the cold room. After dialysis, add sodium acetate (pH 6.5) to a final concentration of 0.3 M and 0.8 volumes 2-propanol. Invert the tube gently and repeatedly. Spool out the precipitate of DNA with a glass pipet with a sealed end and put it into TE (3 ml for each mouse liver or spleen starting material). To dissolve the DNA leave the tube rotating on a wheel (20 rpm) overnight at room temperature.

DNA Preparation from Cell Culture Cells

Cells are grown to a confluent monolayer (750-ml glass or phenol-resistant culture flask), the medium is poured off, and 10 ml of a solution containing TEN9, proteinase K (500 μg/ml), 1% SDS is added. Incubate the flask overnight at 37 or 55°. The lysis solution should cover the cell monolayer. Add 10 ml of phenol to the flask and leave on a rocking platform for 1 hr at room temperature. Pour the solution into a 50-ml plastic tube, and proceed as in the protocol given above.

DNA from Human Blood[3]

To 10 ml of blood plus 1.1 ml 3.8% sodium citrate in a 50-ml plastic tube add 30 ml lysis buffer (155 mM NH$_4$Cl, 10 mM NH$_4$HCO$_3$, 0.1 mM EDTA pH 7.4), leave the sample in ice for 15 min, shaking gently occasionally, and then centrifuge 10 min at 2000 rpm in a table-top centrifuge. The supernatant is discarded. Resuspend the pellet in 10 ml lysis buffer, leave in ice for 5 min, and centrifuge as above. Wash the pellet once with 5 ml lysis buffer and then resuspend in 4.5 ml 75 mM NaCl, 25 mM EDTA (pH 8). To the homogeneous suspension add, at room temperature, 0.5 ml proteinase K (10 mg/ml) and 0.25 ml 20% SDS. The procedure is then continued as described above.

Materials

Proteinase K (Merck)
TEN9: 50 mM Tris–HCl at pH 9, 100 mM EDTA, 200 mM NaCl
TE: 10 mM Tris–HCl at pH 8, 1 mM EDTA

[3] F. Baas, H. Bikker, G.-J. B. van Ommen, and J. J. M. de Vijlder, *Hum. Genet.* **67,** 301 (1984).

Phenol: analysis-grade phenol is melted at 50°; 0.3 g 2-hydroxyquino-line per 250 g phenol is added and dissolved. The phenol is equili-brated with 1 M Tris–HCl (pH 8) and subsequently equilibrated with TE by changing the buffer phase several times. The phenol is kept in a foil-wrapped dark bottle in the cold room.

[16] Digestion of DNA: Size Fractionation

By ANNA-MARIA FRISCHAUF

Fragmentation of DNA

To construct a library with as complete coverage as possible with as few clones as possible, the cloned DNA fragments should be randomly distributed on the DNA. Under these conditions the probability (P) that a unique DNA sequence is represented in N plaques is given by

$$P = 1 - (1 - f)^N \tag{1}$$

where f is the size of the fragment expressed as fraction of the genome,[1] or

$$N = \ln(1 - P)/\ln(1 - f) \tag{2}$$

If the distribution of the inserted DNA is not completely random, that is, if some sequences are preferentially found in the library, the total number of plaques to be screened until another unique DNA sequence is found has to be bigger. The DNA that is to be cloned should therefore be fragmented in as random a fashion as possible to obtain a representative library with the minimum number of clones.

The most frequently used strategy to fragment DNA to a clonable size is cutting with restriction enzymes. If blunt ends are produced by the restriction enzyme, ligatable sequences must be attached to the DNA ends to facilitate cloning in a vector that has been cut to give compatible overhanging ends.[2]

Many libraries have been constructed using partial digestion with *Eco*RI to give DNA fragments of clonable size, but this choice has mainly been dictated by the availability of vectors with other desired characteristics. Cutting with an enzyme that recognizes a 4-bp sequence is generally

[1] L. Clarke and J. Carbon, *Cell* (*Cambridge, Mass.*) **9**, 91 (1976).
[2] T. Maniatis, R. C. Hardison, E. Lacy, J. Lauer, C. O'Connell, D. Quon, G. K. Sim, and A. Efstratiadis, *Cell* (*Cambridge, Mass.*) **15**, 687 (1978).

METHODS IN ENZYMOLOGY, VOL. 152

preferable because the higher density of sites on the DNA makes it less likely that there are long stretches completely without these sites. In addition, the enhancing or inhibiting effects of sequences surrounding a particular site are less important because of the vicinity of many other sites which very likely will not be subject to the same special conditions. *Sau*3A or *Mbo*I are the most frequently used enzymes for partial digestion for genomic libraries. They produce ends that can be ligated into *Bam*HI sites. There are several very useful *Bam*HI λ replacement vectors available, e.g., EMBL3 or 4,[3] Charon 34 or 35,[4] and λ2001 [5]

Even in a library produced by fragmentation with *Sau*3A the distribution of cloned sequences is not completely random. Some sequences may not be present at all, either because of the reasons mentioned above, or because they are unstable in standard cloning hosts (see this volume [14]). In general, however, a *Sau*3A library will contain a reasonable representation of the majority of sequences. For a detailed discussion of DNA sequence representation in a partial digest library, see Seed *et al.*[6]

The protocol given below describes the conditions for partial digestion with *Mbo*I or *Sau*3A. Other enzymes that produce ends that can be ligated into cloning sites of λ (or cosmid) vectors can be used in an analogous manner.

How to Prevent the Random Association of Genomic DNA Fragments during Cloning

When insert DNA is ligated into the phage vector it is important to prevent ligation of fragments that are not contiguous in the genome. There are two ways to achieve this, size fractionation[2] and/or treatment with phosphatase.[7]

Since only DNA 78–105% of the length of wild-type λ (38–55 kb) can be packaged efficiently by the λ packaging system,[8] there is an inherent lower and upper limit on the DNA length that can be inserted into a given vector. It depends on the part taken up by the vector arms, e.g., for EMBL3, inserts can range from 8 to 23 kb. If only DNA of a certain size class, e.g., 15–20 kb, is used in the ligation mixture with one of the vectors mentioned before,[3–5] more than one insert cannot be packaged in a

[3] A.-M. Frischauf, H. Lehrach, A. Poustka, and N. Murray, *J. Mol. Biol.* **170,** 827 (1983).

[4] W. A. M. Loenen and F. R. Blattner, *Gene* **26,** 171 (1983).

[5] J. Karn, H. W. D. Mathes, M. J. Gait, and S. Brenner, *Gene* **32,** 217 (1984).

[6] B. Seed, R. C. Parker, and N. Davidson, *Gene* **19,** 201 (1982).

[7] D. Ish-Horowitz and J. F. Burke, *Nucleic Acids Res.* **9,** 2989 (1981).

[8] M. Feiss and A. Beckerd, *in* "Lambda II" (R. Hendrix, J. Roberts, F. Stahl, and R. Weisberg, eds.), p. 305. Cold Spring Harbor Lab., Cold Spring Harbor, New York, 1983.

single λ head; a double-sized insert plus vector DNA would exceed the maximum packagable size.

An advantage of using sized DNA is that the average size of the insert can be determined by the size of the cut taken, and it is therefore possible to make a library with close to the maximum insert length in the majority of clones. This is especially true for DNA size fractionated in gels and eluted. However, this is also the most costly way in terms of amounts of DNA that have to be processed, even though per microgram of finished DNA preparation the yield can be very high. Size fractionation in sucrose (see below) or salt gradients (this volume [18]) or in gels is time-consuming and usually leads to some loss of material.

An alternative protocol is based on treatment with phosphatase. This leads to removal of the 5'-phosphate groups of the DNA and so prevents ligation of the insert molecules to each other while permitting ligation to the vector arms. The advantages of this method are its speed and the negligible loss of material. It is the method of choice if very small amounts of DNA are available. The disadvantage is that molecules close to the lower size limit of the vector capacity will also be cloned and therefore the average insert size will be lower. This can be partially compensated by using DNA that is digested to a lesser extent than is required for optimal cloning efficiency.

Although it has not been systematically checked, the background of insert to insert ligation is very low, if the phosphatase has cleaved efficiently. It should not be higher than after a good gradient size fractionation.

The choice between size fractionation, phosphatase treatment, or a combination of both methods depends on the problem at hand. If a large amount of DNA can easily be obtained, only one or a small number of libraries need to be constructed, and it is important that the insert should be as large as possible, then clearly size fractionation, possibly in a gel, is advantageous. If the amount of DNA is limiting or if several libraries are to be constructed with minimal effort, phosphatase treatment would seem to be the method of choice.

Methods

To establish the optimum enzyme concentration and digestion time for an unknown DNA preparation, an analytical reaction is performed.

Dependence on Enzyme Concentration

Mix
DNA   0.005–0.2 μg
10× Med   1 μl

10 mM Spermidine (optional)	2 μl
Water to	9 μl
MboI (or another enzyme of choice)	1 μl

 in five concentrations for five digests: 0.3
 units/μl, or 0.06 units/μl, or 0.01 units/μl, or
 0.002 units/μl, or 0.0005 units/μl, or water.

Incubate 20 min at 37°. Add 1 μl 0.5 M EDTA, heat 10 min at 68°, apply the 5 digests and the mock digestion to a 0.3% agarose gel (this volume [8]).

This procedure is only necessary if there are unknown inhibitors in the preparation or if it is not possible to determine the DNA concentration. The latter may be the case for very dilute, precious DNA. In this case the DNA can be visualized by blotting the gel and hybridizing with a repetitive DNA probe; often total nick-translated species-specific DNA is adequate.

Dependence on Time. If the DNA has been prepared by standard procedures (e.g., see preceding chapter) and the concentration is known, an analytical time course of the reaction is usually sufficient.

Mix

DNA	1–2 μg
10× Med	3 μl
10 mM Spermidine	10 mM
Water to	30 μl

Take a 5-μl aliquot *before* adding 0.1 unit MboI enzyme and then 5-μl aliquots at 5, 10, 20, 40, and 80 min of incubation at 37°. Immediately add 1 μl 0.1 M EDTA to each sample, heat 10 min at 68°, and apply to a 0.3% agarose gel.

Figure 1 shows an ethidium bromide-stained gel with aliquots taken at different times. The desired conditions would be between the time points 20 and 40 min (lanes 4 and 5). If the DNA solution is not completely homogeneous some DNA will migrate at the limiting mobility as a band even if the rest of the DNA is digested (see lane 6). This does not indicate, in the case of MboI digestion of mouse DNA, the absence of MboI sites in the DNA; it is dependent on the DNA preparation.

Agarose Gel. Agarose gels, 0.3% in Loening buffer are run at 0.7–1 V/cm overnight. The DNA is visualized under ultraviolet light if more than 0.1 μg DNA/lane is used. Otherwise the DNA is blotted according to standard procedures (this volume [61]) onto nitrocellulose or a nylon membrane and visualized by hybridization to labeled repetitive DNA.

Preparative Digestion. The optimal enzyme concentration and digestion time are estimated from the analytical experiment. All amounts are scaled up proportionally.

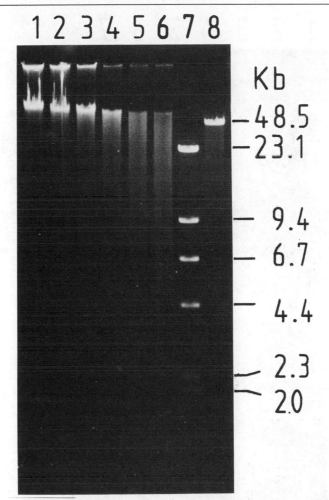

FIG. 1. Agarose gel, 0.3%, showing aliquots taken at 0, 5, 10, 20, 40, and 80 min of digestion with *Mbo*I (lanes 1–6). Size markers: λ *Hin*dIII digested (lane 7), λ uncut (lane 8).

Mix

DNA	20 μg
10× Med	30 μl
10 mM Spermidine	60 μl
Water to	300 μl

*Mbo*I to give optimal digestion after 10 min at 37°.

Take 120-μl aliquots of the above mixture at 5 min and 10 min and the rest at 20 min. Immediately add EDTA to a final concentration of 15 mM.

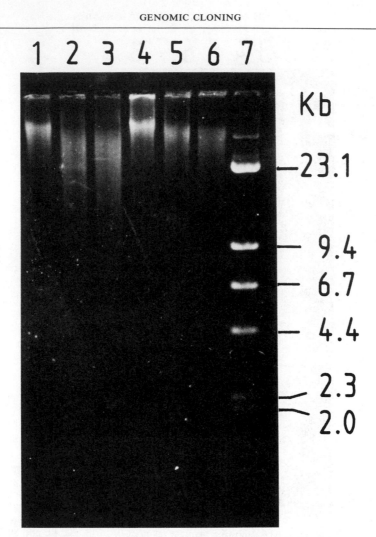

FIG. 2. Control ligations of DNA to test for the efficiency of phosphatase treatment. Sample A: DNA not treated with phosphatase (lane 1), DNA treated with phosphatase and ligated (lane 2), DNA treated with phosphatase, unligated (lane 3). Sample B: like lanes 1–3 (lanes 4–6), λ DNA digested with *Hin*dIII (lane 7).

Inactivate the enzyme by heating 15 min at 68°. (Alternatively the sample may be extracted with phenol.) Remove a 5-µl aliquot from each sample and apply it to a 0.3% agarose gel. Precipitate the rest by adding 0.07 volume 3 *M* sodium acetate (pH 6), 2 volumes ethanol. Leave in ice for 5 min, centrifuge 30 sec in a microfuge, take off the supernatant as completely as possible, and leave the tube open for 10 min at room temperature to dry the pellet. Dissolve the pellet overnight in TE to a final concentration of 0.25 µg DNA/µl.

Sucrose Gradient Size Fractionation. Prepare a 10–40% (w/v) sucrose gradient in a Beckman SW40 polyallomer tube. Load 50 μg of *Mbo*I-digested DNA on top. Centrifuge the gradient for 15 hr at 25,000 rmp at 20°. Collect approximately 30 fractions. Check aliquots of the fractions in a 0.3% agarose gel together with λ size marker fragments. Pool the fractions containing 15–25 kb DNA, and dialyze against TE. Add 0.1 volume 3 M sodium acetate, 2.5 volumes ethanol and leave at −20° overnight. Dissolve in TE to a final concentration of 0.25 μg/μl. If there is still sucrose and salt in the pellet, repeat the ethanol precipitation step using less than the standard amount of sodium acetate (<0.1 volume 3 M sodium acetate).

Note that the separation from higher molecular weight DNA is not very important since fragments that are too large to be packaged will not interfere with the library construction unless they are present in great excess.

Phosphatase Treatment. To 40 μl *Mbo*I-digested DNA (0.25 μg/μl), add 4.5 μl 500 mM Tris–HCl (pH 9.5), 10 mM spermidine, 1 mM EDTA, and 4 units alkaline phosphatase from calf intestine and incubate the samples 30 min at 37°. After addition of 5 μl 100 mM trinitriloacetic acid (pH 8) the samples are heated 15 min at 68°. The DNA is precipitated by addition of 0.1 volume 3 M sodium acetate and 2.2 volumes ethanol, centrifuged, and dissolved overnight in TE to a concentration 0.25 μg/μl.

Phosphatase Control. It is important to dephosphorylate insert DNA completely since random ligation of DNA will result in a library with artifacts. Therefore, this step should be checked. Relevant controls are self-ligation of phosphatase-treated and untreated DNA and comparisons of both samples with unligated DNA. Control reactions should contain 1 μl DNA at 0.25 μg/μl (either treated or untreated with phosphatase), 1 μl 10× L buffer, 2 μl 1 mM ATP, 5 μl water, and 1 μl T4 DNA ligase (80 units/μl). The results of such a test are shown in Fig. 2. While one sample (lanes 1–3) is completely dephosphorylated as shown by the inability to ligate it (lanes 2 and 3 are identical), phosphatase treatment seems incomplete in the second case (lanes 4–6) and should be repeated. Here the presence of high molecular weight material (lane 5) compared with the unligated control (lane 6) is indicative of 5′-phosphates that have not been removed.

Solutions

   10× Mcd: 0.1 M Tris–HCl (pH 7.5), 0.1 M MgCl$_2$,
      10 mM dithiothreitol, 0.5 M NaCl.
   10× L buffer: 0.4 M Tris–HCl (pH 7.6), 0.1 M MgCl$_2$,
      10 mM dithiothreitol.
   Loening buffer: 36 mM Tris, 30 mM NaH$_2$PO$_4$, 1 mM EDTA,
      ethidium bromide at 0.4 μg/ml. The pH is 7.7.

[17] Construction and Characterization of a Genomic Library in λ

By ANNA-MARIA FRISCHAUF

Preparation of Vector

Only about 60% of the DNA of the wild-type λ genome is necessary for the lytic life cycle of the phage. The packaging system of the phage, however, does not package molecules smaller than 78% (or larger than 105%) of the wild-type genome.[1] λ replacement vectors can make full use of the replaceable DNA space and are also able to propagate themselves, because the vector contains a middle, stuffer fragment that is not present in the recombinant phage clone[2] and is not required.

When preparing a library[3] it is helpful to minimize the number of plaques to be screened. Therefore the background should be kept reasonably low. To reduce the level of background due to religated vector phage either the middle fragment can be physically removed in gradients or gels, or the ligatable ends of the middle fragment can be removed, or, for some vectors, genetic selection against the presence of the middle fragment can be applied.

Physical isolation of vector arms is possible for all vectors.[4-7] There are no special constraints on the bacterial host to be used. The disadvantage is that it is laborious and time-consuming and it is rather difficult to remove the middle fragment completely. Biochemical selection or cutting off the ligatable ends from the middle fragment depends on the presence of symmetrical polylinker sequences flanking the middle fragment. Unique sites are necessary on the inner side of cloning sites sufficiently far removed to be completely cut by the second enzyme but still close enough to permit easy removal of the small fragment by 2-propanol precipitation or column chromatography. This is the case for one or more enzyme combinations, e.g., in the vectors λ2001 and EMBL3 or 4.[5,7]

[1] M. Feiss and A. Becker, in "Lambda II" (R. Hendrix, J. Roberts, F. Stahl, and R. Weisberg, eds.), p. 305. Cold Spring Harbor Lab., Cold Spring Harbor, New York, 1983.
[2] N. Murray, in "Lambda II" (R. Hendrix, J. Roberts, F. Stahl, and R. Weisberg, eds.), p. 395. Cold Spring Harbor Lab., Cold Spring Harbor, New York, 1983.
[3] T. Maniatis, R. C. Hardison, E. Lacy, J. Lauer, C. O'Connell, and D. Quon, Cell (Cambridge, Mass.) 15, 687 (1978).
[4] J. Karn, S. Brenner, and L. Barnett, this series, Vol. 101, p. 3.
[5] J. Karn, H. W. D. Mathes, M. J. Gait, and S. Brenner, Gene 32, 217 (1984).
[6] W. A. M. Loenen and F. R. Blattner, Gene 26, 171 (1983).
[7] A.-M. Frischauf, H. Lehrach, A. Poustka, and N. Murray, J. Mol. Biol. 170, 827 (1983).

METHODS IN ENZYMOLOGY, VOL. 152

Genetic selection is based on the use of host strains on which only recombinant phages are able to grow. One such system is provided by the fact that phages containing *red* and *gam* genes (Spi phenotype) cannot grow on a host lysogenic for bacteriophage P2.[8] If those genes are present on the middle fragment, recombinant phages can be selected on a P2 lysogen. (See this volume [13] for a discussion of the Spi selection system.) The background of genetic selection is extremely low and there are no purification steps involved. The disadvantage is that usually a *rec*[+] host is required[2] which may increase the probability of rearrangement or deletion of unstable DNA sequences (see this volume [14]).

Ratio of Insert to Vector DNA

In theory, the yield of plaques per microgram of insert DNA should increase with increasing excess of vector. In practice, the disadvantage of higher background due to religated vector must also be considered. An excess of vector will also reduce any background due to insert ligation caused by either incomplete phosphatase treatment or incomplete sizing. If the insert DNA has been sized, the calculation is easy and a molar ratio of 2 : 1 (vector to insert) may be a good compromise. If the DNA has not been sized but only treated with phosphatase the calculation is more difficult, because an average size must be determined from a gel picture. It is advantageous to use genetic selection against religated vector, so that a large excess of vector does not greatly increase the background. The reaction is relatively insensitive to changes in the vector-to-insert ratio since the background in genetic selection is usually due to nonadsorbed phage rather than to leaky selection.

Packaging

The establishment of *in vitro* packaging systems has been the decisive step in making phage λ such a highly efficient, easy to use vector for cloning DNA.[9,10]

If a packaging efficiency of 1×10^9 plaques/μg λ DNA is obtained, it means that 1 in 20 λ molecules gives rise to a plaque. These yields are, however, only achieved for long concatemeric molecules or molecules annealed intermolecularly by their cohesive ends. Packaging of circular

[8] J. Zissler, E. Signer, and F. Schafer, *in* "The Bacteriophage Lambda" (A. D. Hershey, ed.), p. 469. Cold Spring Harbor Lab., Cold Spring Harbor, New York, 1971.

[9] B. Hohn and K. Murray, *Proc. Natl. Acad. Sci. U.S.A.* **74**, 3259 (1977).

[10] N. Sternbeg, D. Tiemeier, and L. Enquist, *Gene* **1**, 255 (1977).

molecules containing *cos* sites is several orders of magnitude less efficient[11] (see also this volume [14]).

Extracts are prepared from two strains each containing an integrated λ phage with a different amber mutation interrupting the phage assembly in different steps.[9,10] Both phages are heat inducible and contain a deletion that reduces excision of the phage DNA from the chromosome. On mixing of the two extracts and the exogenous DNA, the latter is packaged into phage particles and can be used to infect bacterial cells. An alternative system is based on the use of lysogens in which the *cos* sequence of the phage has been deleted during lysogen formation. Since the phage DNA cannot compete with added DNA a single strain can be used to prepare packaging extracts.[11]

Amplification

When a library has been constructed a choice has to be made whether to screen the primary library directly or to amplify the library first and then to screen aliquots of the amplified library. The advantage of amplification is immediately obvious: there is a library to go back to for each experiment. This library is stable over a long period of time and there is an almost unlimited supply of it. The problem with an amplified library arises from the fact that different recombinant phage clones may grow at very different rates, resulting in unequal representation of the recombinants in the amplified library and if no precautions are taken, repeat isolates of the same clones will unavoidably be characterized. Therefore the primary library should be used if enough starting material is available and a limited number of screens is to be made. Especially if there are several copies of a sequence, as in a gene family, repeated isolation of one overamplified recombinant phage clone is probable and several genome equivalents have to be screened for the other copies. If amplification cannot be avoided it is useful to amplify fractions and to screen the amplified fractions separately. This reduces the problems created by some clones that are amplified much more than others.

Selection for Amber-Containing Phages That Have Incorporated *supF* Sequences

An alternative to the hybridization methods in this volume [45] is to screen a library by homologous recombination in *Escherichia coli*.[12] In

---

[11] S. M. Rosenberg, M. M. Stahl, I. Kobayashi, and F. W. Stahl, *Gene* **38,** 165 (1985).
[12] B. Seed, *Nucleic Acids Res.* **11,** 2427 (1983).

this case the library is constructed in a vector containing two amber mutations and the probe sequence is cloned in a *supF*-containing plasmid. Recombination can occur between the homologous sequences on the probe plasmid and the recombinant phage in the library, leading to covalent integration of the *supF* plasmid into the phage. Since this phage can now grow on suppressor-free hosts, recombination products can be selected in a 10^8-fold excess of amber mutation-carrying phages. If this screening method is to be used it is very important to choose the right vector since recombination frequencies can vary by orders of magnitude in different vector systems.[13]

Since the reversion frequency of double amber mutations is expected to be very low ($\sim 10^{-12}$) and the selection for *supF* sequences in amber-carrying phages is very stringent, *supF* is frequently used for genetic selection of a very small number of recombinant phage in a very large background, e.g., in screening by homologous recombination,[12] tagging of transfected DNA sequences,[14] and the constructing of jumping and linking libraries.[15] The mixture is plated on a suppressor-free host (e.g., MC1061)[16] and only phages that contain the *supF* sequence give rise to plaques. The background due to reversion of the double amber mutation is negligible. We have, however, seen some background for Aam Bam phages that is probably due to recombination between the double amber phage and the phage contained in the packaging extract. In our experience this problem can be eliminated by the use of the S amber mutation that is also carried by the packaging phages.

Characterization of a Library

There is no quick way to be completely sure that a library is perfect. The most disturbing artifact is the ligation of random pieces of DNA during library construction. Therefore size separation or phosphatase treatment should be checked carefully. Once clones are isolated the possibility of rearrangement, deletion, or random ligation can usually be checked by comparison with fragments on genomic Southern blots. If the library construction has been very inefficient, the background, due to vector phage, etc., is usually higher because it is not proportional to the number of plaques.

[13] H. V. Huang, P. F. R. Little, and B. Seed, *in* "Vectors: A Survey of Molecular Cloning Vectors and Their Uses" (R. Rodriguez, ed.), Butterworth, London (in press).
[14] M. Goldfarb, K. Shimizu, M. Perucho, and M. Wigler, *Nature (London)* **296,** 404 (1982).
[15] A. Poustka, T. Pohl, D. P. Barlow, G. Zehetner, A. Craig, F. Michiels, E. Ehrich, A.-M. Frischauf, and H. Lehrach, *Cold Spring Harbor Symp. Quant. Biol.* (in press).
[16] M. J. Casadaban und S. N. Cohen, *J. Mol. Biol.* **138,** 179 (1980).

Methods

This section first gives the procedure for preparation of two-component packaging extracts according to Scherer *et al.*[17]

Strains

For sonic extract (SE): BHB2690 = N205 *recA*$^-$ (λimm434 *c*Its, *b2*, *red3*, *Dam15*, *Sam7*)λ

For freeze–thaw lysate (FTL): BHB2688 = N205 *recA*$^-$ (λimm434 *c*Its, *b2*, *red3*, *Eam4*, *Sam7*)λ

Take a single colony of either strain, touch it to an LB agar plate (LB, see this volume [13]) prewarmed to 42°, and distribute the rest onto a plate at room temperature. Incubate the first plate at 42°, the other at 30° overnight. Nothing should grow at 42°.

Sonic Extract. Scrape the cells (BHB2690) from the 30° plate to inoculate 400 ml prewarmed LB (see [13]) in a 2-liter flask. Shake the flask at 30° until the absorbance at 600 nm reaches 0.3, transfer it to a 43° shaking water bath, induce for 20 min, shake vigorously for 2 hr at 38° and chill rapidly in ice water. Centrifuge the suspension 2 × 200 ml for 10 min at 4000 rpm at 2°. Decant the supernatant very carefully. Keep the pellet in ice. Resuspend each pellet in 1.5 ml cold 20 mM Tris–HCl (pH 8.0), 3 mM MgCl$_2$, 10 mM mercaptoethanol, 1 mM EDTA.

Transfer the suspension to a clear plastic tube. Sonicate while cooling within ice–salt mixture, approximately 15 times for 5 sec at full power (microtip). The solution has to be ice cold at all times and there should be no foaming. Sonication is finished when the solution is translucent. Centrifuge the solution 10 min at 6000 rpm in an SS-34 (Sorvall) rotor. There should be a very small pellet; otherwise the sonication has been insufficient. To the supernatant add 0.6 ml of M1 buffer, mix, distribute aliquots of 20, 50, and 100 μl into cold tubes, freeze them in liquid nitrogen, and store at −70°. The extract is stable for many months.

Freeze–Thaw Lysate. Start an overnight culture as described for Sonic Extract using BHB2688 and inoculating 3 × 400 ml LB. After induction at 43°, continue shaking for 3 hr at 38°. Chill the suspension in ice water and centrifuge 6 × 200 ml for 10 min at 4000 rpm at 20°. Decant the supernatant very well, and keep the pellets in ice. Resuspend each pellet in 0.3 ml cold 10% sucrose, 50 mM Tris–HCl (pH 7.6). The suspension is distributed into cold Oak Ridge centrifuge tubes (Beckman) (1 pellet/tube), 30 μl lysozyme in suspension buffer (2 mg/ml) is added and mixed, and the tube is dropped into liquid nitrogen. At this point it can be stored at −70°. Thaw the samples in ice. This takes approximately 1 hr. Add 125 μl M1 buffer. Shake gently. Centrifuge 30 min at 35,000 rpm in a precooled Ti50 or Ti70.1 rotor at 2°. Pool the supernatants; then distribute

[17] G. Scherer, J. Telford, C. Baldari, and V. Pirrotta, *Dev. Biol.* **86**, 438 (1981).

50-, 100-, and 250-μl aliquots into cold tubes, freeze in liquid nitrogen, and store at $-70°$. The extract is stable for many months.

Packaging. Incubate 1–2 μl DNA solution containing up to 0.2 μg DNA, 2 μl sonic extract, 6 μl freeze–thaw lysate at room temperature for 3 hr. Dilute with 40 μl λ diluent (see end of chapter) or, for preparative packaging reactions, the mixture may be directly applied to CsCl gradients.

The optimum ratio of sonic extract and freeze–thaw lysate may vary somewhat from one preparation to the next. The same is true for incubation times. It is frequently possible to increase the yield by incubating 6 hr or more at room temperature.

Plating Cells. Pick a single colony of the appropriate bacterial strain into L broth (LB, this volume [13]), and shake overnight at the correct temperature. Spin down the cells and resuspend in 0.5 volume 10 mM MgSO$_4$. The cells are usable for 2–5 weeks.

Though the plating efficiency of *in vivo* packaged phage particles is usually independent of the preparation of plating cells, this is not true for *in vitro* packaged phages or cosmids. Differences up to an order of magnitude are possible, especially for cosmids.

There are many different ways of preparing plating cells, frequently involving use of exponentially growing cells and growth in the presence of maltose (see also this volume [13]). We have found it most convenient to prepare several cultures, usually saturated overnight cultures, that are then individually tested for plating efficiencies. The best preparation is used for the plating of libraries.

Preparation of the Vector DNA

For digestion at the cloning site (for *Bam*HI, use the appropriate buffer for any other enzyme), mix

Vector DNA	10 μg
10× Hi	10 μl
100 mM spermidine	2 μl
*Bam*HI	20 units
Water to	100 μl

Incubate 1 hr at $37°$, add EDTA to 15 mM, put in ice, and analyze an aliquot in an agarose gel to determine whether cleavage has occurred. If the vector-DNA is completely digested, either extract the sample with phenol–chloroform–isoamyl alcohol (25 : 24 : 1) or heat at $68°$ for 10 min. Add 8 μl 3 M sodium acetate (pH 6) and 220 μl ethanol, cool to $-70°$ (dry ice 10 min) or $-20°$, centrifuge, wash pellet with 70% ethanol, and dissolve in TE to a concentration of 0.25 μg/μl. Take an aliquot for controls.

If physical separation of the vector arms is planned, more vector DNA should be digested. Heat inactivation is a good alternative to phenol extraction, if digestion with a second enzyme for biochemical removal of the middle fragment is to be performed (if the enzyme can be inactivated by heat, e.g., *Eco*RI, *Bam*HI).

Gradient Purification of Vector Arms.[3] Add $MgCl_2$ to 5 μg vector DNA to a final concentration of 10 mM. Incubate 1 hr at 42° to reanneal the cohesive ends. Apply to a 10–40% (w/v) sucrose gradient in 1 M NaCl, 20 mM Tris–HCl (pH 8.0), 5 mM EDTA; centrifuge in an SW27 rotor for 24 hr at 26,000 rpm at 15° and proceed as described for sucrose gradient separation of *Sau*3A-cut DNA (preceding chapter).

To cut the digestable ends off the middle fragment (using *Eco*RI in EMBL3 as an example), mix

EMBL3 cut with *Bam*HI (0.25 μg/μl)	40 μl
10× Hi	10 μl
100 mM spermidine	2 μl
*Eco*RI	50–100 units
Water to	100 μl

Incubate 1 hr at 37°. Add EDTA to 15 mM. Extract with phenol–chloroform–isoamyl alcohol (25 : 24 : 1), and reextract the organic phase once with 100 μl, 100 mM NaCl, 10 mM Tris–HCl (pH 7.6). To the combined aqueous phases add 0.12 volume 3 M sodium acetate (pH 6), and 0.6 volume (of solution plus sodium acetate) 2-propanol. Mix slowly. Leave in ice for 5 min. Centrifuge 5 min in a microfuge. Wash the pellet once with a mixture of 200 μl 0.45 M sodium acetate plus 120 μl 2-propanol and once with 100 μl 0.3 M sodium acetate plus 220 μl ethanol. Be careful not to lose the pellet at this point. Remove the supernatant completely. Leave the tube open at room temperature for 10 min to dry the pellet. Dissolve the DNA to a concentration of 0.25 μg/μl.

If the vector is to be digested with two enzymes it is necessary to check whether the distance between the two sites is sufficient to allow cutting with the second enzyme. Especially if the sites are close it is advisable to cut first with the cloning enzyme, then with an excess of the second enzyme. Ligation and packaging of the doubly cut vector is the best control to test whether the ligatable site has been removed.

Controls for Ligations

1. Mix

Vector cut with *Bam*HI	1 μl
10× L buffer (preceding chapter)	1 μl
1 mM ATP	3 μl
Water	4.5 μl
T4 DNA ligase (80 units/μl)	0.5 μl

2. Mix

Vector arms separated on gradient	1 μl

10× L buffer	1 μl
1 mM ATP	3 μl
Water	4.5 μl
T4 DNA ligase (80 units/μl)	0.5 μl

3. Mix

Vector cut with BamHI, EcoRI	1 μl
10× L buffer	1 μl
1 mM ATP	3 μl
Water	4.5 μl
T4 DNA ligase (80 units/μl)	0.5 μl

Incubate overnight at 15°.

Package 1 μl of each ligation reaction, 1 μl vector (0.1 μg/μl) cut with BamHI but not religated, and 1 μl vector (0.1 μg/μl) uncut. Then mix 1 μl of each of these DNA solutions, 2 μl sonic extract, and 6 μl freeze–thaw lysate. Incubate 2 hr at room temperature; add 190 μl λ diluent. Plate different dilutions on the appropriate host.

If genetic selection is used, the ligation reactions should be plated on both the selective and the nonselective host. For uncut vector the difference between P2 lysogen and nonselective host should be approximately 10^{-5}. The cut vector should give a thousand-fold fewer plaques than the uncut vector. Phage cut with one enzyme (two sites) and religated should give approximately 10% of the original plating efficiency. The background of Spi⁻ phages (see this volume [13]) should not increase significantly on cutting and religating. Different kinds of media in the plates will give rise to different plaque size. Rich media such as LB (this volume [13]) give smaller plaques. Since the plaque size on a P2 lysogen is small it is important to plate libraries on BBL agar.

Ligation

Mix

Vector DNA	10 μg
Insert DNA, 15–20 kb, or phosphatased	4 μg
10× L buffer	20 μl
10 mM ATP	6 μl
T4 DNA ligase at 400 units/μl	2 μl
Water to	200 μl

Incubate overnight at 15°, then store at 4° (days to weeks).

Packaging

Mix

Ligation mixture	200 μl
Sonic extract	200 μl
Freeze–thaw lysate	600 μl

Incubate at room temperature for 3–8 hr. Then the reaction mixture can be stored in the refrigerator, either diluted with λ diluent or undiluted.

If the reaction has been efficient, the titer will be very high and the library can be plated immediately with the appropriate host strain (see this volume [13] for titering methods). If the titer is quite low it may not be possible to add enough *in vitro* packaged material to the plating cells to give a high-density plate. This may also be the case if few *supF*-containing phages are to be selected in a very large excess of amber phages. In this case, and also for cosmid libraries, it can be helpful to concentrate and purify the phage particles over a CsCl step gradient.

CsCl Step Gradient. Layer 0.5 ml each of 56, 42, and 31% CsCl in λ diluent (w/w) into polyallomer tubes, layer packaged DNA mixture on top, and fill up with λ diluent. Spin in an SW60 rotor for 3 hr at 35,000 rpm at 20°. Collect the fractions from the top with a Pasteur pipet. The phage band is invisible and is located a little under the visible protein band. Test the 200-μl fractions under the protein band by diluting 1 : 50 and plating on the appropriate host. Pool the phage-containing fractions, and dialyze for 6 hr against 2 changes of λ diluent in the cold room. Plate.

Plating the Library. Mix 10–100 μl of packaged DNA mixture or 0.1–3 ml of CsCl purified, dialyzed phage (1–2 × 10⁵ pfu), and 1–2 ml plating cells (2-fold concentrated saturated overnight culture in 10 mM MgSO$_4$). Incubate 15 min at 37°, and add 30 ml BBL top layer agar (for amplification) or BBL top layer agarose (for direct screening). Plate on 22 × 22 cm BBL agar plates. Incubate overnight at 37°.

Large numbers of bacterial cells result in smaller plaques.

Preparation of Amplified Library. Scrape the top agar off the 22 × 22 cm plate, add 30 ml of λ diluent and 0.3 ml chloroform, stir gently at room temperature for 20 min, centrifuge, take off supernatant, titer (this volume [13]), and store in a sterile plastic tube in the refrigerator in the presence of no or very little chloroform.

If there are several plates, the fractions should be kept separate to avoid finding the same overamplified clones many times. Excess of chloroform should be avoided since it seems to accelerate the loss of viable phage in the course of time. The library can be stored for a few years with slow decrease of the titer.

Materials.

BBL agar: 10 g Baltimore Biological Laboratories trypticase, peptone (11921), 5 g NaCl, 10 g agar (Difco) with water added to 1 liter. Adjust to pH 7.2

BBL top layer agar: like BBL agar, but 6.5 g agar/liter, 10 mM MgSO$_4$

BBL top layer agarose: like BBL agar, but 5 g agarose/liter, 10 mM MgSO$_4$

λ diluent: 10 mM Tris–HCl (pH 7.6), 10 mM MgSO$_4$, and 1 mM EDTA

TE: 10 mM Tris–HCl (pH 7.6), 1 mM EDTA

M1 buffer is made by adding the following in order: 0.11 ml water, 1 μl mercaptoethanol, 6 μl 0.5 M Tris–HCl (7.6), 0.3 ml of a mixture of 0.05 M spermidine and 0.1 M putrescine (pH 7), 9 μl 1 M MgCl$_2$, and 75 μl 0.1 M ATP

10× Hi: 0.1 M Tris–HCl (pH 7.6), 0.1 M MgCl$_2$, and 1 M NaCl

Restriction endonucleases were purchased from Bethesda Research Laboratories, New England BioLabs, or Boehringer-Mannheim. T4 DNA ligase was from New England BioLabs and alkaline phosphatase, intestinal, from Boehringer-Mannheim.

[18] Cloning Large Segments of Genomic DNA Using Cosmid Vectors

By ANTHONY G. DiLELLA and SAVIO L. C. WOO

The lambda (λ) bacteriophage cloning system has been used most frequently in the past to generate genomic libraries. The λ DNA in a phage particle is a linear duplex molecule of about 49 kb. At each end are 12-nucleotide single-stranded 5′-projections which are complementary in sequence; these cohesive termini are known as the *cos* site. During normal λ replication, linear concatamers are formed containing tandem arrays of the λ genome with *cos* sites spaced at ~49 kb (genome sized) intervals. These sites are essential for efficient packaging into λ heads *in vivo* and *in vitro*. The λ bacteriophage packaging system selects DNA molecules about the size of the λ genome (37–52 kb).[1] The cosmid cloning system makes use of the *cos* sites but replaces the λ genome with the desired insert DNA. Since cosmid vector molecules are small (i.e., 5 kb), nearly all of the packaging capacity of the λ bacteriophage particle can be used to package genomic DNA. Therefore, one should consider cloning with cosmids whenever it is advantageous to handle large genomic regions.

Cosmids are small plasmid cloning vectors that contain (1) a λ bacteriophage *cos* site, (2) a drug resistance marker and a plasmid origin of replication, and (3) one or more unique restriction sites for cloning.[2] Plas-

[1] M. Feiss, R. A. Fisher, M. A. Crayton, and C. Enger, *Virology* **77**, 281 (1977).

[2] F. Collins and B. Hohn, *Proc. Natl. Acad. Sci. U.S.A.* **75**, 4242 (1979).

mids with a large variety of cloning sites and prokaryotic and eukaryotic selection markers[3] can be converted to cosmids by insertion of the λ *cos* region[2] (usually contained in a 1.7-kb λ *Bgl*II fragment). Since the presence of the *cos* site enables *in vitro* packaging into phage heads,[4] hybrid cosmids can be transduced into *Escherichia coli* with high efficiency. In contrast to λ bacteriophage, however, cosmids replicate in *E. coli* as large plasmids; therefore, cosmid libraries are screened and maintained as populations of transformed bacteria.

Problems involving gene structure and expression can more easily be approached if large DNA segments are isolated. Several advantages of cloning large DNA segments are (1) larger genes can be isolated on a single recombinant clone[3]; (2) several linked genes can be isolated on the same recombinant molecule[5]; (3) genes can be isolated with large stretches of surrounding sequences[6]; (4) fewer colonies need to be screened to isolate the clone of interest[7]; and (5) in chromosome "walking" experiments larger segments of the genome can be covered to facilitate analysis of gene linkage.[8] In this chapter we describe one of the methods available for constructing cosmid libraries.[9]

Priniciple of the Method

The basic scheme for the cosmid cloning technique is shown in Fig. 1. The cosmid DNA vector containing the λ *cos* site and a selectable ampicillin resistance marker (*amp*) is cut with *Bam*HI at a single site and dephosphorylated with calf intestinal alkaline phosphatase to prevent vector–vector ligation. Large genomic DNA fragments (30–50 kb) are generated by partial *Mbo*I digestion of high-molecular-weight DNA (>150 kb), followed by size fractionation and ligation to linearized, dephosphorylated vector. Long concatemeric hybrid structures are formed. Ligations performed with a 10-fold molar excess of cosmid vector over insert DNA inhibits ligation of noncontiguous genomic DNA segments. Molecules containing two *cos* sites in tandem about 37–50 kb apart can be packaged into λ bacteriophage particles and subsequently transduced into

[3] F. G. Grosveld, T. Lund, E. J. Murray, A. L. Mellor, H. H. M. Dahl, and R. A. Flavell, *Nucleic Acids Res.* **10,** 6715 (1982).

[4] B. Hohn and K. Murray, *Proc. Natl. Acad. Sci. U.S.A.* **74,** 3259 (1978).

[5] Y. F. Lau and Y. W. Kan, *Proc. Natl. Acad. Sci. U.S.A.* **80,** 5225 (1983).

[6] G. Scangos and F. H. Ruddle, *Gene* **14,** 1 (1981).

[7] A. G. DiLella, S. C. M. Kwok, F. D. Ledley, J. Marvit, and S. L. C. Woo, *Biochemistry* **25,** 743 (1986).

[8] M. Steinmetz, A. Winoto, K. Minard, and L. Hood, *Cell (Cambridge, Mass.)* **28,** 489 (1982).

[9] A. G. DiLella and S. L. C. Woo, *Focus (Bethesda, Md.)* **7,** 1 (1985).

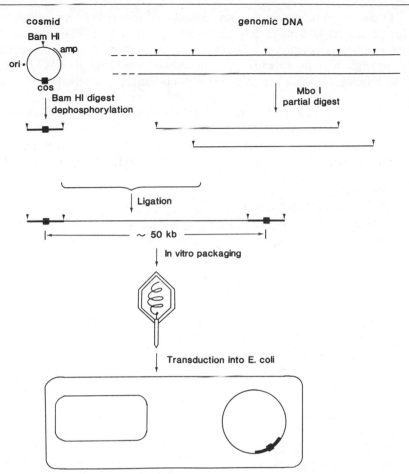

FIG. 1. Gene cloning in cosmids. Cosmid DNA containing the λ *cos* site (*cos*), an origin of replication (*ori*), and an ampicillin resistance marker (*amp*) is linearized with *Bam*HI at a single site (arrow) and dephosphorylated. Genomic DNA is partially digested with *Mbo*I (arrows), producing compatible sticky ends. The dephosphorylated vector and genomic DNA are ligated, packaged into λ bacteriophage particles, and then transduced into *E. coli*. The recombinant cosmid DNA is propagated in *E. coli* as an autonomous replicon (circle) and distinct from the *E. coli* chromosome.

recA⁻ E. coli, thereby reducing recombination and consequently increasing stability of the cosmid DNA. The recombinant cosmid is then propagated as a large plasmid without causing subsequent cell lysis. The ideal cosmid hybrid consists of one cosmid vector molecule and a contiguous unrearranged insert of foreign DNA. The genomic library is then grown under antibiotic selection and individual bacterial colonies are screened using a probe to identify the corresponding genomic DNA sequences.

[*Editors' Note*. Recently, new cosmid vectors (pWE) have been developed that contain phage RNA polymerase promoters. The presence of these promoters permits the radiolabeling of cosmid insert sequences to high specific activity *in vitro*. The properties and applications of these vectors including their use in genomic walking are described in this volume [65].]

Materials. Cosmid vectors can be purchased from Amersham or Boehringer-Mannheim. *Escherichia coli* strains ED8767 (*supE*, *supF*, *hsdS*⁻, *met*⁻, *recA56*) and HB101 (*supE*, *hsdS20*, (r_B^-, m_D^-), *recA13*) can be obtained from the American Type Culture Collection, Rockville, MD.

Solutions

10× Buffer B: 500 mM Tris–HCl (pH 8.0), 500 mM NaCl, 100 mM MgCl$_2$, 10 mM dithiothreitol (DTT), and BSA 1 mg/ml

TBE: 108 g Tris base, 55 g boric acid, and 9.3 g Na$_2$EDTA per liter of water

Phenol–chloroform–isoamyl alcohol: phenol (25 volumes), chloroform (24 volumes), isoamyl alcohol (1 volume) saturated with STE

STE: 0.1 M NaCl, 1 mM Na$_2$EDTA, and 50 mM Tris–HCl (pH 7.5)

Chloroform–isoamyl alcohol: 24 volumes chloroform, 1 volume isoamyl alcohol

10× CIAP buffer: 500 mM Tris–HCl (pH 9.0), 10 mM MgCl$_2$, 1 mM ZnCl$_2$, and 10 mM spermidine

10× Ligation buffer: 0.66 M Tris–HCl (pH 7.5), 50 mM MgCl$_2$, 50 mM DTT, 10 mM ATP

TE: 10 mM Tris–HCl (pH 8.0), 1 mM Na$_2$EDTA (pH 8.0)

0.1× TE: 10 mM Tris–HCl (pH 8.0), 0.1 mM Na$_2$EDTA (pH 8.0)

10× *Mbo*I buffer: 100 mM Tris–HCl (pH 7.5), 750 mM NaCl, 100 mM MgCl$_2$, 10 mM DTT, and 1 mg/ml BSA

*Mbo*I dilution buffer: 10 mM potassium phosphate (pH 7.4), 0.1 mM Na$_2$EDTA, 7 mM 2-mercaptoethanol, and 50% glycerol

SM buffer: 50 mM Tris–HCl (pH 7.5), 100 mM NaCl, 10 mM MgSO$_4$, 0.01% gelatin

20× SSC: 3 M NaCl, 0.3 M sodium citrate (pH 7.0)

50× Denhardt's solution: 1% Ficoll, 1% polyvinylpyrrolidone, 1% BSA in water

Media

LB media: 10 g Bacto-tryptone, 5 g yeast extract, and 5 g NaCl per liter of water. Autoclave.

LB agar: LB media containing 1.2% Bacto-agar. Autoclave.

Procedure I: Linearization of Cosmid Vector DNA

1. Restrict the cosmid vector at 200 μg/ml with BamHI (2 units/μg) in 1× buffer B. We include 4 mM spermidine in the reaction to enhance digestion and maintain the integrity of the BamHI cohesive termini.

2. Incubate the reaction at 37° for 2–5 hr.

3. Put the reaction in ice and check for complete restriction by analyzing a 2.5-μl sample (0.5 μg DNA) in a 0.8% agarose/TBE gel. The time required for complete linearization of the DNA ultimately depends on DNA purity, DNA and enzyme concentration, and enzyme quality.

4. Extract the DNA once with an equal volume of phenol–chloroform–isoamyl alcohol and once with one-half volume of chloroform–isoamyl alcohol. Mix gently for 10 min and centrifuge in an Eppendorf microfuge for 10 min.

5. Remove the upper aqueous layers and reextract and organic phases with 100 μl of STE. Centrifuge and pool the aqueous layers.

6. Adjust the DNA solution to 0.3 M sodium acetate (pH 5.5) and precipitate the DNA with 2.5 volumes of ethanol. Incubate in dry ice for 10 min.

7. Centrifuge the sample in an Eppendorf microfuge for 15 min at 4° and wash the DNA pellet once with 70% ethanol and once with absolute ethanol.

8. Dry the DNA pellet under vacuum for 5 min and solubilize DNA at 1 mg/ml in water.

Procedure II: Phosphatase Treatment of Linearized Vector DNA

1. The linearized DNA (0.2 μM) is dephosphorylated with calf intestine alkaline phosphatase (2.5 units CIAP/50 pmol of 5′ DNA termini) in 1× CIAP buffer at 37° for 1 hr.

2. Adjust the reaction to 15 mM Na₂EDTA and incubate at 68° for 10 min. Cool the sample in ice.

3. Extract the DNA once with an equal volume of STE-saturated phenol for 10 min and once with chloroform–isoamyl alcohol as described in Procedure I.

4. Precipitate the DNA by adding one half volume of 7.5 M ammonium acetate and 2 volumes of ethanol. Incubate in dry ice for 10 min.

5. Centrifuge the DNA sample as above and wash the DNA pellet once each with 70% and absolute ethanol.

6. Dry the DNA pellet under vacuum for 5 min and solubilize the DNA sample in 0.1× TE at 1 mg/ml. Store the DNA at −20°.

Procedure III: Test Ligation of Dephosphorylated Vector DNA

This experiment was designed to test the efficiency of the phosphatase reaction as well as the integrity of the *Bam*HI cohesive termini. The following ligation reactions are performed using pBR322, linearized with *Bam*HI but not dephosphorylated, as the insert DNA.

Vector–Insert Ligation Reaction. Mix with 16 μl water 1.5 μl (1 pmol) pBR322/*Bam*HI (1 mg/ml), 0.5 μl (0.2 pmol) vector/*Bam*HI/dephosphorylated (1 mg/ml), 2.0 μl 10× ligation buffer.

Vector–Vector Ligation Reaction. Mix with 16 μl water 2.0 μl vector/*Bam*HI/dephosphorylated (1 mg/ml), 2.0 μl 10× ligation buffer.

1. Remove an aliquot (1 μl) from each reaction and add to 9 μl of TE. Store these samples at 4° for later analysis by gel electrophoresis.
2. Add T4 DNA ligase (0.5 μl: 1–2 Weiss units) and incubate the reaction at 13° overnight.
3. Add a 1-μl aliquot from each of the ligation reactions to 9 μl of TE and analyze these samples (in addition to those in step 1) by electrophoresis through an 0.7% agarose/TBE gel.

Proceed to the next step if the ligated sample from the vector–insert ligation contains neither of the starting materials (i.e., the linearized cosmid vector or the linearized 4.3-kb pBR322 insert). The ligated DNA should be >20 kb in length, the result of extensive ligation. The sample from the second reaction should be monomeric cosmid vector regardless of whether ligase was added.

Procedure IV: Preparation of High-Molecular-Weight Chromosomal DNA for Cloning

Since large fragments of DNA are required for cloning with the cosmid system, the eukaryotic DNA must be very large (>150 kb) before digestion. This requirement is one of the most critical steps in the cosmid cloning procedure; therefore, precautions are taken not to shear the genomic DNA. Any mixing or sampling of the DNA solution should be done by gentle inversion or gentle pipetting through a wide-tip pipet. The following describes a method successfully used in our laboratory for DNA isolation from tissue culture or white blood cells.

1. Cells from 15 ml of whole blood or a 176-cm^2 confluent cell culture monolayer are gently suspended in 15 ml of STE and adjusted to a 100 μg/ml concentration of proteinase K and 0.5% sodium dodecyl sulfate (SDS).
2. Mix the DNA solution by gentle inversion a few times and incubate about 6 hr in a 50° water bath without shaking.

3. Extract the DNA (by gentle inversion) with an equal volume of phenol–chloroform–isoamyl alcohol for 10 min at room temperature. Carry out the extractions in large tubes or flasks to ensure complete mixing of the aqueous and organic phases.

4. Incubate the sample in ice for 30 min.

5. Centrifuge the DNA solution at 4000 rpm for 20 min at 4°.

6. Remove the upper aqueous layer with a pipet, being careful not to collect any protein interphase.

7. Adjust the DNA sample to 0.2 M sodium acetate (pH 5.5), and gently layer 2 volumes of ethanol ($-20°$) on top of the DNA solution at room temperature. Spool the DNA at the aqueous–ethanol interphase onto a sterile glass rod.

8. Rinse the spooled DNA with 70% ethanol and then touch the glass rod to the side of a sterile plastic tube to drain out the ethanol from the DNA.

9. Dissolve the DNA overnight in 5 ml of TE at 4° without shaking. Occasional pipetting through the wide end of a sterile 1-ml plastic pipet will facilitate dissolving the DNA.

10. Add DNase-free RNase (10 mg/ml) (this volume [13]) to the DNA solution at a concentration of 100 μg/ml and incubate the sample at 37° for 1 hr without shaking. The RNase step may be optional.

11. Add SDS to 0.5% and proteinase K to 100 μg/ml, and incubate at 50° for 1 hr.

12. Repeat steps 3 through 8. Air dry the spooled DNA briefly (about 5 min) and solubilize the DNA in 1 ml of TE at 4° without shaking. Usually the time required to solubilize the DNA is 1–2 days.

13. Check the quality of 100 ng of genomic DNA by electrophoresis through a 0.3% agarose/TBE gel at 2 V/cm gel length. The gel is cast and run in the cold room (8°) using TBE electrophoresis buffer. T4 DNA (160 kb) (Sigma) and λ DNA (50 kb) are used as size standards. Good-quality genomic DNA will comigrate with the T4 DNA marker (Fig. 2).

Procedure V: Analytical *Mbo*I Partial Digestion of Chromosomal DNA

For cloning into the *Bam*HI site of the cosmid vector, chromosomal DNA is digested with *Mbo*I (or *Sau*3A) for a series of increasing times. Three digestion times, a short time (DNA mostly >45 kb), an intermediate time (DNA mostly 35–45 kb), and a long time (DNA mostly <35 kb), are determined from an analytical digestion.

1. Perform analytical *Mbo*I partial digests at 37° in 100 μl reaction volume containing 10 μg of genomic DNA in 1× *Mbo*I buffer. The reaction is preequilibrated at 37° for 5 min prior to the addition of 0.5–1.0 units

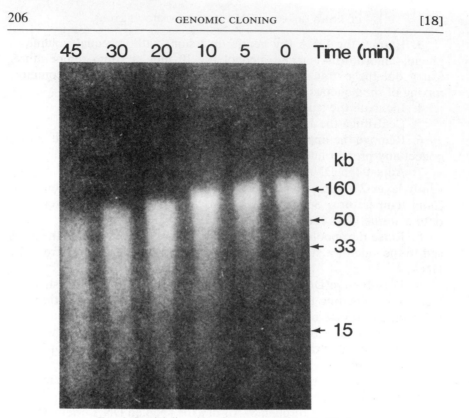

FIG. 2. Partial digestion of genomic DNA with *Mbo*I.

of *Mbo*I (*Mbo*I may be diluted in *Mbo*I dilution buffer immediately before use).

2. Remove aliquots (10 μl) at 0, 5, 10, 20, 30, 45, and 60 min and immediately add each aliquot to 10 μl of 40 mM Na$_2$EDTA (pH 8.0) in ice.

3. Analyze samples by electrophoresis through a 0.3% agarose gel as described above for 36 hr in the cold room. T4 DNA (160 kb), λ DNA (50 kb), and λ digested with *Sal*I (33 and 15 kb) are used as size markers. The samples will appear as smears of decreasing size with duration of digestion (Fig. 2).

Procedure VI: Large-Scale Preparation of Partially Digested DNA

1. Using optimized time intervals determined above, digest 100–200 μg of chromosomal DNA with *Mbo*I in 1–2 ml reaction volume. Scale up the reaction so that the enzyme concentration, time, temperature, and DNA concentrations are identical to those used for the pilot reactions.

2. Remove equal aliquots (333–666 μl) for the three optimal time points discussed above and immediately add 13.5–25 μl of 0.5 M Na₂EDTA (pH 8.0) to each sample and incubate in ice.

3. Pool and mix the three samples from step 2 and analyze a 10-μl aliquot by agarose gel (0.3%) electrophoresis to ensure that the appropriate size distribution has been attained.

4. Meanwhile, gently extract the digested DNA with phenol–chloroform–isoamyl alcohol. Precipitate the DNA with ethanol as discussed in Procedure I.

5. Dissolve the DNA (100–200 μg) in 500 μl of TE and fractionate the DNA in a 13-ml linear NaCl gradient (1.25–5 M NaCl in TE). Centrifuge for 3.5 hr at 39,000 rpm in an SW40.1 Beckman rotor at 18°.

6. Collect 1-ml fractions (12 total), dilute each of the middle six fractions with 1 ml of TE, and ethanol precipitate the DNA in each fraction separately overnight at −20° in SW40.1 polyallomer tubes.

7. Centrifuge the samples in the SW40.1 Beckman rotor for 1 hr at 20,000 rpm at 4°. (The tubes need not be completely filled.)

8. Solubilize each DNA pellet (not visible) in 360 μl of TE, add 40 μl of 3 M sodium acetate (pH 5.5), and ethanol precipitate DNA samples in 1.5-ml microcentrifuge tubes in dry ice.

9. Wash the DNA pellets with ethanol (see above) and dissolve each pellet in 20 μl of 0.1× TE. Store DNA samples at 4°.

10. Check the molecular size of DNA (200 ng) from each fraction by 0.3% agarose/TBE gel electrophoresis.

11. Choose fraction(s) which have a molecular size of approximately 35–50 kb and check the packaging efficiency of each fraction as described below.

Procedure VII: Preparation of Packaging Extracts

The preparation of packaging extracts using *E. coli* lysogens BHB2688 and BHB2690 is described in this volume [17]. Using Charon 4A as a control, packaging efficiencies of 5×10^8 transformants/μg of DNA are obtained. Packaging extracts are also available commercially from Amersham and Promega Biotec.

Procedure VIII: Ligation and Packaging of DNA

A ligation reaction is performed for each salt gradient fraction(s) (Procedure VI, step 11) as follows. Mix, in water to total of 20 μl, 1.5 μg size-fractionated DNA, 3.0 μl (3 μg) vector DNA (1 mg/ml), 2.0 μl 10× ligation buffer.

1. Add 1 μl of each ligation reaction to 9 μl of TE and store at 4° for gel electrophoresis.

2. Add 1 μl of T4 DNA ligase (1–2 Weiss units) to the remaining 19 μl of each reaction and incubate at 14° overnight. Store the ligation reaction at 4°.

3. Remove another 1-μl aliquot from each reaction, and analyze these samples, in addition to those in step 1, by electrophoresis through an 0.3% agarose/TBE gel. If ligation is complete, the eukaryotic DNA should be converted to high-molecular-weight concatamers.

4. For test packaging, place the freeze–thaw lysate (FTL) and sonicated extracts (SE) on ice. The FTL will thaw first.

5. Add the FTL (10 μl) to the still-frozen SE (15 μl) and then immediately add 2 μl of the ligation reaction (from step 2). Mix the packaging reaction at room temperature with a sealed capillary pipet. It is important that the SE does not thaw completely prior to the addition of the DNA since thawing increases the viscosity, which decreases the initial packaging efficiency.

6. The packaging reaction is incubated for 1 hr at room temperature.

7. Dilute the reaction 10-fold with SM buffer and store at 4°.

Procedure IX: Titering the Cosmid Library

1. Inoculate a single colony of *E. coli* ED8767 into 25 ml of LB medium containing 0.4% maltose in a 250-ml flask.

2. Incubate the culture overnight at 37° with vigorous shaking.

3. Centrifuge the culture at 4000 rpm for 20 min at 4°.

4. The cells are resuspended in 12.5 ml of 10 mM MgCl$_2$ by pipetting and stored at 4° until needed. Although the cells are stable for at least 5 days, it is desirable to use a fresh overnight culture.

5. Add 5-μl aliquots of the packaged cosmid DNA (Procedure VIII, step 7) to 45 μl (10-fold dilution) and 245 μl (50-fold dilution) of SM buffer.

6. Mix 25 μl of each dilution (step 5) with 25 μl of cells (step 4) in a 1.5-ml microcentrifuge tube and incubate at room temperature for 30 min.

7. Add 200 μl of LB medium to each sample and incubate at 37° for 1 hr with intermittent mixing to allow expression of the ampicillin resistance gene.

8. Centrifuge the culture for 30 sec in a microcentrifuge and resuspend each pellet in 50–100 μl of fresh LB medium.

9. Plate the cells on LB agar plates (15 cm) containing ampicillin at 50 μg/ml.

10. Incubate plates at 37° until colonies are big enough to count (usually 18 hr). Titers usually range from 100,000 to 800,000 transformants per microgram of size-fractionated chromosomal DNA. Note: At this point the remainder of each ligation reaction (Procedure VIII, step 2) may be packaged and titered as described above.

Procedure X: Screening the Cosmid Library

1. In a series of culture tubes, mix packaged recombinant DNA (20,000–50,000 colony forming units per tube) with an equal volume of *E. coli* ED8767 (Procedure IX, step 4) and incubate at room temperature for 30 min.

2. Add 4 volumes of LB medium to each tube and incubate the culture for 1 hr at 37° as described above.

3. Centrifuge the culture at 4000 rpm for 10 min at room temperature.

4. The cells are resuspended in LB medium (1.5 ml/cell pellet per filter) and spread onto 137-mm nitrocellulose filters (HATF 137-50, Millipore) which are on LB agar plates containing 50 μg/ml ampicillin.

5. Incubate the plates at 37° until the colonies are about 0.2 mm in diameter (usually 8–10 hr).

6. Place blank nitrocellulose filters on LB agar plates containing 50 μg/ml ampicillin for replica plating.

7. In a Biogard Hood, place a sterile Whatman 1 filter on a glass plate, followed by the wetted blank replica filter, the master filter (colony side down), and another Whatman 1 filter. Place another glass plate on top and then press the plates together.

8. Remove the top glass plate and Whatman filter. Mark the nitrocellulose filters in several places using a needle so you will be able to orient one filter with respect to the other. See also this volume [44] for marking filters.

9. Separate the nitrocellulose filters and place them back on the agar plates.

10. Repeat steps 6–9 if additional replica filters are to be made.

11. After replica plating, incubate the master and replica filters until the colonies are approximately 0.5 mm in diameter.

12. Seal the master plates with Parafilm and store at 4° in an inverted position.

13. Remove the replica filters from the LB/ampicillin plates and incubate at 37° on LB agar plates containing 250 μg/ml chloramphenicol for 20 hr to amplify the cosmid DNA.

14. The colonies are then lysed and the DNA is fixed on the filters by sequential 30-min incubations on Whatman 3 MM paper, saturated with 0.5 M NaOH, 1.5 M NaCl followed by 1 M Tris–HCl (pH 7.5) and then 1 M Tris–HCl (pH 7.5), 1.5 M NaCl. Between each incubation the filters are blotted on dry Whatman 3 MM filters.

15. Bake the filters at 68° for 4 hr. A vacuum oven is not required. Store desiccated at room temperature until ready for use.

16. Prehybridize the filters in 6× SSC, 2× Denhardt's solution in baking dishes at 68° overnight.

17. Rinse the filters in 6× SSC and then hybridize in a solution containing 6× SSC, 2× Denhardt's, 1 mM Na₂EDTA (pH 8.0), 100 μg of sheared and denatured salmon sperm DNA per milliliter, 0.5% SDS, and 2–5 × 10⁶ cpm of nick-translated probe/ml of solution. Hybridizations are carried out overnight in petri dishes (2 ml of hybridization solution per filter; 10–15 filters/dish) which are placed in sealable bags and incubated in a 68° oven. Alternatively, hybridizations can be carried out at 42° in 5× SSC, 50 mM sodium phosphate (pH 6.5), 2 mM EDTA, 0.2% SDS, 2× Denhardt's, 45% formamide, and 100 μg salmon sperm DNA per milliliter. Hybridizations can be performed in the presence of 10% dextran sulfate. (See also this volume [9] and [45].)

18. Wash the filters 3 times (1 hr each time) at 68° in baking dishes with a solution containing 2× SSC, 0.5% SDS. Note: More stringent washes can be carried out at 68° in 0.2× SSC.

19. Rinse the filters in 2× SSC at room temperature.

20. Dry the filters on Whatman paper at room temperature and expose to RX-Fuji film using a Dupont intensifying screen for 24 hr at −70° (see this volume [7]).

Procedure XI: Colony Purification

1. Positive colonies are surrounded and frequently in contact with negative colonies. After an area on the master filter has been identified, the positive colony and an approximately 5-mm disk of surrounding colonies are cut from the master plate using a sterile scalpel.

2. Elute (by pipetting) the colonies from the filter disk with 150 μl of LB medium containing 15% glycerol and 50 μg/ml ampicillin and store at −70°.

3. Dilute aliquots of the bacterial stock from step 2 with 1.5 ml of LB medium and plate onto nitrocellulose at a low density (100–1000 colonies per filter) as described above.

4. Repeat the hybridization (Procedure X) to identify a positive isolated colony.

5. Inoculate 2 ml of LB media containing 50 μg/ml ampicillin with an isolated positive colony and incubate overnight with vigorous shaking.

6. Adjust an 0.5-ml aliquot of the overnight culture to 15% glycerol and store at −70°. The remaining culture (1.5 ml) is used for minilysate analysis of the cosmid DNA using the alkaline lysis procedure (see this volume [13]). Large quantities of cosmid DNA can be obtained from a 1-liter culture grown in LB medium overnight by a scaled-up alkaline lysis procedure and purification of the DNA by centrifugation in CsCl gradients (see this volume [13]).

Discussion

The cosmid cloning system is well suited for isolating large DNA segments. The high cloning capacity reduces the number of clones to be screened when unique sequences are isolated from complex genomes. This technology facilitates chromosome walking and the establishment of gene linkage and analysis of large genes or genomic regions. However, several parameters to be considered when dealing with the construction of cosmid libraries are (1) the size of the genomic DNA (>150 kb), (2) the production of a random population of overlapping fragments by partial digestion with enzymes that cut DNA frequently, (3) the generation of cosmid vector–genomic insert concatemers, and (4) the abrogation of intermolecular ligation of vector molecules.

Problems can be encountered which usually result from bad packaging components (<10^8 pfu/μg λ DNA) or degraded genomic DNA (50–100 kb) used for the partial digest. The probe used for screening the library may also be a problem. Due to the large size of cosmids (50 kb), the copy number is much lower than that of small plasmids or λ grown in *E. coli*. Consequently, it is essential that the hybridization probe have a specific activity greater than 10^8 cpm/μg to maximize the signal-to-background ratio. Since cosmid vectors contain pBR322 sequences, the hybridization probes used for library screening should be free of pBR322 contamination. To alleviate this problem, we routinely purify our probes (cDNA inserts) twice by low melting temperature agarose or polyacrylamide gel electrophoresis. Prior to the second gel electrophoresis step, we routinely digest the purified insert with an enzyme that does not cut the cDNA insert but does cut pBR322 many times (i.e., *Hpa*II, *Hha*I). Alternatively, the sensitivity of cosmid library screening can be improved by using high specific activity single-stranded ^{32}P-labeled DNA or RNA probes. (See this volume [10] and [61].)

Because of problems associated with cosmid cloning in the past (i.e., vector concatemerization resulting in cosmids lacking inserted eukaryotic DNA and multiple, noncontiguous genomic DNA inserts ligated into a single cosmid vector), this system has not been widely employed for the isolation of eukaryotic genes. Techniques have been described to alleviate these problems, e.g., phosphatase treatment of vector DNA,[5] and the construction of right- and left-handed cosmid vector ends that are incapable of self-ligation but which accept dephosphorylated genomic DNA fragments.[10] Our procedure is similar to the former method[5] and can be utilized for any available cosmid vector.

[10] D. Ish-Horowitz and J. F. Burke, *Nucleic Acids Res.* **9**, 2989 (1981).

Using this cosmid cloning procedure, we routinely achieve cloning efficiencies ranging from 100,000 to 800,000 (mean 500,000) transformants per microgram of size-selected genomic DNA and have never observed polycosmid or multiple-insert-containing clones. Using only 75 μg of genomic DNA from human blood, this system enabled us to isolate the entire human phenylalanine hydroxylase gene in four overlapping cosmid clones spanning 160 kb of genomic DNA.[7] To isolate single-copy human genes, we recommend screening 500,000 colonies. A quantitative evaluation of the number of colonies to be screened to obtain a specific sequence whose fractional abundance is known can be found in this volume [16,46].

Acknowledgments

We wish to thank Ms. Kelly Porter for typing the manuscript. This work was partially supported by National Institutes of Health Grant HD17711 to S. L. C. W., who is also an investigator of the Howard Hughes Medical Institute. A. G. D. is the recipient of NIH Postdoctoral Fellowship HD-06495.

Section VII

Preparation and Characterization of RNA

[19] Preparation and Characterization of RNA: Overview

By SHELBY L. BERGER

The isolation of bulk RNA and the determination of the properties of specific molecules within that heterogeneous population are the first serious steps in the path toward a cDNA clone. The preparation of messenger RNA also plays a role in the identification and characterization of sequences already cloned, and in some cases, already expressed in cells of a higher organism. Thus, techniques in this section can be used not only at the commencement of cloning but also near its completion. The aim, then, is to explain the organization of this section and to view it within the context of the volume as a whole.

Choice of an RNA Extraction Technique. When investigators not familiar with RNA set out to isolate mRNA, the first question is not which technique to use but whether any technique can succeed. Perhaps more than any other procedure, the preparation of RNA requires near-compulsive elimination of ribonucleases from glassware, pipets, tips, and solutions, indeed, everything that touches RNA. Even implements that do not come into intimate contact with the sample should be scrutinized. Micropipets belong in this category, because they can become contaminated and in turn contaminate solutions. "Finger nucleases" must be barred from the sample by wearing gloves. Hair, dust, sneezes, and coughs are also likely sources of nucleases. When the requirements in chapter [2] have been met, and when the novice is fervently convinced that shortcuts spell failure, only then should a method be selected.

In this section, techniques are presented for isolating RNA from intact cells and tissues [20], nuclei [22], cytoplasm [21], and polysomes [21,23, 24]. The techniques described by MacDonald *et al.* [20] are the most widespread, versatile, and safest in use today. If intact RNA cannot be obtained by blasting whole cells or tissues apart in chaotropic agents, the chances for success with more subtle methods are not great. The main strength of this approach, then, is also the major disadvantage. Subcellular fractionation is not possible under conditions of complete cell disruption.

The methods in chapters [4] and [21] can also be used for preparing RNA from whole cells. Phenol extraction coupled with ethanol precipitation [5] to recover the sample is by far the most common technique of RNA isolation. It is faster and simpler than the guanidine-based methods but may place RNA molecules at slightly greater risk. A second popular method which makes use of sodium dodecyl sulfate (SDS) as a detergent to disrupt cells and proteinase K to degrade proteins,[1] among them ribonucleases, has not been featured here; the success rate with this approach does not equal that in chapter [20] and it, too, is incompatible with the preparation of subcellular fractions.

When subcellular fractionation is selected to enrich for a particular type of mRNA, the problem of endogenous nucleases becomes profound. There is no guarantee of success regardless of the choice of inhibitor(s). The greater the extent of subcellular fractionation that is required, the more difficult it is to find suitable inhibitors. For example, methods for displaying polysomes in a sucrose gradient [21], for isolating membrane-bound and free polysomes [23], and for isolating specific polysomes with antibodies [24] depend on heparin, placental ribonuclease inhibitor, or ribonucleoside–vanadyl complexes for protecting RNA, although none is ideal. Heparin, unlike the others, cannot be removed by phenol extraction, whereas vanadyl complexes result in aggregation and loss of resolution in displays of polysomes. The inhibitor from placenta is expensive and has not been thoroughly tested in this context. With nuclease-rich cells, it is unlikely that mRNA can be isolated reproducibly in a fully intact state from any subcellular organelle although some intact molecules may be evident. As an illustration, the RNA obtained from polysomes in chapter [24] is clearly degraded but, nevertheless, large enough to be suitable as a hybridization probe.

Characterization of Bulk mRNA. Once RNA has been isolated, its quality must be assessed. The simplest method makes use of an agarose gel [8] to fractionate crude cytoplasmic or whole-cell RNA electrophoretically with a commercial preparation of rRNAs serving as a source of markers. However, since rRNAs and some mRNAs have pronounced secondary structure, denaturing gels are essential. Regardless of the quality of the rRNA in a preparation, a direct picture of mRNA is ultimately required.

To separate polyadenylated RNAs from all others, the methods in chapter [25] are effective. Oligo(dT)-cellulose or poly(U)-Sepharose chromatography is a technique used throughout the cloning process for purifying poly(A)$^+$ RNA. Not only can rRNA be reduced and tRNA be removed from messenger RNA preparations, but also polyadenylated RNA can be separated from DNA in the form of oligomers, probes, or primers and from ribonuclease inhibitors, such as heparin or vanadyl complexes. Short columns of these materials can be used to concentrate dilute solutions of mRNA.

[1] H. Hilz, U. Wiegers, and P. Adamietz, *Eur. J. Biochem.* **56,** 103 (1975).

Since mRNA is usually precious, many investigators do not wish to commit microgram quantities for electrophoretic analysis in gels. To conserve material, a small amount of mRNA can be labeled as in chapter [26] before electrophoretic fractionation or used for a Northern blot [61]. (The term Northern blot refers to RNA that has been fractionated in a gel, transferred to a solid support, and hybridized with a labeled nucleic acid.) Upon electrophoretic fractionation, the labeled molecules should be distributed between approximately 0.6 and 4 kb in size. The "RNA ladder" available from Bethesda Research Laboratories (Gaithersburg, MD) provides convenient size markers from 0.3 to 7.5 kb. For Northern blots, a hybridization probe of known size that is common to most cells—actin, for example, or one of many major histocompatibility complex probes—serves as an indication of the quality of the population. The assumption is made that the desired mRNA is intact if the mRNA homologous to the probe is undegraded.

Quantification of Polyadenylated RNA. Methods for measuring the amount of bulk RNA and the amount of a specific mRNA are both presented. Bulk RNA or partially purified poly(A)$^+$ RNA can be quantified in units of mass by measuring A_{260} [6]. However, to assess rRNA contamination in preparations of poly(A)$^+$ RNA, the methods in chapter [26] must be employed. The results are obtained in molar units rather than in microgram amounts.

Specific mRNAs may be accurately quantified only when hybridization probes are available (see below). Then, synthetic mRNAs, constructed as described in chapter [59,61] and evaluated on the same Northern blot, can be used for visual comparisons with the sample. If the investigator is interested in the relative amounts of a specific mRNA in different tissues or in different developmental stages, for example, the dot blot technique in chapter [62] for either whole cells or isolated RNA is useful. Results should always be confirmed with Northern blots to rule out putative spurious cross-hybridization of the probe with rRNA or other mRNAs. If the different types of cells cannot be physically separated, but can be distinguished microscopically, the technique of *in situ* hybridization [68] may be a viable alternative; the results are semiquantitative or qualitative.

Characterization of mRNA by Translation. Since the translation efficiency of an mRNA in cell-free systems or in oocytes is virtually always unknown, translation provides a qualitative rather than a quantitative means for identification of a specific mRNA. In general, the protein produced is completely unprocessed after translation in cell-free systems [27], partly processed by the addition of microsome fractions to cell-free systems [27], or fully processed and directed to a specific subcellular or extracellular location by intact cells [28–30] such as the oocytes of *Xenopus laevis*. To aid in the identification process, techniques for subcellular fractionation of oocytes are included [30]. By far the most common method for distinguishing proteins, and indirectly the mRNA in which they are encoded, is the binding of specific antibodies. The use of antibodies for identification of translation products is described in chapter [31] for cell-free translation but is also applicable to oocytes. Since oocytes supply a full range of protein processing machinery radioactive compounds other than amino acids, e.g., carbohydrate or methyl donors, may be used to indicate posttranslational modification of the nascent chain.

Special Cases. If a hybridization probe is available before the commencement of cloning, a wider variety of techniques becomes available. The probe can be a related RNA, an RNA with sequences in common with the desired RNA, or an RNA from a different tissue or species. Occasionally an RNA with the same repetitive element can be used to advantage. Such cross-hybridizing molecules sometimes are a shortcut to the desired clone: they can be used as hybridization probes for the screening of libraries [45].

Messenger RNA can be purified to near-homogeneity and studied free of other mRNAs once a clone is in hand. The technique of hybrid selection [60], the purification of specific mRNAs by binding to immobilized homologous DNA with concomitant formation of a DNA–RNA hybrid, can be used to obtain abundant mRNAs in high yields. With rare mRNAs such as γ-interferon, losses are usually heavy and the result may be enrichment rather than isolation of the desired species.

The existence of a cloned full-length cDNA makes possible the production of its mRNA in large quantities. By cloning [58,59] or subcloning [55] into a vector with a strong, highly specific bacteriophage promoter, a capped, synthetic transcript can be synthesized *in vitro* in microgram quantities [30]. Although such mRNAs can be translated efficiently *in vitro* into the correct protein, they are not necessarily replicas of their natural counterparts. They lack such refinements as poly(A) tails and the methyl groups found on the bases or the sugars of RNA synthesized *in vivo*. Whereas the absence of a poly(A) tail is easily corrected [26], other structural defects cannot usually be remedied. In contrast, heterologous mRNA synthesized *in vivo* in yeast [70] or cells of a higher organism [71], subsequent to transformation with the desired clone, may reflect the structure of the natural mRNA more closely. Usually, however, transformation of *Escherichia coli* [69] or other cells is performed with truncated cDNAs in which 5' and 3' noncoding sequences have been intentionally removed in order to facilitate the isolation of significant quantities of the protein product.

Other special situations for which methods are described in this volume are the isolation of newly synthesized mRNA by reason of its lengthy poly(A) tail [27], and the "run-on" synthesis of nuclear RNA [26]. Run-on synthesis, also referred to as *in vitro* transcription, can be taken as an indication of which RNA transcripts are expressed or repressed following a change in cell culture conditions or developmental state. The technique is independent of pulse–chase techniques. In eukaryotic cells, the pulse–chase approach is rarely sufficiently sensitive for following rapid fluctuations in the types or amounts of specific mRNAs. Since a cell is distinguished by the messenger RNA it makes, a thorough grounding in the techniques for isolating and characterizing this material is essential for progress in both molecular and cell biology.

[20] Isolation of RNA Using Guanidinium Salts

By RAYMOND J. MACDONALD, GALVIN H. SWIFT,
ALAN E. PRZYBYLA, and JOHN M. CHIRGWIN

The isolation of undegraded ribonucleic acid from cells and tissues involves three steps: (1) inhibition of endogenous nucleases, (2) deproteinization of the RNA, and (3) physical separation of the RNA from the other components of the homogenate. Procedures for the preparation of RNA with strong protein denaturants effectively combine the first two steps. A variety of denaturants are available. Phenol-based procedures have been reviewed extensively.[1–3] This chapter summarizes protocols that utilize guanidinium salts which are more efficient denaturants than phenol and facilitate the isolation of intact, functional RNA from a wide range of biological sources, even those with extraordinarily high levels of nuclease. Inactivation of nucleases parallels the kinetic efficiency of protein denaturation, so that the relative efficacies are guanidinium thiocyanate > guanidine hydrochloride > urea. Denaturation is enhanced by including a reductant, such as 2-mercaptoethanol or dithiothreitol, to break intramolecular protein disulfide bonds. The addition of either competitive inhibitors of ribonuclease or chemical inactivators such as diethyl

[1] J. H. Parish, "Principles and Practice of Experiments with Nucleic Acids." Halsted Press, New York, 1972.
[2] R. D. Palmiter, *Biochemistry* **13**, 3606 (1974).
[3] D. M. Wallace, this volume [4, 5].

pyrocarbonate does not enhance denaturation, and the latter's reactivity toward nucleic acids[4] renders it undesirable.

As an alternative to phenol extraction for the isolation of RNA, homogenization in guanidine hydrochloride solutions is an inexpensive and convenient method. For those samples with very high RNase content or for irreplaceable samples with unknown RNase content, guanidinium thiocyanate becomes the deproteinizing agent of choice. Efficient protein denaturation with consequent elimination of nucleolytic damage to the RNA requires rapid and complete homogenization. For instance, homogenization of a rat pancreas in guanidinium thiocyanate solution by means of a motorized pestle results in extensively degraded RNA, which is avoided by the use of an apparatus such as a Tissumizer or Polytron. Frozen tissue should be pulverized in liquid nitrogen, then homogenized immediately (without thawing) after adding to the guanidinium solution.

Once the RNA is dispersed in the guanidinium solution, steps (1) and (2) have been completed, and only step (3) remains: physical separation of the RNA from the other macromolecular components of the homogenate. A number of separation methods have been described in the literature, and these can be divided into two main categories: selective precipitation based on solubility, and selective sedimentation (by ultracentrifugation) based on buoyant density.

Selective precipitation based on solubility has been described using ethanol[5,6] or lithium chloride[7,8] (see also this volume [5]). These procedures work well for many tissues and they all yield equivalent results when applied to liver, from which RNA also can be isolated readily by conventional phenol extraction followed by ethanol precipitation. Selective precipitations from guanidinium solutions can be nonquantitative and are affected in unforeseeable ways by components of the homogenate. This problem is generally more serious when the RNA concentration in the homogenate is very low. For example, homogenization of fetal bovine ligament in guanidinium solutions results in viscous, very dilute solutions of RNA not precipitable by either ethanol or lithium chloride (J. Chirgwin and R. Mecham, unpublished).

In the cases of intractable tissues such as ligament, irreplaceable sam-

[4] L. Ehrenberg, I. Fedorcsak, and F. Solymosy, *Prog. Nucleic Acid Res. Mol. Biol.* **16**, 185 (1974).

[5] R. A. Cox, this series, Vol. 12B, p. 120.

[6] J. M. Chirgwin, A. E. Przybyla, R. J. MacDonald, and W. J. Rutter, *Biochemistry* **24**, 5294 (1979).

[7] C. Auffray and F. Rougeon, *Eur. J. Biochem.* **107**, 303 (1980).

[8] G. Cathala, J.-F. Savouret, B. Mendez, B. L. West, M. Karin, J. A. Martial, and J. D. Baxter, *DNA* **2**, 329 (1983).

ples such as surgically removed tumors, or very dilute RNA solutions, the method of choice for quantitative recovery of RNA is centrifugation through a dense solution of cesium chloride.[6,9,10] This method depends solely on the high buoyant density of RNA compared with other cellular macromolecules. The pellets of RNA obtained after sedimentation through cesium chloride solutions are often difficult to redissolve. Ways to overcome this problem are discussed in the protocol below.

Once RNA has been recovered from a guanidinium homogenate, particularly after cesium chloride pelleting, it should be free of protein. The addition of a phenol extraction step does not increase the purity of the nucleic acid. Proteinase K digestion would only be useful if peptides were covalently linked to the RNA. However, either of these additional treatments would remove residual nucleases if the guanidinium steps were performed ineffectively.

Three recent demonstrations of the power of the guanidinium thiocyanate/cesium chloride method are the isolation of an approximately 20-kb mRNA for apolipoprotein B,[11] the detection of mouse *H-2D^d* transcripts which were undetectable when the RNA was isolated by conventional means,[12] and the isolation of intact RNA from pulverized cockroach heads.[13]

The Guanidine Hydrochloride Method

1. Prepare a solution of 7.5 M guanidine–HCl by dissolving 72 g (Sigma G4630, practical grade) in 10 ml of 1 M Tris–HCl (pH 7.0), and water to 100 ml. After filtration through Whatman 1 MM filter paper or equivalent, this solution can be stored indefinitely at room temperature. Just before use, add dithiothrietol to 10 mM.

2. Homogenize up to 2 g of tissue in 10 ml of the 7.5 M guanidine–HCl solution at room temperature with a Tissumizer (Tekmar Ind., Cincinnati, OH) or Polytron (Brinkmann, Westbury, NY) at top speed for 60 sec at room temperature.

3. Add 0.5 ml of 10% sodium lauryl sarcosinate (Sigma L-5125) and mix well by vortexing. The presence of sodium lauryl sarcosinate im-

[9] V. Glisin, R. Crkvenjakov, and C. Byus, *Biochemistry* **13**, 2633 (1974).

[10] A. Ullrich, J. Shine, J. Chirgwin, R. Pictet, E. Tischer, W. J. Rutter, and H. M. Goodman, *Science* **196**, 1313 (1977).

[11] L.-S. Huang, S. C. Bock, S. I. Feinstein, and J. L. Breslow, *Proc. Natl. Acad. Sci. U.S.A.* **82**, 6825 (1985).

[12] P. M. Brickell, D. S. Latchman, D. Murphy, K. Willison, and P. W. J. Rigby, *Nature (London)* **316**, 162 (1985).

[13] T. H. Turpen and O. M. Griffith, *BioTechniques* **4**, 11 (1986).

proves the purity of the initial RNA precipitate; its addition after the initial homogenization avoids excessive foaming and obviates the use of antifoam. Centrifuge at 5000 g for 5 min to remove insoluble debris.

4. Add 0.5 ml of 2 M potassium acetate (pH 5.5) and 0.8 ml of 1 M acetic acid and mix. Then slowly add 5 ml of absolute ethanol while vortexing. The slow addition of ethanol while vortexing minimizes precipitation of DNA. Place in an ethanol bath kept in a $-20°$ freezer for at least 2 hr. For tissues with low RNA levels (e.g., kidney or brain), precipitation overnight at $-20°$ increases yields noticeably.

5. Collect the precipitated RNA by centrifugation in a swinging bucket rotor at 10,000 g (e.g., Sorvall HB-4 rotor at 8000 rpm) for 10 min. [If the RNA pellet is difficult to dissolve, centrifuge slower (e.g., 5000 g) and check the efficiency.]

6. Dissolve the pellet at room temperature in 5 ml of unbuffered 7.5 M guanidine–HCl that has been neutralized to pH 7 with NaOH, supplemented with 10 mM dithiothreitol, and filtered. Add 0.25 ml of 2 M potassium acetate (pH 5.5) and mix. Add 2.5 ml of ethanol, mix well, and place at $-20°$ for at least 2 hr.

7. From this point on, all procedures are performed under sterile, RNase-free conditions to prevent RNA degradation. All glassware must be heat-treated at least 8 hr at 180° or treated with 0.2% (v/v) diethyl pyrocarbonate (DEP) (Sigma D-5758) for 0.5 hr at room temperature and rinsed thoroughly with DEP-treated water. Use sterile disposable plasticware whenever possible. For centrifugations use heat-treated 15-ml and 30-ml Corex centrifuge tubes or 12 × 75 mm and 17 × 100 mm disposable polypropylene tubes (Falcon 2005 and 2006). Water is treated with DEP by adding DEP to 0.05% and boiling in a fume hood for 1 hr to remove all traces of DEP odor. All DEP must be destroyed because DEP modifies RNA. DEP is a potent alkylating agent and must be used with caution.

8. Collect the final RNA precipitate out of guanidine–HCl by centrifugation at 10,000 g for 10 min. Extract excess guanidine–HCl by dispersing the pellet in ethanol at room temperature and centrifuging briefly at 10,000 g.

9. Add 1 ml of 20 mM EDTA (pH 7) to the RNA pellet from 1–2 g of tissue. Vortex briefly; then add 2 volumes of chloroform–n-butanol (4 : 1, v/v) and vortex extensively (5 min). Centrifuge briefly to separate the phases. Remove the upper (aqueous) phase, and reextract the interface (which will contain much of the RNA) and organic phase with 1-ml aliquots of 20 mM EDTA until the size of the interface pad stops decreasing (generally only two additional extractions).

10A. Combine the aqueous phases and precipitate the RNA with 2 volumes of 4.5 M sodium acetate (pH 7) at $-20°$ overnight. Collect the

RNA by centrifugation at 10,000 g for 30 min. This salt precipitation removes residual DNA and glycogen, but is not necessary for most applications.

10B. If glycogen or residual DNA is not a problem, precipitate the RNA from the combined aqueous extracts of step 9 by the addition of 0.1 volume of 2 M sodium acetate (pH 7) and 2.5 volumes of ethanol at $-20°$ for at least 2h. Collect the RNA by centrifugation.

11. Dissolve the RNA pellet in DEP-treated water, and estimate the yield by measuring the absorbance at 260 nm (an A_{260} of 1 equals an RNA concentration of 40 $\mu g/ml$). The RNA should have an A_{260}/A_{280} ratio of 1.9–2.2. Reprecipitate the RNA by adding 0.1 volume of 2 M sodium acetate (pH 7) and 2.5 volumes of ethanol. RNA is best stored in this state (precipitated in ethanol at $-20°$).

Guanidinium Thiocyanate Method

For tissues or cells with high RNase levels, the quality of the RNA obtained is enhanced by an initial homogenization and precipitation with guanidinium thiocyanate. As an example, the following procedure describes the preparation of RNA from rat pancreas.[13a]

1. Prepare a solution of 4 M guanidinium thiocyanate by dissolving 50 g of guanidinium thiocyanate (Fluka 50990) in 10 ml of 1 M Tris–Cl (pH 7.5) and water to 100 ml. Filter through Whatman 1 MM filter paper or equivalent. This solution is stable indefinitely. Just before use add 2-mercaptoethanol to 1%.

2. Remove the pancreas from a rat quickly and place the pancreas (about 0.7 g) on the lip of the homogenization vessel containing 10 ml of guanidinium thiocyanate solution at room temperature. Force the tissue into the solution with the tip of the homogenizer probe and immediately homogenize (Tissumizer or Polytron) at full speed for 60 sec at room temperature. Add 0.5 ml of 10% sodium lauryl sarcosinate and mix well. Centrifuge briefly to remove insoluble debris.

3. Add 0.5 ml of 2 M potassium acetate (pH 5.5), 0.8 ml of 1 M acetic acid and mix well. Slowly add 7.5 ml of absolute ethanol while vortexing. The slow addition of ethanol while vortexing minimizes the precipitation of DNA. Place in an ethanol bath kept in a $-20°$ freezer for at least 2 hr.

Three-quarters volume of ethanol (rather than a half volume) is required to prevent the guanidinium thiocyanate solution from freezing at

---

[13a] Additional specialized refinements that may be applicable to the isolation of RNA from pancreas have been described by J. H. Han, C. Stratowa, and W. J. Rutter, *Biochemistry* **26**, 1617 (1987).

$-20°$. However, this amount of ethanol causes the precipitation of some DNA. Removal of this contaminating DNA requires subsequent precipitations out of guanidine hydrochloride with a half volume of ethanol. The additional precipitations also remove residual RNase.

4. Collect the RNA precipitate by centrifugation at 10,000 g for 10 min. Dissolve the RNA pellet in 5 ml of guanidine–HCl solution and proceed with the RNA isolation as in step 6 of the procedure for the Guanidine Hydrochloride Method. For pancreas do two precipitations out of guanidine–HCl.

Guanidinium/Cesium Chloride Centrifugation Method

A useful modification of the guanidinium procedure[6,10,14] separates RNA from the guanidinium homogenate by ultracentrifugation through a cesium chloride cushion.[9] This method is less labor intensive, but requires an overnight centrifugation in an ultracentrifuge.

1. Homogenize tissue at room temperature for 60 sec in 5 volumes of 4 M guanidinium thiocyanate containing 0.1 M Tris–HCl (pH 7.5) and 1% 2-mercaptoethanol using a Tissumizer or Polytron homogenizer. Thorough homogenization shears the nuclear DNA and prevents the formation of an impenetrable DNA mat which blocks the sedimentation of RNA. Add sodium lauryl sarcosinate to 0.5% and mix. Centrifuge briefly (e.g., 5 min at 5000 g) to remove insoluble debris. Some tissues with high levels of mucopolysaccharides and glycosaminoglycans (e.g., ligament) form an occlusion within the gradient, which can be prevented by adding 0.5–1.0 g of solid cesium chloride to each milliliter of homogenate. (Note that centrifugation speeds then must be reduced to stay within safe limits.)

2. Layer about 5.8 ml of the homogenate onto a 2-ml pad of 5.7 M CsCl (BRL 5507UB) in 4 mM EDTA (pH 7.5) in a Beckman SW30.1 tube. [*Note:* The use of less pure CsCl may require treatment with 0.05% DEP and the use of higher concentrations of EDTA (as high as 0.1 M) to chelate the potentially high level of divalent cations.] Table I lists additional ultracentrifugation rotors commonly used and the conditions for centrifugation. When using ultracentrifuge rotors not listed in Table I, calculate the equivalent clearing time using the clearing factor of the rotor at the maximum speed allowed for a tube one-quarter filled with 5.7 M (1.7 g/ml) CsCl.[15] Fixed-angle rotors allow shorter centrifugation times,[13]

---

[14] T. Maniatis, E. F. Fritsch, and J. Sambrook, "Molecular Cloning: A Laboratory Manual." Cold Spring Harbor Lab., Cold Spring Harbor, New York, 1982.

[15] "Rotors and Tubes for Preparative Ultracentrifuges: An Operator's Manual." Beckman Instruments, Palo Alto, California, 1982.

TABLE I

CENTRIFUGATION CONDITIONS FOR GUANIDINIUM THIOCYANATE/CESIUM
CHLORIDE GRADIENTS WITH COMMON ULTRACENTRIFUGE ROTORS

| Rotor | Volume | | Rotor speed[a] (rpm) | Minimum run length[b] (hr) |
	CsCl pad (ml)	Homogenate (ml)		
SW50.1	1.3	3.6	47,000	8
			or 42,000	12
SW60	1.1	3.2	41,000	10
SW30.1	2.0	5.8	29,000	15
SW41	3.3	9.7	30,000	23
SW28	9.5	28.5	27,000	26

[a] Maximum rotor speed for tubes one-quarter full of 5.7 M CsCl.
[b] Time required for pelleting, based upon the clearing factor of each rotor at
the given speed.

but the RNA will be pelleted along the side of the centrifuge tube and thus
will come back into contact with the upper, protein-containing protion of
the gradient when the liquid contents of the tube reorient at the end of the
run.

3. Centrifuge the SW30.1 rotor at 29,000 rpm for 15 hr at 20°. Carefully aspirate the supernatant solution and dry the walls of the tubes with
Kimwipes (without introducing finger RNase).

Low yields of RNA generally involve the difficulty of dissolving recal-
citrant RNA pellets. The efficacy of solubilization can be checked by
dissolving any insoluble remainder in 0.1 N NaOH and estimating the
RNA by absorbance at 260 nm. The correct approach to dissolving the
RNA pellet depends on the tissue source and the size of the pellet:

a. *Small pellets* (less than 1 mg of RNA). Dissolve by vortexing extensively in SET buffer (10 mM Tris–HCl, pH 7.4, 5 mM EDTA, 0.1% SDS).

b. *Large pellets* (more than 1 mg of RNA). The larger the pellets, the
more intractable they become. Vortex the RNA pellet in 4 ml of 70% 2-
propanol–30% 0.2 M sodium acetate to extract CsCl. Carefully remove
the alcohol solution leaving the pellet behind. Solubilization also can be
facilitated by one cycle of freezing and thawing, or by homogenizing the
pellets with a Tissumizer or Polytron using a small probe. Dissolve each
pellet in 1–2 ml of SET buffer with extensive vortexing. Traces of RNase
reintroduced during this step may be removed by chloroform–n-butanol
extraction (see step 9 in Guanidine Hydrochloride Method); beware of
losing RNA at the interface.

c. *Pellets from tissues with high RNase content.* Cut off and discard the top part of the centrifuge tube that was in contact with the homogenate. Dissolve the RNA by extensive vortexing (or homogenize) in unbuffered 7.5 M guanidine–HCl that has been neutralized to pH 7 with NaOH, supplemented with 10 mM dithiothreitol, and filtered. Recover the RNA by precipitating at −20° with 0.05 volume of 2 M potassium acetate (pH 5.5), and 0.5 volume of ethanol. Then reprecipitate with ethanol as in step 6 in Guanidine Hydrochloride Method.

4. If glycogen is present in the RNA pellet, it can be removed by selectively precipitating RNA with a high concentration of salt. Add 2 volumes of 4.5 M sodium acetate (pH 7), mix well, and store for several hours (preferably overnight) at −20°. Collect the RNA precipitate by centrifugation in a swinging bucket rotor at 10,000 g for 30 min. Dissolve the RNA in DEP-treated water.

5. Precipitate the RNA by adding 0.1 volume of 2 M sodium acetate (pH 7) and 2.5 volumes of ethanol. Store at −20° for at least 2 hr.

6. Collect the RNA precipitate by centrifugation in a swinging bucket rotor at 10,000 g for 30 min. Dissolve the pellet in DEP-treated water, estimate the yield by measuring the absorbance at 260 nm, and reprecipitate by adding 0.1 volume of 2 M sodium acetate and 2.5 volume of ethanol.

Guanidine Hydrochloride Method for Cultured Cells

A simplified procedure with guanidine–HCl permits the rapid and efficient isolation of RNA from cultured cells.[16] This procedure is effective for the isolation of RNA from a wide variety of tested cell lines. To ensure precipitation of RNA from guanidine solutions, volumes must be kept as small as possible.

1. Aspirate medium from plates of cells grown in monolayer culture. Rinse once with phosphate-buffered saline at room temperature.

2. For a pair of confluent 100-mm plates use 1 ml of guanidine lysis solution (19 ml 7.5 M guanidine hydrochloride plus 1 ml 2 M potassium acetate, pH 5.5) at room temperature. Pipet across the plate repeatedly until most cells lyse. Combine the lysis solutions from all treated plates. A few cell lines (e.g., macrophage lines) require the use of guanidinium thiocyanate in the initial lysis step.

3. Rinse each set of four plates with an additional 1 ml of guanidine

[16] R. C. Strohman, P. S. Moss, J. Micou-Eastwood, D. Spector, A. Przybyla, and B. Paterson, *Cell (Cambridge, Mass.)* **10**, 265 (1977).

lysis solution. Use a rubber policeman to scrape the remaining lysis solution off the plate.

4. Shear the nuclear DNA with a Tissumizer or Polytron homogenizer at high speeds or pass the lysate through a 25-gauge needle five times to reduce the viscosity.

5. Precipitate the RNA by adding 0.5 volume of ethanol while vortexing. Store overnight at $-20°$ to maximize yields. Collect the precipitate by centrifugation at 10,000 g for 10 min in a swinging bucket rotor.

6. Resuspend the RNA pellet in 0.5 volume of 7.5 M guanidine–HCl containing 10 mM EDTA (pH 7). Add 0.05 volume of 2 M potassium acetate (pH 5.5) and 0.5 volume of ethanol. Keep at $-20°$ for at least 2 hr, or overnight to enhance yields.

7. Collect the RNA precipitate by centrifugation in a swinging bucket rotor at 10,000 g for 10 min. Resuspend the pellet in 0.05 the volume of the original lysate of 10 mM EDTA (pH 7).

8. Extract with three volumes of chloroform–n-butanol (4:1, v:v). Remove the aqueous phase and reextract the interface and organic phase twice with 10 mM EDTA (see step 9 in Guanidine Hydrochloride Method).

9. Combine the EDTA extracts and add 2 volumes of 4.5 M sodium acetate (pH 7). Store at $-20°$ overnight. Collect the RNA precipitate by centrifugation in a swinging bucket rotor at 10,000 g for 30 min.

10. Dissolve the pellet in DEP-treated water (about 0.05 the volume of the original lysate), and add 0.1 volume of 2 M sodium acetate (pH 7) and 2.5 volumes of ethanol. Precipitate at $-20°$.

Generally yields are about 100 μg of total RNA per 100-mm plate of confluent cells.

[21] Isolation of Cytoplasmic RNA: Ribonucleoside–Vanadyl Complexes

By Shelby L. Berger

The isolation of cytoplasmic messenger RNA, polysomes, and ribonucleoprotein particles depends on methods for inhibiting cellular ribonucleases in the presence of native proteins, functional enzymes, and intact organelles. An extensive survey of ribonuclease inhibitors indicated that ribonucleoside–vanadyl complexes were unique in their ability to suppress ribonuclease activity completely in the resting lymphocyte, a cell

METHODS IN ENZYMOLOGY, VOL. 152

renown for its high levels of endogenous ribonuclease.[1] Subsequent experience with these complexes has shown that they are gentle, potent, broad in their inhibitory spectrum, and compatible with a wide variety of experimental protocols.[2] For preparative work, it is also significant that they are easily synthesized from inexpensive, commercially available substances.[1] The complexes themselves may be purchased [Bethesda Research Laboratories (BRL) or New England BioLabs] but, upon making their debut in the business world, an inversion in the name occurred; the term ribonucleoside–vanadyl complexes, first used to describe them, became vanadyl–ribonucleoside complexes (VRC) or vanadyl–ribonucleoside complex, often shortened to vanadyl complexes.

Although vanadyl complexes may be used to inhibit nucleases during phenol extractions of whole cells or lumps of tissue, or together with the chaotropic agents described in this volume [20] for isolating RNA, they are best suited for work with preparations of separated cells. Such cells can be obtained by dispersing attached cells with enzymes (collagenase or trypsin), from culture in spinner flasks, or from tissues such as blood in which individual cells circulate *in vivo*. Theoretically, one can also obtain organized structures containing intact RNA from tissue pulverized in liquid nitrogen with vanadyl complexes as inhibitors, but in practice, I know of no one who has attempted it. Regardless of the protocol, it is essential that the complexes be present throughout the procedure until proteins are quantitatively removed. Intact cells are lysed in their presence, nuclei are removed, and cytoplasmic components are either analyzed or deproteinized with vanadyl complexes standing guard.

Principle of the Method

During enzyme catalysis, an activated complex or transition state forms which is an altered state of a substrate on its way to becoming product. This transition state has no independent existence but requires the surface of the enzyme to survive. Dissociation constants for the hypothetical process of separating the enzyme from its transition state are extremely low, namely, 10^{-7} lower than the dissociation constant of the enzyme–substrate complex.[3] Therefore, stable analogs of transition state intermediates are extremely powerful inhibitors. Vanadyl complexes are transition state analogs of the activated $2',3'$-cyclic phosphates believed

[1] S. L. Berger and C. S. Birkenmeier, *Biochemistry* 18, 5143 (1979).
[2] R. S. Puskas, N. R. Manley, D. M. Wallace, and S. L. Berger, Biochemistry 21, 4602 (1982).
[3] R. Wolfenden, this series, Vol. 46, p. 15.

to occur during catalysis by ribonuclease.[4] The phosphate in the intermediate is replaced by a 2′,3′-linked oxovanadium ion but, unlike the phosphate, the ion is held noncovalently. Nevertheless, the binding is extremely tight; the dissociation constant of the uridine–vanadyl complex from the active site of pancreatic ribonuclease A is 10 μM. Because EDTA can bind the oxovanadium ion, and Mg^{2+} can replace it in the complex, high concentrations of either should be used with caution. Guidelines are presented in Concluding Remarks.

Preparation of Vanadyl Complexes

With the instructions that follow, one produces 20 ml of 200 mM vanadyl complexes.

1. Mix 267 mg adenosine, 243 mg cytidine, 283 mg guanosine, and 244 mg uridine with 17 ml water in a 50-ml tube. Heat in a boiling water bath to dissolve the ribonucleosides and to degas the preparation.

2. Make several milliliters of 2 M vanadyl sulfate (VOSO$_4$, 398 mg/ml) and place in the boiling water bath. For best results weigh quickly and ignore the moisture.

3. Arrange a pH meter, a nitrogen tank, and a magnetic stirring hot plate so that a boiling water bath made in a beaker on the hot plate is accessible to both the N_2 and the electrode of the pH meter.

4. Drop a small stirring bar into the 50-ml tube containing the now fully dissolved ribonucleosides. Transfer the tube to the makeshift water bath in the beaker. Bubble N_2 into the ribonucleoside solution from the bottom of the tube and begin stirring. Insert the electrode. Be sure there is space in this crowded arrangement for delivery of subsequent liquid additions with a hand-held pipet.

5. Add 2 ml 2 M VOSO$_4$. The pH of the mixture should be between 1 and 2.

6. With dropwise addition, add 10 N NaOH (or 10 N KOH) until the pH reaches 6. Continue titrating with 1 N base until the solution is pH 7. Formation of the complex is indicated by a change from bright blue, the color of VOSO$_4$, to green-black, the color of the complexes. Do not be alarmed if, during the titration, the solution becomes cloudy. It will become clear, albeit optically dense, at pH 7.

7. Remove the electrode and N_2 delivery tube, rinsing each with a small amount of boiled water, and bring the complexes to 20 ml with

---

[4] G. E. Lienhard, I. I. Sccemski, K. A. Koehler, and R. N. Lindquist, *Cold Spring Harbor Symp. Quant. Biol.* **36**, 45 (1971).

additional boiled water. Distribute the vanadyl complexes in small portions and store at $-70°$. The complexes are stable for years.

8. When ready for use, thaw at $37°$ and vortex briefly. Avoid extensive aeration, as the oxygen introduced can destroy the inhibitor. If the thawed solution has lost its green cast, discard it. Note that $VOSO_4$ titrated with base in the absence of ribonucleosides gives a dull black suspension. It is the greenish tinge in a clear solution that is diagnostic for vanadyl complexes.

Large-Scale Preparation of Cytoplasm mRNA

The procedure described here was designed for isolating mRNA from lymphocytes and lymphoblastoid cells. It is presented as an illustration of how vanadyl complexes can be used, with the understanding that the exact composition of buffers, times of centrifugation, and the like will be modified for specific applications. The amounts suggested are sufficient for 1 liter of 1.8 to 2.2×10^6 cells/ml of lymphoblastoid cells or actively growing normal lymphocytes. For 1 liter of resting lymphocytes (cells with scanty cytoplasm) the volumes should be reduced to one-quarter of those indicated. The entire preparation may be scaled up or down proportionally. A version suitable for characterizing polysomes on an analytical scale will also be presented.

Reagents for Cell Lysis

Low-salt Tris I: 20 mM Tris–HCl at pH 7.4, 10 mM NaCl, and 3 mM magnesium acetate

Lysing buffer I: low-salt Tris I containing 5% (w/w) sucrose and 1.2% (w/w) Triton N-101 (Rohm and Haas) or Nonidet P-40 (BRL)

Low-salt Tris II: 10 mM Tris–HCl at pH 7.4, 8 mM magnesium acetate

Lysing buffer II: low-salt Tris II containing 5% w/w sucrose and 1.2% (w/w) Triton N-101

Reagents for Display of Polysomes

Low-salt polysome buffer–low-sucrose: 10 mM Tris–HCl at pH 7.4 and 12 mM magnesium acetate.[5] Add 0.05 volume 200 mM vanadyl complexes, and make 10% (w/w) in sucrose

---

[5] The interaction of vanadyl complexes, Mg^{2+}, and ribosomes has not been well studied. Therefore, the Mg^{2+} concentration may have to be adjusted for each lot of vanadyl complexes and for lysates from different types of cells to prevent artifactual aggregation of particles in the gradient.

Low-salt polysome buffer–high-sucrose: 10 mM Tris–HCl at pH 7.4 and 12 mM magnesium acetate.[5] Add 0.05 volume 200 mM vanadyl complexes and make 40% (w/w) in sucrose

High-salt polysome buffer–low-sucrose: 10 mM Tris–HCl at pH 7.4, 50 mM magnesium acetate, 0.5 M NaCl, 1.75 mg/ml heparin, 10% (w/w) sucrose

High-salt polysome buffer–high-sucrose: 10 mM Tris–HCl at pH 7.4, 50 mM magnesium acetate, 0.5 M NaCl, 1.75 mg/ml heparin, 30% (w/w) sucrose

Reagents for Isolating RNA

Ace buffer: 50 mM sodium acetate, 10 mM EDTA at pH 5.1

Phenol: only redistilled phenol, stored frozen at $-20°$ in the presence of 0.1% 8-hydroxyquinoline, should be used. Before the start of an RNA extraction the phenol should be thawed at 60–65°, saturated with 0.3 volume of Ace, and maintained at 50°.

Chloroform–isoamyl alcohol (94 : 6 v/v)

Sodium acetate, 2 M

Sodium dodecyl sulfate (SDS), 10%

All buffers are treated with diethyl pyrocarbonate (this volume [2]).

Procedure for Lysing Cells and Preparing Cytoplasm

1. Recover leukocytes from 1 liter of culture by centrifuging at room temperature for 10 min at 600 and 1600 g for resting lymphocytes or lymphoblastoid cells, respectively. Remove the culture fluids by aspiration or by careful decantation.

2. Resuspend the cells in 250 ml cold Eagle's minimal essential medium or RPMI 1640 and centrifuge again at 4°.

3. Suspend the washed cells in 12 ml ice-cold low-salt Tris I, 0.8 ml 200 mM vanadyl complexes, and add 4 ml ice-cold lysing buffer I. Vortex to mix and lyse cells.

4. Remove nuclei by sedimentation at 1200 g for 5 min at 4°.

5. Recover the supernatant fluid, which contains the cytoplasm, with a pipet. The cytoplasm may be used for analysis of polysomes (see below) or for isolation of RNA.

Procedure for Isolation of RNA

6. Discharge the lysate (~16 ml) immediately into a 250-ml glass bottle fitted with a ground-glass stopper containing 60 ml Ace buffer and 4 ml 10% SDS. Add 80 ml Ace-saturated phenol at 50° and shake vigorously at room temperature for 5 min. The phenol phase, initially bright yellow

owing to the dissolved 8-hydroxyquinoline, will turn gray-black on contact with vanadyl complexes.

7. Recover the aqueous (upper) phase by centrifugation at 4000 g for 10 min at 4° in glass or phenol-resistant plastic tubes.

8. Reextract the aqueous phase in a clean 250-ml glass bottle with 40 ml Ace-saturated phenol and 40 ml chloroform–isoamyl alcohol by shaking at room temperature for 5 min. Separate the phases by centrifugation at 4° as specified in step 7, using Maxiforce, polypropylene bottles (VWR Scientific). Work quickly, because the bottles are not entirely resistant to chloroform. Alternatively, centrifuge at 10,000 g in 30-ml Corex tubes.

9. If a prominent interphase remains, substitute 80 ml chloroform–isoamyl alcohol for the phenol–chloroform–alcohol mix and reextract the aqueous phase as described in step 8.

10. To the final aqueous phase containing cytoplasmic RNA, add 0.05 volumes 2 M sodium or potassium acetate and 2 to 3 volumes of ethanol. Incubate at −20° overnight and recover the RNA by centrifugation at 10,000 g at 4° for 30 min.

Purification of poly(A)-bearing molecules may be carried out using the techniques in this volume [25]. We have obtained 2–3 pg of cytoplasmic RNA per cell, of which ~1% was poly(A)$^+$ RNA. The poly(A)$^+$ RNA was intact and functional.[6]

Note that vanadyl complexes inhibit cell-free protein synthesis. If the RNA is to be translated without first selecting polyadenylated molecules, the complexes must be removed by multiple phenol extractions (this volume [4]). When the phenol phase supplemented with 8-hydroxyquinoline remains bright yellow, complexes have been removed.

Preparation of Cytoplasm for Analysis of Polysomes

The isolation of mRNA from membrane-bound and free polysomes (this volume [23]) and the use of antibodies to obtain specific polysomes (this volume [24]) are both large-scale, batch procedures. Since it is often useful to characterize polysomes on an analytical scale, before committing the time and energy for these more demanding procedures, two such techniques are presented here. Both are compatible with the use of vanadyl complexes as nuclease inhibitors. They can therefore be used as a starting point for those who might need to modify the techniques in chapters [23] or [24] for use with nuclease-rich cells.

---

[6] S. L. Berger, M. J. M. Hitchcock, K. C. Zoon, C. S. Birkenmeier, R. M. Friedman, and E. H. Chang, *J. Biol. Chem.* **255,** 2955 (1980).

Sucrose Gradient Preparation

Prepare a 10–40% linear, low-salt sucrose gradient as follows: Place 5.8 ml of the low-salt polysome buffer–high-sucrose in the well of a gradient maker containing the exit to a 12-ml cellulose nitrate centrifuge tube compatible with a Beckman SW41 rotor; place 5.8 ml of the low-salt polysome buffer–low-sucrose in the other well. Place a stirrer or a vibrator in the denser sucrose solution. Begin stirring. Open the stopcock between the wells and pump the sucrose solution from the exit toward the centrifuge tube using a Pharmacia peristaltic pump. The gradient is formed by allowing the densest aliquot to flow to the bottom of the tube and layering solutions of lower density above it.

Prepare a 10–30% linear high-salt sucrose gradient, similarly, using the high-salt polysome buffer solutions.

Procedure for Lysing Cells

Lyse $\sim 10^8$ washed, pelleted resting lymphocytes or $2.5–5 \times 10^7$ washed pelleted lymphoblastoid cells by resuspending them in 0.3 ml ice-cold low-salt buffer II, 20 μl 200 mM vanadyl complexes, and adding 0.1 ml ice-cold lysing buffer II. Vortex to mix. Remove nuclei by sedimentation as for the large-scale procedure. Recover the supernatant containing the cytoplasm and layer it carefully over the desired sucrose gradient. Centrifuge at 195,000 g at 4° for 165 min and collect in 0.3-ml fractions by upward displacement with 50% (w/w) sucrose using an ISCO Model 185 density gradient fractionator equipped with a Model UA 5 absorbance monitor, a Model 568 fraction collector, and a Model 613 recorder, or the equivalent equipment.

Polysomes should be readily visible with the 80 S ribosome about halfway down either type of gradient. In high-salt gradients, inactive 80 S ribosomes are dissociated into inactive subunits which, for the most part, cosediment with their active counterparts. In low-salt gradients, inactive 80 S ribosomes are not dissociated and cosediment with active 80 S monosomes. As described here, in the low-salt gradient, the included vanadyl complexes which absorb in the ultraviolet make it impossible to obtain an optical density profile of the polysomes with the continuous-flow ISCO equipment detailed above. Therefore, gradient fractions are analyzed by the isolation and characterization of their RNA (this volume [23] and [8]). Cells with labeled RNA can be employed. Using resting lymphocytes, the mRNA recovered from a low-salt gradient was intact. High-salt gradients, with heparin rather than vanadyl complexes as the nuclease inhibitor, yielded partially degraded molecules. It is possible, however, that cells with lower endogenous nuclease levels can be handled successfully using either technique. For some applications, intact RNA is not essential (this volume [30]).

Concluding Remarks

Since cellular nucleases are not well characterized, the inhibitory spectrum of vanadyl complexes is unknown. Nevertheless, it is clear that pancreatic ribonuclease A, ribonuclease T_1, poly(A) polymerase, and micrococcal nuclease are inhibited whereas ribonuclease H and deoxyribonuclease I (DNase I) are not inhibited. Thus, DNA in pulverized tissue fractionated in vanadyl complexes can be digested, without fear of degrading RNA. It is interesting to note that when 7 units of DNase I and 0.4 units of RNase A were incubated at 37° for 30 min at pH 4.5 in the presence of 10 mM vanadyl complexes in a 20-μl reaction, the DNA substrate was completely degraded while the 10 μg of crude tRNA remained intact. Apparently the 10 mM Mg^{2+} in the reaction was sufficient to activate DNase I without destroying the inhibitory function of the vanadyl complexes at the same concentration. Using RNase A with tRNA as the substrate, the function of the complexes can be tested in the presence of suspected interfering substances, such as Mg^{2+}, by visualizing the products in a 12% polyacrylamide gel (this volume [8]). If tRNA does not survive in the desired solution, a change in protocol is required.

[22] Isolation and Analysis of Nuclear RNA

By Joseph R. Nevins

Although in the majority of cases the isolation of total cellular RNA is sufficient for cloning purposes, it is nevertheless sometimes true that a subsequent analysis requires the preparation of RNA from subcellular compartments. In particular, the analysis of nuclear RNA is very often necessary for an understanding of either the transcription of the gene in question or the pathway of processing of the primary transcript. In this chapter, we detail the procedures used in our laboratory for the isolation of nuclear RNA from cultured cells. These procedures have been successful in the analysis of the splicing products of a late adenovirus nuclear RNA whose initial length was 27 kb.[1] In addition, we include procedures that are used for particular cell lines where certain differences may influ-

[1] J. R. Nevins, *J. Mol. Biol.* **130**, 493 (1979).

ence the results. Finally, we describe the methods for transcription rate analysis in isolated nuclei as an assay for the relative rates of *in vivo* transcriptional activity.

Solutions

PBS: $Na_2HPO_4 \cdot 7H_2O$, 2.16 g/liter; KH_2PO_4, 0.2 g/liter; NaCl, 8 g/liter; KCl, 0.2 g/liter

Modified RSB: 10 mM Tris, pH 7.9; 10 mM NaCl; 5 mM $MgCl_2$

NP-40 lysis buffer: 10 mM Tris, pH 7.9; 140 mM KCl; 5 mM $MgCl_2$; 1 mM dithiothreitol

HSB: 10 mM Tris, pH 7.4; 0.5 M NaCl; 50 mM $MgCl_2$; 2 mM $CaCl_2$

NET: 10 mM Tris, pH 7.4; 100 mM NaCl; 10 mM EDTA. For NETS buffer add sodium dodecyl sulfate (SDS) to 0.2%.

Extraction buffer: 0.01 M Tris, pH 7.4; 1.0% SDS; 20 mM EDTA

Ribonuclease (RNase): pancreatic RNase (10 mg/ml) and T1 RNase (5000 U/ml) dissolved in 50 mM sodium acetate (pH 5.1). Heat the solution to 80° for 10 min to destroy any deoxyribonuclease (DNase) activity. Store at 4°.

DNase I: Can be purchased RNase free. $CaCl_2$ at 5 mM can be added as a stabilizer to aqueous solutions. Residual ribonuclease, if present, can be inhibited with 10 mM ribonucleoside–vanadyl complexes[2] (this volume [21]). Store DNase frozen in small aliquots.

Reaction buffer: 20 mM Tris or HEPES, pH 7.9; 20% glycerol; 140 mM KCl; 10 mM $MgCl_2$; 1 mM dithiothreitol; 1 mM each ATP, CTP, GTP; 1–10 μM final concentration UTP; 250 μCi [^{32}P]UTP/ 100 μl reaction. Optional: 10 mM creatine phosphate and 20 U/ml (100 μg) creatine kinase added separately just before incubation.

Prehybridization buffer: 50 mM Tes or HEPES, pH 7.4; 0.3 M NaCl; 10 mM EDTA; 0.2% SDS; 1 mg/ml yeast RNA; 0.5 mg/ml poly(A); 1% sodium pyrophosphate; 5× Denhardt's solution[3] without BSA

Hybridization buffer: 50 mM Tes or HEPES, pH 7.4; 0.3 M NaCl; 10 mM EDTA; 0.2% SDS; 100 μg/ml yeast RNA; 100 μg/ml poly(A); 0.1% sodium pyrophosphate; 1× Denhardt's solution

Preparation of Nuclei

Cells are collected from the culture by centrifugation and then washed two to three times with chilled PBS. All subsequent procedures prior to

[2] R. S. Puskas, N. R. Manley, D. M. Wallace, and S. L. Berger, *Biochemistry* **21**, 4602 (1982).

[3] D. Denhardt, *Biochem. Biophys. Res. Commun.* **23**, 641 (1966).

phenol extraction are performed at 4°. In the various procedures described below, we give the standard conditions for the preparation of nuclei from cells grown in culture that have been used in our laboratory. These procedures were initially employed for HeLa cells and work well for many other cell types used in our laboratory. We also give variations of these standard procedures that have been developed for certain other cell types where the standard procedures were not optimal. Procedures for the preparation of nuclei from animal tissue are not discussed here but have been described elsewhere.[4]

Cells are resuspended in modified RSB at a concentration of $3-5 \times 10^7$ cells/ml and allowed to swell on ice for 5–10 min. They are then broken in a Dounce homogenizer with a tight pestle. Usually, 10–15 strokes are used and then KCl is added to give 100 mM. For Friend erythroleukemia cells, one stroke of the Dounce is used, KCl is added to 50 mM, and then 9–12 more strokes are given in the Dounce. Then KCl is added to 100 mM final concentration. For other cell types, including primary mouse hepatocytes, mouse myelomas, and a mouse B cell lymphoma, it has often been found useful to include detergent during the homogenization.[5] After resuspending cells in RSB, Triton X-100 is added to a final concentration of 0.1% (v/v). The cells are allowed to swell on ice for 5–10 min and then broken with 15–20 strokes. After homogenizing, the nuclei are collected by centrifugation at 1000 g for 3 min.

Alternatively, the nuclei can be liberated from the cells by lysis of the cytoplasmic membrane with the nonionic detergent NP-40. Washed cells are resuspended in NP-40 lysis buffer at a concentration of $3-5 \times 10^7$ cells/ml. For HeLa cells, NP-40 is add to a final concentration of 0.5% (v/v). For Friend cells, NP-40 is added to 0.2% (v/v) final concentration. The cells are gently mixed and kept on ice for 5 min. The optimal concentration of NP-40 for lysis of other cell types should be determined empirically. The nuclei are then recovered by centrifugation at 1000 g for 3 min. With any of the procedures for isolating nuclei, the efficiency of cell lysis and the quality of the nuclei should be checked by examination in a phase-contrast microscope.

Extraction of Nuclear RNA

Hot Phenol Extraction. This procedure is a modification of the original hot phenol extraction procedure[6] which gives good recovery of nuclear

[4] E. Derman, K. Krauter, L. Walling, C. Weinberger, M. Ray, and J. E. Darnell, *Cell* (*Cambridge, Mass.*) **23,** 731 (1981).
[5] D. F. Clayton and J. E. Darnell, *Mol. Cell. Biol.* **3,** 1552 (1983).
[6] R. Soeiro and J. E. Darnell, *J. Mol. Biol.* **44,** 551 (1969).

RNA relatively free of DNA. Nuclei prepared by one of the methods described above are washed once with RSB containing 100 mM KCl and are then resuspended in HSB containing DNase, at a concentration of 5 × 10^7 nuclei/ml. The nuclei will lyse when exposed to the high salt and as a result the viscosity of the solution will greatly increase due to the liberated chromatin. Resuspension is best accomplished by pipetting up and down in a Pasteur pipet. Warm the tube to room temperature in a beaker of water and continue pipetting. The viscosity should disappear within 30 sec. If it takes longer than this, it is essential to add more DNase to the HSB. A longer period of incubation will result in degradation of the nuclear RNA. The purpose of this digestion is not to degrade the DNA totally but rather to reduce it in size such that the solution is physically more manageable. The subsequent hot phenol extraction is efficient in removing the DNA fragments. When homogenization is complete, add an equal volume of extraction buffer which brings the solution to a final composition of 0.25 M NaCl, 0.5% SDS, and 10 mM EDTA. Add an equal volume of phenol–NETS (phenol saturated with NETS buffer) and an equal volume of chloroform. Heat to 65° in a water bath for 5–10 min with periodic shaking. Chill, and then centrifuge for 5 min at 1000 g. Remove the organic layer (bottom phase), leaving the aqueous phase and interphase in the tube, and repeat the phenol and chloroform extraction. Remove the organic phase and add 2–3 volumes of chloroform. Extract and centrifuge. The chloroform will shrink the interphase and remove phenol from the aqueous phase. Remove the aqueous phase to a clean tube and add 2.5 volumes of cold absolute ethanol to precipitate the RNA. The RNA precipitate is then recovered by centrifugation, and the pellet is washed once with 70% ethanol, dried, and dissolved in an appropriate buffer (this volume [4,5]). RNA prepared by this method is relatively free of contaminating DNA and is satisfactory for most purposes of further analysis such as Northern or S_1 analysis.[7] For filter hybridization of labeled RNA, additional DNase treatment is necessary to eliminate background. This procedure is described below.

Guanidinium Isothiocyanate Extraction. As an alternative to the phenol extraction procedure, nuclear RNA can be recovered by the guanidinium procedure usually used for the preparation of total cell RNA.[8] In some cases this may be preferable if nucleases are a severe problem. Basically, after the preparation of nuclei, one can proceed in the same manner as with whole cells. For details, see chapter by MacDonald *et al.* [20] in this volume.

[7] For nuclease-rich cells, these procedures are a starting point. Nuclease inhibitors (this volume [2,4,19,21]) may improve the quality of the isolated RNA.

[8] J. M. Chirgwin, A. E. Przybyla, R. J. MacDonald, and W. J. Rutter, *Biochemistry* **18**, 5294 (1979).

Transcription Rate Analysis in Isolated Nuclei

The analysis of RNA by hybridization with labeled DNA probes, either a Northern or an S_1 protection, is by definition an analysis of steady-state RNA and thus reflects the final accumulation of the transcript. A determination of transcription rate necessitates a brief pulse labeling of the RNA so as to nullify any contribution due to turnover of the RNA. The method of choice is an *in vivo* pulse label, usually with [³H]uridine, for a short period of time, no more than 5 min.[9] However, *in vivo* labeling is only practical when the transcription rate of the gene is sufficiently high such that meaningful incorporation can be obtained. For most genes, certainly any single-copy gene, this will not be the case. Furthermore, one is limited with this procedure to cells growing in culture. An alternative is the use of isolated nuclei, incubated *in vitro* with nucleoside triphosphates, to measure transcription. Under the *in vitro* conditions very little new initiation occurs but previously (*in vivo*) initiated transcripts do elongate.[10-12] Thus, the incorporation of label *in vitro* is believed to be an accurate measure of the number of active transcription complexes that existed prior to breaking the cells. Evaluation of data offered as proof is beyond the scope of this volume. The method has the advantage that a high specific activity pool of nucleotide precursor can be established. Furthermore, nuclei can be prepared from a variety of sources, including animal tissue, thus allowing transcription rate measurements not otherwise possible with metabolic labeling.

Preparation of Nuclei. Nuclei are prepared by one of the procedures described above and washed in reaction buffer minus triphosphates (20 mM Tris or HEPES, pH 7.9; 20% glycerol; 140 mM KCl; 10 mM MgCl$_2$; 1 mM dithiothreitol). The nuclei can then be used directly for transcription measurement or alternatively can be frozen and saved for subsequent analysis.

Freezing Nuclei for Subsequent Analysis. The washed nuclei are resuspended in a small polypropylene tube (or other tube that can be frozen) in 2× reaction buffer minus the triphosphates and creatine kinase. Use 50–100 μl of nuclei suspension per tube. Quickly freeze the nuclei in liquid nitrogen and store in a −70° freezer. Nuclei frozen in this manner can be kept a few months and retain near full transcription activity.

Labeling and Phenol Extraction. Nuclei, washed with reaction buffer, are resuspended in reaction buffer containing triphosphates and label, and creatine kinase is added to a final concentration of 20 U/ml. If frozen nuclei are to be assayed, thaw the nuclei slowly on ice and when thawed,

[9] J. R. Nevins and J. E. Darnell, *Cell (Cambridge, Mass.)* **15,** 1477 (1978).
[10] J. Weber, W. Jelinek, and J. E. Darnell, *Cell (Cambridge, Mass.)* **10,** 612 (1977).
[11] E. Hofer and J. E. Darnell, *Cell (Cambridge, Mass.)* **23,** 585 (1981).
[12] M. Groudine, M. Peretz, and H. Weintraub, *Mol. Cell. Biol.* **1,** 281 (1981).

add [^{32}P]UTP, cold triphosphates, creatine kinase, and water to make 1×
complete reaction buffer. Do this directly in the tube the nuclei were
frozen in. An example of quantities for nuclei frozen in 50 μl 2× reaction
buffer would be

Nuclei	50 μl
5′-[α-^{32}P]UTP (3000 Ci/mmol)	25 μl
10 mM ATP, CTP, GTP (10×)	10 μl
Creatinine kinase (5 mg/ml)	4 μl
Water	11 μl

Proceed directly with nuclear transcription assay as usual (see below).

Caution: The frozen and thawed nuclei may partially lyse during the
procedure but will still incorporate label. *Do not wash* the nuclei after
thawing; just proceed with the reaction in the tube they were frozen in.

The nuclei are incubated, with occasional mixing, for 10 min at 37° or
at 30° for 15 min. Typical reaction conditions are 2–10 × 10^7 nuclei/250
μCi [^{32}P]UTP per 100 μl reaction. The reaction is stopped by the addition
of 1–2 ml of HSB to lyse nuclei. The nuclear RNA is then extracted with
phenol as described above.

In order to remove the unincorporated label, either employ repeated
ethanol precipitations (this volume [5]) or a TCA precipitation step according to the procedure of Groudine *et al.*[12] as follows. Dissolve RNA in
100 μl of TE (NET minus the NaCl) and put on ice. Add 4 ml of cold 5%
TCA, 60 mM sodium pyrophosphate, and 75 μl of tRNA solution (3 mg/
ml). Mix and leave on ice for 30 min. Collect precipitate on a nitrocellulose
filter and rinse with 2–3 ml of TCA solution. Place filter in glass vial with
0.75 ml ETS (NETS minus the NaCl) and incubate at 65° for 10 min.
Count an aliquot of the solution in a scintillation counter and use the
remainder for hybridization. If a DNase step is used (see below), precipitate the RNA with ethanol. On average, 10–20% of the input label is
incorporated into RNA.

Preparation of RNA for Hybridization. It is usually advisable to digest
the extracted RNA with DNase prior to hybridization. RNA is dissolved
in 0.02 M Tris (pH 7.4), 1 mM MgCl$_2$ and DNase is added to a final
concentration of 50 μg/ml. Incubation is at 37° for 20 min. An equal
volume of extraction buffer is added and the sample is extracted with
phenol–chloroform as described before except that all extractions are at
room temperature. The final aqueous phase is made 0.2 M NaCl and the
RNA is ethanol precipitated.

RNA to be used for hybridization can be denatured and partially broken[13] which often improves the efficiency of hybridization. The RNA
solution is brought to 0.2 N NaOH and incubated on ice for 20–30 min.
The solution is then neutralized with HEPES buffer to pH 7.4.

[13] W. Jelinek, *J. Mol. Biol.* **82,** 361 (1974).

Hybridization of Labeled RNA to DNA on Nitrocellulose Filters.
Most commonly, the amount of labeled RNA produced in the isolated
nuclei reaction is quantified by hybridization to specific DNA sequences
immobilized on nitrocellulose filters. Individual filters bearing a single
probe can be prepared or an array of DNA probes can be bound to a single
filter. To a 25-mm² filter circle a maximum of 100 μg of DNA can be
bound. For dot hybridization, 5–8 μg of DNA can be bound to a 5-mm²
area using a minifold filtration apparatus. Alternatively, the DNA can be
loaded directly onto the filter with a capillary pipet.[14] Load the DNA at 1
μg/μl with a 5-μl pipet to a maximum of 3 μg/2-mm² dot. For this procedure, wet the nitrocellulose paper with water, then soak in 20× SSC, and
dry under lamp on 3 MM paper. Denatured DNA, neutralized with
HEPES to pH 7, is dotted onto the paper by capillary action and the filter
allowed to air dry, then rinsed in 20× SSC, and baked at 80° under
vacuum for 2–4 hr. For the manifold apparatus, the nitrocellulose filters
are prewashed with 6× SSC. Single-strand DNA such as an M13 cloned
fragment is loaded in 6× SSC directly to a prewashed filter. Double-
stranded linear DNA is denatured by boiling for 10 min in 0.1× SSC or by
incubation in 0.1 N NaOH for 15–20 min at room temperature. Double-
stranded circular DNA is denatured in 0.1 N NaOH by boiling for 6–8 min
and then neutralized with HEPES to pH 7. After denaturation, the DNA
is brought to an SSC concentration of 6× and then loaded onto the filter at
a flow rate of about 1 ml/min. Filters are washed with 6× SSC, dried, and
baked in a vacuum oven at 80° for 2–4 hr.

Filters are first incubated in the prehybridization buffer at 65° for 3–6
hr. This fluid is removed and replaced with hybridization buffer containing the labeled RNA. Hybridization is at 65° for 36–40 hr in a minimal
amount of hybridization buffer, preferably with agitation.

After the completion of hybridization, the fluid is removed and filters
are washed with agitation in 2× SSC at 65° for 30 min. This is repeated 2–
3 times. The filters are then digested with T1 RNase (5 U/ml) and pancreatic RNase (2.5 μg/ml) in 2× SSC at 37° for 30–60 min. After digestion,
the fluid is removed and filters are washed 2 times with 2× SSC at room
temperature. At this point a further digestion with proteinase K often
improves the background. The filters are incubated with proteinase K
(100 μg/ml) in 2× SSC containing 0.5% SDS at 37° for 30 min. The filter is
then washed again several times with 2× SSC, dried, and used for exposure of X-ray film.

Quantitation of the data can be obtained by cutting out the appropriate
areas of filters and counting directly in scintillation fluid. Alternatively,
when the counts of the hybridized material are too low to give reliable
numbers in the scintillation counter, autoradiographic films can be

[14] F. C. Kafatos, C. W. Jones, and A. Efstradiadis, *Nucleic Acids Res.* **7**, 1541 (1979).

scanned with a densitometer, first ensuring that the signal is within the linear range of the film. It may be necessary to include standards of a range of intensities and different film exposure times to obtain accurate quantitation of hybridizations that vary more than 10-fold in intensity. Finally, it is essential to be certain that the hybridizations are carried out in DNA excess. This can be checked by varying the ratio of input RNA to bound DNA and verifying that the response is linear.

Acknowledgments

I thank Marianne Salditt-Georgieff, who developed several of these procedures, for advice.

[23] Isolation of Messenger RNA from Membrane-Bound Polysomes

By BERNARD M. MECHLER

The proteins of animal cells are synthesized on polysomes that are either free in the cytoplasm or associated with the endoplasmic reticulum. Soluble cytosolic proteins are synthesized on free polysomes whereas secretory or integral membrane proteins are synthesized on membrane-bound polysomes. However, some mitochondrial and chloroplast membrane proteins are synthesized on free polysomes, released in the cytosol, and transported across the organelle membrane by a posttranslational mechanism. Membrane-bound polysomes are present in all nucleated cells except sperm and are especially abundant in cells engaged in protein secretion or extensive membrane synthesis. In such cases as many as half of the total ribosomes are bound to the endoplasmic reticulum. However, in most cells, particularly in cells maintained in culture and actively dividing, the membrane-bound polysomes represent only 5–15% of the total cell ribosomal population. Because the membrane-bound polysomes synthesize specific proteins and contain the corresponding mRNAs, their purification can be a useful enrichment step in the cloning of these sequences and in their further identification and characterization. This method has been recently used to achieve the cloning of the mouse T-lymphocyte antigen receptor.[1]

For good recovery of membrane-bound polysomes, it is essential to arrest protein synthesis in the tissues or cells prior to their homogenization. This is particularly important for cells maintained in liquid culture.

[1] S. M. Hedrick, D. I. Cohen, E. A. Nielsen, and M. M. Davis, *Nature (London)* **308**, 149 (1984).

For relatively small volumes of growth medium (up to about 100 ml), inhibition of protein synthesis can be promptly achieved by diluting the cells in 4 volumes of ice-cold isotonic saline solution. However, with larger volumes a rapid cooling is more difficult to achieve. Partial inhibition of protein synthesis occurs at the level of initiation. This leads to polysome runoff and release of membrane-bound polysomal mRNAs as free mRNP particles. Complete polysome recovery can be achieved by first concentrating the cells in a smaller volume of fresh prewarmed growth medium and then allowing protein synthesis to proceed for a further 30–45 min.[2] The cells are then rapidly cooled, sedimented, and homogenized.

A good recovery of membrane-bound polysomes requires a procedure of cell fractionation which should satisfy the following criteria: (1) high recovery of cytoplasmic polysomes, (2) preparation of a membrane fraction devoid of contaminating free ribosomes, and (3) preservation of the integrity of the polysomal structure. This chapter describes a method for rapidly separating the free and membrane-bound ribosomes of the cytoplasm that fulfills these criteria.[3,4]

Principle of the Method

Membrane-bound cell organelles, including the membrane-bound polysomes, are separated by isopycnic centrifugation in a discontinuous sucrose density gradient which causes the flotation of the membrane vesicles and the partial sedimentation of the free ribosomal particles as illustrated in Fig. 1. This is achieved by adjusting the cytoplasmic extract to a concentration of 2.1 M sucrose which is then loaded over a layer of 2.5 M sucrose in a centrifuge tube. Two successive layers of sucrose solutions, one with 2.05 M sucrose and the second with 1.3 M sucrose, are then layered over the sample. During the centrifugation, all the membrane-containing cell organelles float above the 2.05 M sucrose layer due to the low density of the membranes, whereas the free polysomes, ribosomes, and mRNP particles sediment due to their high density.

Nuclease Precautions

In order to minimize exogenous RNase contamination during the isolation of membrane-bound polysomes and subsequent purification of

[2] B. Mechler, *J. Cell Biol.* **88,** 42 (1981).
[3] B. Mechler and P. Vassalli, *J. Cell Biol.* **67,** 1 (1975).
[4] B. Mechler and T. H. Rabbitts, *J. Cell Biol.* **88,** 29 (1981).

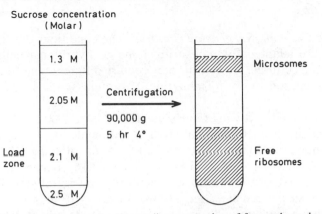

FIG. 1. Diagrammatic outline for the gradient separation of free and membrane-bound ribosomes. The cytoplasmic extract is adjusted to 2.1 M sucrose and loaded at the bottom of the centrifuge tube onto a 2.5 M sucrose layer and covered by layers of 2.05 and 1.3 M sucrose. At the end of the centrifugation the microsomes float at the interface between the 2.05 M and 1.3 M sucrose layers, whereas the free ribosomal particles remain in the 2.1 M sucrose layer.

mRNA, all glassware is washed in chromosulfuric acid, then extensively rinsed in deionized water, and finally baked in an oven at 160° for at least 2 hr. All solutions are treated with 50 μl of diethyl pyrocarbonate per 100 ml of solution. Following vigorous shaking, the solutions are placed in a boiling water bath for 30 min to decompose the remaining diethyl pyrocarbonate. The solutions are shaken while hot to remove any released CO_2 and ethanol.

Endogenous RNase activity varies considerably among different tissues and cells. Mouse myeloma cells, for example the MOPC21-P3K cell line, or hybridomas are almost completely devoid of endogenous RNase. This is also the case with small solid myeloma tumors (MOPC41, 46B, and 315) provided that they are freed of necrotic tissue. Other tissues, for example primary chicken embryonic fibroblasts and *Drosophila* KC or Schneider's L-2 cells, contain endogenous RNase activity which necessitates the use of RNase inhibitors such as heparin, bentonite, polyvinyl sulfate, rat liver supernatant, yeast RNA, or vanadyl–ribonucleoside complexes in order to prevent polysome degradation. For a comprehensive discussion of most of these inhibitors, and others, consult Berger and Birkenmeier.[5]

[5] S. L. Berger and C. S. Birkenmeier, *Biochemistry* **18**, 5143 (1979).

Cell Harvest

Cells Grown in Suspension Cultures. The conditions described in this section are adapted for mouse myeloma or hybridoma cells grown in suspension cultures. Other cells may require different conditions depending on their sensitivity to inhibition of protein synthesis at a high cell density.

Exponentially growing mouse myeloma cells, 1000 ml, from cultures not exceeding 5×10^5 cells/ml are first sedimented in 4×250 ml sterile conical polypropylene tubes (Corning) for 5 min at 300 g (1000 rpm) in a DPR-6000 (IEC-Damon), TGA-6 (Kontron), J-6 (Beckman), Cryofuge 5000 (Heraeus), or RC-3B (Sorvall) centrifuge refrigerated at 4°. The cells are then resuspended at a concentration of 2.5×10^6 cells/ml in 200 ml 37° fresh growth medium and further incubated for 45 min at 37°.

After incubation, the cells are diluted with 800 ml of ice-cold phosphate-buffered saline solution (according to Dulbecco, per 1000 ml: 8.0 g NaCl, 0.2 g KCl, 1.15 g Na_2HPO_4, 0.2 g KH_2PO_4, 0.132 g $CaCl_2 \cdot 2H_2O$, and 0.1 g $MgCl_2 \cdot 6H_2O$) and centrifuged at 4° for 5 min at 300 g (1000 rpm). The cells are then washed twice in 400 ml of ice-cold phosphate-buffered saline solution, resuspended at a concentration of 2.5×10^8 cells/ml in ice-cold hypotonic buffer medium RSB (10 mM KCl, 1.5 mM $MgCl_2$, and 10 mM Tris–HCl at pH 7.4), and transferred into a 7-ml Dounce homogenizer (Kontes Glass Co., Vineland, NJ) on ice. If nuclease inhibitors are necessary, the cell suspension can be made 10 mM in vanadyl complexes[5] or 0.2 to 2 mg/ml in heparin before lysis (see this volume [21] and [24]).

Cells Grown Attached to Plastic Petri Dishes. At the end of the growth period, the growth medium is removed from the plates and the plates are immediately placed on crushed ice. The cells are washed twice with semifrozen phosphate-buffered saline solution and scraped from the dish with a rubber policeman into a 50-ml conical polypropylene centrifuge tube. The cells are then sedimented at 300 g (1000 rpm) for 5 min at 4° in a refrigerated centrifuge and resuspended at a concentration of 2.5×10^8 cells/ml in ice-cold hypotonic buffer medium RSB with the optional ribonuclease inhibitors.

Cell Homogenization

The cells are allowed to swell for 5 min in ice-cold hypotonic buffer medium RSB and are ruptured mechanically with 10 strokes of a 7-ml tight-fitting (B) Dounce glass homogenizer. Cell lysis can be monitored by diluting 1 μl of the homogenate in 100 μl 0.05% trypan blue in saline solution (130 mM NaCl, 5 mM KCl, 7.5 mM MgCl$_2$) and examining the cells under the phase-contrast microscope. The intact cells will exclude the dye.

The homogenate can be directly used for the separation of free and membrane-bound ribosomes, but, if the nuclei are not dense enough to sediment through a 2.5 M sucrose solution, it will be necessary to eliminate the nuclei first by centrifuging the homogenate at 2000 g (3000 rpm) for 2 min at 4°. The supernatant which contains the cytoplasmic extract is retrieved. The nuclear pellet is resuspended with RSB in the same volume as the homogenate and constitutes the nuclear fraction. The cytoplasmic extract is used for the separation of free ribosomal and microsomal fractions. The nuclear fraction can be used to obtain a supplementary microsomal fraction which is pooled with the microsomal fraction prepared from the cytoplasmic extract.

Separation of Free Ribosomal and Microsomal Fractions

The cytoplasmic extract or the nuclear fraction (2 ml each for 5×10^8 myeloma cells) is diluted with 11 ml of 2.5 M sucrose TK$_{150}$M (150 mM KCl, 5 mM MgCl$_2$, 50 mM Tris–HCl at pH 7.4) until the sucrose concentration reaches 2.1 M and is layered onto 4 ml of 2.5 M sucrose TK$_{150}$M in a 38-ml centrifuge polyallomer tube. Two layers of sucrose TK$_{150}$M solution are successively added, one of 13 ml 2.05 M sucrose and a second of 6 ml 1.3 M sucrose. The gradients are then centrifuged for 5 hr at 4° in a Spinco SW27.1 rotor (Beckman Instruments, Inc.) or any other equivalent rotor at 90,000 g (25,000 rpm). After centrifugation, the bottom of the tube is punctured with an 18-gauge needle, 30-drop fractions are collected into 1.5-ml microcentrifuge polypropylene tubes, and the absorbance at 260 nm is measured. Under these conditions, the free ribosomes are present in the load zone whereas the nuclei of the myeloma cells sediment to the bottom of the tube into the 2.5 M sucrose layer. The microsomes are located at the interface between the 2.05 and 1.3 M sucrose layers. When only the microsomal fraction is needed for further investigations, the membrane layer can be collected directly from the top of the gradient with a siliconized Pasteur pipet.

The microsomal fraction contains membrane-bound polysomes and some 80 S ribosomes and 60 S subunits as well as mitochondrial ribosomes. When total polysomal mRNA is required, it is not necessary to purify the membrane-bound polysomes further and the RNA can be directly extracted after diluting the microsomal fraction with an equal volume of TE (10 mM Tris–HCl at pH 7.4, 1 mM EDTA).

The free ribosomal fraction contains the free polysomes, the inactive

80 S ribosomes, and the recycling ribosomal subunits as well as untranslated mRNP particles. When no further purification is required, the total RNA of this fraction can be extracted at this step after addition of two volumes of TE.

Extraction of RNA from Free Ribosomal and Microsomal Fractions

After dilution of the fractions with TE, they are transferred to a large tube and an equal volume of hot SDS buffer (1% SDS, 200 mM NaCl, 20 mM Tris–HCl at pH 7.4, 40 mM EDTA) is added and the tube is incubated in a boiling water bath. After 120 sec, the tube is plunged into ice water and the contents cooled to approximately 30°, after which the contents are digested with proteinase K (50 μg/ml) for 15 min at 37°. Proteinase K is used because it retains its proteolytic activity in the presence of SDS. Then the digest is adjusted to 0.1 M in Tris–HCl (pH 9.0) and 1% SDS and extracted three times at room temperature with an equal volume of phenol–chloroform (1 : 1) saturated with 0.1 M Tris–HCl (pH 9.0). The RNA is precipitated from the aqueous phase by addition of 0.1 volume of 2 M sodium acetate (pH 5.2) and 2 volumes of cold ethanol, and recovered after overnight storage at −20° by centrifugation for 10 min at 12,000 g (10,000 rpm) in a Sorvall SS-34 rotor (see this volume [4,5]). After the precipitate is dissolved in distilled water, poly(A)$^+$ RNA can be isolated by affinity chromatography (see this volume [25]).

Polysome Analysis by Sucrose Gradient Centrifugation

The sedimentation characteristics of the ribosomal particles in the free ribosomal and microsomal fractions can be analyzed on 15–55% linear sucrose gradients in TK$_{80}$M (80 mM KCl, 5 mM MgCl$_2$, 50 mM Tris–HCl at pH 7.4). The membranes are dissolved by adding sodium deoxycholate and Brij 58 (or Nonidet P-40), each to a final concentration of 0.5% to the microsomal fraction. The detergent solutions are freshly made as 10% (w/v) stock solutions. The detergent-treated fraction is diluted with two volumes of TK$_{80}$M buffer to decrease the sucrose concentration to less than 15%. The free ribosome fraction is similarly diluted with 4 volumes of TK$_{80}$M buffer, and 4 ml of each ribosomal fraction is loaded on top of a 34-ml 15–55% linear sucrose density gradient made up in TK$_{80}$M. Centrifugation is carried out for 8.5 hr at 4° in a Spinco SW27 rotor at 95,000 g (23,000 rpm). The gradients are collected in fractions of 1 ml by aspiration from the bottom of the tubes with a peristaltic pump and the absorbance at 260 nm continuously monitored with a Gilford spectrophotometer.

Fractions from the sucrose gradients are then diluted with 0.5 volume of TE and the RNA is extracted at room temperature with 0.1% SDS and an equal volume of phenol–chloroform (1 : 1). RNA is precipitated from the aqueous phase by the addition of 0.1 volume of 2 M sodium acetate (pH 5.2) and 2 volumes of cold ethanol, and recovered after overnight storage at $-20°$ by centrifugation for 20 min at 4000 g (6,000 rpm) in a Sorvall SM-24 rotor. After the precipitate is dissolved in distilled water, RNA can be either translated in a messenger-dependent reticulocyte lysate as described by Pelham and Jackson (see this volume [27]) or immobilized on nitrocellulose filters and hybridized to a specific probe (see this volume [61]).

Distribution of RNA in Mouse Myeloma Cells

On the basis of the separation technique described in this chapter and the subsequent analysis of the sedimentation characteristics of the free and membrane-bound ribosomal particles, the following estimates of RNA in free and membrane-bound ribosomal fractions were obtained.[4] The total cytoplasmic RNA content was as high as 25 pg per cell, of which 14 pg were contained in free polysomes and 3.5 pg in membrane-bound polysomes. The poly(A)$^+$ RNA content of the free and membrane-bound polysomal fractions was 0.18 pg per cell or about 1%. On the basis of the template activity in a messenger-dependent reticulocyte lysate system the poly(A)$^-$ mRNA was estimated to be about 0.05 pg and was essentially present in free polysomes.

Recovery and Purity of the Fractions

For cultured cells, the fractionation procedure allows the recovery of 95% of the ribosomes present in mouse myeloma cytoplasm and gives a yield of membrane-bound polysomes similar to that of free polysomes. The purity of both fractions was high as 98% in several controls as determined by labeling of the membrane lipids with [^{3}H]oleic acid, addition of [^{3}H]adenosine-labeled free polysomes, or [^{3}H]adenosine-labeled microsomes to unlabeled cell extracts followed by an analysis of the radioactivity distribution in the resulting free ribosomal and microsomal fractions.

Separation of Free Ribosomal and Microsomal Fractions from Solid Tissues

This method can also be adapted to the isolation of membrane-bound polysomes from tissues. This section describes briefly the adaptations made for the homogenization of mouse myeloma tumors or mouse liver, the preparation of the cytoplasmic extract, and the loading of the discontinuous sucrose density gradient.

Freshly excised mouse myeloma tumors or mouse liver are rinsed in ice-cold saline and minced thoroughly at 0–4°. Connective tissues and necrotic portions of the tumors are discarded. Two and a half volumes of 0.8 M sucrose in $TK_{150}M$ containing 6 mM 2-mercaptoethanol are added and the tissues are gently homogenized first in a Teflon-glass Potter homogenizer with five strokes at 100–200 rpm followed by six strokes in a 15-ml Dounce homogenizer. The homogenate is then transferred to a 30-ml sterile Corex glass centrifuge tube and centrifuged at 12,000 g (10,000 rpm) for 15 min at 4° in a SS-34 Sorvall rotor to sediment the nuclei. The supernatant or cytoplasmic extract is retrieved. For the separation of free ribosomal and microsomal fractions, 2 ml of the cytoplasmic extract is mixed with 6 ml of 2.5 M sucrose $TK_{150}M$ and layered over 4 ml of 2.5 M sucrose $TK_{150}M$ per 38-ml polyallomer tube. Two layers of 2.05 M and 1.3 M sucrose $TK_{150}M$ are successively added and the separation of the free ribosomal and microsomal fractions is achieved as previously described.

Acknowledgments

This work was supported by the Swiss National Science Foundation, Grant 831.186.73 and the Deutsche Forschungsgemeinschaft, SFB 302-B2 and Me 800-1/1.

[24] Use of Antibodies to Obtain Specific Polysomes

By DENNIS C. LYNCH

In principle, immunoisolation of polysomes is a general technique to enrich for specific mRNAs. The mRNA may then be reverse transcribed to generate probes for cDNA libraries, or cDNAs may be generated directly from the enriched mRNA. In the initial applications of this approach, immunoprecipitation was employed to isolate polysomes whose nascent peptides represented a significant portion of cellular protein synthesis.[1,2] Immunoprecipitation did not prove to be as successful when used for proteins of lower abundance. An important advance in the technique was described by Shapiro and Young,[3] who employed a column of

[1] R. Palacios, R. D. Palmiter, and R. T. Schimke, J. Biol. Chem. 247, 2316 (1972).
[2] I. Schecter, Proc. Natl. Acad. Sci. U.S.A. 70, 2256 (1973).
[3] S. Z. Shapiro and J. R. Young, J. Biol. Chem. 256, 1495 (1981).

protein A-Sepharose to remove the antibody–antigen complex from solution. This avoided the nonspecific trapping of other polysomes in the precipitate and allowed for superior washing of the specifically bound polysomes. The original report described isolation of mRNA for an abundantly synthesized protein (~7–10% of total mRNA), but Korman et al.[4] and Kraus and Rosenberg[5] soon thereafter used protein A-Sepharose columns to immunoisolate successfully low abundance mRNAs (0.01–0.05% of total mRNA).

General Considerations

There are practical and theoretical limitations to this technique which must be considered before it should be applied in a given project. In particular, because the method relies on the physical separation of polyribosomes based on their large size, it is not likely to be useful for isolating mRNAs coding for very small proteins. There are no data which indicate how small is too small, but peptides of 30 amino acids, for example, are not likely to have enough ribosomes attached to their mRNA to allow sedimentation into the proper fraction. In light of the extreme sensitivity of mRNA to ribonucleases, it is somewhat surprising that polysomes can be isolated at all. Other than the addition of heparin (at best, a moderately effective inhibitor of ribonuclease), there is no reasonable means of inhibiting endogenous nucleases in the polysome preparations. It seems likely, then, that the nascent polypeptides themselves must provide protection for the mRNA which binds the ribosomes together. Such protection must inevitably vary from one protein to another, suggesting that the technique may work for some mRNAs from a given tissue and not for others. In this regard, smaller proteins might provide less protection than larger proteins. In addition, mRNAs with long 3' untranslated regions could be susceptible to nuclease cleavage in this region, removing the poly(A) tail and thereby preventing their subsequent binding to oligo(dT)-cellulose. Finally, it might prove difficult to immunoisolate polysomes from tissues with a high endogenous ribonuclease level (e.g., pancreas).

One must also consider if the necessary resources and expertise are available. Clearly, a high-affinity specific antibody is necessary for this technique to be successfully applied. Both polyclonal and monoclonal antibodies have been shown to be useful, but it is desirable to demonstrate that the particular antibody preparation is capable of isolating nascent

[4] A. J. Korman, P. J. Knudsen, J. F. Kaufman, and J. L. Strominger, Proc. Natl. Acad. Sci. U.S.A. 79, 1844 (1982).

[5] J. P. Kraus and L. E. Rosenberg, Proc. Natl. Acad. Sci. U.S.A. 79, 4015 (1982).

polypeptide chains before attempting to isolate polysomes. One major attraction of this technique is that it allows the antibody-based selection of clones to be applied directly to genomic libraries and/or to preexisting, nonexpression cDNA libraries. If a cDNA library is already available in a vector with a high efficiency of expression (e.g., λgt11) it would probably be easier to screen such a library directly with the available antibody rather than to generate a probe via immunoisolation of polysomes.

It is also necessary to have an adequate source of cellular material from which to prepare polysomes. The difficulty in meeting this criterion varies with the source of the material and the abundance of the mRNA. Proteins synthesized in organs that may be obtained from laboratory animals or in readily cultured cells represent the best targets for this approach. Difficulties in generating sufficient polysomes may be encountered if the site of synthesis is in cells that are difficult to culture, or if working with human materials from autopsy specimens where samples may not be made available rapidly enough to prevent autolysis. However, the effort required to prepare polysomes from difficult sources may be justified in some instances by the fact that it is possible to use the same polysome preparation for the sequential enrichment of two or more mRNAs.[4,5] Finally, it is necessary to consider the adequacy of the polysome source in relation to the proposed use of the immunoisolated mRNA. To make an adequately specific probe to obtain an initial positive selection from a cDNA or genomic library requires substantially less mRNA than would be required to generate clones directly from the enriched sample. In addition, a probe would still be useful if only a portion of the mRNA is represented (e.g., only the 3' end), whereas intact mRNA would be required to obtain full-length clones from a library made with immunoisolated mRNA. Thus, it may be feasible to obtain adequate material from some sources (e.g., a fresh surgical sample of human liver) to make a probe which could then be used to screen a preexisting cDNA library, while it might be impractical to attempt to clone an mRNA directly from such a sample.

Preparation of the Antibody

Since, as noted above, the immunobinding step is carried out in the presence of a crude subcellular fraction enriched for polysomes, it is clearly impossible to make the system "ribonuclease free." However, those who have used this method have purified their antibody preparations in order to reduce the level of ribonuclease introduced.

A column of protein A-Sepharose (Sigma or Pharmacia) is prepared and washed with 10 column volumes of sterile phosphate-buffered saline

(20 mM sodium phosphate at pH 7.4, 150 mM NaCl). The size of the column is determined by the amount of antibody to be purified and the binding capacity of the given lot of protein A-Sepharose. The antibody preparation is diluted to ~2.5 mg protein per milliliter in phosphate-buffered saline and passed twice over the column. If the antibody employed does not bind directly to protein A, it should first be incubated with the appropriate rabbit antiserum. For example, a mouse monoclonal IgG which does not bind to protein A could be incubated with an equal amount of rabbit anti-mouse IgG (Cappel) for 4 hr at 4° in phosphate-buffered saline and then applied to the protein A-Sepharose column. The column is then washed with 10 volumes of sterile phosphate-buffered saline and the antibody fraction eluted with 0.1 M glycine (pH 3). The eluate is promptly neutralized with 0.1 volume of 1.0 M Tris. The purified antibodies may be used in this buffer[6] or dialyzed against phosphate-buffered saline.[3-5] In either case, the antibody fraction should be distributed into aliquots and stored at −70° and, prior to use, passed through a 0.2-μm filter to remove aggregates and possible bacterial contamination. The protein A-Sepharose should be reequilibrated with phosphate-buffered saline, stored at 4°, and may be reused to bind immunoisolated polysomes at a later step. Sepharose should not be stored frozen.

Preparation of the Polysome Fraction

The optimal method to lyse cells or homogenize tissues for the purpose of obtaining a postnuclear supernatant will vary widely with the tissue or cells being used. However, in all methods employed, the lysis buffer should contain 5 mM MgCl$_2$, cycloheximide (Sigma) at 1 μg/ml and heparin (Sigma) at 0.2 mg/ml. If tissue culture cells are being used, it is possible to preincubate them with cycloheximide for 30–60 min. This treatment may increase polysome yield somewhat, but is probably not necessary. Cells may be disrupted by any method (hypotonic lysis, freeze–thaw, mild detergent lysis, etc.) which will allow separation of nuclei from cytoplasm without significant nuclear lysis. Solubilization of nuclear DNA would likely increase the viscosity of the ultimate polysome preparation and, thereby, make it more difficult to achieve a clean, specific immunoisolation. The lysate should be centrifuged at 10,000 g for 10 min to remove nuclei and debris and the supernatant made 1% in Nonidet P-40 and 0.5% in sodium deoxycholate to disrupt microsomal membranes. This material is then sedimented into a 65% sucrose cushion by centrifu-

---

[6] D. C. Lynch, T. S. Zimmerman, C. J. Collins, M. Brown, M. J. Morin, E. H. Ling, and D. M. Livingston, *Cell (Cambridge, Mass.)* **41**, 49 (1985).

gation at 4° (~200,000 g for 2 hr or its equivalent). It is convenient to place an intermediate layer of 30% sucrose between the cushion and the loaded sample to facilitate removal of the cushion and pelleted material at the end of centrifugation. Both sucrose solutions are made up in standard polysome buffer (25 mM Tris at pH 7.5, 150 mM NaCl, 5 mM MgCl$_2$, cycloheximide at 1 μg/ml, and heparin at 1 mg/ml). For example, tubes for the SW50.1 rotor could be loaded with 1 ml of 65% sucrose, 0.5 ml of 30% sucrose, and 3.7 ml of lysate, and then centrifuged at 50,000 rpm for 2 hr. After centrifugation, the overlayered material and 30% sucrose are removed and the polysomes resuspended in the 65% sucrose cushion. The tube is then rinsed with standard polysome buffer and the polysomes diluted to a total of five times the volume of the cushion with standard polysome buffer. Alternatively, the undiluted pellet and sucrose cushion may be dialyzed against standard polysome buffer to remove sucrose. Measure the optical density to determine the yield of polysomes at this point. The polysomes can then be frozen at −70° for many months or used directly.

Immunoisolation of Polysomes

Appropriate amounts of antibody and polysomes are mixed and incubated at 4° for 4–16 hr as necessary. The ratio between antibody and polysomes that is optimal for each system must be estimated by the investigator. Assuming all antibodies are of equally high titer, the amount of antibody required is clearly proportional to the prevalence of the antigen. Thus, proteins of high prevalence will require more antibody than those of low prevalence. Others have used ratios of 1 mg purified polyclonal antibody to 160 A_{260} units of polysomes[5] and 0.5 mg of purified monoclonal antibody per gram of tissue culture cells.[4] At the end of the incubation, the solution is diluted with an equal volume of standard polysome buffer (to lower the viscosity and improve the flow rate) and passed over a column of cyanogen bromide-activated Sepharose (Pharmacia) which has been "quenched" with glycine as described by the supplier.[6] The column is equilibrated with standard polysome buffer and is the same size as the protein A-Sepharose column which follows. The purpose of this step is to remove any polysome material which might nonspecifically bind to the succeeding protein A-Sepharose column. The flow-through and first several milliliters of rinse from the preadsorption column are loaded directly onto a protein A-Sepharose column. This column should be made "ribonuclease free" by washing with 10 volumes of 1 M acetic acid, 10 volumes of 0.1 M sodium phosphate (pH 7.5), and then equilibrated with 10 volumes of standard polysome buffer. The minimum column volume needed may be calculated from the amount of antibody used in the incubation. The flow-through is reapplied to the protein A-Sepharose column, which is then washed with 20 volumes of standard polysome buffer. The flow-through fractions may be frozen at −70° for future isolation of other specific polysomes.

Isolation of mRNA

The polysomes bound to the protein A-Sepharose are dissociated by washing the column with two volumes of 25 mM Tris (pH 7.5), 20 mM EDTA, and 0.2 mg/ml heparin. The eluate is then adjusted to 0.5 M LiCl and 0.5% in sodium dodecyl sulfate and applied twice to an oligo(dT)-cellulose column, prepared in the usual fashion (see this volume [25]). This column does not need to be very large (~0.2 ml), because it is unlikely that measurable amounts of specific mRNA will be present. The oligo(dT) column is then washed with ~50 volumes of 0.5 M LiCl, 1 mM EDTA, 10 mM Tris (pH 7.5), 0.5% sodium dodecyl sulfate and allowed to drain as completely as possible. mRNA is then eluted with a small amount of water (0.5 ml for a 0.2-ml column) and ethanol precipitated (see this volume [5]). The material obtained can be characterized in several ways depending on the amount available and the intended use. Previous workers have generally utilized *in vitro* translation to assess purity and identity[3-5] (see this volume [27]). In our experiments,[6] which involved very limited amounts of starting material, the isolated mRNA was reverse transcribed to a high specific activity cDNA (see this volume [33]) and used as a probe without specific characterization.

[*Editors' Note*. Methods for maximizing polysomes, particularly membrane-bound polysomes, while minimizing degradation of RNA by nucleases have not been adequately studied. Intact RNA can be isolated from polysomes prepared in sucrose gradients containing ribonucleoside–vanadyl complexes, provided cells are lysed in the presence of Nonidet P-40 (or Triton N-101) and these complexes. Heparin is ineffective (see this volume [21]). Since the compatibility of vanadyl complexes with many detergents, such as the deoxycholate used here, has not been determined, investigators seeking intact RNA might consider omitting deoxycholate from the above procedure. Concomitantly, a concentration of 10 mM vanadyl complexes (rather than heparin) should be maintained in cell solutions containing polysomes. This approach is a suggestion only; refinements may be required to make the method workable.]

[25] Purification and Fractionation of Poly(A)$^+$ RNA

By ALLAN JACOBSON

Posttranscriptional polyadenylation is a common feature of the biogenesis of most eukaryotic mRNAs. In general, newly synthesized mRNAs have long poly(A) tracts which shorten, as mRNAs age, to steady-state lengths of 40–65 adenylate residues.[1-6] In mammals, the initial size of the poly(A) tract is approximately 200 nucleotides,[1-3] whereas in lower eukaryotes its initial size is considerably shorter.[5,6] There is no significant accumulation of mRNA with poly(A) tracts shorter than steady-state length, either because that length is maintained by cytoplasmic readenylation[7] or because mRNAs with poly(A) tails shortened below 40–65 residues have increased rates of degradation.[8] These properties have made it possible (1) to purify mRNA by virtue of its base pairing with immobilized oligo(dT)[9,10] or poly(U),[11] (2) to recognize a small, but distinct subclass of mRNA which lacks poly(A),[12,13] and (3) to separate functionally different mRNAs by virtue of differences in their poly(A) tract length.[5,6,14] Methods to accomplish these ends constitute the scope of this chapter.

Purification of Poly(A)$^+$ RNA

This section addresses the use of oligo(dT)-cellulose and poly(U)-Sepharose for the purification of poly(A)$^+$ RNA. Other methods, such as

[1] J. R. Greenberg and R. P. Perry, *Biochim. Biophys. Acta* **287**, 361 (1972).

[2] D. Sheiness and J. E. Darnell, *Nature (London), New Biol.* **241**, 265 (1973).

[3] W. R. Jeffery and G. Brawerman, *Biochemistry* **13**, 4633 (1974).

[4] D. S. Adams and W. R. Jeffery, *Biochemistry* **17**, 4519 (1978).

[5] C. M. Palatnik, R. V. Storti, and A. Jacobson, *J. Mol. Biol.* **128**, 371 (1979).

[6] C. M. Palatnik, R. V. Storti, A. K. Capone, and A. Jacobson, *J. Mol. Biol.* **141**, 99 (1980).

[7] G. E. Brawerman and J. Diez, *Cell (Cambridge, Mass.)* **5**, 271 (1975).

[8] U. Nudel, H. Soreq, U. Z. Littauer, G. Marbaix, G. Huez, M. Leclerq, E. Hubert, and H. Chantrenne, *Eur. J. Biochem.* **64**, 115 (1979).

[9] M. Edmonds, M. H. Vaughn, and H. Nakazoto, *Proc. Natl. Acad. Sci. U.S.A.* **68**, 1336 (1971).

[10] H. Aviv and P. Leder, *Proc. Natl. Acad. Sci. U.S.A.* **69**, 1408 (1972).

[11] U. Lindberg and T. Persson, this series, Vol. 34, p. 496.

[12] M. Adesnik, M. Salditt, W. Thomas, and J. E. Darnell, *J. Mol. Biol.* **71**, 21 (1972).

[13] J. R. Greenberg and R. P. Perry, *J. Mol. Biol.* **72**, 91 (1972).

[14] C. M. Palatnik, R. V. Storti, and A. Jacobson, *J. Mol. Biol.* **150**, 389 (1981).

adsorption to and elution from poly(U) filters[15] or nitrocellulose membrane filters,[16] can be used, but they will not be discussed here. Although it is feasible and straightforward to prepare chromatography resins with immobilized oligo(dT)[17] or poly(U),[11] it is assumed that the reader will take advantage of the readily available and inexpensive commercial products.

The following comments pertain to the use of either resin. (1) It is assumed that the starting material will be ethanol-precipitated nucleic acids isolated from cells, tissues, or subcellular fractions by phenol–chloroform extraction or other methods (this volume [4,5,20–22]). Use of crude cellular lysates is not recommended since proteins will remain attached to the mRNA and, in the case of poly(U)-Sepharose, there is a reasonable probability that nucleases in the lysate will degrade the poly(U). (2) All solutions should be sterile (see this volume [2]) and, with the exception of sample preparation, all procedures are carried out at room temperature. (3) Fractions of the column washes should be collected and monitored for A_{260} or, if the sample has been radioisotopically labeled, for total cpm. (4) A second passage through either column is recommended because a single passage yields mRNA that may be only 50% pure. The yield of mRNA should approximate 2–5% of the input, unfractionated RNA. Higher yields indicate contamination with non-poly(A)-containing RNAs, in particular rRNA. Presence of the latter can be readily confirmed either by gel electrophoresis (this volume [8]) or by direct measurements (this volume [26]).

Comparison of Oligo(dT)-Cellulose and Poly(U)-Sepharose

The two resins have their respective advantages and disadvantages. Oligo(dT)-cellulose has a higher binding capacity per gram and a noncollapsible matrix. This permits the use of small columns and small-volume batch techniques for binding, washing, and elution. The major disadvantage of oligo(dT)-cellulose is the short length of oligomers generally used to prepare the resin (usually a maximum of 18–30 dT residues). This leads to inefficient binding of mRNA molecules with relatively short poly(A) tails and, hence, contamination of the nonbound fractions [putative poly(A)$^-$ mRNA] with some true poly(A)$^+$ mRNAs. This resin is also disqualified from fractionation procedures designed to take advantage of length differences in the poly(A) tracts of mRNA (see below) because the oligo(dT) tail is too short. Poly(U)-Sepharose suffers from a lower binding

[15] A. Jacobson, *Methods Mol. Biol.* **8**, 161 (1976).
[16] J. Mendecki, S. Y. Lee, and G. Brawerman, *Biochemistry* **11**, 792 (1972).
[17] P. T. Gilham, *J. Am. Chem. Soc.* **86**, 4982 (1964).

capacity per gram and a collapsible matrix. This means that larger columns (and hence larger volumes) must be used, that more time is required to purify a given amount of mRNA, and that batch procedures which make use of centrifugation are precluded. On the other hand, poly(U)-Sepharose has molecules of poly(U) which are approximately 100 nucleotides long, providing more efficient binding of poly(A)$^+$ RNA than oligo(dT)-cellulose does, giving a more legitimate poly(A)$^-$ fraction, and permitting the fractionation of mRNAs based on the length of their poly(A) tails. Under normal conditions, neither resin sheds its bound oligomer, and both resins yield mRNA that can be used for *in vitro* translation, cDNA synthesis, or microinjection. However, it should be noted that the long poly(U) tracts of poly(U)-Sepharose generally call for elution buffers containing high concentrations of formamide and extra precautions must be taken to ensure that no traces of formamide contaminate the purified RNA.

Procedure

Reagents. Oligo(dT)-cellulose (type 3) is purchased from Collaborative Research, Lexington, MA. It has an advertised binding capacity of 110–120 A_{260} units (or approximately 4.5 mg) of poly(rA) per gram of resin. If we assume that poly(A) comprises 5% of the nucleotides in mRNA, then a gram of oligo(dT)-cellulose would appear to have the capacity to bind 90 mg of poly(A)-containing RNA. However, since the resin is standardized with poly(rA), and since commercial preparations of poly(rA) are often 400–500 nucleotides long, it is probably more accurate to estimate the binding capacity to be approximately 10 mg of mRNA per gram of resin. Poly(U)-Sepharose is purchased from Pharmacia Inc., Piscataway, NJ. It has an advertised binding capacity of 600 μg of mRNA per gram of resin, i.e., the mRNA binding capacity per gram is approximately 15-fold lower than that of oligo(dT)-cellulose. Both resins are stored dry at $-20°$. Formamide (spectral grade) is purchased from Matheson, Coleman and Bell (East Rutherford, NJ) and is deionized prior to use by stirring for 1 hr at room temperature with 5% (w/v) analytical-grade mixed-bed resin [AG501-X8(D), Bio-Rad Laboratories, Richmond, CA] followed by filtration through Whatman 1 or Whatman 50 filter paper.

Buffer Solutions

Oligo(dT) binding buffer: 0.01 *M* Tris–HCl at pH 7.5, 0.5 *M* NaCl, 1 m*M* EDTA, 0.5% SDS

Oligo(dT) wash buffer: 0.01 *M* Tris–HCl at pH 7.5, 0.1 *M* NaCl, 1 m*M* EDTA

Oligo(dT) elution buffer: 0.01 *M* Tris–HCl at pH 7.5, 1 m*M* EDTA

Poly(U) swelling buffer: 0.05 M Tris–HCl at pH 7.5, 1 M NaCl

Poly(U) sample buffer: 0.01 M Tris–HCl at pH 7.5, 1 mM EDTA, 1% SDS

Poly(U) binding buffer: 0.05 M Tris–HCl at pH 7.5, 0.7 M NaCl, 10 mM EDTA, 25% (v/v) formamide

Poly(U) elution buffer: 0.05 M HEPES at pH 7.0, 10 mM EDTA, 90% (v/v) formamide

Binding to and Elution from Oligo(dT)-Cellulose.[10,18] Oligo(dT)-cellulose (0.1–1.0 g) is suspended in 1–5 ml of elution buffer, poured into a 1- to 5-ml disposable column, and washed with 5 column volumes of binding buffer. RNA is resuspended at 1–5 mg/ml in elution buffer, heated to 65° for 5 min, quickly cooled in ice, diluted with an equal volume of 2× binding buffer, and applied to the column. The flow-through is collected and reapplied and then the column is washed with 5–10 volumes of binding buffer and 5 volumes of wash buffer.

The bound RNA is eluted with 2–3 column volumes of elution buffer; subsequently, it is adjusted to a final concentration of 0.5 M NaCl (with 5 M NaCl or 2× binding buffer), rebound, rewashed, and reeluted as before. The RNA in the final eluate is recovered by addition of 0.1 volumes of 3 M sodium acetate and precipitation with 2.5 volumes of ethanol. Sodium dodecyl sulfate (SDS) is deliberately excluded from the wash and elution buffers to avoid precipitating it from ethanol along with the RNA.

The columnn is regenerated by washing with 2–3 volumes of 0.1 M NaOH, 3–5 volumes of distilled water (or until the effluent has a pH < 8.0), and 5 volumes of binding buffer. For storage, 0.2% sodium azide is included in the binding buffer. For prolonged storage, the manufacturer recommends that the regenerated resin be washed with water followed by absolute ethanol, then dried under vacuum overnight, and stored at −20°.

For batch applications,[18,19] oligo(dT)-cellulose is first equilibrated with binding buffer, then aliquots of 25–300 mg are added to 1.5-ml microfuge tubes. RNA dissolved in 1 ml of binding buffer is added and allowed to bind for 15 min with gentle shaking. The suspensions are centrifuged at 1500 g for 3–5 min and washed 3–5 times with 1 ml of binding buffer, and then 3 times with 400 μl of elution buffer. The eluates are pooled, readjusted to 0.5 M NaCl, rebound, and reeluted with three washes of elution buffer. The final three elution buffer washes are pooled and mRNA is recovered by ethanol precipitation.

Binding to and Elution from Poly(U)-Sepharose.[5] Poly(U)-Sepharose is suspended in swelling buffer (20 ml/g) for 1–2 hr, and then washed on a

---

[18] T. Maniatis, E. F. Fritsch, and J. Sambrook, "Molecular Cloning: A Laboratory Manual," p. 198. Cold Spring Harbor Lab., Cold Spring Harbor, New York, 1982.

[19] R. Shapiro, D. Herrick, R. Manrow, D. Blinder, and A. Jacobson, in preparation.

sintered glass filter with 100 ml of swelling buffer per gram of resin. The gel is poured into an appropriately sized column [1 g of swollen poly(U)-Sepharose will have a volume of 4 ml], and washed with 20 column volumes of elution buffer, and then with 20 column volumes of binding buffer. The larger volumes warrant the use of a peristaltic pump (a flow rate of 1 ml/min is recommended).

RNA is resuspended at 1–5 mg/ml in sample buffer, heated to 65° for 5 min, quickly cooled in ice, and diluted 5-fold with binding buffer. The RNA is applied to the column and washed with 5–10 volumes of binding buffer (or until the absorbance or radioactivity in the effluent has reached a baseline). Binding is efficient and there is usually no need to recycle the flow-through. The bound RNA is then eluted with 1.5–2.0 volumes of elution buffer. The RNA elutes with the buffer front which is readily recognized by the change in translucence of the gel (the high formamide concentration of the elution buffer makes it difficult to track the A_{260} of the eluate reliably). The eluate can be either ethanol precipitated or recycled by dilution (1 : 2.5) with a buffer containing 0.05 M Tris–HCl at pH 7.5, 2.5 M NaCl, 10 mM EDTA, and subsequently rebound, rewashed, etc. To avoid losses caused by high formamide concentrations, samples in elution buffer should be diluted 4- to 5-fold with 0.2 M sodium acetate before ethanol precipitation. Moreover, to avoid the possible contamination of the final RNA sample with formamide it is recommended that the RNA be reprecipitated from ethanol and that the aqueous phase be extracted with ether (see this volume [4,5])

After washing with elution buffer, poly(U)-Sepharose is reequilibrated by washing with 3–5 volumes of binding buffer. Storage is at room temperature with no further additions.

It is important to monitor carefully the pH of all solutions containing formamide. Decomposition products of formamide are acidic and may lead to the depurination and subsequent hydrolysis of the RNA in solution. These effects are especially pronounced in the high temperature elution conditions used for the fractionation of mRNAs differing in poly(A) tail length (see below).

Detection of Poly(A)⁻ mRNA

A careful examination of bona fide poly(A)⁻ RNA has revealed the existence of a small amount of mRNA activity.[5,20-22] This activity appears

[20] G. E. Sonenshein, T. E. Geoghegan, and G. Brawerman, *Proc. Natl. Acad. Sci. U.S.A.* **73**, 3088 (1976).
[21] T. Hunter and J. I. Garrels, *Cell (Cambridge, Mass.)* **12**, 767 (1977).
[22] Y. Kaufmann, C. Milcarek, H. Berissi, and S. Penman, *Proc. Natl. Acad. Sci. U.S.A.* **74**, 4801 (1977).

to be restricted to a very limited number of mRNA species, the most well known being the histone mRNA family. The major obstacle in identifying such poly(A)⁻ mRNAs has been the isolation of RNA samples which are truly poly(A)⁻, rather than just poly(A) deficient.

The use of poly(U)-Sepharose, subsaturating amounts of RNA, and modest flow rates is at present the most reliable way to prepare poly(A)⁻ mRNA. We have found that the flow-through fraction from poly(U)-Sepharose contains no detectable poly(A) tracts and has an extremely low translational capacity.[5] Rosenthal et al.[23] have estimated that the lower limit of the binding capacity of poly(U)-Sepharose is 6–10 adenylate residues. On the other hand, oligo(dT)-cellulose has been reported to have a requirement for a minimum poly(A) length of approximately 15 residues[23] and has been shown to have significant contamination of its nonbinding fractions with polyadenylated mRNAs.[5,23,24]

A reliable assay for the presence of mRNA activity in the poly(A)⁻ fraction is *in vitro* translation.[5] The poor translational activity of individual poly(A)⁻ mRNAs[25] and the preponderance in this fraction of rRNA and tRNA tend to reduce its translational specific activity (cpm of protein synthesized *in vitro* per microgram of RNA) to background levels, but gel electrophoresis of the translation products readily reveals the presence of specific proteins, such as histones.

Fractionation of mRNA as a Function of Poly(A) Tail Length: Separation of Newly Synthesized and Decaying mRNAs

This section describes a method for fractionating mRNAs as a function of their poly(A) tail lengths.[5] Messenger RNA is bound to poly(U)-Sepharose in the usual manner and then, instead of being eluted in a single step, is eluted in a series of sequential washes of increasing temperature or formamide concentration. This method is based on the fact that the thermal stability of base-paired nucleic acids is, in part, directly related to the length of the base-paired region.[26] Since molecules with multiple non-contiguous base paired regions will have a T_m which approximates that of the longest base-paired region, this method only works with resins having nucleic acid "tails" significantly longer than the poly(A) length of mRNA in the steady state, e.g., poly(U)-Sepharose, and is not directly applicable to resins capable of forming only short base-paired regions, e.g., oligo(dT)-cellulose.

[23] E. T. Rosenthal, T. R. Tansey, and J. V. Ruderman, *J. Mol. Biol.* **166**, 309 (1983).
[24] F. H. Wilt, *Cell (Cambridge, Mass.)* **11**, 673 (1977).
[25] M. T. Doel and N. H. Carey, *Cell (Cambridge, Mass.)* **8**, 51 (1976).
[26] R. J. Britten, D. E. Graham, and B. R. Neufeld, this series, Vol. 29, Part E, p. 363.

Since most newly synthesized mRNAs contain long, posttranscriptionally added poly(A) tails which shorten as the mRNAs age in the cytoplasm,[5] this method can be used to identify and purify mRNAs differing in their stabilities[6,19] or their differential rates of synthesis.[14] Furthermore, since nuclear pre-mRNA has the longest poly(A) tracts in the cell (approximately 200 nucleotides in mammals) and since mitochondrial mRNA has the shortest (approximately 25 nucleotides), the two are readily separable from cytoplasmic mRNA with its steady-state distribution of 40–65 adenylate residues using this technique.

Procedure

Buffer Solutions. Poly(U) wash buffer: 0.05 M Tris–HCl at pH 7.5, 0.1 M NaCl, 10 mM EDTA, 25% formamide. All other buffer solutions are as listed above.

Thermal Elution from Poly(U)-Sepharose. To avoid contamination of the different mRNA fractions with rRNA, it is advisable (but not essential) to start with an RNA sample which has been purified once by binding and step elution from either oligo(dT)-cellulose or poly(U)-Sepharose (see above). When *Dictyostelium* total cellular RNA which has not been previously purified on oligo(dT)-cellulose or poly(U)-Sepharose is fractionated by thermal elution, rRNA is found to comprise 78% of the RNA in the 25° eluate, 68% of the RNA in the 35° eluate, 22% of the RNA in the 45° eluate, and 1% of the RNA in the 55° eluate.[27]

All of the initial procedures, including swelling and washing of the resin, RNA sample preparation, binding of the RNA, and subsequent washing with binding buffer, are the same as described above (Binding to and Elution from Poly(U)-Sepharose), except that a water-jacketed column is used and that the chromatographic steps are performed at 25°. Messenger RNA fractions of different poly(A) content are then eluted as follows. The flow of the column is halted, binding buffer is replaced with wash buffer, and flow is resumed. The mRNA with the shortest poly(A) tracts (average ~25 A residues) elutes with 1.5–2.0 volumes of wash buffer (25° eluate). After this material is collected, the temperature of the column is raised in 10° increments to 55°, halting the flow during each temperature adjustment and reinitiating the flow as soon as the target temperature is reached. The reservoir of wash buffer is also heated simultaneously to minimize temperature fluctuations. With each temperature increment a reproducible fraction of the bound RNA elutes from the

[27] R. Shapiro, Ph.D. Thesis, Worcester Polytechnic Institute, Worcester, Massachusetts (1985).

column (35° to 55° eluates). In a final wash at 55°, the remaining bound RNA is eluted with elution buffer (EB eluate). Fractions are collected by ethanol precipitation as described above. As a general rule, the size of the poly(A) tail increases by 20–30 residues with each 10° increase in the elution temperature.

Examples of the use of this fractionation procedure with mRNA isolated from *Dictyostelium*,[5,6,14,27,28] *Xenopus*,[29] and yeast[30] have been published. Collectively, these experiments suggest the following general patterns for mRNA distribution in thermal eluates. Newly synthesized mRNA, containing the longest poly(A) tails, first appears in the EB eluate and subsequently accumulates primarily in the 45° eluate. In the steady state, over 50% of the total cellular mRNA is found in the 45° eluate, approximately 15% in the 55° eluate, 5–7% in the EB eluate, and the remainder equally divided between the 25° and 35° eluates. Pre-mRNA is restricted to the EB eluate and mitochondrial mRNA is restricted to the 25° eluate. Unstable mRNAs are degraded before their poly(A) tails have shortened significantly and hence accumulate preferentially in the long poly(A) fractions (55° and EB eluates). Stable mRNAs, on the other hand, have decay rates which are longer than poly(A) shortening rates and, thus, accumulate preferentially in the 35° and 45° eluates.

Acknowledgments

Many of the methods described here evolved from experiments performed in my laboratory by Dmitri Blinder, Anne Capone, David Herrick, Cheryl Mabie, Richard Manrow, Carl Palatnik, Robert Shapiro, Laura Steel, and Carol Wilkins. Their contributions are gratefully acknowledged, as is the constructive advice of Laura Steel and Richard Manrow with the manuscript, the word processing assistance of Linda Dill and Susan Longwell, and the financial support of the NIH.

[28] C. M. Palatnik, C. Wilkins, and A. Jacobson, *Cell (Cambridge, Mass.)* **36,** 1017 (1984).
[29] H. V. Colot and M. Rosbash, *Dev. Biol.* **94,** 79 (1982).
[30] C. A. Saunders, K. A. Bostian, and H. O. Halvorson, *Nucleic Acids Res.* **8,** 3841 (1980).

[26] Determination of the Molar Concentration
of Messenger RNA

By MARC S. KRUG and SHELBY L. BERGER

Purification of messenger RNA and mRNA precursors with columns of oligo(dT)-cellulose or poly(U)-Sepharose results in a highly enriched but nevertheless impure population of polyadenylated molecules. Since the contaminating molecules, principally rRNA, are present in variable but often significant amounts, methods for quantifying the polyadenylated component are required. This chapter presents a technique for determining the mole fraction poly(A)$^+$ RNA and its molar concentration in the presence of roughly equivalent amounts of poly(A)$^-$ RNA.[1] The method consumes about 75–100 ng of total RNA and takes about 45 min to perform.

Principle of the Method

The procedure depends on two reactions: (1) introduction of a radioactive label at the 3′ end of bulk RNA in a manner which cannot distinguish among the possible termini and (2) selective removal of that label only from molecules with legitimate (preexisting) poly(A) tails. The first reaction is catalyzed by poly(A) polymerase, an enzyme which processively adds adenylate residues donated in this case by [α-^{32}P]ATP, to the 3′-hydroxyl ends of RNA.[2,3] In the presence of cordycepin triphosphate, an analog of ATP which serves as a chain terminator,[4] only oligo(A) tails four residues or less in length are synthesized. The second reaction is catalyzed by ribonuclease H (RNase H), an enzyme which attacks the RNA strand of a DNA–RNA hybrid. In the presence of oligo(dT)$_{12-18}$, RNase H cleaves genuine poly(A) tails together with their radioactive tags. All other radioactive molecules remain intact because the synthetic oligo(A) tails are too short to be recognized as substrates by the enzyme.[5] The fraction of the total radioactivity introduced into macromolecules that

[1] M. S. Krug and S. L. Berger, *Anal. Biochem.* **153,** 315 (1986).

[2] A. Sippel, *Eur. J. Biochem.* **37,** 31 (1973).

[3] H. Sano and G. Feix, *Eur. J. Biochem.* **71,** 577 (1976).

[4] W. R. Beltz and S. H. Ashton, *Fed. Proc. Fed. Am. Soc. Exp. Biol.* **41,** 1450 (1982).

[5] J. G. Stavrianopoulos, A. Gambino-Guiffrida, and E. Chargaff, *Proc. Natl. Acad. Sci. U.S.A.* **73,** 1081 (1976).

survives RNase H treatment is therefore a measure of the mole fraction poly(A)$^-$ RNA in the preparation. Focusing on poly(A)$^+$ RNA, we have

$$F_1 = 1 - \frac{\text{radioactivity remaining in RNA sample after RNase H treatment}}{\text{total radioactivity incorporated into RNA sample}} \quad (1)$$

where F_1 denotes the mole fraction of polyadenylated termini in the impure RNA population.

In order to determine the molar concentration of the sample RNA, measured in moles of 3′ termini per liter, it is necessary to add to the sample a measured quantity of standard RNA whose molar concentration (also measured in moles of 3′ termini per liter) and mole fraction of poly(A)$^+$ RNA are known. For convenience, the standard is a poly(A)-negative species such as 18 S rRNA. The amount of standard RNA used must be sufficient to shift the mole fraction of polyadenylated RNA such that F_1 is clearly different from F_{obs}, the fraction of polyadenylated molecules in the mixture. Then

$$F_{obs} = 1 - \frac{\text{radioactivity remaining in RNA mixture after RNase H treatment}}{\text{total radioactivity incorporated into RNA mixture}} \quad (2)$$

$$F_{obs} = \frac{V_1C_1F_1 + V_2C_2F_2}{V_1C_1 + V_2C_2} \quad (3)$$

where F_1, C_1, and V_1 are, respectively, the mole fraction of polyadenylated RNA, the molar RNA concentration, and the volume of the sample RNA included in the mixture and F_2, C_2, and V_2 are the corresponding values for the standard RNA. Since F_2 is zero for rRNA, the expression in this case simplifies to

$$C_1 = \frac{V_2C_2F_{obs}}{V_1(F_1 - F_{obs})} \quad (4)$$

It follows that F_1C_1 is the molar concentration of poly(A)$^+$ RNA in the original sample.

Reagents

Poly(A) polymerase buffer (10×), 500 mM Tris–HCl at pH 8.0, 2.5 M NaCl, 100 mM MgCl$_2$, 10 mM EDTA, 10 mM dithiothreitol

MnCl$_2$, 25 mM

Cordycepin triphosphate, 200 μM (dilute from a concentrated stock of 10 mM cordycepin triphosphate just before use)

p(dT)$_{12-18}$, 100 μg/ml

Ribonucleoside–vanadyl complexes, 200 mM

Poly(A)⁻ RNA, 0.55 μM (a solution of 18 S rRNA)

Combine the reagents into the solutions listed below.

Labeling Mix[6]

Poly(A) polymerase buffer (10×)	2.5 μl
MnCl₂, 25 mM	2.5 μl
[α-³²P]ATP (3000 Ci/mmol)	2.5 to 12.5 μCi
Cordycepin triphosphate, 200 μM	2.5 μl
Poly(A) polymerase	1 to 2 units
Water to	15 μl

Deadenylation Mix

Poly(A) polymerase buffer (10×)	2.5 μl
p(dT)₁₂₋₁₈ (100 μg/ml)	2.5 μl
Vanadyl complexes, 200 mM	2.5 μl
Escherichia coli RNase H	2.5 to 5 units
Water to	25 μl

Procedure

Approximately 0.05 to 1.0 μg of sample RNA (0.5 μl of a solution 0.1 to 2 mg/ml) based on the A_{260} of the sample should be used in the assay. (See this volume [6] for a discussion of A_{260} measurements.) The amount of standard rRNA should be chosen such that $F_{obs} \neq F_1$. In practice, given the large size of rRNA molecules relative to the average size of the molecules in a preparation of bulk mRNA, the amount of standard rRNA (in micrograms) should be 2-fold higher than the amount of mRNA (in micrograms).

1. Prepare the following samples in Eppendorf tubes placed in ice:
   Tube 1:   3 μl labeling mix; 1.5 μl water; 0.5 μl sample RNA.
   Tube 2:   3 μl labeling mix; 1.5 μl water; 0.5 μl standard RNA.
   Tube 3:   3 μl labeling mix; 1.0 μl water; 0.5 μl sample RNA; 0.5 μl standard RNA.

---

[6] Poly(A) polymerase labeling results in the incorporation of <1 [³²P]ATP residue per molecule of RNA. To label RNA to higher specific activities, omit the cordycepin triphosphate, increase the [α-³²P]ATP to several hundred microcuries, and add as much enzyme as possible without exceeding 5% glycerol in the reaction. Under these conditions, some molecules may acquire tails of hundreds of labeled A residues; averages are ~20 residues per molecule. DO NOT DEADENYLATE!

TABLE I
DETERMINATION OF THE MOLAR CONCENTRATION OF mRNA

Time after RNase H treatment (min)	Radioactivity (cpm)		
	Sample, tube 1	Standard, tube 2	Mixture, tube 3
0 (unwashed)	198,000	192,000	184,000
0 (washed with TCA)	42,200	33,300	21,300
20 (unwashed)	202,000	178,000	142,000
20 (washed with TCA)	9,740	30,000	11,600

Some investigators also include a fourth tube similar to tube 1 containing globin mRNA, instead of sample RNA, to serve as a positive control.

2. Transfer all tubes to a 37° water bath; incubate 10 min.

3. To tube 1, add 5 µl ice-cold deadenylation mix, vortex, and immediately withdraw 5 µl. Spot it onto a GF/C glass fiber filter. Repeat for tubes 2 and 3. The radioactivity on these filters will be used to determine the total incorporation into RNA. Note that the vanadyl complexes in the deadenylation mix inhibit poly(A) polymerase[7] and prevent further incorporation of radioactive ATP into RNA during the cleavage process.

4. Incubate all tubes at 37° for 20 min. During this interval, deadenylation goes to completion. Spot 5 µl from each tube onto a separate GF/C glass fiber filter. The volume applied is not critical.

5. Process the filters as detailed in this volume [6]. Briefly, the six filters generated in this experiment are dried, assayed for radioactivity, washed with trichloroacetic acid (TCA) to remove [^{32}P]ATP from the filters, dried again, and reassayed for radioactivity. Table I contains values from an illustrative experiment.

As we will show, the volumes spotted at zero time and after 20 min of RNase H treatment are irrelevant, if the calculations are performed with normalized values. In this case, the normalization is most easily accomplished by converting the acid-insoluble radioactivity (filters washed with TCA) to a percentage of the total on the filter (unwashed filters). Then, F_1 can be calculated from the data for tube 1 as

$$F_1 = 1 - \frac{9,740/202,000}{42,000/198,000} = 0.77$$

Thus, 77% of the termini in the sample are polyadenylated.

[7] R. S. Puskas, N. R. Manley, D. M. Wallace, and S. L. Berger, *Biochemistry* **21**, 4602 (1982).

It is always prudent to calculate the value of F_2 for the standard rRNA assayed in tube 2. This is done by substituting F_2 for F_1 and RNA standard for RNA sample in Eq. (1). In this case, the value, $0.02 \approx 0$, indicates that the labeled oligo(A) tails are indeed very short and that putative nucleases introduced as artifacts have not reduced the rRNA to TCA-soluble fragments.

The term F_{obs} is determined from the results for tube 3 using the approach detailed for F_1:

$$F_{obs} = 1 - \frac{11,600/142,000}{21,300/184,000}$$

The result obtained, 0.30, is used to calculate C_1F_1, the molar concentration of poly(A)$^+$ RNA in the sample. Substituting into Eq. (4) with $V_2 = 0.5\ \mu l$, $C_2 = 0.55\ \mu M$, $V_1 = 0.5\ \mu l$, and F_1 and F_{obs} as measured, a value of $0.35\ \mu M$ is obtained for C_1, the concentration of total RNA in the sample. Then, C_1F_1, the concentration of polyadenylated RNA, is $0.27\ \mu M$.

Concluding Notes

This technique targets 3'-hydroxyl termini in order to distinguish poly(A)$^+$ RNA from poly(A)$^-$ RNA. Molecules with 3'-phosphates or 2',3'-cyclic phosphates which are not substrates for poly(A) polymerase cannot be measured. Furthermore, small oligonucleotides which are not acceptors for poly(A) polymerase-mediated polymerization are similarly undetectable. The assay is unaffected by the presence of nucleoside mono-, di-, and triphosphates with the exception of ATP.

Unlike the poly(U) hybridization technique of Bishop et al.[8] which measures total poly(A) in micrograms or moles of nucleotide without regard for its distribution, our approach tabulates the number of polyadenylated molecules. The length of the poly(A) tail is irrelevant here, since each polyadenylated molecule is counted once. Internal oligo(A) stretches, proven substrates for RNase H in the presence of oligo(dT), will also not be measured provided the 3'-terminal fragment bearing the label remains acid precipitable. In contrast, internal oligo(A) together with poly(A) tails contribute to the results using the Bishop technique.

[8] J. O. Bishop, M. Rosbash, and D. J. Evans, J. Mol. Biol. 85, 75 (1974).

[27] Translation in Cell-Free Systems

By ROSEMARY JAGUS

The simplest, unambiguous identification of a particular mRNA is the identification of its protein product. This can be established by translation of the mRNA of interest in a cell-free protein-synthesizing system. Messenger RNA protein product identification is important in the isolation of a particular mRNA species for cDNA cloning (see this volume, Section VIII) and in the identification of positive cDNA clones (see this volume [60]). The two high-activity translation systems in common use are those prepared from rabbit reticulocytes[1] and from wheat germ.[2,3] Both systems are easy to prepare, and both are available commercially. Each has advantages and disadvantages over the other and a choice between the two will depend on the type of mRNAs to be translated, the prejudices of experience, and availability.

The main disadvantage of the reticulocyte system is that it requires removal of endogenous mRNA. However, this is a relatively simple procedure,[1] involving incubation with a Ca^{2+}-dependent endonuclease, which is subsequently inactivated by EGTA, a chelator of Ca^{2+}. A related disadvantage is that significant levels of mRNA fragments remain after enzymatic digestion, including sequences that contain initiation sites. Added mRNAs have to compete with these fragments for ribosome binding and weakly initiating species may not do so efficiently.

The wheat germ system does not require removal of endogenous mRNA and may translate weakly initiating mRNAs more efficiently. However, ionic optima for translation in the wheat germ system are more sensitive to the nature and concentration of mRNA and may need to be determined for each template. The biggest problem with the use of the wheat germ system is its tendency to produce incomplete translation products due to premature termination. The system is really unsuitable for the translation of mRNAs coding for polypeptides larger than 60,000 Da.

Although both systems may be obtained from a number of commercial sources, which are undoubtedly convenient, such sources represent an extremely expensive alternative. Since one rabbit will yield 20–30 ml of reticulocyte lysate, if the use of any significant quantity is anticipated, it

[1] H. R. B. Pelham and R. J. Jackson, *Eur. J. Biochem.* **131**, 289 (1976).

[2] B. E. Roberts and B. M. Paterson, *Proc. Natl. Acad. Sci. U.S.A.* **70**, 2330 (1973).

[3] C. W. Anderson, J. W. Straus, and B. S. Dudock, this series, Vol. 101, p. 635.

makes good economic sense to make it. In addition, the activity of commercially available translation systems is variable and never specified for a given batch. The problem is that even poor lysates have some translational activity which will give rise to an impressive level of [^{35}S]methionine incorporation. However, if the experimental aim is to monitor a rare or weakly initiating mRNA, it becomes important to have an optimally active system, which may necessitate the production of cell-free translation systems in the home laboratory.

A procedure for the preparation of the wheat germ system is not included in this chapter and the interested reader is referred to refs. 2 and 3.

Solutions, Reagents, Paraphernalia

Phenylhydrazine hydrochloride, 2.5% (w/v): 5 g phenylhydrazine hydrochloride (Sigma) is dissolved in 200 ml deaerated 0.85% sterile NaCl and neutralized to pH 7 with 1 N NaOH. The solution should be a pale straw color, not brown. Single-use aliquots are snap-frozen in airtight, lightproof containers and stored at $-20°$. Once thawed for use, residual solution should be discarded.

Heparin, 500 units/ml in sterile distilled water

Reticulocyte wash buffer: 140 mM NaCl, 5 mM KCl, 7.5 mM magnesium acetate, autoclaved and made 1 mM with respect to glucose before use

Glucose, 100 mM: 100 mM solution made up in sterile distilled water, filter sterilized, and stored in single-use aliquots at $-20°$

Reticulocyte wash buffer (low Mg^{2+}): 140 mM NaCl, 5 mM KCl, autoclaved and made 1 mM with respect to glucose before use

CaCl$_2$: 100 mM, autoclaved

EGTA: 200 mM (pH 7), made up in sterile distilled water

Creatine phosphokinase (Boehringer-Mannheim): 200 units/ml in 100 mM KCl, 20 mM HEPES (pH 7.5), 50% glycerol. This enzyme solution is stable at least 1 year at $-20°$, but is inactivated by freezing.

Phosphocreatine: 0.5 M phosphocreatine, made up in sterile distilled water. Single-use aliquots are snap-frozen and stored at $-20°$ for a maximum of 2 weeks. Once thawed for use, residual solution should be discarded.

Hemin hydrochloride: 1 mM hemin solution is prepared in ethylene glycol by adding 65 mg hemin hydrochloride (Sigma) to 2 ml 100 mM KOH. Sufficient 100 mM Tris–HCl is added to adjust the pH to 7.8 and the volume is adjusted to 20 ml with sterile distilled water. Ethylene glycol, 80 ml, is added and the solution is stored in a lightproof bottle at $-20°$.

K/Mg: 2 M KCl, 10 mM magnesium acetate, autoclaved. Aliquots of 1 ml are stored at $-20°$ to prevent bacterial growth and for convenient use.

Potassium acetate: 0.85–1.85 M, adjusted to pH 7 with acetic acid, autoclaved, and stored in 1-ml aliquots at $-20°$.

Magnesium acetate: 50–100 mM, adjusted to pH 7 with KOH, autoclaved, and stored in 1-ml aliquots at $-20°$.

HEPES: 1 M, made with diethyl pyrocarbonate-treated water (this volume [60]) adjusted to pH 7.5 with KOH. Not stable to autoclaving. Store in 1 ml aliquots at $-20°$.

19-Met or 19-Val: Amino acids, 3 mM are made up in sterile, distilled water. Aliquots of 1 ml are snap-frozen and stored at $-20°$.

Dithiothreitol (DTT), 1 M, made up in deaerated diethyl pyrocarbonate-treated water (see this volume [60]). Store 250-μl aliquots at $-20°$. Once thawed for use, residual solution should be discarded.

ATP/GTP, 50 and 10 mM solutions, respectively, made up in diethyl pyrocarbonate-treated water. Aliquots of 1 ml are stored at $-20°$.

Spermidine tetrachloride, 1 M, made up in diethyl pyrocarbonate-treated water, adjusted to pH 7.

Nuclease S7, Ca^{2+}-dependent (from *Staphylococcus aureus*) (Boehringer-Mannheim): 15,000 units/ml, made up immediately before use in water.

Calf liver tRNA (Boehringer-Mannheim): 5 mg/ml in sterile distilled water. Aliquots of 1 ml are snap-frozen and stored at $-20°$.

Procedures

Preparation of Reticulocyte Lysate

Immature female New Zealand white rabbits (2–3 kg) are injected subcutaneously with 2.5% (w/v) phenylhydrazine hydrochloride (0.25 ml/kg) for 5 days. This schedule allows the reticulocyte count to rise to 85–95%. The rabbits will become weak, but should remain healthy, although one in five may die. The rabbits are allowed to recover for 2 days and then bled on day 8. Before bleeding, the rabbits are injected intramuscularly with 10–15 mg ketamine hydrochloride per kilogram and 0.5 mg acepromazine (10-[3-(dimethylamino)propyl]phenothiazin-2yl-methyl ketone) per kilogram. The rabbits are bled by cardiac puncture with a spring-loaded 60-ml disposable syringe, containing 0.1 ml heparin (500 unit/ml) fitted to a No. 16 Huberpoint needle.

The blood is added to 5 volumes of wash buffer containing 1 unit ml/heparin. The cells are collected by centrifugation at 4000 g for 10 min at 4°. The plasma is discarded and the reticulocytes are washed three times, with the Mg^{2+}-free buffer used for the final rinse. Cells are lysed in 2 volumes sterile distilled water. Lysis is maximized by stirring the cells in ice for 5 min. Cell membranes and mitochondria are removed by centrifugation at 16,000 g for 20 min at 4°. The supernatant is decanted and either snap-frozen in 1-ml aliquots or processed immediately for mRNA-dependent lysate (MDL). Lysate is stored in liquid nitrogen. Translation systems from reticulocyte lysate retain activity for years when stored in liquid nitrogen. At −70°, loss of activity is discernible and the lysate can only be used for 3–4 months. Storage at −20° is not recommended.

Preparation of Messenger RNA-Dependent Lysate

The original procedure of Pelham and Jackson[1] is still the best available. Because of the progressive inactivation of reticulocyte lysate by freeze–thawing, nuclease treatment is performed immediately after lysate preparation. To each 25 ml lysate (~1 rabbit equivalent) is added 250 μl of 100 mM $CaCl_2$, 500 μl of 1 mM hemin, and 250 μl creatine phosphokinase. The Ca^{2+}-dependent nuclease S7 from *S. aureus* is added to a concentration predetermined by pilot studies of the batch used, and the mixture is incubated at 20° for 15 min.

Optimization of endonuclease concentration is important because of batch variability. The range lies between 25 and 100 units/ml. Use of too little nuclease results in high endogenous backgrounds, too much leads to poor stimulation by added mRNA. After incubation, 250 μl of 200 mM EGTA is added to chelate the Ca^{2+} and inactivate the nuclease. Calf liver tRNA (250 μl; 5 mg/ml) is also added, not because endogenous tRNA is hydrolyzed by endonuclease, but because the composition of reticulocyte tRNA is related to the amino acid content of hemoglobin. Single-use aliquots are snap-frozen and stored in liquid nitrogen. Once thawed for use, residual lysate is discarded, since freeze–thawing results in dramatic loss of activity.

Characterization of Lysate

Characterization of Parent Lysate. It is important to check the activity and other characteristics of the parent lysate to ensure that subsequent time and expense are not wasted on inactive preparations and to provide a basis for comparison in assessing the activity of the MDL.

The following characteristics are easily checked: (1) the hemoglobin concentration should be between 200 and 250 mg/ml, which corresponds to an A_{415} of 800–1000. (2) The lysate should contain between 20 and 25 A_{260} units or 200 pmol/ml polysomes. This may be determined after pelleting the ribosomes by centrifugation at 10,000 g for 4 hr. (3) Incorporation of, for example, [^{14}C]valine or [^{35}S]methionine into trichloroacetic acid (TCA)-precipitable product should be linear for 45–60 min and continue for about 2 hr. Incubation with 50 μM [^{14}C]valine (280 mCi/mmol), under the conditions outlined below, should give 50–80 $\times$ 10^3 TCA-precipitable cpm in a 5-μl aliquot, after a 30-min incubation. Incubation with 0.5 μM [^{35}S]methionine (800–1200 Ci/mmol) should give 3–4 $\times$ 10^6 TCA-precipitable cpm in a 5-μl aliquot, after a 30-min incubation. Given endogenous valine and methionine concentrations of 30 and 10 μM, respectively, this corresponds to absolute rates of globin synthesis of ~1 mol globin/mol ribosome/minute, a rate which approaches that found *in vivo*.

Since the MDL will only be as good as the original extract, the MDL should only be used if the above values are found for the parent extract.

Characterization of Messenger RNA-Dependent Lysate. Once the activity of the parent lysate has been determined, the activity of the MDL should be assessed with and without a mRNA of known activity. Incorporation of radioactive amino acids into TCA-precipitable material should be low in the absence of added mRNA; about 5% of the activity of the parent lysate after a 30-min incubation. Globin mRNA is the mRNA of choice for testing the system, since at saturation it should yield exactly the same incorporation as the parent lysate. Globin mRNA saturates at ~20 μg/ml.

Translation of Exogenous mRNAs

Source of mRNAs. Messenger RNAs from a variety of sources may be used for translation. A cytoplasmic RNA preparation which is 90–95% rRNA translates poorly, usually no better than 20–30% of maximum. If enriched fractions of the mRNA of interest cannot be used, then poly(A)$^+$ RNA should be used.

Messenger RNA isolated by agarose gel electrophoresis (this volume [8]), by hybrid selection (this volume [60]), or by *in vitro* synthesis from SP6 promoter-containing constructs (this volume [30]) may all be translated in cell-free translation systems.

Incubation Conditions for Reticulocyte Translation System. The standard assay should contain the following in a 100-μl incubation: 5 μl K/Mg, 5 μl (19-radioactive amino acid), 5 μl (radioactive) amino acid, 2 μl 0.5 M phosphocreatine, 1 μl creatine phosphokinase (200 units/ml), 2 μl 1 mM hemin, 75 μl MDL, 5–10 μl mRNA. All reagents should be kept in ice before commencement of the incubation.

Incubation should be at 30° for up to 2 hr. At higher temperatures, e.g., 37°, the lysate does not maintain a linear incorporation rate as long. This is probably due to the slow activation of an eIF-2-specific kinase, which phosphorylates the α subunit of eIF-2 and leads to the inhibition of protein synthesis initiation. There are occasions when the use of 37° may be advantageous, such as the direct translation of mRNAs from low melting temperature agarose gel fractions.

The incorporation of radioactive amino acids into TCA-precipitable material is determined by diluting small aliquots, 2–10 μl into 1 ml water, to which is added 3 ml of 10% TCA. The mixture is heated at 90° for 15 min to discharge aminoacyl-tRNAs and then cooled in ice for 10 min. Precipitated radioactive protein is collected by vacuum filtration using glass fiber filters (Whatman GF/C or equivalent). After a rinse with 5% TCA the filters are dried under an infrared lamp and assayed in a non-aqueous scintillant with a scintillation counter.

A single globin peptide is synthesized in 1–1.5 min. Using this value, transit times of approximately 80 amino acids/min can be calculated, allowing the estimation of production times. Using these figures, it can be calculated that it will take approximately 10–15 min to synthesize a 100,000-Da peptide and 20–30 min to synthesize a 200,000-Da peptide.

Incubation Conditions for Wheat Germ Translation System. The standard assay should contain the following in a 100-μl incubation: 4 μl potassium acetate, 4 μl 19-Met, 10 μl [^{35}S]methionine (1 μM 800–1200 mCi/ml), 2 μl magnesium acetate, 2 μl phosphocreatine (0.5 M), 1 μl creatine phosphokinase (200 units/ml), 1 μl DTT, 2 μl ATP/GTP, 1 μl spermidine tetrachloride, 65 μl wheat germ extract with an A_{260} of 100, 5–10 μl mRNA. All reagents should be kept in ice before commencement of the incubation. Incubation should be at 25° for up to 2 hr.

Optimization of Ion Concentrations. The optimal concentrations of K^+ and Mg^{2+} ions are dependent on the mRNA being translated and should be determined for each mRNA species. Optimal K^+ ion concentrations range from 100 to 135 mM (endogenous K^+ ion concentration in reticulocyte lysate is ~25 mM). Optimal Mg^{2+} ion concentrations range from 2.5 to 3.5 mM (endogenous Mg^{2+} ion concentration in reticulocyte lysate is ~1 mM). Substitution of potassium acetate for potassium chloride allows a higher K^+ ion concentration to be used since high chloride ion concentrations inhibit initiation.

Ribonuclease Inhibitors. The inclusion of a ribonuclease inhibitor is a wise precaution, particularly for larger mRNAs. The ribonuclease inhibitor from placenta has no inhibitory effect on translation and can be used up to 300 units/ml. It is available commercially from a wide variety of sources, including Promega Biotec (Madison, WI) and Bethesda Research

Laboratories (Gaithersburg, MD). Unpublished results from the author's laboratory show slight degradation of α-globin mRNA in reticulocyte lysate during a 10-min incubation at 30°, as measured by Northern blot analysis. Inclusion of RNasin (Promega Biotec) at 300 units/ml prevents this.

Protease Inhibitors. If *in vitro* translation is used to determine the initial translation product of a mRNA, it is advisable to include protease inhibitors. Inhibitors which work and have no deleterious effects on translation are soybean trypsin inhibitor and ovomucoid at 1 and 0.5 mg/ml, respectively.

Choice of Radioisotope. When analysis of translation products is required, particularly of low abundance mRNAs which are not available in large quantities, [^{35}S]methionine is the radioactive amino acid of choice because of the high specific activities available (800–1200 Ci/mmol). The main disadvantage of radiolabeled methionine for this purpose is that it is a relatively uncommon amino acid in proteins, representing only 1% of the amino acid composition. In choosing a source of [^{35}S]methionine, factors to be considered are its specific activity, the packaging buffer used, and its demonstrated activity in cell-free translation systems.

High specific activity [^{35}S]methionine of high chemical purity is available from a variety of commercial sources. However, different batches of [^{35}S]methionine can contain an inhibitor of translation and some caution in the choice of product is recommended. ICN and New England Nuclear have attempted to remedy this problem by offering "translation-grade" products, that is, products tested for their ability to support activity in cell-free translation systems. The ICN product, Tran^{35}Slabel, has a slightly lower specific activity (800 Ci/mmol) than the New England Nuclear product (800–1200 Ci/mmol), but has the additional advantage that it contains cystine in addition to methionine, which should give rise to additional labeling of polypeptides.

[^{35}S]Methionine is sensitive to oxidation and is susceptible to accelerated decomposition at higher temperatures. Repeated freezing and thawing should be avoided. On arrival, single-use aliquots should be prepared and stored at −70°. The product can be obtained in 50 mM tricine (pH 7.4), which reduces its rate of deterioration on storage and eliminates an artifactual pattern on polyacrylamide gels.

Reticulocyte lysate has an endogenous methionine concentration of 10–15 μM, so when translating at low mRNA concentrations, it is possible to increase the final specific activity by using more isotope.

^{14}C-Labeled amino acids such as leucine or valine are often used. Since these amino acids constitute 7–10% of the amino acid composition of proteins, care must be taken that their concentration is not limiting for protein synthesis, and most commercially available preparations require concentration to 1 mM before use.

Sensitivity of Assay. Viral and globin mRNAs are good templates in the reticulocyte translation system, saturating at concentrations between 5 and 20 μg/ml, although detectable at lower levels. Cellular mRNAs may be translated less efficiently. If only dilute mRNA solutions are available, large volumes of RNA dissolved in water can be lyophilized in the incubation tubes.

For detection of specific products by gel electrophoresis, in combination with fluorography (see this volume [31]), much lower quantities are required than for detection by incorporation of TCA-precipitable radioactivity. If [^{35}S]methionine (800–1200 Ci/mmol) is the isotope, at 0.5 μM, it should be possible to detect the product of 100–200 ng/ml of a particular mRNA species, after 1–2 hr incubation, on fluorographs of products separated by polyacrylamide gel electrophoresis. The limit of detection by this method is determined by the low amount of reticulocyte lysate that can be applied to a gel lane (2–5 μl). If antibodies to the product of the mRNA are available, it is possible to immunoprecipitate from a 100-μl translation mixture and detect the translation of as little 5–20 ng/ml of mRNA (see this volume [31]).

Maximizing Production of a Specific Product. In addition to increasing the specific activity of the radioactive amino acid, increasing the volume of translation mix, and including ribonuclease and protease inhibitors, it is possible to enhance translation of mRNA by treatment of the RNA with methylmercury hydroxide (2.5 mM) prior to use and by inclusion of 2.5 mM methylmercury hydroxide in the incubation mix.[4] The enhancement of translation by methylmercury hydroxide is probably caused by a reduction in mRNA aggregation. The magnitude of the effect is mRNA specific, but can be as high as 5- to 7-fold.

Procedures for Use of Microsomal Membranes

Translation of mRNAs normally found on membrane-bound polysomes can be increased by the addition of dog pancreas microsomal membranes.[5–7] The reason for this is not entirely clear,[8] but may be because reticulocyte lysate contains excess free signal recognition particle (SRP)[9]

[4] F. Payvar and R. T. Schimke, *J. Biol. Chem.* **254,** 7636 (1979).
[5] P. Walter, I. Ibrahimi, and G. Blobel, *J. Cell Biol.* **91,** 545 (1981).
[6] D. Perlman and H. O. Halvorson, *Cell (Cambridge, Mass.)* **25,** 525 (1981).
[7] B. Nash and S. S. Tate, *J. Biol. Chem.* **259,** 678 (1984).
[8] D. I. Meyer, *EMBO J.* **4,** 2031 (1985).
[9] D. I. Meyer, E. Krause, and B. Dobberstein, *Nature (London)* **297,** 647 (1982).

which may inhibit the translation of membrane or secreted proteins and which can be sequestered by microsomes. Rough microsomes and/or salt-washed microsomes may also be required by investigators wishing to study *in vitro* processing of primary translation products and a recipe for their preparation is given. The method is adapted from Shields and Blobel.[10]

Preparation of Rough Microsomes. Probably the most important part of the procedure is to remove and process the dog pancreas immediately after sacrifice. After removal of connective tissue, fat, and large blood vessels, the pancreas should be chopped and the pieces passed through a stainless-steel tissue press (1 mm diameter). The resulting suspension is homogenized in 2 volumes ice-cold TKM buffer (50 mM Tris–HCl at pH 7.5, 50 mM KCl, 5 mM MgCl$_2$) containing 0.25 M sucrose, and homogenized in a stainless-steel homogenizer. Mitochondria and cell debris are removed by centrifugation at 13,000 g for 15 min at 4°. Supernatant, 20 ml, is layered over a discontinuous gradient of 1.5 M, 1.75 M, and 2.2 M sucrose (5 ml of each) TKM. After centrifugation at 140,000 g for 32 hr at 100,000 g and 4°, recover the 1.75 M sucrose layer which contains rough microsomes. Dilute with an equal volume TKM and layer over a cushion of 1.3 M sucrose in TKM. Centrifuge at 100,000 g for 30 min at 4°. Decant the supernatant. The pellet contains rough microsomes which should be resuspended in 0.25 M sucrose in TK to 50 A_{260} units/ml and stored in liquid nitrogen. Rough microsomes can be used in translation systems at concentrations up to 5 A_{260} units/ml.

Preparation of Stripped Microsomes. Rough microsomes contain significant levels of mRNA which can be removed, along with the ribosomes, by treatment with low concentrations of EDTA and nuclease S7. EDTA treatment alone does not remove all mRNA. The use of higher EDTA concentrations gives microsome preparations with lower mRNA contamination, but significantly lower processing activity.[10]

KCl, 1 M and EDTA, 0.2 M are added to the resuspended rough microsomes to give a final concentrations of 100 mM and 1.5 μmol/10 A_{260} units,[10] respectively.[10] Aliquots of 8 ml are layered over a 10–50% sucrose gradient in 100 mM KCl, 20 mM Tris–HCl (pH 7.5) and centrifuged at 60,000 rpm for 2 hr at 4°. A turbid band of EDTA-stripped microsomes is visible in the lower third of the gradient. CaCl$_2$ (100 mM) and nuclease S7 (15,000 units/ml) are added to give final concentrations of 1 mM and 12.5 units/ml, respectively, and incubated for 10–15 min at room temperature. The reaction is terminated by the addition of EGTA (200 mM) to a final concentration of 2 mM. The suspension is diluted 2-fold with TK contain-

---

[10] D. Shields and G. Blobel, *J. Biol. Chem.* **253**, 3753 (1978).

ing 2 mM EGTA and centrifuged at 100,000 g for 30 min at 4°. The resulting pellet is resuspended in 0.25 M sucrose in TK to 50 A_{260} units/ml and stored in liquid nitrogen.

Alternatives to Membrane Preparations. If microsomes are only used to maximize translation, it may be possible to substitute 0.1–0.5% (v/v) Triton X-100.[7] The mechanism by which detergents exert their stimulatory effect and substitute for microsomes is poorly understood, but implies that translation is inhibited in the absence of membranes by some sort of hydrophobic interaction.

Cautions in the Use of Microsomes. Although much data in the literature suggest that the signals and components involved in the translation and processing of membrane, secretory, and lysosomal proteins are universal, significant differences have been noted in the responses of the reticulocyte and wheat germ translation systems to the addition of microsomes and more highly purified components.[10] The investigator studying *in vitro* translation of mRNAs coding for such proteins may have to find out what works for him.

[28] Microinjection into *Xenopus* Oocytes: Equipment

By MICHAEL J. M. HITCHCOCK, EDWARD I. GINNS, and
CAROL J. MARCUS-SEKURA

The microinjection of nucleic acids into *Xenopus* oocytes requires accurate delivery of volumes in the nanoliter range. The capablity of performing injections from a single filling of the needle facilitates processing of large numbers of samples. This chapter describes the construction and operation of a microinjection device fulfilling these criteria.[1]

The reproducibility of the apparatus depends on the fluid transmission of a mechanical impulse using bubble-free oil. A glass microinjection needle (Figs. 1b and 2b) is attached to the tip of a syringe needle which is completely filled with oil (Fig. 2b). Movement of the syringe plunger is controlled by a stepping motor (Fig. 2b), and the oil provides positive displacement for moving the sample fluid in the glass injection needle (Fig. 2b). The sample only makes contact with the glass injection needle and the oil. The stepping motor is mounted directly onto a micromanipulator (Fig. 1c) in order to eliminate the need for flexible tubing connections between the injection needle and the fluid drive.

[1] M. J. M. Hitchcock and R. M. Friedman, *Anal. Biochem.* **109**, 338 (1980).

Construction of the Microinjector

The microinjector is composed of two main parts: a micromanipulator (Fig. 1a, part B) and a microprocessor-controlled pipettor (Fig. 1, b and c). Fluid delivery is accomplished with a Micro Lab P programmable microprocessor-controlled pipettor (Fig. 1c; Hamilton, Reno, NV) equipped with an external speed controller (Fig. 1c, part H) and a foot-operated switch. A MM33 micromanipulator (Fig. 1c, part F) without the fine thrust drive and spacer (Brinkmann Instruments, Inc., Westbury, NY) controls movement of the needle. The coarse thrust drive of the micromanipulator is disconnected and positioned on the solid half of the case as shown in Fig. 1a. For assembly of the device, the two halves of the plastic case of the hand pipettor of the Micro Lab P are disassembled. The positions of the screw holes are marked and after the holes are drilled in the plastic, the two pieces (coarse thrust drive and half of the plastic case) are bolted together (Fig. 1a). The hand pipettor is then reassembled and the thrust drive is reattached to the micromanipulator (Fig. 1b). The micromanipulator holds the needle at an oblique angle to a microscope stage for injection (Fig. 1c). The manipulator controls permit the smooth vertical and horizontal movements needed to fill the needle and inject oocytes.

The pipettor is fitted with a Hamilton 7000 series syringe (7105N, 5 μl; or 7107N, 1 μl) using the 1705-1750 adapter supplied with the Micro Lab P (Fig. 1b, part C). The plunger buttons of these syringes are dome shaped and must be filed flat for them to fit. Polyethylene tubing (0.58 mm i.d., 0.965 mm o.d.) is used to cover all except 6 mm of the wire needle tip of the syringe (Fig. 2b). A 14-gauge cannula (4 in. original length, Becton Dickinson Co., Rutherford, NJ) lined with another piece of polyethylene tubing (1.14 mm i.d., 1.57 mm o.d.) and cut to leave about 1 cm of the wire needle exposed is fitted over this tubing and wire needle to provide more rigidity (Fig. 2). During the cutting of the cannula, care must be taken to avoid burrs and distortion of the cannula.

Two centimeters of silicon tubing (0.79 mm i.d., 3.18 mm o.d.) is pushed over the end of the cannula to cover the needle, polyethylene tubing, and tip of the cannula, leaving about 1 mm of the wire needle exposed (see Figure 2b).

The hand pipettor attached to the micromanipulator is then mounted on a stand, the foot-controlled switch is connected to the external speed controller, and the leads from the hand pipettor and external speed controller are connected to the main control unit (Fig. 1c).

Microinjection needles (3.5–4.0 cm long) are prepared by pulling out glass capillaries (20 μl; Microcaps, Drummond Scientific Co., Broomall, PA) to approximately 25 μm outside diameter using a micropipet puller

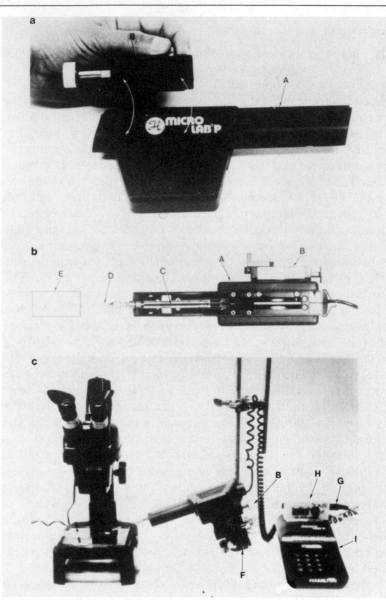

FIG. 1. The microinjection apparatus. (a) Approximate locations of the holes that are drilled in one side of the plastic case of the pipettor (part A) so that they align with the corresponding holes in the coarse thrust drive of the micromanipulator (part B). (b) View of the reassembled hand pipettor (part A) with the micromanipulator drive (part B) bolted in place. A syringe (part C) is fitted with a needle covered by a cannula (part D). The boxed area (labeled E) around the needle tip is shown diagrammatically in more detail in Fig. 2.

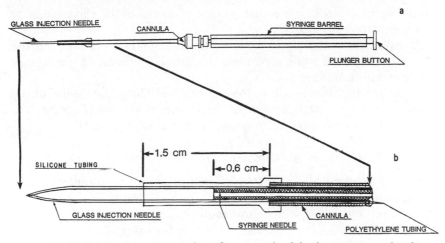

Fig. 2. (a) Schematic representation of oocyte microinjection apparatus showing syringe–injection needle assembly. (b) Expanded view of the glass injection needle assembly.

(J. B. Kefe, Islington, MA). The needles are conveniently stored point up in a piece of styrofoam and placed in a plastic container to prevent contamination and injury.

Operation of the Microinjection Apparatus

Fitting the Glass Needle

1. The syringe must be empty (see steps 1–5 of the following two sections).

2. Break off the old glass needle tip and then remove the remaining glass barrel of the old needle. Removing the old glass needle without breaking off the tip will introduce bubbles into the syringe needle.

3. Hold a new glass needle against a contrasting background and with ethanol-flamed tweezers touch the tip to open the seal.

4. A 3–6 ml syringe is filled with paraffin oil and fitted with a 0.5-in. 24-gauge needle. Bubbles are removed from the needle of this syringe,

(c) Illustration of the microinjection apparatus during injections. The micromanipulator thrust drive (part B) has been reattached to the rest of the micromanipulator (part F) and then mounted on a ring stand. The wire for the foot-operated switch (part G) is connected to the external speed controller (part H) which is taped onto the microprocessor control unit (part I).

and the syringe needle is inserted into the wide end of the glass needle to fill it completely with oil.

5. After filling the glass needle with oil, maintain pressure on the filling syringe barrel while separating the needles in order to prevent air bubbles from forming.

6. Attach the glass needle to the pipettor by sliding it inside the silicon tubing until it rests over the wire of the Hamilton syringe (Fig. 2b). Too much pressure will cause air to appear in the glass needle.

Clearing Bubbles from the Glass Needle

1. Switch on the Micro lab P [the C button on the speed controller is in the off (out) position].

2. Enter # (syringe volume in nanoliters), E.

3. Press *, SPEED, 1. (Display shows E for external.)

4. Press *, MAN.

5. Press START twice.

6. Fit needle (if not already on).

7. Turn the pipettor vertically so that needle points up.

8. Set speed controller at slowest speed (Fig. 1c, part H).

9. Press C button on speed controller (syringe plunger withdraws for approximately 3 min).

10. Release C button.

11. Set speed controller to fastest speed.

12. Press C on speed controller until syringe plunger returns (approximately 3 sec).

13. Release C button. If bubbles are visible in the needle tip, repeat steps 8–13 until no bubbles are produced during three cycles of the above procedure.

Injection Procedure

1. Switch on Micro Lab P main controller [C button on the speed controller must be in the off (out) position].

2. Enter # (syringe volume in nanoliters), E.

3. Press *, SPEED, 1. (Display shows E for external.)

4. Press *, RDIS, # (injection volume in nl), E, # (number of injections + 2), E. Note the volume shown in the volume display.

5. Press START twice.

6. Place a sterile plastic plate on the microscope stage (Falcon lid 3041) and place a sterile glass microscope slide on it (Figs. 1c and 3).

7. Set the speed controller to 40 nl/sec.

8. Fit a glass injection needle (see previous section) if needed.

9. Line up the needle at an angle sloping toward the stage (Figs. 1c and 3) and within the microscope field of view.

10. Maneuver the needle straight up, without changing its angle to the stage, with the vertical micromanipulator control.

11. Place the sample of DNA or RNA to be injected as a droplet on the plastic plate (use 500 nl or the total volume programmed in step 4 plus 200 nl, whichever is larger).

12. While viewing the sample droplet through the microscope bring the needle, still angled at the stage, to the center of the sample droplet, close to, or just touching the plastic lid.

13. Press foot pedal (same as START) to fill the syringe with the sample.

14. After filling is complete, change the speed controller setting to approximately 100 nl/sec.

15. Press foot pedal (or START) to dispense one injection aliquot.

16. Place the oocytes to be injected at the center of the plastic lid in a small amount of modified Barth's medium. The oocytes must not dry out during the injection procedure. The number of oocytes injected at a time is determined by the experience of the operator—initially 5, later 20.

17. Slide the glass microscope slide toward the oocytes until the media is pulled back under the slide. Using blunt, sterile forceps, position the oocytes along the edge of the slide with the dark animal poles toward the needle (see Fig. 3). Although we suggest injection into the dark pole, some investigators inject into the light vegetal pole, while others inject at

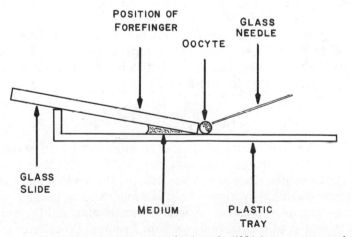

FIG. 3. Diagram illustrating the microinjection of mRNA into oocyte cytoplasm.

the equator. Injection at any of these sites results in translation of mRNA, but regardless of the site chosen one must avoid damage to the nucleus.

18. Move the plastic plate, holding the slide in place, such that the first oocyte is at the bottom of the field of view.

19. Bring the needle, still angled to the stage, into the field of view and adjust its position so that it will pierce the dark pole of the oocyte at 20–30°.

20. Insert the needle into the dark pole of oocyte using the micromanipulator controls.

21. Inject the sample by pressing the foot pedal (or START). If the injected volume is 20 nl or greater, the oocyte should transiently swell.

22. Remove the needle from the oocyte using the micromanipulator.

23. Move the needle to the next oocyte.

24. Continue injection by repeating steps 21–24.

25. When all oocytes have been injected, raise the needle using the micromanipulator controls.

26. Lift the glass slide off the plastic lid and pipet medium onto the oocytes. With a wide-mouth Pasteur pipet transfer the injected oocytes into modified Barth's medium.

27. Repeat steps 17–26 until the number of steps indicated on the Micro Lab P main controller display is 1. This should coincide with the final oocyte injection.

28. Bring the injection needle down to the plastic lid and press start to remove the final aliquot of the sample. If the operation has been carried out successfully, the oil–water interface can be seen moving to the tip of the needle. A small oil bubble is often seen in the aqueous droplet on the plastic lid.

29. Repeat steps 7–28 for subsequent injections of samples of the same volume using the same number of oocytes, or steps 4–28 (step 6 can be omitted) if the volume and/or the number are to be changed.

30. See next section for shutdown.

Storage of Microinjection Apparatus

1. After the final injection, proceed through steps 4–10 of the second section, Clearing Bubbles from the Glass Needle.

2. Switch off the main controller, leaving the syringe filled with distilled water. Bubble formation is kept to a minimum by keeping the injector tip up. The microinjection apparatus can be prepared for use by following steps 1–5 of the previous section, Injection Procedure, using the fastest setting on the speed controller.

Microinjection of DNA into the Germinal Vesicle

The micropipettor assembly described above can readily be adapted for microinjection of DNA into the oocyte germinal vesicle (nucleus). As described in the next chapter [29], oocytes are prepared so that the germinal vesicle is floating at the top surface of the oocyte. The oocytes are positioned on a Spectramesh grid in a small plastic petri dish (Fig. 4). Injection into the nucleus required a vertical angle of injection rather than the oblique angle used for cytoplasmic injections. To accomplish this, the Hamilton pipettor is positioned at the same angle used for cytoplasmic injections, but the glass capillary needle is modified (Fig. 4). Capillaries to be used for these needles are initially pulled with a slightly longer length of barrel tubing (approximately 4.5 cm) than those needles used for cytoplasmic injection. The glass needle tip is broken off by touching it gently with a pair of sterile forceps. Then, while the needle is held with a pair of forceps on the needle barrel close to the tip, the midpoint of the needle is briefly positioned horizontally in a Bunsen burner flame. When the end of the needle away from the tip bends to form an angle of approximately 120°, the needle is pulled out of the flame. After cooling, the needle is filled with paraffin oil and inserted onto the Hamilton syringe as described in Fitting the Glass Needle. The addition of a small amount of an oil-soluble dye, such as O-cresol red to the paraffin oil, makes it easier to visualize the needle in the subsequent microinjection manipulations.

Plasmid DNA at a concentration of 1 mg/ml has been successfully

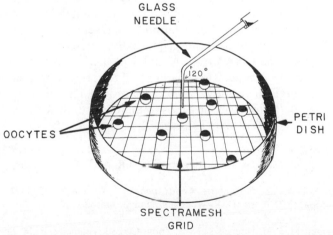

FIG. 4. Diagram illustrating the microinjection of DNA into the germinal vesicles of oocytes.

used in protein expression experiments.[2] The DNA should be centrifuged in a microfuge for 2 min immediately prior to loading into the needle to precipitate particulates which might block the needle. After centrifugation, an aliquot is removed from the DNA solution (avoiding any pellet), placed onto a plastic lid, and an amount sufficient to inject 20–30 oocytes with 15 nl each is loaded into the needle (as described in the Injection Procedures). It is preferable to inject an entire batch of prepared oocytes rapidly because oocyte debris collects on the tip of the needle and, when dry, blocks the needle. Injection of 50 oocytes per DNA sample should provide a detectable level of protein expression.

[2] R. J. Watson, A. M. Colberg-Poley, C. J. Marcus-Sekura, B. J. Carter, and L. W. Enquist, *Nucleic Acids Res.* **11,** 1507 (1983).

[29] Preparation of Oocytes for Microinjection of RNA and DNA

By Carol J. Marcus-Sekura and Michael J. M. Hitchcock

The oocytes of the African clawed toad, *Xenopus laevis*, have been used extensively to test the biological activity of purified macromolecules. The *Xenopus* oocyte provides a unique unicellular test system for the study of transcription,[1–3] translation,[4,5] and secretion[6] of injected molecules. The oocyte will posttranscriptionally process proteins it synthesizes from injected DNA or RNA and will also secrete those proteins normally secreted *in vivo*. This chapter will describe a protocol by which oocytes are surgically removed and prepared for microinjection of RNA into the oocyte cytoplasm or DNA into the oocyte nucleus (germinal

[1] J. E. Mertz and J. B. Gurdon, *Proc. Natl. Acad. Sci. U.S.A.* **74,** 1502 (1977).
[2] S. L. McKnight, E. R. Garvis, R. Kingsbury, and R. Axel, *Cell (Cambridge, Mass.)* **25,** 385 (1981).
[3] E. M. de Robertis and J. E. Mertz, *Cell (Cambridge, Mass.)* **12,** 175 (1977).
[4] J. B. Gurdon, C. D. Lane, H. R. Woodland, and G. Marbaix, *Nature (London)* **233,** 177 (1971).
[5] S. L. Berger, M. J. M. Hitchcock, K. C. Zoon, C. S. Birkenmeier, R. M. Friedman, and E. H. Chang, *J. Biol. Chem.* **255,** 2955 (1980).
[6] A. Colman, C. D. Lane, R. Craig, A. Boulton, T. Mohun, and J. Morser, *Eur. J. Biochem.* **113,** 339 (1981).

METHODS IN ENZYMOLOGY, VOL. 152

vesicle). More extensive information about *Xenopus* and its oocytes can be found in the literature.[7–9]

Surgical Removal of Oocytes. Adult female *Xenopus laevis* can be obtained from Carolina Biological Supply Company, Burlington, NC (catalog no. L1570) and maintained in the laboratory in plastic animal cages in deionized water at room temperature (18–24°). Dried food pellets can be obtained from the same source (catalog no. L1592). Frogs should be fed twice weekly and the water changed several hours after feeding. Frogs can be kept for months before use, and oocytes can be surgically removed from a single frog three to five times before its oocyte population is depleted. [Before proceeding with surgery, have someone with experience in the technique instruct you, to spare the animals as much trauma as possible.]

To anesthetize the frog, add ice to the water in which it is maintained, wait approximately 30 min until movement has stopped, and then transfer the frog to an ice bucket, placing it on its back and covering most of the body but not its head with ice, for an additional 30 min. When the frog can be handled without movement, put it on its back on the ice, and proceed with the surgery. You will need sterile scissors, forceps, a scalpel with a no. 11 blade, an alcohol swab (Tomac Prep Packet) and a surgical needle for sewing (Ethicon 4-0 Vicryl suture, catalog no. J-315H, Ethicon, Somerville, NJ). Begin on one side of the frog to the left (or right) of the midline just above the leg. Swab the area with the alcohol swab to sterilize. Holding the forceps in one hand, pull up a pinch of skin until it is taut, and holding the scalpel in the other hand make a small transverse incision. Use the scissors to widen the incision until it is about 1.5 cm in size. This will expose a second layer of muscle underneath which should be opened in the same manner. The ovary with its black oocytes should now be visible. Using forceps, carefully reach in and pull out a section of ovary (perhaps 2–3 cm in length) and then use the scalpel to separate it from the remaining ovarian tissue. Place the tissue in a petri dish containing modified Barth's medium[10] (see below) at room temperature. Remove additional portions of tissue in this manner. When sufficient tissue has been obtained, sew up the frog using two or three stitches on the inner muscle layer and three or four stitches on the outer skin layer, double-knotting each stitch. The frog can now be returned to its ice-containing animal cage

[7] J. B. Gurdon, *in* "Methods in Developmental Biology" (F. H. Wilt and N. K. Wessells, eds.), p. 75. Crowell-Collier, New York, 1967.
[8] J. B. Gurdon and D. A. Melton, *Annu. Rev. Genet.* **15**, 189 (1981).
[9] J. B. Gurdon, *J. Embryol. Exp. Morphol.* **36**, 523 (1976).
[10] J. B. Gurdon, *J. Embryol. Exp. Morphol.* **20**, 401 (1968).

where it will awaken as the ice melts and the water returns to room temperature.

Preparation of Modified Barth's Solution. Solution A contains 1.76 M NaCl, 20 mM KCl, 48 mM NaHCO$_3$, 16.4 mM MgSO$_4$, and 0.15 M Tris adjusted to a pH of 7.6 with HCl and autoclaved. Solution B contains 3.3 mM Ca (NO$_3$)$_2$ and 4.1 mM CaCl$_2$ and is autoclaved. Combine 25 ml of solution A, 50 ml of solution B, 425 ml of autoclaved water, 50 mg each of penicillin and streptomycin, 25 mg of gentamicin, and 125 μg of fungizone.

Separation of Oocytes from Debris. The oocytes are now gently separated from the ovarian tissue. This is most easily done using a flat spatula and forceps which have been dipped in ethanol and flamed to sterilize just prior to use. Dip them in the Barth's solution in the petri dish to cool before touching the oocytes. Now, using the forceps to hold a section of tissue, gently shake to loosen and then scratch gently with the edge of the spatula to loosen individual oocytes. Very tiny immature oocytes will remain attached to the ovarian tissue. When all the larger oocytes have been freed, discard the remaining tissue.

Selection of Mature Oocytes for Injection. It is now necessary to separate the mature (stage 5 or 6)[11] oocytes suitable for injection from the immature oocytes. These are the large oocytes greater than 1000 μm in diameter which have a distinct boundary between the hemispheres. Break off the tip of a 9-in. sterile Pasteur pipet so that the orifice is approximately 3 mm wide and about 1 in. of constricted tip remains. Insert the tip into a flame of a Bunsen burner until it has become smooth so that it will not tear the oocytes. Fit a Pasteur pipet into a 10-ml Pi Pump (Bel-Art, Fisher Scientific Catalog, no. 13-683C). Draw up about 20–30 oocytes along with about 1.5 ml of Barth's solution into the pipet, hold the pipet vertically, and allow the heavy large oocytes to fall to the bottom and back into the petri dish. Discard the lighter immature oocytes remaining in the pipet. Repeat the procedure multiple times and then transfer the remaining larger oocytes into a fresh petri dish of Barth's solution. Finally, pick up each oocyte individually and examine it by eye to see that it is round and has one yellow hemisphere (yolk) and one dark brown hemisphere (animal pole), and that no bits of tissue are left attached. These oocytes are suitable for microinjection, and oocytes not meeting these criteria should be discarded. It is also possible to inject the mature members of washed clumps of oocytes which are still attached to ovarian tissue. However, in some situations this approach leads to needless complications during analysis of the translation products.[12]

[11] J. N. Dumont, *J. Morphol.* **136,** 153 (1972).
[12] A. Colman, *in* "Transcription and Translation: A Practical Approach" (B. D. Hames and S. J. Higgins, eds.), p. 271. IRL Press, Washington, D.C., 1984.

Storage of Oocytes until Use. Oocytes can be stored in Barth's solution at room temperature for up to 1 week before use. It is recommended that the Barth's solution be changed every few days.

Preparation of Oocytes for Microinjection of RNA. Immediately prior to microinjection, individual oocytes should be removed from the bulk preparation and each oocyte should be examined to be sure it has not lost color or turgor. For detection of protein expression, 20 oocytes each injected with 50 nl of a poly(A)$^+$ mRNA solution should be sufficient. Rabbit globin mRNA which can be purchased commercially (Bethesda Research Laboratories 8103SA) might be used as a control to test that the system is working. Instructions for the injection of the mRNA into the oocytes can be found in this volume [28].

Preparation of Oocytes for Microinjection of DNA. The procedure to be described is a modification of already established microinjection techniques for DNA.[13,14] The principle of the technique is to centrifuge oocytes on a plastic grid so that the germinal vesicle floats to the surface. DNA can then be microinjected into the germinal vesicle using the apparatus described in chapter [28] of this volume.

Polyethylene mesh with a mesh opening of 925 μm (Spectramesh from Spectrum Industries, Los Angeles, CA, catalog no. 146372; Fisher Scientific Catalog no. 08-670-176) is cut using a 35-mm petri dish as a guide, placed in the bottom of a 60-mm plastic petri dish, and fused to the dish with chloroform. All chloroform should be allowed to evaporate before use.

Oocytes prepared as described above are pipetted onto the grid and oriented so that their animal poles (brown hemispheres) are facing up. A small amount of Barth's solution must be present to prevent the oocytes from losing their turgor, but too much liquid will allow them to roll and change their orientation. The petri plate is covered and placed in a Sorvall GLC-4 or RT6000B table-top centrifuge containing an H-1000B rotor and PN11053 buckets with one-place adapters (Sorvall catalog no. 00186). Rubber stoppers (size 11) placed in the bottom of the adapters facilitate placing the petri plate into the rotor and the subsequent removal of the petri plate after centrifugation with minimal jarring of the oocytes. It is important that the centrifuge be located close to the microinjection apparatus. The oocytes are centrifuged at 2600 rpm for 5 min at room temperature. The DNA to be injected is loaded into the needle (this volume [28]) while centrifugation is in progress, so that the petri plate containing the oocytes can be placed directly under the microscope and injected. The

[13] A. Kressmann, S. G. Clarkson, J. L. Telford, and M. L. Birnstiel, *Cold Spring Harbor Symp. Quant. Biol.* **42,** 1077 (1977).
[14] D. Rungger and H. Turler, *Proc. Natl. Acad. Sci. U.S.A.* **75,** 6073 (1978).

position of the germinal vesicle after centrifugation is indicated by a clear area in the animal pole. Oocytes which have changed their orientation or lost most of their turgor during centrifugation should be discarded. Approximately 15 nl of a 1 mg/ml solution of viral or plasmid DNA should be injected per oocyte. After injection, the oocytes are loosened from the grid by spraying gently with Barth's solution. This technique has been used successfully to microinject plasmid clones containing the herpes simplex virus glycoprotein D gene and to detect protein expression after incubation with [^{35}S]methionine followed by immunoprecipitation with monoclonal antibodies.[15] Oocytes can also be simultaneously injected with DNA and radiolabeled ribonucleotide triphosphates for analysis of RNA transcripts.[13] (For further information on postinjection procedures see [30], this volume.)

[15] R. J. Watson, A. M. Colberg-Poley, C. J. Marcus-Sekura, B. J. Carter, and L. W. Enquist, *Nucleic Acids Res.* **11,** 1507 (1983).

[30] Translation of Messenger RNA in Injected Frog Oocytes

By DOUGLAS A. MELTON

This chapter describes the use of injected frog oocytes for the identification and characterization of proteins encoded by messenger RNAs. The main interest in using injected oocytes to study translation derives from the fact that this assay offers a very close approximation to the *in vivo* situation in which mRNAs are normally translated. This method for studying translation has both advantages and disadvantages as compared to *in vitro* systems, such as wheat germ extracts or reticulocyte lysates, and these various points will be considered here.

The translation of exogenous mRNAs injected into frog oocytes was first demonstrated by Gurdon and colleagues.[1] They showed that *Xenopus* oocytes injected with red blood cell poly(A)$^+$ RNA will synthesize globin protein. Subsequent studies have shown that all eukaryotic mRNAs, plant, animal, and viral, will direct the synthesis of the appropriate protein when the RNA is injected into living oocytes.

[1] J. B. Gurdon, D. C. Lane, H. R. Woodland, and G. Marbaix, *Nature (London)* **233,** 177 (1971).

The biology of *Xenopus* oocytes makes it understandable why these meiotic cells are capable of translating exogenous messenger RNAs. During oogenesis, immature eggs or oocytes accumulate a vast store of enzymes, organelles, and other precursors for use during the early development of the fertilized egg. For example, an oocyte contains about 200,000 more ribosomes and an excess of 10,000 tRNA molecules compared to somatic cells. In addition, an abundance of histone protein, nucleoplasmin, and RNA polymerases are accumulated during oogenesis. These materal stockpiles are not stored in an inactive or unavailable form. Indeed, DNAs injected into the nucleus or germinal vesicle are assembled into chromatin and transcribed (reviewed by Gurdon and Melton[2]) and unspliced mRNAs or pre-tRNAs are accurately processed into mature mRNAs or tRNAs.[3,4] And most relevant to the present discussion, exogenous mRNAs are accurately translated into proteins when injected into growing or mature oocytes.

Injected mRNAs compete with endogenous oocyte mRNAs for the translational machinery.[5] Thus, injected oocytes synthesize proteins encoded by endogenous as well as injected mRNAs. Unlike *in vitro* systems translation systems that can be driven exclusively by added mRNA, injected oocytes will always have a "background" of endogenous oocyte proteins. Nevertheless, it is possible to inject large amounts of mRNA into an oocyte and thereby effectively compete with the endogenous mRNAs for the translational machinery. For example, when a high concentration of globin mRNA (100 ng/oocyte) is injected, as much as 50% of the newly synthesized oocyte protein is globin.[6]

The proteins translated from injected mRNAs are correctly modified or matured by the oocyte's enzymes. These posttranslational modifications can include cleavage of precursor proteins, phosphorylation, and glycosylation.[7–9] In addition, injected oocytes will direct proteins into the correct intracellular membrane and will export proteins that are normally secreted from cells.[8,10] Oocytes injected with mixtures of mRNAs will also assemble multisubunit proteins into functional complexes. For exam-

[2] J. B. Gurdon and D. A. Melton, *Annu. Rev. Genet.* **15**, 189 (1981).

[3] D. A. Melton, E. M. De Robertis, and R. Cortese, *Nature (London)* **284**, 143 (1980).

[4] M. R. Green, T. Maniatis, and D. A Melton, *Cell (Cambridge, Mass.)* **32**, 681 (1983).

[5] R. A. Laskey, A. D. Mills, J. B. Gurdon, and G. A. Partington, *Cell (Cambridge, Mass.)* **11**, 345 (1977).

[6] J. Richter and L. D. Smith, *Cell (Cambridge, Mass.)* **27**, 183 (1981).

[7] M. V. Berridge and C. D. Lane, *Cell (Cambridge, Mass.)* **8**, 283 (1976).

[8] A. Colman, C. D. Lane, R. Craig, A. Boulton, T. Mohun, and J. Morser, *Eur. J. Biochem.* **113**, 339 (1981).

[9] J. Matthews, J. Brown, and T. Hall, *Nature (London)* **294**, 175 (1981).

[10] A. Colman and J. Morser, *Cell (Cambridge, Mass.)* **17**, 517 (1979).

ple, oocytes injected with the appropriate mRNAs will assemble functional acetycholine receptors.[11] Thus, unlike most *in vitro* translation systems, injected oocytes provide a complete system for studying protein synthesis, posttranslational modification, assembly, and secretion.

Synthetic mRNAs, such as those prepared by *in vitro* transcription of cloned cDNA with SP6 RNA polymerase,[12] are also translated in injected oocytes.[13] A particularly nice example of the utility of this method has been provided by the experiments of Numa and colleagues in their demonstration that a functional acetylcholine receptor is produced in oocytes injected with synthetic mRNAs.[14]

Messenger RNAs injected into oocytes are unusually stable and are efficiently translated for a relatively long time. The half-life of injected globin mRNA, for example, is estimated to be more than a week.[15] Obviously, not all mRNAs are as stable as globin mRNA and the stability of a given mRNA will depend at least in part on its sequence. Studies with synthetic mRNAs have shown that a 5' cap is essential for stability in injected oocytes,[13,16] confirming earlier work done with natural mRNAs.[17] The sequence at the 3' end of the mRNA may not be as critical with respect to stability and consequently translation. For instance, the role of a poly(A) tail is somewhat controversial in that some RNAs are apparently destabilized by deadenylation[18-20] whereas the stability of other mRNAs is not affected.[21,22] Recent studies by Drummond *et al.*[16] indicate that a poly(A) tail can enhance the stability of mRNAs in long-term (>48 hr) incubations in injected oocytes, though mRNAs lacking a poly(A) tail are efficiently translated.

[11] K. Sumikawa, M. Houghton, J. Emtage, B. Richards, and E. Barnard, *Nature (London)* **292,** 862 (1981).

[12] D. A. Melton, P. A. Krieg, M. Rebagliati, T. Maniatis, K. Zinn, and M. Green, *Nucleic Acids Res.* **12,** 7035 (1984).

[13] P. A. Krieg and D. A. Melton, *Nucleic Acids Res.* **12,** 7057 (1984).

[14] M. Mishina, T. Tobimatsu, K. Imoto, K. Tanaka, Y. Fujita, K. Fukuda, M. Kurasaki, H. Takahashi, Y. Morimoto, T. Hirose, S. Inayama, T. Takahashi, M. Kuno, and S. Numa, *Nature (London)* **313,** 364 (1985).

[15] J. B. Gurdon, J. Lingrel, and G. Marbaix, *J. Mol. Biol.* **80,** 539 (1973).

[16] D. Drummond, J. Armstrong, and A. Colman, *Nucleic Acids Res.* **13,** 7375 (1985).

[17] Y. Furiuchi, A. LaFiandra, and A. Shatkin, *Nature (London)* **266,** 235 (1977).

[18] G. Huez, G. Marbaix, E. Hubert, M. Leclercq, U. Nudel, H. Soreq, R. Salomon, B. Lebleu, M. Revel, and U. Z. Littauer, *Proc. Natl. Acad. Sci. U.S.A.* **71,** 3143 (1974).

[19] G. Huez, G. Marbaix, D. Gallwitz, E. Winber, R. Devos, E. Hubert, and Y. Cleuter, *Nature (London)* **271,** 572 (1978).

[20] G. Marbaix, G. Huez, A. Burny, Y. Cleuter, E. Hubert, M. Leclercq, H. Chantrenne, H. Soreq, U. Nudel, and U. Z. Littauer, *Proc. Natl. Acad. Sci. U.S.A.* **72,** 3065 (1975).

[21] H. Soreq, A. D. Sagar, and P. B. Sehgal, *Proc. Natl. Acad. Sci. U.S.A.* **78,** 1741 (1981).

[22] A. K. Deshpande, B. Chatterjee, and A. K. Roy, *J. Biol. Chem.* **254,** 8937 (1979).

It is difficult to find numerical data on the efficiency of translation (protein molecules per mRNA per time) for injected mRNAs. The rates of protein synthesis in injected and uninjected oocytes have been most carefully measured by Smith and colleagues.[6,23] Their studies show, for example, that oocytes injected with globin mRNA synthesize 5–10 ng of globin protein per hour. It is interesting to note that injected oocytes translate globin mRNA about 100–1000 times more efficiently than is observed with reticulocyte lysates. Indeed, oocytes translate injected globin mRNA about one-fourth as efficiently as do normal reticulocytes.[1]

Isolation of mRNA from Cells

RNA can be isolated from living or frozen cells by virtually any standard procedures (see elsewhere in this volume). To increase the concentration of a particular mRNA in injected oocytes it is often helpful, but not necessary, to isolate poly(A)$^+$ RNA first. In some cases the RNA to be injected is pure, for example if it has been synthesized *in vitro*. It is a good policy to store RNAs as a suspension in ethanol before use. The RNA is pelleted by centrifugation and dissolved in water, MBS-H (see below), or a low-salt buffer for injection. Detergents such as sodium dodecyl sulfate (SDS) must be avoided because they are toxic to oocytes.

Preparation of Synthetic mRNA

As mentioned above, functional messenger RNA can be synthesized *in vitro* from cloned cDNAs.[13] *In vitro* transcription of cloned cDNAs with SP6, T7 or T3 RNA polymerase allows one to prepare relatively large quantities (tens of micrograms) of pure mRNAs of known structure. The cDNA of interest is first inserted into a transcription vector downstream of a bacteriophage promoter, e.g., an SP6 promoter. (For a summary of various transcription vectors, see Krieg and Melton[24]; see, also, this volume [58,59].) This hybrid plasmid is used to prepare a template for *in vitro* transcription by cutting the DNA with a restriction enzyme downstream (3') of the inserted cDNA sequence. A variety of promoters and polylinkers in the transcription vectors make this possible. Details on transcription reaction can be found elsewhere.[12] Here we outline a typical reaction mix for the preparation of 10–50 μg of 5' capped mRNA for oocyte injection. A trace amount of [α ^{32}P]UTP is included in the reaction to label the mRNA so that the exact amount of RNA synthesized can be

[23] M. A. Taylor and L. D. Smith, *Dev. Biol.* **110**, 230 (1985).
[24] P. A. Krieg and D. A. Melton, this series, Vol. 155, in press.

determined and to facilitate following the mRNA through the purification steps. The reaction described below can be easily scaled up or down.

For a final reaction volume of 50 μl the following reagents are added to an autoclaved microcentrifuge tube at room temperature, in the order stated. First, 5 μl of linearized DNA template solution (2.5–5 μg) is added, then 16 μl of diethyl pyrocarbonate-treated water (see this volume [2]). To this mix add 5 μl of 0.1 M dithiothreitol, 2.5 μl (2 mg/ml) bovine serum albumin, 2.5 μl of RNasin (25 units/μl), 5 μl of rNTP stock (CTP, ATP, and UTP each at 5 mM and GTP at 500 μM), 5 μl of 5' cap analog (GpppG or its methylated derivatives at 5 mM), 5 μl of 10× transcription buffer (400 mM Tris–HCl at pH 7.5, 60 mM MgCl$_2$, 20 mM spermidine, 50 mM NaCl), 1 μl (1 μCi) of [α-^{32}P]UTP, and finally 2–3 μl of SP6 RNA polymerase (15–25 units). The components are mixed and incubated at 40° for 1 hr after which time more enzyme can be added to boost the synthesis. Note that all of the reagents are prepared using water that has been treated with diethyl pyrocarbonate and autoclaved to remove ribonucleases.

Following the *in vitro* transcription reaction the DNA template can be removed by addition of ribonuclease-free DNase (see this volume [21]) to a final concentration of 20 μg/ml and further incubation at 37° for 10 min. The mix is then diluted with water to 100 μl and phenol–chloroform extracted and ethanol precipitated (see this volume [4,5]). Two rounds of ethanol precipitation with 0.7 M ammonium acetate and 2.5 volumes of ethanol will remove nearly all of the unincorporated rNTP. Alternatively, the RNA can be purified by chromatography through a small Sephadex G-100 column in 10 mM Tris–HCl (pH 7.5), 0.1% SDS followed by ethanol precipitation. The RNA is dissolved in water, MBS-H, or a low-salt buffer for injection.

[*Editors' Note*. Plasmids pGEM3 and pGEM4 (Promega Biotec), often used for the purpose of synthesizing mRNA, have ATG codons in the *Sph*I site of the polylinker region on both strands. Depending upon where the desired sequence is located, these nucleotides may result in an unsuspected initiation codon upstream from the first ATG codon of the insert.]

Injecting RNA

Details of the preparation and injection of oocytes are described elsewhere (see this volume [29]). Briefly, large fully grown oocytes (stage V and VI according to Dumont[25]) are kept singly or in groups of about five oocytes in MBS-H solution at 18–22°, usually 19°, before injection. The

[25] J. Dumont, *J. Morphol.* **136,** 153 (1972).

oocytes can be conveniently kept in small petri dishes containing sterilized MBS-H solution. If the MBS-H solution is changed every 24 hr,
oocytes can be kept in this manner for at least several days and sometimes
weeks. Immediately before injection, oocytes are placed on a microscope
slide and carefully blotted dry with a paper towel and the slide is placed
on the stage of a standard dissecting microscope. If the follicle cells have
been removed from the oocytes, the injections must be done under a
solution in a dish, i.e., not dry on top of a microscope slide. The oocytes
can be easily manipulated before and during injection with a pair of
blunted watchmaker's forceps.

After the injection pipet (see this volume [28,29]) has penetrated the
oocyte, a solution of 5–50 nl of the RNA is delivered using a micrometer
syringe. The injection volumes can be measured approximately by marking the shaft of the injection pipet with ink and following the movement of
the meniscus. With automatic equipment (as in [28]), this is unnecessary.
Typically, the concentration of poly(A)$^+$ RNA used for injection is 1–10
mg/ml and up to 100 ng of RNA is injected per oocyte. For many mRNAs,
100 ng of mRNA per oocyte is a saturating concentration.[5,6] Following
delivery of the RNA, the pipet is carefully withdrawn and the oocytes are
transferred to MBS-H solution at 19° for incubation.

When injected oocytes are to be incubated in the presence of radioactive amino acids (see below) it is often helpful to begin the labeling period
about 10–24 hr *after* injection. In this way, it is possible to remove any
damaged oocytes and conserve radioactive label. Moreover, damaged
oocytes can release proteases that may adversely affect the incubation
solution shared with other injected oocytes.

Radioactive Labeling

Oocytes will take up exogenous isotopes from the incubation medium.
Indeed, isolated oocytes will even take up [^{3}H]vitellogenin and continue
to grow in culture.[26] Oocytes can be cultured in a variety of saline solutions, but MBS-H is most commonly used. The MBS-H should be supplemented with penicillin and streptomycin (see below). In cases where secreted proteins are being studied, it can be helpful to add gentamicin (50
μg/liter) and/or fungizone (200 μg/liter). Moreover, it has been suggested
that follicle cells may secrete proteases that can digest secreted proteins.
This problem can be minimized by adding 5% fetal calf serum to the MBS-
H solution.[27]

[26] R. Wallace, Z. Misulovin, and L. Etkin, *Proc. Natl. Acad. Sci. U.S.A.* **78**, 3078 (1981).
[27] H. Soreq and R. Miskin, *FEBS Lett.* **128**, 305 (1981).

Since oocytes are permeable to amino acids the injected oocytes can be incubated directly in the radioactive medium. Radioactive amino acids such as histidine, leucine, proline, valine, or methionine are commonly used. The methionine pool is relatively small (about 30 pmol/oocyte) and [^{35}S]methionine is readily available at a high specific activity. In a typical experiment, oocytes labeled with [^{35}S]methionine (1 mCi/ml; 400 Ci/ mmol; 10 hr) each oocyte will synthesize about 1×10^6 cpm of protein. Consequently, only a few oocytes are required for analysis. It is also possible to label proteins with other radioactive precursors. For example, inorganic [^{32}P]phosphate can be added to the media (at 20 mCi/ml) for studies on phosphorylation,[28] [^{3}H]acetate (50–100 mCi/ml) can be used to study acetylation,[29] and [^{3}H]mannose (1 mCi/ml) or [^{14}C]mannose (10–20 μCi/ml) can be added to examine glycosylation.[30]

The radioactive amino acids are typically dried *in vacuo* and redissolved in MBS-H at 0.5–5.0 mCi/ml. The exogenous radioactive amino acids will equilibrate with the internal pool in less than 1 hr. Injected oocytes are most conveniently incubated in the presence of this radioactive medium using microtiter 96-well plates. For studies on secreted proteins that may stick to the plastic, the dishes can be pretreated by soaking them in 0.1% bovine serum albumin in MBS-H for 30 min before rinsing with MBS-H. Injected oocytes (5–12) can be placed into a well and 2–10 μl of radioactive medium is used per oocyte. Routine incubation periods in the presence of radioactive label are 5–24 hr.

Extraction and Identification of Proteins

Proteins can be extracted from injected by a variety of methods, though it is usually helpful to take care to separate the yolk granules from the proteins of interest because the yolk is so abundant. A convenient method is to homogenize a group of oocytes gently in 50 mM Tris–HCl (pH 7.5), 1 mM phenylmethylsulfonyl fluoride $\pm$1% Nonidet P-40 (depending on the protein to be analyzed) in a loose-fitting, small glass homogenizer using about 200 μl of buffer per five oocytes. After centrifugation for 5 min at room temperature in a microfuge (about 10,000 g), two distinct phases will be obvious. The upper phase contains the labeled proteins of interest and the lower phase contains yolk and pigment granules. In most instances the upper phase can be diluted with sample loading buffer and analyzed directly by electrophoresis in SDS–acrylamide

[28] H. R. Woodland, *Dev. Biol.* **68,** 360 (1979).
[29] R. Shih, C. O'Connor, K. Keem, and L. D. Smith, *Dev. Biol.* **66,** 172 (1978).
[30] J. Mous, B. Peeters, and W. Rombatus, *FEBS Lett.* **122,** 105 (1980).

gels. If fractionation of the injected oocytes is not required (see below), oocytes can be frozen on dry ice and stored at −20° before homogenization.

Secreted proteins can be recovered from the incubation medium by precipitation. The incubation medium is first removed and clarified by 5 min centrifugation in a microfuge (10,000 g). The secreted proteins can be directly precipitated by the addition of 0.2 volumes of ice-cold 50% (w : v) trichloroacetic acid in the presence of carrier protein (bovine serum albumin) or precipitation with an equal volume of acetone–ether. The precipitates are collected by centrifugation, washed with ice-cold acetone, dried, and dissolved in sample loading buffer for gel electrophoresis. Alternatively, when antibodies for the proteins of interest are available, these can be used for immunoprecipitation (see this volume [31]). Typically, 50 μl of oocyte extract is diluted 10-fold by the addition of phosphate-buffered saline (recipe below) containing 0.1% Nonidet P-40 1 mM phenylmethylsulfonyl fluoride (pH 7.6), and a dilution of the antibody.

Several subcellular fractions of injected oocytes can be prepared to examine the intracellular location of newly synthesized proteins. The oocyte nucleus can be easily separated from the cytoplasm by placing the injected oocytes in ice-cold MBS-H and pricking the top of the animal pole to allow the large nucleus an exit point. This process is facilitated if the oocytes are first removed from their investing follicle by dissection. Gentle squeezing with watchmaker's forceps often helps the nucleus or germinal vesicle emerge. Alternatively, oocytes can be fixed by placing them in ice-cold 5–20% TCA for 10 min. Each oocyte can be dissected with watchmaker's forceps and the desired number of nuclei are pooled and centrifuged in a microfuge for 30 sec. The pellet is then washed with acetone and prepared for electrophoresis.

Oocytes can also be divided into soluble cytosol and vesicle fractions to analyze the location of newly translated proteins.[8,31] Oocytes are gently homogenized in 50 mM NaCl, 10 mM magnesium acetate, 20 mM Tris–HCl (pH 7.6) using 20 μl per oocyte. The homogenate is layered onto a 20% sucrose pad made in the same buffer and centrifuged for 30 min at 4° at about 15,000 g. The supernatant contains the soluble proteins or the cytosol. The pellets contain the yolk and vesicles and can be further fractionated by extraction with 0.2 ml of phosphate-buffered saline (130 mM NaCl, 7 mM Na$_2$HPO$_4$, 3 mM NaH$_2$PO$_4$) containing 1 mM phenylmethylsulfonyl fluoride and 1% Nonidet P-40 followed by centrifugation at 10,000 g in a microfuge. The supernatant now represents the vesicle fraction.

[31] C. Lane, S. Shannon, and R. Craig, *Eur. J. Biochem.* **101,** 485 (1979).

Reagents

MBS-H: 88 mM NaCl, 1 mM KCl, 0.33 mM Ca(NO$_3$)$_2$, 0.41 mM CaCl$_2$, 0.82 mM MgSO$_4$, 2.4 mM NaHCO$_3$, and 10 mM HEPES (pH 7.4). This can be made as a 10× stock. Antibiotics are often added to suppress bacterial growth, for example, benzylpenicillin and streptamycin sulfate each at 10 mg/liter.

[31] Characterization of *in Vitro* Translation Products

By ROSEMARY JAGUS

Earlier in this volume [27], procedures are described for the translation of mRNA in cell-free systems, as a method for the unambiguous identification of a particular mRNA species. This chapter describes the characterization of *in vitro* translation products by the most commonly used techniques. The methods include SDS–polyacrylamide gel electrophoresis (SDS–PAGE), combined with immunoprecipitation and/or fluorography of [^{35}S]methionine-labeled translation products. The other frequently used characterization tool, translation of hybrid-selected mRNA or hybrid-arrested translation, is treated separately in this volume [60]. Methods are also given for the recognition of mRNAs coding for secreted or membrane proteins.

Analysis by SDS–Polyacrylamide Gel Electrophoresis

The most widely applicable and versatile method for analysis of cell-free translation products, synthesized from mixtures of mRNAs, is polyacrylamide slab gel electrophoresis in the presence of 0.1% sodium dodecyl sulfate (SDS) and a discontinuous buffer system.[1,2] Combined with radiolabeling, using [^{35}S]methionine (2–5 mCi/ml, specific activity 800–1200 Ci/mmol) (this volume [27]), and fluorography of dried gels, this method can resolve a large number of radioactive proteins of varying molecular weights and can be used to monitor many samples simultaneously. This method can be used to detect the products of a particular mRNA species present in the translation mix at 100–200 ng/ml.

[1] U. K. Laemmli, *Nature (London)* **277**, 680 (1970).
[2] C. W. Anderson, P. R. Baum, and R. F. Gestetland, *J. Virol.* **12**, 241 (1973).

METHODS IN ENZYMOLOGY, VOL. 152

Solutions. Electrophoresis-grade chemicals should be used where appropriate.

Acrylamide: 45% polyacrylamide (w/v) (Polysciences, ultra-pure, electrophoresis grade) in water. Stored at 4° in a brown bottle. Useful shelf-life is approximately 3 months. Acrylamide is a cumulative neurotoxin. To reduce the possibility of contaminating the laboratory with acrylamide dust, it is convenient to buy acrylamide in 100 g bottles and make up the whole amount at once in the bottle. The quantity of acrylamide in the bottle can be checked by weighing the bottle full and again after the acrylamide solution has been removed.

N,N'-Methylenebisacrylamide (usually abbreviated to bisacrylamide or bis): 1.6% bisacrylamide (w/v) (Bethesda Research Laboratories, BRL, electrophoresis grade) in water. Solution must be made in warm water; otherwise solution remains cloudy. Many investigators suggest filtering bisacrylamide solutions for this reason. However, this results in a preparation of unknown concentration which can vary considerably from batch to batch. Stored at 4° in brown bottle. Useful shelf-life is approximately 3 months.

Tris–HCl: 1.5 M at pH 8.8 and 0.5 M at pH 6.8. Both are stored at 4°.

Sodium dodecyl sulfate: 4% (w/v) (electrophoresis grade). The entire migration pattern and apparent molecular weights of peptide mixtures can be changed with the source of SDS, reflecting variability in the amount of contaminating higher molecular weight forms. Consequently, it is very important to use electrophoresis grades, e.g., BRL, Bio-Rad, Pharmacia. Stored at room temperature.

Ammonium sulfate: 0.56% (w/v) (electrophoresis grade). Solution stored at 4°, in brown bottle. Solution is relatively unstable but may be used for 3–4 weeks it kept in ice while in use.

TEMED: used as supplied (electrophoresis grade)

Tris–glycine reservoir buffer: 25 mM Tris, 250 mM glycine (electrophoresis grade) at pH 8.3, 0.1% SDS

Sample buffer: 60 mM Tris–HCl (pH 6.8), 5% (w/v) SDS, 1% (w/v) dithiothreitol, 20% glycerol, 0.001% bromphenol blue. Aliquots of 1 ml are snap-frozen immediately after buffer preparation and stored at −20°. Once thawed, residual solution should be discarded.

Procedure for Pouring Gels. Table I gives recipes for the preparation of two 13 × 13 cm resolving slab gels, 0.075 cm thick, such as those accommodated by the cooled vertical slab gel apparatus manufactured by Hoefer Scientific Instruments, Bio-Rad, or LKB. Gels of 0.075–0.15 cm

TABLE I
FORMULATIONS FOR SDS–POLYACRYLAMIDE SEPARATING GELS

| | Volume for different percentages of acrylamide | | | |
Component	10%	12.5%	15%	17.5%
Acrylamide (45%)	6.6 ml	8.3 ml	10 ml	11.6 ml
Bisacrylamide (1.6%)	2.4 ml	1.9 ml	1.6 ml	1.4 ml
Tris–HCl, pH 8.8 (1.5 M)	7.5 ml	7.5 ml	7.5 ml	7.5 ml
SDS (4%)	0.8 ml	0.8 ml	0.8 ml	0.8 ml
Water	11.2 ml	10 ml	8.6 ml	7.2 ml
Ammonium persulfate (0.56%)	1.5 ml	1.5 ml	1.5 ml	1.5 ml
TEMED	40 μl	40 μl	40 μl	40 μl

may be used. The thinner gels give tighter polypeptide bands, but the thicker gels allow the loading of more sample.

After the selection of an appropriate resolving gel concentration, the gel mixture is prepared. Although polymerization of the gels is impeded by dissolved oxygen, the high concentration of ammonium persulfate recommended obviates the need to degas. The resolving gel mix is poured into the assembled gel plates, leaving sufficient space at the top for the stacking gel to be added later. The gel mix should be overlayed gently with 0.1% SDS. After polymerization (10–20 min), as evidenced by the appearance of a sharp interface between the gel and the overlay, the overlay is removed and the surface of the resolving gel is rinsed first with water to remove any unpolymerized acrylamide and then with a small volume of stacking gel mix. The remaining space is filled with stacking gel mix and the comb is inserted immediately.

The recipe for the stacking gel mix is given in Table II. The pH of 6.8 and the ionic strength are chosen to given maximum zone sharpening.[3,4] The concentration of acrylamide in the stacking gel is 5%, a concentration too low to allow any significant molecular sieving effect. The bis/acrylamide ratio is higher than in the resolving gel to give better handling properties. As a general rule, the stacking gel should not be less than twice the height of sample to be applied.

Choice of Acrylamide Concentration for Separating Gel. Recipes are given in Table I for gels of 10–17.5% acrylamide, adapted from Anderson et al.[2] The acrylamide concentration should be chosen to allow the polypeptide of interest to migrate approximately halfway along the length of

[3] L. Ornstein, *Ann. N.Y. Acad. Sci.* **121**, 321 (1964).
[4] B. J. Davis, *Ann. N.Y. Acad. Sci.* **121**, 404 (1964).

TABLE II
FORMULATION OF STACKING GEL

Component	Volume
Acrylamide (45%)	0.9 ml
Bisacrylamide (1.6%)	0.7 ml
Tris–HCl, pH 6.8 (0.5 M)	2.5 ml
SDS (4%)	0.5 ml
Water	4.7 ml
Ammonium persulfate (0.56%)	0.7 ml
TEMED	15 μl

the gel. The most useful concentration is 15%. This concentration gives the maximum (although not linear) separation of peptide mixtures between 20,000 and 100,000 molecular weight, with peptides between 55,000 and 60,000 migrating halfway down the length of the gel.

The proportions of acrylamide and bisacrylamide vary with the acrylamide concentration, such that percentage bis/percentage acrylamide equals 1.333. This ratio was determined empirically to give gels of the optimal combination of pore size, stiffness, light scattering, and swelling properties.[2]

Sample Preparation. For analysis of translation products after radiolabeling with [^{35}S]methionine the translation system should be made 5% with respect to 100 mM cold methionine and "drop" dialyzed against 60 mM Tris–HCl (pH 6.8) for 15 min, using a Millipore V series membrane (0.025 μm, 13 mm diameter).[5,6] An aliquot of 5–100 μl translation mix is dropped onto the center of the membrane, floating on dialysis buffer. After 15 min, samples are recovered with a micropipet. Recovery of protein by this technique is >98%.

Overloading of gel slots causes distortion of the peptide bands. For 3-mm gel slots, the absolute maximum recommended sample load is 250 μg for 0.75-mm gels and 500 μg for 1.5-mm gels. The rabbit reticulocyte lysate translation system contains up to 100 mg/ml hemoglobin which limits the amount of translation mix that can be analyzed by gel electrophoresis. Not more than 2.5 or 5 μl of translation mix can be analyzed directly on 0.75- and 1.5-mm gels, respectively.

The stacking effect which occurs with the SDS discontinuous system[3,4] means that the separation is independent of sample volume, but in practice, sample volume is limited by the size of the sample wells. Using

---

[5] R. Marusyk and A. Sergeant, *Anal. Biochem.* **105**, 403 (1980).
[6] Millipore Application Report No. AR524.

3-mm slots, the maximum desirable sample volume is 30 and 60 μl for a 0.75- and 1.5-mm gel, respectively.

The protein mixture is denatured by heating at 100° for 2–3 min in the presence of excess SDS to denature proteins, and dithiothreitol to cleave disulfide bonds. Most polypeptides bind SDS in a constant weight ratio of 1.4 g SDS/g polypeptide. Consequently, an SDS to polypeptide ratio of at least 3 : 1 is desirable. This means that 2.5 μl of reticulocyte translation mix should be solubilized in 25 μl SDS–sample buffer.

Gel Running Conditions. For 13 × 13 cm gels, 0.075 cm thick, using a cooled vertical slab apparatus, electrophoresis should be at 25 mA/gel, with a voltage maximum of 250 V. Immersion of the gel in the lower reservoir buffer is sufficient to provide the necessary cooling. Under these conditions, gels take approximately 4 hr to run.

Gel Processing. Gels to be used for autoradiography or fluorography (see following section), using [^{35}S]methionine as the radiolabel, should not be stained, since this quenches the emission of radioactivity. Gels should be dried prior to exposure to X-ray film.

Immunoprecipitation of *in Vitro* Translation Products

To increase the sensitivity of detection of *in vitro* translation products from 5–20 ng/ml mRNA, specific products can be immunoprecipitated from larger translation volumes.

Solutions

Immunoprecipitation buffer: 0.1 *M* NaCl, 1 m*M* EDTA, 10 m*M* Tris–HCl (pH 7.5), 1% Nonidet P-40. Store at 4°.

Protein A-Sepharose: 2 mg/ml protein A-Sepharose (Pharmacia), equilibrated in immunoprecipitation buffer.

Antibody directed against protein of interest: this may be a monoclonal antibody or antiserum. Immunoglobulins should be separated from whole serum before use by ion-exchange chromatography.[7] Store at −70°.

Procedure. Numerous examples of immunoprecipitation procedures for *in vitro* translation products may be found in the literature. A generally applicable method is described here, although the most suitable method will depend on the characteristics of the antigen and antibody in question.

Monoclonal antibodies are ineffective in precipitating unless the corresponding antigen has multiple identical determinants. Polyclonal antibodies only precipitate quantitatively at the equivalence point. It is therefore simpler in both circumstances to recover antibody–antigen complexes

[7] A. L. Hurn and S. M. Chantler, this series, Vol. 70, p. 104.

with protein A-Sepharose. Protein A, from *Staphylococcus aureus,* recognizes the Fc region of many, though not all, IgGs. For instance, rabbit, rat, and human IgGs are recognized by protein A, but mouse, sheep, and chicken IgGs are not.

If using a first antibody not recognized by protein A, it is necessary to use a second, "sandwich" antibody. For instance, if the available antibody is a mouse monoclonal, then it is necessary to use, for instance, rabbit anti-mouse IgG as the second antibody.

Immunoprecipitation of *in vitro* translation products should be performed immediately after the translation incubation to lessen nonspecific precipitation of denatured protein. A 100-μl portion of translation mix should be drop dialyzed against immunoprecipitation buffer without detergent. The dialyzate is transferred to a 1.5-ml microfuge tube and 400 μl of immunoprecipitation buffer containing 10 mM methionine is added. An appropriate volume of antibody (determined by pilot experiments) is added to this mixture, which is then allowed to immunoprecipitate at room temperature for 2 hr to overnight. It is advisable to perform preliminary experiments to determine what conditions give maximum immunoprecipitation with minimal nonspecific contamination.

Protein A-Sepharose (10 μl; 2 mg/ml) (final concentration 40 μl/ml) is added to the mixture, which is then incubated at room temperature for a further 30 min to 1 hr. The antibody–antigen complex, bound to protein A-Sepharose, is recovered by centrifugation at 12,000 g for 3 min. The pellet is washed three times with 1 ml of immunoprecipitation buffer. The detergent is omitted from the final rinse as a precaution against the generation of abnormal SDS–polyacrylamide gel electrophoretic migration patterns.

The final pellet is resuspended in 25–60 μl SDS–sample buffer and heated at 100° for 3 min to release the radiolabeled polypeptides. The protein A is removed by centrifugation and the solubilized antigen is analyzed by SDS–polyacrylamide gel electrophoresis.

Fluorography

Low-energy β particles, such as those emitted by ^{3}H, ^{14}C, or ^{35}S, are quenched within gels. Conversion of the emitted energy to visible light by an organic scintillant infused into the gel increases the proportion of the emitted energy which may be absorbed by X-ray film and so enhances the signal.[8,9] The original and excellent procedure of Laskey[8] uses the organic scintillant 2,5-diphenyloxazone (PPO) in dimethyl sulfoxide (DMSO).

---

[8] R. A. Laskey and A. D. Mills, *Eur. J. Biochem.* **56,** 335 (1975).
[9] R. A. Laskey, this series, Vol. 65, p. 363.

Solutions

Protein fixing solution: 10% (w/v) trichloroacetic acid, 10% (v/v) glacial acetic acid, 30% methanol

Dimethyl sulfoxide (DMSO): as purchased

Enhancing solution: 22% (w/v) solution of PPO in DMSO

Procedure. After electrophoresis, place gel in polyethylene tray containing protein fixing solution and agitate gently for 1 hr. After fixing, discard fixing solution and soak in 60–100 ml DMSO for 30 min (volume of DMSO should be at least 20-fold greater than volume of gel). Discard DMSO and repeat. It is essential to remove all water from the gel at this stage or PPO will not enter the gel.

After equilibration, the gel is immersed in 4 volumes 22% PPO in DMSO (approximately 60 ml). The gel is soaked with gentle agitation for 3 hr, after which the impregnated gel is soaked in excess water for 1 hr to remove the DMSO and finally dried under vacuum. The dried gel is exposed to blue-sensitive X-ray film, e.g., Kodak XK-1 or XRP-1, at −70°.

For maximum sensitivity and quantitation, it is necessary to use preexposure of the film.[8,9] Preexposure is achieved immediately prior to use with a single flash from an electronic photographic flash unit such as the Vivitar 283. It is essential that the duration of the flash is short (<1 msec) and the intensity of the flash adjusted to increase the absorbance of the film to $A_{540} = 0.1$–0.2 above the absorbance of unexposed film. Three filters are needed for the window of the flash unit to reduce and diffuse the light output: (1) an infrared-absorbing filter, (2) a colored filter to reduce light output, e.g., Kodak Wratten 21 or 22, (3) translucent diffusing filter to give even exposure (Whatman 1 filter paper is good for this).

Sensitivity. Using preexposed blue-sensitive X-ray film, with exposure at −70°, this method will detect 400 dpm/cm² after 24 hr. This is 15-fold more sensitive than standard autoradiography.[8,9] An alternate method using sodium salicylate as the scintillant has been described.[10] The procedure is more rapid and much less expensive, but does not give quite the same sensitivity as the PPO–DMSO method and gives bands which are somewhat more diffuse.

An acetic acid-based enhancer solution, called EN³HANCE, is available commercially from New England Nuclear. Gel preparation for fluorography using EN³HANCE is very rapid (two-thirds of an hour instead of 5 hr) and less hazardous since DMSO is not used. Use of EN³HANCE gives even higher sensitivity than PPO–DMSO or sodium salicylate; however, it is more expensive than either of the above methods. New England Nuclear markets another enhancer called Enlightening which is also

[10] J. P. Chamberlain, *Anal. Biochem.* **98**, 132 (1979).

very efficient, very rapid (gel processing time of 15–30 min), but very expensive.

Recognition of mRNAs Coding for Secreted or Membrane-Bound Proteins

Occasionally an investigator is pleased just to know whether the mRNA for a cloned gene codes for a secreted or membrane-bound protein. The following section suggests two methods that may alert the investigator to this possibility.

Size of de Novo Translation Product. With only rare exceptions (e.g., ovalbumin), the N-termini of secreted, lysosomal, and membrane proteins consist of a cleavable leader of 16–30 amino acids, which act as signals for the insertion of proteins into or through the membranes of the endoplasmic reticulum.[11] Signal sequences are cleaved by a membrane enzyme before the polypeptide chain is completed.

Primary translation products of mRNAs coding for such proteins, synthesized in the absence of microsomal membranes, are slightly larger than the authentic proteins or translation products synthesized in the presence of membranes. The presence of an additional 16–30 amino acids is sufficient for it to be resolved from the normal-sized product by SDS–PAGE, unless the protein is very large. Electrophoretic mobility can therefore be used as a simple assay for the presence of a signal sequence. It should be noted that protease inhibitors recommended for inclusion in cell-free translation assays (this volume [27]) not be included when cleavage of the signal sequence is required.

If using such a method, it should be noted that there are differences to be found between the reticulocyte and wheat germ cell-free systems. In the reticulocyte cell-free translation system, rough microsomes or KCl-washed rough microsomes are both active in cleaving the signal sequence.[12] In the wheat germ system, only rough microsomes are.[12] Note that a method for the preparation of rough and salt-washed dog pancreas microsomes is given in this volume [27].

Translocation of Polypeptide Can Confer Protease Resistance. Unfortunately, not all membrane, lysosomal, or secreted proteins possess signal sequences. In addition, electrophoretic mobility can be changed by additional modifications of the polypeptide such as glycosylation. However, cleavage of signal sequences is a cotranslational process that occurs

[11] C. Milstein, G. G. Brownlee, T. M. Harrison, and M. B. Mathews, *Nature (London) New Biol.* **239**, 117 (1972).
[12] D. I. Meyer, E. Krause, and B. Dobberstein, *Nature (London)* **297**, 647 (1982).

along with translocation of the nascent polypeptide across the microsomal membrane. This means that in cell-free translation systems such proteins become sequestered within the lumen of the microsomal vesicles and are protected from attack by proteases.

The addition of pronase (100 μg/ml) or a mixture of trypsin and chymotrypsin (each 50 μg/ml) at the end of cell-free translation in the presence of rough microsomes, followed by incubation for 60–90 min at 20°, should destroy any products not protected within the membrane vesicles. The appropriate positive control for such experiments would be to monitor the effect of inclusion of 0.1% Triton X-100 in the translation assay to solubilize the microsomal vesicles. The sample preparation for such incubations should include precipitation with 10% trichloroacetic acid before solubilization in SDS–sample buffer to prevent residual protease activity from digesting the translation products once SDS is added. If immunoprecipitation is required before analysis, an excess of appropriate protease inhibitors should be added to the translation mix, followed by Triton X-100 (to 0.1%), prior to immunoprecipitation.

Section VIII

Preparation of cDNA and the Generation of cDNA Libraries

[32] Preparation of cDNA and the Generation of cDNA Libraries: Overview

By ALAN R. KIMMEL and SHELBY L. BERGER

The conversion of mRNA into cDNA for the purpose of cloning is a complex, interrelated series of enzyme-catalyzed reactions. These reactions together with their substrates are presented in pictorial form in Fig. 1. The scheme consists of six basic approaches (routes), each of which is discussed in condensed form in the first section entitled, Description of the Scheme for cDNA Cloning. The experienced reader will be able to see graphically and verbally how cDNA is constructed, starting with single-stranded mRNA and ending with double-stranded, clonable DNA that is closely related to the original mRNA population. The beginner may have difficulties, since this chapter is not a simplified introduction to cloning. To help the novice progress toward an understanding of the strategies involved in the selection of cloning methods, several suggestions are offered. First, the reader should assimilate chapter [11], "Properties and Uses of Restriction Endonucleases." Fundamentally, cDNA can be defined as a group of molecules possessing the information contained in the various mRNAs, but with termini equivalent to or related to those produced by restriction endonucleases. Cloning is preoccupied with termini because the finished DNA product must be either ligated (covalently attached) or annealed (hybridized by means of complementary polymerized "tails") to a vector with compatible termini. Second, an understanding of the reactions catalyzed by the enzymes is essential; clearly, these enzymes dictate what is possible and, depending on their individual specificities, how well the possible will actually work. This information is concentrated in chapter [10] but some enzymes are discussed only in other chapters; consult the Process Guide and Index for their locations. Finally, chapter numbers (in square brackets) indicate where each technique mentioned in the overview is discussed in detail in the book, usually with an appropriate introduction. Taken together, a multichapter assault should help to reduce the complexity.

The overview also contains a second section, Strategy for Selection of a Cloning Route, which attempts to direct the reader quickly to those aspects of Fig. 1 of specific value. If the novice has become thoroughly confused by the first section, it may be advisable to read the strategy section first. Different investigators, depending on their objectives, will wish to focus on different routes. Once oriented, the reader can target the

METHODS IN ENZYMOLOGY, VOL. 152

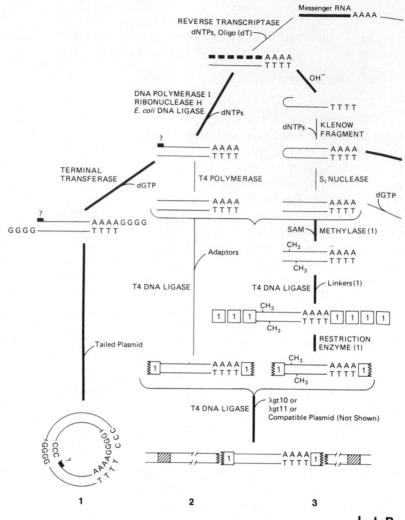

L I B

FIG. 1. Scheme for cDNA cloning. The scheme consists of six routes labeled at the bottom of the figure. (Routes 2 and 3 converge on the same set of vectors.) The following symbols have been used: messenger RNA, heavy horizontal line with consecutive As representing the poly(A) tail; mRNA fragments hybridized to first-strand cDNA, heavy dashed line; first or second strand cDNA, light line with consecutive Ts or As, respectively; putative RNA fragments covalently linked to second-strand cDNA, short heavy line underneath a question mark; homopolymeric dG tails, consecutive Gs; homopolymeric dC tails, consecutive Cs; oligonucleotide linker with a restriction site for enzyme 1 or 2, box surrounding 1 or 2, respectively; cleaved oligonucleotide linker, box with jagged or curvy edge; plasmid, concentric circles; λ, parallel lines with —//— representing sequences not shown and hatched box representing the *cos* sites; cleaved restriction site of plasmid or bacteriophage

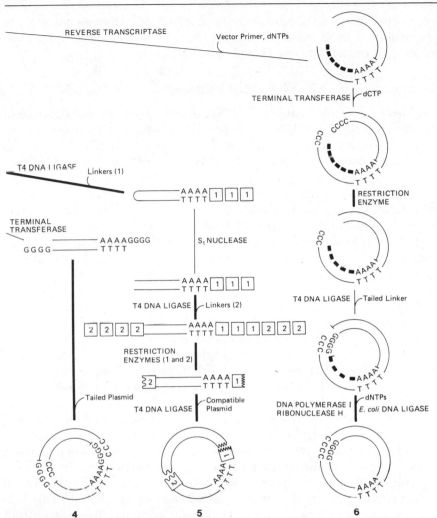

RARY

compatible with cleaved oligonucleotide linker, curvy or jagged edge; and gap or nick, either space between bases or ⌐. Double-stranded cDNAs in brackets are functionally equivalent and therefore interchangeable. Nucleic acid substrates have the first letter capitalized. The vector–primer of route 6 consists of a double-stranded plasmid with a T tail at one 3' terminus. The tailed linker of the same route has dG tails at one 3' terminus and a cleaved restriction site compatible with that of the vector at the other end of the molecule. Enzymes are completely capitalized. The thickness of the connecting lines reflects the relative efficiency of the reaction; high yields are indicated by heavy lines whereas low yields or tricky reactions are shown by light lines. No meaning should be attached to the length of the lines. Abbreviations: SAM, S-adenosyl-L-methionine; dNTPs, deoxyribonucleoside triphosphates.

techniques that need to be mastered and ignore those that do not apply to the project at hand. After reading the chapters that describe those techniques, an overall strategy should become clearer.

Description of the Scheme for cDNA Cloning

There are many routes to a cDNA library. They are not equivalent. Rather they were developed to improve the quality and the quantity of clones, to streamline the process of cloning, and to broaden the possibilities for screening. Here, we compare six basic methods (routes) for constructing a cDNA library.

The first step in cDNA cloning, regardless of pathway, is the synthesis of a DNA strand complementary to the mRNA sequence. The reaction requires template RNA, a complementary primer, reverse transcriptase, and deoxyribonucleoside triphosphates. Routes 1–5 outlined in Fig. 1 make use of a primer composed solely of thymidylate residues to initiate this reaction [33]. Oligo(dT) is the most frequent choice because it can hybridize to 3′ poly(A) tails of template mRNA and theoretically give rise to full-length copies. The yield of such cDNA depends upon the quality of the reverse transcriptase, the length of the mRNA to be copied, and the degree to which mRNA secondary structure interferes with extension of the oligo(dT) primer. The product of this reaction is single-stranded DNA of varying lengths that is base paired with fragments of the original mRNA template. Yields of full-length cDNA are less than 50% relative to input mRNA.

Other molecules used for cDNA priming include random oligomers of DNA [46] as well as specific oligomers [66]. Random hexamers of DNA as primers ensure equal representation of all mRNA sequences in the cDNA library but only at the expense of full-length cDNA copies. In contrast, specific oligomers, able to recognize a known sequence of interest, can direct transcription to one or to a limited number of mRNAs. This process, called primer extension, intentionally generates fragments. Both approaches are frequently employed as analytical tools rather than as a prelude to cloning.

Two methods are presented for second-strand synthesis. One method makes use of the fragments of the original RNA molecules as primers to generate complementary cDNA strands by a process analogous to nick translation (route 2) [35]. Alternatively, a hairpin loop occurring as a transient structure at the 3′ end of the newly synthesized first strand can serve as a primer to generate self-complementary molecules [34] as shown in route 3. In route 3, the hybridized RNA fragments must first be removed and hydrolyzed with alkali. To improve the efficiency of transcribing naked DNA, Klenow fragment is often supplemented with reverse transcriptase. Apparently, the combination of enzymes is more effective for transcribing regions with secondary structure. The conclusion must be drawn that the method in route 3 is still less efficient than the approach dependent on RNA fragments as primers shown in route 2.

With the completion of second-strand synthesis, a method must be selected for coupling the insert (double-stranded cDNA) to the vector (plasmid or bacteriophage DNA). There are two basic options: either the cDNA is tailed and annealed to a cleaved plasmid possessing complementary homopolymeric tails [37], or symmetric, double-stranded oligomers containing a recognition site for a restriction endonuclease (linkers) are covalently attached to the cDNA [38, 39]. The latter approach requires cleavage of the modified cDNA with a restriction endonuclease to generate termini capable of being ligated to a compatible vector. A simpler variation in which asymmetric adaptors substitute for linkers (see below) will also be discussed [38].

For cloning cDNA by tailing, route 1 is far superior to route 4. It is faster and more efficient and usually yields longer cDNA clones because it obviates the requirement for cleavage of the hairpin loop with a single-strand-specific nuclease such as mung bean or S_1. Either nuclease can eliminate unpaired sequences at the hairpin, nucleotides that are derived originally from the 5' ends of mRNA. Furthermore, both enzymes, particularly S_1, can attack double-stranded termini as they "breathe" and introduce nicks into double-stranded molecules. Thus, not only is it impossible to produce full-length cDNA by the method outlined in route 4, but also the ability of terminal transferase to tail at nicks or gaps creates branched structures that interfere with cloning [37]. Although, for historical reasons, many recombinant plasmid libraries exist which were constructed via route 4, it is no longer recommended. A similar warning applies to the cloning of tailed molecules in λ phage; it is extremely difficult to prepare and characterize very large tailed vectors.

The alternative to tailing is the attachment of linkers or adaptors to blunt-ended cDNA molecules. Creation of blunt cDNA ends entails one or more of the following procedures: (1) Klenow fragment to fill in putative 3'-recessed termini [10], (2) S_1 or mung bean nuclease to digest overhangs of either strand [34], or (3) T4 polymerase to fill in 3'-recessed termini [10], and to remove 3' overhanging sequences [35] by means of the 3'-exonuclease activity. Routes 2, 3, and 5 make use of these techniques.

Routes 2 and 3 share intermediates and contain functionally equivalent molecules in brackets (Fig. 1) at two locations. One can switch pathways where such brackets occur.

An analysis of route 2 compared with 3 (starting with the bracketed double-stranded cDNA at the top of Fig. 1) reveals that route 3 is labor intensive yet results in a larger number of clonable molecules. The difference between the methods stems from the requirements of linker as contrasted with adaptor technology [38]. Linkers used in large molar excess relative to inserts help to drive the ligation reaction of insert with linker. In the process, concatamers of linkers at both termini of the cDNA, as well as free polymerized linkers, are synthesized. The high concentration of linkers competitively inhibits self-ligation of inserts. In order to create the staggered cDNA ends needed for efficient ligation into vectors, the linker concatamers are cleaved by a restriction endonuclease. To protect insert DNA from the activity of this enzyme, the double-stranded cDNA in the upper bracket connecting route 2 and 3 is methylated [38]. Methylation prevents specific sequences from serving as substrates for restriction endonucleases.[1] In contrast, the use of adaptors (route 2) does away with the requirement for such cleavage [38]. Adaptors are small, asymmetric, double-stranded molecules with one flush end suitable for ligation to blunt-ended double-stranded cDNA and one staggered end complementary to the staggered ends of a cleaved vector. Ligation of adaptors to cDNA is inefficient because concentrations must be regulated and 5'-terminal phosphate groups must be manipulated in order to prevent self-ligation of adaptors. Thus, three highly efficient steps of route 3—methylation, ligation of linkers, and cleavage of linkers—replace one inefficient step of route 2—ligation of adaptors. The products of routes 2 and 3 can be cloned into plasmid [39][2] or bacteriophage [40] vectors with equal ease.

When the orientation of the insert with respect to the vector must be predetermined, route 5 should be considered [39]. Such directional cloning is made possible by methods that differentiate 5' ends of sense strands (strands equivalent in sequence to mature mRNA) from 3' ends. The use of cDNA with self-complementary hairpins achieves that distinction. In this pathway, the first type of linker is attached to the blunt ends of the cDNA, the hairpin loop is cleaved, and the second type of linker is added to both termini of the now double-stranded cDNA. By cleaving with the two restriction endonucleases whose recognition sequences appear in the two types of linkers, clonable cDNAs are produced. When inserted into plasmids cleaved with the same two restriction endonucleases, the orientation of the insert is fixed relative to that of the vector [39]. Like route 4,

---

[1] Methylation prevents cleavage of insert DNA by certain restriction endonucleases. Once the DNA is propagated as a recombinant plasmid or phage *in vivo*, the final methylation pattern will be a function of the restriction modification system of the host (see this volume [12]).

[2] Chapter [39] contains technical guidelines for cloning cDNA molecules that terminate with a single type of linker or with two different types of linkers. Methods are also presented for optimizing ligation of inserts to plasmid vectors and transformation of competent bacteria in chapter [13].

this route precludes full-length molecules. Unlike route 4, however, the result is more favorable, since linkers cannot be attached to nicks. Furthermore, since this approach is used virtually exclusively for expressing the sequences encoded in the inserts, the absence of some 5' untranslated cDNA sequences from the cloned product is sometimes an advantage; noncoding sequences usually inhibit synthesis of heterologous proteins [69, 71].

Route 5 uses a different approach from route 3 for manipulating inserts. In route 5, the cDNA is not methylated but is nonetheless subjected to digestion with restriction endonucleases. If cleavage within the insert were to occur, depending on the location of the restriction site, more efficient expression of at least a portion of the desired protein could be the result. To maximize the possibility of producing expressible fragments after cleavage, the method requires that a second library be constructed in which the positions of the linkers are reversed. Concomitantly, in this library, the position of the restriction sites of the vector are also reversed with respect to the heterologous promoter used for expression. The result is two sublibraries which can be combined. The inserts in this library may be flanked by the following: two types of linkers; one linker and one internal site in cDNA; and, if sites recognized by both restriction enzymes occur in insert DNA, no linkers may appear.

Route 6 departs from standard oligo(dT) priming of cDNA with the aim of maximizing the percentage of full-length clones [41]. Toward this end, the primer and the vector are incorporated into a single molecule. The steps in cloning here are similar to elements found in routes 1–3 but the order is rearranged. Chronologically, the following reactions are carried out: (1) synthesis of first-strand cDNA primed with a T-tailed vector; (2) polymerization of homopolymeric tails both on single-stranded cDNA (with mRNA fragments still hybridized) and on the 3' terminus of the plasmid; (3) removal of tails from the plasmid by cleavage with a restriction enzyme; (4) ligation of an oligomer (usually called a tailed linker) to the plasmid; (5) cyclization by means of the annealed complementary homopolymeric tails on linker and plasmid as in route 1; and (6) concomitant synthesis of second-strand cDNA as described in route 2.

One step requires attention. The removal of tails from the plasmid (step 3 in route 6) can be accomplished without first protecting the insert DNA because single-stranded cDNA and DNA–RNA hybrids are usually not substrates for restriction enzymes. In addition, the tailing of nicks in the plasmid is probably not deleterious here. The inclusion of DNA ligase together with a polymerase with 5'-exonuclease activity during second-strand synthesis probably results in degradation of tails before the recombinants are transfected into the host.

Like route 5, route 6 results in directional cloning. The vector–primer can include either prokaryotic or eukaryotic promoters situated so as to direct expression of the encoded sequences in appropriate cells.

Route 6 is usually thought of as a means for increasing the number of full-length molecules. This is correct in one respect. The percentage of full-length molecules is enhanced, but only at the expense of total clones. Because incomplete first-strand cDNA produced in step 1, with its accompanying 3′ recessed terminus, cannot be tailed efficiently in step 2, incomplete molecules are at a disadvantage. Furthermore, since this approach, unlike all other routes, requires a 3-fold molar excess of mRNA relative to the vector–primer, it is actually a highly inefficient process.

Strategy for Selection of a Cloning Route

The most important decision an investigator must make before constructing a cDNA library is how the library will be screened. If only nucleic acid probes are to be used, any of the procedures outlined in Fig. 1 will suffice [43, 45]. However, if antibodies will be used to identify clones capable of synthesizing specific peptides, expression vectors such as the pUC family of plasmids [50] employed in route 5 or λgt11 phage [51] mentioned in route 2 must be chosen.

The second factor that must be considered is the number of recombinants needed to ensure that a specific RNA sequence is represented in the library [40, 46]. Minimally, 5-fold more individual recombinants need to be generated than the number expected, given the relative abundance of the desired mRNA in the population to be cloned. That is, one needs to construct a recombinant library composed of >5000 different cDNAs to have a >99% probability of finding a specific sequence that represents ~0.1% of the mRNA population. For most expression libraries constructed in λgt11, the equivalent number is at least 6-fold greater, since only one of every six clones is a template for the appropriate peptide. This last value is based on the possible orientations of the clone and its frame relative to the initiation codon. One may need to screen >10^6 clones to identify a rare mRNA using an antibody. Although phage [40] and plasmid [39] cDNA libraries can be constructed and handled with approximately equal efficiency and facility, phage vectors are often the choice for large libraries of >100,000 recombinants whereas libraries of 10,000 to 50,000 members are often prepared in plasmid vectors.

The most direct method for estimating the abundance of mRNA requires a nucleic acid probe. Probes can be synthetic oligomers based on an amino acid sequence or a homologous nucleotide sequence from a related cloned gene. Methods for quantifying mRNA in solution [67] or bound to a solid support [61, 62] by hybridization with labeled probes are detailed. Estimates of the relative abundance of an mRNA made without a nucleic acid probe are indirect and usually inaccurate. If an antibody exists, one can calculate the relative abundance of a specific radiolabeled protein by immunoprecipitation after translating bulk mRNA [31]. However, in order to relate the amount of protein to the amount of a specific mRNA in the population used to direct protein synthesis, assumptions (often sheer guesswork) about the translation efficiency of the given mRNA and relative abundance of the radiolabeled amino acid in the given protein are made.

Several general recommendations are offered. Route 1 [33, 35, 36, 37, 13] is the simplest, most efficient procedure for cDNA cloning particularly for libraries of 10,000 to 50,000 clones. An amount of 10–40 μg of poly(A)$^+$ RNA will be needed to generate this library. It should be screened exclusively with nucleic acid probes. Although in the past the dG-dC tails made sequencing or preparing hybridization probes problematic, many of these difficulties have now been resolved [57–59]. Any of a variety of vectors is suitable. The pUC plasmids[3] have largely supplanted pBR322 for the following reasons: (1) they have a polylinker (multiple cloning site); (2) they can be propagated at a higher copy number; (3) they are smaller; (4) they contain a promoter for β-galactosidase which can direct attention exclusively to clones with inserts; and (5) they do not contain sequences that inhibit transformation of eukaryotic cells.

λgt10 is the obvious choice of vector for large libraries that are to be screened with nucleic acid probes [33, 35, 36, 40]. For this purpose the cDNA should be prepared by using the beginning steps of route 2, up to the first switching point, and completing the construct using the steps outlined in route 3. Once again, 10–40 μg of starting mRNA may be necessary to generate a library of $>10^6$ clones.

For small or large libraries that are to be screened with antibodies, use λgt11 as the vector and prepare the cDNA as suggested above for λgt10. Toward the same end, one can also use route 5 and a plasmid expression vector such as pUC18 or pUC19 [39]. There are more steps in route 5, but some investigators prefer it because characterization of cDNA inserts in plasmids is easier than in λ. Thus, one recoups some of the labor involved in constructing the plasmid expression library when the desired clone is

---

[3] Recently, new varieties of cloning vectors have been developed. They incorporate many of the advantageous characteristics of pUC vectors. In addition, they contain sequences recognizable by phage RNA polymerases [30, 59]. Although these new vectors have not been widely used for the construction of cDNA libraries, there are no obvious features to disqualify them.

analyzed. Nevertheless, the λ vector is by far the most common method for constructing expression libraries.

Finally, if mRNA is not limiting and if full-length clones are the major objective, route 6 is recommended [41]. Nucleic acid probes are used for screening. The expression elements that can be included in the vector primer are usually not used for initial screening; however, they can become important when the cloned sequence is transfected in order to select for phenotypic expression. One should keep in mind that it is always possible and often preferable to isolate partial cDNAs constructed by routes other than 6 and, subsequently, using specific fragments as primers, to repeat the process of cloning to enrich for the missing segments of desired cDNAs [66].

[33] First-Strand cDNA Synthesis Primed with Oligo(dT)

By Marc S. Krug and Shelby L. Berger

The quality of a cDNA library depends on the integrity of the messenger RNA and the fidelity with which it can be reverse transcribed. RNA cannot be cloned directly; in a reaction catalyzed by reverse transcriptase, the RNA, together with a suitable primer and a supply of deoxyribonucleoside triphosphates (dNTPs), must be converted to a double-stranded molecule. The product contains a complementary strand (first, antisense, or minus-strand cDNA) that is hybridized to what remains of the original RNA template. Such DNA–RNA hybrids can be cloned (see this volume [41]) albeit often with lower efficiency than their double-stranded DNA counterparts. Usually the hybrid molecules are treated as intermediates in a scheme aimed at replacing the fragmented RNA with continuous DNA to form a double-stranded cDNA molecule (this volume [34, 35, 41]). Upon modification with homopolymeric tails (this volume [37]) or linkers (this volume [38]), a well-constructed double-stranded cDNA can be inserted into a plasmid (this volume [13,39]) or phage (this volume [40]) vector that is capable of replicating in bacteria and, in some cases, capable of giving rise to mRNA and protein.

From this necessarily brief summary of cDNA cloning, it should be obvious that, regardless of the strategy, reverse transcriptase is a key element. An understanding of what reverse transcriptase does and how it does it *in vitro* is essential. Like many polymerases, reverse transcriptase catalyzes more than one reaction. The enzyme embodies two functions *in vitro,* a polymerase activity and an associated ribonuclease H (RNase H) activity.[1] The polymerase, the desired activity, requires a template RNA molecule hybridized to a DNA primer with a 3'-hydroxyl group, and all four deoxyribonucleoside triphosphates, in order to synthesize, processively, a DNA molecule which is a faithful complement of the RNA.[1] During this process, the RNase H, an exonuclease that degrades RNA in DNA–RNA hybrids, is fully active. The products of the exonucleolytic cleavage reaction are oligomers 2 to 30 bases in size. When poly(A) RNA is mixed with an oligo(dT) primer, RNase H can deadenylate the RNA in the presence or absence of dNTPs.[2] Without its poly(A) tail, template RNA cannot bind to primer DNA and, therefore, cannot contribute to the yield of full-length cDNA. As a further complication, RNA fragments produced by RNase H *after* the polymerase has synthesized complementary DNA can also act, inefficiently, as primers; owing to their small size, oligomers of RNA dissociate from the newly synthesized cDNA at moderate temperature. When such fragments are unwittingly reverse transcribed, the resultant DNA oligomers are excellent primers. Both RNA and DNA fragments represent a mixture of great complexity that is able to hybridize to internal sequences of the RNA; the priming of RNA molecules internally precludes production of full-length cDNA. Clearly, the RNase H activity of reverse transcriptase is counterproductive during first-strand cDNA synthesis.

Much effort has been expended in attempting to inhibit, selectively, the RNase H activity of reverse transcriptase from avian myeloblastosis virus. Although claims of success have been reported,[3–5] they did not withstand more careful screening. For example, pyrophosphate,[6] transient low pH,[6] and fluoride[7] do not selectively inhibit RNase H activity. Neither does RNasin[6] (ribonuclease inhibitor from placenta) nor do ribonucleoside–vanadyl complexes[6]; both nuclease inhibitors are without effect on either function of reverse transcriptase when high levels of dNTPs are included. The presence of high concentrations of deoxyribonucleoside triphosphates does not inhibit RNase H.[6] The triphosphates have no effect on RNase H activity, but rather enhance polymerase activ-

[1] I. M. Verma, *in* "The Enzymes" (P. D. Boyer, ed.), 3rd ed., Vol. 14, p. 87. Academic Press, New York, 1981.

[2] S. L. Berger, D. M. Wallace, R. S. Puskas, and W. H. Eschenfeldt, *Biochemistry* **22,** 2365 (1983).

[3] J. C. Meyers and S. Spiegelman, *Proc. Natl. Acad. Sci. U.S.A.* **75,** 5329 (1978).

[4] M. Gorecki and A. Panet, *Biochemistry* **17,** 2438 (1978).

[5] L. Brewer and R. D. Wells, *J. Virol.* **14,** 1494 (1974).

[6] M. S. Krug and S. L. Berger, unpublished.

[7] S. K. Srivastava, E. Gillerman, and M. J. Modak, *Biochem. Biophys. Res. Commun.* **101,** 183 (1981).

ity by supplying limiting substrates. In fact, there is no additive that is capable of specifically inhibiting the ribonuclease H activity of reverse transcriptase under conditions in which full polymerase activity is maintained. Nevertheless, it *is* possible that preparations of reverse transcriptase differ in the ratio of the two activities. Such dissimilarities may be a consequence of partial protease degradation of the subunits or holoenzyme, dissociation of subunits, or the presence of unidentified components in the enzyme solution.

Several approaches to cDNA synthesis are presented. Methods for synthesizing cDNA with vector–primers (T-tailed plasmids) and conditions for using cDNA synthesis as a means for generating labeled probes (see also this volume [41, 46]) are also detailed.

The reader is enjoined to create a ribonuclease-free environment in the laboratory (reviewed in this volume [2]) and to prepare poly(A)$^+$ RNA of high quality (this volume [20, 21, 25]) before proceeding.

Reverse Transcriptase. Commercially available reverse transcriptases are of two types: cloned mouse Moloney leukemia virus (MMLV) reverse transcriptase (Bethesda Research Laboratories, BRL); and reverse transcriptase isolated from avian myeloblastosis virus (AMV) [Life Sciences; Molecular Genetic Resources; Pharmacia; Takara Shuzo (available from Amersham)]. The origin of reverse transcriptases from other sources is often proprietary information. Nevertheless, most companies market enzyme from one of the above sources. Each of these enzymes has been tested and conditions for their use have been optimized in our laboratory using a synthetic 7.5-kb mRNA (BRL) as the template. Approximately 55% of these RNA molecules have poly(A) tails (based on the assay in this volume [26]) and none has a cap structure. The enzymes have also been evaluated with a genuine capped mRNA, namely rabbit globin mRNA, but because of its small size (600 nucleotides) and apparent stability, globin mRNA cannot be used to distinguish among the enzymes. Note that any particular mRNA may respond differently from the 7.5-kb test RNA. The reader should also be advised that the choice of enzyme is probably irrelevant for templates of <2 kb *provided the reaction conditions detailed here* and not those of the manufacturer are employed. In many cases, suppliers seem to recommend suboptimal protocols for synthesizing cDNA.

The criteria for evaluating reverse transcriptases are based on maximizing the yield of full-length first-strand cDNA, minimizing the yield of small fragments, either single stranded or double stranded, and inhibiting second-strand synthesis, a reaction in which reverse transcriptase uses DNA rather than RNA as the template by forming a hairpin (partial DNA duplex) at what was the 5′ end of the mRNA. (When hairpins are desired for priming second synthesis as in this volume [34], they may be generated as needed regardless of whether they appeared during synthesis of the antisense strand.) Our criteria do not include maximizing the incorporation of radioactive precursors into trichloroacetic acid-insoluble cDNA because such measurements are unreliable. Fortunately, those preparations that best satisfy the requirement for full-length cDNA also best satisfy the others; it is not necessary to compromise. We routinely use AMV reverse transcriptase from Molecular Genetic Resources.

In general, high concentrations of reverse transcriptase (and dNTPs) promote synthesis of full-length cDNAs. AMV reverse transcriptase is supplied as a concentrated solution in buffered 50% glycerol. Although this solvent stabilizes the enzyme, it is not advantageous for activity. Upon dilution of the enzyme 10-fold into 10% glycerol, 10 mM potassium phosphate at pH 7.4, 0.2% Triton X-100, and 2 mM dithiothreitol, and incubation in ice for 30 min with occasional stirring, the AMV enzyme from all sources tested increased in activity.[6] The diluted enzyme at 250–500 units per milliliter in the reaction mixture performed about twice as well (based on the above criteria) as the undiluted enzyme at 1000 units per milliliter added directly to the reaction.[8] Since enzymes are known to undergo conformational changes when transferred from fully aqueous solutions to solutions containing high concentrations of glycerol,[9,10] this observation suggests that a conformational change, which may involve the subunits, may have occurred. Note that dilution of the enzyme is not simply a method for obtaining the same amount of cDNA with less enzyme; it is a means for improving the yield of full-length cDNA, a result which excessive amounts of undiluted enzyme cannot equal. The cloned enzyme from mouse Moloney leukemia virus is highly unstable and should not be diluted.

Other Reagents. Reverse transcriptases should be stored at −70° in small aliquots. Reactions are performed in buffers prepared from concentrated stock solutions: AMV reverse transcriptase 10× buffer (500 mM Tris–HCl at pH 8.3,[11] 500 mM KCl, 100 mM MgCl$_2$, 10 mM dithiothreitol, 10 mM EDTA, 100 μg/ml bovine serum albumin; and MMLV reverse transcriptase 10× buffer (500 mM Tris–HCl at pH 8.3,[11] 750 mM KCl, 30

[8] Since unit measurements undoubtedly differ among manufacturers, even when unit definitions are identical, the maximum activity for each enzyme was determined. In all cases, the diluted AMV enzyme at 250–500 manufacturer's units per milliliter in the final reaction mixture gave maximum yields of cDNA.

[9] S. L. Bradbury and W. B. Jakoby, *Proc. Natl. Acad. Sci. U.S.A.* **69**, 2373 (1973).

[10] J. S. Myers and W. B. Jakoby, *J. Biol. Chem.* **250**, 3785 (1975).

[11] Adjust pH at the temperature at which the reaction is run: 42° for AMV enzyme and 37° for the cloned enzyme from MMLV.

mM MgCl$_2$, 100 mM dithiothreitol, 1 mg/ml bovine serum albumin. A concentrated solution (100 μg/ml) of the primer oligo(dT)$_{12-18}$, is prepared; it is brought to 0.1% diethyl pyrocarbonate, stirred vigorously at room temperature, and heated at 70° for 1 hr. Deoxyribonucleoside triphosphates (dNTPs) are dissolved at high concentration (>40 mM) and diluted to stock solutions of 10 mM each. It is also useful to prepare a mixture containing 10 mM of each dNTP. These solutions can be filter sterilized, a procedure which appears to remove nucleases. Use the extinction coefficients listed in this volume [6], together with the absorbance of the solution at the recommended wavelength, to determine the concentration of the stock solutions. Do not store dilute solutions of dNTPs. Other required solutions are 80 mM sodium pyrophosphate [treated with diethyl pyrocarbonate as for oligo(dT)], 10 mM spermidine–HCl, and actinomycin D at 1 mg/ml. The concentration of actinomycin can be ascertained by measuring absorbance at 441 nm and computing the precise value based on a molar extinction coefficient of 21,900 in water and a molecular weight of 1255. Actinomycin D is light sensitive and toxic and should be handled with gloves at all times regardless of whether or not RNA is involved. It can be stored in a foil-wrapped vessel at −20°.

The construction of T-tailed plasmids for use in the vector–primer approach to cDNA synthesis is detailed elsewhere in this volume [41].

Synthesis of cDNA: Oligo(dT) as Primer

These methods were developed based on the criteria, summarized above, for evaluating reverse transcriptases. Reaction mixtures contain either pyrophosphate or actinomycin D to inhibit second-strand synthesis, spermidine, and a ribonuclease inhibitor.

Regardless of the source of the AMV enzyme, the yield of full-length cDNA is enhanced by the addition of pyrophosphate. Although the mode of action of this substance remains obscure the practical result of including it is clear: an increase in large cDNA, a decrease in small cDNA, and greatly reduced formation of hairpins. Actinomycin D is qualitatively similar but quantitatively less effective. However, as we have not found conditions under which the enzyme from MMLV functioned in the presence of pyrophosphate, actinomycin D was used as a substitute. Either substance inhibits incorporation of radioactive precursors into acid-insoluble cDNA, probably by reducing synthesis of fragments of the second, or sense, strand. Apparently, these materials do not act synergistically because addition of both to the reaction results in inhibition of the synthesis of both large and small cDNA.

The presence of spermidine results in as much as a 2-fold increase in full-length cDNA in the presence or absence of pyrophosphate or actinomycin D using the AMV enzyme. Amounts greater than those recommended are inhibitory. (Spermidine inhibits the MMLV enzyme.)

Strictly speaking, a ribonuclease inhibitor should have no effect on cDNA synthesis if all materials are free of nucleases. Nevertheless, either RNasin or ribonucleoside–vanadyl complexes sometimes increase yield. RNasin is the recommended inhibitor because it is unaffected by other components of the reaction. Ribonucleoside–vanadyl complexes, at 2 mM, inhibit cDNA synthesis unless the concentrations of all four deoxynucleoside triphosphates exceed ~1 mM. Mixtures of the inhibitors have not been tested.

Template poly(A)⁺ RNA is diluted to 0.25 μg/ml and heated to 70° for 3–5 min immediately before use. When a specific RNA is believed to have extensive secondary structure, the RNA solution can be treated with methylmercury hydroxide[12] (Alfa Division, Danvers, MA) to destroy base pairing, although this procedure reduces the overall yield of cDNA. Since this compound is extremely dangerous, all manipulations are carried out in a hood with gloved hands. The RNA, at a maximum concentration of 1 μg/μl, is mixed in a total of 10 μl with 1 μl 0.1 M methylmercury hydroxide and incubated for 10 min at room temperature. (If this relatively gentle treatment does not improve the qualify of the *desired* cDNA, the reaction mixture with methylmercury hydroxide should be boiled for 1–3 min and quickly cooled in ice.) Methylmercury hydroxide is inactivated by the addition of 1 μl 0.35 M 2-mercaptoethanol. The reaction is probably complete virtually instantaneously, but 5 min at room temperature are usually allowed. The total amount of RNasin to be included during reverse transcription may be added once the sulfhydryl compound is present. The "denatured" RNA is transferred to a tube, in ice, containing the remaining components. Optimum final concentrations of poly(A)⁺ RNA during cDNA synthesis are between 50 and 200 μg/ml. (See this volume [26] for methods to determine the amount of rRNA contamination in polyadenylated RNA.)

Procedure for Using Either AMV or MMLV Reverse Transcriptase

Before assembling the reaction mixture, dilute the AMV reverse transcriptase into buffered 10% glycerol as described above. Denature

12 F. Payvar and R. T. Schimke, *J. Biol. Chem.* **254,** 7636 (1979).

RNA using either heat or methylmercury. Then, in ice, mix the following in a total volume of 40 μl:

	AMV enzyme	MMLV enzyme
AMV reverse transcriptase (diluted)	10 units	
MMLV reverse transcriptase		200 units
AMV reverse transcriptase 10× buffer	4 μl	
MMLV reverse transcriptase 10× buffer		4 μl
Oligo(dT)$_{12-18}$, 100 μg/ml	4 μl	4 μl
dNTP mixture (10 mM each dNTP)	4 μl	2 μl
Spermidine–HCl, 10 mM	2 μl	2 μl
RNasin	40–80 units	40–80 units
Sodium pyrophosphate, 80 mM	2 μl	
Actinomycin D, 1 mg/ml		2 μl
Template poly(A)$^+$ RNA	4 μg	4 μg

1. Transfer 2–4 μl of the complete reaction mixture to a separate tube containing 1–2 μCi of [α-^{32}P]dCTP or dGTP (it is best to avoid TTP and dATP). Incubate both reaction mixtures, labeled and unlabeled, at 42° (AMV reverse transcriptase) or 37° (MMLV enzyme) for 45 min. The tube containing the labeled sample may be sealed with Parafilm and submerged completely in order to avoid evaporation of the small volume. The labeled sample serves for analysis; the unlabeled sample is used for second-strand synthesis.

2. Characterize the labeled sample in a 1.0–1.5% alkaline agarose gel (this volume [8]). Since sample volume is small, it is usually not extracted with phenol, but, instead, is mixed with an equal volume of 0.1 N NaOH, 10 mM EDTA and boiled for 5 min. A fraction of a microliter is diluted with a convenient volume of loading buffer, usually ~10 μl or more (depending on the size of the wells), and analyzed electrophoretically with appropriate DNA markers. (See this volume [8] for choice of markers, and [10] for labeling methods.) The finished gel should be simultaneously neutralized and fixed by soaking in 400 ml 10% acetic acid–10% methanol for 30 min. Subsequently the gel is dried on Whatman 3 MM paper (this volume [8]). Subject the dried gel to autoradiography as described in this volume [7]. Eukaryotic cDNA molecules should range in size from ~600 bases to 2 kb with small amounts of larger material. In general, the closer the cDNA is in size to the mRNA, the better the cDNA preparation. The remainder of the labeled reaction should be diluted and stored. It will be needed for comparison when characterizing the synthesis of the second strand (this volume [34, 35]).

If the synthesis of second-strand cDNA is contemplated by the classical method (this volume [34]), RNA must be hydrolyzed (step 3). If the

DNA–RNA hybrid itself is the starting material for second-strand synthesis (this volume [35]), omit step 3 and proceed immediately to step 4.

3. Hydrolyze RNA by adding 40 μl 0.3 N NaOH, 30 mM EDTA and boiling for 5 min. Neutralize with ~10 μl 1 M Tris–HCl at pH 8. An alternative procedure, in which the DNA–RNA hybrid is denatured by boiling while the RNA is partially degraded, is presented in this volume [34]. The latter approach avoids the need for handling purified single-stranded cDNA. Single-stranded DNA adheres to plastic and glass tubes, pipets, and tips and *requires* the use of siliconized materials. Losses are inevitable!

4. Extract the sample once with phenol (saturated with 50 mM Tris–HCl at pH 8) and once with a 1 : 1 mixture of phenol and chloroform–isoamyl alcohol as described in this volume [4]. Precipitate with ethanol using ammonium acetate as the salt (this volume [5]). For samples of less than 5 μg, the addition of 5 μg tRNA is recommended. Recover the sample as a pellet by centrifugation (this volume [5]). Discard the supernatant fluid and go on to the second strand ([34] or [35]).

Vector-Primed cDNA Synthesis

The preparation of cDNA libraries with combined vector–primers is detailed in this volume [41]. Here we present a method for synthesis of first-strand cDNA which makes use of the same components found to be advantageous with oligo(dT) as the primer. Unlike methods for oligo(dT)-primed cDNA synthesis which require excess primer in order to maximize utilization of mRNA, vector–primer techniques require a 3-fold molar excess of poly(A)$^+$ RNA.[13] By maximizing utilization of the vector–primer, the number of empty clones, namely, tailed vector-primers without inserts, is reduced. For vectors of 2.7 kb, and a mRNA population with an average size of 1.5 kb, 2–3 μg of vector per microgram mRNA is suggested.

In ice, mix, in a total volume of 30 μl:

AMV reverse transcriptase 10× buffer	3.0 μl
dNTPs mix (10 mM each dNTP)	1.5 μl
Spermidine–HCl (10 mM)	1.5 μl
Sodium pyrophosphate (80 mM)	1.5 μl
RNasin	30 to 60 units
AMV reverse transcriptase (diluted)	8 units
Linearized plasmid, T-tailed at one end (described in [41])	8 to 12 μg
Template poly(A)$^+$ RNA	4.0 μg

[13] H. Okayama and P. Berg, *Mol. Cell. Biol.* **2,** 161 (1982).

1. Transfer 1.5–3.0 μl of the reaction to a tube containing 1–2 μCi [α-^{32}P]dCTP as described for oligo(dT)-primed synthesis of cDNA.

2. Incubate both the labeled reaction and the unlabeled reaction at 42° for 45 min.

3. Analyze the labeled sample electrophoretically in a 1% alkaline agarose gel as described above. Take note of the fact that the cDNA is attached to vector DNA. Therefore, T-tailed vector should also be analyzed in the same gel. T-tailed vector may be labeled either by the phosphatase–kinase method in this volume [10] or by the addition of a few residues to the 3′ end with terminal transferase (this volume [37]). A shift to larger size of the plasmid bearing the cDNA should be observed.

4. Extract sequentially with phenol and phenol–chloroform–isoamyl alcohol and ethanol precipitate as described above. *Do not treat with alkali!*

Synthesis of [^{32}P]cDNA

Occasionally it is useful to prepare radioactive cDNA for use as a hybridization probe. Since low concentrations of dNTPs necessitated by use of labeled compounds reduce full-length cDNA production, the protocol in chapter [46], which makes use of random hexamers rather than oligo(dT) as a primer, should be considered as an alternative. In the method presented here, the preponderance of label is located in sequences complementary to the 3′ end of mRNA even when longer molecules of cDNA are synthesized.

In a sterile tube, dry 10 μl [α-^{32}P]dCTP (10 μCi/μl, 400 Ci/mmol). Add, in a total volume of 10 μl:

AMV reverse transcriptase 10× buffer	1.0 μl
Oligo(dT)$_{12-18}$, (100 μg/ml)	1.0 μl
dNTPs mix (2 mM each of dATP, dGTP, and TTP)	0.5 μl
Spermidine–HCl, 10 mM	0.5 μl
Sodium pyrophosphate, 80 mM	0.5 μl
RNasin	10 to 20 units
AMV reverse transcriptase (diluted)	2.5 to 3 units
Template poly(A)$^+$ RNA	0.2 to 1 μg

1. Incubate at 42° for 10 min.

2. Add 1 μl chase mix (chase mix contains all four dNTPs, each at 10 mM) and continue incubating for 20 min.

3. Monitor the reaction using a very small aliquot and the filter assay described in this volume [6].

4. Treat with alkali, boil, neutralize, extract sequentially with phenol and phenol–chloroform–isoamyl alcohol and ethanol precipitate with

added carrier tRNA as described earlier in this chapter (steps 3 and 4 of Procedure for Using Either AMV or MMLV Reverse Transcriptase).

5. Recover the probe by centrifugation. It should contain $\sim 10^8$ cpm/μg and can be hybridized without further denaturation.

[*Editors' Note.* Methods for the synthesis of cDNA using primers other than oligo(dT)$_{12-18}$ can be found in this volume [66] under the heading Primer Extension.]

[34] Second-Strand cDNA Synthesis: Classical Method

By UELI GUBLER

The "classical" scheme for the synthesis of double-stranded cDNA as it was reported in 1976[1,2] is as follows: reverse transcription of mRNA with oligo(dT) as the primer generates first strands with a small loop at the 3' end of the cDNA (the end that corresponds to the 5' end of the mRNA). Subsequent removal of the mRNA by alkaline hydrolysis leaves single-stranded cDNA molecules again with a small 3' loop. This loop can be used by either reverse transcriptase or Klenow fragment of DNA polymerase I as a primer for second-strand synthesis. The resulting products are double-stranded cDNA molecules that are covalently closed at the end corresponding to the 5' end of the original mRNA. Subsequent cleavage of the short piece of single-stranded cDNA within the loop with the single-strand-specific S$_1$ nuclease generates "open" double-stranded molecules that can be used for molecular cloning in plasmids or in phage. Useful variations of this scheme have been described.[3]

Technical Remarks

Before describing actual laboratory protocols, I would like to address four important points relevant to the procedure. (1) The quality of the first-strand cDNA (mass yield, length, purity) is obviously a crucial parameter for successful second-strand synthesis. (2) Reverse transcriptase and Klenow fragment of DNA polymerase I should both be used for second strand synthesis. The best approach seems to be a stepwise syn-

---

[1] A. Efstratiadis, F. C. Kafatos, A. Maxam, and T. Maniatis, *Cell (Cambridge, Mass.)* **7,** 279 (1976)

[2] F. Rougeon and B. Mach, *Proc. Natl. Acad. Sci. U.S.A.* **73,** 3418 (1976).

[3] H. Land, M. Grez, H. Hauser, W. Lindenmaier, and G. Schuetz, *Nucleic Acids Res.* **9,** 2251 (1981).

METHODS IN ENZYMOLOGY, VOL. 152

thesis starting with Klenow fragment followed by reverse transcriptase. It has been found that the two enzymes have different abilities to "plough" through secondary structures in a piece of single-stranded cDNA, a fact that has also been exploited in enzymatic DNA sequencing. Use of both Klenow fragment and reverse transcriptase will give a high yield of full-length second strands. One should obtain direct proof that the second strand has been made. The following principle can be adopted: run a first-strand reaction without any label in order to follow accurately how much label is subsequently incorporated into the second strand. The point cannot be stressed enough that both overall mass yield (incorporation) and size under truly denaturing conditions (alkaline or methylmercury hydroxide gels) are essential control steps for evaluating the quality of the second strand. Another point to keep in mind is that the products made should not be labeled to high specific activities since highly labeled molecules are rather unstable and cannot be kept indefinitely before they are actually cloned. One alternative to circumvent that problem is to synthesize approximately 95–98% of the second-strand cDNA with unlabeled nucleotides. The remaining 2–5% of the complete reaction is then transferred to a second vessel (usually a 0.5-ml Eppendorf tube) in which 2–20 μCi of $5'[\alpha\text{-}^{32}P]dCTP$ or dGTP has been dried. Both samples are incubated in parallel; the labeled reaction is analyzed while the unlabeled reaction is cloned. However, if one is confident that all the steps in the cDNA cloning system are working properly, it is possible to work with moderately radioactive cDNA ($\sim 10^6$ cpm/μg) for about 2–3 weeks. (3) The step that needs the most attention in this approach for second-strand synthesis is the S_1 nuclease-mediated cleavage of the hairpin loops. S_1 nuclease is a very aggressive enzyme that needs proper attention and titration. If used in excess, it will tear even perfect double-stranded cDNA to shreds, leaving only small amounts of short cDNAs ready for the cloning. The best way to titrate the S_1 enzyme for proper hairpin loop cleavage is to perform a series of digestions on analytical amounts of sample with different amounts of S_1 nuclease and to look at the products by denaturing gel electrophoresis (shift down after cleavage). After complete digestion of the loops, the products need to be separated according to size before cloning in order to remove the small fragments created by the S_1 treatment. (4) Last, but not least, we have found that two ethanol precipitations out of 2.5 M ammonium acetate with chilling the reaction and subsequently warming it up to room temperature prior to centrifuging[4] eliminate the need for any column-based separation of free nucleotides from cDNA products.

4 H. Okayama and P. Berg, *Mol. Cell. Biol.* **2,** 161 (1982).

Second-Strand Synthesis with Klenow Fragment

The protocols presented here follow the approach published in the Cold Spring Harbor Manual[5] on molecular cloning. Mix the following components in a total volume of 100 μl:

Single-stranded cDNA	20 μl
2× Buffer	50 μl
25 mM each deoxynucleoside triphosphate (dNTP)	2 μl
Water	18 μl
Klenow fragment	10 μl

Incubate the reaction at 15° for 3–4 hr. Add EDTA to 20 mM to stop the reaction. Extract with buffered phenol twice. To the aqueous phase, add 0.5 volume of 7.5 M ammonium acetate and 2 volumes of ethanol. Chill the mixture in dry ice–ethanol for 10 min, let warm up to room temperature for 10 min and centrifuge at 4° for 10 min in a microfuge. Redissolve the pellet in half the original volume (50 μl) of sterile water and repeat the precipitation as described. Wash the final pellet once with 80% ethanol at room temperature, centrifuge again, and dry *in vacuo* for 10 min. Redissolve the final pellet in 20 μl of sterile water. Saving a small aliquot of the products for later gel analysis is optional at this point.

Comments. (1) The maximum amount of cDNA that can be processed in a 100-μl reaction has to be determined by the amount of Klenow fragment needed to run the reaction and by the concentration in which the enzyme is supplied. About 20–50 units of enzyme per microgram of single-stranded cDNA should be used, and the total glycerol concentration in the reaction (due to the enzyme storage buffer) should not exceed 10–15%. (2) The composition of the 2× second-strand buffer is 0.2 M HEPES (pH 6.9), 20 mM MgCl$_2$, 5 mM dithiothreitol, 0.14 M KCl. (3) We usually keep a stock of a mixture of all four dNTPs, each nucleotide at 25 mM, with the pH adjusted to approximately 7.5. Stocks at concentrations less than 10 mM are not kept. If radioactive nucleotide is to be incorporated into the second strand it should be added at this point. (4) Single-stranded cDNA is sticky. To avoid handling pure single-stranded material, the following alternative adapted from Wickens *et al.*[6] may be used. At the end of the first-strand synthesis, boil the reaction for 3 min to denature the DNA–RNA hybrids. Quickly cool in ice and remove insoluble material by brief centrifugation. The reaction is reconstituted for second-strand syn

[5] T. Maniatis, E. F. Fritsch, and J. Sambrook, eds., "Molecular Cloning: A Laboratory Manual." Cold Spring Harbor Lab., Cold Spring Harbor, New York, 1982.
[6] M. P. Wickens, G. N. Buell, and R. T. Schimke, *J. Biol. Chem.* **250**, 2483 (1978).

thesis by adding an equal volume of 0.2 M HEPES at pH 6.9, 90 mM KCl, 0.5 to 1 mM each dNTP, and Klenow fragment as suggested, to the supernatant fluid. Incubate as described above.

Completion of the Second Strand with Reverse Transcriptase

Mix, in a total volume of 50 μl:

Klenow-treated cDNA from above	20 μl
1 M Tris–HCl (pH 8.3)	5 μl
1 M KCl	7 μl
0.25 M MgCl$_2$	2 μl
25 mM dNTPs	2 μl
0.1 M dithiothreitol	5 μl
Water	7 μl
Reverse transcriptase	2 μl

Incubate the reaction at 42° for 1 hr. Add EDTA to 20 mM to stop the reaction and extract with buffered phenol twice. Precipitate with ethanol as described above. Redissolve the final pellet in 20 μl of sterile water and measure the Cerenkov radiation in a 1-μl aliquot (see this volume [6]) to determine overall recovery of material. The cDNA is now ready to be treated with S$_1$ nuclease. (When directional cDNA cloning is contemplated, as in this volume [39], S$_1$ nuclease treatment is postponed. See [39] for instructions.)

Comments. (1) The pH of the Tris–HCl stock has to be adjusted at 42°, i.e., the temperature at which the reaction is run. (2) The solution of the dNTPs should be prepared as described above. (3) The amount of avian myeloblastosis virus reverse transcriptase added to the reaction should be about 40–60 units; the volume added should not exceed 10% of the total reaction volume. (For a discussion of the relative merits of reverse transcriptases, see this volume [33].)

S$_1$ Nuclease-Mediated Cutting of the Hairpin Loop

As discussed in the introduction to the chapter, S$_1$ nuclease needs proper attention. One needs to titrate the optimal amounts of S$_1$ nuclease to be used per mass of double-stranded cDNA by running small-scale analytical reactions. Such a titration should be run for every new reaction, since every preparation of cDNA might be just different enough from the previous one to require a different optimal amount of S$_1$ enzyme. Once that amount has been determined, the reaction has to be scaled up exactly based on the total amount of cDNA to be treated. The conditions given here represent *general guidelines*. Certain reactions might need amounts of enzyme different from those suggested here.

Prepare five tubes with the following general composition in a total volume of 5 μl:

Double-stranded cDNA	1 μl
(See comment 1 below)	
5× Buffer	1 μl
(See comment 2 for composition)	
Water	2 μl
S₁ nuclease	1 μl

(See comment 3 on different activities in the five tubes)

Incubate the tubes at 37° for 30 min (see comment 4, below). Add EDTA to 20 mM to stop the reactions and analyze the products in an alkaline agarose gel (see this volume [8]). Include a sample of the first-strand cDNA that was saved prior to the second-strand synthesis reaction. Dry the gel down and expose to X-ray film (see this volume [7]). Determine the concentration of S₁ nuclease that cuts the double-stranded cDNA to a size distribution identical to that of the single-stranded cDNA. This is the amount of S₁ nuclease that should be used in an exactly scaled-up reaction to cut the hairpin loops in the large remaining portion of the cDNA. The scaled-up reaction should be extracted with phenol and the products should be precipitated with ethanol as described before. The cDNA is now ready to be sized (see this volume [36]) prior to cloning in either plasmids or bacteriophage λ (see this volume [39,40]).

Comments. (1) The cDNA should have approximately 2000 cpm per reaction, equivalent to about 2–5 ng. (2) The composition of the 5 × buffer is 1 M NaCl, 0.25 M sodium acetate (pH 4.5), 5 mM ZnSO₄, 2.5% glycerol. (3) The five different tubes should contain 0, 0.5, 1, 2, and 4 units of S₁ nuclease,[7] respectively. The enzyme should be diluted in the enzyme storage buffer containing 50% glycerol, as recommended by the supplier. Dilutions of the enzyme should not be kept. (4) Incubation at temperatures lower than 37° will also help to control the reaction. (5) A note on the scale-up: as mentioned above, find the optimal amount of S₁ enzyme that cuts the loops in the cDNA present in the pilot reaction (say 5 ng). If the large remaining portion of cDNA represents 1 μg, the reaction should be scaled up 200-fold. This pertains to the mass of cDNA, volume of the reaction, and amount of S₁ enzyme. (6) See this volume [35] for the use of mung bean nuclease to cleave hairpin loops as an alternative to S₁ nuclease.

[7] V. M. Vogt, *Eur. J. Biochem.* **33**, 192 (1973).

[35] Second-Strand cDNA Synthesis: mRNA Fragments as Primers

By Ueli Gubler

The basic principle of this method for second-strand synthesis was first described in 1982[1] and later modified to some extent.[2] The method uses standard conditions of reverse transcription for the first-strand synthesis; upon completion of this reaction, the RNA–DNA hybrid is, in contrast to the classical second-strand synthesis method, left intact. *Escherichia coli* RNase H is used to nick the RNA in the hybrids, leaving only small RNA primers with free 3'-OH groups attached to the cDNA. These 3'-OH groups can subsequently be used by DNA polymerase I to synthesize efficiently a second strand all along the length of the first strand (original cDNA). The resulting double-stranded cDNA molecules are essentially ready for the cloning steps at this point. There is in general no need for either sizing or hairpin loop cleavage by S_1 nuclease as required in the classical cDNA synthesis method. Longer cDNA molecules can be obtained in higher yields in this way, facilitating the cloning of low-abundance mRNAs considerably.

Technical Remarks

Before detailing actual laboratory protocols, I would like to discuss four relevant technical aspects in greater detail. (1) As stated in the previous chapter the quality of the first-strand cDNA made (mass yield, length, and purity) is crucial for good second-strand synthesis. (2) It is often advisable to obtain direct proof that the second strand has been synthesized. Methods for accomplishing this are described in the previous chapter. (3) With the development of cDNA cloning techniques in bacteriophage λ (using blunt-end ligation of EcoRI linkers to the cDNA for its subsequent cloning at EcoRI sites of λ DNA), an additional treatment of the cDNA with T4 DNA polymerase to create blunt ends has been introduced.[3] This step will assure efficient ligation of the linkers to the cDNA inserts prior to cloning. If the cDNA is to be tailed, blunt ends are not essential. (4) After completed second-strand synthesis and creation of

---

[1] H. Okayama and P. Berg, *Mol. Cell. Biol.* **2,** 161 (1982).
[2] U. Gubler and B. Hoffman, *Gene* **25,** 263 (1983).
[3] J. J. Toole, J. L. Knopf, J. M. Wozney, L. A. Sultzman, J. L. Buecker, D. D. Pittman, R. J. Kaufman, E. Brown, C. Shoemaker, E. C. Orr, G. W. Amphlett, W. B. Foster, M. L. Coe, G. J. Knutson, D. N. Fass, and R. M. Hewick, *Nature (London)* **312,** 342 (1984).

blunt ends, we have found it practical to test for possible residual hairpin-looped molecules within the double-stranded cDNAs. Our experience has been that in some cases, even though the first strand has been synthesized under conditions that suppress second-strand synthesis by reverse transcriptase (i.e., high mRNA template concentration and inclusion of actinomycin D or sodium pyrophosphate), loops at the 3′ ends of the first strands can remain. These loops can be used by the DNA polymerase I as primers for second-strand synthesis, resulting in molecules that are covalently closed at the ends corresponding to the 5′ ends of the original mRNAs. Since the RNase H-mediated method for second-strand synthesis will in general not involve any nuclease step (but see below), such hairpin molecules cannot accept either linkers or tails at *both* ends and consequently cannot be cloned.

There are two easy ways to determine the presence and extent of such loops. First, denaturing gel electrophoresis of products after second-strand synthesis compared to first-strand products will show a shift of the looped second strands to higher molecular sizes, because the hairpin molecules upon denaturation will be twice the size of their non-hairpin-looped counterparts. Second, S_1 nuclease digestion of labeled double-stranded cDNA with and without prior heating and quick cooling of the DNA will determine the percentage of such loops relative to a control (labeled nick-translated plasmid DNA). The rationale behind the heating and cooling step is that properly double-stranded molecules without hairpin loops will denature when heated and stay single-stranded upon cooling, thus remaining susceptible to S_1 nuclease attack. Hairpin-looped molecules on the other hand will quickly snap back to (at least partial) double-strands upon cooling and will thus remain S_1 resistant to a larger extent. A nick-translated plasmid DNA heated and quickly cooled in this manner will yield approximately 10–15% of the input radioactivity as S_1-resistant counts; these are acceptable background values. If there is indication of hairpin-looped molecules by gel shift analysis and S_1 digestion, two things can be done: go back to the first strand reaction and try to minimize loop formation during first strand synthesis; or treat the double-stranded cDNA molecules with mung bean nuclease, a single-strand-specific nuclease that is similar in activity to S_1 nuclease, i.e., it cleaves the hairpin loops and renders the molecules clonable. The advantage of using mung bean nuclease as an alternative to the procedure with S_1 nuclease, detailed in this volume [34], is that it is a much less aggressive enzyme. Thus it needs less titration work and gives higher yields of longer cDNAs (D. Colman, personal communication).

Editors' Note: A one-tube reaction containing the buffers specified in this volume might contain 40 μl first strand cDNA reaction which should be diluted to 250 μl for second strand synthesis. Following an incubation at 70° for 10 min, blunt ends should be formed with 2 units of T4 polymerase/μg mRNA.

Second-Strand Synthesis

Mix, in a total volume of 100 μl:

cDNA–mRNA hybrid	20 μl
(See comment 1 below for concentration)	
2× Buffer	50 μl
Water	25.7 μl
4 mM each deoxynucleoside triphosphate (dNTPs)	1 μl
RNase H	1 μl (1.8 units)
DNA polymerase I	2.3 μl (23 units)

Incubate the reaction at 12° for 1 hr and subsequently at 22° for 1 hr. Stop the reaction by adding EDTA to 20 mM and extracting with buffered phenol twice. To the aqueous phase, add 0.5 volume of 7.5 M ammonium acetate and 2 volumes of ethanol. Chill the mixture in dry ice–ethanol for 10 min, let warm up to room temperature for 10 min, and centrifuge at 4° for 10 min in a microfuge. Redissolve the pellet in half the original volume (50 μl) of sterile water and repeat the precipitation as described. Wash the final pellet once with 80% ethanol at room temperature, centrifuge again, and dry *in vacuo* for 10 min. Redissolve in 20 μl sterile water. The product is now ready for blunt ending with T4 DNA polymerase. If any label was incorporated into the reaction, estimation of incorporation by trichloroacetic acid (TCA) precipitation should be done at this point (see this volume [6]). If nonradioactive cDNA was synthesized, a reaction containing radioactive precursors run in parallel must be used for quantification (see this volume [34]). The Cerenkov radiation in a small aliquot of the latter can be measured to determine the recovery of material (see this volume [6]).

We usually assume that 100% of the cDNA–mRNA hybrid is converted to double-stranded cDNA and base all the further calculations on that assumption, keeping in mind that the mass of cDNA doubles during the reaction.

Comments. (1) The maximum amount of single-stranded cDNA that should be added to a reaction of 100 μl is 500 ng (i.e., 1 μg of hybrid). If more material needs to be made double-stranded, scale the reaction up appropriately. Small amounts of cDNA can be processed in smaller volumes, e.g., 50 ng in 50 μl or less than 50 ng in 20 μl. (2) The buffer that we use for the reaction can be made as a 2× buffer and safely stored in aliquots at −20° for periods of up to a year. The composition of the 2× buffer is as follows: 40 mM Tris–HCl (pH 7.5), 10 mM MgCl$_2$, 20 mM

$(NH_4)_2SO_4$, 200 mM KCl, 100 μg/ml bovine serum albumin. We recommend the use of standard reagent-grade chemicals, using autoclaved HPLC-grade water, and filtering the final solution through a nitrocellulose (0.45 μm pore size) membrane filter. (3) We usually keep a stock of a mixture of all four dNTPs, each nucleotide at 25 mM, with the pH adjusted to approximately 7.5. A dilution of that stock is made to 4 mM immediately before the reaction, used, and then discarded. If the entire second-strand synthesis reaction is to be run with label, add 10–20 μCi of 5'[α-^{32}P]dCTP (3000 Ci/mmol) at this point. (4) We use *Escherichia coli* RNase H from Bethesda Research Laboratories; the final concentration of enzyme should be 18 units/ml. (5) We use DNA polymerase I from New England BioLabs, at a final concentration of 230 units/ml. We have not tried to substitute DNA polymerase I with Klenow fragment at this point.

Blunt Ending with T4 DNA Polymerase

Once the cDNA has been rendered double stranded, this additional blunt-ending step will assure that the ends of all the molecules are really flush. This is important if the cDNA molecules are to be cloned by the addition of linkers and subsequent ligation and packaging in λ DNA. It is not necessary to perform this step if the cDNA is to be tailed. A typical reaction is as follows.

Mix, in a total volume of 50 μl:

1 M Tris–HCl (pH 8.3)	2.5 μl
0.1 M MgCl$_2$	5 μl
0.1 M dithiothreitol	5 μl
1 M NaCl	2.5 μl
2.5 mM each deoxynucleoside triphosphate	1 μl
Double-stranded cDNA	20 μl
Water	13.5 μl
T4 DNA polymerase I	0.5 μl (5 units)

Incubate the reaction for 30 min at 37°. Add EDTA to 20 mM to stop the reaction, extract with buffered phenol twice, and precipitate with ethanol as described above. Wash with 80% ethanol once, dry, and redissolve the final pellet in 20 μl of sterile water. Determine the Cerenkov radiation in a small aliquot (1–2 μl) to estimate the recovery. The cDNA is now ready for either sizing (see this volume [36]) or cloning via the addition of homopolymeric tails or linkers (see this volume [37–40]).

Analysis for Residual Hairpin Loops by Analytical Digestion with S$_1$ Nuclease

This analysis can be performed on approximately 2000 cpm of double-stranded cDNA per tube, with a total of four tubes. A nick-translated plasmid DNA serves as the control (two tubes). All volumes are given in microliters.

		Tube				
	1	2	3	4	5	6
cDNA (in water)	1	1	1	1	—	—
Control DNA	—	—	—	—	1	1
Water	28	29	28	29	28	29
Boil for 5 min	+	+	−	−	+	−
Quench in ice water	+	+	−	−	+	−
5× Buffer (comment 1, below)	20	20	20	20	20	20
Carrier DNA (comment 2)	50	50	50	50	50	50
S_1 nuclease (comment 3)	1	—	1	—	1	—

Incubate the reactions at 37° for 90 min. Precipitate two parallel aliquots of 45 μl per tube with 10% TCA on glass fiber filters (see this volume [6]) and determine the amount of acid-insoluble radioactivity. Calculate the percentage of S_1-resistant counts. Background levels (tube 5 compared to tube 6) should be on the order of 10%. A loop-free cDNA preparation should give similar values (compare tube 1 to tube 2).

Comments. (1) The composition of the 5× buffer is as follows: 1 M sodium acetate (pH 4.5), 2 M NaCl, 12.5 mM ZnCl$_2$. (2) Carrier DNA (calf thymus, herring sperm, or the like at 1 mg/ml and undenatured) is added to "buffer" the excess S_1 nuclease added. (3) Add 5–10 units of S_1 nuclease defined according to Vogt.[4] Dilute the enzyme in 1× digestion buffer immediately before using it and discard the diluted solution.

Hairpin Loop Cleavage with Mung Bean Nuclease

The following protocol was developed and communicated by D. Colman. It can be used as the standard procedure after RNase H-mediated second-strand synthesis. Blunt ending with T4 DNA polymerase becomes unnecessary, but the cDNA products must be fractionated according to size before cloning (see this volume [36]).

Mix the following components:

Double-stranded cDNA	20 μl
5× Buffer	20 μl
Water	59 μl
Mung bean nuclease	1 μl (10 units)

Incubate the reaction at 37° for 15 min. Neutralize by adding 10 μl of 1

[4] V. M. Vogt, *Eur. J. Biochem.* **33,** 192 (1973).

M Tris–HCl (pH 8), extract with buffered phenol twice, and precipitate with ethanol as described above.

Comments. (1) One can safely process 0.5 to 2 μg of double-stranded cDNA in 100 μl. (2) The 5× buffer has the following composition: 150 mM sodium acetate (pH 5), 250 mM NaCl, 5 mM ZnCl$_2$, 25% (v/v) glycerol. (3) We use mung bean nuclease from Pharmacia. The enzyme should be handled and diluted according to the manufacturer's specifications.

[36] Purification of Large Double-Stranded cDNA Fragments

By WILLIAM H. ESCHENFELDT and SHELBY L. BERGER

During the construction of a cDNA library, polyadenylated RNA is converted to double-stranded DNA using a series of enzyme-catalyzed reactions. The enzymes, mainly polymerases, are capable of directing the synthesis of a complementary DNA when provided with an appropriate primer and template. Because the polymerase activities occupy center stage in the process, it is easy to overlook the degradative activities that are an intrinsic part of these proteins. For example, reverse transcriptase has an associated ribonuclease H which can cleave RNA in heteroduplex molecules; DNA polymerase I and its large fragment have exonuclease functions. These enzymes can also displace both DNA and RNA from a template during synthesis of the second strand. To complicate matters further, certain properties of the enzymes, such as lack of processivity, and properties of the template, such as secondary structure and sequences that are difficult for the enzymes to traverse (so-called "strong stops"), contribute to the production of incomplete chains; the cDNA remains nicked or gapped. The end result is a plethora of both single-stranded and double-stranded fragments which can interfere with the cloning of full-length molecules.

The debris generated during cDNA cloning is often a small percentage of the total mass of double-stranded DNA. When the labeled products are examined in gels, high-molecular-weight material predominates. Nevertheless, if the number of *moles* of cDNA is considered, rather than mass, the undersized component represents a substantial fraction of the total. It is, therefore, always advantageous to fractionate cDNA molecules according to size before the addition of homopolymeric tails and it is sometimes useful to discard smaller fragments before ligating cDNA to linkers. This is best accomplished by gel filtration in columns of Sepharose CL-4B.

Procedure

1. Pour a column, 20 × 1 cm (diameter), of Sepharose CL-4B: add ~1 ml of water to an Econo column (Bio-Rad) or the equivalent, dilute the Sepharose slurry with water, and pour it into the column until a bed height of ~19 cm is obtained. Wash with water followed by several column volumes of water saturated with 0.05% diethyl pyrocarbonate. Wash again with water to remove the diethyl pyrocarbonate and equilibrate the column with 50 mM NH$_4$HCO$_3$. The pH value of the buffer and of the eluate from an equilibrated column should be between 8.0 and 8.5.

2. Calibrate the column. This is most easily accomplished by chromatographing 100 μl of *Hae*III-digested ϕX174 DNA at a concentration of 20 μg/ml. Upon loading the sample, immediately commence collecting 0.20- to 0.25-ml fractions (seven drops) in a Gilson fraction collector (Medical Electronics, Inc., Middleton, WI) or any convenient substitute. A pressure head of 30–35 cm assures an adequate flow rate. Transfer the individual fractions to 1.5-ml Eppendorf tubes and reduce to dryness in a Speed Vac (Savant Instruments, Inc., Hicksville, NY). The samples can also be lyophilized. Analyze each entire fraction in a 2% agarose gel (see this volume [8]). Double-stranded DNA molecules, 600 bp and larger, are found in the void volume whereas smaller molecules, ~300 bp, appear after about 5.5 ml of column effluent have been collected.

3. Prepare double-stranded cDNA for gel filtration. If the DNA has been ethanol precipitated, recover the pellet by centrifugation (this volume [5]) and dissolve it in 100 μl of water. If the cDNA is unlabeled, it is helpful to add a small amount of tracer cDNA to monitor the column. Since only a few thousand cpm are needed, an aliquot of an analytical reaction used to characterize the synthesis of the second strand will suffice (see this volume [34,35]). However, if the calibration step has been carried out carefully, labeled material is not required. Charge the column with the sample and activate the fraction collector. When the sample has entered the gel, wash with 100 μl of buffer, connect the reservoir of NH$_4$HCO$_3$, and collect fractions as for step 2.

4. Pool the high-molecular-weight material found in the fractions on the ascending limb front, and two or three fractions on the descending limb (fractions 20 to 23). If the sample is labeled, radioactivity may be quantified by transferring each individual fraction in its entirety to a 1.5-ml Eppendorf tube and measuring Cerenkov radiation in the ^{3}H channel of a scintillation counter. This is done by placing the Eppendorf tube on the rim of a scintillation vial as detailed in chapter [6]. The sample is not consumed.

5. Concentrate the pooled material by reducing to dryness. Rehydrate with 0.5 ml of water and dry again. Finally, dissolve in 20 μl water, add 2–

5 μg poly(A) as carrier, if necessary, and ethanol precipitate using potassium acetate as the salt (see this volume [5]). It may be necessary to precipitate the sample with potassium acetate and ethanol a second time to remove residual NH_4HCO_3.

Terminal transferase and T4 polynucleotide kinase are extremely sensitive to ammonium ions. Therefore, it is essential to exclude NH_4HCO_3 from column-purified products if use of either enzyme is contemplated.

Sepharose CL-4B columns may be regenerated and reused for years. The column is run and stored at room temperature.

[*Editors' Note.* Other chromatographic procedures for purifying cDNA can be found in this volume [39].]

[37] Homopolymeric Tailing

By WILLIAM H. ESCHENFELDT, ROBERT S. PUSKAS, and
SHELBY L. BERGER

The insertion of cDNA into vector DNA to construct recombinant plasmids can be accomplished by either of two methods: ligation of "bridge" molecules such as linkers or adaptors discussed elsewhere in this volume [38] and polymerization of homopolymeric tails. Tailing, as the latter process is commonly called, is a reaction catalyzed by terminal deoxynucleotidyltransferase in which nucleoside monophosphate residues donated by deoxyribonucleoside triphosphates (dNTPs) are polymerized onto 3'-hydroxyl termini of either single-stranded or double-stranded DNA. The reaction has been extensively studied using DNA with blunt ends, 3'-protruding ends, and 3'-recessed ends as acceptors, and the triphosphates of each of the four canonical bases as donors.[1,2] A gapped plasmid capable of transforming bacteria can be fabricated by annealing tailed cDNA molecules with vector molecules bearing complementary tails. In general, 3'-protruding ends are most reactive, and thus tailing at a cleaved *Pst*I site is the preferred approach for vectors. Double-stranded cDNA, however, is a complex population of termini and successful tailing of all molecules requires a detailed understanding of the reaction. In both cases, the length of the tails must be rigorously controlled in order to clone efficiently.

[1] G. D. Deng and R. Wu, *Nucleic Acids Res.* **9**, 4173 (1981).
[2] A. M. Michelson and S. H. Orkin, *J. Biol. Chem.* **257**, 14773 (1982).

METHODS IN ENZYMOLOGY, VOL. 152

The length and size distribution of homopolymeric tails are functions of the concentration of enzyme, the concentration of the chosen dNTP, and the concentration of termini to be tailed. Furthermore, since the enzyme acts processively, a high concentration is essential to avoid synthesizing long tails on selected molecules while others acquire no tail. Under conditions of enzyme excess (in moles), the concentration of termini exerts little influence on the rate of the reaction, and the reaction is driven by enzyme. Then the major factors governing the average tail length are the choice of triphosphate, its concentration, and the temperature. The breadth of the distribution of tail lengths, however, is a function of the relative rates of initiation (usually rate limiting regardless of the enzyme concentration) and elongation.

Under specified conditions, with dGTP as the substrate, the tailing reaction is self-limited[3,4]; apparently, the secondary structure of homopolymers of dG inhibits polymerization. By limiting the concentration of dGTP, the rate of elongation can be reduced further. Since these conditions favor a narrow distribution of sizes, regardless of the type of terminus to be tailed, we recommend tailing cDNA with dG. As a consequence, polymerization of dC residues, a reaction inherently more difficult to control, is performed with the less valuable DNA, namely, the vector.

There are several advantages to this technique when compared with linkers and adaptors: (1) multiple ligation steps are unnecessary; (2) the need for methylation and restriction of DNA which accompanies the use of linkers is avoided; (3) there are fewer purification steps; and (4) the approach takes less time. There are also disadvantages: (1) GC stretches are detrimental when screening libraries with GC-rich oligomers; (2) special sequencing methods are needed to "read through" the tails flanking the insert DNA; (3) gaps and nicks of poorly constructed cDNA may be tailed, with resultant loss of cloning efficiency; (4) tailing is not recommended when the identification of clones of interest depends on expression of fusion proteins able to bind to antibodies.

A procedure for homopolymeric "tailing" of insert cDNA and vector DNA will be described. The method yields inserts with tails of defined length and vectors with complementary tails having a broader size distribution.

The technique of homopolymeric tailing can also serve as a means for labeling oligomers. Methods for obtaining radioactive molecules of high specific activity will be presented, together with an evaluation of the factors governing their use in hybridization experiments.

[3] A. Dugaiczyk, D. L. Robberson, and A. Ullrich, *Biochemistry* **19,** 5869 (1980).
[4] A. Otsuka, *Gene* **13,** 339 (1981).

Reagents

Terminal Transferase. The concentration of terminal transferase and the composition of its storage buffer are considerations when choosing a commercial source of the enzyme. Because Co^{2+} is a cofactor in the reaction and because cobalt phosphate is insoluble, it is best to avoid enzyme preparations containing phosphate buffer. Further, the investigator should be aware that cobalt pyrophosphate is also insoluble and that pyrophosphate is a product of terminal transferase-catalyzed polymerization reactions. At high substrate concentrations, the reaction can terminate spontaneously owing to depletion of the cofactor as a precipitate.

Buffer. Tailing buffer is composed of 0.1 M potassium cacodylate at pH 7.2, 2 mM cobalt chloride, and 0.2 mM dithiothreitol. The preparation of 5× tailing buffer is as follows. Equilibrate 5 g Chelex 100 (Bio-Rad, Richmond, CA) with 10 ml of 2–5 M potassium acetate at room temperature. After 5 min, remove excess liquid using a Millipore sterile filtration unit or a suction flask fitted with a Büchner funnel containing Whatman 1 filter paper. Wash the Chelex with 10 ml deionized water. Prepare a solution of 1 M potassium cacodylate by adjusting cacodylic acid (IBI, New Haven, CT) to pH 7.2 with KOH. Equilibrate the potassium cacodylate with Chelex at room temperature for 2–3 min. Recover the cacodylate solution by filtration, and discard the Chelex. To the purified potassium cacodylate add, in order, water, dithiothreitol, and cobalt chloride to make final concentrations of 0.5 M cacodylate, 1 mM dithiothreitol, and 10 mM cobalt chloride.

Procedures

Tailing Double-Stranded cDNA with dG

Under the conditions recommended for tailing cDNA, namely, 37°, 5 μM dGTP, and excess enzyme, the homopolymer tails terminate spontaneously at ~20 residues regardless of the concentration of blunt, 3'-recessed or 3'-protruding ends to be tailed. The tail length is so uniform that the radioactive dGTP incorporated into the cDNA can be used to calculate the concentration of termini and therefore the concentration of cDNA. Since the cDNA is already labeled, in some cases it is important to monitor the reaction at zero time as well as at completion. (See this volume [6] for methods for measuring trichloroacetic acid-precipitable radioactive material using very small volumes.) It is also imperative to limit the radioactivity in the untailed cDNA so that the increase upon tailing is significant. The amount of cDNA estimated from the assay is used to determine the amount of vector needed for optimal transformation of bacteria.

The components of a typical reaction are as follows:

cDNA	dried pellet
5× Tailing buffer	5 μl
[α-^{32}P]dGTP (3000 Ci/mmol, 10 μCi/μl)	2 μl
20 μM dGTP	6 μl
Water	10 μl
Terminal transferase (25 units/μl)	2 μl

1. Incubate at 37° for 30 min. To stop the reaction, inactivate the enzyme by heating at 65° for 5 min.

2. Measure the incorporation of dGTP into trichloroacetic acid-precipitable material using the filter assay described in this volume [6].

3. Phenol extract and ethanol precipitate the product, using ammonium acetate as the salt, as detailed in this volume [4,5].

4. Calculate the amount of cDNA from the net acid-precipitable radioactivity per unit volume at 30 min, taking into account the total volume of the reaction and the specific activity of the dGTP. Assume each terminus (two per molecule) has 20 dG residues affixed to it. Then

$$\text{Amount of cDNA (pmol)} = \frac{(\text{dpm}/\mu\text{l at end} - \text{dpm}/\mu\text{l at start}) \times \text{ total volume } (\mu\text{l})}{40 \times \text{ specific activity of dGTP (dpm/pmol)}}$$

Note that dGTP at 3000 Ci/mmol has a specific activity of 6.6×10^6 dpm/pmol and that dpm may be replaced by cpm provided the specific activity of the dGTP is expressed with respect to the same counting efficiency.

If the net incorporation of dGTP is much lower than expected from even the crudest estimate of the concentration of cDNA, phenol extract the sample, ethanol precipitate with potassium acetate twice (this volume [4,5]), and tail with dG again. (Avoid ammonium ion. It inhibits terminal transferase!) Poly(A) or poly(C) (2 μg) can be used as carrier. Since the dG-containing adducts cannot become too long under these tailing conditions, there is no danger in re-tailing the sample.

Tailing Linearized Plasmid DNA with dC

The following conditions are suitable for tailing 3'-protruding ends with dCTP as the substrate. If the termini are the result of cleavage at a *Pst*I site, that site will not be regenerated. However, the multiple cloning sites available in most vectors provide adequate means for recovering inserts regardless of the destruction of any specific site. (Conditions for T-tailing vectors are found in this volume [41].)

Mix:

pUC9 cut with *Pst*I	75 to 80 μg
5× Tailing buffer	20 μl

100 μM dCTP	30 μl
Terminal transferase (25 units/μl)	8 μl
Water to	100 μl

1. Transfer 5 μl of the complete reaction mix to a separate tube containing 10 μCi (dried) [α-^{32}P]dCTP. Spot an aliquot on a GF/C filter as described for dG tailing (and in this volume [6]).

2. Incubate at 15° for 20 min. Stop the reaction by placing the tube in ice. Spot an aliquot and, based on the known concentration of termini, calculate the average tail length. Continue incubating at 15° if necessary. When the average reaches ~20 residues, stop the reaction by heat inactivation of the enzyme (65° for 5 min).

3. Phenol extract and ethanol precipitate the tailed vector as described for tailed cDNA.

4. Characterize the product electrophoretically in a 2.5% agarose gel. Digest 0.5 μg of tailed vector with HaeII as specified by the supplier. In a parallel reaction, digest PstI-cut untailed vector. Run both sets of fragments in adjacent lanes of the gel.

Using this technique, the four fragments generated by HaeII digestion of PstI-cut pUC9 were evident. The two smallest, 170 and 250 bp, in the untailed sample, were virtually absent from the tailed sample. Since two bands of material slightly larger in size (190 and 270 bp) appeared in the digest of the tailed vector, apparently both ends had tails of approximately 20 residues.

Annealing of Double-Stranded cDNA to Vector[5]

Prepare 10× annealing buffer: 100 mM Tris–HCl (pH 7.5), 1 M NaCl, 10 mM EDTA. Then combine

10× Annealing buffer	1 μl
dC-Tailed vector	5.5 fmol (10 ng)
dG-Tailed cDNA	2 to 5 fmol
Water to	10 μl

1. Incubate at 65° for 5 min.

2. Reduce the temperature to 57° and incubate for 60 min.

3. Transform Escherichia coli HB101 competent cells (Bethesda Research Laboratories or the equivalent) with the annealed DNA, using 5 μl per 100 μl cells, according to the manufacturer's instructions or those found in this volume [13 and 39].

[5] T. Maniatis, E. F. Fritsch, and J. Sambrook, "Molecular Cloning: A Laboratory Manual," p. 242. Cold Spring Harbor Lab., Cold Spring Harbor, New York, 1982.

Transformation efficiencies are routinely one-tenth to one-hundredth those obtained with supercoiled pBR322.

Oligonucleotide Labeling

Tailing with terminal transferase is a convenient way to introduce radioactivity into oligomers for use as hybridization probes. A modification of the method of Collins and Hunsaker,[6] which follows, results in the addition of 10–20 dA residues and a specific activity of $10–25 \times 10^9$ dpm/μg. Combine in 10 μl: 2 μl 5× tailing buffer, 10 ng oligomer, 200 μCi [α-^{32}P]dATP (5000 Ci/mmol), 1 μg bovine serum albumin (restriction endonuclease grade from BRL or the equivalent), and 20 units of terminal transferase. Incubate at 37° for 60 min. Remove unincorporated label with a spun Sephadex G-25 column[7] or precipitate in the presence of 50 μg carrier tRNA with ethanol using ammonium acetate as the salt (this volume [5]). Monitor the reaction using DE-81 filters as described in this volume [6].

The specific activity of the product is calculated by assuming that the tail contributes radioactivity, but no mass. The tailed probe has the same dissociation temperature (T_d) as its untailed counterpart; the dA polymer has no effect on the specificity of hybridization. Hybridization signals obtained with such probes are nearly 10-fold greater than with the same mass of 5'-^{32}P probe prepared by "kinasing" with [γ-^{32}P]ATP and polynucleotide kinase. Longer tails are associated with a decrease in signal.

The following conditions have been recommended for prehybridizing and hybridizing blots with tailed probes. Prehybridize, in 6× SSC containing 0.5% sodium dodecyl sulfate (SDS), cleaved denatured salmon sperm DNA, 0.5 mg/ml (150–200 bp) (see this volume [46]), and proteinase K, 0.1 mg/ml, at 65° for 2 hr. Continue the prehybridization at 10–20° below T_d for 2 hr in 6× SSC, 0.5% SDS, 70 mM Tris–HCl (pH 7.4), 0.1 mg/ml poly(A), 5 μg/ml (dpA)$_{50}$, 0.1 mg/ml proteinase K, and 0.3 mg/ml hydrolyzed cleaved salmon sperm DNA. (DNA is hydrolyzed at 100° for 1 hr in 0.5 M NaOH; the products are 1–20 bases.) Hybridize by adding tailed probe. Hybridization in general and hybridization with oligomers, in particular, are detailed in this volume in [43,45] and [47], respectively.

---

[6] M. L. Collins and W. R. Hunsaker, *Anal. Biochem.* **151**, 211 (1985).
[7] A spun G-25 column is prepared by substituting Sephadex G-25 (Pharmacia) for Sephadex G-50 in the technique described in this volume [9]. Oligomers tailed with dA may also be purified by heat denaturing at 100° for 5 min, binding to oligo(dT)-cellulose at 20°, and eluting at 60° using the column and buffers described in this volume [25].

[38] Adaptors, Linkers, and Methylation

By Ray Wu, Tiyun Wu, and Anuradha Ray

One way to clone blunt-ended cDNA molecules is to add synthetic linkers or adaptors and thereby introduce restriction enzyme sites to their ends, which can be ligated to compatible termini in a vector. The synthetic duplex molecules used for this purpose are of two kinds: (1) duplex molecules which are blunt at both ends, more commonly termed *linkers*,[1] and (2) ready-made *adaptors* which form short duplexes with one blunt end and one sticky end.[2,3] With linkers, the cloning efficiency can be as much as 5-fold higher than with adaptors, but, as will become evident, adaptors are safer and require fewer manipulations.

Use of synthetic oligodeoxynucleotides (oligomers) for cloning blunt-ended DNA fragments involves two ligation reactions. The first is the ligation of the linker or adaptor to the fragments to be cloned. The second is the ligation of these tailored fragments to suitable vectors either of phage[4,5] or plasmid origin.[6,7] There are two additional steps involved in linker ligation that are not required in adaptor ligation. Unlike adaptors, linkers can be ligated to one another to form long blunt-ended concatamers consisting of linkers alone or linkers attached to cDNA. It is, therefore, necessary to generate cohesive ends flanking each DNA fragment by cleaving with the required restriction enzyme. During the step, the DNA molecules may also be cleaved at their internal restriction enzyme sites. This problem can be overcome, in some cases, by enzymatically methylating the internal sites beforehand, so that they become resistant to subsequent cleavage.[8] The last step before ligation to the vector molecules is the removal of excess linkers and adaptors by gel chromatography[9] using either Ultrogel AcA 34 (LKB) or the Sepharose CL-4B column described in detail in this volume [36].

[1] C. P. Bahl, K. J. Marians, R. Wu, J. Stawinsky, and S. A. Narang, *Gene* **1,** 81 (1976).
[2] C. P. Bahl, R. Wu, R. Brousseau, A. K. Sood, H. M. Hsiung, and S. A. Narang, *Biochem. Biophys. Res. Commun.* **81,** 695 (1978).
[3] R. J. Rothstein, L. F. Lau, C. P. Bahl, S. A. Narang, and R. Wu, this series, Vol. 68, p. 98.
[4] N. E. Murray, W. J. Brammar, and K. Murray, *Mol. Gen. Genet.* **150,** 53 (1977).
[5] R. A. Young and R. W. Davis, *Proc. Natl. Acad. Sci. U.S.A.* **80,** 1194 (1983).
[6] U. Ruther and B. Müller-Hill, *EMBO J.* **2,** 1791 (1983).
[7] K. R. Stanley and J. P. Luzio, *EMBO J.* **3,** 1429 (1984).
[8] T. Maniatis, R. C. Hardison, E. Lacy, J. Lauer, C. O'Connell, D. Quon, G. K. Sim, and A. Efstratiadis, *Cell (Cambridge, Mass.)* **15,** 687 (1978).
[9] S. Anderson, *Nucleic Acids Res.* **9,** 3015 (1981).

Use of linkers, then, requires sequential methylation of the insert, ligation of linkers, cleavage at an internal site of the linker, purification of the cDNA (to remove excess linker molecules), and ligation of the tailored insert to a suitably cleaved vector. In contrast, the use of adaptors requires only the initial ligation of the adaptor, purification of the cDNA, and the final ligation to cleaved vector DNA. The investigator must decide whether to risk the extra steps that accompany the linker technique in order to achieve a higher cloning efficiency.

Solutions and Buffers

TE buffer: 10 mM Tris–HCl (pH 8.0), 1 mM EDTA (pH 8.0)

Standard T4 ligase buffer: 66 mM Tris–HCl (pH 7.6), 10 mM MgCl$_2$, 10 mM dithiothreitol (DTT), 1 mM spermidine–HCl, 0.3 mM ATP, and 0.1 mg/ml bovine serum albumin (BSA)

10× kinase buffer: 660 mM Tris–HCl (pH 7.6), 10 mM ATP, 10 mM spermidine–HCl, 100 mM MgCl$_2$, 100 mM DTT, and 2 mg/ml BSA

DNA methylase buffer: 50 mM Tris–HCl (pH 8.0), 100 mM NaCl, 10 mM EDTA, 0.2 mg/ml BSA, and 0.1 mM S-adenosylmethionine

EcoRI digestion buffer: 100 mM Tris–HCl (pH 7.5), 50 mM NaCl, 10 mM MgCl$_2$, and 0.1 mg/ml BSA

Phenol–chloroform mixture: The mixture contains phenol and chloroform in equal volumes. The phenol is previously equilibrated with 100 mM Tris–HCl (pH 8.0), and the chloroform is a mixture of chloroform and isoamyl alcohol in a ratio of 24 : 1 (by volume).

Linker Ligation

Phosphorylation of Linkers by Polynucleotide Kinase Followed by Self-Annealing

Unphosphorylated single-stranded oligomers (8–12 bases in length) are first heat denatured in water at 90° for 2 min and quickly cooled in ice, and then phosphorylated at their 5′ ends using T4 polynucleotide kinase [Bethesda Research Laboratories (BRL)]. The reaction mixture (25 μl) contains 4–6 μg of linkers (approximately 1.5 nmols), 10 μCi of [γ-^{32}P]ATP, 2.5 μl of 10× kinase buffer (contains 2.5 nmol unlabeled ATP), and 15–20 units of T4 polynucleotide kinase. After incubation at 37° for 1 hr, the tube is heated to 70° for 5 min and transferred to a beaker containing water at 65°. The water is allowed to cool slowly to room temperature to promote annealing of the oligomers to give duplex linkers. The tubes are centrifuged briefly, and the reaction mixture is divided into small aliquots and stored at −20°. Duplex linkers should not be heated over 25° during thawing when used later. Yields of phosphorylated linkers are 60–

80%, depending on the nature of the 5′ terminus. The reaction may be monitored and the yield assessed by using the methods for oligomers described in this volume [6].

Testing of Linkers by Blunt-End Ligation

Before using the linkers to tail cDNA molecules, it is advisable that the linkers be checked by self-ligation and redigestion with the appropriate restriction enzyme as follows. Incubate 200 ng of 5′-^{32}P-phosphorylated linkers (colloquially known as "kinased" linkers) in T4 ligase buffer with 2 units of T4 DNA ligase (BRL) in a final reaction volume of 10–15 μl for 6 hr at 15°. After heating to 65° for 10–15 min, 5 μl of the ligated linker is removed and digested with 10–15 units of the restriction enzyme after appropriate adjustments of buffer conditions. The products are analyzed by electrophoresis in a 10% polyacrylamide (this volume [8]) gel along with unligated "kinased" linkers and ligated "kinased" linkers. The ligation reaction should produce a ladder of linker oligomers, and restriction digestion should produce a single band of monomer length.

Methylation of DNA

Protection of internal restriction enzyme sites (e.g., EcoRI sites) may be achieved by treating the DNA with EcoRI methylase as follows. Between 0.5 and 1 μg of DNA is methylated in 20 μl DNA methylase buffer using 20 units of EcoRI methylase (New England BioLabs). Incubation is carried out for 60–90 min at 37°. The mixture is extracted with a phenol–chloroform mixture. The organic phase is reextracted with TE buffer; the aqueous phases are pooled and extracted with ether. The DNA is precipitated by adding 0.1 volume of 3 M sodium acetate (pH 5.0), 2 volumes of ethanol and chilling to −70° in a dry ice–ethanol bath for 30–60 min. The methylated DNA is recovered by centrifugation (this volume [5]).

To verify that methylation has indeed occurred, control DNA, such as pBR322 linearized with an irrelevant enzyme (not EcoRI in this example), is methylated (with EcoRI methylase) and restricted (with EcoRI). When the methylated sample is compared by electrophoretic analysis in a 1% agarose gel, with a similarly restricted unmethylated sample, the latter should yield two fragments whereas the former should yield the intact linearized plasmid used as the starting material.

The type of restriction site represented in the linker must be chosen based on the availability of the required methylase. If the insertion of cDNA is to be accomplished by means of a site for which no methylase is available, the investigator *must* use the adaptor technique detailed elsewhere in this chapter.

Ligation of Linkers to Blunt-End DNA

The next step is to ligate the "kinased" linkers to blunt-ended DNA molecules. Usually blunt-end ligation is much less efficient than sticky-end ligation. A basic requirement for efficient cloning of cDNA molecules is thus an improvement in the ligation efficiency. We have manipulated various parameters (temperature, concentration of ligase and ATP, linker concentration, etc.) to obtain an optimal condition for linker ligation. We have also examined the effects of polyethylene glycol 6000[10] and hexamminecobalt chloride.[11] Polyethylene glycol failed to improve linker ligation (at concentrations 5–15%) but hexamminecobalt chloride at a 1 mM concentration did stimulate ligation.

Conditions for linker ligation were optimized by taking two factors into consideration: (1) efficient ligation of linkers to target DNA, and (2) high yield of transformants. To check the effects of various factors on ligation efficiency, blunt-ended fragments of pBR322 generated by digestion with HaeIII were chosen as substrates for the ligation of a short (8-mer) EcoRI linker 5′d(G-G-A-A-T-T-C-C), and a longer (12-mer) EcoRI linker, 5′d(C-C-G-G-A-A-T-T-C-C-G-G).

Optimal Conditions for Linker Ligation

Taking into consideration the various conditions reported in the literature for blunt-end ligation, and based on the results of a number of experiments carried out by us recently, the reaction conditions for optimal linker ligation have been found to be 66 mM Tris–HCl (pH 7.6), 10 mM MgCl$_2$, 10 mM DTT, 0.3 mM ATP, 1 mM spermidine–HCl, 1 mM hexamminecobalt chloride, 200 μg/ml BSA, 20-fold molar excess of the 8-mer or the 12-mer linker over total 5′ ends of target DNA (keeping in mind that 1 mol of DNA equals 2 mol of 5′ ends), and T4 DNA ligase at 200 units/ml in a total volume of 15–20 μl. Incubation is at 15° for 6 hr.

Ligation of cDNA to Vector DNA and Transformation of E. coli

To check the transformation efficiency of fragments ligated to linkers under optimal conditions, a 2.5-kb PvuII fragment of λ DNA containing an EcoRI site was methylated with EcoRI methylase according to the methylation conditions mentioned before. The reaction mixture was extracted once with a phenol–chloroform mixture, once with ether, and precipitated by the addition of 0.1 volume of 3 M sodium acetate and 2 volumes of ethanol. An amount of 0.05–0.5 μg of the DNA was ligated in separate reactions with either the 8-mer or the 12-mer linker under opti-

[10] B. H. Pheiffer and S. B. Zimmerman, Nucleic Acids Res. 11, 7853 (1983).
[11] J. R. Rusche and P. Howard-Flanders, Nucleic Acids Res. 13, 1997 (1985).

mal reaction conditions. After incubation for 6 hr at 15°, the mixture was successively extracted with phenol–chloroform, chloroform, and ether and the DNA precipitated with salt and ethanol. The DNA was digested with 15 units of *Eco*RI and purified in a 1% low melting temperature agarose gel. The DNA was next ligated to *Eco*RI-cleaved 5'-dephosphorylated pUC19 under standard ligation reaction conditions for 6 hr at 22° using a vector-to-insert ratio of 3 : 1 (100–200 ng of vector and 30–60 ng of insert DNA). The ligation mixtures were diluted with TE buffer, and 5–20 ng (total DNA) was used to transform competent *E. coli* HB101 cells.[12] We found that the number of transformants was 5×10^5 to $2 \times 10^6/\mu g$ of insert DNA. As a positive control, undigested pUC19 DNA gave 4×10^7 colony forming units/μg DNA. As a negative control, the transformation efficiency of ligated dephosphorylated vector (pUC19) was only $10^3/\mu g$ of DNA. We have not used the highest efficiency methods for transformation reported in the literature[13] and expect to get higher yields of transformants if such methods are used.

When a complex mixture of double-stranded cDNA is used to generate a cDNA library, the concentration of termini must be estimated in order to choose a vector concentration. Unlike the well-characterized *Pvu*II fragment of λ DNA in the example, cDNA contains a heterogeneous mixture of fragment lengths, each at an unknown concentration. To determine the molar cDNA concentration it is useful (1) to select only the larger size classes by means of a Sepharose CL-4B column[14] (this volume [36]) *before* methylation and ligation of linkers, and (2) to calculate the concentration of termini *after* ligation, restriction, and chromatography (a second time) in the Sepharose CL-4B column.[14] The second passage through the column results in a population with a single cleaved linker at each terminus. (Spermine precipitation, this volume [5], is an alternative to chromatography.) Given the specific activity of the "kinased" linkers in cpm/mol of linker, the number of moles of cDNA can be obtained. Remember, when determining the specific activity of the linkers, that those linkers that did not become phosphorylated do not play a role in the reaction: neither do they contribute to the numerator because they are not radioactive nor should they be included in the denominator because they are incapable of ligation. (Conditions for ligating cDNA to plasmid and phage vectors appear in this volume in [10,39] and [10,40], respectively.)

[12] T. Maniatis, E. F. Fritsch, and J. Sambrook, "Molecular Cloning: A Laboratory Manual," p. 250. Cold Spring Harbor Lab., Cold Spring Harbor, New York, 1982.

[13] D. Hanahan, *J. Mol. Biol.* **155,** 557 (1983).

[14] When purifying cDNA before the addition of linkers or adaptors or when repurifying it afterward to eliminate excess linkers or adaptors from the final cDNA product, select only the material excluded by Sepharose CL-4B.

Ligation of Adaptors to Blunt-Ended DNA

As discussed in the introduction to this chapter, linker tailing of DNA molecules involves protection of internal sites, generally by methylation. The choice of linkers is thus restricted to those for which methylases are available. An alternative approach is to ligate duplex molecules with preformed cohesive ends (ready-made adaptors[2]) to the ends of the molecules to be cloned.

Phosphorylation of Adaptors by Polynucleotide Kinase Followed by Annealing

In forming ready-made adaptor molecules, for example a *Bam*HI–*Sma*I adaptor,

$$5' \text{ pC-C-C-G-G-G}$$

$$3' \text{ G-G-G-C-C-C-C-T-A-G-OH}$$

only one of the oligomers is phosphorylated at the 5' end to create a 5'-phosphate group at the blunt end of the molecule. The sticky end is kept unphosphorylated to prevent self-ligation during ligation to DNA molecules to be cloned. The hexamer is phosphorylated according to the conditions mentioned for phosphorylation of linkers. It is extracted with phenol–chloroform, and the organic phase reextracted with TE. The aqueous phases are pooled and the DNA precipitated using 0.1 volume of 3 *M* sodium acetate, 3 volumes of ethanol, and chilling at $-70°$ for 90–120 min. (Recovery of the phosphorylated hexamer after performing these steps was greater than 90%.) To form the ready-made adaptor, equimolar concentrations of the phosphorylated and unphosphorylated oligomers are suspended in TE buffer at a concentration of 0.5–1 μg/μl and heat denatured at 70° and annealed as mentioned before.

Efficiency of Adaptor Ligation

We have used the optimal conditions for linker ligation and a 20 to 50-fold molar excess of adaptor molecules, and we have utilized *Hae*III-cleaved pBR322 as the test system to check for adaptor tailing (*Bam*HI–*Sma*I adaptor). We found that although a major proportion of DNA molecules are tailed by adaptors, a smaller fraction receives a *Sma*I hexamer linker at either one or both ends as judged by the mobility shifts of the bands on the gels.

Transformation Efficiency of Adaptor-Tailed Fragments

The same 2.5-kb *Pvu*II fragment of λ DNA was used to check the transformation efficiency when tailed with *Bam*HI–*Sma*I adaptors. For the transformation assay, the adaptor was first ligated to DNA ends, and the fragment purified in an agarose gel and then phosphorylated at the 5'-OH groups of the sticky ends (using the same kinase buffer mentioned before). The DNA was extracted with phenol–chloroform, ether and precipitated with salt and ethanol. It was next ligated under standard ligation conditions to dephosphorylated BamHI-cleaved pUC19 vector. The transformation efficiency of adaptor-tailed fragments was approximately 4 to 5 × 10^5 cfu/μg of insert DNA.

When cloning cDNA, purification by column chromatography before and after ligation of adaptors, together with calculation of the cDNA concentration from the specific activity of the adaptors, may be accomplished as described above for cloning cDNA with linkers.

[39] Directional cDNA Cloning in Plasmid Vectors by Sequential Addition of Oligonucleotide Linkers

By DAVID M. HELFMAN, JOHN C. FIDDES, and DOUGLAS HANAHAN

Expression vectors make possible the synthesis of foreign proteins in E. coli by providing bacterial promoters, signals for translation, and restriction sites located downstream for insertion of foreign DNA (Fig. 1 and [69]). To be expressed as a bacterial fusion protein, the foreign coding sequences must be in the correct orientation and must be in frame with the vector initiation codon. Foreign 5' noncoding sequences located between the AUG codon and the foreign coding region are potential problems because in frame stop codons can interrupt protein synthesis. When subsequent screening methods depend on antibody recognition of the fused product as in [50], cloning methods to maximize correct expression must be chosen. In this chapter we show how to insert cDNA into expression plasmids using two different linkers,[1] an approach which concomitantly increases the cloning efficiency.

J. C. Fiddes and D. Hanahan (unpublished results) developed a simple improvement of the double-linker method such that the linkers are added sequentially. One set of linkers is ligated to the end of the double-stranded cDNA (this volume [34]) corresponding to the 3' end of the poly(A)+ RNA prior to cleavage of the hairpin loop with nuclease S_1. After the nuclease S_1 reaction the second set of linkers is ligated to the end of the cDNA that was previously protected by the hairpin loop. As a result, the majority of cDNA molecules contain different linkers at each end, which has been

[1] D. T. Kurtz and C. F. Nicodemus, Gene 13, 145 (1981).

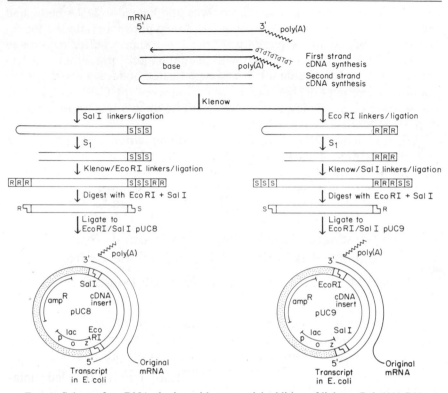

FIG. 1. Scheme for cDNA cloning with sequential addition of linkers. Poly(A)$^+$ RNA is copied into single-stranded DNA by reverse transcriptase by using oligo(dT) primers. *Sal*I linkers are marked S and *Eco*RI linkers are marked R. While the figure presents the use of *Eco*RI and *Sal*I double-cut pUC8 and pUC9 (stippled), other linkers or vectors may be used. P, O, and Z refer to the promoter, operator, and part of the β galactosidase gene, elements needed for expression of foreign DNA (open).

found to increase the efficiency of inserting the cDNA into the vector, and hence for producing clones of the cDNAs. In addition, sequential addition of linkers permits the cDNA to be inserted in a defined orientation into the vector. The latter property is particularly useful in the construction of cDNA expression libraries using plasmid vectors.[2,3] Furthermore, double-cut vectors produce head-to-head or tail-to-tail dimers upon self-ligation, and these are unable to transform *Escherichia coli*. This provides a selection for transformation of recombinant plasmids. The synthesis of double-stranded cDNA is presented elsewhere in this volume [33,34] and will not be described. What follows is a description of the steps used following

[2] D. M. Helfman, J. R. Feramisco, J. C. Fiddes, G. P. Thomas, and S. H. Hughes, *Proc. Natl. Acad. Sci. U.S.A.* **80,** 31 (1983).

[3] D. M. Helfman, J. R. Feramisco, J. C. Fiddes, G. P. Thomas, and S. H. Hughes, *Genet. Eng.* **7,** (1985).

synthesis of double-stranded cDNA containing intact hairpin loops (i.e., not treated with S_1 nuclease). Figure 1 shows the steps in preparation of a cDNA library by sequential addition of linkers. The advantages of adding linkers in reverse order to the cDNA and using two vectors (e.g., pUC8 and pUC9) that differ in the orientation of their polylinker are also discussed. The example described below uses EcoRI and SalI linkers and pUC plasmids, but in principle any combination of nonhomologous linkers can be employed with any compatible double-cut vector.

Oligonucleotide Linker Preparation

5'-Phosphorylated and unphosphorylated EcoRI and SalI octanucleotide linkers can be obtained from PL Biochemicals, New England BioLabs, or Collaborative Research, etc. (most commercially available linkers work well, although it is advisable to test each lot for ligatability). EcoRI or SalI nonphosphorylated linkers (8 μg) are incubated with 2.5 units of T4 polynucleotide kinase (PL Biochemicals) in a 40-μl reaction volume containing 50 mM Tris–HCl (pH 8.0), 7 mM MgCl$_2$, 1 mM dithiothreitol, and 100 μCi [γ-^{32}P]ATP at 37° for 15 min. The reaction is diluted to 80 μl with the same buffer, nonradioactive ATP is added to a final concentration of 1 mM, and 10 additional units of T4 polynucleotide kinase added. The reaction is incubated at 37° for an additional 45 min, heated to 68° for 15 min, and chilled to 0°. The radiolabeled linkers are added to the unlabeled linkers in a molar ratio of 1 : 100 (labeled : unlabeled). The resulting solution is used as a linker stock with a concentration of 100 ng/μl. The use of radiolabeled linkers allows one to follow linker ligations and subsequent restriction enzyme digestions. It is very important that the linkers are tested prior to ligation to the cDNA by blunt-end self-ligation and endonuclease digestion. This test has been described in this volume [38].

Linker Ligation to Double-Stranded cDNA

A preparation of double-stranded cDNA (with intact hairpin loops at the 5' end) is treated with the Klenow fragment of E. coli DNA polymerase I to produce flush double-stranded ends. The cDNA is incubated with 5 units of DNA polymerase I (Klenow fragment, New England BioLabs) in a 20-μl reaction containing 50 mM Tris–HCl (pII 8.0), 7 mM MgCl$_2$, 1 mM dithiothreitol, and 30 μM each dATP, dCTP, dGTP, and dTTP at 15° for 4–16 hr. The reaction is then divided in half for the ligation of either EcoRI or SalI linkers to the end of the cDNA corresponding to the 3' end of the mRNA [poly(A) tract]. Each reaction is adjusted to a 20-μl volume containing 70 mM Tris–HCl, (pH 8.0), 7 mM MgCl$_2$, 1 mM dithiothreitol, 1 mM ATP, 800 ng EcoRI or SalI linkers, and 600 units of

T4 DNA ligase (New England BioLabs). The ligation is carried out at 4° for 12–16 hr. The extent of ligation can be monitored by electrophoresis of a small sample on a 10% polyacrylamide gel, and observing a characteristic linker ladder. The 800 ng of linkers should correspond to at least a 100× molar ratio of linkers to cDNA. In practice it is advisable to increase the linker concentration even further to a very high excess of linkers to cDNA (approximately 1000×). This is to minimize the possibility of cDNA molecules ligating to each other. Such "junk" cDNAs present a problem in subsequent analyses of a cloned cDNA, as they contain fragments of unrelated genes. We have found it advantageous to fractionate the cDNA preparation by size (this volume [36]) prior to the addition of linkers. This tends to minimize the chance of obtaining such "junk" cDNA molecules, and facilitates achieving an appropriate ratio of linkers to cDNA ends.

The two cDNA preparations (which carry either $EcoRI$ or $SalI$ linkers at their 3' ends) are extracted with phenol–chloroform, precipitated with ethanol, washed with 70% ethanol, and dried under vacuum (this volume [4,5]). In order to remove the hairpin loops each cDNA preparation is then digested with S_1 nuclease (Bethesda Research Laboratories) in a 25-μl volume containing 300 mM NaCl, 30 mM sodium acetate (pH 4.5), 3 mM ZnCl$_2$, at 42° for 5 min. The appropriate amount of S_1 nuclease to use is determined as described elsewhere[4] and in this volume [34]. The reaction is stopped by the addition of 1 μl 1 M Tris base (pH unadjusted) and 1 μl 250 mM EDTA (pH 8.0). The cDNA samples are extracted with phenol–chloroform, and fractionated in a 1-ml Sephadex G-75 column in 10 mM Tris–HCl (pH 7.6), 1 mM EDTA. The excluded peak is collected and the DNA precipitated with ethanol, washed with 70% ethanol, and dried under vacuum.

Each set of cDNAs is then treated with 5 units of DNA polymerase I (Klenow fragment), in a 10-μl reaction containing 50 mM Tris–HCl, (pH 8.0), 7 mM MgCl$_2$, 1 mM dithiothreitol, and 30 μM each dATP, dCTP, dGTP, and dTTP at 12° for 30 min. The second oligonucleotide linker is then ligated to the blunt-ended cDNA molecules (see Fig. 1). Each reaction is adjusted to a 20-μl volume containing 70 mM Tris–HCl (pH 8.0), 7 mM MgCl$_2$, 1 mM dithiothreitol, 1 mM ATP, 800 ng $EcoRI$ or $SalI$ linkers, and 600 units of T4 DNA ligase. The ligation is carried out at 4° for 12–16 hr. Following the ligation reaction, the ligase is inactivated by heating at 68° for 15 min. The reactions are then diluted to about 500 μl with the appropriate restriction enzyme buffer, and digested with $EcoRI$ and $SalI$

---

[4] T. Maniatis, E. F. Fritsch, and J. Sambrook, "Molecular Cloning: A Laboratory Manual." Cold Spring Harbor Lab., Cold Spring Harbor, New York, 1982.

(at least 250 units each) for several hours. In order to monitor for complete digestion of the linkers, a small aliquot is removed and fractionated by electrophoresis in a 5% polyacrylamide gel (this volume [8]). If the linkers are not digested, one will still observe the presence of a linker ladder. Accordingly, additional restriction enzymes may be added for complete digestion. After complete digestion of the linkers (no linker ladder in the gel), the cDNA preparations are extracted with phenol–chloroform, ethanol precipitated, washed with 70% ethanol, and dried briefly under vacuum. The cDNAs are then size-fractionated with concomitant removal of linker debris in a 1-ml Sepharose CL-4B column in a buffer containing 300 mM NaCl, 10 mM Tris–HCl (pH 7.6), 1 mM EDTA. Fractions (approximately 75 μl each) are assayed by electrophoresis of 3 μl aliquots in 5% polyacrylamide gels or 1.4% agarose gels, followed by autoradiography. Those fractions containing cDNA greater than 500 base pairs in length are pooled and precipitated with ethanol.

Preparation of Double-Cut Plasmid Vectors

The use of two heterologous restriction enzyme sites in cDNA cloning requires the preparation of a suitable double-cut plasmid vector. The pUC8 and pUC9 vectors[5] are useful for this purpose. Both vectors contain a polylinker that includes sites for the enzymes EcoRI and SalI (and several others). The vectors differ in the orientation of the polylinker. In addition, the pUC plasmids are useful for expression cloning.[2,3] It is worth noting that many other plasmid vectors are available that contain suitable cloning sites and may be used in the construction of a cDNA library. Starting with 25–50 μg of pUC8 or pUC9, the plasmid is first digested with SalI and a small aliquot of the digest is tested by agarose gel electrophoresis to determine that all the plasmid has been linearized. The plasmid is then digested with EcoRI. Since the distance between the EcoRI and SalI sites is very small, it is not possible to monitor the EcoRI digest by agarose gel electrophoresis. It is advisable to remove a small aliquot of the double-digested DNA, and end label the fragment using [^{32}P]dATP or [^{32}P]dTTP and the Klenow fragment of DNA polymerase I (this volume [10]). This is then separated in a 15% polyacrylamide gel to determine if the small EcoRI–SalI fragment has been liberated (this volume [8]). In addition, this test may be used to determine if the liberated fragment is completely removed in the subsequent purification of the double-cut vector. The double-cut plasmid is then isolated. Any one of a number of techniques may be used to remove the small EcoRI–SalI fragment, in-

[5] J. Vieira and J. Messing, Gene 19, 259 (1982).

cluding gel electrophoresis, sucrose gradient centrifugation, or passage over a Sephacryl CL-6B or Sepharose 4B column. For double-cut vectors derived by digesting "polylinker" clusters (e.g., pUC8 and pUC9), the small fragment can be removed by passing the digestion products over a small column of Sepharose CL-4B or Sepharose 4B (bed volume 1–2 ml) equilibrated in 300 mM NaCl, 10 mM Tris–HCl (pH 7.4), 1 mM EDTA. The vector is then precipitated with ethanol, washed with 70% ethanol, dried briefly under vacuum, and resuspended in a final concentration of 1 mg/ml in 10 mM Tris–HCl (pH 7.4), 1 mM EDTA. It is advisable to wash the column extensively (5–10 bed volumes) prior to chromatographing the vector DNA to remove compounds that may inhibit subsequent ligations of the cDNA to the plasmid.

Ligation of cDNA to Vector and Bacterial Transformation

Before ligation of cDNA to the vector, the purified vectors are first tested in transformation assays. This involves transforming competent bacteria (this volume [13]) with the following preparations: (1) the unligated double-cut vector, (2) self-ligation of the double-cut vector, and (3) ligation to an equimolar amount of a compatible EcoRI–SalI insert. The results of such an experiment are shown in Table I for an EcoRI–SalI double-cut vector derived from pBR322. The unligated vector fragment (line 2) does not transform $E.\ coli$, while the self-ligated vector (line 3) transforms very poorly, about a thousandfold less efficiently than when the vector is ligated to an EcoRI–SalI fragment (line 4). The use of a

TABLE I
CHARACTERISTIC TRANSFORMATION EFFICIENCIES OF DOUBLE-CUT VECTORS[a]

DNA	Condition	Colonies/ng
pBR322	Supercoiled	5×10^5
pBR322-RS vector	Unligated	5×10^{-2}
pBR322-RS vector	Ligated	25
pBR322-RS vector + RS small fragment	Ligated	4×10^4

[a] pBR322 was digested with EcoRI and SalI, and the vector fragment (pBR322-RS vector) containing the origin of replication and the ampicillin resistance gene was purified by sucrose gradient centrifugation or by agarose gel electrophoresis. The small fragment obtained from this digest (RS small fragment) was purified by agarose gel electrophoresis. Ligations were at 4° for 12–24 hr with DNA concentrations of about 20 ng/μl. The reconstitution of pBR322 was done with approximately equimolar amounts of the two fragments. Transformations were performed on numerous occasions using 1–10 ng of DNA and colonies were selected for ampicillin resistance.

double-cut vector provides an intrinsic selection for the acquisition of an insert to form a recombinant plasmid. It is therefore not necessary to prevent self-ligation of the vector by treating the cleaved plasmid with alkaline phosphatase. The vector itself cannot cyclize efficiently through the nonhomologous sticky ends and the vector dimer (vector-to-vector ligation) cannot give rise to stable transformants. Thus the background of nonrecombinant plasmids in a transformation can be negligible. In practice we find that greater than 95% of the recombinants obtained contain a cDNA insert.

Once the vector has been tested it is generally desirable to optimize the colony-forming potential of a cDNA preparation. First, a series of ligations are carried out using varying amounts of cDNA (approximately 0.1–2 ng per ligation) and a constant amount of the vector (10 ng) to determine the optimal ratio of cDNA to vector. It is possible to estimate the amount of cDNA based on the specific activity of the linkers. Following ligation of the cDNA to appropriate plasmid vectors, E. coli is transformed with the recombinant plasmids. In our hands we find that ligase dramatically inhibits the efficiency of transformation. Accordingly, following the ligation reactions the ligase is removed by precipitating the DNA with two-thirds volume 5 M ammonium acetate and 2 volume ethanol, followed by a 70% ethanol wash. The pellet is dried briefly under vacuum and resuspended in 10 mM Tris–HCl (pH 7.5), containing 0.1 mM EDTA. It is desirable to optimize the conditions for transformation to determine the amount of the ligated vector–insert to use per transformation. A series of parallel transformations are carried out using varying amounts of the ligated sample and a constant amount (e.g., 200 μl) of competent E. coli. The colony-forming potential of a cDNA preparation can thus be significantly extended by remaining in the linear range during transformation and optimizing the ratio of cDNA ligated to vector. Figure 2 shows the results of such an experiment. Three independent cDNA preparations were used: one was synthesized from embryonic chicken smooth muscle RNA and was ligated to EcoRI–SalI pUC9, and two were synthesized from human term placental mRNA and ligated to EcoRI–SalI pBR322 vector. Each cDNA was employed in a standard transformation with about 1.5×10^8 cells.[6,7] The amounts of total DNA ranged from 0.25 to 64 ng, increasing by factors of 2. Figure 2 shows the results of these transformations. The response is linear from 0.25 ng to 4–8 ng. Above this

[6] D. Hanahan, J. Mol. Biol. 166, 557 (1983).
[7] D. Hanahan, in "DNA Cloning" (D. M. Glover, ed.), Vol. 1, p. 109. IRL Press, Washington, D.C., 1985.

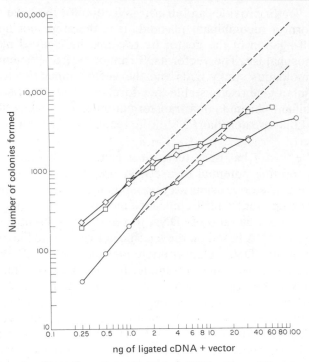

FIG. 2. Linearity of transformation with ligated cDNAs. Three different preparations of cDNAs were used in a series of transformations of *E. coli* DH1 in which varying levels of DNA were combined with a fixed volume of competent cells. Two of the cDNA preparations (□ and ○) were derived from human term placenta mRNA and were ligated to *Eco*RI–*Sal*I pBR322 vector at a 25:1 mass ratio of vector to cDNA (approximately a 5-fold molar excess of vector). The third (◇) cDNA preparation was from embryonic chicken smooth muscle mRNA and was ligated to pUC9 at a 20:1 mass ratio. In all cases the noted amount of ligated DNA was used in a standard transformation, in which supercoiled pBR322 gave 5×10^8 transformants/μg DNA. The dotted lines are extrapolations based on the expectation that a doubling of the amount of DNA should double the number of colonies. Both scales on the graph are logarithmic.

level the response becomes increasingly nonlinear. The chicken cDNA saturated the transformation at 8 ng, while the human cDNAs saturated at 30–50 ng. It is important not to saturate the transformation competent bacteria, e.g., no more than 10–20 ng of total DNA in a standard bacterial transformation[6,7] using 200 μl of bacteria. Once the optimal ratio of cDNA to vector has been established the ligation can be scaled up to prepare a cDNA library. By optimizing the ligation and transformation conditions, we routinely obtain 5000–15,000 transformants per nanogram of size-selected cDNA, corresponding to about 250,000 transformants per microgram of mRNA. If larger libraries are required, more mRNA can be used in the initial stages of cDNA synthesis.

Discussion and Alternative Strategies

The sequential addition of linkers ensures that a majority of the cDNA molecules have nonhomologous ends (Fig. 1), which greatly improves the efficiency of cloning. One advantage of adding the linkers sequentially is that the orientation of the cDNA within the vector is known. In addition to being very useful in the analysis of cDNA clones, this property is useful in the construction of cDNA expression libraries in *E. coli*.[2,3] Furthermore, cDNA may be prepared for expression in eukaryotic cells (e.g., yeast or vertebrates) using suitable vectors. Another benefit of directional cloning is the ability to produce the cDNA library in a dual riboprobe vector (e.g., containing a polylinker flanked by T3 and T7 polymerase promoters; see this volume [58,59] for an example). This would allow one to prepare sense RNA with one promoter and antisense RNA with the other promoter. The antisense RNA could then be used for subtractive hybridization to produce enriched probes or cDNA libraries.

One problem in cloning using sequential linker addition is the presence in some cDNAs of restriction sites that are the same as those of the linkers. This will give either an unclonable DNA insert or, in the case of expression libraries, an insert that is always in the same, possibly incorrect, translational frame. The pUC8/pUC9 plasmid expression system has been used in an attempt to overcome this problem. As described, both of these vectors have a polylinker with multiple cloning sites inserted in the amino-terminal region of the *E. coli lacZ* gene but differ in the orientation of the polylinker.[5] Therefore, it is possible to take half of a cDNA preparation, add linkers in temporal order, and clone in the appropriate orientation in pUC8 and then take the other half of the same cDNA preparation, add the same set of linkers in the opposite order, and clone in the correct orientation in pUC9 (Fig. 1). This will overcome most of the problems caused by internal restriction sites in the cDNA. Of course if one is not interested in preparing an expression library, both cDNA preparations can be ligated to the same double-cut vector (e.g., pUC8).

There are other approaches which could be used to overcome the problems caused by internal sites. One way would be to methylate the cDNA with the appropriate specific methylase before adding the oligonucleotide linkers (see [38] this volume). The cDNA would then be resistant to cleavage during the digestion of the linkers. However, it is our experience that methylation tends to inhibit transformation of *E. coli*. An alternative possibility to minimize the problem of internal restriction sites is to prepare separate libraries using different sets of oligonucleotide linkers, especially ones that cut DNA infrequently. Finally, it should also be possible to ligate one set of linkers corresponding to a rare cutter (e.g., *Sal*I or *Xho*I) to the poly(A) end of the cDNA, and then to ligate a sticky-ended phosphatase-treated adaptor to the other end, one which does not require digestion with a restriction enzyme in order to produce an end ready for ligation to the vector (see [38] this volume). This would reduce the potential problems, as the cDNA would then only have to be digested with a single restriction enzyme with an infrequently observed recognition site.

The sequential addition of linkers increases the number of steps in an already difficult cDNA synthesis. One might argue that a single linker would suffice, when used in conjunction with a phosphatase-treated vector, so as to reduce vector self-ligation. In our experience, this alternative is significantly less efficient, and suffers both from the inability to orient the cDNA insert, and from the empirical observation that phosphatase-treated ends do not compete equally with phosphorylated ends. This latter point means that the cDNA would rather self-ligate than ligate to the vector. In contrast, a double-linkered cDNA cannot self-ligate, nor can a double-cut vector, thus the reaction requires that intermolecular ligation occurs (as intramolecular or self-ligation cannot). This fact is borne out in comparative ligation/transformation experiments. For example, a double-cut vector will typically show relative transformation efficiencies of 10^{-7} when unligated, 10^{-4} when self-ligated, and 10^{-1} when ligated to an insert (this compared to an equal mass of supercoiled vector, $10^0 = 1$). In contrast, a single-cut vector will give relative efficiencies of 10^{-3} unligated, 10^{-1} self-ligated, and 10^{-1} ligated to insert (no difference with or without insert). A single-cut phosphatase-treated vector will give 10^{-4} colonies per unit mass when unligated, 10^{-3} after self-ligation, and 10^{-2} when ligated to an insert. These values are characteristic of the alternative methods using plasmid vectors. If, for some reason, a single linker is desired on both ends of the cDNA, one would simply begin with the S_1 step followed by end filling (Klenow reaction), or prepare cDNA or described in chapter [35] with careful attention to generating blunt ends, and then ligate to the single linker. In conclusion then, the persistent result is that sequential addition of linkers is more efficient at producing cDNA clones. When the consequential benefit of oriented insertion is also considered, the extra steps required in this procedure can be seen to merit their inclusion.

[*Editors' Note.* If expression is not important, tailing as a means for insertion of cDNA into a vector (this volume [37]) should be considered because it eliminates the need for blunt ends. Furthermore, since there is no need for cleavage of polymerized linkers attached to the termini of cDNA, the double-stranded cDNA is never at risk from restriction endonuclease cleavage at internal sites.]

Acknowledgments

This work was supported in part by grants from the National Institutes of Health, the Muscular Dystrophy Association, and the American Cancer Society. We are grateful to Madeline Szadkowski for preparation of this manuscript.

[40] Cloning cDNA into λgt10 and λgt11

By JERRY JENDRISAK, RICHARD A. YOUNG, and
JAMES DOUGLAS ENGEL

This chapter provides instructions for generating cDNA libraries with the λ bacteriophage vectors gt10 and gt11.[1-8] Briefly, double-stranded cDNA containing *Eco*RI cohesive termini (this volume [38]) is ligated into the unique *Eco*RI cloning site present in λgt10 or λgt11 DNA. Recombinant DNA is then packaged into viable phage particles which are plated on appropriate *Escherichia coli* hosts for amplification and screening. When only nucleic acid probes are available for library screening, λgt10 is the vector of choice. When antibody probes are available for screening, λgt11 is used since it is an expression vector (meaning that a fusion protein is formed between *E. coli* β-galactosidase and eukaryotic protein from the cDNA inserts). Advantages of λgt10 and gt11 over the use of plasmid vector cDNA cloning include (1) the high efficiency of introduc-

[1] T. V. Huynh, R. A. Young, and R. W. Davis, *in* "DNA Cloning" (D. M. Glover, ed.), Vol. 1, p. 49. IRL Press, Washington, D.C., 1985.

[2] R. A. Young and R. W. Davis, *Proc. Natl. Acad. Sci. U.S.A.* **80,** 1194 (1983).

[3] R. A. Young and R. W. Davis, *Science* **222,** 778 (1983).

[4] R. A. Young and R. W. Davis, *Genet. Eng.* **7,** 29 (1985).

[5] R. A. Young, V. Mehra, D. Sweetser, T. Buchanan, J. Clark-Curtiss, R. W. Davis, and B. R. Bloom, *Nature (London)* **316,** 450 (1985).

[6] M. Yamamoto, N. S. Yew, M. Federspiel, J. B. Dodgson, N. Hayashi, and J. D. Engel, *Proc. Natl. Acad. Sci. U.S.A.* **82,** 3702 (1985).

[7] N. R. Landau, T. P. St. John, I. L. Weissman, S. C. Wolf, A. E. Silverstone, and D. Baltimore, *Proc. Natl. Acad. Sci. U.S.A.* **81,** 5836 (1984).

[8] S. P. Leytus, D. W. Chung, W. Kisiel, K. Kurachi, and E. W. Davie, *Proc. Natl. Acad. Sci. U.S.A.* **81,** 3699 (1984).

ing recombinant DNA into *E. coli* by *in vitro* packaging followed directly by infection, and (2) the efficiency of screening bacteriophage at high plaque densities using either nucleic acid or antibody probes.

The subject of constructing and screening cDNA libraries in λgt10 and λgt11 has been reviewed recently by Huynh, *et al.*[1] The reader is referred there for details on vector construction, genetic maps, and for a more detailed discussion on the rationale for using either λgt10 or λgt11 for a particular application.

It is the intent of this chapter to serve as a practical, step-by-step guide beginning with double-stranded cDNA in hand (replete with *Eco*RI cohesive termini) and ultimately ending up with a cDNA recombinant library in either λgt10 or λgt11. Many methods used in this process are the subjects of other more detailed chapters in this volume; the reader is referred to chapters on microbiological techniques [13], the purification of phage λ DNA [13], electrophoresis of nucleic acids [8], the use of restriction enzymes [11], ligation of DNA [10] (see Process Guide), and the preparation and use of bacteriophage λ DNA *in vitro* packaging extracts [17].

Generating cDNA Libraries in λgt10

λgt10 is an insertion vector with a cloning capacity of up to 7 kb. The unique *Eco*RI cloning site is located in the λ repressor (*c*I) gene. Insertion of foreign DNA at this restriction site interrupts the *c*I coding sequence and causes the phenotype of the phage to change from $c\mathrm{I}^+$ (wild type) to $c\mathrm{I}^-$. Since $c\mathrm{I}^-$ phage are unable to lysogenize the host, clear plaques are produced by recombinants. When plated on mutant bacteria which produce lysogeny (bacteriophage integration) at a high frequency, only recombinant $c\mathrm{I}^-$ phage produce plaques. Nonrecombinants (e.g., λgt10 without an insert) are effectively suppressed from plaque formation. This phenomenon serves as the basis for the biological selection for recombinant phage during λgt10 library amplification.

Step 1. Plaque Purify λgt10 Phage and Prepare a Stock

a. Prepare plating bacteria: grow a 50-ml overnight culture in LB of *E. coli* C600 (*hsdR*⁻, *hsdM*⁺, *supE44*, *lacYI*, *ton A21*) starting from a single colony. Sediment the cells (5000 *g* for 5 min) and resuspend them in 0.5 volume of 10 m*M* MgSO₄. Store the cells at 4°. The plating bacteria can be used for up to 1 week, but highest plating efficiencies are obtained when cells are 0–2 days old.

b. Streak λgt10 phage (λimm434, b527) from a phage stock or a plaque using a sterile wire loop onto the surface of an LB plate. Mix 200 μl of

C600 plating bacteria with 2.5 ml of top agar (at 47°) and pour on the streaked plate. Incubate the plate (inverted) for 8 hr to overnight.

c. Pick a well-isolated plaque (they should all be "turbid" in appearance) with a sterile glass capillary or Pasteur pipet (1–3 mm diameter). Suspend the plug containing the single plaque in 1 ml of λ diluent (10 mM Tris–HCl at pH 7.5, 10 mM MgSO$_4$) and add a single drop of chloroform (to kill any residual live bacteria); allow the phage to diffuse out of the agar plug for several hours at 4°. The titer is generally 10^6 phage/ml (do not vortex, and allow the chloroform to settle before the next step).

d. Mix 100 μl of phage (0.1 of the plaque) with 100 μl of *E. coli* C600 plating bacteria and incubate at 37° for 20 min. Transfer to 50 ml of LB containing 10 mM MgSO$_4$. Incubate overnight at 37° with shaking (the culture should be completely lysed after 12 hr). Add 1 ml chloroform and continue shaking at 37° for 20 min. Sediment the lysis debris for 15 min at 5000 g. Collect the supernatant into a sterile screw-capped tube (preferably glass), and add 50 μl of chloroform. Titer the primary bacteriophage stock on *E. coli* C600 plating bacteria; the titer should be >10^9 pfu/ml. (See this volume [13] for a more detailed description of these techniques.)

Step 2. Make a Large-Scale λgt10 Bacteriophage Preparation

a. Mix 1 ml of C600 plating bacteria with 10^7 λgt10 phage and incubate for 15 min at 37°. Transfer to 1 liter of LB containing 10 mM MgSO$_4$ in a 4-liter flask. Shake at 37° overnight. Add 5 ml of chloroform and shake for 15 min. Centrifuge 15 min at 5000 g for 5 min and collect the supernatant. Save a small sample for titer; it should be >10^{10} pfu/ml.

b. Precipitate the phage by adding 54 g of NaCl and 63 g PEG 8000 to the supernatant and stir gently until dissolved.

c. Chill in an ice bath for 2–3 hr.

d. Sediment the phage by centrifugation at 5000 g for 20 min at 4°. Discard the supernatant.

e. Resuspend the pellet in 5 ml of λ diluent by scraping and stirring (not strong pipetting) and transfer to a 15-ml polypropylene centrifuge tube.

f. Extract with 5 ml chloroform using agitation (shake gently by hand). Centrifuge 5000 g for 15 min. Save the opalescent aqueous (phage-containing) upper phase.

Step 3. Purify the λgt10 Phage

Purify the phage by two sequential CsCl density gradient centrifugations.

a. Measure the volume of phage. Add 0.5 g CsCl/ml.

b. Make a step gradient in Beckman SW41 centrifuge tubes. One 11.4-ml nitrocellulose or cellulose acetate centrifuge tube will be enough for one preparation. Add the phage solution to the tube first (~5.75 ml), and then insert a long Pasteur pipet to the bottom of the tube. First add 2.5 ml of ρ = 1.5 g/ml CsCl (in λ diluent) to the Pasteur pipet, and when that solution has completely underlayed the phage solution, add another 2.5 ml CsCl solution of ρ = 1.7 g/ml (in λ diluent) to the pipet. Carefully withdraw the pipet, and add mineral oil to the rim of the tube.

c. Centrifuge at 30,000 rpm for 16 hr in an SW41 rotor or an equivalent g force × time in another rotor.

d. Remove the opalescent phage band (~1 ml) through the side of the tube using a syringe and a 20-gauge needle.

e. Add an equal volume of CsCl solution (ρ = 1.5 g/ml in λ diluent) and centrifuge overnight at 30,000 rpm.

f. Collect the phage (as described above) in a volume of approximately 0.5 ml.

Step 4. Purify λgt10 DNA

a. Dialyze the CsCl from the phage suspension against λ diluent for several hours with two dialysis changes.

b. Add one-fourtieth volume of 0.5 M EDTA, pH 8.0 (EDTA chelates the Mg^{2+} inside the phage head, and thus disrupts the compacted bacteriophage DNA) to the dialyzed phage solution and extract with phenol–chloroform 3–4 times. Phage DNA must be handled gently to avoid random shearing; do not vortex (see this volume [13]).

c. Extract with chloroform one additional time.

d. Dialyze versus DNA buffer (10 ml Tris–HCl at pH 7.9, 0.1 mM EDTA) for several hours to overnight with at least three buffer changes.

e. Determine the DNA concentration by absorbance measurement at 260 nm (this volume [6]). The yield should be about 1 mg.

Step 5. Prepare λgt10 Phage Vector DNA for Cloning

a. Starting with λgt10 DNA at a concentration of approximately 200 μg/ml, add 0.1 volume of 10× ligase buffer (0.5 M Tris–HCl at pH 7.8, 0.1 M MgCl$_2$, 0.2 M dithiothreitol). Incubate at 50° for 15 min to anneal the *cos* sites.

b. Cool to 16°, add 0.1 M ATP to 1 mM (final concentration) and T4 DNA ligase to a final concentration of 100 Weiss units per ml, and incubate for 2–3 hr at 16°.

c. Heat at 70° for 10 min (to inactivate the ligase).

d. Add 0.1 volume of 10× *Eco*RI digest buffer (1 M Tris–HCl at pH

7.5, 0.5 M NaCl, 50 mM MgCl$_2$) and *Eco*RI restriction enzyme to 200 units/ml. Incubate 1 hr at 37°.

e. Add an additional 200 units/ml *Eco*RI and incubate 1 additional hr at 37°.

f. Add DNase-free calf intestinal alkaline phosphatase to a final concentration of 5 glycine units/ml to the remainder and incubate at 37° for 30 min. (One glycine unit is that amount of alkaline phosphatase that will hydrolyze 1 μmol of *p*-nitrophenyl phosphate in 1 min at pH 10.4 at 37° in a reaction containing 0.1 M glycine, 1 mM ZnCl$_2$, 1 mM MgCl$_2$, and 6 mM *p*-nitrophenyl phosphate.)

g. Add one-thirtieth volume 0.5 M EDTA and gently extract DNA with phenol–chloroform three times (this volume [4] and [13]). Extract once with chloroform. Precipitate with 2 volumes of absolute ethanol.

h. Microcentrifuge for 5 min at 4°; aspirate the supernatant and then rinse the pellet once with 70% ethanol. Allow the rinsed pellet of DNA to air dry until the residual supernatant ethanol evaporates. Resuspend the damp pellet at a final concentration of approximately 500 μg/ml in DNA buffer.

i. Check the DNA concentration by absorbance measurement at 260 nm.

j. Check the quality of the vector DNA by self-ligating a small amount in the absence or presence of T4 polynucleotide kinase (PNK) (kinase is fully functional in ligase buffer). Package the DNA and titer the phage on *E. coli* C600 (see this volume [13,17]; also step 7a below). The *Eco*RI-digested and phosphatase-treated vector should produce less than 10^5 pfu/μg. Self-ligation in the absence of PNK should not change this background number, but ligation in the presence of PNK (0.1 Weiss units) and 0.1 mM ATP should produce greater than 10^7 pfu/μg. If the vector passes these quality control tests it is suitable for use in subsequent cloning steps. (Note: alkaline phosphatase-treated λgt10 vector DNA is commercially available, e.g., from Promega, which allows you to proceed directly to step 6.) If it does not pass these tests, start over from purified phage, step 4.

Step 6. Prepare Recombinant λgt10 Phage DNA

a. Mix double-stranded cDNA containing cohesive *Eco*RI termini with an equimolar amount of prepared λgt10 vector DNA. (For cDNA of average length of 1 kb, this is a 50 : 1 weight ratio of vector to insert.)

b. Add 0.1 volume of 10× ligase buffer, 0.01 volume of 0.1 M ATP, and T4 DNA ligase to a concentration of 100 Weiss units/ml. Incubate for 2 hr at 16°. (See this volume [10] and the Process Guide for protocols for ligation.)

Step 7. In Vitro Package Recombinant λgt10 Phage DNA

a. Package the ligated DNA using *in vitro* packaging extracts prepared by any of several methods listed in another chapter ([17]) or obtained commercially from a variety of suppliers. Use a wide-bore microcapillary pipet for transferring the viscous ligated DNA to the packaging extract. (Packaging efficiency should be greater than 10^8 pfu/μg ligated λ DNA. Most extracts are capable of packaging from 1 to several micrograms of DNA.)

b. Dilute packaging extract to 0.5 ml with λ diluent and add a drop of chloroform and gently mix. This diluted packaging extract is termed "the library." Titer the library on *E. coli* C600hfl A150.

Step 8. Plate the Packaging Mixture to Produce an Amplified λgt10 Library

a. Mix aliquots of the packaging mixture containing 10^4 to 2×10^4 bacteriophage with 200 μl of C600 *hfl* plating bacteria. Incubate at 37° for 20 min.

b. Mix 6.5 ml of melted top agar (46°) with each aliquot of infected bacteria and spread onto a freshly poured 150-mm LB plate.

c. Incubate at 37° for a maximum of 8 hr (until plaques are ~0.5 mm in diameter).

d. Overlay with 12 ml λ diluent and store plates at 4° overnight to allow the phage to desorb.

e. Remove the overlay solution containing the phage with a Pasteur pipet. Rinse the plate with 4 ml λ diluent and pool with initial overlay solution. Add chloroform to 5%. Allow the phage solution to equilibrate for 15 min at room temperature with occasional shaking.

f. Remove debris by centrifugation at 4000 *g* for 5 min at 4°.

g. Remove the supernatant to tight screw-capped glass tubes, vials, or bottles. Add chloroform to 0.3% and store aliquots at 4°. An amplified library stored in this manner should be stable for several years without significant loss of titer after the first month of storage.

(Note that the primary, unamplified library may be screened with nucleic acid probes directly during the initial plating for amplification. See chapters [44,45] on screening bacteriophage λ with hybridization probes.)

Generating cDNA Libraries in λgt11

λgt11 is also an insertion vector with a cloning capacity of up to 7 kb. The single *Eco*RI site for cloning is located at the carboxy-terminal end of the β-galactosidase (*lacZ*) gene present in the vector molecule. Insertion of foreign DNA (up to 7 kb) into this site causes the phenotype of the phage to change from $lacZ^+$ to $lacZ^-$. When plated on an appropriate host (which is deleted for $lacZ$ function), parental λgt11 phage produce blue plaques in the presence of isopropyl-β-D-thiogalactopyramoside (IPTG; a synthetic inducer of the lac repressor) and X-Gal (a chromogenic substrate for β-galactosidase activity). Under the same conditions, recombinant λgt11 phage produce colorless plaques due to the interruption of the coding sequence in $lacZ$ by inserted DNA, which generally produces an enzymatically inactive β-galactosidase fusion protein. Since there is no biological selection against the growth of parental λgt11 phage, background reduction relies solely on the use of alkaline phosphatase-treated vector molecules in the vector preparation step. Properly functional λgt11 vector should result in obtaining greater than 60% recombinant (colorless) plaques on X-Gal IPTG plates. As discussed by Huynh *et al.*, expression of foreign DNA as a β-galactosidase fusion under transcriptional control improves the probability of detecting foreign DNA gene products by antibody screening procedures (see this volume [51]).

The methods for use of λgt11 are very similar to those used for λgt10. The only major differences are that the parent phage are produced by lysogen induction rather than infection, and the use of special host strains for library amplification and screening which were designed to optimize insert protein expression and detection.

Step 1. Colony Purify the λgt11 Lysogen BNN97

a. Streak the λgt11 (λcI857, $lac5$, S100) lysogenic strain BNN97 for single colonies on an LB plate and incubate overnight at 32°.

b. Pick a colony and restreak it onto two LB plates. Incubate one plate at 32° and the other at 42° to be sure that this colony harbors a temperature-sensitive lysogenic phage (i.e., there should be very many fewer colonies of BNN97 at 42 than at 32°).

Step 2. Produce λgt11 Phage

a. Grow a 50-ml overnight culture of BNN97 at 32° with shaking.

b. Use 10 ml of this stock to inoculate 1 liter of prewarmed LB (pH 7.5) in a 4-liter flask. Grow at 32° with shaking until the $A_{600 \, nm}$ is 0.6 (3–4 hr).

c. Shift the temperature rapidly to 42–43° and hold at that temperature with shaking for 15 min.

d. Grow at 38° for 3 more hr (if need be, adjust the pH to 7.5–8 with NaOH) with vigorous shaking.

e. Add 5 ml of chloroform and continue shaking at 38° for 10 min. Lysis should occur.

f. Sediment the cell debris at 7000 *g* for 10 min at 4°.

g. Proceed with steps 2b–f of the section on λgt10.

Step 3. Purify the λgt11 Phage. Proceed as described in step 3 of the section on λgt10.

Step 4. Purify the λgt11 Phage DNA. Proceed as discussed in step 4 of the section on λgt10.

Step 5. Prepare λgt11 Phage DNA for Cloning. Proceed as described in step 5 of the section on λgt10. Again, one has the option to obtain commercially available alkaline phosphatase-treated, *Eco*RI-digested λgt11 DNA and proceed from step 6.

Step 6. Prepare λgt11 Recombinant Phage. Proceed as described in step 6 of the section on λgt10.

Step 7. In Vitro Package Recombinant λgt11 Phage. Proceed as described in step 7 of the section on λgt10 (save 10 μl of the diluted packaging mixture for a later determination of the number of recombinants in the initial λgt11 library).

Step 8. Plating to Amplify the Recombinant λgt11 Phage Library

a. Prepare *E. coli* Y1090 (R⁺M⁺) and Y1088 (R⁻M⁺) for plating bacteria (first section, step 1), or *E. coli* Y1090 (R⁻M⁺). [If you have the latter strain (available from Promega Biotec, Madison, WI) you can plate and screen the phage library directly. In order to screen λgt11 recombinants using the original strains, you must amplify the library first on *E. coli* Y1088 (to methylate host specifically inserts containing a recognition site for *E. coli* K restriction endonuclease) and then screen, using antibodies, by plating the Y1088-amplified library on the original (R⁺M⁺) strain Y1090].

b. Mix aliquots of the packaging mixture containing 10,000–20,000 bacteriophage with 200 μl of host strain Y1090 (R⁻M⁺) or Y1088. Incubate at 37° for 15 min.

c. Mix the bacteriophage solution with 6.5 ml of top agarose (46°) and pour onto a fresh 150-mm LB plate.

d. Incubate at 37° for ≤8 hr.

e. i. If you used strain Y1090 (R⁻M⁺), you can either screen the plate directly using antibodies (this volume [51]) and prepare the amplified library (after isolation of the positive plaques), or prepare the amplified library directly (prior to screening) at this time (step ii, below).

ii. If you used strain Y1088, overlay the plates with 12 ml λ diluent and store in a level position at 4° overnight.

f. Repeat steps 8, e–g, in the section on λgt10 but using Y1090 (R⁺M⁺).

Step 9. Determination of the Number of Recombinant λgt11 Bacteriophage in the Phage Library

In order to estimate the number of recombinant bacteriophage which must be screened for isolation of the recombinants of interest, one must first determine the number of recombinants in the libraries. When using λgt10 as a vector, if the packaging efficiencies of the vector alone (when compared to the vector plus cDNA inserts) are roughly the levels as outlined in the section on λgt10, one may assume that 95–99% of the recovered bacteriophage (after amplification on an *hfl* strain) are indeed recombinant. Thus the λgt10 library can be titered directly on an *hfl* strain to determine the number of recombinant bacteriophage.

In order to determine the number of recombinant phage in λgt11 libraries, we make use of the property of the (wild-type) phage to catabolize a chromogenic β-galactosidase analog (X-Gal) from an uncolored precursor to a blue product. Since cDNA inserts are ligated into an *Eco*RI site within the β-galactosidase structural gene, inactivation of the β-galactosidase protein leads to production of a colorless (clear) plaque in recombinant bacteriophage and production of a blue plaque in nonrecombinants.

a. Prepare *E. coli* Y1090 (R⁻M⁺) for plating bacteria.

b. Mix 100-μl dilutions of the amplified library (this section, step 8) or of the original library (this section, step 7) with 100 μl of Y1090 strain plating bacteria. Incubate at 37° for 20 min.

c. Add 2.5 ml of melted LB top agar containing 20 μl of IPTG stock solution (20 mg/ml) and 50 μl of X-Gal stock solution (20 mg/ml in dimethylformamide.) Pour onto a fresh, dry LB plate.

d. Incubate (inverted) for 6–12 hr at 37°.

e. Score the plates for number of clear plaques versus total number of plaques (this equals the percentage of recombinants).

Screening Considerations

Screening λgt libraries for any particular sequence of interest follows relatively straightforward statistical probabilities on the one hand; on the other hand, problems arising in attempting to clone a particular cDNA sometimes obey nonpredictable criteria. These are both addressed in the two following sections.

Statistical Probabilities

A typical vertebrate cell contains between 10^4 and 3×10^4 different mRNA sequences.[9] As discussed previously,[10] in order to ensure cloning of any particular sequence statistically, one can calculate the strict probability of the number of unique clones which must be screened in order to ensure isolation of a particular sequence using the equation

$$N = \frac{\ln(1 - P)}{\ln(1 - n)}$$

where N is the number of clones required; P, the probability of isolating the clone (usually set at a desired value of 99%); and n, the fractional proportion of the total mRNA population represented by a single mRNA species (the clone you want).

A priori estimation of the abundance of the mRNA species encoding the gene of interest is difficult without a direct (e.g., nucleic acid probe) assay. Most often, rough estimates have been made using immune precipitation of radiolabeled *in vitro* translation reactions. Estimations of the relative mRNA levels based on such data can be quite misleading, and vary according to the isotope(s) used in *in vitro* translation reactions, the integrity of the mRNA used, and the size of the primary translation product. While all of these factors detract from the ability to afford an accurate assessment of mRNA concentration of a particular species, without nucleic acid probes such estimates sometimes afford the only possible avenue for gaining any knowledge of mRNA abundance prior to cloning.

As a specific example, suppose we wish to clone a "rare" mRNA (10 copies/cell) from chicken red blood cells.[11] Rare mRNA sequences represent about 10% of the total polysomal message of an immature chick RBC (in which there exist only ~100 different mRNA species). Therefore the actual fractional representation of any mRNA of this class in the total mRNA of a red cell is ~1 in 10^3 molecules. Therefore if $n = 10^{-3}$, to ensure cloning of this sequence to a 99% statistical probability, we need theoretically to construct and screen a library of only 5000 members to ensure that this sequence is adequately represented (Table I).

A more common example might be encountered in examination of the population of mRNAs in a transformed human fibroblast.[12] Williams has

[9] J. G. Williams, *in* "Genetic Engineering" (R. Williamson, ed.), Vol. 1, p. 1. Academic Press, New York, 1981.

[10] L. Clarke and J. Carbon, *Cell* (Cambridge, Mass.) **9**, 91 (1976).

[11] L. Lasky, N. D. Nozick, and A. J. Tobin, *Dev. Biol.* **67**, 23 (1978).

TABLE I
THEORETICAL CONSIDERATIONS FOR CLONING cDNA SEQUENCES

A, Cell type	B, Abundance class	C, Percentage representation	D, Number of mRNAs	E, mRNA copies/cell	F, n (fractional representation of each mRNA)	G, N (total number of clones required)
Chick RBC	Abundant	90	3	1,500	0.30	13
	Rare	10	100	10	10^{-3}	4.6×10^3
Fibroblast	Abundant	20	30	3,500	6.7×10^{-3}	6.8×10^2
	Moderately abundant	50	1,000	200	5×10^{-4}	9.2×10^3
	Rare	30	11,000	10	2.7×10^{-5}	1.7×10^5
Hypothetical	Abundant	15	10	10,000	1.5×10^{-2}	3.1×10^2
	Moderately abundant	40	1,000	100	4×10^{-4}	1.2×10^4
	Rare	40	10,000	10	4×10^{-5}	1.2×10^5
	Very rare	5	10,000	1	5×10^{-6}	9.2×10^5

determined the number of clones necessary to ensure successful cloning of a single mRNA in the "rare" abundance class of message in this cell type.[9] These values are also shown in Table I. Note that simply because there are many more total mRNA species in fibroblasts than in red cells, in order to clone a message present at 10 copies per cell in fibroblasts one would need a library some 35-fold greater in total to isolate the same 10 mRNA/cell species as is present in red blood cells!

Finally, we have included the probability of cloning an mRNA represented once per cell in a hypothetical cell type which contains 21,000 different mRNA species, all at relatively different levels, an example which might well represent a "typical" mammalian cell. As shown in this hypothetical case, one would need to have an initial library of 10^6 recombinants to ensure statistically that a very rare mRNA was contained within that library.

The numbers shown in Table I are the theoretical considerations for cloning of a particular cDNA sequence. Thus, those considerations would be straightforwardly applied to cloning of a cDNA into λgt10 and screening with (e.g.) oligonucleotide probes. When cloning into λgt11, however, one usually assumes faithful expression of a cDNA fusion of β galactosidase and a eukaryotic mRNA copy to be only one-sixth the simple cloning

12 J. G. Williams, M. M. Lloyd, and J. Devine, Cell (Cambridge, Mass.) 17, 903 (1979).

efficiency deduced from Table I, since only one-third of the cDNA clones will be in the proper reading frame to produce a fusion gene with the correct eukaryotic amino acid sequence, and only half of the cDNA clones will be in correct reading orientation when cloning cDNA containing EcoRI linkers at either end into an EcoRI site.

Nonstatistical Considerations

By far the most common problem encountered in isolating clones from cDNA libraries is the overall quality of the initial double-stranded cDNA inserted into prokaryotic λ or plasmid vectors. If, in comparison of an ethidium bromide-stained denaturing gel of the mRNA which you are using when run side by side with the cDNA you have prepared, the two are not at least comparable in average molecular weight, it is unlikely that any cDNA cloning method recommended will be successful. Several previous chapters in this volume deal with minimizing such initial technical problems in cDNA size.

Outside of this most common problem, several other considerations might be applicable in particular cases when screening λgt11 libraries with monoclonal antibodies and monospecific antisera. In general, if both are available, clean monospecific antisera are preferable to using monoclonal antibodies for λgt11 screening. This is because an antiserum may react with several epitopes in the fusion polypeptides, whereas (by nature) the monoclonals can only recognize a single epitope. One should then expect to find a particular clone more easily with a good antiserum if the epitopes are spread from the amino to the carboxy termini. Furthermore, one might also expect greater signal intensities by recognition of multiple epitopes in one cloned sequence.

In general, if a monoclonal recognizes an antigen on Western blots, it can be successfully used in λgt11 library screening, since a reasonable rule of thumb is that because since the antibody recognizes an SDS-denatured form of the protein on Western blots, the antigen is likely to be a simple (primary structural) epitope. The same does not hold true for antigens reactive with monoclonal antibodies after fixation for immuno-fluorescence. Since, in theory, antigens are "fixed" in place by aldehyde, acetone, or methanol treatment of cells, in many cases such complex epitopes (including appropriate secondary folds or tertiary interactions of the protein of interest with other cellular components) may only infrequently be present in λgt11 fusion polypeptides.

Of course, the monoclonal specificity may include a posttranslational modification of the protein (such as glycosylation or phosphorylation) as part of the epitope, and this "Western rule" may not always hold true.

However, such problems can usually be eliminated by testing immune precipitates of *in vitro* translation products from the mRNA to be used in initially forming the library.

Another theoretical problem which is a potential pitfall in the immunological approach in screening λgt11 libraries is the secondary structure of the mRNA substrate. If a strong secondary structure is encountered, the initial cDNA may be invariably prematurely truncated at a single nucleotide flanking the 3′ end (in mRNA sense) of a strong secondary structure, and therefore always leading to a β-galactosidase fusion product truncated at that point. Such a "strong stop" secondary structure has been observed in c-*abl* mRNA, and was overcome (as recommended in this volume [33]) by reverse transcription in the presence of methylmercury hydroxide (O. Witte, personal communication).

[41] Full-Length cDNA Clones: Vector-Primed cDNA Synthesis

By Prescott L. Deininger

This chapter is concerned with methods which are of particular use for the formation of cDNA libraries containing high proportions of long or full-length inserts, specifically the use of vector-primed cDNA synthesis. Many methods exist which aid in the preparation of long cDNAs and do not require the use of vector priming. However, most of these methods can be, and are, incorporated into the vector-primed cDNA protocols.

Scheme for Vector-Primed cDNA Cloning

The preparation of the vector–primer and linker molecules needed for these cDNA cloning procedures is illustrated in Fig. 1 and is essentially that of Okayama and Berg.[1] There are a number of variations on these protocols and, as discussed below, several procedures now do away with the preparation of the linker molecule altogether. The vector can be any number of plasmids that have the proper arrangement of restriction sites, and choice of vectors will be discussed in more detail later. It is helpful if the restriction enzyme cleavage site at Z (Fig. 1A) generates a 3′ overhanging end (e.g., *Kpn*I or *Pst*I) which will facilitate the tailing reactions,

---

[1] H. Okayama and P. Berg, *Mol. Cell. Biol.* **2,** 161 (1982).

METHODS IN ENZYMOLOGY, VOL. 152

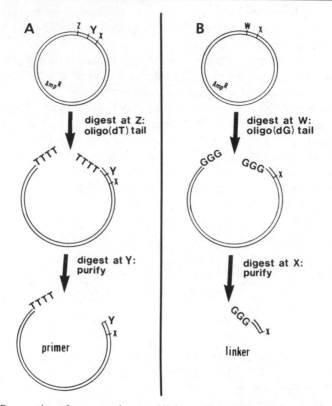

FIG. 1. Preparation of vector–primer and linker molecules. (A) Besides the requirement for selection (typically ampicillin resistance) and replication within a bacterial host, the vector must have three single restriction cleavage sites. Site Z should create a 3' overhang upon digestion (*Kpn*I in this protocol for the use of pUC18 as the vector). This site is cleaved and tailed with dT for the vector–primer. The tailed molecule is cleaved at restriction site Y (*Bam*HI in this protocol) and the large vector–primer fragment purified away from the small fragment by gel electrophoresis. This molecule is then ready for use to prime cDNA synthesis as shown in Fig. 2. (B) The linker preparation is similar to that of the vector–primer. Site W varies depending on the vector system being used, but can be the *Kpn*I site again for pUC18. Tailing is with dG (although dC can be used if the proper modifications are made later in the protocol). The linker is then cleaved from the vector using enzyme X (*Hin*dIII in this protocol) and purified by gel electrophoresis.

but otherwise any enzymes which generate "sticky" ends are acceptable for the other cleavage sites. The protocols described later in this chapter utilize the multilinker of the plasmid pUC18 to supply these sites. Once the plasmid is linearized, oligo(dT) tails are created on both ends of the plasmid. These tails will eventually be used as primers for cDNA synthesis starting at the poly(A) end of an mRNA molecule. The T-tailed vector is cleaved with a second enzyme (site Y, Fig. 1A) to create a large vector–primer molecule with oligo(dT) at only one end and a small molecule that must be removed before use of the vector. If the small tailed molecule is not removed from the vector–primer, it, too, will be able to prime cDNA synthesis but will not function as a vector. This would result in an approximately 2-fold lower yield in the cloning procedures. For many applications this might be found to be acceptable and avoiding this step would make vector–primer preparation much easier. There is a special dimerized vector (pARC7)[2] which allows the use of vector which is T-tailed at both ends without any loss of yield. This vector is discussed in more detail later.

The scheme for generation of cDNA clones using the T-tailed vector–primer and the linker molecule is shown in Fig. 2. The first step is the annealing of the vector–primer to the mRNA followed by cDNA synthesis. This cDNA is then tailed with dC (or dG in variants of the protocol discussed below) and the dC is cleaved from the vector at site X. In theory this digestion could also cleave within the cDNA–mRNA duplex region if the appropriate restriction site were present. This would make recyclization of any such cDNAs impossible and could potentially eliminate specific cDNAs from the resulting library. In practice, cleavage of DNA–RNA hybrids with most restriction enzymes is very inefficient.[1] Thus, as long as digestions are not carried to extremes, this digestion step should not lead to serious depletion of specific cDNAs. Addition of the linker molecule, which consists of a short DNA fragment with one end generated by cleavage at site X and the other end dG tailed (Figs. 1 and 2), will then allow cyclization of the vector–cDNA by interaction with the X enzyme ends and C tails of the vector–cDNA. After ligation and replacement of the mRNA strand with DNA, the synthesized cDNA library is ready for introduction into bacteria. There are a number of variations on these procedures which are also discussed below.

Why Use Vector-Primed cDNA Synthesis Techniques?

In many ways the vector-primed cDNA synthesis approaches require a more difficult and time-consuming initial preparation than traditional cDNA cloning protocols. There are several advantages to the vector priming approach which make them well worth any additional effort required for many experiments. Among these advantages are (1) relatively low backgrounds of clones containing no inserts, (2) the ability to obtain high yields of cDNA clones from small amounts of mRNA, and (3) the

[2] D. C. Alexander, T. D. McKnight, and B. G. Williams, *Gene* **31**, 79 (1984).

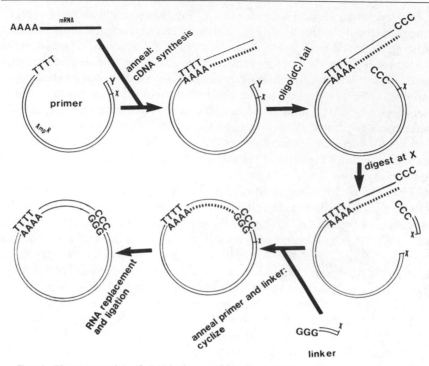

FIG. 2. The preparation of cDNA clones. This schematic shows the steps in the standard protocol presented in this chapter. The primer (from Fig. 1A) and mRNA (heavy solid line) are mixed and cDNA synthesis carried out. During the course of the reverse transcription the mRNA is nicked by the RNase H activity and is then represented as a heavy dashed line. The cDNA is tailed with dC (which also tails the other end of the vector). The tails at the other end are removed by cleavage at site X (*Hin*dIII in these protocols). The dG-tailed linker (from Fig. 1B) is added and ligated, resulting in circularization of the recombinants. The mRNA can then be replaced with DNA using *E. coli* RNase H, DNA polymerase I, and *E. coli* DNA ligase.

ability to prepare quite long cDNAs with a high proportion of full-length clones.

The first two advantages are derived directly from the vector priming. Low backgrounds can be obtained because the vector–primer is purified on an oligo(dA) affinity column. This removes any uncleaved or untailed vectors and eliminates background due to simple religation of the vector molecules. By adding the mRNA in a reasonable excess, the majority of the vector molecules are then involved in cDNA priming. The fact that every cDNA synthesized is already attached to the vector and need only be circularized effectively then helps contribute to the yields. Also important to the high yields is the fact that all of the manipulations can be carried out without the need for purification (e.g., column chromatography) of any of the intermediates. Using the most efficient preparations of competent cells (about 10^8 recombinants from 1 μg of supercoiled plasmid DNA) we have been able to exceed 10^5 recombinants/μg input mRNA. We have found that traditional methods can exceed this yield severalfold, however.

The reasons for the ability of these procedures to make long cDNAs are somewhat complex and some aspects are a little speculative. The tailing of the cDNA to prime the second-strand synthesis and the use of RNase H and DNA polymerase I to replace the RNA mean that no steps are necessary which directly degrade the cDNA. This is in contrast to the use of S_1 nuclease[3] which can remove several bases from the cDNA and also lead to occasional internal nicking, resulting in fragmented cDNA. As a matter of fact, traditional approaches can sometimes lead to cDNA clones from which large portions of the 3' region are missing, as well as short clones which do not extend near the 5' end. Both tailing for the priming of second-strand synthesis and the RNA replacement strategy used here have been reported to improve the length of cDNAs prepared by methods other than vector priming as well.[4,5]

One of the reasons suggested for the success of these procedures is that the tailing may preferentially select those molecules which are full length.[1] Certainly, in the presence of Mg^{2+}, tailing is very inefficient on duplex molecules with recessed 3' ends. Therefore, cDNAs that have synthesized all the way to the 5' end of the mRNA could be tailed much more effectively. However, using terminal transferase in the presence of either Co^{2+} or Mn^{2+}, as is done in all of these protocols, it is not clear how much of an effect the exact nature of the end has on tailing efficiencies. In studies with different restriction fragment ends, dC tails were synthesized onto 3' recessed ends with higher efficiency than on blunt ends.[6] The differential polymerization of dG tails on a blunt end relative to a recessed end was minor in the presence of Mn^{2+}, but did show a 4-fold effect under Co^{2+} conditions. Whether a 4-base recessed end will tail more or less effectively than one recessed by 500 bases has not been demonstrated. Since most of the tailings in these protocols are carried out under conditions that optimize tailing efficiencies on all types of ends, it is not clear whether this is an important factor in selecting for long cDNAs or not. Certainly any selection for long cDNA by these procedures would have

[3] U. Gubler, this volume [34].

[4] U. Gubler and B. J. Hoffman, Gene 25, 263 (1983).

[5] H. Land, M. Grez, H. Hauser, W. Lindenmaier, and G. Schutz, Nucleic Acids Res. 9, 2251 (1981).

[6] G.-R. Deng and R. Wu, this series, Vol. 100, p. 96.

an adverse effect on the cDNA cloning yield at the expense of the shorter cDNAs. Although these methods are very good at the generation of full-length cDNAs, it is still quite possible to create libraries with very short cDNAs if care is not taken to maintain the integrity of the mRNA and in synthesizing the first strand of the cDNA.

Variations and Alternate Approaches

Most of the variations on the vector-priming approach are designed to simplify the preparation of the vector–primer and linker components. Consider the pARC7 plasmid for use as the vector–primer (Fig. 3)[2]; it is a specially dimerized plasmid which can be thought of as analogous to Siamese twins. It can be used as vector–primer after restriction digestion, T-tailing, and selection by oligo(dA)-cellulose chromatography. However, the necessity for removing the unwanted T-tailed fragment generated by cleaving at Y in the classical scheme (Fig. 1A) is eliminated because no such fragment is produced (Fig. 3). One simply works both ends of the T-tailed Siamese twins until their surgical separation at site X is indicated (this would be at the same point as is shown in the standard scheme in Fig. 2). There is also a companion plasmid, pARC5, which is the source for the linker used with pARC7. In fact, pARC5, is designed to yield two linker molecules per plasmid. The authors report a cloning yield of about 10^5 recombinants/μg of input mRNA. This is the same cloning efficiency as that reported for the original procedure, but apparently without some of

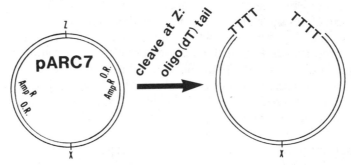

FIG. 3. Vector–primer prepared from pARC7. The vector pARC7 is a dimerized vector, each half of which carries essential elements for DNA replication (O. R., origin of replication) and ampicillin selection (Amp^R). Thus, when the vector is cleaved at Z (*Sst*I in this particular vector) and oligo(dT) tailed, each tail is attached to a functional vector. Eventually cleavage at site X (*Bam*HI in this case) as is shown in Fig. 2 will liberate two vector molecules attached to separate cDNAs. There are also a number of useful sites flanking the tailing site (Z) in this vector which can be used later for excising the insert. A separate plasmid pARC5 is available for preparation of the linker molecule as in Fig. 1B.

the labor-intensive steps. Since I have not actually used this vector, I will not present detailed protocols. However, it could easily be incorporated into a slightly simpler version of these protocols.

Several other variations on this technique are designed to eliminate the tailed linker molecule altogether (Fig. 4).[7,8] One of these procedures uses oligo(dG) to prime second-strand synthesis after tailing the cDNA with dC.[7] The unwanted dC tail at one end is cleaved with a restriction endonuclease, leaving a staggered end. Oligo(dG) is used to prime second-strand synthesis of cDNA in conjunction with replacement synthesis of the mRNA while concurrently the restriction site of the opposite end of the plasmid is filled in to create a blunt end. The molecules are then circularized by blunt-end ligation. Yields for this approach are reported at about the same level as the linker approach (about 10^5 recombinants/μg of input RNA) and it does avoid the rather cumbersome procedures of preparing and calibrating the linker molecule. Since I do not have direct experience with this variation I have not included it in the protocols described, but it may well be worth incorporating into a modified protocol. A second variation which avoids the linker molecule is also presented in Fig. 4.[9] In this variation, the dG-tailed cDNA has a dideoxyG (ddG) added at the end of the tail. This makes further lengthening of these tails by terminal transferase impossible. After a secondary restriction cleavage at site X, dC tails can be synthesized on the new restriction end, but will not be added onto the dideoxyG-terminated dG tails. This molecule is then ready to self-circularize and have the RNA replaced by DNA. This approach has reported lower yields (about 10^4 recombinants/μg of input mRNA) of cDNA clones than some of the others. Since none of these modified procedures has been carried out in parallel, it is difficult to determine whether the lower reported yields are due to inherent difficulties in the procedure, such as failure to incorporate ddG, or simply variations in some of the other experimental techniques (e.g., less efficient bacterial transformation). Although I have not seen it reported, it should also be possible to replace the linker molecule in the original protocols with a synthetic DNA fragment. Synthesis of the linker fragment would allow vast quantities of linker to be made fairly easily.

Last, there is another vector-priming protocol that varies significantly from the others.[10] In this protocol (Fig. 5) the vector–primer is prepared

[7] S. Falkenthal, V. P. Parker, W. W. Mattox, and N. Davidson, *Mol. Cell. Biol.* **4**, 956 (1984).

[8] H. Okayama and P. Berg, *Mol. Cell. Biol.* **3**, 280 (1983).

[9] R. J. Milner, C. Lai, K.-A. Nave, D. Lenoir, J. Ogata, and J. G. Sutcliffe, *Cell (Cambridge, Mass.)* **42**, 931 (1985).

[10] G. Heidecker and J. Messing, *Nucleic Acids Res.* **11**, 4891 (1983).

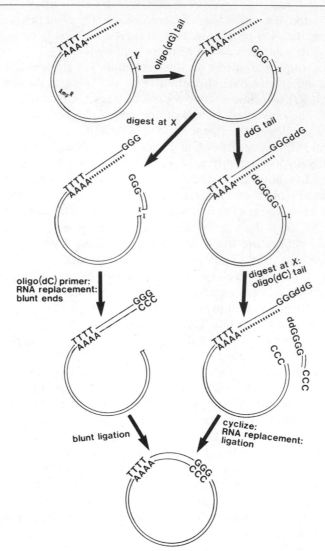

FIG. 4. Alternate recircularization strategies. Two variations on the strategy presented in Fig. 2 are presented which use alternative procedures for cyclizing the recombinants. Both procedures are identical in the priming of cDNA synthesis and the addition of tails. In the procedure on the left, the tails are cleaved from one end of the vector at site X. Second-strand synthesis is then primed using oligo(dC) to prime on the dG tail and also including the RNase H and DNA polymerase I to replace the mRNA strand. The molecules are then blunt-ended and then cyclized by a blunt-end ligation. In the variation on the right, dideoxyG is added to the tails on the molecule. After cleavage at X, dC tails are added to site X, but are not able to add to the dideoxyG-terminated end. The molecule is then able to circularize by annealing of the dG and dC tails, followed by replacement synthesis of the mRNA and ligation.

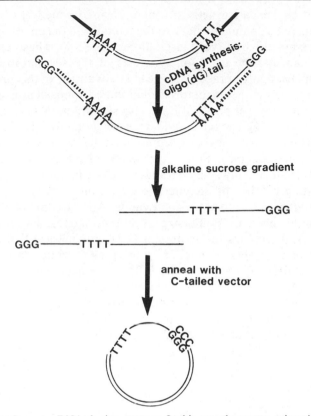

FIG. 5. An alternate cDNA cloning strategy. In this case the vector–primer is left with dT tails on both ends. This primer then primes synthesis with two different mRNA molecules. These cDNAs are both tailed with dG and then the sample is denatured and fractionated by size on an alkaline sucrose gradient. One cDNA strand stays with each strand of the vector and the mRNA is destroyed by the alkali. This gradient allows removal of vector with no cDNA or very short cDNA. dC-tailed vector (the same vector as is used for priming, but with dC instead of dT tails) is then denatured and annealed to the size-fractionated cDNAs. The vector strands anneal and the dC tails of the added vector hybridize with the dG tails on the cDNA to allow cyclization. The single-stranded region can then be readily filled in with DNA polymerase.

by cleavage and tailing of both ends of the plasmid. This vector–primer is then used to prime cDNA synthesis and the cDNA is tailed much as in the other procedures. At this point the vector–cDNA is then denatured and applied to an alkaline sucrose gradient for size fractionation. Concomitantly, remaining mRNA is destroyed. Both strands of the vector can prime cDNA synthesis and the size fractionation allows the elimination of vector with no insert or very small inserts. These molecules are then circularized by annealing with the same plasmid whicn has been tailed with dC tails to be complementary to the dG tails on the end of the cDNA. This acts as essentially a very large linker molecule. These circularized molecules can then be used for synthesis of the second strand of the cDNA with DNA polymerase. The greatest advantage of this procedure is that the vector components require a minimum of steps in preparation and do not require any gel purifications. The size fractionation can also be useful for eliminating background and for selecting for cDNAs of a specific size. The greatest drawback seems to be the sensitivity of the procedure to nicking enzymes. Since the vector–cDNA must go through a single-stranded form, any single-stranded nicks will destroy it. Thus some care in testing batches of enzymes (e.g., terminal transferase) for any nicking activities is essential for good yields. Another drawback, possibly serious, is the need to synthesize second-strand cDNA using single-stranded naked DNA as the template. DNA polymerase I, the required enzyme, cannot traverse sequences known as "strong stops," making this approach less efficient than the mRNA replacement synthesis technique employed in the other procedures. The alkaline gradient is also a tedious procedure if the size fractionation is not needed. Yields for this procedure have been reported at about 2×10^4 recombinants/μg of input mRNA,[10] but without using the highest efficiencies for bacterial transformation.

Choice of Vector

There is a wide range of vectors that will work with most of these vector-primed cDNA synthesis methods. The only requirement for most of the approaches described above is to have several appropriate restriction sites available. It is also convenient to have other single-cut restriction sites flanking those used for the cDNA cloning to allow cDNA inserts to be readily excised from the vector for further characterizations. Because of this, any of the vectors now available with a multilinker cloning site would work well. The following protocols describe the procedures as we use them on the vector pUC18. However, essentially identical cloning procedures could be carried out with specialized vectors, such as the pARC vectors described above or any of the Okayama–Berg vectors.[1,8]

Besides the basic requirements for these vectors, it may also be useful to carry out the cloning in any one of a number of more specialized vectors which may make the characterization of the resulting clones easier. For instance, expression of the resulting cDNAs in eukaryotic cells is possible with the Okayama and Berg expression system.[8] All of the relevant eukaryotic expression signals are found in the linker and vector fragments. This can be very useful both for proving the identity of a cDNA clone and also for later studies on the biology of a cDNA product. Another method which is helpful in screening cDNA libraries is to use a vector that allows the preparation of large amounts of RNA from the cDNAs *in vitro* (see this volume [30, 61]).[11] These RNAs (produced using the SP6 polymerase in this case) can then be used in *in vitro* translations followed by immune precipitation to help identify clones producing the gene product.

Methods

Vector Preparation

It is best to start by preparing fairly large quantities of both the vector–primer and the linker. These components store well frozen in aliquots, and preparation of larger amounts of material requires only a little more work than small batches. Steps are described here for preparing 200 μg of vector–primer as well as the linker from 200 μg of vector. It would not be unreasonable to double all of the amounts and recipes described below, for those who expect to use these procedures extensively. Prepared vector–primer and linker components are also available commercially for this procedure[1] as well as for the procedure of Heidecker and Messing.[10]

Begin by digesting 400 μg of pUC18 (200 μg for vector–primer and 200 μg for linker) with *Kpn*I. This digestion is carried out in 800 μl of the *Kpn*I buffer [6 m*M* Tris–HCl at pH7.5, 6 m*M* NaCl, 6 m*M* MgCl$_2$, 1 m*M* dithiothreitol, and bovine serum albumin (BSA) at 100 μg/ml] with 800 units of *Kpn*I for 2 hr. This should result in complete digestion, but should be checked by running 200 ng of the cut vector adjacent to uncut vector on a 1% agarose minigel. If the digestion is incomplete, add more enzyme and incubate further. When the digestion is complete, extract the reaction with 600 μl of phenol, add 80 μl of 3 *M* sodium acetate to the aqueous phase, and ethanol precipitate with 2.5 volumes of ethanol.[12]

After precipitating the DNA and washing the pellet once with 70% ethanol, resuspend the dried pellet in 400 ul of water. Half (200 μl) of this linearized vector is used in the vector–primer preparation and 200 μl saved for the linker preparation. The tailing reaction is carried out as described below, with a small aliquot being tailed in the presence of radioactive dTTP to act as a tracer for later steps. The actual times pre-

[11] Y. Noma, P. Sideras, T. Naito, S. Bergstedt-Linquist, C. Azuma, E. Severinson, T. Tanabe, T. Kinashi, F. Matsuda, Y. Yaoita, and T. Honjo, *Nature (London)* **319,** 640 (1986).

[12] D. M. Wallace, this volume [5].

sented for the tailing reactions are approximate and are best calibrated as described.

Tailing Calibrations

The exact time for the tailing reactions should be carefully calibrated with a scaled-down version of the reaction conditions that will be used for the larger preparations. For oligo(dT) tailing, mix

*Kpn*I-cleaved pUC18	5 μg
10× CoCl$_2$ tailing buffer (10× = 250 mM potassium cacodylate at pH 7.2, 20 mM CoCl$_2$, and 10 mM dithiothreitol)	2.5 μl
5 mM dTTP	2.5 μl
5'-[α-^{32}P]dTTP (or [^{3}H] dTTP if preferred)	10 μCi
Water to	25 μl

Incubate this mix at 37° for 10 min to prewarm the solution. Prepare six tubes with 50 μl of 10 mM EDTA as stop tubes. Take 4 μl of the reaction mix and add it to one of the stop tubes as a 0 time point. Add 50 units of terminal transferase to the remainder and begin timing the reaction. Quench 4 μl of the reaction at 5, 10, 15, 20, and 30 min. Each of the time points can be quickly assayed by trichloroacetic acid (TCA) precipitation for the extent of reaction as described in this volume [6]. From this measurement and the molar concentration of dTTP in the reaction, the approximate length of the tails can be calculated as discussed elsewhere.[13] The tail lengths can be measured directly as described below.

For direct measurement, first phenol extract each of the quenched reactions with 50 μl of phenol. Add 5 μl of 3 M sodium acetate to the aqueous phase and 125 μl of ethanol. After chilling and spinning down the precipitate, wash with 500 μl of 70% ethanol and dry the pellet. Redissolve each of the pellets in 20 μl of *Hin*dIII buffer (10 mM Tris–HCl at pH 7.5, 50 mM NaCl, 10 mM MgCl$_2$, and 100 μg/ml BSA) and incubate for 1 hr with 10 units of *Hin*dIII. This will cleave off a fragment which is 37 bases long plus the length of the added tails. Add 10 μl of formamide loading dye, place the samples in a boiling water bath for 3 min, and load aliquots on a 6% sequencing gel as discussed elsewhere.[14-16] Also load a radioactive marker, either sequencing reactions or end-labeled restriction fragments. Autoradiography of this gel allows a direct visualization of the lengths of the tailed fragments. The time which produces most of the radioactivity 40–90 bases longer than the original fragment size is then chosen for the larger-scale preparation.

[13] W. H. Eschenfeldt *et al.*, this volume [37].
[14] W. M. Barnes, this volume [57].
[15] B. Ambrose and R. Pless, this volume [56].
[16] R. C. Ogden and D. A. Adams, this volume [8].

Both oligo(dG) and oligo(dC) tailing can be calibrated in the same manner. It is only necessary to replace the dTTP (both radioactive and nonradioactive) with dCTP or dGTP. In these cases decrease the nucleotide triphosphate concentration somewhat, and only use a stock of 2 mM in making up the reaction mix (200 μM final concentration). Time points are also typically shortened to 0, 2, 4, 8, 12, and 20 min. It is best to obtain tails of about 10–20 bases in length.

Vector–Primer Preparation

To prepare the oligo(dT)-tailed vector–primer, make up the following mix:

(200 μg) KpnI-cleaved pUC18	200 μl
10× CoCl$_2$ tailing buffer	100 μl
100 mM dTTP	5 μl
Water to	950 μl

Incubate this mixture at 37° for 10 min before adding 1000 units of terminal transferase. Immediately remove a 25-μl aliquot of this reaction mix and add it to 2 μCi of [α-^{32}P]dTTP. Allow both the radioactive and nonradioactive mixes to incubate at 37° for the appropriate time (about 20 min) as determined in the tailing calibrations. Add 10 μl of 0.5 M EDTA (pH 8.0) to the main reaction, quench both reactions on ice, and mix them. The radioactive reactants are now used to follow the sample through further steps. Phenol extract the reaction with 500 μl of phenol, add 100 μl of 3 M sodium acetate to the aqueous phase, and precipitate with 2.5 ml of ethanol. If desired, an aliquot of this tailed material can be directly checked for the length of the tails produced as described in the tailing calibration section.

After ethanol precipitation and washing with 70% ethanol, the pellet is then dissolved in 400 μl of BamHI buffer (6 mM Tris–HCl at pH 7.5, 150 mM NaCl, 6 mM MgCl$_2$, and 100 μg/ml BSA) and digested with 400 units of BamHI for 2 hr. This reaction is then phenol extracted, and 40 μl of 3 M sodium acetate and 1 ml of ethanol are added to precipitate the DNA. The pellet is washed with 70% ethanol and dried as above.

It will increase the yield of the cloning reactions to separate the small fragment freed by the BamHI digestion from the large, vector–primer fragment. This is a relatively easy separation using this vector, as the small fragment is on the order of 100 bases with tails and the large fragment is on the order of 3000 bases. We accomplish this by dissolving the DNA in a neutral gel loading buffer and applying it to a 0.8% low gel temperature (LGT) agarose minigel. We overload the gel heavily, with 200 μg of sample being applied to about 5 cm^2 of well cross section. Electrophoresis need only be carried out until the major band has migrated into the gel 1–2 cm. This will allow ample separation from the small fragment and minimize the amount of gel to be excised for eluting the sample. The vector–primer band is then excised using ethidium bromide fluorescence and eluted by the phenol extraction method described elsewhere.[16]

[*Editors' Note*. Because the vector–primer is several thousand bases in length and the small fragment separated from it is less than 50 bases in length plus the length of the tails, it may be possible to use one of several alternatives to an agarose gel for purification of the vector–primer. Of particular interest might be the use of PEG precipitation in the presence of NaCl,[16a] which can selectively precipitate large nucleic acid molecules or also the possible use of gel exclusion chromatography.]

The last step in the purification of the vector–primer is affinity purification on an oligo(dA) column. This step selects for molecules with tails long enough to prime cDNA synthesis effectively and keeps the background contributed by untailed vector to a minimum. We carry out this affinity purification essentially as described in the original method.[1] The oligo(dA)-cellulose is washed in buffer (10 mM Tris–HCl at pH 7.3, 1 mM EDTA), 1 M NaCl) and the fines decanted. A 0.6-cm diameter column is then packed 2.5 cm deep and cooled to 0° (this column can effectively handle twice this amount of vector). The vector–primer, which was ethanol precipitated after elution, is suspended in 500 μl of the same buffer, chilled to 0°, and slowly loaded onto the chilled oligo(dA) column. The column is then washed extensively with the same loading buffer (about 20 ml). Elution was carried out with water at room temperature. Since this vector–primer will be used in the presence of RNA, we treat the water with diethyl pyrocarbonate and handle the vector–primer from this point with full precautions to maintain the stability of the RNA.[17,18] Fractions of about 0.5 ml are collected and monitored for radioactivity. The peak fractions are pooled, the concentration of the DNA measured spectrophotometrically (this volume [6]), and the sample is ethanol precipitated with the addition of 0.1 volume of 3 M sodium acetate and 2.5 volumes of ethanol. After washing with 70% ethanol and drying the pellet, the vector is then dissolved in water at a concentration of 1 mg/ml and is ready for cDNA synthesis.

Linker Preparation

The preparation of an oligo(dG)-tailed linker is presented here as in Fig. 1. However, the use of an oligo(dC)-tailed linker with dG tails being placed on the cDNA would work just as well.[2] As a matter of fact, since

---

[16a] J. T. Lis, this series, Vol. 65, p. 347; G. F. Hong, *Biosci. Rep.* **1**, 243 (1981).

[17] D. D. Blumberg, this volume [2].

[18] S. L. Berger, this volume [19].

dG tailing on blunt ends in the presence of Mn^{2+} is reported to be more effective than dC tailing under either Mn^{2+} or Co^{2+} conditions, it might improve yields to use the dG tailing with Mn^{2+} to tail the cDNA.[6]

The linker molecule is a *Hind*III/*Kpn*I fragment from pUC18 that has oligo(dG) or oligo(dC) tails of about 10–20 bases in length at the *Kpn*I end. It is prepared by tailing 200 μg of *Kpn*I-cleaved pUC18 with dGTP followed by *Hind*III cleavage and purification of the small linker fragment. Many protocols tail with oligo(dC) at this point, instead of oligo(dG). To prepare the oligo(dG)-tailed linker, we prepare the following tailing reaction mix:

*Kpn*I-cleaved pUC18 (200 μg)	200 μl
10× $CoCl_2$ tailing buffer	100 μl
100 m*M* dGTP	2 μl
Water to	1 ml

This mixture is preincubated and then incubated as for the dTTP reactions except that a small portion is mixed with labeled dGTP instead of dTTP and the reaction times are generally shorter (1–5 min). The reactions are then stopped as described above for the vector–primer and cleaved with *Hind*III. After phenol extraction and ethanol precipitation the primer is resuspended in a neutral gel loading buffer and applied to a 12% neutral polyacrylamide gel.[16] This amount of DNA should be loaded on about 0.2 cm² of gel surface. As a marker, 5′-end-labeled *Hind*III/*Kpn*I-cleaved pUC18 should be run in an adjoining well (see this volume [10]). Autoradiography of the gel will allow detection of the untailed, labeled marker and also the tailed material. It is possible to count the bands in the tailed sample so that the material with tails of six and longer is easily detected. The region of the gel containing tails of 6–20 in length is excised and the DNA eluted by crush elution as described elsewhere.[16]

After elution, the concentration of the linker is too low to be measured accurately. We, therefore, simply estimate the yield at this point and carry out a practical titration of this material at a later point. From 200 μg of pUC18 the maximum theoretical yield of linker should be 3 μg. A good preparation may yield as much as 1.5 μg after losses, but may be as little as half that. We dissolve the linker preparation at an estimated 20 μg/ml.

First-Strand Synthesis

The conditions we employ are essentially those of Okayama and Berg.[1] It is essential to use an excess of RNA over vector. Any excess vector that does not anneal to RNA is likely to result in background recombinants with no cDNA inserts. For total poly(A)⁺ mRNA, if we estimate that the average length is 1500 bases, then 1 μg of mRNA will represent approximately 2 pmol, and 1 μg of the vector–primer will represent about 0.5 pmol. Therefore for priming 1 μg of mRNA, about 2–3 μg

of vector–primer would result in a sfae excess of mRNA. If size fractionated or partially purified specific mRNA populations are to be used it is
important to consider these molar ratios and not only the mass.

One of the most important considerations in obtaining long and full-
length cDNA clones is that the mRNA must be of high quality. Preparations of mRNA populations are discussed elsewhere.[19] It is well worth the
trouble to look at the mRNA size distribution on a denaturing gel prior to
cDNA synthesis and even to carry out Northern blot analysis with a
known probe to detect any signs of degradation within the mRNA preparation. All preparations of solutions, glassware, and handling of the reactions for the first stages of the cDNA cloning should also be carried out
with all precautions relevant to maintaining the integrity of the mRNA.[17]

The size of the first-strand synthesis reaction also depends to some
extent on the quantities of mRNA available and the size of the library
desired. We will present the protocol for the use of 4 μg of poly(A)$^+$
mRNA. This should be capable of generating a fairly representative
cDNA library with as many as 4×10^5 recombinants. However, these
reactions can be readily scaled up or down.

The first-strand synthesis reaction mix should be made up of

poly(A)$^+$ mRNA (8 pmol)	4 μg
Vector-primer (8–12 μg, 4–6 pmol)	2 μl
10× first-strand buffer (10× = 500 mM Tris–HCl at pH 8.3, 80 mM MgCl$_2$, 300 mM KCl, 3 mM dithiothreitol)	3 μl
20 mM of each of the four dNTPs	3 μl
[α-^{32}P]dNTP (use either dCTP or dATP)	30 μCi
RNasin (placental RNase inhibitor)	20 units
Water to	30 μl

Add 40 units of reverse transcriptase. See this volume [33] for a complete discussion of the reaction conditions for first-strand synthesis. The
reaction is then incubated at 42° (Okayama and Berg[1] carry out the reaction at 37°) for 30 min. The reaction is stopped by the addition of 1 μl of
0.5 M EDTA (pH 8.0) and 1 μl 10% sodium dodecyl sulfate and the mix is
extracted with 20 μl of phenol. After mixing and centrifugation, the aqueous phase is mixed with 30 μl of 4 M ammonium acetate and 60 μl of
ethanol. After quick freezing (5 min) in dry ice, the mix is warmed to room
temperature with gentle shaking before centrifugation. The pellet is redissolved in 30 μl of TE (10 mM Tris–HCl at pH 7.3, 1 mM EDTA) and
reprecipitated after addition of 30 μl of 4 M ammonium acetate and 60 μl
of ethanol. After quick chilling, warming to room temperature, and cen-

[19] A. Jacobson, this volume [25].

trifugation as above, the solution is decanted and the pellet washed with 70% ethanol. This pellet should now be free of nucleoside triphosphates and represents the second step in Fig. 2 and is ready for oligo(dC) tailing.

Tailing the cDNA

A calibration of the rate of incorporation of dCTP (dGTP in the presence of Mn^{2+} reaction buffer if the linker has dCTP tails) into tails should be carried out using 4 μg of BamHI or similarly cleaved pUC18 in the tailing mix described below, but with the addition of labeled dCTP and the use of time points as described in the tailing calibrations above. The tailed plasmid can then be cleaved with EcoRI and the size of the tails measured as above. Conditions should be determined for obtaining tails of 10–20 bases predominately.

The vector–cDNA is suspended and oligo(dC) tailed in the following mix:

10× CoCl$_2$ tailing buffer	2.5 μl
5 mM dCTP	1 μl
Water to	25 μl

This mixture was preincubated for 5 min at 37° before addition of 25 units of terminal transferase for the appropriate time to add 10–20 Cs (about 5 min). The reaction is terminated by the addition of 1 μl of 0.5 M EDTA (pH 8.0), extracted with 20 μl of phenol and twice precipitated in the presence of 2 M ammonium acetate, as above. The pellet, after washing with 70% ethanol, is ready for cleavage with HindIII.

[Editors' Note. The reason many investigators prefer to tail cDNA with dG is because the reaction stops spontaneously when the tails are about 20 bases long (see procedure in this volume [37]). Thus, improper tail lengths are encountered only when preparing the dC-tailed linker.]

Restriction Cleavage

The tailed vector–cDNA must now be cleaved with HindIII to remove the tail from the other end of the vector (see Fig. 2) and to leave a fragment end which is compatible with ligation to the previously prepared linker molecule.

The dry DNA pellet is dissolved in 25 μl of HindIII buffer and incubated with 12 units of HindIII at 37° for 1 hr. It is important that the enzyme be active and cleave the DNA to completion, but too extensive an overdigestion can also decrease the ligatability of the restriction fragment end. This sample is phenol extracted with 20 μl of phenol, precipitated with ethanol, and washed with 70% ethanol.

Cyclization and RNA Replacement

It is necessary to introduce the linker to allow recircularization of the recombinants. This involves a ligation of the HindIII sites followed by annealing of dG tails on the linker with the dC tails on the recombinants. It is necessary to add an excess (molar) of linker because it must not only react with every recombinant molecule, but also compete with the small dC-tailed fragment cleaved from the recombinants for religation back together at the HindIII site. An annealing step between the dG tails and dC tails before the ligation at the HindIII ends may help the linker compete more effectively by allowing for intramolecular ligations to occur.

At least the first time a batch of linker is used some calibration is required to determine the optimal amount to be added. In general it will be advantageous to add a 2- to 3-fold molar excess. Therefore, set up three test circularizations. Each should consist of 0.1 μg of vector DNA (0.05 pmol) and either 1.5, 3.0, and 6.0 ng of the linker (assuming that the overall yield was 1.5 μg from the preparation described above, this will yield 0.05, 0.10, and 0.20 pmol, respectively).

The cDNA–linker mixtures above are placed in a 10-μl volume with 10 mM TE, 0.1 M NaCl, heated at 65° for 5 min, and then annealed at 42° for 30 min. After quenching at 0°, 10 μl of 5× E. coli ligase buffer [5×: 100 mM Tris–HCl at pH 7.4, 20 mM MgCl$_2$, 50 mM (NH$_4$)$_2$SO$_4$, 500 mM KCl, 0.5 mM β-NAD, and 250 μg/ml BSA] is added and the mixture made up to 50 μl with water. This ligation mixture is incubated overnight with 2 units of E. coli DNA ligase at 12–15°. This results in the ligation of the HindIII ends of the linker and vector molecules.

To complete the synthesis of the cDNA molecules, the RNA strand, which is now no longer completely intact, is generally replaced with DNA by a nick translation procedure and the molecule is ligated. In several control experiments, we have found that this replacement of the RNA strand is unnecessary and it does not affect the yield if it is eliminated. Other researchers find that its replacement increases the yield several-fold, so that we still include this step in the protocol. In order to replace the RNA strand, add to the ligation mixture above 1 μl of a solution containing 10 mM of each of the four deoxynucleoside triphosphates, 1 μl of 5 mM β-NAD, 1 more unit of E. coli DNA ligase, 2 units of E. coli DNA polymerase, and 1 unit of E. coli RNase H. This mixture is incubated for 1 hr at 12° and then 1 hr at room temperature to allow repair synthesis, nick translation of the RNA strand, and ligation. The reaction is terminated by the addition of 2 μl of 0.5 M EDTA, and frozen.

An aliquot of the above cDNA consisting of 10 ng of vector DNA (10 μl of the above mixture) should be transfected onto competent cells[20] and the cloning efficiency of each of the three test mixes calculated. A yield of 500 clones from 10 ng of vector is good ($10^5/\mu$g of mRNA). It may also prove valuable to prepare minipreps[20] of the DNA from a number of the resulting clones and analyze them for the size of the insert. This will allow an assessment of the size of the inserts and the relative background containing no inserts before carrying on with the library.

The rest of the vector–cDNA mix can then be mixed with linker and cyclized through a scaled-up version of the above procedure using the least amount of linker which produced a maximal yield. The plating of the library and any amplification steps desired are discussed in detail elsewhere.[20,21]

---

[20] H. Miller, this volume [13].
[21] G. Vogeli and P. S. Kaytes, this volume [44].

Section IX

Selection of Clones from Libraries

[42] Selection of Clones from Libraries: Overview

By Alan R. Kimmel

There are several procedures for selecting cloned sequences of interest from recombinant DNA libraries. Individual recombinants within libraries can be screened for homology with a nucleic acid sequence, for expression of an antigen (antibody recognition), or for expression of a phenotype. In addition, populations can be screened as mixtures of recombinants. Selection using only a single approach is rarely proof that a clone has been identified correctly, because even the most stringent criterion for screening may select false positives. Therefore, a combination of selection schemes is often needed and further characterization is usually essential for corroboration. Such analyses must rely on information which supplements that originally used for screening [54] (numbers in brackets refer to chapters in this volume).

Nucleic Acids Probes

Screening with nucleic acid probes is dependent upon molecular hybridization, the annealing of single-stranded nucleic acid molecules to form duplex structures stabilized by sequence-specific hydrogen bonds [43]. Only nucleic acids of related sequence organization will base pair or hybridize with each other. The stability of hybrids is a function of the relatedness of the two nucleic acid sequences. Theoretical considerations and practical applications for screening with nucleic acid probes are critical for discriminating sequence-specific hybridization from nonspecific interactions. Empirical procedures are described which relate the effect of probe length, probe base composition, reaction temperature, reaction time, organic solvent concentration, cation concentration, probe concentration, base-pair mismatch, and differences in reactivity of RNA and DNA on the ability to form stable hybrids [43].

Before screening, recombinant libraries must be organized to permit the isolation of one or a few positive colonies or plaques from the $>10^4$ clones probed. Libraries are distributed on master agar plates; clones are transferred to filter matrices and oriented to permit alignment of master plates with the pattern of recombinants replicated on filters [44]. Often duplicate filters are made. DNA from these clones is then immobilized *in situ* and hybridized with denatured or single-stranded radiolabeled nucleic acid probes [45]. Under appropriate conditions, only clones containing DNA sequences that share homology with the probe will hybridize. Positive recombinants can be identified by autoradiography of the filter replicates. Appropriate clones can be isolated and characterized further (see [54]). If small plasmid libraries (<1000 recombinants) are to be screened it may be advantageous to transfer the clones to agar plates in a grid pattern which matches the arrangement of wells in microtiter dishes. Devices (Falcon) exist which permit the transfer in either direction of colonies between agar plates and microtiter wells; clones can be grown in a sterile manner in the dishes in the presence of glycerol and then frozen at $-70°$ for long-term storage [13].

Nucleic acid probes are generally of three types. Probes can be homogeneous cloned sequences, heterogeneous sequences corresponding to mRNA populations, or synthetic oligonucleotides.

Cloned sequences are usually derived from a gene of interest that is isolated from a different species or a related gene that is isolated from the same species. Hybridization conditions for heterologous probes must permit imprecise matching of DNA sequences [43]. It is advisable to establish appropriate conditions of hybridization prior to screening by probing genomic DNA or mRNA on Southern or Northern blots, respectively [61]. Stringent conditions which exclude imprecise hybrids can be identified, in this way. Similarly, relaxed conditions which allow hybridization to sequences evolutionarily distant from those under investigation can also be ascertained. In addition, cloned sequences complementary to repetitive DNA are easily recognized; on genomic DNA blots hundreds of restriction fragment bands hybridize to a repetitive DNA element [63]. If a library were screened with this repeat sequence, potentially thousands of clones with only the repetitive element in common would be identified. By isolating specific restriction fragments from the original clone it should be possible to identify quickly sequences complementary to repetitive or to low-copy DNA and to eliminate the former from the strategy of screening. A final caveat about using cloned DNA to screen libraries concerns potential contamination with *Escherichia coli* genomic or vector DNA sequences. Procedures described for purifying plasmid or phage DNA eliminate contaminating *E. coli* DNA [13]. However, extreme care must be taken in purifying insert from vector DNA if plasmid- or phage-derived sequences are used to screen plasmid or phage libraries, respectively. Since cross-reactivity between plasmid, M13, and λ vectors is usually low, probes should be transferred to a vector type different from that of the library in order to reduce background hybridization [55].

Tissue-specific expression may be exploited to generate nucleic acid probes [46]. These probes are usually heterogeneous mixtures of cDNAs complementary to mRNA populations which reflect cell-specific and/or gene-specific attributes. For example, large numbers of genes have been isolated based on their specific patterns of expression in different tissues, cell types, developmental stages, or physiological environments. Ideally, related cell "types" express all genes at similar levels except for one. In practice, expression patterns are more complex since related cells rarely express most genes at identical levels. Marked differences in mRNA abundance are essential to distinguish between tissue-specific expression and normal variation in hybridization.

Screening for cloned DNAs complementary to mRNAs which exhibit tissue-specific expression is usually performed with duplicate sets of filters. One set is hybridized with RNA-derived probes (often single-stranded cDNA) from one cell type while an identical set of filters is screened with probes from a different cell type. A collection of recombinant clones which hybridizes to probes derived from one cell type but not from the related type is likely to include the clone for which the investigator is searching. This method of screening, often referred to as plus/minus (+/−), is usually not sufficiently sensitive to identify cDNA clones complementary to mRNAs which represent less than 0.1% of total mRNA.

Procedures also exist to enrich populations of RNA sequences that are present in certain cells but not in other physiologically related cells [46]. Messenger RNA from the former cells is reverse transcribed into cDNA. The cDNA is hybridized to an excess of RNA from the latter cells and the nonhybridizing cDNA purified by chromatography. This cDNA is derived from cell-specific mRNAs and can be used to construct an enriched cDNA library or to probe for complementary cloned sequences in an already existing library. This method, called subtracted hybridization, is clearly more technically demanding than that of the +/− procedure. The advantage of the subtraction approach is that clones complementary to mRNAs representing only 0.01% of total mRNA can often be detected.

Specific mRNAs can also be enriched by isolating RNA from nuclei [22], membrane-bound polysomes [23], polysomes associated with specific nascent proteins [24], and other subcellular fractions; in addition newly synthesized mRNA can be separated from older molecules [25]. These enriched populations can be used to focus on recombinant clones which encode RNAs of certain classes. However, if a specific gene within this population is desired, screening by other methods will be required [54].

It should also be noted that clones complementary to abundant mRNAs can be selected directly [46]. By controlling hybridization conditions, only clones which hybridize to mRNA-derived cDNA probes representing greater than 1% of the input mRNA will be radiolabeled sufficiently to be detected by autoradiography.

Synthetic oligonucleotides constitute the final class of nucleic acid probes [47–49]. Most often these are constructed based on codon sequences predicted by an amino acid sequence of a protein. Several parameters need to be considered before choosing an oligonucleotide sequence. Among these are length, GC composition, and codon usage by the organism. There are advantages and disadvantages for each choice and possibly the single most logical one may not be sufficiently specific to assure success. Multiple choices of oligonucleotide probes are therefore recommended. The oligonucleotide should be hybridized to genomic DNA blots to assess the specificity of the probe and to determine reaction conditions. Similarly, mRNA blots should be hybridized, and therefore the probes synthesized should be complementary to the RNA strand, i.e., they ought to be antisense. If amino acid sequence information exists from different regions of a protein, oligonucleotide probes from several regions should be synthesized. These may hybridize to identical clones or to different clones with overlapping sequences.

Antibody Recognition

Manipulation of expression libraries in *E. coli* prior to screening with antibodies to specific proteins is similar but not identical to that described for nucleic acid probes [50,54]. Characterization of antibody probes is as critical as the characterization of nucleic acid probes. Antibodies should have specificity for a single protein, bind to that protein immobilized on solid matrices (Western blots), and interact with an amino acid sequence rather than with products of post translational modifications such as glycosylation or phosphorylation [50,51]. Polyclonal antibodies have generally been used for such screening since they recognize many epitopes. The broad specificity of polyclonal antibodies guarantees that many regions of a protein can be recognized, but backgrounds can be unacceptably high. In contrast, monoclonal antibodies recognize one epitope, which if specific to a particular protein can give very high signal-to-noise ratios. A disadvantage is that monoclonal antibodies often recognize only a short epitope or an epitope not present in the partly denatured, partly folded antigen expressed in *E. coli*.

Phenotypic Expression

Another method of screening is dependent on expressing DNA sequences to produce an identifiable phenotype, usually in eukaryotic cells. Functional selections rely on numerous approaches which are specific for each gene and species. However, there are some general classes. In one, genomic DNA covalently linked *in vitro* to a heterologous DNA "tag" is transfected into eukaryotic cells and clones selected which express a new phenotype [53].[1] Genomic libraries are then constructed from the DNA

[1] I. Lowy, A. Pellicer, J. F. Jackson, G.-K. Sim, S. Silverman, and R. Axel, *Cell (Cambridge, Mass.)* **22,** 817 (1980).

isolated from the selected cells. By screening these new libraries with radiolabeled tag sequences, clones bearing DNA responsible for the altered phenotype can be identified. For example, DNA from a cosmid library can be transfected and the new libraries subsequently screened with cosmid vector DNA. Another variation on this technique introduced genomic DNA into cells of a different species. Since species-specific repetitive DNA may be associated with many, if not most genes, screening the newly constructed genomic libraries with this repetitive element obviates the requirement for the heterologous tag. For example, repetitive Alu sequences can identify human DNA sequences in mouse cells.[2]

A second approach is often described as insertional tagging. Cells are transfected with a heterologous DNA sequence capable of integrating into genomic DNA of the host and an alteration of phenotype selected. If the heterologous DNA can insert randomly at any DNA sequence in the genome, potentially any gene with a selectable phenotype can be identified. In practice, most insertional sequences exhibit some site preference for integration. In mammalian cells, specifically engineered retroviral DNA can be used as the tag [52]. In *Drosophila* transposable P elements are agents of insertional mutation.[3] A related approach has been applied to *Caenorhaditis elegans*.[4] In haploid genomes, the application of the method is conceptually straightforward; however, most species are diploid. In the absence of a heterozygous genotype, insertional inactivation can only succeed if a dominant mutation is generated.

A third approach makes use of a vector which promotes the expression of cloned cDNA sequences in eukaryotic cells (see [32,41]). Again the vector provides a tag to identify associated transfected cDNA sequences.

Regardless of the selection scheme implemented, it should be emphasized that all of the tagged clones generated from DNA obtained from cells with the desired phenotype are rarely correct. Difficulties with the technique are elaborated in considerable detail [53]. Usually many different donor DNA sequences integrate in the host genome following transfection. Similarly, multiple insertional events often occur with retroviral vectors or P elements. Since genes other than the one encoding the selectable phenotype are tagged, many incorrect clones can be isolated. These clones, taken together, however, potentially represent a population enriched for the gene of interest. Thus, with a series of retransfections and reselections, it should be possible to eliminate all of these incorrect se-

[2] L. C. Kuhn, A. McClelland, and F. H. Ruddle, *Cell (Cambridge, Mass.)* **37,** 95 (1984).
[3] M. G. Kidwell, *Drosophila* (in press).
[4] I. Greenwald, *Cell (Cambridge, Mass.)* **43,** 583 (1985).

quences and to isolate a DNA sequence responsible for the observed phenotype.

Population Screenings

If all of the aforementioned procedures fail or are inapplicable, a "brute force" approach may be considered in which recombinants are manipulated in mixed populations rather than as individual clones. Initially, a recombinant library is divided into a series of clonal populations that represent portions of the original library. Each population is then screened for a specific characteristic. Once a population containing a positive recombinant has been identified, it is subdivided and each of these subdivisions is similarly analyzed. The process is repeated until an individual clone is selected that exhibits the specific characteristic.

The method of analysis that is chosen should be sufficiently specific and sensitive to detect a few positive recombinants within a large population of negatives and should be based on characteristics that differ from the methods of screening already discussed. Obviously, the more sensitive the assay, the greater the chance of success. For example, it is sometimes possible to detect an activity following the translation *in vitro* [27] of total cellular mRNA. A screening procedure can be devised in which recombinant DNAs that are complementary to an mRNA are selected [60]. A series of populations are grown containing (e.g.) 500 recombinants per population; the number of populations to be grown is a function of the library size and the estimated abundance of the mRNA (see [32]). An aliquot of cells from each population is saved and DNA purified from the remainder of the culture [13]. The DNAs are then immobilized on filter matrices and hybridized to total mRNA [60]. The mRNA selected by hybridization would be complementary to all of the recombinants in each of the populations. Each heterogeneous mixture of mRNAs is then translated *in vitro* and assayed. Recombinant populations containing cloned DNAs that are complementary to the specific mRNA are thus identified. The aliquot of cells which had been saved from this population is used to establish small subpopulations of recombinants and the screening process is repeated. Eventually a specific clone can be identified which can hybridize to an mRNA that directs the *in vitro* synthesis of the activity being assayed.

Choice of Library

Finally the investigator must decide what kind of library to screen—a genomic or a cDNA library. This is mostly of concern when screening by nucleic acid hybridization. Selections using phenotypic expression can be based on cDNA expression libraries or genomic DNA. Different selections may require different approaches.

The choice of genomic or cDNA library for nucleic acid screening is dependent on the relative abundance of a particular clone within the entire recombinant library. In λ libraries derived from mammalian DNA an average genomic clone (~15 kb) is present approximately 10-fold less frequently than is a clone corresponding to a rare mRNA (representing 0.01% of total mRNA), in a cDNA library. Thus it is usually more convenient to screen cDNA libraries. If a cDNA library from a specific cell type were difficult to construct, screening a genomic library should be attempted. In organisms with genomes sufficiently less complex (e.g., *Saccharomyces, Dictyostelium, Caenorhaditis, Drosophila*) than those of mammalian cells, each rare mRNA sequence is represented with approximately equal frequency in genomic or cDNA libraries. Thus, similar numbers of genomic or cDNA recombinants need to be screened to obtain clones complementary to rare mRNAs. Clones complementary to more abundant mRNAs may still be represented more frequently in cDNA libraries than in comparable genomic libraries.

There are obvious differences between clones obtained from genomic or cDNA libraries. Entire genes can be isolated from genomic libraries including nontranscribed flanking sequences, mRNA coding sequences, and introns, whereas only transcribed sequences can be identified in cDNA libraries. Eventually clones of both types may be required to complete the analysis of the genes. It should be clear that probes from one can be used to isolate clones from the other.

[43] Molecular Hybridization of Immobilized Nucleic Acids: Theoretical Concepts and Practical Considerations

By Geoffrey M. Wahl, Shelby L. Berger, and Alan R. Kimmel

Molecular hybridization is the formation of double-stranded nucleic acid molecules by sequence-specific base pairing of complementary single strands.[1,2] The development of methods for immobilizing DNA on nitro-

[1] R. J. Britten and E. H. Davidson, *in* "Nucleic Acid Hybridization: A Practical Approach" (B. D. Hames and S. J. Higgins, eds.), p. 3. IRL Press, Oxford, 1985.
[2] J. Meinkoth and G. Wahl, *Anal. Biochem.* **138**, 267 (1984).

cellulose paper[3,4] and for detecting the fixed nucleic acid with radioactive probes by molecular hybridization[5,6] (which we will refer to as mixed-phase hybridization) forms the cornerstone of gene (and gene product) detection methods. These techniques have revolutionized our understanding of gene structure, genome organization, and the control of gene expression. The sensitivity (<1 pg of complementary sequence), speed (<24 hr), and convenience (simple machines and inexpensive materials are required) of the nucleic acid hybridization procedures have enabled them to be applied not only to basic research problems but also to the diagnosis of heritable diseases and to the detection of a wide variety of microbial and viral pathogens.

Our aim is to describe methods for detecting sequences in unfractionated nucleic acids applied directly to solid supports. Recombinant cDNA and genomic libraries generated in plasmid or phage vectors are amenable to this technique, as are electrophoretically fractionated nucleic acids either after transfer to solid supports (blots) or directly in the gel matrix.

In all cases, the probe is a radiolabeled nucleic acid (either DNA or RNA) capable of hybridizing to complementary sequences found among the molecules bound to the solid support. Although each case is different, and depends on the information one has and what one is seeking to learn, nucleic acid probes are usually one or more of the following: a related gene or cDNA; a viral sequence; an oligonucleotide based on a known amino acid sequence (see this volume [47–49]); bulk cDNA probes derived from mRNA isolated from related cell types such as uninduced and induced populations or from cells at different developmental stages (see this volume [46]); or any other nucleic acid capable of hybridizing to the sequence of interest. The principles specified here apply to the screening of DNA and RNA blots (see this volume [61,62]), dried gels [47], and recombinant DNA libraries [45]. We will also emphasize the most sensitive and rapid techniques, simplifying the procedures on the basis of published observations or where our experience and that of many colleagues permit.

Hybridization Parameters

The kinetics of hybridization of RNA or DNA probes with RNA or DNA tethered to nitrocellulose or free in solution are very similar,[1–4,6,7]

[3] A. P. Nygaard and B. D. Hall, *Biochem. Biophys. Res. Commun.* **12,** 98 (1963).
[4] A. P. Nygaard and B. D. Hall, *J. Mol. Biol.* **9,** 125 (1964).
[5] D. Denhardt, *Biochem. Biophys. Res. Commun.* **23,** 641 (1966).
[6] D. Gillespie and S. Spiegelman, *J. Mol. Biol.* **12,** 829 (1965).
[7] G. B. Spiegleman, J. E. Habe, and H. O. Halvorson, *Biochemistry* **12,** 1234 (1973).

suggesting that parameters which influence nucleic acid reannealing in solution will have similar effects in mixed-phase systems. Here, we discuss the parameters which affect hybrid stability and hybridization rate. An understanding of these parameters enables one to derive hybridization conditions which should yield optimal signal-to-noise ratios.

Hybrid Stability

The formation of nucleic acid hybrids is a reversible process and an understanding of the parameters which affect their stability enables one to derive the optimal conditions for discriminating between perfect and imperfect hybrids. The melting temperature (T_m) is defined as the temperature when half the duplex molecules have dissociated into their constituent single strands. It is affected by the monovalent cation concentration (M, in moles per liter), the base composition expressed as mole fraction of G and C residues, the length in nucleotides of the shortest chain in the duplex (L), and the concentration of helix-destabilizing agents such as formamide.[8,9,10] The following equation, valid from pII 5 to 9, has been derived from analyzing the influence of these factors on hybrid stability of probes longer than ~50 nucleotides:

$$T_m = 81.5° + 16.6 \log M + 41 \text{ (mole fraction G + C)}$$
$$- 500/L - 0.62 \text{ (\% formamide)} \quad (1)$$

From this expression it follows that there is a 16° increase in T_m with each log increase in cation concentration. For example, the T_m of mammalian DNA (GC mole fraction ~0.4) is ~85° under standard salt conditions, namely 0.18 M Na$^+$, and ~97° in 1 M Na$^+$. Hybrids between oligonucleotides (14–20 bp) and immobilized DNA show decreased stability and an empirical formula has been determined[11] to define the temperature at which 50% of these short duplexes dissociate (T_d) when the denaturation is performed at 0.9 M NaCl:

$$T_d \text{ (°C)} = 4(G + C) + 2(A + T) \quad (2)$$

where G, C, A, and T indicate the number of the corresponding nucleo-

8 B. L. McConaughy, C. L. Laird, and B. J. McCarthy, *Biochemistry* **8**, 3289 (1969).
9 C. Schildkraut and S. Lifson, *Biopolymers* **3**, 195 (1965).
10 R. J. Britten, D. E. Graham, and B. R. Neufeld, this series, Vol. 29, p. 363.
11 R. B. Wallace, J. Shaffer, R. F. Murphy, J. Bonner, T. Hirose, and K. Itakura, *Nucleic Acids Res.* **6**, 3543 (1979).

tides in the oligomer. Similar expressions have been derived for RNA probes hybridizing to immobilized RNA[12]:

$$T_m = 79.8° + 18.5 \log M + 58.4 \text{ (mole fraction GC)}$$
$$+ 11.8 \text{ (mole fraction GC)}^2 - 820/L - 0.35 \text{ (\% formamide)} \quad (3)$$

For DNA–RNA hybrids the expression differs from Eq. (3) in the last term: 0.35 (% formamide) is replaced by 0.5 (% formamaide).[13] Note that the T_m for RNA–RNA duplexes is about 10° higher than that of comparable DNA–DNA hybrids in the absence of formamide and RNA–DNA duplexes are intermediate in stability. RNA–DNA hybrids denature at greater than 10° higher than comparable double-stranded DNA structures in 80% formamide. This fact accounts for the observation that under appropriate conditions RNA can displace a homologous DNA sequence in a double-stranded molecule.

The stability of duplexes formed between strands with mismatched bases is decreased according to the number and location of the mismatches and is especially pronounced for short (e.g., <14 bp) oligonucleotides. For hybrids longer than 150 bp, the T_m of a DNA duplex decreases by ~1° with every 1% of base pairs which are mismatched.[14] For hybrids shorter than 20 bp, the T_m decreases by approximately 5° for every mismatched base pair.[10,11,15] In order to minimize the hybridization of probe to related but nonidentical sequences, hybridization reactions must be performed under the most stringent conditions possible. From the discussion above, hybridization stringency can be altered by adjusting the salt and/or formamide concentrations and/or by changing the temperature. The stringency can be adjusted either during the hybridization step, or in the posthybridization washes. It is often convenient to perform the hybridization at low stringency and wash at increasing stringencies, analyzing the results after each wash. Once achieves maximum stringency at $T_m - 5°$ for probes >150 bp in length. Hybridization at $T_m - 25°$ (where the maximum hybridization rate is observed, see below) then washing under conditions of higher stringency enables the detection of related sequences and the monitoring of the effectiveness of the washes. This strategy also enables one to obtain an estimate of sequence relatedness.

[12] D. K. Bodkin and D. L. Knudson, *J. Virol. Methods* **10,** 45 (1985).
[13] J. Casey and N. Davidson, *Nucleic Acids Res.* **4,** 1539 (1977).
[14] T. L. Bonner, D. T. Brenner, B. R. Neufeld, and R. J. Britten, *J. Mol. Biol.* **81,** 123 (1973).
[15] R. B. Wallace, M. J. Johnson, T. Hirose, T. Miyake, E. H. Kawashima, and K. Itakura, *Nucleic Acids Res.* **9,** 879 (1981).

Hybridization Rate

Kinetics of Hybridization of Single-Stranded Probes. The rate of hybrid formation for single-stranded probes in mixed-phase hybridizations should follow pseudo-first-order kinetics since the concentration of probe is almost always in vast excess over that of target sequences. (Complex cDNA probes derived from mixed mRNA populations described in this volume [46] are notable exceptions.)

The time required for half of the tethered DNA to anneal with the probe is

$$t_{1/2} = \ln 2/kC \qquad (4)$$

where k is the first-order rate constant (expressed as liters $\times$ mol nucleotides^{-1} seconds^{-1}) for formation of a hybrid molecule and C is the probe concentration (mol nucleotides $\times$ liters^{-1}). Under standard conditions, defined as $T_m - 25°$, 0.18 M cation concentration, and a single-stranded fragment length of 500 nucleotides, the rate constant k can be approximated as

$$k = 10^6/N \qquad (5)$$

where N is the molecular sequence complexity of the probe or the total number of unique nucleotide sequences.[16] Under conditions of hybridization different from the standard case, k, and thus the rate of hybridization, will be affected.

In probe excess, the most important parameters governing the rate of hybridization are the concentration of DNA in solution (C) and its sequence complexity (N). If there are no repeating units, a probe consisting of fragments 1 kb in length generated by cleaving a cloned insert of 5 kb has a sequence complexity of 5000 nucleotides and a length of 1000 nucleotides. It follows that the concentration of complementary sequences decreases as the sequence complexity increases at a fixed concentration of DNA. As a consequence, the rate of hybridization is inversely proportional to the sequence complexity and directly proportional to nucleic acid concentration.

Conditions which affect stability of double-stranded nucleic acids often influence the relative rate of hybrid formation. Thus, temperature, cation concentration, base mismatch, probe strand length, solvent, pH, and base composition affect both stability of the hybrid and the rate at which it forms.

---

[16] Note that for double-stranded probes, sequence complexity is defined in terms of unique base pairs.

The maximum rate of hybridization in solution has been determined empirically to occur at 25° below the T_m of DNA–DNA duplexes and 15° below T_m for RNA–DNA duplexes. There is a broad temperature optimum with little change (20%) in rate occurring between $T_m - 32°$ and $T_m - 18°$ for DNA–DNA duplexes. This observation also pertains to mixed-phase hybridizations which utilize probes longer than approximately 150 nucleotides.[17] For shorter nucleic acid probes, <20 nucleotides in length, a temperature 5° below the T_d is used for hybridization experiments.[10,18] Hybridization with oligonucleotides is considered in depth in this volume [47, 48].

Cation concentration (M) has little effect on the rate constant for DNA–DNA hybridization as long as it is kept above 0.4 M (e.g., k increases 2-fold between 0.4 and 1.0 M NaCl and then reaches a plateau). At low values of ionic strength the rate of hybridization is affected significantly. Hybridization in 1.0 M NaCl occurs 7 times more rapidly than at 0.18 M NaCl. Decreasing the salt concentration to 0.09 M NaCl reduces the rate an additional 5-fold.

The effects of salt on DNA–RNA duplex formation are somewhat different. At 0.18 M NaCl, the rate is equivalent to that of DNA–DNA duplexes. However, at 1 M NaCl, the rate of DNA–RNA duplex formation is only twice that measured at 0.18 M M. This contrasts with a 7-fold increase in relative rates of DNA–DNA hybridization under the same conditions.

There is a 2-fold reduction in rate for each 10% base mismatch of probes > 150 nucleotides assuming hybridization is occurring at ~25° below the T_m for the mismatched duplexes.

Probe length also affects the rate of hybridization. For fragments that are 50 to 5000 nucleotides long, the rate is proportional to the square root of the length of the smallest single-stranded fragment. Since most probes vary in length between a few hundred base pairs up to ~2 kb, the change in k does not exceed 3-fold and can often be ignored.

The concentration of formamide has a slight affect on k. Less than a 2-fold reduction in the rate of hybridization is observed in 50% formamide.

The effects of pH and base composition on k are also small. In the range pH 5 to pH 9, when the salt concentration is above 0.4 M NaCl, k changes less than 1.3-fold[19]; a change in mole fraction (GC) from 0.2 to 0.6 causes an increase in k of less than 1.5-fold.

[17] K. H. Cox, D. V. DeLeon, L. M. Angerer, and R. C. Angerer, *Dev. Biol.* **101,** 485 (1984).
[18] S. V. Suggs, R. B. Wallace, T. Hirose, E. H. Kiwashima, and K. Itakura, *Proc. Natl. Acad. Sci. U.S.A.* **78,** 6613 (1981).
[19] J. G. Wetmur and N. Davidson, *J. Mol. Biol.* **31,** 349 (1968).

Kinetics of Hybridization of Double-Stranded Probes. The rate of hybridization of denatured double-stranded probes follows second-order kinetics. Furthermore, unlike single-stranded probes, double-stranded molecules can hybridize to the immobilized nucleic acids on the filter or renature in solution, concomitantly. Since the probe strands, in general, are in vast excess over complementary bound molecules, reassociation of the probe to itself limits the time during which useful hybridization to the filter can occur. No increase in hybridization signal can be expected by hybridizing longer than the interval required for complete renaturation of the probe in solution. The time required for half of the double-stranded fragments to renature is

$$t_{1/2} = 1/kC \qquad (6)$$

where k is the same rate constant defined in Eq. (5) and C is the concentration of double-stranded probe in moles of nucleotide per liter, taking into account both strands. In general, conditions which affect the rate of hybridization of single-stranded probes similarly affect the rate of hybridization of double-stranded probes. However, there are some expectations which will be discussed below.

Difficulties can arise when high molecular weight double-stranded probes are used in hybridization experiments. Renaturation in solution can proceed more rapidly than hybridization to nucleic acid sequences immobilized on filters because of decreased accessibility of the immobilized target sequences to long probe molecules and decreased diffusion of these probes from solution to the filters.

Random overlapping double-stranded probes have the ability to form duplex molecules in which the single-stranded "tails" of one duplex are available for further hybridization. This secondary hybridization which can form long networks occurs with a rate that is 5-fold slower than the rate of hybridization of a single-stranded probe to the immobilized nucleic acids. Since the networks can make a significant contribution to the signal, both rates must be considered when designing hybridization experiments (reviewed in Meinkoth and Wahl[2]).

The rate of probe reannealing can be enhanced in solution[20] and in mixed-phase hybridizations using anionic dextran polymers (e.g., dextran sulfate 500)[21] or polyethylene glycol 6000. Dextran sulfate acts to exclude DNA from solution, thus raising the effective DNA concentration and the hybridization signal in a given unit of time. In mixed-phase hybridization, the effect of dextran sulfate is most pronounced for polynucleotides

[20] J. G. Wetmur, *Biopolymers* **14,** 2517 (1975).
[21] G. M. Wahl, M. Stern, and G. R. Stark, *Proc. Natl. Acad. Sci. U.S.A.* **76,** 3683 (1979).

longer than about 250 nucleotides (G. Wahl, unpublished observations) and has no effect on oligonucleotides 14 bases long (R. B. Wallace, personal communication). The increase in the apparent rate of hybridization is approximately 3-fold for mixed-phase hybridizations which utilize single-stranded probes and up to 100-fold for hybridizations utilizing random, overlapping, nick-translated probes.[21] This effect has been attributed to the accelerated rate of formation of probe networks or "hyperpolymers" between the partially overlapping sequences of the molecules generated by nick translation.[21] It is important to emphasize that the ability to form such networks is critically dependent on probe size since the overlaps in small probes are not likely to be able to initiate or maintain the formation of stable networks. It should also be noted that inclusion of dextran sulfate in filter hybridizations can result in high nonspecific backgrounds, although the background can generally be attributed to the probe preparation.

Empirical Considerations

In this section, the practical aspects of hybridization reactions are considered. These observations are based on experiments performed in solution and are applicable to both single-stranded and double-stranded probes in excess. A term used to describe hybridization, C_0t, must be defined. The product of C_0, the initial concentration of the probe in moles of nucleotides per liter, and t, time in seconds, determines the progress of hybridization. A similar term R_0t, defines the progress of RNA hybridization. C_0t can be redefined in terms of the units usually encountered in the laboratory as

$$C_0t \approx 1/100 \ [\text{hours} \times \text{DNA concentration } (\mu g/ml)]$$
$$\approx (A_{260}/2) \times \text{hours} \tag{7}$$

With these parameters alone, however, one cannot estimate when, for a given C_0, the hybridization reaction will have gone to completion.

In order to estimate how long a probe in solution should be hybridized to immobilized nucleic acids, the sequence complexity must be considered. A term $C_0t_{1/2}$ can be defined which describes that combination of time (seconds) and concentration (moles of nucleotide per liter) when half a double-stranded probe has renatured at $T_m - 25°$ in 0.18 M salt. Clearly, the more complex the probe at a given C_0, the longer it will take for renaturation to occur. $C_0t_{1/2}$, then, is a measure of sequence complexity N (measured in base pairs). It is independent of the nature of the sequence, that is all nick-translated unique 5-kb fragments have the same sequence complexity. A useful approximation relating these terms is

$$C_0 t_{1/2} \approx N \times 10^{-6} \qquad (8)$$

where C_0 is expressed as the molar concentration of nucleotides and $t_{1/2}$ is in seconds. In practical terms, hybridization is virtually complete (75%) in 3 times the number of hours required to reach $C_0 t_{1/2}$. An empirical formula that relates these parameters is

$$t_{1/2} \approx \frac{(F)(N)(\text{ml})}{(\mu\text{g DNA})10{,}000} \qquad (9)$$

where the units of $t_{1/2}$ are hours, the units of N are base pairs, and F is the ratio of the rate of hybridization under the conditions used to the rate under standard conditions. (Standard conditions refer to $T_m - 25°$ at 0.18 M cation concentration for a fragment of 500 bp in length.) For example, at a concentration of 0.1 μg/ml (3×10^{-7} mol nucleotides/liter) of a mammalian DNA fragment with a length and complexity equal to 1000 bp and a GC mole fraction equal to 0.4, hybridization at a temperature of 50° in 0.18 M salt will have a $C_0 t_{1/2}$ of 0.001. From either Eq. (8) or (9) we calculate that $t_{1/2} \approx 1$ hr. [From Eq. (8), $C_0 t_{1/2} = 1000 \times 10^{-6} = 10^{-3}$; $t_{1/2} = 10^{-3}/(3 \times 10^{-7}) = 3.3 \times 10^3$ sec = 0.92 hr; from Eq. (9) $(1 \times 1000 \times 1)/(0.1 \times 10{,}000) = 1$ hr.]

At an identical DNA concentration in 50% formamide and 1 M Na$^+$ at 42° the rate of hybridization will increase 4-fold (~7-fold increase as a result of the increase in salt and ~2-fold decrease as a result of adding formamide) and the reaction will approach the $t_{1/2}$ in 0.25 hr, and completion in ~0.75 hr. Note that these expressions do not include network formation. Therefore, longer hybridization times *may* increase signal strength without affecting noise. Many investigators routinely hybridize for 20 hr when shorter times would seem to be indicated.

[44] Amplification, Storage, and Replication of Libraries

By GABRIEL VOGELI and PAUL S. KAYTES

The successful isolation of clones from cDNA or genomic libraries is dependent in part on the ease of handling the library. For small libraries, such as cDNA libraries in plasmid vectors, methods such as stabbing into microtiter wells or the storage of the initial filters at −80° are feasible.[1]

[1] D. Hanahan and M. Meselson, *Gene* **10**, 63 (1980).

METHODS IN ENZYMOLOGY, VOL. 152

For example, a few hundred to a few thousand colonies can be picked up from the initial transformed plates and stabbed into 200 μl of LB broth in the wells of a microtiter plate. They are grown at 37° and are stored at −20° after the addition of an equal volume of glycerol. Alternatively, the initial filters can be incubated on LB plates with glycerol and then stored at −70°; copies are made from these filters whenever needed. Similarly, methods for the convenient handling of phage libraries with cDNA or genomic inserts containing up to 10^8 independent colonies have been worked out[2,3] (also this volume [17]). However, with the availability of methods that can create plasmid cDNA libraries with up to 10^6 initial transformants[4] (this volume [39]), and the development of genomic cosmid libraries capable of carrying large inserts in a plasmidlike vector[5] (this volume [18]), simple methods for the storage and replication of these libraries are important.

Four problems arise during the use of a library. First, during the growth (amplification) of the initial transformants, there should be a minimal change in the relative distribution of the different recombinant molecules. Second, the library must be stored in such a manner that the cells or phage particles can be recovered with good viability if the library is to be used at a later date for screening with other probes. Third, one has to be able to replicate the stored library at a very uniform colony or phage density of up to 10,000 plaques or colonies per plate. Fourth, the screening of the library should yield only specific hybridization signals with as few false positives as possible and it should also be possible to screen the same filters several times with different probes. Here we describe methods for the handling of plasmid and phage libraries which address these four problems.

Libraries in Plasmid Vectors

A plasmid library may be screened directly after the initial transformation, or it may be amplified first and stored for screening at a later time. Direct screening eliminates the problem of preserving the initial distribution of recombinants. Direct screening of colonies grown on the surface of agar plates allows an easy alignment of the autoradiogram signals with the clear plates. However, a library stored on agar plates can be screened

---

[2] T. Maniatis, R. C. Hardison, E. Lacy, J. Lauer, C. O'Connel, D. Quon, G. K. Sim, and A. Efstratiadis, *Cell (Cambridge, Mass.)* **2,** 687 (1978).

[3] R. A. Young and R. W. Davis, *Proc. Natl. Acad. Sci. U.S.A.* **5,** 1194 (1983).

[4] H. Okayama and P. Berg, *Mol. Cell. Biol.* **2,** 161 (1982).

[5] J. Collins and B. Hohn, *Proc. Natl. Acad. Sci. U.S.A.* **9,** 4242 (1978).

only a few times since the bacteria on plates are viable for only a limited period of time, and the preparation of replicas smears such a library with accompanying loss of resolution (see also this volume [13]). In addition, storage in the refrigerator may result in the plates becoming contaminated, a process that will destroy the library. In contrast, if the library is amplified first, aliquots may be stored frozen and grown at will on nitrocellulose filters which then act as masters for the preparation of replicas. Replica plating from master filters is believed to result in more efficient transfer than replica plating from plates, and allows one to plate a large number of colonies on a filter (approximately 3000 colonies per 88-mm filter). The main disadvantages are the added steps involved in amplification, the need for screening a minimum of five times the number of recombinants in the initial library in order to be sure that very rare clones will be represented (see this volume [16,46] for calculations), and the potential difficulty in aligning autoradiogram signals with the opaque master filter.

Procedures for both of these methods are presented in this chapter. In the first method, we describe the amplifications of a library, the subsequent storage, and the plating of the library on filters for the preparation of a master template and replicas. In the second method we describe the growing of bacteria on master plates and the transfer of colonies to filters for screening. Both methods are suitable for maintaining live bacteria during the screening process. (See this volume [45] for details of screening.)

Procedures for Amplifying Plasmid Libraries

The following method is useful for preparing transformants from a cosmid or plasmid library for storage, or for the amplification of an already existing library.[6] The growth of the initial transformants during the construction of a cDNA or genomic library is an amplification step of individual ligation products. However, because colony size is frequently unequal, such amplification will distort the relative frequency of clones in the stored library. In contrast, when initial transformants are plated in soft agarose the colony sizes vary less within the soft agarose layer. Thus, amplification of a plasmid library in soft agarose (in much the same manner as a phage library is plated) does not markedly alter the distribution of recombinant plasmids in the library.[6] A method is presented below. We will assume that double-stranded DNA suitable for insertion into a vector is in hand, and that competent *Escherichia coli* are available for transformation.

---

[6] G. Vogeli, E. Horn, M. Laurent, and P. Nath, *Anal. Biochem.* **151**, 442 (1985).

1. Titer the recombinant library. This step is designed to find out how many recombinant bacteria can be produced from the DNA to be cloned. If you have double-stranded cDNA prepared, optimize the yield of ligation by trying several vector-to-insert ratios. Transform *E. coli* with several dilutions of ligation mixture and plate several dilutions of each transformation. (The techniques are described in detail in this volume [13,39]). For each plate, mix the transformed bacteria with 2.5 ml 0.7% top agarose in LB broth at 45° and pour on a 100-mm LB agar plate supplemented with the appropriate antibiotic. If an existing library is plated, determine the titer of the library using a series of 10-fold serial dilutions in LB broth and plate 100 μl of each dilution as described above.

2. Grow overnight, inverted at 37°, and count bacteria in order to choose the most efficient ligation procedure and to determine the titer of the library.

3. Ligate the remaining DNA and transform *E. coli;* scale up proportionally. From the amount of ligated DNA available for the transformation you will now have a good estimate of the total number of transformants your library will contain.

4. Plate the library and grow it as described above in step 1. In some cases[7] the soft agar technique results in more recombinants than direct plating of bacteria. Aim for approximately 3000 colonies per 100-mm dish.

5. Since the colonies on the surface of the agar spread rapidly and are not the standard size, remove them by rinsing with 3 ml LB and rubbing gently with a glass rod. These can be maintained separately or discarded.

6. Scrape the soft agarose layer with the majority of the colonies embedded in it off the agar plates. Combine the agarose layers from all plates and make it into a paste by adding 1 volume of LB broth and passing the agarose through needles of decreasing size until it can pass through a 22-gauge needle.

7. Mix the paste with 0.5 volume of Sephadex G-25 slurry (autoclaved in LB broth). Prepare several Econo columns (Bio-Rad, no. 731 1550, or equivalent) by adding a 2-ml bed of Sephadex G-25. Subsequently load 12 ml of the agarose-Sephadex G-25 slurry into the reservoir of each column.

8. Place each column in a 15-ml plastic tube and centrifuge in a tabletop centrifuge with a swing-out rotor to collect the cells in the centrifuge tube (for example, in the Sorvall GLC-4 centrifuge, H1000 rotor, centrifugation for 5 min at 1000 rpm, equivalent to about 100 *g*, is adequate). The agarose remains with the Sephadex G-25 in the reservoir of the column and can be removed with a spatula so that the same column can be reused.

[7] M. W. Norgard, K. Keem, and J. J. Monahan, *Gene* **4,** 279 (1978).

9. Resuspend the cell pellet in the supernatant (or a smaller volume as desired). Make 12% in glycerol and store in aliquots at −70°.[8]

10. Titer the frozen library using the conditions under which it will be plated later (a typical titer is 10^9 colony forming units/ml).

11. To screen the library, thaw an aliquot of the frozen library in a water–ice bath and withdraw a small sample (5 μl). Freeze the aliquot again in dry ice. The thawing and freezing step can be repeated at least five times without loss of cell viability.[6] Dilute the aliquot in LB broth to a concentration of around 3000 colony forming units/ml and plate as described below.

12. Mark a nitrocellulose filter, by labeling it asymmetrically with a black ballpoint pen and number the filter to correspond with the plate. The nitrocellulose filter need not be sterilized. Lay it on an LB agar plate supplemented with the appropriate antibiotic. One milliliter of the properly diluted cell suspension (3000 colony forming units/ml) can be distributed evenly over an 82-mm nitrocellulose filter by rotating the agar plate. Keep the plates open until the 1 ml of liquid has dried.

13. Incubate this master filter, colony side up, on its plate of agar at 37° in an inverted position until the colonies are visible (around 8–10 hr at 37°, colony size around 1 mm). (Note, for libraries constructed in X 1776, choose only detergent-free nitrocellulose.)

14. Remove the filters and replicate them singly by placing the filter, colony side up, on Whatman 1 paper. Prewet a filter by applying it to an LB agar plate, with the side of the filter up that will later face the colonies of the master filter. Lay the filter, either a nylon hybridization membrane or a nitrocellulose filter (Millipore HAWP or the equivalent), on top of the master filter (see the manufacturers instructions for the orientation of the nylon-based filters). Sandwich the filters between Whatman 1 paper and press together with finger strokes to transfer the colonies. Also transfer the orientation marks on the master copy by puncturing with a 19-gauge needle into the copy filter. A drop of India ink in the needle facilitates later identification of the punctures. Several copies can be made from one master filter. However, it may be necessary to regenerate the colonies for a few hours at 37° before preparation of each replica in excess of four copies.

15. Put the master filter, colony side up, on a fresh LB agar plate and incubate for a few hours at 37° to regenerate the colonies before storage at 4°. (Stored plates should be sealed with Parafilm and kept inverted.) Incubate the filter copies overnight at 37° on fresh LB plates with the appropri-

---

[8] D. A. Morrison, *J. Bacteriol.* **1**, 349 (1977).

ate antibiotic. If desired, the plasmid DNA in the colonies can be amplified before screening by transferring the copy filters, colony side up, onto LB agar plates with 250 μg/ml chloramphenicol. These plates can be supplemented with 1 mg/ml uridine.[9,10] After overnight incubation at 37°, remove the filters, air dry for 1 hr, and treat for hybridization either by following the procedures recommended by the manufacturer or by using the directions for nitrocellulose filters in step 16 below.

In order to hybridize the filters, colonies must be lysed; cellular debris, protein, and RNA must be removed; and the plasmid DNA must be denatured and immobilized.[11] This is done, assembly-line fashion, by setting up four trays each large enough to accommodate four filters. Each tray is fitted with a piece of Whatman 3 MM filter paper at the bottom. Tray 1 should contain a 10% sodium dodecyl sulfate (SDS) solution used as a prewash for removing debris; tray 2, alkali (0.5 N NaOH, 1.5 M NaCl) for lysing bacteria and denaturing DNA; tray 3, a neutralization solution (0.5 M Tris–HCl at pH 8, 1.5 M NaCl); and tray 4, 2× SSC (see this volume [45] for recipe). In each case the Whatman 3 MM filter paper is made thoroughly wet with the appropriate solution. The idea is to expose the colonies on nitrocellular filters to the various solutions *through* the nitrocellulose. These filters are not to be submerged, that is the colony side should not be wetted directly.

16. Treat the filters, colony side up, as follows: tray 1, 3–5 min; tray 2, 5–15 min, tray 3, 5–15 min; tray 4, 5 min. The alkali step is critical; it should be continued until the colonies become smeared and gooey in appearance, indicating that they have lysed. A few minutes or so more or less in the other solutions will not alter the results. Finally, dry the filters in a single layer on dry Whatman 3 MM paper at room temperature for about an hour and bake at 80° between layers of Whatman 3 MM filter paper in a vacuum oven for 20 min. (Many investigators bake for 2 hr but as detailed in this volume [45], this is unnecessary.)

Note, after about 10 150-mm nitrocellulose filters have been processed, the Whatman filter paper in each tray and the liquid should be discarded in favor of fresh filter paper saturated with the appropriate solution.

17. Alternatively, one can float filters, colony side up, on a thin film of liquid (2.5 ml per 88-mm filter, 7.5 ml per 150-mm filter) directly on the lab bench instead of using Whatman 3 MM paper. This allows each filter to be treated with fresh solution. For nylon-based filters we float the filters on

[9] J. P. Gergen, R. H. Stern, and P. C. Wensink, *Nucleic Acids Res.* **8**, 2115 (1979).
[10] M. V. Norgard, K. Emigholz, and J. J. Monahan, *J. Bacteriol.* **1**, 270 (1979).
[11] W. D. Benton and R. W. Davis, *Science* **196**, 180 (1978).

0.5 M NaOH (2 × 2 min), followed by 1 M Tris–HCl at pH 7.4 (2 × 2 min). Between each treatment the filters are blotted on dry 3 MM paper for 1 min and at the end filters are air dried.

18. Because of the unusual amount of debris which continues to remain associated with the filters, the baked filters can be incubated for 2 hr at 42° in 50 mM Tris–HCl (pH 8), 1 M NaCl, 1 mM EDTA, and 0.1% SDS. Loose material may be gently removed with gloved hands. Prehybridize and hybridize colony-bearing filter DNA as described in this volume [45].

Alternative Procedure for Screening an Unamplified (or Amplified) Plasmid or Cosmid Library—Direct Screening

1. Titer recombinant bacteria by transforming *E. coli* with a small portion of the DNA to be cloned. Use several vector-to-insert DNA ratios. Plate several dilutions directly on LB plates with appropriate antibiotics. This step is equivalent to step 1 in the pervious section with the omission of soft agarose in the plating procedure.

2. Invert plates and grow the bacteria at 37° overnight. Count clones to obtain a titer. Choose the most efficient ratio of vector to insert and use the remaining DNA to transform *E. coli*. Scale up proportionally. Plate the recombinant bacteria directly on LB plates, aiming for 10,000 colonies per 150-mm plate. Grow the bacteria overnight at 37°, plates inverted. This is the master plate.

3. Replicate the master by placing a prewetted (Millipore) HAWP numbered nitrocellulose filter or the equivalent (see step 14 of the previous procedure) on the master plate. Avoid bubbles. Press the two together with a bent sterile glass rod.

4. Mark the filter by stabbing several times in an asymmetric pattern with a needle containing India ink. The ink left in the agar will aid in aligning plates and filters.

5. Peel off the replica. Incubate the master plate at 37° for a few hours to regenerate the colonies. Store sealed with Parafilm, inverted, at 4°. Place the replicas, colony side up, on fresh LB plates and incubate plates upside down at 37° overnight.

6. Process replica filters as in step 16 and its preceding paragraphs in the previous section.

Note, if contamination is a major problem in your locale, use only sterile nitrocellulose. Sterilize filters between sheets of Whatman 3 MM filter paper moistened in water. Wrap in aluminum foil to autoclave. Filters may be labeled with ballpoint ink before or after sterilization, after the final baking, or even when wet.

Once filters have been baked at 80° (step 16) they may be stored at room temperature, or if molds are prevalent in your area, desiccated at room temperature. If filters have been in contact with liquid, as in the note following step 16, they should be stored at −20° between the layers of plastic wrap with a hard surface, such as used X-ray film, serving as a support.

A similar procedure appears in this volume [18] for cosmids.

Amplification and Storage of Phage Libraries

Because phage infection of host bacteria is an extremely efficient procedure, whereas plasmid transformation of competent bacteria is grossly inefficient, phage libraries are stored as isolated phage particles in a bacteria-free suspension. The phage are viable for years in a chloroform-saturated buffer.

Chapter [17] in this volume contains a discussion of when to amplify a phage library and when to screen directly. Methods for amplification and storage of the library are also presented.

Preparation of Filters for Screening a Bacteriophage Library

The procedure for plaque screening that follows is similar to methods described previously for colony screening. We assume that a suspension of phage particles is available and that the titer has been determined.

1. Plate 10,000 to 20,000 plaques per 13.5-cm Petri dish as described in this volume [17].

2. Before transferring plaques to filters, cool the plates at 4° for 1 hr. Cold helps to prevent transfer of the soft agarose layer. (If agarose does stick to the filters, it can be removed later.)

3. Place a dry nitrocellulose filter on each plate. (Sterile filters are not necessary.) When the filter has become wet, the plaques have transferred.

4. Mark the filters as in step 4 of the alternate procedure and process them, plaque side up, as in step 16 of the first procedure, omitting tray 1 (10% SDS treatment). If soft agarose remains, it can be floated away during the denaturation treatment in tray 2.

5. Store plates, inverted, sealed with Parafilm at 4°. Filters are ready for prehybridization (this volume [45]).

Amplification of Plaques *in Situ*

Several filters may be prepared from the same plates without "regenerating" fresh plaques. However, it is occasionally useful to amplify plaques *in situ* to replace those that have been removed during the preparation of filters or to increase subsequent hybridization signals.

1. Prepare an overnight culture of plating bacteria (this volume [13 or 17]). You will need 400 ml of culture for 20 filters.

2. Collect bacteria by centrifugation and resuspend them in an equal volume of fresh LB + 10 mM MgSO$_4$. General methods for handling λ are found in this volume [13].

3. Label the filters and mark them asymmetrically, with a black ballpoint pen, on the side that will be in contact with the plaques.

4. Dip the filters in the bacterial cell suspension and allow them to air dry briefly.

5. Lay the filters on the surface of the plates containing plaques. Transfer the orientation marks to the agar plate. The techniques in step 4 of the alternate procedure can be employed.

6. Prepare additional filter copies, if desired. Be sure to transfer orientation marks from agar to filter. A light box is a useful aid here.

7. Lay copy filters, phage plaque side up, on fresh LB + Mg^{2+} plates and incubate, inverted, at 37° overnight.

During the overnight growth at 37° the plaques infect the growing *E. coli,* leading to a substantial amplification of phage DNA. After this amplification, it is usually not necessary to hybridize two sets of filters to avoid false positives.

8. Remove the filters from the plates, air dry for at least an hour and process filters as in step 16, omitting the 10% SDS treatment as described above. Store plates, inverted, sealed in Parafilm at 4°. The filters are ready for prehybridization (this volume [45]).

[45] Screening Colonies or Plaques with Radioactive Nucleic Acid Probes

By Geoffrey M. Wahl and Shelby L. Berger

Colony or plaque hybridization is a technique for screening replicated material *In situ* on filters with labeled probes.[1-5] The probes most com-

[1] M. Grunstein and D. S. Hogness, *Proc. Natl. Acad. Sci. U.S.A.* **72**, 3961 (1975).
[2] M. Grunstein and J. Wallis, this series, Vol. 68, p. 379.
[3] W. D. Benton and R. W. Davis, *Science* **196**, 180 (1978).
[4] D. Hanahan and M. Meselson, *Gene* **10**, 63 (1980).
[5] D. Hanahan and M. Meselson, this series, Vol. 100, p. 333.

monly used are nucleic acids or antibodies. Here we will describe techniques for using nucleic acids to analyze libraries generated in either phages or plasmids. The use of antibodies for screening libraries can also be found in this volume [50, 51].

A library is a mixture of clones constructed by inserting either cDNA or fragments of genomic DNA into a suitable vector. The term *library* implies the existence of large numbers of different recombinants, only one or a few of which are of immediate interest to the investigator. The desired clone is located by performing the following steps: (1) transfected bacteria or phage are grown on master plates (or filters) and replica plated; (2) the original plates called *master plates* are preserved while the replicas, hereafter called *filters,* are processed; (3) phage are disrupted or bacteria are lysed *in situ* on filters; (4) DNA is bound to the filter while RNA is hydrolyzed; (5) the resulting partially denatured DNA is hybridized to sequences able to bind specifically to the desired insertions. (6) Because the configuration of DNA on the filter replicas matches the configuration of live bacteria or phage on the master plates, DNA on replicas which binds to the probe (so-called positive signals) can direct the investigator to the bacterial colony or phage plaque from which the DNA was derived; (7) the positive colony or plaque is then purified and grown in quantity for further analysis.

Chapters [44] and [18] describe steps 1–4 for plasmid or λ libraries and cosmid libraries, respectively. Here we will focus on steps 5–7.

Colony hybridization is a rapid but inexact procedure aimed at calling attention to clones worthy of serious consideration. Falsely positive clones are therefore not uncommon. To some extent these can be reduced by the following: (1) use both negative control filters and, if possible, positive control filters; (2) screen duplicate filters of each master plate; and (3) prepare probes carefully.

To satisfy the requirements of point 1, it is advisable to include clones containing the vector without an insert or containing an irrelevant insert. The latter is particularly important when fragments bearing homopolymer "tails," usually composed of dG on one strand and dC on the other, are screened; the GC-rich regions on either end of the insert can hybridize to GC-rich probes and cause spurious positive signals. Thus, the use of known negative recombinants acts as a means for detecting unwanted cross-hybridization of the probe to vector and host DNA (which are also present) and also serves to establish the intensity of a background signal, one that should be ignored. Since intensities are relative, a genuine positive signal is needed for comparison. If there are no known positive clones, one can always clone the probe itself and create a positive recombinant. Such engineered positive colonies or plaques are rarely perfect because they may give more intense signals than the positive constructs in the library. For example, when using probes from one animal source to screen libraries generated from another, a genuine positive signal in the library will be less intense than the "synthetic" control owing to sequence divergence between species. Nevertheless, the synthetic positive can still be used to detect cross-reacting probes: there should be a dramatic difference between the signal obtained from the positive control and that obtained from an irrelevant clone.

The requirement stated in point 2 is designed to distinguish true positive signals from "spontaneous" spots, speckles, and smears that sometimes appear on the film used to identify positive signals on the DNA-bearing filters. Since whatever caused the false spot, be it a cosmic ray, a contaminated cassette or screen, or the static electricity released by crushing plastic wrap (see below), it is highly unlikely that both filters of a pair will have such artifacts in the identical location. Genuine positive clones should appear on both members of the duplicate pair. Clones that are positive on only one filter should be ignored.

Probe Preparation

The ratio of specific to nonspecific hybridization is directly related to the purity of the probe, hence the requirement stated in point 3. In this chapter, we will discuss ^{32}P-labeled nucleic acid probes exclusively. Nonradioactive probes, which can eliminate some of the disadvantages of radioactive probes, e.g., short half-life, high cost, hazardous nature, inconvenience, are not yet sufficiently sensitive for screening libraries for rare genes or cDNAs. Nevertheless, nick-translated DNA containing biotin-labeled nucleotides[6–8] and enzyme-linked nonradioactive assays[9] for detection may well be the technology of the future.

In this volume you will find a myriad of techniques for introducing ^{32}P into DNA or RNA. Any of these methods can be used to prepare probes for screening libraries. They are as follows, together with their location in this volume, respectively: end labeling RNA with ATP [6] or pCp [10]; tailing [10, 37]; introducing 5'-phosphates, filling in the staggered DNA ends created by digestion with restriction endonucleases or exonucleases [10]; nick translation (by far the most common) [9]; primer extension on

[6] D. J. Brigati, D. Myerson, J. J. Leary, B. Spalholz, S. Z. Travis, C. K. Y. Fong, G. D. Hsiung, and D. C. Ward, *Virology* **126**, 32 (1983).

[7] P. R. Langer, A. A. Waldrop, and D. C. Ward, *Proc. Natl. Acad. Sci. U.S.A.* **78**, 6633 (1981).

[8] P. R. Langer and D. C. Ward, *ICN-UCLA Symp. Mol. Cell. Biol.* **23**, 647 (1981).

[9] J. J. Leary, D. J. Brigati, and D. C. Ward, *Proc. Natl. Acad. Sci. U.S.A.* **80**, 4045 (1983).

RNA templates [33,46,66] or DNA templates [74]; preparation of enzymatically synthesized RNA [61]; preparation of M13 single-stranded and partially double-stranded probes [10]; and synthesis of double-stranded probes using random hexamers as primers [10]. Oligonucleotide probes are discussed separately [47–49].

Since nick-translated double-stranded probes are the most widely used, they deserve special attention. Best results are usually obtained with labeled fragments having a single-strand length of 500–1500 bases; they run as compact smears on alkaline agarose gels. Larger probes correlate with high backgrounds. It is therefore prudent to characterize all probes, particularly when valuable libraries are to be screened (see also this volume [9]).

Some forms of noise can be removed by filtering the hybridization mix containing the probe through a 0.45-μm nitrocellulose membrane. If the membrane removes impurities which bind to nitrocellulose strongly, background will be reduced. However, if the undesirable impurities are contaminant DNA molecules, the more drastic measures outlined below will be needed to remove them.

Regardless of the nature of the probe, the purity of the starting DNA or RNA from which it was made is of paramount importance. (The use of RNA probes is described elsewhere in this volume [61].) Consider the case of a nick-translated insert, excised with a restriction endonuclease and gel purified to remove plasmid sequences. Unless one has been exquisitely careful, the insert will *always* be contaminated with plasmid sequences which will also be nick translated. If one screens a library constructed in the same or a related plasmid vector, all the colonies in the library will hybridize to the probe or to the contaminants or to both. A large irrelevant colony may hybridize better, by virtue of the large amount of plasmid it contains, than a small colony containing a small amount of the desired recombinant plasmid. This problem can be alleviated by purifying the insert at least 2 times in a gel and, if possible, by using restriction enzymes which do not cleave the insert to cut the offending DNA into tiny fragments small enough to be removed by differential precipitation such as spermine precipitation (this volume [5]). It is sometimes useful to characterize the insert DNA by performing a Southern blot (this volume [61]) in which the DNA in question serves as both the tethered sample and the labeled probe. Contaminants appear as low-intensity bands or smears different in size from the intense signal derived from the insert hybridizing to itself. Attempts to "compete out" these DNA molecules by adding large amounts (0.1 to 1 mg/ml) of unlabeled plasmid sequences, in this example, to all prehybridization and hybridization solutions are usually not entirely satisfactory.

Reduction of Background Hybridization

Prehybridization of filters containing immobilized DNA for 2–24 hr with substances designed to preempt nonspecific nucleic acid binding sites on the solid supports was prescribed early on as an effective means of reducing background hybridization.[5] We have found that prehybridization times as short as 5 min (a convenient time to wet the filter completely with the prehybridization mixture) are usually sufficient. It is also possible to use the prehybridization mixture for the hybridization itself in Southern and Northern blots, avoiding the necessity of preparing two separate solutions. However, there is often sufficient bacterial debris in colony hybridizations to justify discarding prehybridization fluid in favor of a fresh solution. We have found that the most frequent cause of background is the probe itself and not the time of prehybridization. Background is usually not significantly reduced by minor modifications of the prehybridization solution as discussed above.

Methods for Hybridization

1. Prepare prehybridization mix (50–100 μl mix/cm^2 filter). For hybridization at 42° use 50% formamide, 1× to 5× Denhardt's reagent, 100–200 μg/ml denatured sheared heterologous DNA or tRNA, 5× SSPE or 5× SSC, 0.1% sodium dodecyl sulfate (SDS), and (optional) 10% dextran sulfate. It is often convenient to perform hybridizations in a warm room at 37° rather than at the slightly more stringent 42° when using formamide. For hybridization at 68° use 5× SSPE or 5× SSC, 1× to 5× Denhardt's reagent, 100–200 μg/ml denatured heterologous DNA or tRNA, and 0.1% SDS. SSC (1×) is 0.15 M NaCl, 0.015 M sodium citrate (pH 7.0). SSPE (1×) is 0.18 M NaCl, 0.01 M NaPO$_4$ (pH 7.7), 0.001 M EDTA. Please note that unlike SSPE, SSC has little buffer capacity near pH 7. Denhardt's reagent is 0.02% Ficoll, 0.02% bovine serum albumin, and 0.02% poly(vinylpyrrolidone) and is usually made up as a 100× solution. It is filter sterilized and stored frozen or at 4°. Thaw in a room temperature water bath with occasional agitation. Do not heat Denhardt's reagent concentrates!

2. Put gloves on. Never touch filters with bare hands. Have stored filters at hand. If the filters have not been washed free of debris (see the note following step 16 in the previous chapter [44]), do so at this time. This step is optional, particularly if the density of colonies or plaques on the master plates was low. (Filters are prepared as described in this volume [44 or 18].)

3. Nitrocellulose filters with denatured DNA from colonies or plaques affixed are prehybridized at the desired temperature for 5 min or longer. If

there are only a few filters, prehybridize in heat-sealable bags. Make sure there are no bubbles trapped between the filter and the solution. It is often convenient to prehybridize filters in bags lying flat at the bottom of a water bath with a weight anchoring one corner. Agitation is not necessary. When many filters must be handled, use either the heat-sealable bags specified above or a Petri or cell culture dish slightly larger in diameter than the filters. Place the prehybridization solution in the dish. Stack the filters in the dish by transferring them one at a time, using gloved hands to squeeze out bubbles. Seal the dish with Parafilm. Prehybridize in a warm room (37°) or incubator at an appropriate temperature with slow rocking

4. Denature the probe by adding NaOH to 0.1 N and incubate for 5 min at 37°. Neutralize with HCl and any convenient concentrated, neutral buffer when hybridizing in SSC solutions. For 5 ml of hybridization mix, the probe volume should be 100 μl or less. Alternatively, double-stranded DNA probes in water or Tris–EDTA buffer may be boiled for 5 min and quickly cooled in ice for 2 min to denature them. Single-stranded probes generally do not require denaturation but may be heated briefly (1 min, 70–100°) to melt secondary structure. The amount of probe used ranges from 1×10^5 to 1×10^7 cpm/ml mix depending upon the application. However, if large amounts of immobilized sequences are being probed (e.g., 100 ng of pure plasmid, or colonies containing plasmids), 500–1000 cpm/ml of the probe will hybridize sufficiently to produce a detectable signal in an overnight exposure. Under these conditions the tethered sequences, rather than the probe, are in excess. Extensive libraries with pinpoint-sized colonies require a minimum of 2×10^6 cpm/ml of a probe with a specific activity of 1×10^8 cpm/μg.

5. Hybridize the filters with the probe as follows. For filters in bags, one may simply add the denatured probe directly to the prehybridization mix and filter already in the bag. Reseal, mix well, and return the assembly to the water bath to hybridize. For many filters in a dish, add the denatured probe to fresh prehybridization mix in a new dish to make hybridization mix. Use 50–100 μl of mix/cm^2 filter. Then, with gloves, carefully transfer the filters one at a time to the probe-containing hybridization mix. Use the edge of the old dish as well as a piece of Whatman 3 MM filter paper to blot away or remove the prehybridization mix because it will dilute the probe during hybridization. Squeeze out bubbles between filters, as each is added to the stack. Seal the dish with Parafilm and hybridize, rocking, usually overnight. It is possible to hybridize 20 137-mm filters in less than 50 ml of hybridization mix. However, once in contact with the probe, the filters must never be allowed to dry. If they are not completely submerged, the rocking motion must rewet them continuously. Use precautions consistent with handling radioactive materials at all times (see this volume [3]).

6. Wash the filters. There are two philosophies for this procedure: several short washes at relatively high stringency; or longer washes under conditions at or near those used for hybridization. We present examples of both.

a. Remove the filters from the hybridization solution and wash sequentially in a large volume (200–500 ml) as follows: 2×5 min in $1\times$ SSPE ($1\times$ SSC), 0.1% SDS at 25°; 2×15 min in $0.1\times$ SSPE ($0.1\times$ SSC), 0.1% SDS at 42°; an empirically determined time in $0.1\times$ SSPE ($0.1\times$ SSC), 0.1% SDS at 50–70°.

b. Alternatively, place the hybridized filters in a large volume of $2\times$ SSC, 0.1% SDS. Wash for 15 min at room temperature. Repeat 3 or 4 times until the radioactivity on the filters remains constant and the radioactivity in the wash is undetectable when assayed with a Geiger counter. Transfer the filters to fresh prehybridization mix omitting Denhardt's reagent and heterologous DNA or tRNA from the solution. Wash 6 hr to overnight. Washes at higher stringency in $1\times$ SSC, 0.1% SDS or $0.1\times$ SSC, 0.1% SDS at 50–70° may then be applied.

At any time during the procedure after the room temperature washes, the filters may be examined by autoradiography. This is usually carried out as described below.

1. Wrap a used piece of X-ray film with plastic wrap. Prepare enough sheets of used film to accommodate all filters in a single layer. For orientation, place a small piece of tape in each corner and mark it with a phosphorescent pencil (New England Nuclear) or radioactive ink (India ink containing any convenient radioactive compound).

2. Transfer the wet filters to the used film. They adhere easily to the plastic wrap. Cover them with plastic wrap and autoradiograph with unexposed X-ray film, screens, and a cassette as detailed in this volume [7]. Do not let the filters dry. If, on developing the film, the signal is not sufficiently resolved from noise, continue the more stringent washes and reautoradiograph. Filters can be matched with their respective X-ray images using the signal generated by the phosphorescent markings in the corners of each film.

Filters and blots may be reused by removing probe as follows. Heat $0.1\times$ SSPE, 0.1% SDS to boiling, remove from heat, add to blot, and incubate for 15 min. Repeat. Follow with a brief wash at 25° with $0.1\times$ SSPE, 0.1% SDS to remove residual eluted probe. Alternatively, place blot in 0.1% SDS, 5 mM EDTA in a boiling water bath. Heat 15 min and repeat with fresh solution. Expose the film for the anticipated time of the next exposure to check if hybridized material has been removed.

Methods for Selecting and Purifying Positive Clones

Autoradiography of filters bearing DNA from plaques or transfected bacteria should result in signals which are intense (dark spots) relative to a weak background. If the master plates have a high density of recombinants, the size of each positive signal will correspond to a colony or plaque of the same size and may not be much larger than the head of a pin. If a pair of duplicate filters from each master plate has been made, spurious dark spots, those present on only one filter of a pair, can safely be ignored. All other positive signals should be investigated using the instructions below.

1. Circle positive spots with a red or blue marker.

2. Orient the autoradiograph with respect to the master plate. This is most easily done by transferring the orientation marks from the filter which has just been screened to the autoradiograph, with a red pen. (The red color makes it easy to distinguish an orientation mark from a black colony or plaque on the X-ray film.) Use the transferred red marks to align the film with the master plate. Remember, film has two sides. Therefore you have either the image or the *mirror* image of the master plate before you. If your orientation marks are indeed asymmetric, this difficulty will be easy to resolve; reverse the film if necessary. A light box aids visibility.

3. Once the film and the plate are aligned (film underneath the master plate), you should be able to see the circled spots on the film through the agar. Pick positive areas as detailed below.

For plaques

a. Use a sterile razor and forceps or the wrong end of a Pasteur pipet (the end to which a rubber bulb is usually attached) and remove a plug of agar containing the area of the phage that gave rise to the positive signal.

b. Drop the plug into 1 ml λ diluent (see this volume [17]). Add 100 μl chloroform, vortex, and incubate at room temperature for 30 min. Centrifuge at 3000 g for 3 min. Remove and reserve the supernatant fluid containing the positive phage, among many others.

For plasmids in bacteria

a. Touch a sterile wire loop to the area of the plate defined by the circle surrounding the positive signal. Make sure all bacteria within the circle are represented. (It will be necessary to smear the master plate at each positive signal to do the job.)

b. Transfer bacteria to LB broth containing the antibiotic corresponding to the resistance marker carried on the plasmid vector.

4. Plate bacteria or λ at several dilutions. Aim for 500–1000 colonies or plaques per 150-mm plate. Each positive area must be subjected to

these procedures. Use the methods in chapter [13] for titering both λ and bacteria containing plasmids.

5. Repeat the entire screening procedure using a plate with the required density of clones as the new master. Transfer to new filters, process the filters, and rehybridize. Positives should now represent 5–10% of the total plaques or colonies per plate and should superimpose with spots on film.

At this point, or after repeating the screening procedure one additional time, it should be possible to align a well-separated single colony or plaque with a positive signal. More than 50% should now be positive. This should be possible for each of the positive areas selected after the first screen.

6. When pure cultures have been obtained, they are grown for the purpose of isolating DNA for further analysis. These small-scale cultures are called *minipreps*. When preparing small cultures of phage, use the procedure for "plate lysates" to grow a stock of λ. After concentrating the phage with polyethylene glycol (PEG), phenol extract as described in this volume [13]. It will be necessary to scale down both the PEG method and the phenol extraction method. As a rule of thumb, an 82-mm plate will yield 10^8–10^{10} pfu. Therefore, phenol extract the PEG precipitate using an aqueous volume of 0.5 ml. (Note, the DNA need not be purified in a CsCl gradient.) For plasmid minipreps follow the instructions in this volume [13] for Small-Scale Plasmid Preparations.

[46] Isolation of Differentially Expressed Genes

By THOMAS D. SARGENT

Cell and tissue differentiation is thought to be the result of selective expression of the genome, and therefore genes that are differentially expressed, i.e., in a regionally, temporally, or environmentally specific way, are often of interest. This chapter reviews selection strategies, including "+/− screening" and "subtraction" approaches, which exploit the differential expression of such genes. These methods use probes derived from populations of mRNA molecules, as opposed to gene-specific reagents such as antibodies, oligonucleotides, or evolutionarily related sequences. The "+/−" procedure is designed to detect cDNA clones derived from mRNAs which are present (+) in one cell type and absent (−) or expressed at significantly lower levels in another cell type. The approximate limitations of detection are mRNAs representing 0.1% of the total

METHODS IN ENZYMOLOGY, VOL. 152

mRNA population. Probes derived from differentially expressed genes can be enriched by "subtraction" or "cascade" hybridization. Bulk cDNA from one cell type is hybridized to a sequence excess of mRNA from a related cell type. Thus sequences which are present in both cell types can be removed (subtracted) from the original cDNA preparation. The remaining unhybridized cDNA is highly enriched for differentially expressed gene sequences. RNAs representing only 0.01% of the total mRNA population have been successfully identified by this procedure. A difficulty arising from use of complex cDNA probes is the potential for a very large number of positives, from which a limited number of desirable clones must be selected. If clone verification involves tedious or time-consuming procedures, direct screening with mRNA-based probes may not be practical. This problem can be circumvented by choosing a source of RNA for cloning in which the desired gene is highly abundant, or alternatively by reducing the complexity of either the probe or the library, accomplished by use of subtraction hybridization.

Consideration of the abundance level of the mRNA sequence that is to be cloned is of critical importance in the selection of a cloning strategy. A typical eukaryotic cell contains approximately 1 pg of mRNA, equivalent to about 10^6 molecules, transcribed from about 15,000 different genes.[1,2] Most genes are members of the low abundance class, and are represented by about 20 mRNA copies per cell. A smaller number are expressed at higher mRNA levels (medium abundance), and a few mRNAs accumulate to a few percent of cellular mRNA, or even more in some cases. Genes in this last category (high abundance) are easiest to clone, because a simple duplicate screening approach can be used.

For this reason it is usually worth attempting to identify a tissue, developmental stage, culturing condition, etc. that maximizes the abundance of the desired mRNA sequence. If the relevant protein can be readily detected, then *in vitro* translation of various mRNA samples followed by immunoprecipitation, electrophoresis, and autoradiography or some other detection procedure can provide an estimate of abundance (see this volume [27, 31]). Even if a source of abundant expression is not discovered, such information is helpful in planning alternate strategies.

+/− Screening

If a gene is expressed as a very abundant RNA in one cell type and is rare or absent in another, it can be isolated with minimal effort by +/−

[1] B. P. Brandhorst and E. H. McConkey, *J. Mol. Biol.* **85,** 451 (1974).
[2] H. Latham and J. E. Darnell, *J. Mol. Biol.* **14,** 1 (1965).

screening. After selection of a mRNA source that contains a maximal level of the desired sequence, a cDNA library is prepared from this material. Any type of cDNA library is suitable, although plasmid vectors are probably preferable since these are somewhat more versatile than bacteriophage vectors. This library is then screened in duplicate with labeled cDNA made from mRNA containing the desired sequence and, in parallel, with labeled cDNA made from mRNA where this sequence is absent or much less abundant (see this volume [33]).

It is possible to estimate the number of clones that must be screened in order to be arbitrarily certain of encountering at least one copy of the desired clone. A statistical calculation can be made using the first term of the binomial expansion

$$P_0 = (1 - A)^N \tag{1}$$

and solving for N:

$$N = \log P_0 / \log(1 - A) \tag{2}$$

where P_0 is the probability of missing the desired clone, A is the fractional abundance of the relevant mRNA sequence, and N is the number of clones screened. For example, only about 500 clones need to be screened to be 99% certain of identifying at least one derivative of a gene represented by at 1% of total cellular mRNA. Since relatively few recombinants need to be screened, it is feasible to prepare orderly grids of clones by inoculating wells of microtiter dishes with individual colonies. This array can be replica plated onto nitrocellulose filters on nutrient agar plates, and processed for hybridization (see this volume [44, 45]). Clones that hybridize with positive but not negative probe are isolated.

Cloning Rare mRNAs: Subtraction Techniques

When a gene is differentially expressed but is not represented by an mRNA of abundance much greater than 10^{-3}, it becomes difficult to isolate it by simple $+/-$ screening. In such cases it is possible to increase the effective concentration of the desired sequence by subtraction hybridization.[3–5] This procedure is made possible by the fact that mRNA populations can be readily hybridized to complete kinetic termination with homologous cDNA. Briefly, mRNA from cell type A is used as a template to

[3] W. E. Timberlake, *Dev. Biol.* **78**, 497 (1980).
[4] T. D. Sargent and I. B. Dawid, *Science* **222**, 135 (1983).
[5] S. M. Hedrick, D. I. Cohen, E. A. Nielsen, and M. M. Davis, *Nature (London)* **308**, 149 (1984).

produce a radiolabeled tracer cDNA. The tracer cDNA is hybridized to a sequence excess of driver mRNA isolated from cell type B. The unhybridized tracer cDNA represents an enriched population of sequences expressed in cell A but not cell B. The following is an outline of the procedures used in preparation of subtracted cDNAs for use as probes and for cloning.

RNA Purification

Any RNA purification method that yields undegraded material should be adequate. However, if the subtracted cDNA is to be cloned, it is especially important to remove all traces of DNA from RNA preparations. In most cases contaminating genomic DNA will not be efficiently removed by the hybridization step and would be cloned as though it were cDNA. The result could be quite confusing. Two or three rounds of oligo(dT)-cellulose chromatography effectively eliminate DNA contamination of poly(A)$^+$ RNA. Subsequent treatment with RNase-free DNAse I or DNase in the presence of vanadyl complexes (this volume [21]) may be used in preparation of the driver RNA and is probably advisable when the RNA has been isolated from tissues with larger amounts of DNA relative to poly(A)$^+$ RNA content. It may be desirable to exclude nuclear RNA or nonpolysomal cytoplasmic RNA sequences from the driver and/ or the cDNA template RNA preparations. This will vary depending upon the circumstances, but should be carefully considered; extremely rare and highly complex RNA sequence subpopulations may be substantially enriched in the selective library. Poly(A)$^+$ RNA, which must be used in order to carry out the subtraction hybridization (see below), may be selected by three cycles of binding to oligo(dT)-cellulose (Bethesda Research Laboratories) in 1 M NaCl, 0.1 M Tris–HCl (pH 7.4), 1 mM EDTA at room temperature and elution with 1 mM Tris–HCl (pH 7.4), warmed to 50°. RNase-free buffers can be easily prepared using new bottles of reagents and deionized water, followed by two cycles of filtration through a nitrocellulose membrane (e.g., Millipore), which binds RNase and other trace proteins. Finally, it is very helpful to have undegraded RNA, both for the usual reasons and also because degraded RNA will be relatively impoverished in 5'-proximal sequences after oligo(dT)-cellulose chromatography, and this could result in false enrichment of such sequences in cDNA subtracted with such material.

Synthesis of High Specific Activity cDNA

All of the procedures for synthesizing tracer cDNAs can be found in this volume [33, 66]. If cDNA is to be used only as a hybridization probe, it should be synthesized to very high specific activity, $>10^8$ cpm/μg. If the cDNA will be used for cloning, lower specific activities are appropriate, e.g., 10^6 cpm/μg. When preparing high specific activity tracer cDNA, it is necessary to use rather low concentrations of dNTPs in the synthesis reaction. This usually results in cDNA that is much smaller than the average template mRNA, due to premature termination of reverse transcription. It is possible that an artificial bias for 3'-proximal sequences may occur if oligo(dT) is used as the synthesis primer. This bias can be circumvented by priming instead with random oligodeoxynucleotides.[6] The resulting cDNA will not necessarily be much longer, but it will contain 5' mRNA sequences as well as those closer to the 3' ends. Random primer cDNA can be synthesized by the following conditions: 5 mCi/ml [^{32}P]dCTP (3000 Ci/mmol), 20 μM unlabeled dCTP, 200 μM each dATP, dGTP, and TTP, 20 mM dithiothreitol, 50 mM Tris–HCl (pH 7.5), 75 mM KCl, 3 mM MgCl$_2$, 10 μg/ml poly(A)$^+$ RNA, 300 μg/ml random primer,[6] and 6000 U/ml MMLV reverse transcriptase (MMLV, mouse mammary leukemia virus, from Bethesda Research Laboratories). Following incubation at 37° for 1 hr, the reaction is halted and RNA hydrolyzed by addition of an equal volume of 0.6 N NaOH, 20 mM EDTA and incubation at 65° for 30 min, followed by chromatography in Sephadex G-50 in NETS buffer [0.1 M NaCl, 10 mM Tris–HCl (pH 7.6), 0.1 mM EDTA, 0.1% sodium dodecyl sulfate (SDS)], to remove unincorporated label. The specific activity of the resulting cDNA is 4×10^8 cpm/μg, and 1–2×10^8 cpm of probe should be derived from 1 μg of mRNA.

Hybridization

WHen driver poly(A)$^+$ RNA is mixed with tracer cDNA and incubated under the appropriate conditions, RNA–cDNA hybrids will form. If the driver poly(A)$^+$ RNA is present in large sequence excess relative to the homologous cDNA, then even when all of the cDNA has been rendered double stranded, most of the homologous RNA will remain single stranded. Since the driver RNA reactant in this instance is not significantly consumed, its concentration remains approximately constant, and the kinetics of such a reaction are effectively first order. This is significant because a first-order hybridization proceeds faster than a second-order reaction with the same reagents. If the driver RNA is very similar, but not identical to the tracer cDNA, it is feasible to carry out a hybridization reaction that will result in virtually all of the homologous cDNA sequences becoming double stranded while the cDNA molecules that are

[6] J. P. Dudley, J. S. Butel, S. H. Socher, and J. M. Rosen, *J. Virol.* **28,** 743 (1978).

not represented in the RNA will remain single stranded. Note that if an mRNA sequence is only a few times more abundant in the positive relative to the negative source, it may not be possible to isolate it by subtraction, since use of a small excess of negative mRNA to positive cDNA would result in deviation from first-order kinetics and failure to terminate. The RNA to cDNA ratio should be at least $10:1$, so the difference in abundance of the desired sequences must be greater than 10-fold, i.e., at least 10 times more of the desired mRNA in positive cells than in negative cells.

The progress of a pseudo-first-order reaction is described by the following equation:

$$D/D_0 = e^{-kR_0 t} \tag{3}$$

where D is the remaining single-stranded cDNA, D_0 is the total cDNA, k is the pseudo-first-order rate constant (see below), R_0 is the RNA concentration, and t is the elapsed time. This equation is pseudo-first order due to the presence in the exponent of a concentration term, which is irrelevant in true first-order reactions, such as radioactive decay.

The rate constant k reflects the probability that the average cDNA molecule will collide with a homologous RNA molecule, and therefore depends on the complexity, i.e., the number of different RNA sequences in the population. Analysis of simple RNA excess hybridization reactions[7] has yielded the following relationship between complexity of the excess RNA in nucleotides (C) and the rate constant k expressed in liters/mol-sec:

$$k = 10^6/C \tag{4}$$

This relationship is defined for hybridization of cDNA ~500 nucleotides long occurring at $T_m - 25°$ in the presence of $0.18\ M$ Na^+. For example, a typical cell with 15,000 different mRNAs, all equally abundant and an average of 2000 nucleotides in length, has a complexity C of $15,000 \times 2000 = 3 \times 10^7$ nucleotides. The rate constant of a reaction "driven" by this RNA population will be $10^6/(3 \times 10^7) = 3.3 \times 10^{-2}$. The low abundance mRNA class, which is the slowest to hybridize, usually comprises about 20% of the total cellular mRNA mass, and therefore the effective rate constant for hybridization of this class is $0.2 \times 3.2 \times 10^{-2} = 6.7 \times 10^{-3}$.

Using this value for k it is possible to calculate the $R_0 t$ required for an arbitrarily complete hybridization reaction. In order to attain 99.9% completion, solve Eq. (3) for $R_0 t$:

$$D/D_0 = 0.001 = e^{-kR_0 t}$$

[7] G. A. Galau, R. J. Britten, and E. H. Davidson, *Proc. Natl. Acad. Sci. U.S.A.* **74**, 1020 (1977).

so

$$\ln 0.001 = -kR_0t$$
$$(-6.91)/(-6.7 \times 10^{-3}) = R_0t = 1030$$

Thus a hybridization of cellular mRNA to cDNA taken to a R_0t of about 1000 should theoretically reach 99.9% completion, and any tracer cDNA remaining single stranded after such a reaction is probably not homologous to the driver mRNA.

R_0t is formally expressed in units of (moles per liter) times seconds. Conversion to the more familiar (micrograms per microliter) times hours is accomplished by dividing by 10.3; R_0t 1000 is equivalent to 97 $(\mu g/\mu l)$-hr. In other words, the above reaction will be terminated in 97 hr at an mRNA concentration of 1 $\mu g/\mu l$, or 19 hr at an mRNA concentration of 5 $\mu g/\mu l$. The rate of hybridization can be accelerated by carrying out the incubation in higher salt concentrations.[8,9] By using as a hybridization buffer 0.12 M NaH$_2$PO$_4$ (pH adjusted to 6.8 with concentrated NaOH), 0.82 M NaCl, 1 mM EDTA, 0.1% SDS, the effective rate is increased about 2- to 3-fold, so in the above example, termination would be reached in 6.5 hr at 5 $\mu g/\mu l$.

In a typical experiment, an RNA to cDNA ratio of 30:1 is utilized, and thus, in theory, any cDNA sequence present at one part per 30 parts or higher level in the driver RNA would be hybridized, and any cDNA sequence absent from the driver RNA or present at a level below one-thirtieth of its concentration in the cDNA would remain at least partially single stranded. The RNA to cDNA ratio should not be calculated using the actual mass of tracer cDNA but rather using the mass of the template RNA used to prepare the tracer cDNA. This is based on the assumption that all poly(A)$^+$ mRNA molecules serve as templates for cDNA synthesis initiation, but not all are copied into full-length cDNA. For highly efficient cDNA synthesis reactions, this is a minor factor, but if most of the cDNA is prematurely terminated, this could lead to a large error in calculating the molar ratio. Hybridization rates are also affected by the size of the reactant molecules. If either the cDNA or the mRNA is degraded to less that 200–300 nucleotides, then appreciable rate reductions will result. Presence in the cDNA of extremely rare sequences, such as nuclear RNA or rare mRNAs from a minor cell type, can delay termination, which will lead to contamination of the subtracted cDNA with sequences that are not necessarily differentially expressed, but may be merely slow to react with their complementary RNA. It is probably prudent to carry out subtraction hybridization to the maximum practical

[8] R. J. Britten, D. E. Graham, and B. R. Neufeld, this series, Vol. 29, Part E, p. 363.
[9] J. Van Ness and W. E. Hahn, *Nucleic Acids Res.* **10**, 8061 (1982).

extent, i.e., increase the incubation time to 20–24 hr. Incubation times longer than this will tend to exacerbate rather than remedy the size retardation problem, due to thermally induced degradation of RNA and, to a lesser extent, cDNA. It is necessary to have high mRNA concentrations in the reaction, requiring the use of poly(A)$^+$ RNA. Highly purified RNA is not difficult to dissolve at several micrograms per microliter, and if a large amount of material is available, a 10 μg/μl stock can be prepared and combined with cDNA and concentrated buffer. If only limited quantities of mRNA can be obtained, setting up the hybridization reaction can be facilitated by coprecipitating the cDNA with the mRNA, and dissolving in water at a few tenths of a microgram per microliter. The nucleic acid mixture is then concentrated to 10 μg/μl by repeated extraction with equal volumes of 2-butanol in a microfuge tube treated with dichlorodimethylsilane, followed by one chloroform extraction to remove residual alcohol. An equal volume of 2× reaction buffer (see above) is then added. This procedure eliminates the difficulties associated with dissolving samples in very small volumes. The most convenient way to carry out the incubation is to place the reactants in a silanized screw-capped microfuge tube which is then topped off with paraffin oil, closed, and submerged in a 65–70° water bath. Following incubation, the paraffin oil is removed, and single- and double-stranded cDNAs are separated chromatographically.

Hydroxylapatite Chromatography

Hydroxylapatite is a form of insoluble calcium phosphate that binds nucleic acids.[8] Double-stranded molecules are bounds more tightly than single stranded, and both can be released by treatment with phosphate buffers. Chromatography is usually carried out in a water-jacketed column maintained at 55–60°, containing approximately 0.5 g hydroxylapatite (HTP grade, Bio-Rad) per 100 μg RNA. A terminated hybridization reaction is diluted into 0.5 ml of 0.12 M NaH$_2$PO$_4$ (pH 6.8), and applied to the column. If necessary, the flow rate can be increased by applying gentle air pressure using a syringe fitted with a rubber stopper inserted into the column top. Single-stranded cDNA will pass through the hydroxylapatite and cDNA–mRNA hybrids will bind. Residual single-stranded material is recovered by washing the column with several volumes of 0.12 M NaH$_2$PO$_4$. Most of the mRNA will remain bound to the column under these conditions, and carrier DNA (see Appendix, this article, for method of preparation) may be added at this step to minimize losses. If desired, the double-stranded material can then be eluted with 0.48 M NaH$_2$PO$_4$. The single-stranded fraction is sometimes contaminated with cDNA degradation products that may have failed to hybridize due to their small size.

Such material can be removed by concentrating the single-stranded material approximately 10-fold by extraction with 2-butanol followed by chromatography on Sepharose CL-6B, in NETS buffer, collecting the excluded peak. After removal of low molecular weight contaminants, the cDNA is ready to be used to probe libraries.

Screening with Subtracted cDNA

Because subtracted cDNA contains only sequences of interest, and usually represents a small portion of the unfractionated cDNA, a few hundred thousand cpm of it can be used to hybridize to plasmid or phage libraries. This reduces background correspondingly, allowing long exposure times and increased sensitivity. By this method it is possible to detect rare mRNAs, on the order of 0.01% or less of the total cellular mRNA content, provided that the subtraction removes >95% of the original cDNA. Note that if the two mRNA populations used in the subtraction differ by more than 5–10%, it may be unproductive to use this procedure. The amount of labeled cDNA needed to screen a library depends on the abundance of the sought mRNA species, which determines the number of clones that need to be screened (see above) and the signal that will be obtained with a given amount of probe. For an mRNA that is 0.01% of the total, 50,000 clones should be screened, which can be distributed on two or three large filters and can be hybridized in 15–20 ml. It would be desirable to start with at least 10^8 cpm of cDNA prior to subtraction. This much cDNA can be obtained from 0.5 μg of poly(A)$^+$ mRNA by using the random primer conditions (see above), and should be subtracted with at least 10 μg of driver mRNA.

Cloning Subtracted cDNA

If the desired mRNA sequences are present at very rare levels, or if numerous (i.e., a few hundred) different genes are to be studied, then it may be advisable to prepare a library of clones from the subtracted cDNA. The single-stranded molecules that result from subtraction and hydroxylapatite chromatography can be cloned by conventional procedures designed for unfractionated cDNA. Serious difficulties can be expected due to the small amount of cDNA mass that is usually available, in most cases only a few nanograms. Special care must be taken to avoid losing such trace quantities. Of course, it is not possible to add carrier DNA to the cDNA as may be done when the cDNA is for probe purposes only. Procedures that minimize handling of the cDNA are desirable. (For details of cDNA cloning procedure, see [32, 34] this volume.)

Preparation of a subtracted library is technically very challenging, but gives a significant advantage regarding the sensitivity of screening. Since only a few different clones exist in the library, a large fraction of its complexity can be screened by preparing DNA from a manageably small number of clones, perhaps only a few dozen. This DNA can be radiolabeled and used to probe RNA gel or dot blots containing samples from various sources. Extremely rare RNAs can be easily detected by such methods. Clones can also be screened by sequence determination, hybrid-selected or arrested translation, or genomic blot analysis.

Appendix: Preparation of Carrier DNA

1. Dissolve 1 g DNA (e.g., salmon testes DNA, Sigma D-1626) in 50 ml 0.1 N NaOH. Add 1.5 ml concentrated HCl, and mix quickly. The DNA will precipitate immediately, and should not be stirred more than a few seconds to prevent formation or a large aggregate.

2. Incubate at room temperature for 20 min to partially depurinate the DNA. Add 2 ml 10 N NaOH (OH$^-$ concentration to 0.1 N) and stir until the DNA redissolves completely.

3. Incubate at 65° for 30 min. This hydrolyzes the DNA to a broad size distribution, from 250 to around 1000 nucleotides. Cool to room temperature, and add SDS to 0.5%.

4. Extract twice with phenol–chloroform (1 : 1).

5. Add 7 ml 3 M sodium acetate (pH 4.5), mix, and then add 60 ml 2-propanol.

6. Chill 15 min in dry ice to precipitate DNA.

7. Centrifuge to recover precipitate, which is then washed with 70% ethanol, decanted (do not dry the DNA), and finally dissolved in 50 ml 25 mM NaOH.

8. Store at 5° in the 25 mM NaOH to prevent renaturation. Do not freeze DNA in NaOH.

[47] Oligonucleotide Probes for the Screening of Recombinant DNA Libraries

By R. Bruce Wallace and C. Garrett Miyada

There are several situations in which the logical screening strategy is to use a synthetic oligonucleotide as a hybridization probe. When a portion of the amino acid sequence of the protein is known, an oligonucleotide probe can be designed based on this information. This idea was first proposed[1,2] when it was determined that oligonucleotides hybridize to their complementary sequence with a high degree of specificity. Under appropriate conditions only duplexes will form in which all of the nucleotides are base paired, while mismatched duplexes will not. Thus, it is possible to design an oligonucleotide probe that is a mixture of all possible coding sequences for a given amino acid sequence. Only one of the oligonucleotides in the mixture is complementary to the coding region of the protein. All of the other sequences in the mixture are only capable of forming mismatched duplexes with this region.

Oligonucleotides of unique sequence are also useful for screening recombinant DNA libraries. The hybridization specificity of oligonucleotide probes allows one to use unique sequence probes to screen for genomic clones or cDNAs encoding a specific member of a multigene family,[3] to screen for a new allele when the sequence of one allele is known, to screen for a specific region of a gene, to screen for specific mutants created by site-directed mutagenesis,[4,5] or to screen libraries with probes whose sequence represents a consensus coding sequence.[6]

This chapter deals with procedures for the use of oligonucleotides as hybridization probes including probe design, labeling, and hybridization to colonies, phage plaques, DNA, and RNA. It does not deal with oligonucleotide synthesis.

Screening Strategy

Oligonucleotides have several unique properties which make them suitable as probes for screening recombinant DNA libraries. The major unique property is the hybridization specificity discussed above. There are also some limitations of oligonucleotides which should be considered when designing experiments which use these probes. The major limitation when using mixed oligonucleotide probes is that often there is no obvious positive control to use. Oligonucleotides have a tendency to bind nonspe-

[1] R. B. Wallace, J. Shaffer, R. F. Murphy, J. Bonner, T. Hirose, and K. Itakura, *Nucleic Acids Res.* **6,** 3543 (1979).

[2] R. B. Wallace, M. J. Johnson, T. Hirose, T. Miyake, E. H. Kawashima, and K. Itakura, *Nucleic Acids Res.* **9,** 879 (1981).

[3] D. H. Schulze, L. R. Pease, Y. Obata, S. G. Nathenson, A. A. Reyes, S. Ikuta, and R. B. Wallace, *Mol. Cell. Biol.* **3,** 750 (1983).

[4] R. B. Wallace, P. F. Johnson, S. Tanaka, M. Schold, K. Itakura, and J. Abelson, *Science* **209,** 1396 (1980).

[5] R. B. Wallace, M. Schold, M. J. Johnson, P. Dembek, and K. Itakura, *Nucleic Acid Res.* **9,** 3647 (1981).

[6] R. Lathe, *J. Mol. Biol.* **183,** 1 (1985).

cifically to noncomplementary DNA sequences. This is probably due to an unavoidable low degree of homology of short oligonucleotides to other DNA sequences[6] and is a particular problem with probes shorter than 14 bases long. Thus, one is often faced with the problem of falsely identifying clones as positive due to the lack of an appropriate control. Negative controls should always be used. When possible, the specificity of the probe should be determined, for example by using it in a Northern blot experiment (see Ito *et al.*[7] for example, and this volume [61]).

When using mixed oligonucleotides as probes for cDNA clones, it is best to choose the longest probe possible, preferably 20 or longer, as the primary probe. If one has additional protein sequence information, it is desirable to synthesize a second probe to another region. This probe can be used to rescreen positive clones obtained with the first probe, either in a colony screening experiment, or by Southern blotting (see this volume [45, 61]) or dried gel hybridization with DNA obtained from miniplasmid or miniphage preparations. Clones positive with the two independent probes are very likely to be the desired clone.

Probe Design

In the design of oligonucleotide probes for any application, several aspects of the DNA sequence must be considered. These include oligonucleotide length, G + C content, self-complementarity, and complexity (in the case of mixed oligonucleotide probes). These aspects shall be considered separately.

Length

Basically, the length of the oligonucleotide probe determines its hybridization specificity. The longer a sequence, the more likely it is to be unique among the collection of sequences and oligonucleotide is used to probe and the less likely it will bind nonspecifically to other sequences. The length of the oligonucleotide is limited more by need and appropriate information than by synthetic considerations, since it is possible to synthesize oligonucleotides in excess of 100 nucleotides in length on most of the modern DNA synthesizers.

For a genome whose nucleotides are distributed randomly, the expected frequency of occurrence (*a*) of an oligonucleotide sequence is

$$a = (g/2)^{G+C} \times [(1 - g)/2]^{A+T} \tag{1}$$

[7] H. Ito, S. Yamamoto, S. Kuroda, H. Sakamoto, J. Kajihara, T. Kiyota, H. Hayashi, M. Kato, and M. Seko, *DNA* **5**, 149 (1986).

where g is the fractional G + C content of the genome, and G, C, A, and T are the number of guanines, cytosines, adenines, and thymines, respectively, in the oligomeric sequence.[8] For a genome of size N (in nucleotide pairs) the expected total number n of oligonucleotide-complementary sites is

$$n = 2Na \tag{2}$$

since the oligonucleotide could be complementary to either DNA strand. As stated by Nei and Li,[8] n follows a Poisson distribution with mean of $2Na$.

In addition to specificity, oligonucleotide length determines duplex stability.[1] This is particularly important for oligonucleotides 11 bases long or shorter. While these sequences might be expected to be unique on statistical grounds, they have been found to give unacceptable results as probes for cloned DNAs.[9] Therefore, oligonucleotide probes should be made as long as possible.

G + C Content

The G + C content of an oligonucleotide is also important in duplex stability.[10] While no detailed thermodynamic study has been done, Suggs *et al.*[10] have determined an empirical effect of oligonucleotide G + C content on duplex stability. By comparing the dissociation of several oligonucleotide DNA duplexes as a function of temperature, a parameter T_d (the temperature at which half of the duplex is dissociated) was determined.[1] When the effect of G + C content was taken into account, an empirical relationship was derived

$$T_d = 2° \text{ (number of A + T residues)} + 4° \text{ (number of G + C residues)} \tag{3}$$

for duplexes 11–23 bases long in 1 M Na$^+$. This relationship is only valid within these ranges of length and is meant to serve as a guide. Smith[11] has presented a more detailed analysis of the effect of length on duplex stability. The above relationship is useful for estimating the effects of length

[8] M. Nei and W.-H. Li, *Proc. Natl. Acad. Sci. U.S.A.* **76,** 5269 (1979).

[9] S. V. Suggs, R. B. Wallace, T. Hirose, E. H. Kawashima, and K. Itakura, *Proc. Natl. Acad. Sci. U.S.A.* **78,** 6613 (1981).

[10] S. V. Suggs, T. Hirose, T. Miyake, E. H. Kawashima, M. J. Johnson, K. Itakura, and R. B. Wallace, *in* "Developmental Biology Using Purified Genes" (D. D. Brown, ed.), p. 683. Academic Press, New York, 1981.

[11] M. Smith, *in* "Methods of RNA and DNA Sequencing" (S. M. Weissman, ed.), p. 23. Praeger, New York, 1983.

and G + C content on duplex stability, as well as for determining an appropriate hybridization temperature (see below).

In addition to effects on the duplex stability, G content of oligonucleotides has an effect on the DNA synthesis itself, particularly on purification of the oligonucleotide. G-rich oligonucleotides are notoriously difficult to purify. Although purification schemes have been devised,[12] it is best to avoid the problem when possible. This is often achieved by simply synthesizing the complementary sequence which would serve the same purpose for many applications. Alternatively, synthesis of an oligonucleotide complementary to a different region is required.

Self-Complementarity

Self-complementarity of oligonucleotides, like G content, creates problems for oligonucleotide purification.[12] Unlike G richness, it is obviously not possible to avoid self-complementarity by synthesis of the complementary sequence. It has not been determined what effect, if any, oligonucleotide self-complementarity would have on the efficiency of hybridization to a complementary DNA or RNA sequence. However, it is possible that self-complementarity would affect 5' labeling and should be avoided whenever possible.

Complexity

When synthesizing a mixed-sequence probe, the complexity (number of different oligonucleotides in the mixture) must be taken into account. Increases in the mixture complexity result in two effects, a decreased hybridization specificity and a decreased abundance of the single correct probe in the mixture. No detailed studies have been done to determine the maximal acceptable complexity of the mixed probes. Wherever possible, the complexity should be minimized. Single mixtures as complex as 384 different sequences have been used successfully.[13] However, complex probe regions can be handled by the synthesis of two or more pools of mixtures for the same region. A recent good example of this approach is the cloning of the cDNA for rabbit tumor necrosis factor (TNF) by Ito *et al.*[7] In this study a region of protein sequence was encoded by 128 different possible nucleotide sequences. Five pools were synthesized, and each pool hybridized to a Northern blot of RNA isolated from rabbit macrophages which had been stimulated to produce TNF. The pool which gave

[12] M. D. Edge, A. R. Greene, G. R. Heathcliffe, P. A. Meacok, W. Schuch, D. B. Scanlon, T. C. Atkinson, C. R. Newton, and A. F. Markham, *Nature (London)* **292,** 756 (1981).
[13] A. S. Whitehead, G. Goldberger, D. E. Woods, A. F. Markham, and H. R. Colten, *Proc. Natl. Acad. Sci. U.S.A.* **80,** 5387 (1983).

the strongest hybridization signal was used as the hybridization probe in the subsequent cloning experiment. Clearly, antisense oligomers must be used for analysis of mRNA.

Hybridization Conditions

The G + C content and the length of an oligonucleotide affect the thermal stability of the duplex formed (above). The parameter T_d is also useful for determining the appropriate hybridization temperature. Suggs et al.[10] have suggested that hybridization should be done at 2–5° below the calculated T_d such that duplexes with one or more mismatched base pairs will not form.

One aspect of oligonucleotide hybridization that is little appreciated is the kinetics of the reaction. Since synthetic oligonucleoides are of low sequence complexity and are available in large amounts, hybridization rates are much higher than those normally encountered in the molecular biology laboratory. In the most typical application of oligonucleotide probes, a labeled oligonucleotide is hybridized with DNA immobilized on a solid matrix. The oligonucleotide is present in vast excess with respect to the immobilized DNA sequences. Hybridization kinetics are essentially first order with respect to oligonucleotide concentration. The rate constant (k) as a function of length and complexity of the probe has been described by Wetmur and Davidson[14] as

$$k = 3 \times 10^5 \, L^{0.5} \, N^{-1} \quad \text{(liters per mole of nucleotides per second)} \quad (4)$$

where L is length and N is complexity of the probe in nucleotides for hybridizations done in 1 M Na$^+$. Since the reaction is first order, then

$$t_{1/2} = (\ln 2)/kC \quad (5)$$

where C is the concentration of probe in moles of nucleotides per liter. As an example, for a 15-base long oligonucleotide hybridized at 0.01 μg/ml, the hybridization is half complete in about 4 min. Wallace et al.[1] have shown, however, that the actual rate of hybridization is about 3–4 times slower than the calculated rate. Thus in the above example, the half-time of hybridization is 12–16 min. (See also this volume [43] for a more general discussion. The rate constant k is defined in Eq. (5) under "standard" conditions.)

Reagents

Denhardt's solution: 0.02% bovine serum albumin, 0.02% poly(vinylpyrrolidone), 0.02% Ficoll

[14] J. G. Wetmur and N. Davidson, J. Mol. Biol. **31**, 349 (1968).

Kinase buffer: 67 mM Tris–HCl (pH 8.3), 10 mM MgCl$_2$, 10 mM dithiothreitol

NET: 0.15 M NaCl, 15 mM Tris–HCl (pH 8.3), 1 mM EDTA

SSC: 0.15 M NaCl, 0.015 M sodium citrate

SSPE: 180 mM NaCl, 10 mM sodium phosphate, 1 mM Na$_2$EDTA (pH 8)

TBE running buffer: 89 mM Tris base, 89 mM boric acid, 2 mM EDTA

TE: 10 mM Tris–HCl (pH 8.0), 1 mM EDTA

Procedures

Determination of Oligonucleotide Concentration

Like other nucleic acids, oligonucleotides are measured spectrophotometrically. However, since the base composition of different synthetic sequences can vary widely, it is necessary to calculate the molar extinction coefficient for the particular sequence to permit determination of the concentration. The molar extinction coefficient at 260 nm (pH 8.0) is calculated by summing the contribution of each base: G, 12,010; A, 15,200; T, 8400; and C, 7050. The concentration of a solution of the oligonucleotide can then be determined by measurement of the absorbance of the solution at 260 nm. Solutions of oligonucleotides should be made in TE and stored frozen.

Labeling of Oligonucleotides

Oligonucleotides synthesized by the phosphotriester or the phosphoramidite approaches contain a free 5'-OH. There are two general methods of radiolabeling these molecules, labeling by phosphorylation of the 5'-OH with [^{32}P]phosphate and by primer extension.[15] The later approach will not be discussed here because it is not generally useful for mixed oligonucleotide probes.

The oligonucleotide which has a 5'-OH is labeled by transfer of the [^{32}P]phosphate from [γ-^{32}P]ATP using T4 polynucleotide kinase. A typical kinase reaction contains in 10 μl of kinase buffer: 18.5 pmol oligonucleotide, 30 pmol [γ-^{32}P]ATP (>7000 Ci/mmol, New England Nuclear), 5–6 units of polynucleotide kinase. The reaction is incubated at 37° for 30 min. An equal volume of deionized 98% formamide containing 0.15% bromphenol blue and 0.15% xylene cyanole is added; the sample is heated 5 min at 95°, and then subjected to electrophoresis on a 14.5% acrylamide, 0.5% bisacrylamide gel (15 cm × 30 cm × 1 mm) containing 7 M urea in

---

[15] A. B. Studencki and R. B. Wallace, *DNA* **3**, 7 (1984).

TBE buffer. Electrophoresis is done at 27.5 mA until the bromphenol blue reaches the gel bottom. The labeled oligonucleotide is located by autoradiography (see this volume [7, 8]), the radioactive band excised, and the oligonucleotide eluted by soaking the gel slice in two 300-μl changes of TE over a period of at least 12 hr at 37°. This procedure removes the excess [γ-^{32}P]ATP and separates the phosphorylated oligonucleotide from any remaining unlabeled probe.

If the oligonucleotide has been previously purified by gel or HPLC, the unincorporated [γ-^{32}P]ATP may be removed by chromatography on DE-52 cellulose (Whatman). After the kinase reaction, the labeled oligonucleotide is diluted 10-fold in TE. The sample is then applied to a small column of DE-52 (bed volume, approximately 0.2 ml), previously equilibrated with TE. The column is then washed with 5 bed volumes of TE followed by 5 bed volumes of 0.2 M NaCl in TE, which removes the unincorporated label. The labeled oligonucleotide is then eluted with 0.5 ml of 1 M NaCl in TE and used directly in hybridization experiments. Kinased oligonucleotides prepared by either method may be used for at least 1 week when stored at −20°.

Colony Screening

A typical hybridization procedure for screening filter replicas from a recombinant DNA library is shown in Table I. Colonies containing recombinant plasmids are transferred to filters as described in this volume [44] or they can be transferred to Whatman 540 filter circles and prepared for hybridization as described by Gergen et al.[16] In the case of nitrocellulose filters a prehybridization step is included as described below for plaque screening. In the case of the cellulose filters, the filters are first rinsed in a solution containing 6× NET, 5× Denhardt's solution, and 0.1% sodium dodecyl sulfate (SDS). The filters are hybridized in the above solution containing 1 × 10^6 cpm/ml labeled oligonucleotide (or up to 1 × 10^7 cpm/ml for mixed probes) at the appropriate temperature, T_d − 5° for unique sequence probes and T_{dmin} − 2° for mixed probes, for 2 hr. The filters are then washed four times for 5 min each in 6× SSC at 30–40° below the hybridization temperature. The filters are then exposed to Kodak XAR-5 film with two Cronex Lightning Plus intensifier screens for 1–4 hr at −70°. The filters can then be washed in 6× SSC for 1 min at the hybridization temperature and reexposed to X-ray film. By comparing the two films the consequence of the higher temperature wash can be assessed and any positive colonies identified. Additional washes in 6× SSC are used as needed to control background. All stringent washes (at T_d − 5°) are done for short times (1–5 min).

---

[16] J. P. Gergen, R. H. Stern, and P. C. Wensink, *Nucleic Acids Res.* **7**, 2115 (1979).

TABLE I
CONDITIONS FOR OLIGONUCLEOTIDE HYBRIDIZATION[a]

Step	Conditions
Prehybridization (for nitro-cellulose)	
Solution	6× NET, 0.1% SDS, 5× Denhardt's solution, 100 μg/ml *E. coli* DNA
Incubation	65° for 5 hr
Hybridization	
Solution	6× NET, 0.1% SDS, 5× Denhardt's solution, 1–10 ng/ml ^{32}P-labeled oligonucleotide
Incubation	$T_d - 5°$ ($T_{dmin} - 2°$ for mixed probes) for 2 hr
Washing	
Solution	6× SSC, 0.1% SDS; ignore SDS precipitate
Initial wash	Four changes at 4° for 5 min each
Initial X-ray film exposure	1–24 hr (typically)
Additional washes	As needed to control background
Final wash	One change at hybridization temperature for 1–2 min
Final X-ray film exposure	Same time as initial exposure

[a] A typical procedure for hybridizing immobilized DNA with labeled oligonucleotide probes.

Plaque Screening

A prehybridization step is beneficial in the screening of recombinant phage by the method of Benton and Davis[17] with oligonucleotide probes. The filters are prehybridized in a solution containing 6× NET, 5× Denhardt's solution, 0.1% SDS, and 100 μg/ml sonicated, denatured *E. coli* DNA at 65° for 4–16 hr. The prehybridization solution is then removed and the filters are hybridized and washed as described in the colony screening section. It is necessary to expose the filters to X-ray film about 10 times longer than for the colony screening due to the lower signals obtained with phage plaques.

Other Oligonucleotide Hybridization Techniques

Hybridization to DNA in Dried Agarose Gels (Unblots)

In our hands, oligonucleotide probes produce a hybridization signal with a genomic restriction digest (in which fragments are >1.5 kb) approximately five times stronger when hybridized directly in the agarose

[17] W. D. Benton and R. W. Davis, *Science* **196,** 180 (1977).

gel matrix than when the same digest has been transferred to nitrocellulose or similar hybridization membrane.

Preparation of DNA Samples for Dried Gel Hybridization.[18] After electrophoresis, gels are stained with ethidium bromide and then photographed. The DNA samples are denatured by soaking the gel in 0.5 N NaOH, 150 mM NaCl for 30 min at room temperature with gentle shaking. The gel is then neutralized by soaking the gel in 500 mM Tris–HCl (pH 8), 150 mM NaCl for 30 min at 4° with occasional shaking. The gel is then placed on two sheets of Whatman 3 MM paper. The gel is trimmed at this point to remove any unused lanes. The gel (still on the Whatman paper) is covered with plastic wrap, placed in a gel dryer (Bio-Rad), covered with only the neoprene rubber sheet, and dried with only the vacuum until the gel is nearly flat (approximately 30 min). Then the gel dryer heater is turned on to 60° and the gel is dried an additional 0.5–1 hr. The vacuum is then released. At this point the gel has been reduced to a thin membrane on the Whatman paper. The gel is now ready for hybridization or it can be stored at room temperature indefinitely on the paper after wrapping it in plastic wrap.

Dried Gel Hybridizations.[19] The following conditions apply to the hybridization of an oligonucleotide probe to a single-copy genomic sequence and allow the discrimination between two genes that differ by a single nucleotide. The dried gel is removed from its paper backing by soaking it in a shallow dish of distilled water. Although the gel is no longer stuck to the paper, the gel is supported by the paper to facilitate its handling. After the gel is blotted to remove excess water, the gel and supporting paper are placed in a Seal-a-Meal (Dazey) plastic bag. The gel is pressed against the Seal-a-Meal bag, which leads to the preferential adherence of the gel to the plastic and allows the paper backing to be removed. Prehybridization is not necessary. The dried gel is hybridized in a solution that contains 5× SSPE, 0.1% SDS, 10 μg/ml sonicated, denatured *E. coli* or salmon sperm DNA, and 2 × 10^6 cpm/ml of labeled oligonucleotide. The hybridization volume for a 10 × 14 cm dried gel is 6 ml. The gel is hybridized at $T_d - 5°$ or 60° whichever is lower for 16 hr. After the hybridization period, the gel is washed with 6× SSC first at room temperature then at a stringent temperature. Usually two 15-min washes at room temperature are done followed by a 4-hr wash at room temperature. The stringent wash is done in 6× SSC at $T_d - 5°$ for 1–3 min. Gels can be reused for additional hybridizations by merely subjecting them to the alkali denaturation and neutralization as described in the section on drying the gels.

[18] S. G. S. Tsao, C. F. Brunk, and R. E. Perlman, *Anal. Biochem.* **131,** 365 (1983).
[19] C. G. Miyada, C. Klofelt, A. A. Reyes, E. McLaughlin-Taylor, and R. B. Wallace, *Proc. Natl. Acad. Sci. U.S.A.* **82,** 2890 (1985).

After the appropriate washes, the hybridized gel is wrapped with plastic wrap and exposed to Kodak XAR-5 X-ray film between two Cronex Lightning Plus intensifying screens at −70° (this volume [7]).

Hybridization to RNA in Dried Agarose Gels (Unblots)

RNA can also be hybridized in a dried gel. After electrophoresis of RNA in formaldehyde-containing gels the gel is washed for 30 min in 0.1 M Tris–HCl (pH 7.5), dried, and hybridized as described for the DNA dried gel. RNA molecules as small as 9 S (globin mRNA) appear to be retained by this procedure. Hybridization conditions are a described for DNA hybridization in dried gels.

Hybridization to DNA and RNA Bound to Membranes

Oligonucleotide hybridizations to restriction fragments less than 1.5 kb in size require the transfer of the DNA to a hybridization membrane since restriction fragments of this size tend to be lost from the agarose gel matrix during the hybridization and subsequent washes. Acid depurination (nicking)[20] of the DNA, which facilitates its transfer out of the gel, should be avoided since oligonucleotides do not hybridize efficiently to DNA treated in this manner. Conditions for electrophoresis of DNA and RNA and transfer to membranes are described in this volume [8, 61]. Blots are prehybridized, hybridized, and washed using conditions described in Table I. (See also this volume [43, 45, 61].)

Acknowledgments

The work described herein could not have been done without the collaboration of Dr. Keiichi Itakura. The work was supported by NIH Grants GM31261. RBW is a member of the Cancer Center at the City of Hope (CA33572).

[20] G. M. Wahl, M. Stern, and G. R. Stark, *Proc. Natl. Acad. Sci. U.S.A.* **76,** 3683 (1979).

[48] Gene Cloning Based on Long Oligonucleotide Probes

By WILLIAM I. WOOD

The most commonly used technique for gene cloning has been to utilize oligonucleotide probes based on protein sequence data. Of course this approach requires characterized and purified protein so that at least a portion of amino acid sequence can be determined and used to infer the corresponding DNA sequence. Based on the amino acid sequence information, either short or long oligonucleotide probes can be synthesized chemically.

Short probes are typically 11–20 bases in length and are pools of 8–32 (or more) sequences including all of the possible codon choices for each amino acid. There are three disadvantages of short probes. (1) They can generally only be used in regions of low codon redundancy; otherwise the pool size becomes unmanageable. (2) The amino acid sequence must be correct. A single mismatch is generally sufficient to prevent hybridization of the probe. (3) Only probes of 17 or longer can be used to screen high-complexity libraries (e.g., a human genomic library). This is because the complexity of the mammalian genome is such that an exact match of any 16-base sequence would be expected at random. When a pool of sequences is used, the number of false positives can be a problem. In some cases this difficulty can be overcome by using two nearby short probes. The advantage of short probes is that if the protein sequence data are correct, the probe should hybridize faithfully as all the codon choices are covered. Also, the exact hybridization conditions used need not be determined empirically when tetramethylammonium chloride is used[1] (see also this volume [49]).

Long probes on the other hand are typically 30–100 nucleotides long and are a single sequence based on a best guess for each codon. The long probe approach was first used to screen for three different genes: bovine tryspin inhibitor,[2] human insulin-like growth factor I,[3] and human factor IX.[4] There are three advantages of long probes. (1) Any stretch of amino acid sequence 10 or longer can be used; regions of low redundancy while

---

[1] W. I. Wood, J. Gitschier, L. A. Lasky, and R. M. Lawn, *Proc. Natl. Acad. Sci. U.S.A.* **82**, 1585 (1985).

[2] S. Anderson and I. B. Kingston, *Proc. Natl. Acad. Sci. U.S.A.* **80**, 6838 (1983).

[3] A. Ullrich, C. H. Berman, T. J. Dull, A. Gray, and J. M. Lee, *EMBO J.* **3**, 361 (1984).

[4] M. Jaye, H. de la Salle, F. Schamber, A. Ballard, V. Kohij, A. Findeli, P. Tolstoshev, and J. P. Lecocq, *Nucleic Acids Res.* **11**, 2325 (1983).

always a help are not especially important. (2) The amino acid sequence need not be absolutely correct. An erroneous amino acid or two can be tolerated.[5,6] (3) These probes can be used to screen high-complexity libraries with fewer false positives. The only disadvantage of long probes is the uncertainty of the codon choice. With the right codon choice even a 30-mer will hybridize very specifically in a high complexity library screen. However, if the codon choices are completely incorrect, the probe will never hybridize under any conditions.

In spite of the uncertainties over codon selection, the long probe approach is currently the method of choice in screening for genes based on protein sequence data. This is not to say that pools of short oligonucleotides do not have utility. The wisest course in screening for any new gene is to pursue all the avenues possible consistent with the available time, energy, and manpower. However, the utility of the long, single sequence probes has been demonstrated repeatedly for the screening of high-complexity libraries starting with any stretch of protein sequence data.

A variety of codon usage information can be used depending on the particular gene to be screened. A number of workers have used the codon usage table for mammalian genes compiled by Grantham et al.[7] It is also possible to use the codon frequency of a gene already cloned from the same or a related family. The most extensive consideration of the codon usage problem is given by Lathe[8] which includes considerations to reduce the number of CG base pairs in adjacent codons and other optimizations. It should also be noted that some workers have used long probes which are pools of 8 or 16 sequences covering several common codon choices at a few of the amino acids.[9,10] In some respects this approach encompasses some of the best features of the long and short probe approaches.

The probe can be labeled by any of the standard techniques including end labeling, filling in with the Klenow fragment of DNA polymerase I (used with two long oligonucleotides that have 12- to 15-base overlap), cloning in M13 and primer extending across the probe region, and even

[5] A. Ullrich, J. R. Bell, E. Y. Chen, R. Herrera, L. M. Petruzzelli, T. J. Dull, A. Gray, L. Coussens, Y.-C. Liao, M. Tsubokawa, A. Mason, P. H. Seeburg, C. Grunfeld, O. M. Rosen, and J. Ramachandran, *Nature (London)* **313,** 756 (1985).

[6] D. Pennica, G. E. Nedwin, J. S. Hayflick, P. H. Seeburg, R. Derynck, M. A. Palladino, W. J. Kohr, B. B. Aggarwal, and D. Goeddel, *Nature (London)* **312,** 724 (1984).

[7] R. Grantham, C. Gartier, M. Gouy, M. Jacobzone, and R. Mercier, *Nucleic Acids Res.* **9,** r43 (1981).

[8] R. Lathe, *J. Mol. Biol.* **183,** 1 (1985).

[9] P. H. Seeburg and J. P. Adelman, *Nature (London)* **311,** 666 (1984).

[10] J. J. Toole, J. L. Knopf, J. M. Wozney, L. A. Sultzman, J. L. Buecker, D. D. Pittman, R. J. Kaufman, E. Brown, C. Shoemaker, E. C. Orr, G. W. Amphlett, W. Barry Foster, M. L. Coe, G. J. Knutson, D. N. Fass, and R. M. Hewick, *Nature (London)* **312,** 342 (1984).

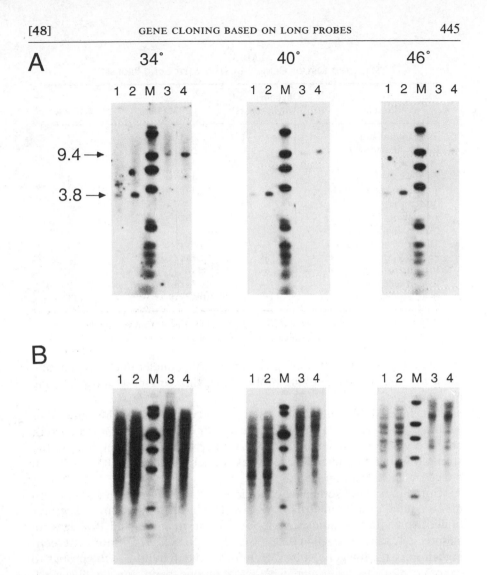

FIG. 1. Long probe hybridization to genomic blots. Two long probes for human factor VIII were hybridized to genomic blots of DNA, containing 1 and 4 X chromosomes.[11] Since factor VIII was known to be on the X chromosome, authentic hybridization is indicated by a relative intensity of 1 : 4 rather than 1 : 1 for the autosomes. Human genomic DNA, 5 μg, was digested with EcoRI or BamHI, separated electrophoretically in an agarose gel, and transferred to nitrocellulose. Lane 1, EcoRI male DNA; lane 2, EcoRI 46,XXXXY DNA; lane M, markers, end-labeled λHindIII and φX174 HaeIII; lane 3, BamHI male DNA; lane 4, BamHI 46,XXXXY DNA. The hybridization was as described in the text. The blots were washed in 1× SSC, 0.1% SDS at the temperature indicated. (A) A 36-mer end-labeled probe; (B) an 81-mer probe, cloned in M13 and labeled by fill-in reaction.

TABLE I
SELECTED LIST OF GENES ISOLATED WITH LONG PROBES

Protein	Oligonucleotide length (bases)	Match[a]	Reference
Trypsin inhibitor	86	...5x11x8...	2
Insulin-like growth factor I	94	...5x11x2x5x8x5...	3
Factor IX	52	...14x5x12	4
Transforming growth factor α	74	...14x2x6...	12
Luteinizing-hormone-releasing hormone	38	...14...	9
Tumor necrosis factor α	42	...8x17...	6
Factor VIII	36	...14x2x10...	11
Insulin receptor α	63	11x8x11x3x10...	5
Insulin receptor β	54	...5x12...	5
β-Adrenergic receptor	38	14x3x19	13

[a] Numbers give the bases in a row which match. Each mismatch is one x. The three dots (...) indicate an extension of the probe with poor match.

end labeling a series of short overlapping oligonucleotides and ligating them together to make a long probe.[5] (See this volume [10] for labeling methods.)

The labeled probes can be used to screen any available cDNA or genomic library in a plasmid or phage vector. The most common current practice would be to screen a cDNA library in λgt10. Depending on the strength of the hybridization, it is best to screen about 10,000 to 20,000 phage on each 150-mm plate.

Commonly used screening conditions are to hybridize in 6× SSC, 50 mM sodium phosphate (pH 6.8), 5× Denhardt's solution, 0.1 g/liter boiled, sonicated salmon sperm DNA, 20% formamide, and 10% dextran sulfate at 42° and to wash in 0.2× SSC, 0.1% SDS at 37°.[3] Prior to screening a library for the gene, it is often informative to hybridize the probes to genomic blots to determine whether a unique band can be observed. These blots can be used for two purposes: (1) to test different wash conditions to find the most stringent wash possible (see also this volume [43,45,61]) and (2) to determine which of several long probes is the most likely to be useful, thus eliminating some probes as having insufficient specificity to be of further use (especially in screening genomic libraries). Figure 1 shows the hybridization of two oligonucleotide probes to human EcoRI- and BamHI-digested DNA.[11] The probes were hybridized under

[11] W. I. Wood, D. J. Capon, C. C. Simonsen, D. L. Eaton, J. Gitschier, B. Keyt, P. H. Seeburg, D. H. Smith, P. Hollingshead, K. L. Wion, E. Delwart, E. G. D. Tuddenham, G. A. Vehar, and R. M. Lawn, Nature (London) 312, 330 (1984).

the conditions given above and washed in 1× SSC, 0.1% SDS at 34°, 40°, and 46°. Probe A, a 36-mer, clearly is much more suitable than probe B, an 81-mer, from another portion of the same protein. A screen of a genomic library with probe A isolated the derived clones easily. A screen with probe B was not attempted because of the large number of hybridizing bands. Another example of genomic blots with long probes has been published.[2]

Table I lists several examples where long probes based on protein sequence data have been used to clone new genes.[12,13] The table is not intended to be complete but purports to show the kinds of sequence match necessary for suitable hybridization. Of primary importance in obtaining adequate hybridization is the number of nucleotides in a row which match rather than the percentage homology, although clearly two regions separated by a single mismatch hybridize better than the longer of the two regions alone.

[12] R. Derynck, A. B. Roberts, M. E. Winkler, E. Y. Chen, and D. V. Goeddel, *Cell* (*Cambridge, Mass.*) **38**, 287 (1984).
[13] R. A. F. Dixon, B. K. Kobilka, D. J. Strader, J. L. Benovic, H. G. Dohlman, J. Frielle, M. A. Bolanowski, C. D. Bennett, E. Rands, R. E. Diehl, R. A. Mumford, F. E. Slater, I. S. Sigal, M. G. Caron, R. J. Lefkowitz, and C. D. Strader, *Nature* (*London*) **321**, 75 (1986).

[49] Hybridization of Genomic DNA to Oligonucleotide Probes in the Presence of Tetramethylammonium Chloride

By ANTHONY G. DiLELLA and SAVIO L. C. WOO

In this chapter we present a powerful method for the hybridization of genomic DNA to AT-rich oligonucleotide probes. The method utilizes tetramethylammonium chloride (TMA), a reagent which binds AT-rich DNA polymers[1] while concomitantly abolishing the preferential melting of AT versus GC base pairs.[2] Thus, hybridization of these probes to genomic DNA (Southern blots) or to recombinant DNA libraries is a function of probe length; it occurs in a manner independent of base composition.[3] The technique is particularly well suited for detecting and dis-

[1] J. T. Shapiro, B. S. Stannard, and G. Felsenfeld, *Biochemistry* **8**, 3233 (1969).
[2] W. B. Melchior and P. H. von Hippel, *Proc. Natl. Acad. Sci. U.S.A.* **70**, 298 (1973).
[3] W. I. Wood, J. Gitschier, L. A. Lasky, and R. M. Lawn, *Proc. Natl. Acad. Sci. U.S.A.* **82**, 1585 (1985).

tinguishing between normal and mutant alleles in the human genome[4-7] since single base-pair mismatches between an oligonucleotide probe and the complementary DNA sequence are sufficient to destabilize the duplex structure of the hybrid.[8-10] This methodology is applicable when genetic mutations do not result in obvious gene deletions or alterations in restriction fragment patterns.

Principle of Method

The temperature at which an oligonucleotide dissociates from a complementary genomic DNA fragment has been determined empirically as $T_d = 2° \times$ the number of AT base pairs $+ 4° \times$ the number of GC base pairs[11] (see also [47] this volume). Stringent hybridization and wash conditions for DNA analysis by Southern blotting or *in situ* gel hybridization using oligonucleotide probes are carried out at $T_d - 5°$. For probes of normal GC content, hybridization signals produced under these conditions are usually strong enough, relative to background, to detect perfect hybrids. However, empirically determined T_d values are unsuitable for oligonucleotide probes rich in AT content; the resulting background is unacceptable. If such probes are hybridized under nonstringent conditions and washed in the presence of TMA to control the hybridization stringency, strong hybridization signals are detected with negligible background. This method eliminates the dependence of T_d on DNA base composition; therefore, T_d (3 M TMA) can be determined as a function of probe length from previously reported values,[3] e.g., 45° (11-mer), 47° (13-mer), 50° (15-mer), 54° (16-mer), 58° (18-mer), and 71° (27-mer). We used a 21-mer with 16 AT residues (5′-AAATTACTTACTGTTAATGGA-3′) corresponding to the border of intron 12/exon 12 in the human phenyl-

[4] B. J. Conner, A. A. Reyes, C. Morin, K. Itakura, R. L. Teplitz, and R. B. Wallace, *Proc. Natl. Acad. Sci. U.S.A.* **80**, 278 (1983).

[5] V. J. Kidd, R. B. Wallace, K. Itakura, and S. L. C. Woo, *Nature (London)* **304**, 230 (1983).

[6] H. H. Kazazian, S. H. Orkin, A. F. Markham, C. R. Chapman, H. Youssoufian, and P. G. Waber, *Nature (London)* **310**, 152 (1984).

[7] D. J. G. Rees, C. R. Rizza, and G. G. Brownlee, *Nature (London)* **316**, 643 (1985).

[8] R. B. Wallace, J. Shaffer, R. F. Murphy, T. Bonner, T. Hirose, and K. Itakura, *Nucleic Acids Res.* **6**, 3543 (1979).

[9] R. B. Wallace, M. J. Johnson, T. Hirose, T. Miyake, E. H. Kawashima, and K. Itakura, *Nucleic Acids Res.* **9**, 879 (1981).

[10] R. B. Wallace, M. Schold, M. J. Johnson, P. Dembek, and K. Itakura, *Nucleic Acids Res.* **9**, 3647 (1981).

[11] S. V. Suggs, T. Hirose, T. Miyake, E. H. Kawashima, M. J. Johnson, K. Itakura, and R. B. Wallace, *ICN-UCLA Symp. Dev. Biol.* **23**, 683 (1981).

alanine hydroxylase gene[12] to illustrate the technique. T_d (TMA) for this oligomer is 63°.[3]

Materials

TMA (Aldrich) is made up as an aqueous 5 M stock solution and filtered through an 0.8-μm nitrocellulose filter to remove impurities. The actual molar concentration (c) must be determined from the refractive index (N) by the formula $c = (N - 1.331)/0.018$.

Methods

1. Digest DNA with the desired restriction endonuclease and separate fragments electrophoretically in an agarose gel. Prepare the gel for *in situ* hybridization (see this volume [47]). Alternatively, after electrophoresis transfer DNA fragments to nitrocellulose by Southern blotting (see this volume [61]).

2. For Southern analysis, prehybridize the nitrocellulose filter in a buffer containing 6× SSC, 50 mM sodium phosphate (pH 6.5), 5× Denhardt's solution, 0.1% sodium dodecyl sulfate (SDS), 2 mM EDTA, and 0.2 mg/ml of denatured salmon sperm DNA prehybridization solution. Note: dried gel membranes for *in situ* hybridization are not prehybridized.

3. Label the oligonucleotide probe by primer extension (see procedure 3 for Klenow fragment in this volume [10]). In this case, a 9-mer complementary to the 21-mer was used as the primer.[12]

4. Hybridize nitrocellulose filters overnight with shaking at 37° in the above prehybridization solution containing 2 × 10⁶ cpm probe/ml solution. Dried gel membranes are hybridized overnight with shaking at 37° in 6× NET buffer (6× NET: 0.9 M NaCl, 6 mM EDTA, 0.5% SDS, 0.09 M Tris–HCl at pH 7.5) containing 0.2 mg/ml denatured salmon sperm DNA and 2 × 10⁶ cpm probe/ml of hybridization solution. These conditions are suitable for oligomers of 19 nucleotides or longer.

5. Wash the dried gel membrane or nitrocellulose filter with shaking, twice at 4° for 30 min in TMA buffer (3 M TMA, 2 mM EDTA, 50 mM Tris–HCl at pH 8.0), once each at room temperature (30 min) and 60° (7 min) ($T_d - 3°$) in TMA buffer containing 0.2% SDS, followed by a room temperature rinse (10 min) in TMA buffer.

6. Autoradiograph gels or filters between two Quanta III intensifier screens (DuPont) at −70° (see this volume [7]).

[12] A. G. DiLella, J. Marvit, A. S. Lidsky, F. Güttler, and S. L. C. Woo, *Nature* (*London*) **322**, 799 (1986).

Discussion

Figure 1 shows the results using TMA (lanes 3 and 4) compared with a standard hybridization protocol (lanes 1 and 2) for oligonucleotides. Clearly, the use of TMA made possible the identification of the human phenylalanine hydroxylase gene by Southern blotting (lane 4) or *in situ* gel hybridization (lane 3) with the AT-rich oligonucleotide probe. This method is applicable to synthetic probes of varying GC composition and can be applied to detect single base-pair mismatches for hybrids

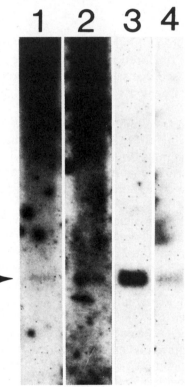

FIG. 1. Southern hybridization of genomic DNA to an oligonucleotide probe using standard (lanes 1 and 2) and TMA (lanes 3 and 4) hybridization methods. *Standard method: Pvu*II digests of genomic DNA were resolved in 1% agarose gels, which were then prepared for direct gel hybridization (see this volume [47]). Using the equation in the text, $T_d = 52°$ for the 21-mer used in this study. Dried gel membranes were hybridized overnight at 47° ($T_d - 5°$) in 6× NET buffer to 5'-end-labeled probe (lane 1) or primer-extended probe (lane 2). The gel membranes were then washed four times (15 min each) in 6× SSC at 4°, followed by a 60-min wash at room temperature in 6× SSC containing 0.5% SDS. Exposure time: 4 days. *TMA method (see text):* The primer-extended probe was hybridized to the dried gel membrane (lane 3) or nitrocellulose filter (lane 4). Exposure time: 2 days.

formed between oligonucleotides and genomic DNA.[12] The negligible background resulting from the TMA procedure obviates the need for a detailed restriction map of the gene, which was previously required to generate small hybridizing DNA fragments using the standard hybridization method, and greatly increases the overall sensitivity of the method compared to the standard hybridization protocol.

Acknowledgments

We wish to thank Ms. Kelly Porter for typing the manuscript. This work was partially supported by National Institutes of Health Grant HD17711 to S.L.C.W., who is also an investigator of the Howard Hughes Medical Institute. A.G.D. is the recipient of NIH Postdoctoral Fellowship HD-06495.

[50] Use of Antibodies to Screen cDNA Expression Libraries Prepared in Plasmid Vectors

By DAVID M. HELFMAN and STEPHEN H. HUGHES

Immunological screening has become a widespread method for the identification and isolation of particular clones from recombinant DNA expression libraries. Such libraries can be constructed using either plasmid[1-3] or bacteriophage[4] (λgt11) vectors. The construction of cDNA expression libraries was described previously in chapters [39] and [40]. In this chapter we describe the identification and isolation of cDNA clones by immunological screening of libraries prepared using plasmid vectors. The following chapter [51] describes the use of antibodies to screen λgt11 libraries.

Plating Bacteria

The library to be screened is plated onto nitrocellulose filters (Millipore Triton-free HATF) and overlaid on LB ampicillin plates (or appropriate selection media; this volume [13]). When plating the library after transformation of *Escherichia coli* with recombinant plasmids we use

[1] D. M. Helfman, J. R. Feramisco, J. C. Fiddes, G. P. Thomas, and S. H. Hughes, *Proc. Natl. Acad. Sci. U.S.A.* **80,** 31 (1983).

[2] A. J. P. Brown, E. A. Leibold, and H. N. Munro, *Proc. Natl. Acad. Sci. U.S.A.* **80,** 1265 (1983).

[3] T. F. Meyer, N. Mlawer, and M. So, *Cell (Cambridge, Mass.)* **30,** 45 (1982).

[4] R. A. Young and R. W. Davis, *Proc. Natl. Acad. Sci. U.S.A.* **80,** 1194 (1983).

Triton-free filters because we find filters containing Triton X-100 give lower transformation efficiencies, presumably due to detergent lysis of some of the bacteria. However, for subsequent replica plating of the original library it is not necessary to use Triton-free filters. If the library is obtained as a frozen liquid stock of *E. coli,* the bacteria should be plated onto Triton-free filters. In both cases, since the library will be replica plated from a master filter, the number of colonies per filter should be carefully considered. We routinely plate at a density of approximately 1000–3000 colonies per 82-mm filter. Larger filters (132 mm) may be substituted at plating densities of approximately 10,000 colonies per plate. In control experiments we found that at relatively high plating densities of nonexpressors (approximately 20,000 colonies on an 82-mm filter) a few colonies (10–15) known to express a portion of tropomyosin could be easily detected.

The bacterial colonies on the master filters are grown to a suitable size (0.1–0.5 mm). The library can then be replicated onto nitrocellulose filters for immunological (or nucleic acid) screening or onto glycerol plates for long-term storage at −70°.[5] Replica plating is carried out as described[5] (also this volume [44]). Either Schleicher and Schuell or Millipore nitrocellulose filters (with or without Triton X-100) are suitable for replica plating and screening using antibody or nucleic acid probes. It is worth noting that for immunological screening we find nitrocellulose filters containing a pore size of 0.45 μm give superior results to nitrocellulose filters containing pore sizes of 1.2 or 3.0 μm.

Bacterial Lysis and Antibody Screening

Bacterial colonies are grown on nitrocellulose filters until they reach a size of at least 1–2 mm. Letting the colonies grow larger than 2 mm does not appear to affect the assay. If an inducible promoter is being used, it will be necessary to place the filters onto a suitable induction medium to optimize expression of the cloned DNA. In the case of cDNA expression libraries prepared using pUC plasmids cloned into bacterial strain DH1 or DH5, expression of the fusion proteins is constitutive and does not require induction. The filters are removed from the media plates and suspended by a binder clip in a chloroform vapor chamber for 15–20 min. Each filter is then placed in an individual Petri dish (100 mm) containing 10 ml of 50 mM Tris–HCl (pH 7.5), 150 mM NaCl, 5 mM MgCl$_2$, 1 μg/ml DNase I, 40 μg/ml lysozyme, and 3% (w/v) bovine serum albumin. The filters are agitated gently overnight at room temperature on a rotatory

[5] D. Hanahan and M. Meselson, *Gene* **10,** 63 (1980).

shaker. When a large number of filters are screened, two filters can be put in each Petri dish, the bottom filter colony side down and the top filter colony side up.

After the filters are incubated overnight, they are washed in Tris-buffered saline (50 mM Tris–HCl at pH 7.5, 150 mM NaCl). We routinely wash all the filters from a particular screen together in square glass baking dishes (20 × 20 cm). The wash removes the bacterial debris. In instances in which the remnants of a colony remain on the filter, remove them by gently scraping with one's finger (while wearing a glove) or with a moist Kimwipe. After these procedures the filters can be incubated with the primary antibody.

It is advisable to preabsorb the primary antibody with bacterial lysates prepared from the host strain of bacteria. The bacterial lysates are prepared by growing 1 liter of *E. coli* overnight to stationary phase. The bacteria are recovered by centrifugation. (See this volume [13] for techniques for handling bacteria.) The bacterial pellet is resuspended in 10 ml of deionized water in a 50-ml plastic conical tube and placed in a boiling water bath for 5–10 min. One milliliter of the resulting bacterial lysate is used to absorb 100 ml of diluted antibody for at least 2 hr at 4°. The antibody is diluted in Tris-buffered saline containing 3% bovine serum albumin (w/v). We routinely use a dilution of primary antibody sufficient to detect 1 ng or less of purified antigen spotted onto nitrocellulose. It is important to be sure that the diluted antiserum used is capable of detecting antigen spotted onto nitrocellulose filters since this is the basis for the immunological screening procedure. It is also advisable to use control filters, i.e., nitrocellulose spotted with the antigen of interest and a few negative controls each time a cDNA library is screened. This control is most useful when the assay fails to detect any antigen-expressing colonies, since it helps to decide whether the antiserum or library is inadequate.

After preabsorbtion of antiserum with bacterial lysate, the bacterial debris is removed by centrifugation for 15 min at 5000 to 10,000 rpm. Each filter is then incubated for 1 hr at room temperature with 8 ml of antiserum in a 100-ml Petri dish with gentle agitation using a rotatory shaker. It is possible to place two filters in one Petri dish in 9 ml of antiserum. When two filters are incubated simultaneously, care should be taken that no bubbles form between the bottom and top filter, preventing the interaction of the antibody with part of a filter. Following incubation with the primary antibody, the filters are placed together in a glass dish and washed in Tris-buffered saline at room temperature (five changes, 10–30 min each wash). As many as 50 filters can be washed in one dish.

The primary antibody can be detected using ^{125}I-labeled second antibody, or ^{125}I-labeled protein A. Horseradish peroxidase available in kit form (Bio-Rad, Richmond, CA) may also be used. What follows is a protocol using ^{125}I-labeled second antibody. Because the primary antibody was raised in rabbit, the second antibody was goat anti-rabbit IgG. Following incubation with a primary antibody and subsequent washes, the filters are then incubated with 5×10^6 cpm of ^{125}I-labeled second antibody (specific activity, approximately 10^7 cpm/μg), diluted in 10 ml of Tris-buffered saline containing 3% (w/v) bovine serum albumin. It is not usually necessary to preabsorb the ^{125}I-labeled second antibody with bacterial lysates. In experiments using the ^{125}I-Fab fragment from New England Nuclear (available in iodinated form), we obtain suitable results without preabsorbing the antibody. ^{125}I-labeled protein A also gives satisfactory results and is used at a concentration of 0.1 μCi/10 ml of Tris-buffered saline containing 3% bovine serum albumin (w/v). After a 1-hr incubation at room temperature, the filters are washed extensively in Tris-buffered saline (five changes), dried, and autoradiographed for 24–72 hr in the presence of Dupont Cronex Lightning Plus X-ray intensifying screens (see this volume [7]).

There is considerable variation between different antisera with respect to the signal from an antigen-producing colony, background from the nonexpressing colonies, and the nitrocellulose matrix itself. While various factors may be responsible for this it is obvious that a high signal-to-noise ratio is desirable. One simple procedure that often improves the signal-to-noise ratio is to wash the filters in detergents. We routinely wash the filters after both the first antibody and second antibody for 30–60 min in a solution of Tris-buffered saline containing 1% Triton X-100, 0.5% deoxycholate, and 0.1% SDS, followed by a rinse in Tris-buffered saline with no detergents.

When screening a library we find it advisable to screen two sets of filters with a given antiserum. This provides immediate confirmation that an antibody-positive colony is in fact reproducible. The same antibody can be used to screen the two sets of filters sequentially. We find the autoradiographs of the second set of filters have a much lower background than the first, making antibody-positive colonies more apparent (better signal-to-noise ratio), although in some cases, the signal from positive colonies is diminished. Alternatively, if one has access to separate preparations of antiserum to a given protein, it may be advisable to screen each set of filters with each independent antiserum. Antibody-positive colonies common to both antisera are chosen. Using two separate antisera, genuine positives unique to a given antibody may exist, since particular regions of a protein can be recognized by one antiserum and not the other. The use of antiserum obtained from two independent sources should decrease the probability of false positives obtained by the presence in a particular serum of antibodies that recognize proteins other than the one of interest.

Isolation of Single Colonies

After identification of positive colonies in the primary screen, the positive clones are isolated as a single colony. There are two points worth mentioning in the isolation of particular clones by immunological screening. First, generally all colonies on a filter are weakly discernible on the autoradiograph, although the antibody-positive colonies clearly give a stronger signal. Although the filters are keyed to line up the autoradiograph with a master copy, the faint background from the negative colonies is beneficial because it simplifies locating and picking a colony of interest. When a short exposure (12–24 hr) shows the antibody-positive colonies but fails to indicate the position of all the colonies, it is useful to get a longer exposure clearly showing the position of all the colonies on a filter. Second, the actual isolation of a colony is made easier, if each of the clones picked are restreaked and five colonies (or more if necessary) are picked and dotted onto nitrocellulose using a grid. The colonies are grown to approximately 1 mm, the filter is replica plated, and the replicas screened with antibody again to obtain a single colony. As many as 100 colonies from as many as 20 separate clones can be spotted onto a single nitrocellulose filter. These filters can now serve as a master copy and can be replica plated numerous times. This is especially useful for reconfirming a screen (i.e., when duplicates are done, or if two different antisera to the same protein are used; the identical clones can be screened and compared). If necessary, these procedures can be repeated until a pure clone is fully characterized. Alternatively, the bacteria from the master filter may be picked and suspended in liquid medium and then plated onto nitrocellulose at a colony density suitable for obtaining single colonies (see this volume [13]). This filter can then serve as the master copy, and can subsequently be replica plated. The copies can be screened with antibody probes to obtain a single antibody-positive colony.

Analysis of Bacterial Fusion Proteins

Once a single clone or group of clones positive with respect to a given antibody has been isolated, the fusion proteins produced in the bacteria can be analyzed. For cDNA libraries constructed using pUC plasmids we routinely grow bacteria overnight or until stationary phase in 3–4 ml of medium (e.g., LB), with or without ampicillin (100 μg/ml). Bacterial culture (1–1.5 ml) is transferred to an Eppendorf centrifuge tube and the bacteria recovered by centrifugation for 1 min in a microfuge. The medium is aspirated off and the bacterial pellet resuspended in 200 μl of Laemmli sample buffer,[6] supplemented with 2 mM EDTA, 2 mM EGTA, and 2 mM PMSF (phenylmethylsulfonyl fluoride). The tube is then placed in a boiling water bath for 3 min, and the proteins analyzed by SDS–polyacrylamide gel electrophoresis (this volume [31]). Usually between 5 and 10 μl of the bacterial lysate is sufficient to visualize the proteins by Coomassie blue stain. In addition, if the solution is viscous after boiling, it may be necessary to shear the DNA by putting the solution through a syringe fitted with a 26-gauge needle 2–3 times. The identity of the fusion proteins can be confirmed by immunoblot analysis.[7]

In experiments using the pUC8/pUC9 expression plasmids we have found that it is critical to harvest bacteria for analysis of their fusion proteins during late log-phase or stationary-phase growth. Virtually no fusion protein can be detected when the bacteria are analyzed during log-phase growth. The reason for this is unclear at this time. In addition, using the pUC8/pUC9 expression vectors and *E. coli* strain DH1, it is not necessary to induce the expression of the cloned cDNA. This is probably a consequence of the high copy number of this plasmid and the relatively low levels of repressor protein in the host bacteria. However, some expression vectors do require induction for maximal expression.

Final Identification of a Clone

The identification of a clone by immunological screening does not in itself prove that a cDNA clone codes for the protein of interest. Obviously, colonies may express a protein product that cross-reacts with an antibody but does not encode a cDNA for a particular clone of interest. Final identifications are usually made by sequence analysis of the cDNA insert and comparison of the derived amino acid sequence with the known protein sequence (if available). Alternatively, the cloned cDNA can be used for hybrid selections or hybrid-arrest translations as a means for identification. These procedures are discussed in this volume (see this volume [56–60]).

Considerations

The two most important components for immunological screening of expression libraries are the cDNA library itself and the antibody prepara-

[6] U. K. Laemmli, *Nature (London)* **227**, 680 (1970).
[7] H. Towbin, T. Staehelin, and J. Gordon, *Proc. Natl. Acad. Sci. U.S.A.* **76**, 4350 (1979).

tion used for screening. With respect to the cDNA library used, a variety of factors need to be taken into account, including the type of library and size of library to be screened. One potential problem is the synthesis of a protein product that is lethal to *E. coli*. Fortunately, all that is required in the immunological screening procedure is that an antigenic fragment be produced. For example, we successfully detected a recombinant expressing less than 50% of the full-sized tropomyosin protein (133 of the 284 amino acids). It is likely that in many cases in which a complete or nearly complete protein would be lethal to *E. coli,* an antigenically active fragment would not. Even if there are problems with lethal synthesis of proteins in some cases, this problem may be overcome using inducible expression vectors. In addition, there may be problems associated with the stability of some expressed proteins in *E. coli*. In many cases the use of expression vectors which contain a large fusion protein encoded by the vector itself tends to give stability to the smaller region encoded by the cloned cDNA.

With respect to the antibody preparation used, this is likely to be the most variable and crucial reagent. From tests with purified proteins spotted onto nitrocellulose, we have found that the sensitivity depends critically on the quality of the antibody preparation used. It is important to be sure that the antibody preparation used is capable of detecting the antigen of interest spotted onto nitrocellulose filters, since this is the basis for immunological screening. In addition, high-titer antibodies tend to give better results than low-titer antibody preparations due to lower background observed during immunological screening of expression libraries. Both polyclonal and monoclonal antibodies have been used successfully for screening libraries. With respect to the use of monoclonal antibodies, it is important to remember that a single monoclonal antibody might detect a site near the amino terminus of the protein, which would mean that only a small number of appropriate cDNA clones would make an antigen that a particular monoclonal antibody would recognize. Ideally, the use of several monoclonal antibodies (if available) which recognize antigenic determinants throughout the length of the protein would be most advantageous, and would also decrease the chances of obtaining false positives.

Acknowledgments

This work was supported in part by grants from the National Institutes of Health, the Muscular Dystrophy Association, and the American Cancer Society. We are grateful to Madeline Szadkowski for preparation of this manuscript.

[51] Gene Isolation by Screening λgt11 Libraries with Antibodies

By ROBERT C. MIERENDORF, CHRIS PERCY, and RICHARD A. YOUNG

Bacteriophage λgt11 and its bacterial host strains were developed to allow the isolation of DNA sequences that encode determinants recognized by antibodies.[1] The system incorporates features that maximize the ability to detect antigenic determinants (epitopes) on polypeptides expressed from DNA of any source. Successful isolation of genomic or cDNA sequences using this system requires both a λgt11 recombinant DNA library and an antibody probe of adequate quality. In addition, the success of the screening can depend on the sensitivity of the method used to detect antibody binding. This chapter describes the principle of the method, criteria for choosing and assessing the quality of the λgt11 library and the antibody probe, methods for sensitive antibody detection, our favorite protocol, and finally, approaches for troubleshooting.

Principle and Practice

Successful gene isolation with antibody probes requires the expression of the antigen of interest at detectable levels. The stability of foreign polypeptides expressed in *Escherichia coli* is variable, and in some cases their accumulation may be harmful to cell viability.[2] The λgt11 expression vector–host system has been designed to minimize these problems. The fusion of foreign sequences to the carboxy terminus of β-galactosidase supplies efficient and controllable signals for bacterial expression machinery and reduces the instability of many foreign polypeptides.[3,4] The host strain Y1090 is deficient in the *lon* protease, which increases the stability of β-galactosidase fusion proteins without affecting cell growth.[5] This strain also contains the *lac* repressor which prevents *lacZ*-directed gene expression until it is derepressed by the addition of isopropyl-β-D-thiogalactopyranoside (IPTG) to the medium. This allows the induction of potentially toxic foreign antigens to be delayed until after the cells have

---

[1] J. Jendrisak and R. A. Young, this volume [40].

[2] R. A. Young and R. W. Davis, *in* "Genetic Engineering: Principles and Techniques" (J. Setlow and A. Hollaender, eds.), Vol. 7, pp. 29–41. Plenum, New York, 1985.

[3] H. Kupper, W. Keller, C. Kurtz, S. Forss, H. Schaller, K. Franze, K. Strommaier, O. Marquardt, V. G. Zaslavsky, and P. H. Hofschneider, *Nature (London)* **289**, 555 (1981).

[4] K. Stanley, *Nucleic Acids Res.* **11**, 4077 (1983).

[5] R. A. Young and R. W. Davis, *Proc. Natl. Acad. Sci. U.S.A.* **80**, 1194 (1983).

undergone early log-phase growth and still contain fully active transcriptional and translational mechanisms. The *supF* gene product is also present to suppress the phage mutation causing defective lysis.

For a screening experiment the λgt11 library is grown on a lawn of *E. coli* Y1090 R⁻ [Δ*lacU169 proA⁺* Δ*lon araD139 strA supF*(*trpC22*::Tn*10*) (pMC9)]. When a large number of infected cells surrounds each plaque (usually 3–4 hr) *lacZ*-directed gene expression is induced by placing an IPTG-soaked nitrocellulose membrane over the plate. After a 3.5-hr incubation at 37°, during which protein from the lytically infected cells becomes immobilized, the membrane is removed and rinsed, and unsaturated protein binding sites are blocked by incubation in a solution containing excess protein, such as bovine serum albumin or calf serum. The membrane is then incubated with the primary antibody against the antigen of interest, followed by a washing step and another incubation with the appropriate anti-IgG ("second antibody") alkaline phosphatase conjugate. After a final wash, the color development substrates are added. Usually within a few minutes, positive plaques turn a dark purple color as a result of alkaline phosphatase activity. The positives are then located and removed from the original plates and the recombinants are plaque purified by repeating the same screening procedure until all plaques are positive.

Other detection methods for antibody screening have involved the use of anti-IgG peroxidase conjugates[6] and ¹²⁵I-labeled protein A.[5] A discussion of these and how they compare with the alkaline phosphatase-based system is presented below.

Recombinant DNA Expression Libraries

The quality of the λgt11 recombinant DNA library is an important factor in the success of screening with antibodies (immunoscreen). It is also one of the most difficult parameters to measure. Two types of library can be constructed, genomic DNA and cDNA. The type of library that is most useful depends primarily on the relative frequency with which a particular coding sequence is expected to exist within the library of recombinant molecules. Thus, in all but prokaryotes and lower eukaryotes, specific protein coding sequences are usually more abundant in cDNA from the appropriate cell types than in total genomic DNA. Protocols for constructing both types of libraries are described elsewhere.[1]

The size of the recombinant DNA library is also important. Because antigen coding sequences are generally expressed only if they are in the

[6] V. C. W. Tsung, J. M. Peralta, and H. R. Simons, this series, Vol. 92, p. 377.

appropriate transcriptional orientation and translation frame, DNA sequences of interest can usually be detected 5- to 10-fold more frequently with nucleic acid probes than with antibody probes. Thus, the recombinant DNA expression library must contain sufficient numbers of individual recombinants to ensure that the DNA sequence of interest appears in multiple copy in the library.

Finally, the manner in which a recombinant DNA expression library is screened affects the ability to detect specific clones with antibody probes. The plaque density should be in the range $2.5-5 \times 10^4$ plaques per 150-mm plate to maximize the efficiency of screening but minimize the number of clones missed through overcrowding.

Antibody Probes

Whether a given antibody preparation is useful for screening depends on several factors. First, as discussed above, the amount of antigen that is deposited during a plaque lift incubation varies with each fusion protein based on its expression and stability in the host cells. Successful screenings have been carried out with positives containing 30–60 pg of immunoreactive material per plaque, whereas some plaques appear to contain as much as 200–800 pg of fusion protein.[7]

Second, the titer and binding characteristics of the antibody are of obvious importance. The best signals are generally produced by high-titer, high-affinity antibodies. Exceptions include antibodies that recognize modifications such as carbohydrate moieties that are not added in *E. coli,* or antibodies (particularly some monoclonals) that recognize specific conformations of the antigen which are either made unavailable upon binding to the nitrocellulose membrane or are not assumed by the recombinant polypeptide. The use of polyvalent antibodies has the potential advantage that a number of different epitopes can be recognized. In general, antibodies that produce good signals on Western blots also produce good signals in the screening procedure at similar dilutions. If little or no signal is obtained or if the background is too high in the screening procedure, the antiserum or ascites should be checked on a Western or dot blot before proceeding. These steps may reveal conformation-dependent antibodies and others that may tend to stick nonspecifically to surfaces such as nitrocellulose, and which therefore are not suitable for immunoscreening.

Third, most crude antisera and ascites fluids contain IgG components that bind to *E. coli* proteins. If the titer and/or binding affinity of these

---

[7] R. Kincaid, personal communication.

endogenous antibodies is high relative to the antibodies of interest, it may be difficult to distinguish true positives from the background of other plaques. The effect of the contaminants can be minimized by optimizing the dilution of primary antibody used for screening. This can be accomplished by dot blotting a series of known amounts of the antigen on a membrane disk which is subsequently used to make a plaque lift, and testing several dilutions of primary antibody for maximum sensitivity with minimum background. In general, a range of dilutions from 1 : 200 to 1 : 10,000 will reveal the optimal levels for most sera, ascites fluids, and purified antibodies, whereas hybridoma tissue culture fluids (which have much lower antibody concentrations) usually require dilutions from 1 : 10 to 1 : 100 for optimal performance.

If the nonspecific plaque signals are still unacceptable, the antibody preparation can be preadsorbed with an extract of Y1090 cells to inhibit or remove the anti-*E. coli* IgG. The extract is prepared by growing a culture of Y1090 to saturation, harvesting the cells by centrifugation, resuspending the cells in a buffer containing 50 mM Tris–HCl (pH 8.0), 10 mM EDTA, breaking the cells by a freeze–thaw cycle followed by sonication, and removing the debris by another centrifugation. The extract is used by incubating an appropriate amount with the diluted primary antibody for 30 min before adding the plaque lift to be screened. The amount of extract needed to be effective depends on the dilution of the antiserum used for screening. As a guideline, we have found that a 1 : 1000 dilution of antiserum requires about 1 mg/ml of extract protein to reduce the background effectively.

Larger amounts of antiserum can be treated at one time by immobilizing the extract on a solid support, such as agarose beads, and performing either a batchwise or column adsorption. To prepare an immobilized extract, first transfer it into 0.1 M NaHCO$_3$ (pH 8.5–9.0), 0.5 M NaCl by dialysis or gel filtration. It can then be bound to cyanogen bromide-activated agarose or other activated matrices following the manufacturers' instructions.

Antibody Detection Methods

The detection methods of choice is based on the binding of species-specific antiimmunoglobulins conjugated with alkaline phosphatase to the primary antibody, followed by a color development reaction with the phosphatase substrate 5-bromo-4-chloro-3-indolyl phosphate (BCIP) in combination with nitro blue tetrazolium (NBT). The enzyme converts the BCIP to the corresponding indoxyl compound, which precipitates and tautomerizes to a ketone. Under the conditions of the assay, the ketone undergoes oxidation and dimerizes to form an indigo, which is blue. The hydride ions released from dimerization are acquired by the NBT salt, and this reduction causes the formation of an intensely purple diformazan, which is deposited along with the indigo at the site of enzyme activity.[8,9]

Other detection methods that have been used for immunoscreening include ^{125}I-labeled protein A (from *Staphylcoccus aureus*),[5] and antiimmunoglobulins conjugated with horseradish peroxidase.[6] The enzymatic procedures have an important advantage over the use of ^{125}I-labeled protein A in that the signals are produced directly on the nitrocellulose membrane, thereby exactly reproducing the pattern of plaques on the plate, with the positives appearing as darkly stained plaques over a faint plaque background. This allows the precise location of the positive plaques and reduces the uncertainty and subsequent work involved in plaque purification. In addition, ^{125}I-labeled protein A produces more false positive signals than the enzymatic methods. The use of antiimmunoglobulins also allows the detection of different antibody isotypes of a variety of species.

The experimental protocols for the two enzymatic methods are extremely similar. In the case of peroxidase, the substrate of choice has been 4-chloronaphthol, which turns purple and precipitates on oxidation by the enzyme.

The relative sensitivities of the different enzymatic assays has been studied. Figure 1 shows a comparison of the sensitivity of detection obtained with alkaline phosphatase- and peroxidase-conjugated second antibodies in a dot blot assay using identical amounts of antigen and the same primary antibody. The results indicate that the alkaline phosphatase conjugate was able to produce a detectable signal with 20–50 pg of antigen, whereas the lower limit of detection was between 200 and 500 pg with the peroxidase system. Because the amount of antigen deposited on the nitrocellulose membrane from some plaques is at or below the limit of sensitivity of the peroxidase, the peroxidase-based assay may not detect some recombinants that would be seen with an alkaline phosphatase conjugate. Figure 2 shows a comparison of the performance of alkaline phosphatase and peroxidase anti-IgG conjugates in detecting positive clones on a plaque lift. In this case the peroxidase-based detection yielded unequivocal positives, but the enhanced signal-to-noise ratio obtained with the alkaline phosphatase system is clearly evident.

[8] J. McGadey, *Histochemie* **23**, 180 (1970).
[9] M. S. Blake, K. H. Johnston, G. J. Russell-Jones, and E. C. Gotschlich, *Anal. Biochem.* **136**, 175 (1984).

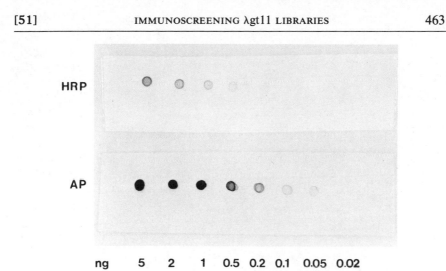

ng   5   2   1   0.5 0.2 0.1 0.05 0.02

FIG. 1. "Dot" blot comparison of the sensitivities of alkaline phosphatase and peroxidase-based detection assays. The indicated amounts of pure β-galactosidase were spotted directly on a nitrocellulose membrane in 1-μl volumes, followed by immunodetection using a rabbit anti-β-galactosidase antiserum (1 : 1000 dilution) as the primary antibody, and either alkaline phosphatase (AP) or peroxidase (HRP)-conjugated goat anti-rabbit IgG for detection. The substrates used in the color development reaction were NBT/BCIP and 4-chloro-1-naphthol for alkaline phosphatase and peroxidase, respectively. Incubation times and buffers were identical, and second antibody dilutions and color development were performed according to the instructions of Promega and Bio-Rad, respectively.

It should be noted that the high sensitivity of the alkaline phosphatase-based method can reveal some types of background which may be less significant with the other methods. For example, contamination of primary antibodies with anit-*E. coli* IgG may appear more pronounced, or minor specificities within a polyclonal serum may reveal more recombinants corresponding to contaminants in the original antigen preparation. However, as discussed earlier, the signal-to-noise ratio can be manipulated by changing the antibody dilutions and using preadsorption steps when appropriate (also see Troubleshooting section).

Alkaline phosphatase conjugates offer two additional advantages over peroxidase conjugates. The phosphatase color development reaction can continue for hours or even overnight, whereas peroxidase is substrate inactivated, resulting in an effective reaction time of only about 30 min. This can create an even larger difference in the relative sensitivities of the two methods. In addition, alkaline phosphatase-developed blots are less susceptible to fading or photobleaching and thus maintain their color for longer periods of time.

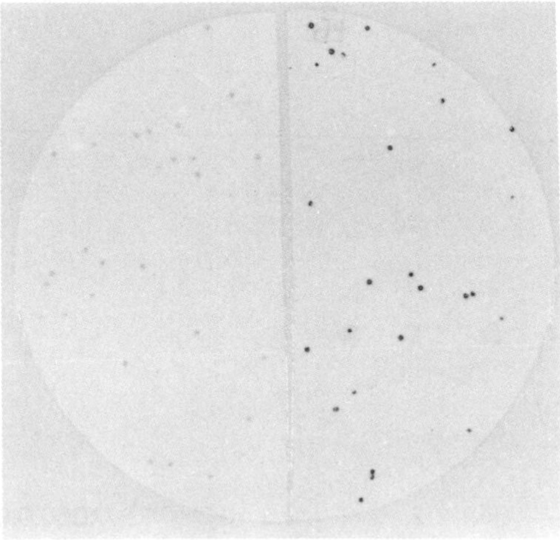

FIG. 2. Comparison of alkaline phosphatase- and peroxidase-based detection assays under immunoscreening conditions. Approximately 50 recombinant λgt11 phage containing an ovalbumin cDNA were mixed with 10^4 λgt10 phage and plated on the Y1090 host. Plaque lifts were made and membranes processed according to the recommended protocol. The primary antibody used was a mouse monoclonal anti-β-galactosidase which detects the fusion protein expressed only by the λgt11 recombinants. Following incubation with the primary antibody and washing, the membrane was cut in two. The half shown on the right was incubated with goat anti-mouse IgG alkaline phosphatase followed by color development as described in the text using NBT/BCIP, and the half on the left was incubated with the corresponding peroxidase conjugate and developed using 4-chloro-1-naphthol according to the manufacturer's instructions.

Reagents and Solutions Recommended for Screening Phage Libraries[10]

λ diluent: 10 m*M* Tris–HCl (pH 7.5), 10 m*M* $MgCl_2$
IPTG: 10 m*M* isopropyl-β-D-thiogalactopyranoside in water
Tris-buffered saline + Tween 20 (TBST): 10 m*M* Tris–HCl (pH 8), 150 m*M* NaCl, 0.05% Tween 20
Blocking solution: 20% calf serum (or other appropriate protein, such as 1% bovine serum albumin or 1% gelatin) in TBST
Affinity purified antiimmunoglobulin alkaline phosphatase conjugate. The specificity of the second antibody should match the species and isotype of the primary antibody, e.g., anti-rabbit IgG if the

[10] Reagents for cloning in λgt11, for immunoscreening libraries using anti-IgG alkaline phosphatase conjugates, and for purifying β-galactosidase fusion proteins with immobilized monoclonal antibodies are commercially available from Promega (Madison, WI).

primary antibody is a rabbit antiserum. The most commonly used antiimmunoglobulins are prepared in goats.

Alkaline phosphatase (AP) buffer: 100 mM Tris–HCl (pH 9.5), 100 mM NaCl, 5 mM MgCl$_2$

Color development substrates: NBT at 50 mg/ml in 70% dimethyl-formamide; BCIP at 50 mg/ml in 100% dimethylformamide

LB broth, LB agar, LB soft or top agar (see this volume [13])

Nitrocellulose membrane disks

Protocol for Immunoscreening Libraries

1. *Grow plating cells.* Streak out *E. coli* Y1090 R$^-$ for single colonies on LB plates (pH 7.5) containing ampicillin at 100 μg/ml. Incubate at 37° overnight. Starting with a single colony, grow Y1090 to saturation in LB (pH 7.5) at 37° with good aeration.

2. *Infect cells with library.* If plating is to be done on 90-mm plates, mix 0.2 ml of the Y1090 overnight with 0.1 ml of λ diluent containing 3 × 10^4 pfu of the λgt11 library for each plate. If plating is to be done on 150-mm plates, use 0.6 ml of the Y1090 culture with up to 5 × 10^4 pfu in λ diluent for each plate. Adsorb the phage to the cells at room temperature for 20 min.

3. *Plate cells.* Add 2.5 ml (for a 90-mm plate) or 7.5 ml (for a 150-mm plate) LB soft agar (pH 7.5) to the culture and pour onto an LB plate (pH 7.5). Use slightly dry plates (i.e., 2 days old) so that the soft agar will stick to the plate. Incubate the plates at 42° for 3.5 hr.

4. *Overlay nitrocellulose membrane.* Remove the plates to a 37° incubator. Carefully (avoiding air bubbles) overlay each plate with a dry nitrocellulose membrane disk which has been saturated previously in 10 mM IPTG in water (dry saturated disks can be stored up to a week at 4°). Incubate for 3.5 hr longer at 37°. Although this amount of time appears sufficient for the deposition of most fusion proteins, some individual proteins (which may be expressed at lower steady-state levels) may require longer incubations. A second membrane may be overlaid after the first has been removed and incubated for an additional 3 (or more) hr at 37° if a duplicate is required.

5. *Prepare antigen-bound membrane for antibody screen.* Remove the plates to room temperature, quickly mark the position of the membrane on the plate with a needle, and carefully remove the membrane. If the top agar begins to stick to the membrane rather than the bottom agar, chill the plates at 4° for 20 min. Place the membranes in TBST (10 mM Tris–HCl at pH 8.0, 150 mM NaCl, 0.05% Tween 20) and rinse briefly to remove any remnants of agar.

Notes: Do not allow the membranes to dry out during any of the subsequent steps. At this point the damp membranes may be wrapped in plastic, then in foil and stored at 4° overnight. Perform all of the following washing and incubation steps at room temperature with gentle shaking. It is convenient to perform the incubations and washes in plastic petri dishes that are slightly larger than the membranes with gentle shaking on a rotating platform. Use one membrane per dish, with the surface that was in contact with the agar facing up.

6. *Block nonspecific protein binding sites.* Float the membranes on the surface of TBST in individual petri dishes until they are evenly wet, submerge them, and shake for 20 min with one buffer change (this wash is optional but appears to help in reducing backgrounds). To saturate nonspecific protein binding sites, decant the buffer and incubate the membranes in TBST + 20% calf serum or other suitable protein blocking agent (e.g., 1% bovine serum albumin, 1% gelatin) for 30 min. Use 7.5 ml per 82-mm membrane and 15 ml per 132-mm membrane.

7. *Bind primary antibody.* Incubate the membranes in TBST + primary antibody for 30 min. The optimal incubation time may be longer depending on the titer and affinity of the antibody. Use 7.5 ml per 82-mm membrane and 15 ml per 132-mm membrane. The diluted primary antibody can be reused several times.

8. *Wash.* Wash the membranes in 15–20 ml TBST three times for 5–10 min each.

9. *Bind anti-IgG alkaline phosphatase conjugate.* Transfer the membranes to TBST containing the appropriate second antibody–alkaline phosphatase conjugate (at the manufacturer's recommended dilution). Allow 7.5 ml per 82-mm membrane and 15 ml per 132-mm membrane. Incubate for 30 min at room temperature.

10. *Wash.* Wash the membranes in TBST, as in step 8.

11. *Start color reaction.* Blot the membranes damp dry on filter paper and transfer to the color development substrate solution, prepared as follows: for 5 ml solution, add 33 μl NBT substrate to 5 ml AP buffer, mix, add 16.5 μl BCIP substrate, and mix again. Protect the solution from strong light and use within 1 hr. Positive clones will appear as purple plaques on the filters. Since the fusion protein is synthesized in infected cells at the periphery of the plaque, positives often appear as "doughnuts" with clear centers. Development of color will continue for at least 4 hr.

12. *Stop color development.* When the color has developed to the desired intensity, stop the reaction by rinsing the membrane with several changes of deionized water. Membranes can be photographed while still moist by placing them on top of a piece of damp filter paper on a glass

plate. For storage, the membranes can be air dried on filter paper. The color will fade slightly upon drying but can be restored by moistening with water. Protect the membranes from light during prolonged storage.

13. *Plaque purification*. To retest a putative positive signal, remove an agar plug containing phage particles from the region of the plate corresponding to the signal on the membrane. Incubate the agar plug in 1 ml of λ diluent for at least 1 hr at room temperature. Replate the phage (see this volume [44]) and repeat the screening procedure with the antibody probe until all the plaques on the plate produce a signal.

Troubleshooting

Plaque Problems

1. *Plaques smear over plate*. The plates were too wet at plating. Remove moisture by leaving the tops off of the plates for 10–15 min. This can be done soon after plating at 42° if excess water "pools" on the surface of the plates.

2. *Plaques are too small*. The plates may be too dry. Alternatively, the cell density may be too high; reducing the number of plating cells will increase the plaque size.

3. *Plaques are small on one side of the plate, large on the other side*. The plating bench is not level.

Signal Weak or Absent

1. *Primary antibody binds weakly or not at all*.

a. Antibody lost activity during storage, or is of low titer. Store antisera/ascites at −20° in aliquots: avoid repeated freeze–thaw cycles. Use some other immunochemical assay, such as ELISA, immunoprecipitation, immunodiffusion, to check reactivity toward the antigen. If positive, repeat the immunoscreening assay using higher concentrations of primary antibody.

b. Antibody is of low affinity. Low-affinity antibodies are more affected by buffer conditions, incubation times, and relative concentrations than are high-affinity antibodies. Eliminate Tween 20 from the buffers and increase the incubation time and concentration of the primary antibody.

c. Antibody is conformation dependent. See above section on antibody probes.

2. *Anti-IgG conjugate activity is too low*. Conjugate may have been improperly stored. Store at 4° or in aliquots at −20°. Avoid repeated freeze–thaw cycles, heat treatment, and bacterial contamination. Test activity by adding 1 ml of diluted conjugate in TBST to 1 ml of color

development solution. Intense purple color should appear within 5 min. (This is also a test for activity of the color development substrates.) Although unlikely, it is possible that some water may contain inhibitors. Use reagents of highest quality available.

3. *Improper blocking agent was used.* Some blocking proteins may have effects on antigen recognition by some antibodies. If this is suspected, try a different blocking protein such as bovine serum albumin, casein, gelatin, or another type of serum to saturate excess binding sites. Include controls omitting the primary antibody, the anti-IgG conjugate, and both antibodies from the procedure to check for possible effects on background.

Background Too High

1. *General purple background throughout the membrane.*

a. Color development reaction was too long. Stop the reaction when the color has reached the desired intensity.

b. Poor quality nitrocellulose.

c. Poor quality alkaline phosphatase conjugates.

d. Improper blocking. The blocking step incubation time can be increased if necessary. Some alternative blocking agents such as nonfat milk may contain alkaline phosphatase activity or IgG that bind conjugates. Perform controls by omitting the primary antibody and the anit-IgG conjugate from the procedure.

e. Anti-IgG conjugate concentration too high.

f. Primary antibody sticks nonspecifically even to blocked membranes. Perform the procedure using just a blocked membrane without plaques. A few IgG and other immunoglobulins (particularly IgM) that may have determinants recognized by the second antibody conjugates are especially "sticky" and may be difficult to use for immunoscreening.

2. *Background localized; unexpected plaques appear to be positive.*

a. The primary antibody contains components that react with *E. coli* proteins. See above section on antibody probes.

b. The host strain, phage library, or culture medium is contaminated with organisms producing alkaline phosphatase. This possibility can be tested by omitting the primary antibody and anti-IgG conjugate incubations from the procedure. If color development is significant, there is contamination. If it is not practical to start over with uncontaminated materials try to inactivate the alkaline phosphatase by either heating the plaque lift at 80° for 20 min, or incubating it in 0.1 M acetic acid for 20 min prior to the blocking step. If successful, test the effect of the treatment on primary antibody binding.

c. The anti-IgG alkaline phosphatase conjugate cross-reacts with bacterial proteins. Preadsorb the diluted second antibody conjugate with an *E. coli* extract as described in the section on antibody probes.

[*Editors' Note*. If alkaline phosphatase contamination cannot be eliminated, use either the horseradish peroxidase screening method[6] or the protein A technique.[5] Reagents for both methods are available in kit form with instructions from Bio-Rad (Richmond, CA).]

Production and Purification of Fusion Proteins

After a recombinant has been detected and plaque purified, it is often desirable to obtain preparative amounts of the fusion protein. The first step in this process is to express the λgt11 recombinant as a lysogen in *E. coli* strain Y1089, which contains *hf1A150,* a mutation that enhances the frequency of phage lysogeny. The lysogen is grown to a high density, *lacZ*-directed fusion protein production is induced by the addition of IPTG to the medium, and the cells are harvested and lysed (see also this volume [13]).

The fusion protein can be purified in several ways. Successful methods have taken advantage of the large size of the fusion protein and have therefore involved preparative SDS–polyacrylamide gel electrophoresis and gel permeation chromatography. Alternatively, β-galactosidase fusion proteins can be purified from crude lysates in one step by immunoaffinity chromatography using anti-β-galactosidase antibodies covalently coupled to agarose beads.[10]

[52] Gene Isolation by Retroviral Tagging

By Stephen P. Goff

Insertional mutagenesis has long been recognized as a powerful method for the inactivation and identification of genes in prokaryotes. The technique of transposon tagging to mutate and subsequently to clone selected genes has been extended to eukaryotes; transposable elements have been demonstrated to generate mutations in yeast[1,2] (see this volume

[1] G. S. Roeder and G. R. Fink, *Cell (Cambridge, Mass.)* **21,** 239 (1980).
[2] S. J. Silverman and G. R. Fink, *Mol. Cell. Biol.* **4,** 1246 (1984).

[53]), and the transposable P elements of *Drosophila* have been used to clone several new genes.[3,4] The analogous elements of choice for the mutagenesis of mammalian cells are the retroviruses: agents which upon infection direct the synthesis of a provirus, a DNA copy of the viral genome, and insert that DNA at randomly selected sites in the host genome.

Numerous studies have shown that the insertion of retroviral DNA can result in mutation of cellular genes. Early studies showed that oncogenesis by the leukemia viruses occurs by insertional activation of cellular oncogenes; examples include the activation of genes known as c-*myc*,[5,6] c-*myb*,[7] and c-*erbB*.[8] Insertion events in tumorigenesis have also allowed the identification of novel putative oncogenes as common target genes for retroviral insertions.[9-14] Infection of early embryos, or microinjection of DNAs into eggs, can result in the formation of germ-line insertions. Surprisingly often these insertions cause mutations with observable phenotypes.[15,16] In tissue cultures, infection has been used to induce mutations in several known loci, including a resident v-*src* gene,[17] and a gene for a nucleotide salvage enzyme.[18] It is reasonable to expect that new genes will soon be identified and cloned by retroviral tagging. This chapter describes procedures for the generation of mutations by retroviral infection and gives some examples of applications.

[3] L. L. Searles, R. S. Jokerst, P. M. Bingham, R. A. Voelker, and A. L. Greenleaf, *Cell (Cambridge, Mass.)* **31,** 585 (1982).

[4] W. Engels, *Annu. Rev. Genet.* **17,** 315 (1983).

[5] W. S. Hayward, B. G. Neel, and S. M. Astrin, *Nature (London)* **290,** 475 (1981).

[6] D. Steffen, *Proc. Natl. Acad. Sci. U.S.A.* **81,** 2097 (1984).

[7] G. L. C. Sheng-Ong, M. Potter, J. F. Mushinski, S. Lavu, and E. P. Reddy, *Science* **226,** 1077 (1984).

[8] Y.-K. T. Fung, W. G. Lewis, L. B. Crittenden, and H.-J. Kung, *Cell (Cambridge, Mass.)* **33,** 357 (1983).

[9] R. Nusse and H. E. Varmus, *Cell (Cambridge, Mass.)* **31,** 99 (1982).

[10] G. Peters, S. Brookes, R. Smith, and C. Dickson, *Cell (Cambridge, Mass.)* **33,** 369 (1983).

[11] C. Dickson, R. Smith, S. Brookes, and G. Peters, *Cell (Cambridge, Mass.)* **37,** 529 (1984).

[12] P. N. Tsichlis, P. G. Strauss, and L. F. Hu, *Nature (London)* **302,** 445 (1983).

[13] G. Lemay and P. Jolicoeur, *Proc. Natl. Acad. Sci. U.S.A.* **81,** 38 (1984).

[14] H. T. Cuypers, G. Selten, W. Quint, M. Zijlstra, E. R. Maandag, W. Boelens, P. van Wezenbeek, C. Melief, and A. Berns, *Cell (Cambridge, Mass.)* **37,** 141 (1984).

[15] A. Schnieke, K. Harbers, and R. Jaenisch, *Nature (London)* **304,** 315 (1983).

[16] E. F. Wagner, L. Covarrubias, T. A. Stewart, and B. Mintz, *Cell (Cambridge, Mass.)* **35,** 647 (1983).

[17] H. E. Varmus, N. Quintell, and S. Oritz, *Cell (Cambridge, Mass.)* **25,** 23 (1981).

[18] W. King, M. D. Patel, L. I. Lobel, S. P. Goff, and M. C. Nguyen-Huu, *Science* **228,** 554 (1985).

Selection of Insertional Mutants: Requirements and Frequencies

General Considerations for Applicable Selection Methods

The key step in the successful recovery of insertional mutants is the selection procedure. Each target gene represents a different case, and so we can only consider general restrictions on the selection procedures here. One approach is to ask for an insertional activation event, i.e., the turning on of a previously silent locus. This approach has not been extensively explored in mammalian cells, but the example of oncogene activation would suggest that powerful selections for a given gene product, beginning with an appropriate parent, could succeed. For gene inactivation, the main problem is to devise protocols which allow the growth only of those cells *lacking* the expression of the target gene. The difficulties are greatly enhanced by the fact that most cultured cells are diploid or aneuploid. This fact means that there are usually two codominantly expressed copies of each autosomal gene, and that both must be mutated before the cell is free of the gene products. Our experience has been that insertions into even a single-copy gene are not trivial to detect (see below). Two options exist. One can attack genes known to be located on the X chromosome and present in only one active copy, or one must devise schemes that will allow selection for an insertion into a single allelic copy of the two present in each cell.

A mammalian cell is usually considered to contain about $2-3 \times 10^6$ kb of DNA per haploid complement. An ordinary gene might span 2–30 kb of DNA, and if we assume that a proviral insertion anywhere in this region will inactivate the gene, then a single-copy gene is a target that represents 10^{-5} or 10^{-6} of the total DNA. If retroviral integrations were truly randomly distributed throughout the mammalian genome, then the introduction of one provirus per cell should cause a mutation in 10^{-5} or 10^{-6} of the cells. In the limited number of cases tested to date, this expectation has not been realized. For example, the insertion of about 25 proviruses per cell generated mutations in the hypoxanthine–guanine phosphoribosyltransferase (HGPRT) locus in slightly less than 10^{-6} of the cells.[18] The most likely reason is that retroviral insertion is not truly random, and that the HGPRT gene is a "cold spot" for integration, since it seems to be affected about $50\times$ less often than would be predicted for random insertions. It is not clear that this is a general phenomenon for all target genes, but it seems appropriate to be prepared for the possibility.

The use of transposon tagging to identify new genes requires that one or at most a very few proviruses be inserted in each cell. If there are too many proviruses, it becomes problematical to determine which one is causing the mutation by being in the desired locus and which ones are unimportant. The number of proviruses present after infection by replication-competent viruses (perhaps five or so; see below) is probably in the right range. If the target gene behaves like the HGPRT locus, then the insertional mutation frequency will be about 10^{-7}. This is a low frequency, placing two important constraints on the selection schemes if the majority of the selected cells are to contain insertional mutations. First, the gene must exhibit a very low spontaneous mutation rate, below the insertional mutation rate; the rate probably must be in the range of 10^{-7} to 10^{-8}. Second, the selection must be very powerful, allowing the survival of a very small fraction of the parent cells; again the fraction of survivors must probably be in the range of 10^{-7} to 10^{-8}. If less favorable conditions are used, the problem is that most of the mutant cells will not bear insertional mutations. In early studies of insertions into a v-*src* gene this was precisely the case: only 2 out of 60 mutants arising after Moloney murine leukemia virus infection were due to insertion at the v-*src* locus.[17] If there were no preexisting *src* probe it would not have been easy to determine that these two mutants were the relevant ones.

Specific Examples of Selective Methods

We will not consider selections for the activation of a silent gene, as these tend to be highly specific for the particular gene and conceptually straightforward. Of more general application, but more difficult, are selections for the *absence* of expression of a particular gene. We will describe two such methods.

Toxic Metabolites (Suicide Selections). In particular cases it may be possible to kill selectively the parent cells with toxic compounds that are metabolized by these cells but not by mutants. Sometimes radioactive, isotopically labeled substrates can be used. The classic drugs in this category are the nucleoside or base analogs, which selectively kill those cells able to convert them to the nucleotide form. Examples include 5-bromodeoxyuridine (BrdU), 8-azaguanine (8-AG), 6-thioguanine (6-TG), and azaadenine (AA). Mutant cells lacking the salvage pathways for the utilization of these compounds are unaffected by exposure. Unfortunately, most of the enzymes of these pathways are encoded by autosomal genes and so are not useful targets. The exception is the HGPRT locus on the X chromosome, which we have studied.[18] Exposure of infected cells to a mixture of two drugs (3 μg/ml 8-AG and 6 μg/ml 6-TG in ordinary medium) kills the majority in a few days and leaves the HGPRT$^-$ survivors unaffected. The spontaneous mutation frequency depends on the cell line but in EC cells it is below 10^{-7}. Introduction of 20–50 proviruses per cell gives a marked stimulation in the mutation frequency above the spontaneous background.

Immunoselection. A powerful solution to the problem of detecting inserts into one of two autosomal alleles of a locus is to mutagenize target cells that are heterozygous at the target locus, and to use allele-specific antibodies to select against cells expressing a particular one of the two alleles. The target cells are killed by exposure to the antibody in the presence of rabbit complement, and the selection is strictly limited to genes whose gene products are cell-surface proteins. The net effect is to counterselect for expression of a single-copy gene, eliminating the ploidy problem inherent in the mutagenesis of most autosomal genes. The utility of the method has been demonstrated by the isolation of insertions into the β_2-microglobulin locus.[19]

Immunoselection depends heavily on the quality and specificity of the antibody preparation. The antibody source will vary with target gene, but monoclonal antibodies will be the most commonly used. There must be available strains of animals, normally mice, which show strong allelic differences in their response to the antibody. Conditions of exposure to antibody and complement must be devised so as to kill nearly all of the sensitive cells of a cell population without effect on cells of the resistant strain. Different dilutions of the antibody are simply tested in parallel on the two cell types to determine the appropriate level. Typical exposure conditions are for 3 hr at 37°, after which the survivors are allowed to grow out and are counted. The ratio of survival rates between the two strains should be in the range of 10^6.

If antibodies of sufficient selectivity are available, F_1 heterozygous animals are prepared by mating the two distinctive strains. Permanent cell lines must then be established from these animals, and the expression of the cell surface marker on these cells must be demonstrated. There are many ways to establish immortal cell lines from mice, including immortalization by exposure to chemical carcinogens; one of the most straightforward is transformation by any of a number of retroviruses, such as the Abelson murine leukemia virus.[20] The method of choice will be most probably determined by the tissue distribution of the surface marker. The sensitivity of the established, heterozygous cell line to antibody and complement and the spontaneous mutation frequency at the sensitive allele should be determined. These cells are then infected with the mutagenic

[19] W. Frankel, T. A. Potter, N. Rosenberg, J. Lenz, and T. V. Rajan, *Proc. Natl. Acad. Sci. U.S.A.* **82**, 6600 (1985).
[20] N. Rosenberg and D. Baltimore, *J. Exp. Med.* **143**, 1453 (1976).

virus construct and exposed to selection, and the survivors cloned and grown into large cultures.

Retroviral Infection

Choice of Virus as Mutagen

The mammalian retrovirus now in widest use as a genetic tool in molecular biology laboratories is the Moloney murine leukemia virus[21] (M-MuLV). The virus grows to high titers in a variety of murine cells, infectious DNA clones are available,[22,23] the entire genome has been sequenced,[24] and numerous variants have been constructed that allow sophisticated manipulation of the life cycle. The standard host cell for the preparation of viral stocks is the NIH 3T3 cell line, a hardy and commonly used line.

Wild-type M-MuLV can be used to induce insertional mutations,[17] but a serious disadvantage of the use of the wild-type virus in murine cells is that the newly inserted proviruses cannot be readily detected by hybridization, e.g., by Southern blots (see this volume [61]). The reason is that the mouse genome contains approximately 50–100 copies of endogenous sequences exhibiting strong homology to the exogenous MuLVs, and the presence of these sequences interferes with the detection of the new proviruses. A more useful virus is a viable variant of M-MuLV, in31SuIII, which carries a bacterial suppressor tRNA gene.[25] This virus was constructed by the insertion of a 220-bp fragment containing the *Escherichia coli SuIII*[+] tyrosine suppressor tRNA gene into the retroviral long terminal repeat (LTR) sequence. The inserted DNA at this position is stably retained through many infectious cycles by the virus. This virus has no disadvantages in comparison with the wild-type M-MuLV: it is replication competent, grows to equally high titers, and can infect the same spectrum of cells. Virtually identical viruses have been constructed in Jaenisch's laboratory.[26] The advantage of the use of these viruses is that the integrated proviral DNAs can be readily recovered in the form of recombi-

[21] J. B. Moloney, *J. Natl. Cancer Inst. (U.S.)* **24**, 933 (1960).

[22] C. Shoemaker, S. P. Goff, E. Gilboa, M. Paskind, S. W. Mitra, and D. Baltimore, *Proc. Natl. Acad. Sci. U.S.A.* **77**, 3932 (1980).

[23] I. Chumakov, H. Stuhlmann, K. Harbers, and R. Jaenisch, *J. Virol.* **42**, 1088 (1982).

[24] T. M. Shinnick, R. A. Lerner, and J. G. Sutcliffe, *Nature (London)* **293**, 543 (1981).

[25] L. I. Lobel, M. Patel, W. King, M. C. Nguyen-Huu, and S. P. Goff, *Science* **228**, 329 (1985).

[26] W. Reik, H. Weicher, and R. Jaenisch, *Proc. Natl. Acad. Sci. U.S.A.* **82**, 1141 (1985).

nant phage (see below). The $SuIII^+$ gene can also serve as a hybridization probe, as there are no homologous sequences in the mammalian genome.

Other virus constructs are also useful for particular purposes. Replication-defective constructs, carrying foreign DNA, are potentially valuable as insertional mutagens. Constructs carrying and expressing a number of marker genes have been described, including resistance to the aminoglycoside G418,[27] the thymidine kinase gene conferring resistance to HAT medium,[28] the gene for hypoxanthine–guanine phosphoribosyltransferase (HGPRT) conferring resistance to HAT medium,[29] and many others. These constructs are as a rule replication defective, and their transmission through virion particles requires that they be introduced by transformation into cells already expressing a "helper" virus genome. Wild-type virus can be used as helpers, but in many ways the ideal helpers are the Psi2[30] or Psi-am viruses,[31] which produce virions but are not themselves transmissible. The constructs have the advantage that successfully infected cells can be isolated by selection for the marker gene on the virus. The Psi2 and Psi-am helpers have the additional advantage that there is no replicating virus in the target cells. Only the defective construct genome is transmitted to the target cell and no helper genome virus is transferred.

Choice of Target Cell Line

Numerous cell lines can be infected by retroviruses and thus can serve as targets for insertional mutagenesis. The majority of murine cell lines in culture can be infected with M-MuLV. Receptor is present on a wide spectrum of cell types, and there are no genetic backgrounds known that can block provirus insertion of the NB-tropic M-MuLV intracellularly. Any cell line, however, that has previously been infected with an ecotropic MuLV or that expresses an ecotropic MuLV envelope protein, will be resistant to infection. The popular L cell line, for example, is resistant because it is chronically infected by and is a producer of the L cell virus (LCV), an ecotropic MuLV related to M-MuLV. Rat cells are generally somewhat less sensitive to infection with MuLVs than mouse cells, but are perfectly usable as targets. These cells have the enormous advantage

[27] C. L. Cepko, B. E. Roberts, and R. C. Mulligan, *Cell* (*Cambridge, Mass.*) **37**, 1055 (1984).
[28] C. J. Tabin, J. W. Hoffmann, S. P. Goff, and R. A. Weinberg, *Mol. Cell. Biol.* **2**, 426 (1982).
[29] A. D. Miller, D. J. Jolly, T. Friedmann, and I. M. Verma, *Proc. Natl. Acad. Sci. U.S.A.* **80**, 4709 (1983).
[30] R. Mann, R. Mulligan, and D. Baltimore, *Cell* (*Cambridge, Mass.*) **33**, 153 (1983).
[31] R. D. Cone and R. C. Mulligan, *Proc. Natl. Acad. Sci. U.S.A.* **81**, 6349 (1984).

that the endogenous retroviral sequences are only very weakly homologous to MuLVs, and the presence of new MuLV proviruses can be readily detected by hybridization with MuLV probes.

One murine cell type deserves particular mention. Embryonal carcinoma (EC) or teratocarcinoma cell lines can be infected with ecotropic viruses like M-MuLV, and proviral DNAs are inserted normally into the DNA of these hosts. The inserted viral DNA, however, is not actively transcribed in these cells to form viral mRNAs or genomic RNAs.[32,33] The result is that there are no viral gene products and no progeny virus formation. Insertional mutagenesis of these cells is particularly attractive because there can be no complications in selection procedures that virus progeny might cause.

Cells of nonrodent species are generally resistant to infection by ecotropic MuLVs. Cells from an enormous spectrum of species, however, can be infected by so-called amphotropic viruses.[34,35] These are mouse viruses which utilize a receptor different from that of the ecotropic viruses, one that is apparently common on cells of most mammalian species. Cells susceptible to amphotropic viruses include human cells, numerous primate cells, canine and feline cells, and most rodent cells. Significant exceptions are the popular Chinese hamster ovary (CHO) cell line and the L cell line. The viruses of choice for the permissive cells are a virus called 292A,[36] and one called 4070A.[34] Biologically cloned isolates of both viruses are available, and the latter virus has been molecularly cloned.[37] The same spectrum of cells that can be infected with these amphotropic viruses can be infected with replication-defective virus constructs based on the M-MuLV genome, if the constructs are transferred with the Psi-am helper virus.[31] This helper was constructed by inserting the 4070A *env* gene into a M-MuLV backbone, conferring on that virus the broad host range of the amphotropic parent.

Infection Protocols

The simplest procedure for the transfer of virus from a producer cell to an adherent target cell is by infection with a cell-free virus harvest. Under

[32] C. L. Stewart, H. Stuhlmann, D. Jahner, and R. Jaenisch, *Proc. Natl. Acad. Sci. U.S.A.* **79**, 4098 (1982).

[33] O. Niwa, Y. Yokota, H. Ishida, and T. Sugahara, *Cell (Cambridge, Mass.)* **32**, 1105 (1983).

[34] J. W. Hartley and W. P. Rowe, *J. Virol.* **19**, 19 (1976).

[35] S. Rasheed, M. B. Gardner, and E. Chan, *J. Virol.* **19**, 13 (1976).

[36] M. B. Gardner, *Curr. Top. Microbiol. Immunol.* **70**, 215 (1978).

[37] S. K. Chattopadhyay, A. I. Oliff, D. L. Linemeyer, M. R. Lander, and D. R. Lowy, *J. Virol.* **39**, 777 (1981).

normal conditions the virus is collected whenever it is convenient, stored at $-70°$, and thawed at the time of infection at room temperature or 37°. Polybrene (Sigma; stock solution of 8 mg/ml in phosphate-buffered saline) is added to a final concentration of 8 μg/ml to enhance the infectivity (usually a 5–10× enhancement). Some cells are rather sensitive to Polybrene and lower concentrations (2 μg/ml) or none at all may be used.

The recipient cells should be subconfluent and rapidly growing at the time of infection; typically they would be at a density of 10^6 cells per 10-cm dish (10 ml of medium). With adherent cells, the infection is simply carried out by removing the medium and replacing it with 0.1 volume of undiluted virus. For suspension cultures, the cells are centrifuged and resuspended in the virus. The virus is allowed to adsorb to the cells for 1–2 hr with occasional mixing (either rocking plates of adherent cells or stirring tubes of suspension cells), and then 1 volume of medium is added to the cells. Normally we leave the virus in the medium but if the cells are sensitive to Polybrene, the virus preparation and the Polybrene it contains are removed before addition of the medium.

The cells are allowed to grow for at least 48 hr before any selection is imposed. Retroviral integration begins in perhaps 12 hr, but circular preintegrative forms persist for several days, and there may be integration continuing during this long period. Integration is thought to require cell division, and it is probably wise to keep the cells rapidly growing for at least 2 days. If low titers of virus are used, or if the cells are rather resistant to infection, it may be desirable to wait several days until all the cells are infected.

When replication-competent virus is used, the multiplicity of virus used is not very critical. If a multiplicity of infection (MOI) of 10 is achieved, the target cells will probably receive about 10 proviral inserts per cell. If an MOI of 1 is used, the cells initially receive one provirus, but such infected cells begin to produce progeny virus that can superinfect the producer cells themselves for a period of about a day or two. At that time envelope protein, the product of the *env* gene, appears on the cell surface in substantial quantity, and the cell becomes resistant to further superinfection. The result is that usually about five proviruses are formed before superinfection resistance appears. If a very low MOI is used, virus will spread through the culture until essentially all the cells are infected, with the result that about five insertions per cell still occur.

An exception to this rule occurs when EC cells are the targets. These cells do not express virus or envelope protein, and they never exhibit superinfection resistance.[32,33] Thus, the MOI directly determines the number of insertions per cell. One can increase the effective MOI by applying virus repeatedly to the same cultures, adding a few proviruses each time.

Cocultivation

Cocultivation of the target cells with the producer cells can result in the efficient transfer of virus. The target cells are simply replated onto a lawn of producer cells. About 10^6 target cells could be plated on 10^6 producer cells in a 10-cm dish. Direct contact seems to promote highly efficient transfer of virus, and extended contact allows continuous infection to generate a large number of inserts in the target cells. Our experience with the F9 EC cell line as recipient showed that cocultivation for 1 day yielded about 5–10 copies per cell, and for 3 days yielded 20–50 copies.[18] These numbers are certain to vary somewhat with choice of producer and target cell lines.

When cocultivation is used, it may be problematical to separate the donor from the target cells. One solution is to kill the donor cells before cocultivation with a brief exposure to ultraviolet light. The medium is removed, the cells are washed once with phosphate-buffered saline, and the uncovered dishes are exposed to the germicidal ultraviolet lights in a tissue culture hood (distance of about 50 cm) for 10–20 sec. Such cells continue to release virus for several days but cannot divide and eventually die. Alternatively, growing producer cells can be killed by treatment with mitomycin C (10 μg/ml for 2–4 hr), washed, and overlaid with target cells.[32]

Analysis and Recovery of Mutant Alleles

The value of insertional mutagenesis in the identification of new target genes is that the inserted DNA provides a physical and genetic tag for the isolation of the target region. It is here that the use of marked retroviral genomes provides a most important advantage over wild-type genomes.

Counting Inserts

If the spontaneous mutation rate at the target locus were zero, if the selection procedure were perfect, and if there were only one retroviral insertion per mutant cell, it would be possible to clone one provirus region from one mutant and be confident that the target gene was in hand. Unfortunately these conditions are never met in reality. Usually, there are more than one provirus per cell, and many of the cell lines surviving the selection procedure are spontaneous mutants, bearing proviruses at irrelevant loci. One must therefore isolate many independent potential insertional mutants, and determine whether there are any insertion sites in common among them. Such common sites must then be the site of the target gene.

Under normal circumstances one should isolate 20–50 independent surviving cell lines after infection. Each clone is grown to about 10^7 cells, and genomic DNA is prepared by SDS–proteinase K lysis and repeated phenol extractions (see this volume [15,4]). The DNAs are cleaved with restriction enzymes that do not cleave within the mutagenic retroviral genome, and analyzed by blot hybridization[38] after electrophoresis on low percentage agarose gels (see this volume [8,61]). For variant $in31SuIII$[25] and related viruses,[26] the $SuIII$ gene of plasmid piVX[39] is the appropriate probe. For constructs like pZIP-NeoSV(X)1, the bacterial neomycin-resistance gene can be used.[27] The number of bands seen by blot hybridization should reflect the number of insertions per cell. As noted above, it is helpful if this number is small: one, or at most a few.

If DNAs of several cell lines, cut with enzymes that cut outside the provirus, show the same pattern of hybridizing bands, then this is indicative of repeated hits of the same region. The provirus is cloned from these cells. The results of such surveys, however, are usually ambiguous, due to the fact that the relevant fragments are large (all are bigger than the virus) and that the gels have limited resolution. If the target gene is large, two independent hits of that gene may not even yield similar patterns. In these cases all the proviruses must be cloned from each of many cell lines.

Cloning the Proviruses and Flanking DNA

ISOLATION OF CLONES. Provirus clones must be isolated from several genomic DNAs. The procedure must be rapid and easy, since clones must be obtained from as many as 20–50 genomic samples. Two methods have been used to facilitate the cloning of these elements.

Selection for $SuIII^+$. If the mutagenic provirus carries the bacterial $SuIII^+$ gene, all the proviruses can be recovered from a genomic library in phage vectors. The procedure is similar to the recovery of transforming genes artificially linked to the $SuIII^+$ gene by cotransformation.[40] A phage library is prepared by conventional means in one of the Charon phage vectors[41] carrying amber mutations in essential phage genes (see this volume [16,17]). We have used Charon 30A, but 4A, 16A, and 21A work as well. Partial cleavage of the genomic DNA with Sau3A, and insertion into vectors cut with BamHI, is a good procedure in that this ensures

[38] E. M. Southern, *J. Mol. Biol.* **98**, 503 (1975).

[39] B. Seed, *Nucleic Acids Res.* **11**, 2427 (1983).

[40] M. Goldfarb, K. Shimizu, M. Perucho, and M. Wigler, *Nature (London)* **296**, 404 (1982).

[41] D. L. Rimm, D. Horness, J. Kucera, and F. R. Blattner, *Gene* **12**, 301 (1980).

good probability of cloning the provirus no matter what the structure of the flanking DNA. The DNA is ligated to the vector, packaged into phage coats, and amplified on a suppressing bacterial host such as LE392. At least 10^6 independent recombinant phage must be formed. About 10^8 of the amplified phage are plated on a nonsuppressing, amber lac^- host, including the indicator dye X-Gal[42] (40 μgml, final concentration) in the plate. Any phage carrying the $SuIII^+$ gene can grow on the nonsuppressing host, and will also suppress the amber lac^- mutation of the host, forming blue plaques. These plaques are simply picked and grown up for further analysis. White plaques also present are phage that have lost the amber markers by recombination with DNA in the packaging extracts, but do not carry the desired $SuIII^+$ gene.

Selection for Plasmid Replicon and Drug Marker. If the retrovirus contains a complete functioning plasmid replicon, the provirus can be cloned directly. The genomic DNA is cleaved to completion with a restriction enzyme that does not cleave in the provirus, and 5–10 μg is ligated at low concentration (5–10 μg/ml; total volume 1 ml) to promote circularization. The ligated DNA is then used to transform *E. coli* to drug resistance, selecting for the marker on the provirus. Standard procedures can be used, so long as the efficiency of transformation is in the range of 10^7 transformants/μg of DNA. No more than about 50 ng can be used per 0.1 ml of competent cells, plated on each 10-cm plate. For example, the pZIP-NeoSV(X)1 factors[27] contain a ColE1 replicon and confer resistance to kanamycin (50 μg/ml in L plates). The resulting colonies are picked and plasmid DNA is isolated from the cultures.

ISOLATION OF TARGET GENE: COMPARISON OF FLANKING DNAS. In the final step, the clones obtained from the genomic DNAs must be compared to determine if any represent insertions into common, shared target loci. One screen to apply to the clones is simply to digest them all with enzymes which cleave often, producing numerous small fragments. Examination on gels might in favorable cases reveal common-size fragments. The fragments from the retrovirus and vector itself, which will be common to all clones, have to be ignored.

The ultimate test for homology between flanking clones is to use each clone as a probe for Southern blot hybridizations against the other clones (see this volume [61]). The DNAs are digested with enzymes so as to produce at least a few fragments free of vector and provirus, and multiple copies of a blot are prepared. Hybridization with individual clones as probes are performed. If one probe cross-hybridizes with a flanking DNA fragment of another, more detailed studies are warranted. The possibility

---

[42] X-Gal is 5-bromo-4-chloro-3-indolyl-β-D-galactoside.

of repetitive elements being responsible for the cross-reaction must be eliminated. To accomplish this, unique sequence (nonrepetitive) probes must be prepared and subclones of short subregions of the flanking DNA must be screened by hybridization with total, labeled, genomic DNA for the absence of such repetitive sequences. If a unique flanking probe still shows cross-reaction, the gene is identified. This probe can then be used to clone the wild-type locus, to prepare cDNAs, and to examine other mutants for alterations in the locus.

[53] Isolation of Genes by Complementation in Yeast

By MARK D. ROSE

The yeast *Saccharomyces cerevisiae* has a unique utility in recombinant DNA technology. *Saccharomyces cerevisiae* is a simple eukaryote that has technically simple methods of transformation while permitting extraordinarily powerful methods of genetic analysis. The isolation of the putative chromosomal origins of replication and the major structural components of the chromosome has allowed an enormous diversity of transformation techniques each suited to a different analytical purpose. This chapter will cover the various strategies that have been used to isolate yeast genes by complementation and will point out some of the novel approaches that are being used to extend these techniques. Detailed protocols have been included so that anyone can begin to isolate a gene by complementation in yeast. A complete review of all of the genetic methods available to the researcher for the analysis of cloned genes is well beyond the scope of this chapter. General methods for handling yeast strains can be found in the excellent methods book used in the yeast course at Cold Spring Harbor.[1]

Background

Advantages of a Small Genome

Saccharomyces cerevisiae has a genome that is only 15,000 kb in size.[2] Therefore any sufficiently large recombinant DNA library having ran-

---

[1] F. Sherman, G. R. Fink, and C. W. Lawrence, "Methods in Yeast Genetics." Cold Spring Harbor Lab., Cold Spring Harbor, New York, 1979.

[2] G. O. Lauer, T. M. Roberts, and L. Klotz, *J. Mol. Biol.* **114**, 507 (1977).

domly generated inserts with size-selected inserts averaging 15 kb or larger should have any region of DNA represented at a frequency approaching 1 in 1000, assuming all other things are equal. Several such libraries currently exist and are freely available.[3-5] Because of the small number of clones that must be screened, many cloning experiments that rely on labor-intensive methods or subtle phenotypes are feasible for yeast but would be unthinkable for higher eukaryotes.

Transformation

Saccharomyces cerevisiae can be transformed readily with DNA either by preparation of spheroplasts[6] or by treatment with alkaline salts such as LiCl.[7] Transformation may utilize any one of several fairly well-defined modes of maintenance of transforming DNA. In essence, DNA may be integrated into the chromosome via homologous recombination or maintained as an episome using plasmids with varying degrees of stability and copy number. In practice therefore, it has become routine to clone different yeast genes based on the ability to complement different mutations. This approach is feasible if the gene of interest has been previously defined by virtue of a useful mutation and if that mutation is recessive. However, current interest is shifting to genes that encode proteins for which no mutations have been isolated.

The great advantage of yeast for long-term genetic analysis is the ability to recombine virtually any cloned DNA with its normal chromosomal locus. To begin with, the novel phenotypes that can arise from the gene overexpression associated with high copy number plasmids mean that isolation of a complementing DNA fragment is not proof of its identity. Therefore, other criteria must be used to verify the identity of a cloned gene. Integration of the cloned DNA at the genetic locus associated with a given mutation via homologous recombination is a priori evidence for (but not absolute proof of) the identity of the DNA. Absolute proof of identity requires construction of mutations *in vitro* and replacement of the wild-type allele with the new mutation. Indeed the construction of mutations in cloned genes followed by observation of the altered phenotype of cells bearing that mutation is the fundamental genetic approach to determining the normal physiological role of any given element. To date, yeasts are the only eukaryotic organisms in which the phenotype of recessive mutations constructed *in vitro* can be assessed in a routine fashion.

[3] M. Carlson and D. Botstein, *Cell (Cambridge, Mass.)* **28,** 145 (1982).
[4] K. Nasmyth and S. I. Reed, *Proc. Natl. Acad. Sci. U.S.A.* **77,** 2119 (1980).
[5] M. D. Rose, P. Novick, J. H. Thomas, D. Botstein, and G. R. Fink, in preparation.
[6] A. Hinnen, J. B. Hicks, and G. R. Fink, *Proc. Natl. Acad. Sci. U.S.A.* **75,** 1929 (1978).
[7] H. Itoh, Y. Fukada, K. Murata, and A. Kimura, *J. Bacteriol.* **153,** 163 (1983).

History of Yeast Cloning

One major strategy of gene isolation has taken advantage of the physical properties of a gene or its product. Examples include the use of specific hybridization probes such as radioactive cDNAs and oligonucleotides or specific antibodies to screen, in *Escherichia coli*, libraries constructed in plasmid or phage vectors. Many of these techniques are described in detail elsewhere in this volume. Those aspects that are not unique to yeast will not be discussed further. Alternatively, the functional properties of the gene can be exploited by complementation of relevant mutations or by otherwise altering the phenotype of the cell.

The first experiments in cloning yeast genes utilized hybridization schemes to isolate the genes for the abundant stable RNAs.[8,9] Schemes that rely on hybridization remain of great import both by allowing selection based on changes in the regulation of interesting genes[10] and by allowing access to interesting genes that have been isolated previously in other organisms. Examples of this approach include the isolation of the genes for β-tubulin[11] and the *ras* oncogenes.[12]

The demonstration that some yeast genes can complement the equivalent mutation in *E. coli*[13] also led to the rapid isolation of several yeast genes. This method has been completely superceded by the direct complementation of genes in yeast.

The advent of methods for yeast transformation[6] has allowed the isolation of many genes that do not complement the equivalent *E. coli* mutation,[14] or for which no comparable *E. coli* mutation exists,[4] or for which the abundance or regulation of the RNA transcript prevented the isolation of the gene by hybridization. The ease and efficiency of yeast transformation are such that one report[15] has demonstrated the feasibility of cloning genes by direct transformation into yeast without the use of *E. coli* as an intermediate to amplify the cloned DNA. In fact, transformation of yeast has allowed the isolation of at least one gene from a higher eukaryote, the

[8] R. A. Kramer, J. R. Cameron, and R. W. Davis, *Cell (Cambridge, Mass.)* **8**, 227 (1976).

[9] T. D. Petes, J. R. Broach, P. C. Wensink, L. M. Hereford, G. R. Fink, and D. Botstein, *Gene* **4**, 37 (1978).

[10] T. St. John and R. W. Davis, *Cell (Cambridge, Mass.)* **16**, 443 (1979).

[11] N. N. Neff, J. H. Thomas, P. Grisafi, and D. Botstein, *Cell (Cambridge, Mass.)* **33**, 211 (1983).

[12] D. Defeo-Jones, E. M. Scolnick, R. Koller, and R. Dhar, *Nature (London)* **306**, 707 (1983).

[13] K. Struhl, J. R. Cameron, and R. W. Davis, *Proc. Natl. Acad. Sci. U.S.A.* **73**, 1471 (1976).

[14] A. Hinnen, P. J. Farabaugh, C. Ilgen, G. R. Fink, and J. Friesen, *ICN-UCLA Symp. Mol. Cell. Biol.* **14**, 43 (1979).

[15] M. C. Kielland-Brandt, T. Nillson-Tillgren, J. R. Litske Peterson, and S. Holmberg, *Alfred Benzon Symp.* **16**, 369 (1980).

ade8 gene of *Drosophila melanogaster*,[16] based on its ability to complement the cognate mutation in yeast. Although in this case the relevant
gene was expressed in yeast, this is very much an exception. It is likely
that this approach will only prove successful in cases where the organism
is closely related to yeast or where the library uses full-length cDNAs
fused to a yeast promoter.

Yeast Vectors

Yeast plasmid vectors[17] can be separated into two basic types depending on their mechanism of replication in the transformed cell. The integrating vectors (e.g., YIp5, *Yeast Integrating plasmid*) have a selectable
gene (in this case the yeast *URA3* gene) and the means for selection and
replication in *E. coli* but no mechanism for replication in yeast. Therefore,
selection after transformation results in cells that contain an integrated
copy of the plasmid as a product of a homologous recombination event
between the plasmid and the genomic DNA. Typically, with closed circular plasmid DNA, the frequency of transformation is on the order of 1–2
transformations/μg of input DNA. Creating linear plasmids by cleaving
the plasmid within the yeast DNA greatly increases the frequency of
transformation from 10- to 100-fold and also has the useful effect of directing the integration event to the site of the cleavage. Therefore the integration event can be directed to one site among several homologous regions
present on the plasmid. In addition, cleavage can allow the isolation of
strains bearing integrated copies of plasmids that would otherwise remain
autonomous.

The second class of yeast plasmids have in common their capacity to
be autonomously replicated in the yeast cell. The YRp series (for *Yeast
Replicating plasmid*) have a so-called *ars* element that has the characteristics expected for a chromosomal origin of replication. Transformation
with these vectors is extremely efficient, approaching several thousand
per microgram of input DNA. In the absence of a mechanism for efficient
segregation of such plasmids into the daughter cells, such transformants
are inherently unstable. Asymmetric segregation causes accumulation of
the plasmid into the mother cell, raising the copy number to very high
levels. After overnight nonselective growth of cultures containing these
plasmids, typically less than 10% of the cells will still harbor the plasmid.

[16] S. Henikoff, K. Tatchell, B. H. Hall, and K. A. Nasmyth, *Nature (London)* **289,** 33
(1981).
[17] D. Botstein and R. W. Davis, *in* "The Molecular Biology of the Yeast *Saccharomyces*;
Metabolism and Gene Expression" (J. N. Strathern, E. W. Jones, and J. R. Broach, eds.),
p. 607. Cold Spring Harbor Lab., Cold Spring Harbor, New York, 1982.

The YEp plasmids (*Yeast Extrachromosomal plasmid*) contain a portion of the endogenous yeast plasmid, the 2-μm circle.[18] The various vectors differ in the particular portion of the plasmid that they contain. Their unique features are derived from the capacity of the endogenous plasmid to amplify its copy number and to be segregated efficiently to both mother and daughter cell. The YEp plasmids typically contain at least one copy of the 560-bp inverted repeat region from the 2-μm plasmid. This sequence is efficiently recombined in yeast by a site-specific recombinase elaborated by the 2-μm plasmid. Some of the YEp plasmids contain other parts of the 2-μm plasmid, usually the origin of replication. However, an origin of replication is not required to be present on the vector to obtain high frequency transformation and high copy number. The plasmid rapidly recombines with endogenous plasmid by the action of *FLP* recombinase on the inverted repeat sequences. Once recombination has occurred, the vector is replicated and amplified as a part of the 2-μm plasmid. The transformation frequencies of these plasmids approach several thousand per microgram. The plasmids tend to be fairly stable; over half of the cells in an overnight culture of the transformed strain will still harbor the plasmid after nonselective growth. The plasmids are generally present at around 10–40 copies per cell, though this estimate varies widely in the literature, possibly because of differences in plasmid structure, method of assay, and the particular selection history of the cultures assayed.

The addition of a chromosomal centromere to an *ars*-containing plasmid produces a YCp vector (for *Yeast Centromere plasmid*) and has two related beneficial effects. The copy number of the plasmid is reduced to 1–2 per cell and the plasmid acquires a much greater stability. Typically an overnight culture of a YCp-containing strain grown without selection will be composed of 90–99% or more plasmid-containing cells. This frequency depends upon the size of the plasmid, with stability increasing with size to a plateau value that is similar to that observed for circular chromosomes.

Linear plasmids (YLp) have been constructed by utilizing the telomeres (ends of chromosomes) from both the naturally linear amplified ribosomal DNA repeats from the macronuclei of ciliates and the normal yeast chromosomes.[19] Although useful for answering questions of chromosome structure and function, the linear plasmids have little utility for the cloning of most genes in yeast. Their great virtue may lie with the

---

[18] J. Broach, *in* "The Molecular Biology of the Yeast *Saccharomyces*; Life Cycle and Inheritance" (J. N. Strathern, E. W. Jones, and J. R. Broach, eds.), p. 445. Cold Spring Harbor Lab., Cold Spring Harbor, New York, 1981.

[19] J. W. Szostak and E. H. Blackburn, *Cell (Cambridge, Mass.)* **29**, 245 (1982).

isolation of telomere-proximal genes, which are likely to be underrepresented in circular plasmid libraries. To date this has not been a problem because the telomere-proximal sequences have been found to be tandem repeats of a long sequence. Therefore the functional genes of interest have proved to be sufficiently distal to the telomere that they have been present in the circular plasmid banks currently available.

Limitations to Cloning by Complementation

No Mutations in Gene

The isolation of yeast genes by the complementation of preexisting mutations has been extremely successful. The method has the obvious limitation that many interesting genes have yet to be identified by mutation. Moreover, it is likely that some genes will prove to be intractable genetically by virtue of being present as members of multigene families (e.g., the two α-tubulin genes in *S. cerevisiae*[20]). Typically such genes have been accessible via dominant mutations, such as those that confer resistance to specific inhibitors. In many cases, however, a specific inhibitor is unknown.

Genes Missing from Libraries

The complementation method has also failed for the technical reason that some genes fail to be represented in the different plasmid libraries. In a few instances the reason for the lack of representation is readily apparent. Some genes are toxic in yeast when present on high copy number plasmids. Examples include the genes for actin (*ACT1*) and β-tubulin (*TUB2*)[21] and the *KAR1* gene.[22] This particular problem has been bypassed by the use of a YCp plasmid library.[5] Some yeast genes appear to have toxic effects when they are present in *E. coli*. Since most yeast genes are expressed only poorly in *E. coli* to begin with, this problem can be effectively solved by use of low copy number plasmids in *E. coli* (e.g., the *KAR2* gene is toxic to *E. coli* on vectors that use the high copy number ColE1 origin of replication such as pBR322 but is not toxic on plasmids that use the pSC101 origin of replication[23]). Alternatively, cloning directly from yeast to yeast is feasible and has been successful for cloning the *HOL1* gene which is adjacent to a gene that is quite toxic to *E. coli*.[24]

[20] P. J. Schatz, L. Pillus, P. Grisafi, F. Solomon, and D. Botstein, *Mol. Cell Biol.* **6,** 3711 (1986).
[21] J. H. Thomas, Doctoral Dissertation, M. I. T., Cambridge, Massachusetts (1984).
[22] M. D. Rose and G. R. Fink, *Cell (Cambridge, Mass.)* **48,** 1047 (1987).
[23] M. D. Rose, unpublished observation.
[24] R. Gaber and G. R. Fink, personal communication.

No one plasmid library can solve all the problems that prevent cloning of a particular gene. For example, whereas the YCp banks contain genes that are not present in the YEp banks, genes that are very close to the chromosomal centromeres will be underrepresented in the centromere bank. The presence of the additional centromere from the insert DNA produces a dicentric plasmid which is subjected to the bridge–breakage–fusion cycle observed in dicentric chromosomes.

Multiple Complementing Genes

Another problem frequently accompanying the use of high copy number plasmid vectors is too many rather than too few clones. For certain genes the presence of a different gene in high copy number causes the suppression of the mutation that is being complemented.[25-29] Typically this phenomenon has been observed in situations where the products of the other genes are thought to interact with the first gene product or where being present in high copy number causes alteration or escape from the regulatory consequences of the original mutation. Although this property is interesting and useful, it nevertheless introduces uncertainty into the identification of the gene of interest based on complementation alone, particularly when YEp or YRp vectors are used.

Extensions of the Method

Cross-Suppression

Even in the absence of a mutation in the gene of interest, the relevant gene can still be isolated by extension of the principle of complementation. Basically, yeast complementation cloning depends on the observation of an altered phenotype in the transformed cell. Several instances have documented the fact that the presence of multiple copies of a wild-type gene can lead to alteration in the phenotype of the cell in spite of a wild-type copy of that same gene on the same chromosome. In the case of general amino acid regulation the genes for *GCN1* and *GCN2* were found to suppress the mutant phenotypes of mutations in *GCD1* as well as each other.[25] Although this cross-suppression complicated the identification of

[25] A. G. Hinnebusch and G. R. Fink, *Proc. Natl. Acad. Sci. U.S.A.* **80,** 5374 (1983).

[26] G. Natsoulis, F. Hilger, and G. R. Fink, *Cell (Cambridge, Mass.)* **46,** 235 (1986).

[27] J. R. Pringle, S. H. Lillie, A. E. M. Adams, C. W. Jacobs, B. K. Haarer, K. G. Coleman, J. S. Robinson, L. Bloom, and R. A. Preston, *ICN-UCLA Symp. Mol. Biol., New Ser.* **33,** 47 (1986).

[28] V. L. MacKay, this series, Vol. 101, p. 325.

[29] C.-L. Kuo and J. L. Campbell, *Mol. Cell Biol.* **3,** 1730 (1983).

the correct genes, the fact of cross-suppression was turned to advantage for the dissection of the regulatory pathway. Similarly the high copy number plasmids that suppress a temperature-sensitive mutation in the gene for the histidyl-tRNA synthetase include the genes for the histidyl-tRNAs as well as the gene for the synthetase.[26] Cases of cross-suppression include examples from the *cdc* genes,[27] the sterile genes,[28] and genes required for DNA synthesis.[29] It is likely that the cross-suppression of mutations in related genes will be common in circumstances where the gene products are part of regulatory pathways or where they physically interact. Moreover, in cases in which a mutation is "leaky" and the phenotype arises from the instability of the gene product (rather than a null mutation for example), it may be common that cross-suppressor genes can be isolated that lead to an increase in the level of the substrate. An example of this is the *hts1* mutation which can be suppressed both by addition of histidine to the medium and by increased numbers of the histidyl-tRNA genes.

Use of Quasi-Mutant Phenotypes

A similar approach can be taken when there is no mutation to suppress as long as there is a sufficiently well-defined phenotype to observe. In some cases, resistance to specific inhibitors can be achieved by overexpression of the wild-type gene product that is the target of the drug. In principle overexpression can be achieved by having the structural gene or genes for positive activators or limiting modifiers on high copy number plasmids. By this route one of the genes for HMG-CoA reductase has been isolated based on the increased resistance to the inhibitor compactin.[30] Resistance to tunicamycin and ethionine also yielded interesting plasmid clones.[31]

Similarly, by using a sensitive assay for chromosome instability, Meeks-Wagner and colleagues[32,33] showed that an imbalance in the ratio of histones created by carrying the genes on high copy number plasmids leads to increased frequencies of chromosome loss. They were then able to isolate two new genes that are involved with the fidelity of chromosome segregation. Both genes caused increased frequencies of chromosome loss when present in high copy number.

---

[30] M. E. Basson, M. Thorsness, and J. Rine, *Proc. Natl. Acad. Sci. U.S.A.* **83,** 5563 (1986).
[31] J. Rine, W. Hansen, E. Hardeman, and R. W. Davis, *Proc. Natl. Acad. Sci. U.S.A.* **80,** 6750 (1983).
[32] D. Meeks-Wagner and L. Hartwell, *Cell (Cambridge, Mass.)* **44,** 43 (1986).
[33] D. Meeks-Wagner, J. S. Wood, B. Garvik, and L. H. Hartwell, *Cell (Cambridge, Mass.)* **44,** 53 (1986).

In essence, the appearance of new phenotypes when certain genes are present in high copy number is akin to the idea of isolating dominant mutations. Dominant mutations may arise from higher level expression or constitutive synthesis of regulated genes. Alternatively, dominant mutations may arise by alteration of gene function. By using high copy number as a mechanism for generating increased gene expression, altered function mutations may be avoided. Of course, any such selection scheme may be prone to the "artifactual" isolation of genes which can cross-suppress a given mutation by virtue of some subtle but unrelated alteration in cell physiology. Therefore the burden remains on the investigator to demonstrate that the gene that has been isolated is relevant. In particular it is essential to demonstrate, if possible, that the gene product of interest is overexpressed in the transformed cells and that a loss of function mutation in the isolated gene leads to the predicted phenotype. In the case of multiple gene families this may be difficult as the loss of function may be masked by other copies of the gene. Nevertheless, in the case of α-tubulin, in both *S. cerevisiae* and *Schizosaccharomyces pombe,* loss of a single copy of one of the genes has a pronounced effect on cell viability which can be suppressed by the presence of extra copies of the other gene.[34,35] In the case of HMG-CoA reductase, loss of one of the two genes does not affect cell viability but does result in increased sensitivity to compactin.[30]

Antigen Screening

A quite different approach to the problem of cloning a gene when no mutation exists is to search for clones that overproduce the antigen, provided the protein has been identified. This method is well described in this volume [46, 54] using the λgt11 vector for isolating genes from yeast and other organisms in *E. coli*. In addition, similar methods of antigen detection have been described for yeast,[36] and in principle, these could be adapted to looking for the overproduction of a yeast antigen based on the overexpression observed from 2-μm plasmid-derived vectors. The problem with the λgt11 system is that rare transcripts may be present at much less than one part in 10^{-5} in a cDNA bank derived from yeast mRNA (particularly if the bank was prepared from cells in an inappropriate regulatory state). By using a genomic plasmid bank with sufficiently large inserts, no gene need be less abundant than 1 per 1000 plasmids. In practice this approach requires extremely sensitive antibody probes cou-

[34] P. J. Schatz, F. Solomon, and D. Botstein, *Mol. Cell. Biol.* **6,** 3722 (1986).
[35] Y. Adachi, T. Toda, O. Niwa, and M. Yanagida, *Mol. Cell Biol.* **6,** 2168 (1986).
[36] S. Lyons and N. Nelson, *Proc. Natl. Acad. Sci. U.S.A.* **81,** 7426 (1984).

pled with scrupulous care in preventing cross-reaction to irrelevant antigens, because the level of overexpression is unlikely to approach that attainable from viral replication and transcription. An extension of this approach, however, might allow the isolation of a gene for which some product of its activity can be immunologically detected. Possibilities include increased synthesis of cell wall constituents or increased levels of a modification when that modification can be specifically probed with a monoclonal antibody.

Limitations of High Copy Number Screens

One problem with these methods is that not all genes are overexpressed when present on high copy number plasmids. For example, although *lacZ* gene fusions to the *URA3* gene present on YEp plasmids are expressed at levels up to 40-fold higher than single copy,[37] *HIS4–lacZ* gene fusions on YEp plasmids are expressed only a few-fold higher[38] than single copy. The reasons for this are obscure. Therefore high copy number screens should not be considered an exhaustive selection for any gene. An alternative method would be to construct plasmid banks in which the genes have been fused to inducible, highly expressed promoters.

Cloning by Regulation

In the absence of any direct information about either the identity of a gene or its gene product, one recourse is to isolate a set of genes whose regulation fulfills some interesting set of criteria. The first example of this was the isolation of the genes for the Leloir pathway of galactose utilization, in which differential plaque filter hybridization detected genes whose RNA levels are controlled by the presence of galactose.[10] Inevitably this set of genes could have been readily isolated on the basis of complementation of the many mutants in these genes. A similar approach used cDNA probes which were specifically prepared on the basis of whether they are synthesized constitutively or are regulated by the addition of the yeast mating hormone.[39]

Quite a different approach was used by Ruby and Szostak who extended the observation that *lacZ* gene fusions can be used to study regulation in yeast.[40] They screened a yeast genomic library of *Sau*3A fragments fused to the *lacZ* gene for those plasmids that showed altered levels of β-

[37] M. Rose and D. Botstein, *J. Mol. Biol.* **170**, 883 (1983).
[38] S. J. Silverman, M. Rose, D. Botstein, and G. R. Fink, *Mol. Cell Biol.* **2**, 1212 (1982).
[39] G. L. Stetler and J. Thorner, *Proc. Natl. Acad. Sci. U.S.A.* **81**, 1144 (1984).
[40] S. W. Ruby and J. W. Szostak, *Mol. Cell Biol.* **5**, 75 (1985).

galactosidase in yeast after treatment of the transformants with DNA-damaging agents.

Cloning by Transposon Mutagenesis

A method that has great potential for both the isolation of useful mutations and the subsequent isolation of the gene is an extension of induced Ty element transposition.[41] The Ty element of yeast is a transposon which shares many structural and functional characteristics with retroviruses[42,43] including transposition via an RNA intermediate. Systems have been developed that increase specifically the frequency of transposition,[41,44] so that many of the mutations arising after a round of transposition are due to Ty insertions. Identification of the Ty element responsible for the mutation thereby defines the approximate physical location of gene of interest. For example, a system in which enhancement of Ty transposition was achieved by incubation of strains at reduced temperature allowed the efficient selection of Ty insertions that made alcohol dehydrogenase constitutive.[44] This yielded Ty insertions in the promoter region of the *ADH1* gene and in a newly identified gene involved with its regulation. In the long range, since the Ty elements can be marked genetically, it may be possible to recover the insertion mutation directly by including within the marked Ty insertion the elements required for replication in *E. coli* and a selectable marker. The difficulty with the Ty mutagenesis system is that to date there is no strain of *S. cerevisiae* that totally lacks Ty elements. Therefore there is no equivalent of the *M* cytotype of *D. melanogaster* that has made the P element transposition mutagenesis system so useful. After mutagenesis the researcher must find the relevant Ty insertion against a background of greater than 30 preexisting Ty elements as well as the set of new insertions caused by the mutagenesis. However, it is likely that physical marking of the induced Ty will greatly reduce the complexity of this problem. A second limitation of the Ty system concerns the specificity of Ty insertion. In most cases the insertions that lead to an observable phenotype are due to insertion into the 5'-noncoding region of the gene. Two cases are known for internal insertion.[45,46] It remains to be seen whether there is a further level of specificity

[41] J. D. Boeke, D. J. Garfinkel, C. A. Styles, and G. R. Fink, *Cell (Cambridge, Mass.)* **40**, 491 (1985).
[42] R. T. Elder, E. Y. Loh, and R. W. Davis, *Proc. Natl. Acad. Sci. U.S.A.* **80**, 2432 (1983).
[43] D. J. Garfinkel, J. D. Boeke, and G. R. Fink, *Cell (Cambridge, Mass.)* **42**, 507 (1985).
[44] C. E. Paquin and V. M. Williamson, *Mol. Cell. Biol.* **6**, 70 (1986).
[45] M. Rose and F. Winston, *Mol. Gen. Genet.* **193**, 557 (1984).
[46] G. Simchen, F. Winston, C. A. Styles, and G. R. Fink, *Proc. Natl. Acad. Sci. U.S.A.* **81**, 2431 (1984).

whereby different 5'-noncoding regions would have greatly different mutation frequencies. It is likely that these problems will be elucidated very rapidly in the coming years.

Constructing a Library for Cloning in Yeast

In some cases, none of the existing yeast genomic plasmid libraries will suffice to isolate a given gene by complementation. In particular, when the only mutation available is dominant to the wild-type allele, the only recourse is to create a plasmid library from the mutant strain. Moreover, the strains from which the current banks have been derived may themselves have been mutant for the gene of interest. Or it may be that the gene from a particular genetic background is required. In this volume and elsewhere there is much guidance on the construction of appropriate libraries. We will confine our discussion to some of the considerations that are unique to the construction of libraries to be used for complementation in yeast.

Strains

The initial concern in creating a library is the strains to be used. The donor strain from which the DNA is to be made may be constrained by genetic considerations. All other things being equal, the researcher should consider the fact that most of the libraries have been freely shared within the yeast research community, and thus select a donor strain corresponding to one of the isogenic genetic backgrounds widely in use, such as S288C.[1] One reason for this is that different strains can vary widely in the restriction map of the same genes. In addition, haploid strains should be used as the source of the DNA to avoid any difficulties about the origin of a given allele.

The recipient strain for isolating the gene by complementation requires careful attention. The strain should carry at least one nonreverting auxotrophic mutation corresponding to the marker to be used on the vector. The standard set now includes the following mutations: *ura3–52*, a Ty insertion within the gene[45]; *leu2–3*, *leu2–112*, a pair of frameshift mutations[6]; *his3–200*, an *in vitro* constructed deletion[47]; and Δ*trp1-901*, an *in vitro* constructed deletion.[48] This is important to prevent the isolation of revertants. Additional markers are particularly useful for ensuring that the candidate clone is not a prototrophic contaminant. The strain should

[47] M. Fasullo, Doctoral Dissertation, Stanford University, Stanford, California (1986).
[48] P. Hieter, C. Mann, M. Snyder, and R. W. Davis, *Cell* (*Cambridge, Mass.*) **40**, 381 (1985).

be checked to determine that it transforms well. A good frequency is above 1000 transformants per microgram of pure supercoiled YEp plasmid DNA. Lower frequencies can be tolerated at the expense of preparation of larger amounts of library DNA. Strains can generally be improved by repeated crosses to some good transforming strain. Because transformability may be a polygenic trait, repeated backcrosses are more likely to be successful. In addition strains that transform well by one method often do not transform well by the other. Therefore it may be worthwhile to test transformation by the spheroplast transformation method also.

Preparing Yeast DNA

The DNA should be prepared in as high molecular weight as possible, to avoid a large population of useless fragments with sheared ends. We have used the following variant of the method of Cryer *et al.*[49] with good success. Addition of a preparative sucrose gradient after cell lysis removes low-molecular-weight shear fragments, RNA, 2-μm circle DNA, and much of the mitochondrial DNA.

1. Grow 1 liter of yeast in YEPD to late log phase (2×10^8 cells/ml). Harvest cells by centrifugation for 5 min at 5000 rpm in a Sorvall GS-3 rotor or equivalent.

2. Wash cells in 200 ml of ice-cold 50 mM Na$_2$EDTA (pH 8.5) and centrifuge as above.

3. Suspend cells in 25 ml 50 mM Tris–HCl (pH 9.5), 2% 2-mercaptoethanol (2-ME). Incubate at room temperature for 15 min. Centrifuge for 5 min at 5000 rpm in a Sorvall SS-34 rotor or equivalent.

4. Suspend cells in 20 ml of 1 M sorbitol, 1 mM Na$_2$EDTA (pH 8.5) and add 1 ml of Zymolyase 60,000 (Miles) (1 mg/ml). Incubate with gentle shaking at 37° until greater than 95% of the cells have been converted to spheroplasts. An adequate conversion to spheroplasts can be demonstrated microscopically by the lysis of cells upon dilution in 1% Sarkosyl. Times of incubation vary between different strains, from 30 min to 2 hr. Harvest spheroplasts by centrifugation as above.

5. Suspend cells by gentle stirring in 5 ml of lysis buffer (0.1 M Tris–HCl at pH 9.5, 0.1 M Na$_2$EDTA, 0.15 M NaCl, 2% 2-ME). Freeze in liquid nitrogen.

6. Thaw cell suspension and gently mix in 10 ml of lysis buffer containing 4% Sarkosyl. Incubate at 45° for 20 min.

7. Add 12.5 ml of lysis buffer made up at pH 8.0 and containing 4% Sarkosyl. Mix by gentle swirling. Heat at 70° for 15 min.

[49] D. R. Cryer, R. Eccleshall, and J. Marmur, *Methods Cell Biol.* **12**, 39 (1975).

8. Add RNase A (boiled in 50 mM potassium acetate at pH 5.5 for 15 min) to 100 μg/ml. Incubate at 45° for 1 hr.

9. Add 1 ml of a freshly prepared stock of Pronase (Sigma, B grade) at 20 mg/ml. Incubate at 45° for 1 hr. Add a second aliquot of Pronase and incubate for another hour.

10. Heat at 70° for 15 min.

11. Add an equal volume of chloroform–isoamyl alcohol (24 : 1). Gently rock tube until a white emulsion forms. Remove debris and separate phases by centrifugation at 20,000 g. Carefully remove supernatant. Incubate at 45° until odor of organic solvents has dissipated.

12. To obtain high-molecular-weight DNA, preparative sucrose gradients are run. Prepare three 24-ml 5–20% sucrose gradients. The gradients are prepared in SW-27 tubes (Beckman) on top of a 3-ml cushion of Angio-Conray made 20% in sucrose. The sucrose gradients are prepared in 20 mM Tris–HCl (pH 8.0), 20 mM Na$_2$EDTA, 0.2 M NaCl, and 0.1% Sarkosyl. An aliquot of 10 ml of sample is layered on top of the gradient (note that sometimes the sample sinks to a density below the top of the gradient, but this has no deleterious effect). Run at 13,500 rpm for 17 hr.

13. The DNA is collected by carefully removing 1-ml samples from the top of the gradient. The DNA should not be passed through tubes with an orifice smaller than a Pasteur pipet. We use plastic 5-ml pipets. The DNA is found in a viscous fraction near the bottom of the gradient. The DNA can be located quickly by mixing 10 μl of each fraction with 10 μl of ethidium bromide (1 μg/ml) and examining with ultraviolet light or by running a small agarose gel.

14. Remove sucrose from the DNA-containing fractions by dialysis against at least two changes of 0.15 M NaCl, 10 mM Tris–HCl (pH 8.0) and 1 mM Na$_2$EDTA.

15. Add CsCl to 10 g per 8 ml of DNA solution, rocking slowly to dissolve. Refractive index should be 1.400. Centrifuge at 50,000 rpm in a type 50 rotor (Beckman) or equivalent for 36 hr. Collect DNA by dripping gradient through 16-gauge needle. DNA is contained in a viscous fraction which can be found as above. Remove CsCl by dialysis against several changes of 10 mM Tris–Cl (pH 8.0), 1 mM Na$_2$EDTA.

16. Typical yields are 300–400 μg of DNA per 10 g of cell pellet.

The purified yeast DNA is then digested partially with a 4-base recognition restriction enzyme, and the appropriate size range of DNA fragments is purified by sedimentation on sucrose gradients or gel electrophoresis. It is important to use partially digested DNA that has been digested with a range of different enzyme concentrations to avoid bias due to differences in the rate of cleavage at different sites. A size range of 15–20 kb ensures that the library contains a complete genome in 1000 insert-containing plasmids.

Vectors

The vector plasmid should also be carefully considered. The criteria to be considered include the selectable marker to be used in yeast and *E. coli*, the mode of replication in yeast, copy number in yeast and *E. coli*, and whether there is a simple screen for the presence of inserts in the vector. Detailed restriction maps for three commonly used yeast plasmid vectors, YIp5, YEp24, and YRp17 (all use the *URA3* gene as a selectable marker and YRp17 also carries *TRP1*), are diagrammed in the New England BioLabs catalog. Figure 1 shows a map for the centromere-based plasmid YCp50 which was constructed in the laboratory of R. Davis. The

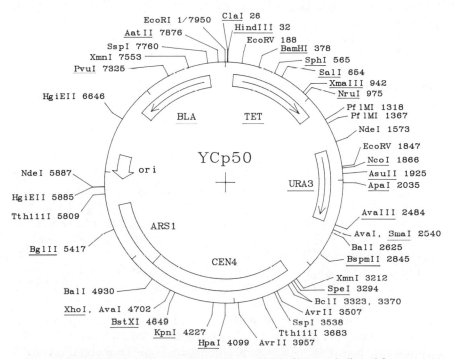

Fig. 1. Restriction map of yeast shuttle vector YCp50. Sites are indicated for enzymes that cleave once or twice. Those that make a single cut are underlined. Type II enzymes that do not cut are *Afl*II, *Bst*II, *Not*I, *Sac*I, *Sac*II, *Sna*BI, *Mlu*I, *Pvu*II, *Sau*I, *Xba*I, *Bss*HII, *Mst*II, and *Sfi*I. Enzymes *Acc*I, *Acy*I, *Afl*III, *Aha*III, *Cfr*I, *Eco*P15, *Gdi*II, *Hae*I, *Hae*II, *Hgi*AI, *Hgi*CI, *Hgi*JII, *Hin*dII, *Mst*I, *Nae*I, *Nar*I, *Nsp*(7524)I, *Nsp*BII, *Nsp*CI, *Sca*I, *Sna*I, *Stu*I, *Tth*111II, *Xho*II, and *Xmn*I all cut three or more times.

plasmid is derived from YIp5 by the insertion of a DNA fragment containing the centromere from chromosome 4 and *ars1* inserted into the *Pvu*I site. YCp50 is available from the American Type Culture Collection under entry 37419. In addition, a large list of yeast plasmids has recently been compiled.[50]

After ligation and transformation the size of the library must be assessed. A good library should be large enough that the probability of complete representation of the genome approaches 1. Formulas for the determination of the required number of inserts have been described in this volume [16, 26]. For practical purposes, a library should have at least 10,000 independent plasmids with an average insert size of 15 kb (i.e., about 10 times the genome). If possible such a library should be separated into smaller libraries each containing perhaps two or three times the total genome represented. This is to avoid the repeated isolation of the same plasmid from a single large library. By breaking up the library into small separate libraries initially, the investigator guarantees that each isolate of a complementing plasmid from different libraries is independent. This is particularly important since the subsequent analysis can be laborious. In addition, isolation of a large number of different plasmids is useful in the localization of the complementing portion of the insert, as different isolates overlap in a way that narrows the area to search.

Media

Media for growing yeast strains[1] will be briefly described in this section. All components are available from Difco. The routine nonselective medium is a complex medium called YEPD (for yeast extract peptone dextrose). Per liter it is composed of 10 g yeast extract powder and 20 g peptone. After autoclaving, sterile glucose is added to a final concentration of 2% (w/v). For petri plates, agar is added before sterilization to a final concentration of 2% (w/v).

The defined medium is called SD (for synthetic dextrose). Per liter it is composed of 6.7 g of yeast nitrogen base (without amino acids). After autoclaving, glucose is added to a final concentration of 2% (w/v). For plates, the solution is made up at 2× concentration and after sterilization it is mixed with an equal volume of sterile molten 4% (w/v) agar after the solutions are below 65°. Add sterile glucose to a final concentration of 2% (w/v).

For most purposes, the SD medium is supplemented with a complete set of the growth supplements commonly required by different auxo-

[50] S. A. Parent, C. M. Fenimore, and K. A. Bostian, *Yeast* **1,** 83 (1985).

tropns. Growth on this rich defined medium is much faster than on minimal medium. A powder mixture is prepared containing 2.0 g each of all of the amino acids except leucine, of which 4.0 g is added. To this add 0.5 g adenine, 2.0 g uracil, 2.0 g inositol, and 0.2 g p-aminobenzoic acid. If selection of a gene for a particular growth supplement is desired, that supplement is left out of the mixture. A couple of clean marbles is added to the powder and it is mixed end over end for at least 15 min . Use 2 g of the powder mixture per liter of SD. Autoclave with the yeast nitrogen base.

Other carbon sources can be used instead of glucose. For a nonfermentable carbon source glycerol or ethanol can be used at 2% (v/v). For a carbon source that does not cause catabolite repression, use nonfermentable compounds or raffinose at 2% (w/v).

Transforming Yeast

Given a plasmid library and a strain to transform, the next step is the introduction of the DNA into the host strain. The following is a general protocol for transformation of S. cerevisiae cells based on the alkaline cation method of Itoh et al.[7]

1. Grow cells to midlogarithmic phase (approximately 2×10^7/ml) in YEPD medium. You will need to grow up 10 ml of cells for each different transformation.

2. Harvest cells by centrifugation; 5000 rpm for 2 min in a Sorvall SS-34 rotor or equivalent is adequate.

3. Resuspend cells in one-half the culture volume of 0.1 M lithium acetate in TE (pH 7.5) (10 mM Tris at pH 7.5, 1 mM Na$_2$EDTA).

4. Incubate at 30° for 1 hr using gentle shaking.

5. Centrifuge cells as above and resuspend at a density of approximately 2×10^9 per ml.

6. Place aliquots of cells into small culture tubes and add DNA. Typically we use about 1 μg of plasmid DNA per transformation of 0.1 ml of concentrated cells. Add 50 μg of carrier DNA per 0.1 ml of concentrated cells. Carrier DNA is prepared by dissolving DNA from calf thymus, chicken blood, or salmon sperm in TE and shearing the DNA by repeated passage through a graded series of syringe needles, finishing with a 26-gauge needle or smaller. The DNA is then deproteinized by repeated phenol extraction, ethanol precipitated, and dissolved in TE at a concentration of around 5 mg/ml.

7. Incubate 30 min, 30°.

8. Add 0.7 ml of 40% PEG 4000, 0.1 M lithium acetate in TE (pH 7.5) to each tube of 0.1 ml of cells. (The 40% PEG stock is made up by

autoclaving a stock of 44% PEG and adding 0.1 volume of a sterile solution of 1 M lithium acetate, 0.1 M Tris–HCl at pH 7.5, 10 mM Na$_2$EDTA after it is cool.) Mix well.

9. Incubate 1 hr, 30°.

10. Place tubes in a 42° water bath for 5 min.

11. Centrifuge cells as above and wash once in sterile distilled water. Resuspend cells in 0.2–0.4 ml of water per tube.

12. Spread 0.2 ml of the resuspended cells onto a selective medium plate. Any remaining cells may be refrigerated for a week if you need more transformants. Transformants will appear after 2 or 3 days at 30° depending on the strain, the selected marker, and the vector.

Screening the Transformants

The transformant ordinarily should be selected by using the selectable marker built into the vector rather than by selecting for the gene of interest directly or by selecting for both together. Usually the mutation available can revert, and direct selection will yield strains containing either no plasmid or random plasmids requiring sorting out at a later step. By far the preferable method is to replica plate the transformants to medium that selects for the desired phenotype. The pattern of colony growth or nongrowth is then diagnostic of whether the clone contains the plasmid of interest. Positives give strong growth over the entire extent of the colony, whereas reversion of the marker yields numerous papillae.

A second advantage of the indirect approach is that it allows the purification of the candidate transformant without requiring the strain to grow up under the selective regime. Selection via cross-suppression may not yield optimal growth rates of transformants, allowing plenty of room for selection of modifying mutations.

After the purification and retesting of the candidates, it is essential to demonstrate that the complementing activity resides on the plasmid. This is usually achieved by showing that the complementing activity and the plasmid marker cosegregate after mitotic loss. A colony of the transformant strain is picked off of a medium plate that selects for the plasmid and is used to inoculate 5 ml of YEPD. After overnight growth the dilutions of the culture are plated onto YEPD plates so as to provide 100–300 colonies per plate. After colonies have appeared they are replica plated to solid media that select for the plasmid marker and for the complementing activity. To prove that the complementing activity is on the plasmid, each colony must have only one of a pair of phenotypes: either the colony grows on both plates or it grows on neither. Appearance of plasmid-bearing cells that do not complement indicate that the transformant contained more than one plasmid only one of which may complement. Appearance of "complemented" cells which do not bear the plasmid marker indicate that the strain has reverted or acquired a suppressor mutation. For this analysis, the lithium acetate transformation protocol is preferred because this tends to give a lower frequency of multiple plasmids per transformant than the spheroplast transformation method. Using the YCp vectors, in which the individual plasmids are relatively stable, it can be difficult to generate a strain that harbors only a single plasmid.

Recovering the Plasmid from Yeast

Next, the plasmid is recovered in a form which permits physical analysis. Usually this means transforming *E. coli* with DNA prepared from the transformant. A protocol for preparing yeast DNA that is usually successful[41] is as follows:

1. Grow 10 ml culture of cells to saturation in YEPD or SD.

2. Harvest cells by centrifugation and wash in 1 ml of 0.9 M sorbitol, 0.1 M Na$_2$EDTA (pH 7.5). All steps can be performed in microfuge tubes as long as cells and spheroplasts are not centrifuged for longer than about 10 sec.

3. Resuspend cells in 0.4 ml of 0.9 M sorbitol, 0.1 M Na$_2$EDTA (pH 7.5), 14 mM 2-mercaptoethanol. Add 0.1 ml Zymolyase 60,000 (Miles) (2 mg/ml) made up in the sorbitol solution. Incubate at 37° for 20–30 min. Monitor spheroplast formation by examination of detergent sensitivity: a small sample of cells is diluted into 1% sodium dodecyl sulfate (SDS), and spheroplasting is sufficient when greater than 90% of the cells burst when examined under the microscope.

4. Centrifuge spheroplasts and resuspend gently in 0.45 ml of 50 mM Tris–HCl (pH 8.0), 50 mM Na$_2$EDTA. Mix in 50 μl of 2% SDS. If necessary stir with the tip of a micropipet.

5. Incubate at 65° for 30 min.

6. Add 80 μl 5 M potassium acetate. Put on ice for at least 60 min.

7. Centrifuge precipitate in microfuge for 15 min. Transfer supernatant to a fresh tube.

8. Precipitate the DNA by adding 1 ml of ethanol at room temperature and mixing. Centrifuge briefly. Rinse pellet with cold 70% ethanol and air dry.

9. Resuspend in 0.5 ml TE (10 mM Tris–HCl at pH 8.0, 1 mM Na$_2$EDTA). Centrifuge insoluble material for 15 min in a microfuge. Transfer supernatant to a fresh tube. Add 25 μl of RNase at 1 mg/ml and incubate at 37° for 30 min.

10. Add an equal volume of 2-propanol, mix gently, and spin in microfuge for 10 min.

11. Discard supernatant, rinse pellet with cold 70% ethanol, and air dry.

12. Resuspend DNA in 50 μl of TE.

The DNA prepared by this protocol is relatively clean and can be cut to completion with most restriction enzymes. Further purity can be obtained by phenol and chloroform extraction. The concentration is approximately 100 μg/ml. For transformation of *E. coli* cells it is best to try a range of DNA concentrations from 0.5 to 10 μl per tube. There is a persistent inhibitor of *E. coli* transformation that is not removed and causes variable transformation frequencies. Frequently more DNA added leads to fewer transformants recovered. Under favorable conditions several dozen transformants can be obtained per tube. Such DNA preparations can also be used to transform yeast directly with an efficiency similar to that for transforming *E. coli*.

Proving Clone Identity

Localizing the Complementing Activity

Having a set of plasmid transformants in *E. coli*, the investigator next must localize the gene and prove its identity. Location of the functional genes is first indicated by examination of the restriction map of the different complementing plasmids. Assuming that a single gene can complement the mutation, all of the plasmids should contain overlapping DNA fragments and the gene should be contained within the overlap region. The presence of more than one gene will be readily apparent, because the plasmids will assort into different groups according to the restriction fragments that they have in common. There are several sophisticated methods for the further localization of the functional gene including insertional inactivation by transposon mutagenesis[51] and linker insertion.[52] A simple method is to subclone portions of the complementing DNA fragment into a yeast plasmid. In all cases the yeast strain is then transformed with the new constructs and the transformants are examined to determine whether the plasmid still alters the phenotype of the cell. In the case of subcloning, a small fragment containing the complementing gene can usually be obtained. Restriction enzymes that produce no functional subclonable fragments must have at least one site within the gene. By this means restriction sites that lie within the gene can be determined.

[51] H. S. Seifert, M. So, and F. Heffron, *Genet. Eng.*, **8**, 123 (1986).
[52] D. Shortle, *Gene* **22**, 181 (1983).

Determining the Genetic Locus of the Complementing DNA

Proving that the complementing gene is in fact the same gene as the one which contains the mutation requires a genetic test. Homologous recombination will integrate a YIp plasmid bearing the complementing DNA at its normal location in the genome (and also frequently at the location of the selectable marker, thus providing an excellent control). A mutant strain is transformed by the standard protocol selecting for a vector-derived selectable marker. Usually this will produce a tandem duplication of the cloned DNA separated by the vector sequences (see Fig. 2). Alternatively, integration at the selectable marker will duplicate that gene; the vector and cloned gene will separate the repeats. Transformants are checked to determine that they have become phenotypically wild type (this is necessary because gene conversion events can result in production of the same mutation in both copies of the duplicated DNA or in conversion of the selectable marker without integration). The transformants are then crossed to two strains; one carries the wild-type allele of the gene of interest and the selected marker gene and the second strain carries mutant alleles for both genes. Subsequent sporulation and analysis of tetrad spores will yield different patterns depending on whether the normal genetic locus of the gene is the same or different from the location of the mutation. If the gene and the mutation are at the same locus, then all (greater than 99%) spores from the cross to the wild-type strain will be wild type for that phenotype. If the plasmid has integrated elsewhere (greater than about 150 kb away or on a different chromosome) then the mutant phenotype will be apparent in roughly one-fourth of the spores. Lesser numbers of mutant spores will appear if the cloned gene has integrated closer than this limit. Likewise, the marker gene will show appearance of mutant spores if integration occurred at the locus of the cloned DNA fragment, but all spores will be wild type if integration occurred at the location of the selectable marker. The key observation that would indicate that the complementing gene is not the same as the mutant gene is the appearance of spores that are mutant for both genes. The cross to the double mutant strain is necessary to demonstrate that the two wild-type genes that come to the strain via the plasmid are actually linked to one another (all spores are either wild-type for both markers or mutant for both markers) and that each segregates equally at meiosis (half of the spores are wild type and half are mutant, for each marker).

Failure to integrate at the genetic locus of the mutation may indicate that the cloned gene is not the same gene as the one that has mutated or that the cloned DNA fragment bears a portion of repeated DNA such as a Ty element. Therefore a Southern blot hybridization analysis using the

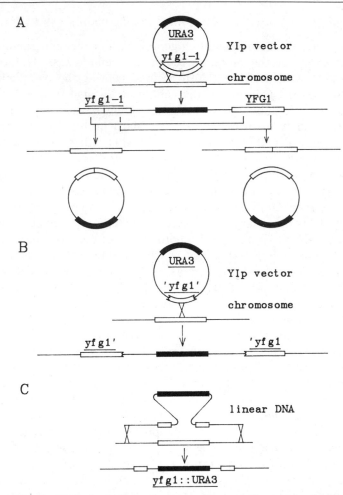

FIG. 2. Methods of inactivating a gene in yeast. See text for details.

cloned gene as radioactive probe should be performed soon after obtaining the gene of interest. If a repeated element is present on the fragment then two recourses are available. A smaller functional subclone can be used that lacks the repetitive DNA. Alternatively, the integration event may be directed to the nonrepetitive DNA sequences by cleaving the plasmid prior to transformation with a restriction enzyme that cuts the plasmid only once and in the unique yeast DNA sequences.[53]

[53] T. L. Orr-Weaver, J. W. Szostak, and R. J. Rothstein, this series, Vol. 101, p. 228.

In the case of genes for which no mutations are known, the integrated plasmid can now be used as a genetic marker to follow through crosses by virtue of the plasmid selectable marker. The first step is to identify the chromosome. Several methods have been developed to do this. The simplest is to integrate a YEp vector bearing the gene, using restriction enzyme cleavage to force recombination. The chromosome containing an integrated YEp plasmid is unstable.[54] If the transformant is crossed to multiply marked strains, then the subsequent appearance of the recessive markers indicates which chromosome has become unstable. The second approach is to use the cloned DNA as a hybridization probe for filters bearing yeast DNA that has been separated into different chromosomes by orthogonal field alternating gel electrophoresis.[55] In either case, the next step is to perform standard crosses against known genetic markers on the identified chromosome.

Inactivation and Gene Replacement

In some cases, proof of the correctness of the cloned gene will come from sequence analysis and demonstration of homology to another previously cloned gene. In other cases, where no mutation is known and no homolog has been identified, judgment of the "correctness" of the cloned gene must rely on the phenotype produced by mutations made *in vitro* and used to replace the wild-type allele.

The simplest kind of mutation that can be produced *in vitro* is a null allele produced by deletions or insertions. In this case, plasmid constructions used to localize the gene include some which define internal restriction sites. Constructions which place a transposon, linker, or gene fragment into the gene can also be used. Three different strategies have been devised for placing these mutations onto the chromosome to replace or destroy the wild-type gene.

The first method of gene replacement was pioneered by Scherer and Davis[56] and makes use of the tandem duplication that forms during integration of a YIp plasmid (Fig. 2A). If the integrating plasmid contains a mutation in the gene then one of the gene copies will harbor a mutant allele. The plasmid can then be lost by a homologous recombination to either side of the mutation. Recombination on one side results in the restoration of the wild-type allele, whereas recombination on the opposite side of the mutation leaves the mutation in the chromosome. With powerful selection for both the presence and the absence of the *URA3* gene, this

[54] S. C. Falco and D. Botstein, *Genetics* **105**, 857 (1983).
[55] G. F. Carle and M. V. Olson, *Proc. Natl. Acad. Sci. U.S.A.* **82**, 3756 (1985).
[56] S. Scherer and R. W. Davis, *Proc. Natl. Acad. Sci. U.S.A.* **76**, 3912 (1979).

is a very efficient method. In addition, sister segregant strains are isogenic.

Two other methods produce null mutations in one step. In the first
case (Fig 2B), a YIp plasmid is constructed that carries a wholly internal
fragment of the gene of interest. When integration of the plasmid occurs
by homologous recombination within this DNA segment, the gene is split
into two overlapping but nonfunctional fragments.[57] The second one-step
method (Fig. 2C) involves first inserting a selectable marker into the gene.
Using appropriate restriction enzymes sites, a linear fragment is excised
such that both ends lie within the yeast DNA sequences and contain the
gene bearing the insertion. Transformation with this fragment, selecting
the marker gene, proceeds normally, and integration occurs by recombination at both ends of the fragment. This procedure is called "omega"
transformation[58] and results in the replacement of the wild-type gene with
the mutant allele.

In all of these procedures for introducing a null allele into the chromosome, the appropriate strain to transform is a diploid. This will allow a
lethal mutation to be covered by the wild-type allele on the other chromosome. The mutation is then observed in spore clones after sporulation and
tetrad dissection.

In the long run, null alleles have limited value for assessing the precise
role of a gene in the physiology of the cell. To that end, conditional alleles
are required to allow the observation of the mutant phenotype in large
populations of cells. The isolated gene can be mutagenized *in vitro* by
chemical mutagens,[22] by misincorporation,[59] and by synthetic oligonucleotide priming of replication.[60] The resulting mutations can be screened
for failure to complement the null allele by transforming a heterozygous
diploid and sporulating or by "plasmid shuffling."[61] Finally the new mutations can be placed onto the chromosome in place of the wild-type allele
by variations in the methods described above.

Acknowledgments

I thank Ron Sapolsky for sharing unpublished data on the sequence of YCp50. I also
thank Gerry Fink and David Botstein, in whose laboratories I learned much of what is
written here.

[57] D. Shortle, J. E. Haber, and D. Botstein, *Science* **217,** 371 (1982).
[58] R. J. Rothstein, this series, Vol. 101, p. 202.
[59] D. Shortle, P. Grisafi, S. J. Benkovic, and D. Botstein, *Proc. Natl. Acad. Sci. U.S.A.* **79,**
1588 (1982).
[60] M. J. Zoller and M. Smith, *DNA* **3,** 479 (1984).
[61] J. D. Boeke, J. Trueheart, G. Natsoulis, and G. R. Fink, this series, Vol. 154.

Section X

Identification and Characterization of Specific Clones

[54] Identification and Characterization of Specific Clones: Strategy for Confirming the Validity of Presumptive Clones

By ALAN R. KIMMEL

Keeping Your Plasmids Circular and Your Arguments Linear

The selection of a clone by any of the procedures discussed (see [42]) is rarely sufficient for proving that the sequences isolated are derived from the gene of interest. Further, the most convenient method of characterization is often just a reidentification of some of the same properties used to select the clone originally. For example, a clone, selected from an expression library by antibody recognition [50,51], obviously encodes amino acid sequences which are cross-reactive with that antibody. Thus, subsequently demonstrating that this clone is complementary to an mRNA which will direct the synthesis *in vitro* [60] of a protein recognized by the antibody is not sufficient for unambiguous identification. As another example, methods have been presented for isolating clones representing genes with particular patterns of expression [46]. Since several or many other genes may share a common characteristic, demonstrating that an isolated clone hybridizes to an mRNA with a predicted pattern of expression or subcellular localization is again insufficient proof of identification. In addition, clones can be artifactually selected by cross-hybridization with nucleic acid sequences [43,45]. Since hybridization conditions for such screenings may be relatively relaxed, incorrect clones are often selected which exhibit only partial sequence homology to the probes.

It should be clear that circular proofs must be avoided. However, if characteristics in addition to those used for the original screen can be included in the evaluation, it is more likely that a clone will be identified correctly. Such auxiliary characteristics might include DNA sequence information [56–58], subcellular localization of mRNA complementary to the clone [61,62,67], developmental- or tissue-specific expression patterns of complementary mRNA [61,62,67], and functional analysis of gene products [69,70,71]. Ultimately, however, unequivocal identification of any clone may require analysis of its nucleotide sequence and a comparison of the derived amino acid sequence with that of the corresponding protein. If an amino acid sequence is not available a functional analysis of the gene product may suffice. Cloned DNA can be introduced into a variety of cell types and the transfected cells assayed for the expression of the corresponding phenotype. Although this approach has been used successfully, it is technically complex, and results may be difficult to interpret (see [53]).

Once a clone has been positively identified a final analysis of the genomic and mRNA sequences can be completed. Gene products can be produced in homologous or heterologous systems for functional studies [69,70,71]. Full-length genes and associated flanking sequences can be isolated [63,64] and sequences corresponding to primary transcripts and to mature mRNA can be precisely defined [66]. Eventually cis and trans acting regulatory elements can be identified [72–74].

Methods for Characterizing Cloned Sequences

Subcloning. Before proceeding with any in-depth characterization of a genomic or λ cDNA library clone it is usually advantageous to subclone into plasmid vectors [11, 55]. Subcloning of genomic fragments permits the analyses of DNA several kilobases apart from the remainder of the clone. Subcloning cDNA from λ vectors into plasmid vectors enriches cDNA-to-vector mass ratios. Clearly, prior to subcloning, a map of the positions of the recognition sequences of various restriction enzymes is essential [11]. By comparing these restriction maps with the cloning sites of vectors the most direct schemes for subcloning become clear. There are many choices for plasmid vectors. Plasmids (see [32]) may possess promoters for phage (SP6, T7) RNA polymerases for *in vitro* transcription, multiple cloning sites (polylinkers), sequences complementary to oligonucleotide primers for DNA sequencing, genes (β-galactosidase) for identifying clones by insertional inactivation, and properties to permit the direct isolation of the recombinant DNA in a single-stranded form (M13).

DNA Sequencing. Technology now exists for sequencing tens of thousands of base pairs of DNA [56–59]. DNA sequences can be used to predict amino acid sequences. If a partial protein sequence is known, a precise match with the predicted amino acid sequence is usually sufficient to establish the identity of a clone. Further, a complete nucleic acid sequence may permit the description of the entire amino acid sequence of a protein by an indirect method that is more efficient and often more accurate than direct protein sequencing. An amino acid sequence can sometimes be used to predict structural features of proteins. Membrane components or secreted proteins may possess hydrophobic amino acids that are organized differently from those of soluble proteins. In the absence of protein sequence, a DNA-derived amino acid sequence is, thus, often useful for evaluating the nature of an isolated clone.

mRNA Distribution. Several methods exist for determining the relative levels of expression of specific mRNAs in various subcellular components or during different developmental stages. Methods that are based on blot hybridizations can be rapid but only approximate mRNA levels. Whereas "fast blots" [62] do not require the purification of RNA prior to hybridization, difficulties with nonspecific binding to abundant (ribosomal) RNA sequences can sometimes present problems. This problem can often be eliminated by hybridizing to Northern blots [61] of electrophoretically fractionated RNAs. Nonspecific hybridization to rRNAs is identified apart from hybridization to specific mRNAs. Furthermore, the size of the mRNA under study can be accurately determined. Quantifying mRNA levels on Northern blots involves a comparison with RNAs of known abundance. Actin mRNA is a widely used control since it is ubiquitous in eukaryotes and is expressed at moderately high levels in most cells. Recently, sense-strand RNAs synthesized *in vitro* have been used in a similar manner. A cloned sequence complementary to the mRNA being studied can be transcribed *in vitro* using [^{3}H]ribonucleoside triphosphates of known specific activity [30]. The mass of RNA synthesized is, thus, a function of the amount of radioactivity incorporated into RNA. This RNA when present on the same Northern blot used for abundance studies serves as a known standard for quantitation by hybridization.

Titration analysis by solution hybridization [67] is the most quantitative and sensitive procedure for determining mRNA levels in different cells. RNAs representing less than 0.001% of total mRNA are readily detectable by this procedure. In contrast to Northern blot hybridization, solution hybridization does not require intact mRNA and problems with nonspecific interactions are rarely encountered. However, no information about transcript size can be obtained. One difficulty with titration hybridization is encountered when analyzing the level of an mRNA encoded by a single member of a multigene family. Probes specific to the individual gene members may be required to discriminate their hybridization from hybridization to all related mRNAs. These probes are usually derived from 5'- or 3'-untranslated regions. If such probes are unavailable, related mRNAs of different size can be quantified by Northern blot hybridization. An alternative approach uses a combination of titration hybridization [67] and the nuclease protection assay [66].[1,2]

In situ hybridization is also used for determining which cells in a population possess a particular mRNA product [68]. It has also been effective for analyzing the relative abundance of a specific mRNA in a wide variety of cell types.

An indirect method for analyzing mRNA levels makes use of antibod-

---

[1] M. McKeown and R. A. Firtel, *Cell* **24,** 799 (1981).
[2] P. Romans, R. A. Firtel, and C. L. Saxe III. *J. Mol. Biol.* **186,** 337 (1985).

ies to detect tissue-specific distributions of proteins. If antibodies to specific proteins do not exist, cloned sequences can be used to generate them. Cloned DNA can be ligated to a gene (e.g., β-galactosidase) such that both sequences are translated in-frame into a single polypeptide in *Escherichia coli* (see [40]); the fusion protein synthesized can then be purified and used as an antigen to generate polyclonal or monoclonal antibodies specific for the protein of interest [50, 51].[3]

Hybrid-Selected Translation. *In vitro* translation experiments [27] can also be used for confirming the identity of a cloned sequence. Cloned DNAs can be used to purify complementary RNA by hybridization and the selected RNA subsequently translated in a cell-free system [60]. The proteins synthesized *in vitro* can be analyzed directly or immunoprecipitated if an appropriate antibody exists. As with mRNA hybridization experiments, previously characterized cloned sequences serve as good controls for hybrid selection of mRNA and its subsequent translation. When analyzing members of multigene families, sequences specific to individual genes or stringent conditions of hybridization must be used to eliminate the selection of multiple, related mRNAs (see above).

Characterization of the size or isoelectric point of the protein translated *in vitro* and a comparison with its bona fide counterpart synthesized *in vivo* is critical since artifacts generated by fortuitous cross-reactivity with heterologous protein products can occur. Difficulties in characterization can arise if the mature protein differs from the primary translation product in size or is modified posttranslationally. Sometimes, these differences can be reduced or eliminated by translating the hybrid-selected mRNA in frog oocytes [28–30]; the oocyte system can process proteins and may translocate them to their correct subcellular or extracellular compartment.

Functional Expression Studies. Cloned DNA can be introduced into various cell types for *in vivo* expression of gene products [69–71]. If expression correlates with a specific phenotype it can be used to identify positively the cloned DNA. An entire gene or a full-length cDNA incorporated into a eukaryotic expression vector [32, 41] can be transfected into a cell line which is deficient for the specific phenotype. Subsequently, cells are analyzed to confirm that they exhibit the predicted phenotype following DNA-mediated transformation solely with the specific cloned DNA. Heterologous and homologous expression systems for producing relatively large quantities of gene products *in vivo* [69–71] make possible a detailed analysis of their structural and functional properties or subcellular localizations.

---

[3] See this series, volumes 100, 101, 153, 154, and 155.

Genomic Organization. Once a cloned sequence has been isolated, specific restriction fragments can be used to screen genomic or cDNA libraries for nucleic acid homology [45]. Full-length genes and associated flanking sequences are easily obtained as are evolutionarily related genes, pseudogenes, and members of multigene families [63, 64]. If full-length cDNA clones are not in the library, specific restriction fragments can be used as primers to construct a new cDNA library enriched for the missing sequences (see [32, 66]). Sequences within the genes which correspond to mRNA sequences can be identified by S_1 nuclease [66] or ribonuclease [67][3] protection studies. Similarly, a comparison of the sequence organization of genomic and cDNA clones can be used to determine regions of discontinuity in mRNA coding regions and to locate splice junctions between introns and exons. Such analyses in conjunction with primer extension experiments [66] can precisely define sites for transcription initiation. Various regions of the genes can be linked to "reporter" genes and transfected into homologous cell systems [72]. Cis acting elements regulating transcription can be putatively identified by determining which fusions are capable of promoting the expression of the reporter gene. Specific nucleotides or sequence organizations responsible for the cis regulation can be further defined by mutating[3] the putative regions *in vitro* and reassaying their activity in transfection experiments. These cis regulatory elements can be used as probes to identify trans acting factors, such as site-specific DNA-binding proteins [73, 74] that would interact with them.

Summary

It should now be apparent that establishing the identity of a particular recombinant is not a simple procedure. In the absence of unambiguous and internally consistent DNA and protein sequence information, analyses become dependent on a series of characteristics which, together, may be unique to a specific gene and/or gene product. The process of identification is, thus, often more a corroboration of predicted characteristics than absolute proof. It is critical that the process of identification is not simply reduced to the circular arguments generated by reconfirming characteristics used during the initial selection scheme.

Note Added in Proof: Recently, anti-sense RNAs and gene constructs have been used to block the expression of specific genes.[4] This approach may be useful for the analysis of certain gene functions in cells that cannot be studied using traditional genetic manipulations. It should be noted that the technique has met with varied success.

[4] D. A. Melton, *Proc. Natl. Acad. Sci.* **82,** 144 (1985).

[55] Subcloning

By JONATHAN R. GREENE and LEONARD GUARENTE

In this chapter we consider methods and rationales employed to cull small defined fragments of DNA of interest from large insert-bearing clones, such as genomic clones, and some uses of defined clones in modern biological studies. The chapter is organized in three sections. First is a general discussion of studies employing small subclones that often contain protein-coding sequences. Within this discussion is an enumeration and description of vectors designed for specific studies. Next is a general description of the art of subcloning. Last is a more detailed discussion of methods employed in the construction of subclones.

Uses of Subclones

Determination of Protein Sequences

Perhaps the most obvious utility of subclones is that they allow the prediction of protein sequences encoded by specific genes. DNA sequences are determined either by partial chemical degradation of end-labeled DNA fragments[1] or by enzymatic chain termination of DNA programmed by M13 derivatives.[2] It is often helpful to insert small DNA fragments into plasmid vectors to provide substrates for chemical sequencing. Several suitable vectors exist including the pBR322-derived pUC series. The pUC plasmids contain a "multiple cloning site," a region of DNA that contains cleavage sites for many restriction endonucleases. This is a great aid in the subcloning and subsequent end labeling of a DNA fragment. Chain termination sequences are most often carried out using single-stranded DNA templates. For the purpose of construction of these templates, derivatives of the male-specific single-strand phage M13 have been developed. Plasmids that contain origins of replication for the single-strand phages f1 or M13 have been described which allow the plasmid DNA to be recovered in single-strand form from defective phage particles.[3,4]

Deduced amino acid sequences are of primary importance in several

[1] A. Maxam and W. Gilbert, this series, Vol. 65, p. 449.
[2] F. Sanger, A. Coulson, B. Barnell, A. Smith, and B. Roe, *J. Mol. Biol.* **142,** 1617 (1980).
[3] L. Dente, G. Cesareni, and R. Cortese, *Nucleic Acids Res.* **11,** 1645 (1983).
[4] R. J. Zagursky and M. Berman, *Gene* **27,** 183 (1984).

contexts. In many cases, genes are cloned on the basis of nucleic acid homology to characterized clones derived from other organisms. Evidence that a bona fide related gene has been cloned is provided by a comparison of the DNA sequence or the deduced protein sequence with those provided by the probe. To cite an example, the sequences of the yeast *RAS*1 and *RAS*2 genes provided compelling evidence that the loci were related to the *Ras* gene borne by Harvey and Kirsten animal viruses.[5,6] In other cases, particularly for loci in which mutations have been identified, the function and indeed even the sequence of the encoded product are unknown. A comparison of the deduced protein sequence with a large number of coding sequences in computer data bases could provide the first clue to a possible function of the gene product. For instance, the first indication that the yeast *SNF*1 gene encodes a protein kinase was provided by the homology of a region of the encoded product with many known protein kinase domains.[7]

Generation and Use of Nucleic Acid Probes

The advent of filter hybridization techniques has allowed progress in studying several areas of gene structure and expression. These include the number of gene copies, linkage to restriction enyzme polymorphisms, structural rearrangements, and the existence of cognate genes in the same organism or homologous genes in other organisms. Hybridization probes that are specific for protein-coding sequences or exons will hybridize to a single region of a Southern gel containing appropriately digested genomic DNA.[8] It is thus possible to determine the number of gene copies in an organism or to identify homologous sequences in other organisms. Such homologs may then be cloned using phage or plasmid genomic libraries.

Hybridization technology also potentiates rapid characterization of mutations that result in deletion, insertion, or rearrangement of DNA or that alter specific restriction enzyme sites. This approach has allowed rapid characterization or mutant alleles involved in several human diseases, including β-thalassemia and familial hypercholesteremia.[9] Mutations that alter restriction enzyme sites are generally useful as genetic markers. For example, it is possible to establish genetic linkage between restriction site polymorphisms and loci involved in disease. The knowl-

[5] M. Wigler, *Cell (Cambridge, Mass.)* **36,** 607 (1984).
[6] D. DeFeo-Jones, E. M. Scolnick, R. Koller, and R. Dhar, *Nature (London)* **306,** 707 (1983).
[7] J. L. Celenza and M. Carlson, *Science* **233,** 1175 (1986).
[8] E. Southern, *J. Mol. Biol.* **98,** 503 (1975).
[9] J. Goldstein, M. Brown, R. Anderson, D. Russel, and W. Schneider, *Annu. Rev. Cell. Biol.* **1,** 1 (1985).

edge of such linkages can be extremely useful in rapidly diagnosing the likelihood of disease.

In addition, filter hybridization is the tool that allows the rapid quantitative analysis and physical mapping of transcriptional units by Northern gel hybridization. Mapping of 5' and 3' ends of RNA is carried out using either double-strand DNA probes labeled at one end or single-stranded probes that are uniformly labeled.[10,11] The latter may be either DNA derived from M13-derived vectors or RNA. Vectors have been constructed that are designed to direct the *in vitro* synthesis of large quantities of radioactively labeled sense or antisense RNA complementary to a given DNA sequence.[12] The insert DNA is placed downstream of a promoter for phage-specific RNA polymerase, usually that of the *Salmonella typhimurium* phage SP6 or the *Escherichia coli* phage T7. The high promoter selectivity and enzymatic activity of these RNA polymerases allow for the synthesis of significant quantities of RNA *in vitro*. Vectors exist which contain promoters for both SP6 and T7 polymerases that converge on a multiple cloning site. Insertion of DNA into such a vector allows for synthesis of RNA of one sense by one RNA polymerase and of RNA of the other sense by the other RNA polymerase.

RNA synthesized *in vitro* that is complementary to mRNA can be microinjected into *Xenopus* eggs or cultured animal cells and has been shown capable of arresting synthesis of the encoded gene product *in vivo*.[13] It is thus possible that null mutant phenotypes for specific cloned genes may be determined in this fashion.[14] RNA synthesized *in vitro* can be used as a template for *in vitro* translation of a specific gene product. In this way, radiochemically pure protein can be synthesized and used in biochemical studies.[15] Subclones of particular segments of DNA can serve as the source of uniquely end-labeled DNA fragments that can be used in the study of the structure of that DNA[16,17] or in the study of interactions between that DNA and proteins that bind to it.[18,19]

[10] A. J. Berk and P. Sharp, *Cell (Cambridge, Mass.)* **12,** 721 (1977).

[11] A. J. Berk and P. Sharp, *Proc. Natl. Acad. Sci. U.S.A.* **75,** 1274 (1978).

[12] D. A. Melton, P. A. Kreig, M. R. Rebaglati, T. Manniatis, K. Kimm, and M. R. Green, *Nucleic Acids Res.* **18,** 7035 (1984).

[13] J. Izant and H. Weintraub, *Cell (Cambridge, Mass.)* **36,** 1007 (1984).

[14] T. Maniatis, E. Fritsch, J. Lauer, and R. Lawn, *Annu. Rev. Genet.* **14,** 145 (1980).

[15] I. A. Hope and K. Struhl, *Cell (Cambridge, Mass.)* **43,** 177 (1985).

[16] H. R. Drew and A. Travers, *Cell (Cambridge, Mass.)* **37,** 491 (1984).

[17] H.-M. Wu and D. M. Carothers, *Nature (London)* **308,** 509 (1984).

[18] D. J. Galas and A. Schuitz, *Nucleic Acids Res.* **6,** 111 (1975).

[19] M. G. Fried and D. Crothers, *Nucleic Acids Res.* **9,** 6505 (1981).

Overexpression of Gene Products

It is often useful to express the product of a cloned gene in either *E. coli* or the yeast *Saccharomyces cerevisiae* in order to simplify its purification. Several vectors for expression of cloned genes in *E. coli* exist. In general, these use strong promoters, whose activity can be induced. One common system employs the phage λ P_L promoter whose activity can be controlled by the presence of a temperature-sensitive repressor in the same cell.[20] Another commonly used promoter is the TAC promoter, which is very strong and can be induced by isopropyl-β-D-thiogalacto-pyranoside (IPTG).[21] A system has also been developed to utilize the phage T7 promoter to direct the expression of a gene *in vivo*.[22] In this system, the T7 RNA polymerase is expressed from an inducible promoter so that expression of the gene of interest can be controlled by growth conditions. Expression of a cloned gene in yeast can be obtained by using vectors that contain well-characterized promoters, such as pLGSD5 which contains a hybrid *GAL1*-10 *CYC1* promoter that is inducible by growth in galactose.[23] Another commonly used yeast promoter is the constitutive *ADH1* promoter.[24] The overexpression of gene products in *E. coli* and higher organisms is considered in greater detail elsewhere in this volume [69–71].

Gene Fusions

It is often desirable to express a portion of a polypeptide as a fusion product. In particular, those that contain a β-galactosidase moiety at the carboxyl terminus can be used to study regulation of synthesis and protein localization, and can be an aid in protein purification.[25,26] Vectors exist for construction fusions to *lacZ in vivo* in *E. coli* or *in vitro* using plasmid or phage vectors.[27,28] Alternatively, strategies have been employed that fuse β-galactosidase to the amino terminus of high-value pro-

---

[20] J. E. Mott, R. A. Grant, Y.-S. Ho, and T. Platt, *Proc. Natl. Acad. Sci. U.S.A.* **82,** 88 (1985).

[21] J. Brosius, *Proc. Natl. Acad. Sci. U.S.A.* **81,** 6929 (1984).

[22] F. W. Studier and B. Moffat, *J. Mol. Biol.* **189,** 113 (1986).

[23] L. Guarente, P. Yocum, and P. Gifford, *Proc. Natl. Acad. Sci. U.S.A.* **19,** 7410 (1982).

[24] A. Ammerer, this series, Vol. 101, p. 192.

[25] T. S. Silhavy, S. Benson, and S. Emr, *Microbiol. Rev.* **47,** 313 (1983).

[26] L. Guarente, *Genet. Eng.* **6** (1984).

[27] M. Casadaban and S. N. Cohen, *J. Mol. Biol.* **138,** 179 (1980).

[28] L. Guarente, G. Lauer, T. M. Roberts, and M. Ptashne, *Cell (Cambridge, Mass.)* **20,** 543 (1980).

teins, assuring a high level of synthesis and apparently stabilizing the foreign protein in *E. coli*.[29] Libraries constructed with vectors such as λgt11 that fuse random sequences to *lacZ* may be screened immunologically for the detection of the expression of a polypeptide of interest.[30] Genes commonly used in fusions other than *lacZ*, especially for studies in animal cells, include *galK*, *cm^R*, *tk*, *neo^R*, and *gpt*.[31–35]

Identifying Coding Sequences

Cloning by complementation is commonly employed to isolate DNA from fungi (this volume [53]) and bacteria. Several strategies may be employed to subclone specific genes from genomic clones (typically 10–20 kb). Most straightforwardly, specific restriction fragments may be subcloned into vectors that can be installed into the host bearing the mutation. Such subcloning is done most systematically if a restriction enzyme map of the primary clone has already been determined. Specific fragments may then be tested for ability to complement the lesion, indicating that they bear a functional gene. In addition, subgenic fragments that do not complement the mutation but cover the site of the lesion can be identified by their ability to recombine with mutation at high frequency and generate a wild-type phenotype (marker rescue).

Alternatively, *in vivo* and *in vitro* methods exist to generate insertion mutations within the cloned sequences. Insertions that abrogate the ability of the clone to complement *in vivo* define the location of the functional sequences. Perhaps the best method of performing random insertional mutagenesis involves treating plasmid DNA with DNase I to generate one break per molecule followed by insertion of oligonucleotide linkers that encode restriction enzyme sites. The position of the insertions may be quickly mapped relative to restriction sites in the genomic sequences and their effect on complementing activity easily monitored.

A second convenient approach in generating insertion mutations is transposon mutagenesis. Since most vectors bearing genomic clones also carry sequences allowing their expression in *E. coli*, it is possible to insert

[29] K. Itakura, T. Hirosc, R. Crea, and A. O. Riggs, *Science* **198**, 1056 (1977).
[30] R. A. Young and R. W. Davis, *Proc. Natl. Acad. Sci. U.S.A.* **80**, 1194 (1983).
[31] D. Shumperli, B. Howard, and M. Rosenberg, *Proc. Natl. Acad. Sci. U.S.A.* **79**, 157 (1982).
[32] C. Gorman, L. Moffat, and B. Howard, *Mol. Cell. Biol.* **2**, 1044 (1982).
[33] M. Wigler, S. Silverstein, L. Lee, A. Pellicer, Y. Cheng, and R. Axel, *Cell (Cambridge, Mass.)* **11**, 223 (1977).
[34] P. Southern and P. Berg, *J. Mol. Appl. Genet.* **1**, 327 (1982).
[35] R. Mulligan and P. Berg, *Science* **209**, 1422 (1980).

transposons such as Tn5 or Tn10 randomly into the plasmid.[36] Analysis may proceed analogous to that described for linker mutagenesis. Recently, recombinant transposons have been described that bear the *lacZ* gene close to one end such that in-frame fused genes can be generated by transposition into a coding sequence. This transposon also bears an antibiotic resistance marker and a yeast marker (the *URA3* gene). Thus, insertion mutations into primary clone sequences may be rapidly crossed into the yeast genome by the method of Rothstein.[37] The location and direction of transcription of the gene of interest may also be deduced by this method. Further, the final product is a yeast strain bearing a *lacZ* fusion which may be used in subsequent studies on gene regulation.

General Subcloning Strategies

In this section we discuss some of the considerations involved in the design of a subcloning strategy. One of the first things to consider is of course the vector. This choice is guided by the type of studies to be carried out, as was discussed in the previous section. In addition to this, it should be noted that many cloning vectors already described in this chapter have been entirely sequenced and hence exhaustive restriction enzyme maps are available. It is advisable to use vectors that have been entirely sequenced so that there will be no uncertainties about a cloning vector and its restriction digestion patterns.

The simplest subcloning strategy would of course be to subclone a restriction fragment from a genomic clone into a unique site recognized by the same restriction enzyme in the vector. Such a subcloning strategy should produce a population of subclones in which half of the plasmids have the insert in one of two possible orientations. However, such a simple strategy may not be applicable if the desired restriction sites are not available. One must then consider other strategies. It is often useful to exploit the fact that different restriction endonucleases produce DNA molecules whose ends are cohesive. For example, DNA molecules with ends generated by *Bam*HI, *Bgl*II, and *Sau*3A can all be annealed to one another. For a complete list of compatibilities among restriction endonuclease-generated ends, see this volume [11]. The ligation of two compatible ends generated by different restriction enzymes often will not regenerate either of the original sites. However, if the site chosen in the vector DNA molecule lies in a multiple cloning site, then convenient unique restriction sites will be placed nearby.

[36] N. Kleckner, J. R. Roth, and D. Botstein, *J. Mol. Biol.* **116**, 125 (1977).
[37] R. J. Rothstein, this series, Vol. 101, p. 202.

Sometimes the fragment that one wants to subclone can be generated by digestion with two different restriction endonucleases. In this case the vector can also be treated with enzymes that generate the appropriate cohesive ends. This strategy should also be used if one must ensure the orientation of the insert in the desired subclones. Another benefit from using two different restriction enzymes in the subcloning strategy is that it impairs the ability of the vector DNA to recircularize and therefore reduces the "background" of insert-deficient vector DNA molecules obtained in the ligation reaction.

Another important element to consider in designing a subcloning strategy is the fact that any blunt-ended DNA molecule can be ligated to any other blunt-ended DNA molecule. Many restriction enzymes generate blunt ends (see this volume [11]). However, any DNA molecule that has 3' or 5' extensions can be modified to have blunt ends and can therefore be made compatible with any other blunt-ended DNA molecule. A 5' extension can be "filled in" using the large fragment of E. coli DNA polymerase I as described in this volume [10]; 3' ends can be degraded by the exonucleolytic activity of bacteriophage T4 DNA polymerase. Both 3' and 5' extensions can both be removed by digestion with single-stranded DNA-specific nucleases such as S1 (this volume [34, 35]).

For some studies, the addition of an oligonucleotide restriction enzyme linker to the ends of an insert fragment may be desirable, but requires additional manipulations (this volume [38]) and is not always necessary. Again, insertion of DNA into a "multiple cloning site" will place restriction sites nearby and often has the same result as adding a linker sequence.

As discussed in this volume [11], it is possible to cleave DNA at any specific location regardless of whether or not a restriction site is available. The DNA is subcloned into M13, and single-stranded DNA is prepared [13] and hybridized to an oligomer bearing a double-stranded hairpin region with a FokI site that is attached to a single-stranded region able to hybridize to the sequence of interest. Since FokI cleaves outside its recognition sequence, by positioning the FokI site with respect to the desired site of cleavage, the oligomer can direct the enzyme to cut at virtually any location. Once cleaved, the oligomer can be dissociated, and the single-stranded DNA is made double stranded with the M13 sequencing primer (among others) and Klenow fragment [10] and subcloned. Careful attention to the orientation of the fragment in M13 is required. This technique is also useful when one wishes to cleave at one restriction site of a DNA fragment that contains many such sites.

There are advantages to treating the vector DNA with alkaline phos-

phatase (either from calf or bacteria) just after restriction enzyme digestion. Since T4 DNA ligase requires that one of the substrate's ends have a
5'-phosphate, removing the 5' phosphates from the vector DNA prevents
it from recircularizing. In a strategy in which the vector is being treated
with two restriction enzymes, phosphatase treatment will prevent the two
vector fragments from being ligated together to reform the original vector.

If one has a large piece of genomic DNA and wishes to subclone each
of its fragments generated by any restriction endonuclease (such as for
"shotgun" dideoxy sequencing; this volume [57]) it is not necessary to
purify the fragments before subcloning. The DNA can be digested with
the desired restriction endonuclease, purified by phenol extraction and
ethanol precipitation, and resuspended in TE (see below) at a concentration of 100–500 μg/ml. The subclones will then be sorted out by analyzing
their structure by either restriction enzyme mapping or sequencing. If,
however, a plasmid construction involves subcloning a particular DNA
fragment from a large clone then that fragment can be purified by gel
electrophoresis. In a subcloning strategy in which the desired clones are a
subset of the potential products, whether to gel purify inserts is the choice
of the investigator. Some prefer to isolate fragments, carry out individual
ligation reactions and transformations, and then find the desired clones by
analyzing plasmids from a small number of transformants from each
group. Others prefer to do one ligation and transformation with a mixture
of fragments and then sort the clones by analyzing plasmids from several
transformants. The products of shotgun cloning, a complex mixture of
restriction fragments, can often be simplified by using two restriction
endonucleases to generate the desired insert. If one of the chosen restriction endonucleases cuts infrequently in the large clone, then very few of
the resulting fragments will be competent to ligate to a vector that has
only one of its ends generated by that enzyme.

To identify and confirm the structure of a desired subclone, plasmid
DNA can be prepared from individual transformants by any available
methods (see this volume [13]). In a simple subcloning strategy, comparison of the restriction digestion pattern of the subclone with that of the
vector and insert DNA should be sufficient to confirm the identity of a
subclone. In more complicated strategies, Southern hybridization using
an appropriate labeled probe can be used to identify the desired subclone.

Methods

The techniques involved in subcloning are straightforward and many
are discussed elsewhere in this volume. The steps involved are (1) to

isolate the vector DNA and the DNA to be subcloned with compatible ends, (2) to ligate the molecules, (3) to transform *E. coli* with the ligated DNA, and (4) to select for transformants.

Preparation of Vector DNA Fragments

Preparation of vector DNA is often quite simple. It is usually sufficient to digest the DNA, remove the restriction enzyme(s) by phenol extraction (this volume [4]), then ethanol precipitate the DNA (this volume [5]), and resuspend it in a low ionic strength buffer such as 10 mM Tris–HCl (pH 8), 1 mM EDTA (TE buffer) at a concentration of about 100–500 μg/ml. Since ligation reactions for subcloning often require ~100–300 ng of vector DNA, it is often sufficient to digest only 1–5 μg of vector DNA which should be enough digested DNA for several ligations.

Both bacterial and calf alkaline phosphatase are active in reaction mixtures used for restriction endonucleases. One unit of enzyme is added 15–30 min prior to completion of the restriction enzyme reaction (except for those carried out at 60–65°). For restriction enzymes that generate blunt ends or recessed 5' ends, the restriction enzyme reaction is allowed to go to completion and then the pH of the reaction mixture is raised to pH 9.5 by adding 0.1 volume at 1 M Tris–HCl (pH 9.5) or 1 M glycine–NaOH (pH 9.5). One unit of phosphatase can be added and allowed to react for 15–30 min at 37°. The DNA is purified by phenol extraction and ethanol precipitation.

Preparation of Insert DNA

There are several available methods for purifying DNA molecules from polyacylamide and agarose gels, and we will present some which are in current use. Others surely exist and are waiting to be devised (see also this volume [8]).

"Crush and Soak" Method. One of the simplest methods for isolating DNA fragments from polyacrylamide gels is the "crush-and-soak" method.[38] DNA (10–30 μg) is digested with the desired restriction endonuclease and is separated electrophoretically in a 3.5% polyacylamide gel. The gel is stained in 0.05% methylene blue until the DNA-containing bands become visible under ambient light, or the gel is stained in ethidium bromide (0.5 μg/ml) for 10 min and the DNA-containing bands are visualized under long-wavelength ultraviolet (uv) light. The region of the gel containing the desired DNA fragment is excised using a razor blade and placed in a tube. The gel slice is then crushed with a pestle and 3 ml of 10

---

[38] A. Maxam and W. Gilbert, this series, Vol. 65, p. 499.

mM Tris–HCl (pH 8), 1 mM EDTA, 150 mM NaCl is added. Sodium dodecyl sulfate can be included in this buffer at a concentration 0.5% or an equal volume of buffer-saturated phenol can be added. The suspension is then incubated overnight at room temperature. The DNA is recovered by filtering the aqueous phase through glass wool and precipitating with ethanol. The resulting pellet is rinsed, dried, and resuspended in 50–100 μl TE.

Electroelution. DNA fragments can be isolated from agarose gels by electroelution. DNA is digested with the desired restriction enzyme(s) and fractionated electrophoretically through an agarose gel that contains ethidium bromide at 0.5 μg/ml. After electrophoresis, the bands are located by ultraviolet fluorescence. An incision is made in the gel ahead of the desired band (toward the positive electrode). A piece of precut DEAE membrane (available from Schleicher and Schuell) is placed in the incision and the DNA is driven electrophoretically onto the paper; its progress can be monitored by uv fluorescence. The DEAE paper is then removed from the gel, shaken in a tube containing 20 mM Tris–HCl (pH 8), 0.1 mM EDTA, 0.15 M NaCl, and then placed in a microfuge tube. The paper is then covered with 100–250 ml of 20 mM Tris–HCl (pH 8), 0.1 mM EDTA, 1.0 M NaCl and incubated at 55–68° for 10–45 min with occasional mixing to elute the DNA. The eluate is transferred to a fresh tube and the paper is then rinsed with another 50 ml of buffer. The DNA can be precipitated directly from the eluate with ethanol.

Low Temperature Melting Agarose Gels. Low temperature melting agarose gels can be used to isolate DNA fragments. The DNA-containing band of interest is excised from the gel after ethidium bromide staining and combined with an approximately equal volume of 10 mM Tris–HCl (pH 8), 1 mM EDTA, 0.1 M NaCl in a microfuge tube and incubated at 65° for 15 min. The suspension is then extracted with phenol. The aqueous phase is incubated on ice for 30 min and centrifuged in a microfuge for 1 min to remove residual particulate matter. The DNA can be precipitated from the supernatant with ethanol.

A method in which DNA is isolated from low temperature melting agarose gels and used directly in ligation reaction has been described.[39] In general, it appears that the quality of the agarose used in this procedure is critical to its success.

Estimating the Concentration of DNA in a Small Volume

It is often desirable to determine the concentration of a DNA fragment preparation. However, spectrophotometry is often not practical due to

[39] K. Struhl, *BioTechniques* **3**, 452 (1985).

the small volume of the fragment preparation. The following "spot test" procedure is well suited for estimating the concentration of a DNA solution.

Standard solutions of known DNA concentrations are prepared in the range of 0–20 μg/ml in TE. On a piece of Parafilm, 2-μl drops of a 5 μg/ml solution of ethidium bromide are arranged and a standard curve is prepared by mixing 2 μl of standard DNA solutions each with an individual 2-μl drop of ethidium bromide. Dilutions of the DNA fragment preparation are made in TE and 2-μl drops are mixed with 2-μl drops of ethidium bromide on the Parafilm. The Parafilm is then placed on a UV transilluminator and photographed under UV light. The fluorescence of the DNA fragment preparation is compared to the standard curve in order to estimate its concentration.

Ligation

Ligations for subcloning are carried out as described (see Process Guide) and with the same theoretical considerations. In general, 100–500 ng of vector DNA is combined with 100–500 ng of insert in a 10- to 20-μl reaction mixture containing 10 mM Tris–HCl (pH 8), 10 mM MgCl$_2$, 1 mM dithiothreitol, BSA at 100 μg/ml, 1 mM ATP; T4 DNA ligase is added (unit definitions vary from manufacturer to manufacturer); and the reaction is allowed to proceed. Incubation times and temperatures vary and appear to be flexible. For ligations involving blunt ends, incubation at 4° for 8–16 hr is adequate. For ligations involving cohesive ends, incubations at 4–18° for 2–16 hr have been reported to be successful. Ligation reaction mixtures can be directly used to transform competent *E. coli* (this volume [13]) or stored frozen for use at some future date.

[56] DNA Sequencing: Chemical Methods

By Barbara J. B. Ambrose and Reynaldo C. Pless

Four-Lane Methods

Limited base-specific or base-selective cleavage of a defined DNA fragment yields polynucleotide products, the length of which correlates with the positions of the particular base (or bases) in the original fragment. Sverdlov and co-workers[1-3] recognized the possibility of using this principle for the determination of DNA sequences. In 1977, Maxam and Gilbert[4] introduced a fully elaborated method, based on this principle, which allowed routine analysis of DNA sequences over distances greater than 100 nucleotide units from a defined, radiolabeled terminus. Six procedures for partial cleavage were described, with the following cleavage specificities: (G > A), (A > G), G, (A > C), (C + T), and C. Simultaneous parallel resolution of an appropriate set of partial cleavage mixtures by polyacrylamide gel electrophoresis, followed by visualization of the radioactive bands by autoradiography, allows the deduction of nucleotide sequence. A further developed version of this method, incorporating several modifications, was described in great experimental detail by Maxam and Gilbert in 1980.[5] An adaptation of five cleavage protocols from their publication is given at the end of this chapter. The five modification reactions, which upon subsequent treatment with hot aqueous piperidine result in base specific cleavage, are (1) methylation of the DNA with dimethyl sulfate in aqueous solution at pH 8, for cleavage at G sites, (2) partial depurination in piperidine formate buffer (pH 2), giving approximately equal cleavage at G sites and at A sites, (3) treatment with 57% aqueous hydrazine for cleavage at C sites and T sites, (4) treatment with 57% aqueous hydrazine containing 1.5 M NaCl, cleaving at C sites only, and (5) treatment with 1.1 M NaOH at 90°, resulting in strong cleavage at A sites and weak cleavage at C sites.

The history of the development of these procedures and their logical framework have been outlined by Maxam.[6] It speaks for the power and general applicability of this method that it is extensively used today, largely in the form described by Maxam and Gilbert in 1980. For some tasks, as for the location of rare bases, the chemical cleavage analysis cannot be replaced by the dideoxynucleotide terminator method,[7] as the latter analyzes the DNA of interest via its complementary sequence; it can, thus, only give sequence information in terms of the four canonical

[1] E. D. Sverdlov, G. S. Monastyrskaya, E. I. Budowsky, and M. A. Grachev, *FEBS Lett.* **28,** 231 (1972).

[2] E. D. Sverdlov, G. S. Monastyrskaya, A. V. Chestukhin, and E. I. Budowsky, *FEBS Lett.* **33,** 15 (1973).

[3] E. D. Sverdlov, G. S. Monastyrskaya, and F. I. Budowsky, *Mol. Biol. (Moscow)* **11,** 116 (1977).

[4] A. M. Maxam and W. Gilbert, *Proc. Natl. Acad. Sci. U.S.A.* **74,** 560 (1977).

[5] A. M. Maxam and W. Gilbert, this series, Vol. 65, p. 499.

[6] A. M. Maxam, *in* "Methods of DNA and RNA Sequencing" (S. M. Weissman, ed.), Chapter 4. Praeger, New York, 1983.

[7] F. Sanger, S. Nicklen, and A. R. Coulson, *Proc. Natl. Acad. Sci. U.S.A.* **74,** 5463 (1977).

TABLE I
ALTERNATIVE OR MODIFIED CHEMICAL CLEAVAGE PROCEDURES[a]

Specificity	Reagent	Temperature (°C)	Time (min)	References
T	5% OsO_4 in H_2O–pyridine (11 : 1, v/v)	0	15	b,c
T ≫ G,C	10^{-4} M $KMnO_4$ in H_2O	20	10	d,e,f
T > G ≫ A,C	1 M Cyclohexylamine in H_2O + UV irradiation	RT[g]	5	h,i
T	1 M Spermine in H_2O + UV irradiation	0	20	j,k,l
G > T	1 M Methylamine in H_2O + UV irradiation	0	20	j,k,l
T	0.5 M $NaBH_4$ in H_2O, pH 8–10	37	30	m,n
T ≫ C	2–3 M H_2O_2 in carbonate buffer, pH 9.6	37	60	o,m,n
C	2–3 M H_2O_2 in carbonate buffer, pH 8.3	37	60	o
C	2–3 M H_2O_2 in Tris–HCl buffer, pH 7.4	37	60	o
C	N_2H_4–H_2O (3 : 1, v/v), 5 M N_2H_4·HOAc	20	10	d
C	N_2H_4–H_2O (3 : 2, v/v), 0.2 M NaOH	p	p	q
C	3 M NH_2OH–HCl in H_2O, pH 6.0	20	10	d,e
G	0.1% Methylene blue + visible light	RT[g]	15	c,r
G	4% Dimethyl sulfate in formate buffer, pH 3.5	20	12	s,q,f
G ≫ C	0.3% Diethyl pyrocarbonate in cacodylate buffer, pH 8	90	5	t
A + G	60–88% Aqueous formic acid	p	p	u,v,f
A + G	Citrate buffer, pH 4	80	5	w
A + G	2–3% Diphenylamine in 66% formic acid	20	40	x,u,q,l
A + G	0.1% Diethyl pyrocarbonate in acetate buffer, pH 5	90	5	t
G	**0.5% Dimethyl sulfate in 50 mM cacodylate buffer, pH 8**	**37**	**15**	**y,z**
A + G	**0.13 M Aqueous piperidine formate buffer, pH 2**	**45**	**40**	**y**
A + G	**2% Diphenylamine in 66% formic acid**	**37**	**12**	**z**
C + T	**N_2H_4–H_2O (7 : 4, v/v)**	**45**	**20**	**y,z**
C	**N_2H_4–H_2O (7 : 4, v/v), 1.1 M NaCl**	**45**	**20**	**y**

[a] Unless otherwise stated in these footnotes, the base-specific reaction given was followed by treatment with hot aqueous piperidine. Conditions given in bold print are optimized specifically for oligonucleotide sequencing.

[b] Osmium tetroxide is highly toxic.

[c] T. Friedmann and D. M. Brown, *Nucleic Acids Res.* **5**, 615 (1978).

[d] C. M. Rubin and C. W. Schmid, *Nucleic Acids Res.* **8.** 4613 (1980).

[e] M. E. S. Hudspeth, W. M. Ainley, D. S. Shumard, R. A. Butow, and L. I. Grossman, *Cell (Cambridge, Mass.)* **30,** 617 (1982).

bases. Accordingly, genomic sequencing[8] for location of 5-methylcyto-sine sites in mammalian DNA is a derivative of the Maxam–Gilbert procedure for sequencing by chemical cleavage.

Reviews of the method have been published,[9,10] and adaptations for the sequencing of short oligonucleotides[11–14] and alternative base-specific DNA cleavage procedures have been described. A selection of such variants is listed in summary form in Table I.

Much information has accumulated on the susceptibility of DNA to cleavage at sites occupied by various rare bases. 5-Methylcytosine posi-

---

[8] G. M. Church and W. Gilbert, *Proc. Natl. Acad. Sci. U.S.A.* **81**, 1991 (1984).
[9] J. Hindley, *Lab. Tech. Biochem. Mol. Biol.* **10**, 230 (1983).
[10] N. L. Brown, *Methods Microbiol.* **17**, 259 (1984).
[11] E. Jay, A. K. Seth, J. Rommens, A. Sood, and G. Jay, *Nucleic Acids Res.* **10**, 6319 (1982).
[12] A. M. Banaszuk, K. V. Deugau, J. Sherwood, M. Michalak, and B. R. Glick, *Anal. Biochem.* **128**, 281 (1983).
[13] K. Rushlow, *Focus (Bethesda, Md.)* **5**, No. 2, 4 (1983).
[14] A. Rosenthal, S. Schwertner, V. Hahn, and H. Hunger, *Nucleic Acids Res.* **13**, 1173 (1985).

---

f A. Rosenthal, S. Schwertner, V. Hahn, and H. Hunger, *Nucleic Acids Res.* **13**, 1173 (1985).
g Room temperature.
h A. Simoncsits and I. Török, *Nucleic Acids Res.* **10**, 7959 (1982).
i A cleavage specificity of (T + G) was found for oligonucleotides (A. Simoncsits, as quoted in ref. *h*).
j H. Sugiyama, I. Saito, T. Matsuura, K. Ueda, and T. Komano, *Nucleic Acids Symp. Ser.* **12**, 103 (1983).
k I. Saito, H. Sugiyama, T. Matsuura, K. Ueda, and T. Komano, *Nucleic Acids Res.* **12**, 2879 (1984).
l Backbone cleavage does not require a separate treatment with hot piperidine.
m E. D. Sverdlov and N. F. Kalinina, *Dokl. Akad. Nauk SSSR* **274**, 1508 (1984).
n Procedure hardly cleaves at T sites having another T as their next 3' neighbor.
o E. D. Sverdlov and N. F. Kalinina, *Bioorg. Khim.* **9**, 1696 (1983).
p No details were given.
q V. N. Dobrynin, V. G. Korobko, I. V. Severtsova, N. S. Bystrov, S. A. Chuvpilo, and M. N. Kolosov, *Nucleic Acids Symp. Ser.* **7**, 365 (1980).
r D. M. Stalker, W. R. Hiatt, and L. Comai, *J. Biol. Chem.* **260**, 4724 (1985).
s V. G. Korobko, S. A. Grachev, and M. N. Kolosov, *Bioorg. Khim.* **4**, 1281 (1978).
t A. S. Krayev, *FEBS Lett.* **130**, 19 (1981).
u K. G. Skryabin, V. M. Zakharyev, and A. A. Bayev, *Dokl. Akad. Nauk SSSR* **241**, 488 (1978).
v Yu. A. Ovchinnikov, S. O. Guryev, A. S. Krayev, G. S. Monastyrskaya, K. G. Skryabin, E. D. Sverdlov, V. M. Zakharyev, and A. A. Bayev, *Gene* **6**, 235 (1979).
w A. Maxam, as quoted in ref. *e*.
x V. G. Korobko and S. A. Grachev, *Bioorg. Khim.* **3**, 1420 (1977).
y K. Rushlow, *Focus (Bethesda, Md.)* **5**, No. 2, 4 (1983).
z A. M. Banaszuk, K. V. Deugau, J. Sherwood, M. Michalak, and B. R. Glick, *Anal. Biochem.* **128**, 281 (1983).

tions have been reported[15,16] to be much more resistant than C sites to the C- or the (C + T)-specific hydrazinolysis conditions of Maxam and Gilbert.[4] A 5'-terminal 5-methylcytosine site in an oligonucleotide was found to be weakly but significantly susceptible to cleavage in (C + T)-specific hydrazinolysis,[17] but resistant to the action of hydroxylamine.[14] N^6-Methyladenine sites react similarly to adenine sites in the (A + G)-specific acid depurinations, but are much more resistant than A sites to the direct action of hot aqueous piperidine[18] or to the action of alkali[19] in the (A > C)-specific cleavage procedure of Maxam and Gilbert.[4,5] In sequence analysis of short synthetic deoxyribooligonucleotides, uracil sites were found to react much more strongly than thymine sites and cytosine sites in (C + T)-specific hydrazinolysis, and 5-bromouracil positions were strongly modified both in (C + T)-specific hydrazinolysis and in (A > C)-specific reaction with alkali.[17] In contrast to T sites, both uracil sites and 5-bromouracil sites are susceptible to the action of hydroxylamine.[14] For one analog base, 2-aminopurine, a highly specific cleavage procedure has been found in simple treatment with hot aqueous piperidine.[20]

Accessory techniques for DNA sequencing have undergone considerable development. Ansorge and Barker[21] have used thin gels (0.1 or 0.2 mm) of up to 1.2 m length, of 4% polyacrylamide, to achieve resolution of sequence ladders for regions more than 500 nucleotide units distant from the labeled end. Use of a thermostated mold plate, kept at 70° during the electrophoresis, minimizes bending of the electrophoretic front and band compression caused by formation of hairpin structures.[22]

An increase in the conductivity of the gel toward the bottom of the mold decreases the local electric field and thus retards the migration of the faster running polynucleotides, allowing more time for the electrophoretic resolution of the longer fragments in the same lane. Such a progressive increase in conductivity can be achieved by use of a steep buffer salt gradient set up in the lower part of the gel as it is poured.[23] A different

[15] H. Ohmori, J. Tomizawa, and A. M. Maxam, *Nucleic Acids Res.* **5,** 1479 (1978).
[16] J. R. Miller, E. M. Cartwright, G. G. Brownlee, N. V. Fedoroff, and D. D. Brown, *Cell (Cambridge, Mass.)* **13,** 717 (1978).
[17] A. Rosenthal, D. Cech, V. P. Veiko, T. S. Orezkaja, E. A. Romanova, A. A. Elov, V. G. Metelev, E. S. Gromova, and Z. A. Shabarova, *Tetrahedron Lett.* **25,** 4353 (1984).
[18] B. J. B. Ambrose and R. C. Pless, *Biochemistry* **24,** 6194 (1985).
[19] N. Okawa, Y. Suyama, and A. Kaji, *Nucleic Acids Res.* **13,** 7639 (1985).
[20] R. C. Pless and M. J. Bessman, *Biochemistry* **22,** 4905 (1983).
[21] W. Ansorge and R. Barker, *J. Biochem. Biophys. Methods* **9,** 33 (1984).
[22] H. Garoff and W. Ansorge, *Anal. Biochem.* **115,** 450 (1981).
[23] M. D. Biggin, T. J. Gibson, and G. F. Hong, *Proc. Natl. Acad. Sci. U.S.A.* **80,** 3963 (1983).

approach uses wedge-shaped gels, 0.15 or 0.2 mm thick in the upper regions and flaring to a thickness of close to 1 mm at the bottom.[24,25]

One-Lane Methods

The traditional methods for DNA sequencing by chemical cleavage are based on the parallel execution of four base-specific or base-selective modification protocols and the parallel electrophoretic resolution of the hydrolysates in four lanes. An interesting alternative is analysis of the DNA sequence based on one chemical modification procedure and electrophoresis in a single lane. Logic dictates that such a method should be capable of elaboration, and defines the features which such a method should exhibit:

(1) The procedure should produce backbone cleavage at *all* bases in the DNA, since the single electrophoretic lane must signal every individual nucleotide position in the sequence.

(2) The procedure should produce cuts in analogous positions in the DNA backbone, such that a uniquely end-labeled, defined DNA segment will be cleaved to a nested set of radioactive fragments which have uniform end configuration and therefore differ by multiples of full nucleotide units.

(3) The rate of cleavage at the four canonical bases in DNA should be clearly different, so that the intensities of the radioactive bands in the electrophoretogram can be interpreted in terms of base sequence. Minor bases in DNA, 5-methylcytosine and N^6-methyladenine, should have cleavage propensities distinct from those of the other bases, or else should be identical in cleavage propensity to the cognate canonical bases, cytosine and adenine, respectively.

(4) The cleavage propensity of a given base should be essentially unaffected by the local sequence surrounding it.

We have developed a procedure[18] involving treatment with 0.5 M aqueous piperidine containing sodium chloride, at high temperature for several hours, which largely fulfills these requirements.

(1) Examination of the time course of the digestion with hot aqueous piperidine indicates that the treatment brings about base modification at A, G, and C bases in a time-dependent fashion, leading ultimately to

---

[24] A. Olsson, T. Moks, M. Uhlén, and A. B. Gaal, *J. Biochem. Biophys. Methods* **10,** 83 (1984).

[25] W. Ansorge and S. Labeit, *J. Biochem. Biophys. Methods* **10,** 237 (1984).

cleavage of the backbone at these positions.[26] In contrast, hot aqueous piperidine appears to attack thymine moieties in DNA only very slowly, if at all. The weak bands seen at thymine positions may be due to initial radiolytic damage to the thymine bases, which are the moieties in DNA most vulnerable to ionizing radiation,[27] followed by backbone cleavage at the damaged sites during the treatment with piperidine.

(2) Densitometric analysis of the autoradiograms indicates virtually complete absence of spurious bands.[18] As in sequence analysis of 5′-end-labeled DNA according to Maxam and Gilbert,[4,5] the radioactive fragments produced have the general formula $d\text{-}\overset{*}{p}(Np)_n$, with a base Z in a generalized sequence $d\text{-}\overset{*}{p}XpYpZp$ ... signaled on the gel by the fragment $d\text{-}\overset{*}{p}XpYp$ (where asterisk denotes the labeled phosphate).

(3) In the presence of 0.3 M NaCl, solvolysis in 0.5 M aqueous piperidine at 90° for several hours produces strong bands signaling A positions, C bands and G bands of medium intensity, and weak T bands. The relative band intensities, as determined by densitometry of the autoradiograms, are approximately 8 : 4 : 4 : 1 for A : C : G : T.[18] While it is ultimately desirable to define conditions providing for clearly distinct cleavage rates at the four canonical bases, we have been able to determine DNA sequences from hydrolyses performed as specified above, by applying an independent criterion to distinguish G bands from C bands: the observation, already made by others,[4,12] that in electrophoretic ladders obtained by chemical cleavage of end-labeled polydeoxyribonucleotides, the bands denoting G positions have a distinctly larger separation from the next higher band in the ladder than do the bands denoting C positions.

(4) The strongly denaturing conditions of the digestion with hot aqueous piperidine (90°, pH >12) minimize local sequence effects on the reactivity of a given type of base in the DNA. For instance, all A sites in a fragment will be modified at a similar rate. Nonetheless, the radioactivity distribution obtained by treating 5′-end-labeled DNA with 0.5 M piperidine, 0.3 M NaCl at 90° for 5 hr shows a gradual increase in the intensity of the A bands (and of G bands and C bands, as well) for progressively shorter polynucleotides. This reflects loss of long-chain products during the course of the reaction, due to repeated cleavage.

Lanes 1–4 in Fig. 1 show sequence ladders for two fragments of mouse thymidylate synthase cDNA, inserted into pBR322 by Johnson, Vanin, and co-workers,[28] and kindly put at our disposition. Lanes 1 and 2 corre-

---

26 B. J. B. Ambrose and R. C. Pless, unpublished observations (1984).
27 R. Uliana, P. V. Créac'h, and A. Ducastaing, *Biochimie* **53,** 461 (1971).
28 S. M. Perryman, C. Rossana, T. Deng, E. F. Vanin, and L. F. Johnson, *Mol. Biol. Evol.* **3,** 313 (1986).

spond to a double-stranded *Dde*I–*Ava*II fragment, 188 nucleotide units long, 5' labeled at the *Dde*I end. This fragment was heated at 90° for 5 hr in 0.5 *M* aqueous piperidine, 0.3 *M* NaCl, and then analyzed by electrophoresis through a 20% polyacrylamide sequencing gel. The first two bands shown at the bottom of lane 2 (A and G), corresponding to positions 3 and 4 counting from the labeled end, are obviously badly distorted by the presence of salt migrating in this region of the gel. From the next band on, the sequence can be read according to the prescription formulated in Ambrose and Pless,[18] namely, A: relatively strong band, relatively short separation to the next higher band in the ladder; G: medium intensity band, relatively large separation; C: medium intensity band, relatively short separation; T: relatively weak band, relatively large separation. Both the band intensity and the band spacing parameters have to be evaluated in a relative manner, i.e., in comparison to neighboring bands in the ladder, because the standard for both parameters shifts along the electrophoretic lane; average band intensities increase for the shorter oligonucleotides due to repeated cleavage in one chain (as explained above), and average band spacings increase toward the bottom of the gel because the band mobilities are approximately inversely related to the fragment length.[29] Lane 1 represents an earlier loading of this sequencing mixture, in which the first 17 bands were run off the gel. For these higher regions of the sequence, displayed in lane 1, the systematic change in average band spacing is moderate, facilitating the reading of the sequence. In this lane, the lettering gives the sequence up to a position 71 nucleotide units removed from the labeled end.

Lanes 3 and 4 in Fig. 1 show two loadings for a double-stranded *Bgl*II–*Pst*I fragment, 5' labeled at the *Bgl*II site, 176 nucleotide units long in the labeled strand, after treatment with 0.5 *M* aqueous piperidine, 0.5 *M* NaCl at 110° for 1.5 hr and electrophoresis through a 20% polyacrylamide gel. These hydrolysis conditions again produce a cleavage pattern readable using the prescription given above, but the increase in solvolysis temperature and in salt concentration allows a reduction of the reaction time by a factor of 3 while still providing satisfactory signal intensities.

Lanes 5–8 in Fig. 1 show four successive loadings of a hydrolysate of a 137-mer fragment of pBR322 DNA (positions 3867–3731, according to Sutcliffe[30]) obtained by restriction with *Dde*I and *Ava*II and 5'-end-labeled at the *Dde*I terminus (position 3867). This DNA was treated with 0.5 *M* aqueous piperidine, 0.3 *M* NaCl at 90° for 5 hr and then fractionated electrophoretically through a 10% polyacrylamide gel. The lowest bands

[29] E. M. Southern, *Anal. Biochem.* **100**, 319 (1979).
[30] J. G. Sutcliffe, *Cold Spring Harbor Symp. Quant. Biol.* **43**, 77 (1979).

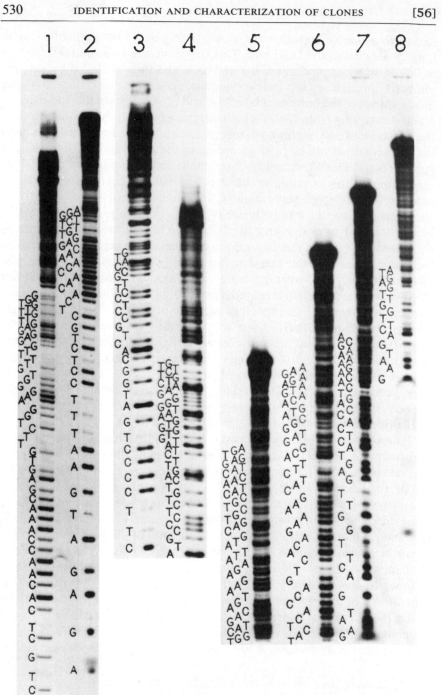

lettered in lanes 8 and 7 correspond to the G in fourth position from the labeled end. In lanes 7, 6, and 5, the first 3, 19, and 51 bands, respectively, were run off the gel. In lane 5, the bands are lettered up to a position 99 nucleotide units removed from the labeled end. Compared to the denser gels, in these 10% polyacrylamide gels a much larger number of bands close to the labeled end are affected in their shape and mobility by comigrating salt.

Lane 6 in Fig. 1 shows that in 10% polyacrylamide gels band spacing follows the same general rules which govern spacing in 20% gels, as documented earlier[18]; there is a systematic increase in average gap size as one moves down in the electrophoretic ladder, and superimposed on this general trend there is a base-specific effect, with band spacings in the order G > T > A > C. In a given region of the gel, band separations are 25–30% larger for G bands than for C bands.

N^6-Methyladenine sites are distinctly less reactive in hot aqueous piperidine, at 0.3 M NaCl or 1.0 M NaCl, than are A sites.[18] Judged by band intensity and spacing, N^6-methyladenine sites appear as C sites in this analysis. Therefore, when DNA cloned in pBR322 is sequenced by this method, reading the series GCTC leaves a doubt whether the actual sequence is not $G^{me}ATC$, as GATC is A methylated in the *Escherichia coli* host. While this issue can be resolved by sequencing the complementary strand, it is desirable to have a procedure which presents methyladenine sites as A sites. We have developed such a method,[31] based on partial depurination of 5'-end-labeled DNA in 0.25 M aqueous piperidine formate buffer (pH 2) at 37° for 25 min, followed by extended heating in 0.5 M aqueous piperidine at 90°. This procedure is very similar to the (A+G) procedure of Maxam and Gilbert,[5] except for the prolonged treatment with hot piperidine, necessary to bring about distinct cleavage at all sites.

In this system, G bands are stronger than A bands, and the two types of sites can also be distinguished by the criterion of band spacing. C bands are weaker, and T bands are the weakest. N^6-Methyladenine sites examined by this analysis were read as adenines. As expected, 5-methylcytosine sites were not cleaved directly by prolonged heating in aqueous piperidine, nor in the variant which includes the acid pretreatment.

The variant procedure is particularly useful in the analysis of short

---

[31] B. J. B. Ambrose and R. C. Pless, unpublished observations (1984).

FIG. 1. Autoradiographed gel electrophoretograms of hydroysates obtained from 5'-end-labeled DNA fragments by treatment with 0.5 M aqueous piperidine, 0.3 M NaCl at 90° for 5 hr (lanes 1, 2, 5, 6, 7, 8) or by treatment with 0.5 M aqueous piperidine, 0.5 M NaCl at 110° for 1.5 hr (lanes 3, 4). Polyacrylamide gel densities were 20% for lanes 1–4 and 10% for lanes 5–8. Band identifications are lettered on the left of the corresponding lane.

oligonucleotides, as it produces a hydrolysate which is essentially salt free, such that the electrophoretic mobility of even the shortest product fragments on the gel is unaffected by salt. Treatment of 5′-end-labeled synthetic oligodeoxyribonucleotides with 0.4 M aqueous piperidine formate buffer (pH 2.0) at 37° for 6 hr, followed by repeated lyophilization to remove the volatile salt, heating in 0.5 M aqueous piperidine at 90° for 6 hr, and electrophoresis of the hydrolysate through a 27% polyacrylamide gel, creates a one-lane pattern of radioactivity in which the band intensities are ordered as G > A > C > T, and band spacings decrease in the sequence G > T > A > C.[32]

Comparison of One-Lane Methods with Four-Lane Methods

The one-lane methods described here derive the nucleotide sequence of a DNA fragment from the radioactivity distribution in a single electrophoretic lane; the informational content of the lane is heavily utilized. The partial hydrolysates which are analyzed in this manner are generated by simple, rapid procedures. By comparison, sequence analysis using four (or more) separate cleavage procedures is more laborious, and there is a considerable measure of informational redundancy in the display of radioactive bands in four (or more) parallel gel lanes. This redundancy is one of the reasons why the four-lane methods are more accurate than the one-lane methods, at least at their present stage of development. A contaminating fragment of DNA would be recognized in multilane analysis by its simultaneous appearance in all lanes, while it may cause misreading in one-lane methods. To the extent that the one-lane methods depend on band spacing for base assignments, they are also more error prone in the case of spacing irregularities due to salt or to hairpin formation.

Use of one-lane sequencing appears warranted when a somewhat higher level of error (about 2%) can be tolerated or when analysis is confirmed by sequencing of the opposite strand. The method will also be useful in confirmatory sequencing of synthetic oligonucleotides,[32] for establishing location in known sequences, and for comparison and identification of mutant sequences.

Procedures

The following contains a close adaptation of five modification and cleavage protocols given by Maxam and Gilbert in 1980,[5] and instructions

[32] B. J. B. Ambrose, M. M. Castro, and R. C. Pless, *Anal. Biochem.* **159,** 24 (1986).

on the two one-lane procedures described in this chapter. An extensive guide for troubleshooting in connection with the sequencing procedures of Maxam and Gilbert is given in Maxam and Gilbert.[5]

General. It is assumed that the DNA fragment of interest, 5'- or 3'-end-labeled with ^{32}P, is available in essentially salt-free aqueous solution, at about 5000 dpm/μl. Unless stated otherwise, reactions are run in 1.5-ml polypropylene snap-cap tubes. Heating is done in thermostated heating blocks, with the wells filled with water (for temperatures up to 90°) or with silicone oil (for 90° or higher). An ice–water mixture is used for chilling to 0°, and a dry ice–ethanol bath for chilling to −70°. Mixing is done by Vortexer, and, for 1-ml volumes, by repeatedly inverting the tube. Centrifugation is carried out in a microcentrifuge (12,000 g) in a cold room. Vacuum evaporations are performed with a Savant Speed Vac. Unless stated otherwise, all additions of solutions are done by mechanical pipet. Supernatant liquids are removed from DNA pellets by Pasteur pipet.

Reagents and Solutions

Dimethyl sulfate: 99%, Aldrich Chemical Company

DMS buffer: 50 mM aqueous sodium cacodylate (pH 8.0); 10 mM MgCl$_2$, 1 mM EDTA

DMS stop solution: 1.5 M aqueous sodium acetate (pH 7.0), 1.0 M mercaptoethanol, 100 μg/ml tRNA

Piperidine formate, 1.0 M (pH 2.0): 4% (v/v) aqueous formic acid, adjusted to pH 2.0 with piperidine

Hydrazine: 95%, Eastman Kodak, stored in aliquots in snap-cap tubes at −20°

Hydrazine stop solution: 0.3 M sodium acetate, 0.1 mM EDTA, 25 μg/ml tRNA

Aqueous sodium acetate, 0.3 M

Ethanol, 95%

Aqueous sodium chloride, 5 M

Aqueous sodium hydroxide, 1.2 M with 1 mM EDTA

Sonicated salmon sperm DNA, 1 mg/ml, as carrier DNA

tRNA, 1 mg/ml

Aqueous acetic acid, 1 M

Aqueous piperidine, 1 M: 1 volume piperidine to 9 volumes water, freshly mixed

Loading solution: 80% (v/v) formamide, 10 mM NaOH, 1 mM EDTA, 0.1% (w/v) xylene cyanole, 0.1% (w/v) bromphenol blue

Waste bottle for dimethyl sulfate: 5 M aqueous sodium hydroxide

Waste bottle for hydrazine: 3 M aqueous ferric chloride

Dimethyl sulfate and hydrazine are toxic. They should be dispensed in a hood. Use above waste bottles for disposal of small amounts of these reagents.

Limited Cleavage at Guanines (G)

Mix 200 μl DMS buffer, 3 μl sonicated salmon sperm DNA, 5 μl end-labeled DNA in water, and chill to 0°. Add 1 μl dimethyl sulfate (delivered from a graduated microcapillary pipet). Cap tube, mix, and heat 4 ± 2 min at 20°. Add 50 μl DMS stop solution at 0° and 750 μl 95% ethanol at 0°. Cap tube, mix, and chill 5 min at −70°. Centrifuge 5 min. Transfer supernatant to the dimethyl sulfate waste bottle. Add 250 μl 0.3 M sodium acetate, 0°. Cap tube; mix to dissolve DNA. Add 750 μl 95% ethanol, 0°. Cap tube, mix, and chill 5 min at −70°. Centrifuge 5 min. Remove supernatant. Add 1 ml 95% ethanol, 0°. Centrifuge 15 sec. Remove supernatant. Vacuum evaporate for 5 min. Preparation is now ready for piperidine treatment.

Limited Cleavage at Purines (A + G)

Combine 3 μl sonicated salmon sperm DNA and 9 μl end-labeled DNA in water. Cap tube and mix. Add 4 μl 1 M piperidine formate (pH 2). Cap tube, mix, and heat 15 ± 7 min at 37°. Vacuum evaporate to dryness. Add 20 μl distilled water and mix. Vacuum evaporate to dryness. Preparation is now ready for piperidine treatment.

Limited Cleavage at Pyrimidines (C + T)

Combine 10 μl distilled water, 3 μl sonicated salmon sperm DNA, and 10 μl end-labeled DNA in water. Cap tube and mix. Add 35 μl hydrazine. Cap, mix gently, and heat 5 ± 2 min at 20°. Add 200 μl hydrazine stop solution at 0° and 750 μl 95% ethanol at 0°. Cap tube, mix, and chill 5 min at −70°. Centrifuge 5 min. Transfer supernatant to hydrazine waste bottle. Add 250 μl 0.3 M sodium acetate at 0°. Cap tube; mix to redissolve DNA. Add 750 μl 95% ethanol at 0°. Cap tube, mix, and chill 5 min to −70°. Centrifuge 5 min. Remove supernatant. Add 1 ml 95% ethanol at 0°. Centrifuge 15 sec and remove supernatant. Vacuum evaporate for 5 min. Preparation is now ready for piperidine treatment.

Limited Cleavage at Cytosines (C)

Combine 15 μl 5 M sodium chloride, 3 μl sonicated salmon sperm DNA, and 5 μl end-labeled DNA in water. Cap tube and mix. Add 35 μl hydrazine. Cap, mix gently, and heat 5 ± 2 min at $20°$. Add 200 μl hydrazine stop solution at $0°$ and 750 μl 95% ethanol at $0°$. Cap tube, mix, and chill 5 min at $-70°$. Centrifuge 5 min. Transfer supernatant to hydrazine waste bottle. Add 250 μl 0.3 M sodium acetate at $0°$. Cap tube; mix to redissolve DNA. Add 750 μl 95% ethanol at $0°$. Cap tube, mix, and chill 5 min to $-70°$. Centrifuge 5 min. Remove supernatant. Add 1 ml 95% ethanol at $0°$. Centrifuge 15 sec and remove supernatant. Vacuum evaporate for 5 min. Preparation is now ready for piperidine treatment.

Limited Cleavage at Adenines and Cytosines (A > C)

Combine 100 μl 1.2 M sodium hydroxide, 1 mM EDTA, 1 μl sonicated carrier DNA (1 mg/ml), and 5 μl end-labeled DNA, in water. Cap tube using Teflon tape (Chemplast, Norton Company) as a cap liner and mix. Heat 10 ± 5 min at $90°$, using a heavy weight to hold down the cap. Add 150 μl 1 M acetic acid, 5 μl tRNA (1 mg/ml), and 750 μl 95% ethanol at $0°$. Cap tube, mix, and chill 5 min at $-70°$. Centrifuge 5 min. Remove supernatant. Add 1 ml 95% ethanol at $0°$. Centrifuge 15 sec. Remove supernatant. Vacuum evaporate several minutes. Preparation is now ready for piperidine treatment.

Piperidine Treatment and Loading of Samples on Gel

To effect cleavage of the DNA backbone, the dried, base-modified DNA samples obtained by the various procedures listed above are immediately treated as follows. Add 100 μl 1 M piperidine. Cap tube using Teflon tape as a cap liner. Mix to redissolve the DNA. Centrifuge 15 sec to collect the liquid at the bottom of the tube. Heat 30 min at $90°$, using a weight to hold down the cap. Vacuum evaporate to dryness. Add 10 μl distilled water, mix to dissolve DNA, and vacuum evaporate to dryness. Again add 10 μl distilled water, mix to dissolve DNA, and vacuum evaporate to dryness. Add 10 μl loading solution. Cap tube, mix to dissolve pellet, and centrifuge 15 sec to collect the solution at the bottom of the tube. Heat 1 min at $90°$. Quickly chill at $0°$. Load 3 μl of this per electrophoretic lane, using a drawn glass capillary tube.

Limited Cleavage (A > G ≃ C > T) for One-Lane Analysis

Combine 3 μl sonicated carrier DNA (1 mg/ml), 3.4 μl 5'-end-labeled DNA, 1.6 μl 5 M sodium chloride, and 8 μl 1 M piperidine. Cap tube, mix, and centrifuge 15 sec to collect solution at bottom of tube. Draw the solution into a pointed glass capillary tube (about 1 mm inner diameter).

Flame-seal the tip; then flame-seal the opposite end. Heat capillary 1.5 hr at 110°, in a silicone oil bath or in an oven. Wipe and wash silicone oil off the capillary with ethanol. Break open capillary. Transfer contents to a snap-cap tube with a drawn glass capillary. Rinse capillary with 10 μl of water, and combine this with the sample. Vacuum evaporate to dryness. Add 10 μl distilled water, mix to dissolve DNA, and vacuum evaporate to dryness. Again add 10 μl distilled water, mix to dissolve DNA, and vacuum evaporate to dryness. Add 10 μl loading solution, cap tube, mix to dissolve pellet, and centrifuge 15 sec to collect the solution at the bottom of the tube. Heat 1 min at 90°. Quickly chill at 0°. Load 3 μl of this per electrophoretic lane, using a drawn glass capillary.

Limited Cleavage (G > A > C > T) for One-Lane Analysis

Combine 3 μl sonicated carrier DNA (1 mg/ml), 3 μl 5'-end-labeled DNA, and 2 μl 1 M piperidine formate (pH 2). Cap tube. Mix. Heat 25 min at 37°. Vacuum evaporate to dryness. Add 20 μl distilled water, mix to dissolve DNA, and vacuum evaporate to dryness. Add 5 μl distilled water and 5 μl 1 M piperidine. Cap tube, mix, and centrifuge 15 sec to collect solution at bottom of tube. Draw the solution into a pointed glass capillary tube (about 1 mm inner diameter). Flame-seal the tip; then flame-seal the opposite end. Heat capillary 5 hr at 90°, in a silicone oil bath or in an oven. Wipe and wash silicone oil off the capillary with ethanol. Break open capillary. Transfer contents to a snap-cap tube with a drawn glass capillary. Rinse capillary with 10 μl of water, and combine this with the sample. Vacuum evaporate to dryness. Add 10 μl distilled water, mix to dissolve DNA, and vacuum evaporate to dryness. Again add 10 μl distilled water, mix to dissolve DNA, and vacuum evaporate to dryness. Add 10 μl loading solution, cap tube, mix to dissolve pellet, and centrifuge 15 sec to collect the solution at the bottom of the tube. Heat 1 min at 90°. Quickly chill at 0°. Load 3 μl of this per electrophoretic lane, using a drawn glass capillary.

Polyacrylamide Gel Electrophoresis and Autoradiography

Commercial gel molds are used, with a gel length of about 40 cm and a thickness of 0.2 or 0.4 mm. The gels are polymerized with a molar ratio of acrylamide to N,N'-methylenebisacrylamide of 19:1, in 8.3 M urea, 100 mM Tris–borate (pH 8.3), 2 mM EDTA.

Stock Solution for Gels. For 20% gels use 95.0 g acrylamide and 5.0 g bisacrylamide; for 10% gels, use 47.5 g acrylamide and 2.5 g bisacrylamide. Gel solutions also contain 250 g urea, 50 ml 1.0 M Tris–borate buffer, (pH 8.3), 20 mM EDTA, and 150 ml water. Mix to dissolve, and pass through filter paper. Make up to 500 ml with water. Store in an opaque bottle.

Polymerization on the 70-ml Scale. Take 70 ml of polyacrylamide stock solution. Degas by stirring 5 min under reduced pressure (water aspirator). Add 420 μl of 10% (w/v) aqueous ammonium persulfate solution, and mix by swirling. Add 20 μl TEMED, and mix briefly and vigorously by swirling. Pour into gel mold immediately. Polymerization should occur in about 10 min.

Gel electrophoreses are performed at an applied voltage of 1500 to 2000 V, adjusted so that the surface temperature of the glass mold is about 40°, as measured with a surface temperature probe. Approximate migration positions for the marker dyes are bromphenol blue, with 8-mers in 20% gels and with 17-mers in 10% gels; xylene cyanole, with 27-mers in 20% gels, with 60-mers in 10% gels, with 75-mers in 8% gels, and with 115-mers in 6% gels. For 20% gels, electrophoresis is carried out after the first loading until the xylene cyanole marker has traversed two-thirds of the lane, then another aliquot of the sample is added to a new well and electrophoresis is resumed until the bromphenol blue dye of the second loading has migrated halfway down the lane. The second loading will display the bands from the very shortest oligomers up to about the 40-mer in readable form, while the first loading allows reading from around the 24-mer, at the bottom of the lane, to about the 80-mer. For 10% gels, the xylene cyanole is allowed to move two-thirds down the lane in both the first and second loadings. Sequences between approximate positions 32 and 100 are read in the lane for the second loading, while 90 to about 160 are read in the first loading. If xylene cyanole is allowed to reach the bottom of a 40-cm gel, one may expect to read from nucleotide 30 to 150 in a 12% gel, 75 to 170 in an 8% gel, and 115 to 225 in a 6% gel.

After electrophoresis, the gel mold is opened by lifting one of the glass plates off the gel. A used piece of X-ray film is placed on the open face of the gel, the gel is turned over, and the second glass plate is lifted off. The gel, adhering to the film support, is covered with a tightly drawn layer of plastic wrap. In the dark room, a new sheet of X-ray film is placed against the plastic-covered face of the gel, and an intensifying screen is placed next to the unexposed film. The entire packet is enclosed in a lighttight film holder and kept at −70°, compressed under a brick, until film development (usually one to several days). Cassettes for this purpose are commercially available (see this volume [7]).

Alternatively, the gel may be dried prior to autoradiography. This often results in increased sensitivity and the ability to read longer distances. Gel drying (see this volume [8]) is especially recommended for ^{35}S labeling during dideoxy sequencing (see this volume [57–59]).

Additional suggestions for increasing the resolution of sequencing gels may be found in this volume [57].

Acknowledgments

The work on one-lane methods described in this chapter was carried out at the Ohio State University, Department of Chemistry, and was supported, in part, by a Public Health Service grant (GM 34450)

[57] Sequencing DNA with Dideoxyribonucleotides as Chain Terminators: Hints and Strategies for Big Projects

By Wayne M. Barnes

The dideoxy DNA[1] sequencing method first demonstrated by F. Sanger[1a] has a claim to be the least labor-intensive method for rapid data acquisition. An important feature of this method is that it can easily be modified to use several strategies designed to gather large amounts of data for large sequencing projects. We here discuss and compare four of these strategies, and also include some recipes and advice for obtaining good data.

Ordered Strategies for Large-Scale Dideoxy Sequencing

Once primer and template are in hand, only a few hours of work and only 1 or 2 days of elapsed time will produce 300–400 nucleotides of sequence data for the region adjacent to the 3' side of the primer, with 4–48 experiments (depending on skill) carried out simultaneously in 1 day by one person. Facing the possibility of more than 1000 nucleotides of data per day, scientists interested in large stretches of DNA sequence determination have employed two basic general strategies which I will call *random* and *directed*. The random approach[2] collects data from randomly cloned fragments. For each experiment, the researcher does not know in advance where in the target DNA the data will be obtained, nor even

---

[1] Abbreviations: dGTP, deoxyguanosine triphosphate; dITP, deoxyinosine triphosphate; ddGTP, dideoxyguanosine triphosphate; dc⁷GTP, deoxy-7-deazaguanosine triphosphate.

[1a] F. Sanger, S. Nicklen, and A. R. Coulson, *Proc. Natl. Acad. Sci. U.S.A.* **74,** 5463 (1977).

[2] S. Anderson, *Nucleic Acids Res.* **9,** 3015 (1981).

which strand of the DNA will be sequenced. A computer is then used to assist in putting the data together into a whole.[3] The largest sequencing projects when they were published (human mitochondrial DNA, 16 kb[4]; phage λ, 49 kb[5]; Epstein–Barr virus, 170 kb[6]) were completed by the laboratory that invented large-scale dideoxy sequencing, and they used mainly the random method.

In the face of this obvious example of success, many people nevertheless prefer a more directed approach since there are psychological, economical, and ecological benefits to having nearly every sequencing experiment generate new data just where it is needed to complete a project or answer a biological question. This chapter discusses various approaches to directed, as opposed to random, large-scale dideoxy sequencing.

Custom Primer-Directed Sequencing—The Look and Leap Approach

An obvious approach to extending a sequence by primer-directed dideoxy sequencing is merely to synthesize a primer at the edge of the known sequence with its 3' end pointing off into the unknown sequence. If the target sequence is available in large (4–14 kb) stretches of single-stranded form (see below for a discussion of suitable vectors for this purpose), within hours 300 nucleotides of new data can be generated at the point of greatest interest.

The recent widespread commercial introduction of automated DNA synthesizers has eliminated the technical obstacle to easy availability of custom, single-use primers. Continuing advances are reducing the cost of each primer to a range that nearly allows this approach to be recommended as the major strategy for large-scale sequencing, especially if mutant variations are contemplated after determination of the initial sequence.

Rules for Primer Design

Some primers with homology to the template sequence do not work well for dideoxy sequencing. They can fail in two ways: either they do not

[3] R. Staden, *Nucleic Acids Res.* **14,** 217 (1986).
[4] S. Anderson, A. T. Bankier, B. G. Barrell, M. H. L. De Bruijn, A. R. Coulson, J. Drouin, I. C. Eperon, D. P. Nierlich, B. A. Roe, F. Sanger, P. H. Schreier, A. J. H. Smith, R. Staden, and I. G. Young, *Nature (London)* **290,** 457 (1981).
[5] F. Sanger, A. R. Coulson, G. F. Hong, D. F. Hill, and G. B. Petersen, *J. Mol. Biol.* **162,** 729 (1982).
[6] R. Baer, A. T. Bankier, M. D. Biggin, P. L. Deininger, P. J. Farrell, T. J. Gibson, G. Hatfull, G. S. Hudson, S. C. Satchwell, C. Seguin, P. S. Tuffnell, and B. G. Barrell, *Nature (London)* **310,** 207 (1984).

prime well enough or they prime in too many places on the template. In this section some rules for custom primer design are suggested that will reduce the number of futile primers to a minimum.

Rule 1. Use the following formula to determine the length of the designed primer:

Primer length = 18 + 1 extra nucleotide for each 2% off of 50% GC

Aim for 50% GC base composition (range 40–60%). Primers with too high an AT composition will not prime effectively in some template situations. Bias toward high GC composition is dangerous in another way: accidental homology with high GC regions on the template may cause background priming which results in more than one sequence appearing on the same ladder. This problem occurs with high GC and not high AT because even short, relatively probable stretches of homology with high GC have a high stability, with a concomitant and significant possibility of undesirable DNA polymerase priming. This rule may be stretched by stretching the length of the primer to compensate.

While primers that break this rule (for instance 90% AT composition and a length of only 12–14 nucleotides) do work well sometimes, in other cases they will not work at all. A possible reason for the lack of consistency may be competition with template structure. Some sequences on the target DNA template or M13 vector may happen to cause the template molecule to fold in such a way as to make the priming site inaccessible to weakly annealing primers. In one case the two strands (the two orientations cloned on M13) of the same target DNA fragment allowed differing performance by the standard *lac* DNA sequencing primers. One orientation of the template (and both orientations of most other templates) produced clean sequences with the 15-mer *lac* primer (TCCCAGTCACGACGT), and the other orientation template gave rise to two confusing sequence ladders. The 17-mer *lac* primer (GTAAAACGACGGCCAGT) performs cleanly on both of these templates and more reliably on all templates.

Rule 2. Use the computer (with a program such as COMPARE[7,8]) to check the primer for accidental base pairing with undesirable locations on the template. If left undiscovered, this priming could cause more than one sequence to appear on the sequencing ladder, rendering the true target impossible to read accurately with confidence. Check for obvious problem homology with the known parts of the sequence of your template molecule.

[7] Described software (for IBM or VAX) and gel devices are available from CompuGene, 223 Renaldo Drive, Chesterfield, MO 63017 (tel: 314-994-1587).
[8] E. W. Myers and D. W. Mount, *Nucleic Acids Res.* **14,** 501 (1986).

Check the last ten. The most important part of the primer to pay attention to for problem homology is the last (3' terminal) 10 nucleotides. A homology of 8/10 in the last nucleotides may prime in the unwanted location, particularly if this part of the primer is high in GC content. With this point in mind, I enter each potential primer twice into the input site file for the program SEARCH[7]: once as full length, then followed immediately in the list as the last 10. This list may be generated automatically.[9] Then I run the program a few times with the fraction-of-match ranging from 1.00 down to 0.80, which would pick up an 8/10 match in the last 10. The problematic *lac* 15-mer mentioned above fails this test by exhibiting a good homology between its 3' region and an unwanted region of M13 vector DNA.

Note that DNA polymerase I (large fragment) is quite capable of removing at least four 3' unpaired nucleotides under sequencing reaction conditions, then synthesizing in the forward direction to generate a dideoxy data ladder that is remarkably readable (or confusing, as the case may be.) Remember that no sequence data can show up on the gel until after the polymerase has incorporated a radioactive nucleotide. Therefore, if dATP is the label, be sure there are some *A*s to incorporate between the primer and a nearby target sequence.

Unpurified primers work just fine. Oligonucleotides straight off the DNA synthesizer machine may be used without size purification. When unpurified primers are 5'-labeled and analyzed, they may appear to be only 90–95% pure, with the contamination as apparently one longer or one to several nucleotides shorter. However, when this mixture is used as a primer, one does not obtain the relative intensity of additional bands expected from the primer heterogeneity—the sequencing gel looks better than expected. This is probably due to the fact that the majority, correct primer sequence anneals better than the contaminating primers, which have mismatches (loop-outs) and therefore compete less well than their concentration would indicate. Counter examples have been reported,[10] but these do not follow the length and base composition rule above.

It is important that the synthetic oligonucleotides be free of organic contamination, however. They should be ethanol precipitated twice from a large volume: resuspend the dried oligonucleotide in 3 ml (6 ml of 0.3 *M* sodium acetate for each micromole of primer), and then add 4 volumes of 95% ethanol.

Assuming the ready availability of a modern automated DNA synthesizer, approximately 2 days must still elapse between ordering a new

[9] P. Lobel, unpublished.
[10] F. C. Strauss, J. A. Kobori, G. Siu, and L. E. Hood, *Anal. Biochem.* **154**, 353 (1986).

primer and applying it to determine the next 300–400 nucleotides of sequence. Such sequence determination and design of the next primer might take another 2 days, so that the minimum sustainable turnaround time for this approach is probably 4 days. In the likely event that 300 nucleotides every 4 days is not fast enough for a large project, several moving fronts of data acquisition should be arranged by the investigator. It is for the rapid achievement of many data fronts that one of the nested deletion strategies should be employed in the initial phase of a sequencing project.

Nested Deletion Strategies

Nested deletions prepared for the purpose of dideoxy sequencing are a set of deletions that all start adjacent to the sequencing primer and then extend various distances across the target DNA (Fig. 1). These deletions thus bring various portions of the target sequence into proximity of the standard primer for sequencing, without the necessity for a recloning event. When combined with a method for sizing the deletions in advance of sequencing,[11,12] the use of nested deletions eliminates most of the randomness from a sequencing project.

Although some of the deletion strategies shorten the target DNA in various ways and then reclone the shortened fragments,[13] this chapter discusses only those nested deletion strategies that employ an intramolecular recircularization event to close the deletion endpoint to result in thousands of independent deletions per microgram of input DNA. Several other similar approaches[14–16] are not discussed.

Kilo-Sequencing

Kilo-sequencing[11,12] was developed in response to disappointment with the use of Bal 31 nuclease for long (greater than 1500 bp) deletions. It was also designed to use a quantitatively complete enzyme reaction at each step so that DNA need not be gel purified away from unreacted (uncleaved) material. This method requires as starting material replicative form (RF) DNA with one unique restriction site adjacent to the 3' side of the primer homology. There are no other restriction site requirements for this method.

[11] W. M. Barnes, M. Bevan, and P. H. Son, this series, Vol. 101, p. 98.
[12] W. M. Barnes and M. Bevan, *Nucleic Acids Res.* **11**, 349 (1983).
[13] A. M. Frischauf, H. Garoff, and H. Lehrach, *Nucleic Acids Res.* **8**, 5541 (1983).
[14] G. F. Hong, *J. Mol. Biol.* **158**, 539 (1982).
[15] M. D. Poncz, D. Solowiejczyk, M. Ballantine, E. Schwartz, and S. Surrey, *Proc. Natl. Acad. Sci. U.S.A.* **79**, 4928 (1982).
[16] A. Ahmed, *Gene* **39**, 305 (1985).

```
calDNA ->                4-10 kb of Target DNA                                    ...


calDNA ->(                  )AGCTTC                                               ...
calDNA ->(                       )TCCGCGG                                         ...
calDNA ->(    Various              )GGACTAG                                       ...
calDNA ->(    extents of              )AGGATCC                                    ...
calDNA ->(    deleted DNA                )CCTCTCA                                 ...
```

Determined DNA sequence: AGCTTCCGCGGACTAGGATCCTCTCA

FIG. 1. Nested deletion strategies. The set of nested deletions for sequencing all have the same boundary adjacent to the sequencing primer, but the other side of the deletions is at many locations within the target sequence. calDNA is the template strand of *lac* DNA from the M13 cloning vector; →, the standard sequencing primer; AGCTTC, 300 nucleotides of new data obtainable by sequencing each particular deletion.

The RF DNA is treated first to create a quantitative, single, randomly located scission across both strands; this scission is the distal and variable side of the nest of deletions desired. It was found that a combination of three enzyme treatments is necessary for this result, but each enzyme is used for only a brief time so that contaminating activities are not a problem, and the result is a 100% conversion to the randomly linear form. These three enzyme treatments (described elsewhere in more detail[11] are as follows.

1. Treatment with pancreatic DNase I in the presence of DNA-saturating amounts of ethidium bromide. A single nick is produced in every supercoiled RF molecule. Mix 5 μg RF DNA, 10 μl of 10× nicking buffer (10×: 1.25 M NaCl, 0.2 M MgCl$_2$, 40 mM Tris–HCl at pH 7.9, 600 μg/ml bovine serum albumin), 10 μl ethidium bromide at 5 mg/ml, and water to 100 μl. Add 50 ng pancreatic deoxyribonuclease I (DNase) (stored as a 1 mg/ml stock in 1 mM HCl and diluted in nicking buffer before use). Incubate at 25–30° for 1 hr. Add 0.05 volume 2 M Tris–HCl at pH 7.9, extract twice with phenol and twice with chloroform and ethanol precipitate (this volume [4] and [5]).

2. Treatment with exonuclease III for 10 min to widen the nick to a gap, since single-strand specific nucleases do not efficiently attack a nick. Resuspend DNA in 45 μl DNA buffer (10 mM Tris–HCl at pH 7.9, 10 mM NaCl, 0.1 mM EDTA), 5 μl 10× exonuclease III buffer (10×: 0.66 M Tris–HCl at pH 7.9, 6.6 mM MgCl$_2$, 100 mM 2-mercaptoethanol), and 10–30 units of exonuclease III (New England BioLabs). Incubate at 25–30° for 10 min. Heat at 70° for 10 min to inactivate the enzyme.

3. Treatment with Bal 31 nuclease that has been diluted so much that only its single-stranded nuclease activity is evident. This is equivalent to a treatment with S$_1$ nuclease to cut across the gap. Use 50 μl of the exonuclease III reaction, 100 μl 2× Bal 31 buffer (2×: 50 mM Tris–HCl at pH

7.9, 24 mM MgCl$_2$, 24 mM CaCl$_2$, 1.2 M NaCl, 2 mM EDTA), 50 μl water, and 1 μl Bal 31 [1/16 unit prepared by diluting Bal 31 (New England BioLabs) with 0.2 M Tris–HCl at pH 7.9, 10 mM 2-mercaptoethanol, 50% glycerol] for this purpose. Incubate at 25–30° for 5 min and transfer 100 μl of the reaction to excess EDTA. Stop the remainder of the reaction similarly after 9 min. Save aliquots from both sections (at least 0.2 μg) for analysis in an agarose gel, mix the samples and extract with phenol once and chloroform twice, and precipitate with ethanol.

Check the progress of the reactions after nicking, widening the gap, and cutting across the gap. In a 1% agarose gel you should see mostly nicked circles (no supercoils) after the first two steps and full-length linear fragments after the third. The nested deletions cannot be seen until the cleavage and recircularization reactions have been completed.

Treatment with the unique restriction enzyme (whose site is adjacent to the sequencing primer) and polishing the ends with the large fragment of DNA polymerase complete formation of linear molecules. Conditions for restricting the DNA depend on the choice of unique site. Ends are polished (made blunt) by resuspending DNA in 45 μl of DNA buffer (see above), 6 μl of 10× Klenow buffer (10×: 0.1 M Tris–HCl at pH 7.9, 0.1 M MgCl$_2$, 0.5 M NaCl, 0.1 M 2-mercaptoethanol), 6 μl deoxynucleoside triphosphates (all four, each at 1 mM), and 3 μl (1.5 units) of Klenow fragments (DNA polymerase large fragment). Incubate at 25–37° for 15–30 min. Extract with phenol once and chloroform twice and precipitate with ethanol. The linear constructs only need recircularization to form a large collection of deletions. Recircularization is best performed with linker tailing (see below). Deletions that go the wrong way are inviable, since the origin of replication of M13 lies nearby on that side.

Run mixtures of plaques in live phage gels. Pour a 0.7% high melting temperature agarose gel in 3× GGB running buffer (20× GGB: 0.8 M Tris–acetate at pH 8.3, 0.4 M sodium acetate, 4 mM EDTA). Make up gel sample at 60° using 0.3 ml of phage stock, 0.2 ml of blue 2 (0.1% bromphenol blue, 0.1% xylene cyanole, 0.1 M EDTA, and 12.5% Ficoll), and 0.3 ml of 0.7% agarose. Load the wide well of a horizontal gel two-thirds full. Load vector and undeleted M13 phage in some side slots, for size standards. Top off with 0.7% agarose, cover with plastic wrap, and run gel at 100 mA (3 V/cm) for 16 hr. Sample gel at 2-mm intervals from 2 to 13 cm migration, using an array of blunt syringe needles. Rinse gel samples into microtiter wells with 2× 0.1 ml of M13 buffer. (See this volume [16] for M13 handling techniques.)

Staining gel: Soak a half-hour in 1 liter of 0.1 N NaOH to strip phage coats from the DNA. Neutralize by soaking for a half-hour in 3× GGB. Then soak for 1–16 hr in ethidium stain (0.5 μg/ml in 0.5× GGB).

The advantages of the kilo-sequencing method are the following. (1) Intramolecular recircularization seals the deletion endpoints. (2) There are no homopolymer tails to sequence adjacent to the primer. (3) Size is selected using live phage gels (this step could usefully be applied in any of these methods). (4) The target sequence is independent of relative size. This size independence is evident at two points. At initial construction of M13-transducing phage carrying target DNA, the M13 vectors (mWB series) introduced with the kilo-sequencing method are more able to carry large inserts without giving rise to unwanted deletions. Then during the enzymatic steps, the deletion-forming protocol does not require any exonucleases to traverse thousands of nucleotides, during which incubation contaminating activities are often found to destroy the biological activity of the subject DNA from the long time points. (5) A restriction site of choice becomes located at the deletion boundary, through the use of a judicious linker (also true for the Exo III method below). (6) Automatic indication of deletions that have gone too far, or of any other aberrant events that correlate with loss of the *lac* operator, is provided by blue plaque color for all desired phage. (7) Kilo-sequencing starts with double-stranded DNA, so it is just as useful for plasmids as for M13. Care should be taken to clone near a selectable marker or origin so that deletions that go the wrong way will still be selected against. (8) Except for blunt-end-reactive T4 ligase, the enzymes used need not be of particularly high quality, since they are each used only in minimal amounts and for short times.

One-Way Digestion by Exonuclease III

Henikoff[17] has demonstrated a strategy based on the observation that exonuclease III will not begin digestion at a 3' overhanging end such as that left by the restriction enzyme *Pst*I. If the substrate molecule can be constructed to contain a unique *Pst*I site between a unique blunt or 5' overhang restriction site and the sequencing primer, then exonuclease III will digest in only the one desired direction. Subsequent treatment with a single-stranded nuclease such as S_1 nuclease, mung bean nuclease, or very dilute Bal 31 nuclease will then remove the other strand of the DNA covered by the desired deletions. After the ends are polished with the large fragment of DNA polymerase I as for the kilo-sequencing method, linker tailing (see below) is recommended for recircularizing the RF DNA.

This method appears to be remarkably fast, simple, and effective. At

17 S. Henikoff, *Gene* **28,** 351 (1984).

times a particular batch of exonuclease III that has contaminating nucleases leads to very low yields from the longer time digestions. Several lots of exonuclease III should be tried. As for all of the deletion methods and each of their enzymatic steps, frequent agarose gel assays should be used to monitor the effectiveness and quality of each processing step, and alternative sources of each enzyme should be tried if necessary.

T4 Polymerase/Terminal Transferase Tailing

Dale *et al.*,[18] regardless of the advantages of the kilo-sequencing method, have devised a way to create a nest of deletions starting with single-stranded DNA. They create the DNA cleavage for the fixed end of the deletions first, by employing a double-strand-specific restriction enzyme and a short oligonucleotide synthesized to match the vector DNA adjacent to the target DNA. This synthetic oligonucleotide has a dual use, however, as it also contains a short homopolymer tail for use in the last (recircularization) step. After cleavage, they exploit the fact that the only activity that T4 polymerase exhibits against single-stranded DNA is a 3′ exonuclease, which they found acts distributively to create a range of deletions up to at least 3 kb over 1 hr of time points. After the T4 polymerase treatment, terminal transferase is used to add a short homopolymer tail complementary to the 3′ end of the oligonucleotide, which is then reintroduced to form a "Band-aid" during ligation.

The advantages of the T4 exonuclease/terminal transferase tailing method are that (1) intramolecular recircularization seals the deletion endpoints; (2) the deletions are constructed on single-stranded DNA; and (3) as shown by Dale *et al.* the deletion procedure can be carried out in one reaction tube without phenol extractions and ethanol precipitations to separate the various enzymatic steps.

Comments. The restriction enzyme used to carry out the initial linearization must be several orders of magnitude more free of single-strand nuclease than is required for most purposes for which restriction enzymes are normally checked for quality control. For this reason, not every source of restriction enzyme will be suitable for this step. The enzyme to be used may easily be tested for suitability, however, by testing for a lack of effect against the biological activity of single-stranded DNA as a substrate. Similarly, the terminal transferase is problematic if not of very high quality.[19] The methods using exonuclease III and T4 polymerase (but not

[18] R. Dale, B. McClure, and J. Houchins, *Plasmid* **13**, 31 (1985).
[19] W. H. Eschenfeldt, Robert S. Puskas, and S. L. Berger, this volume [37].

kilo-sequencing) have been commercialized into kit form, although the deletion-resistant M13 vectors are not in the kits.

Use of Linkers

For the methods of kilo-sequencing and one-way digestion by Exo III for creating deletions, one has the option of introducing any restriction site of choice at the deletion boundary. After the sequence has been determined, this restriction site may be valuable for other research purposes which would help to define the active sites of the DNA in question, or which would use the DNA under study for expression experiments. The original standard methods of introducing this restriction site[11] were limited to sites that did not appear elsewhere in the molecule, and were in addition fraught with potential inefficiencies at the final restriction enzyme recleavage step. These difficulties have been overcome by the introduction of "linker tailing."[20]

We have found that this procedure can be even more convenient than originally presented. It is not necessary to remove excess linkers, nor is it necessary to reanneal before transformation.

Simplified Linker Tailing

For a total volume 30-μl reaction, mix together 5 μg of any linear, blunted plasmid or RF DNA (such as that resulting from the restriction enzyme cleavage which creates the fixed end of the deletions in the kilo-sequencing method or the Exo III method), 3 μl of 10× ligase buffer (stored as 300 mM Tris–HCl at pH 7.9, 40 mM MgCl$_2$, 400 μM rATP, 10 mM EDTA, 100 mM dithiothreitol), 3 μl 10 mM rATP, 3 μl 10 mM hexamminecobalt chloride, 0.05 OD$_{260}$ of linker that has *not* been kinased, and 3 μl T4 ligase, sufficient to carry out blunt-end ligation. Incubate overnight at 4–14°. It is not necessary to phenol extract the reaction.

Heat the DNA to 70° for 10 min amd then chill. The purpose of this step is not to inactivate enzyme (although it does) but rather to melt off the unligated strand of linker at each end of the molecule. The DNA is immediately ready for transformation[21] and nearly all of the clones contain the desired linker. Apparently recircularization and ligation are

---

[20] R. Lathe, M. P. Kieny, S. Skory, and J. P. Lecocq, *DNA* **3**, 173 (1984).
[21] No more than 3 μl of ligase reaction should be introduced into each 200 μl of competent cells, since various components of the buffer can interfere with transformation.

accomplished intracellularly, as in phage λ infection and λ DNA transfection.

Dideoxy Sequencing Reactions

Many detailed recipes have been published for dideoxy sequencing. We prefer the one previously described in detail in this series, but another excellent protocol and discussion, and many references to others, may be found in Williams et al.[22] Not least among the available recipes is one that can be carried out for 48 primings at once using 96-well microtiter dishes.[23] A more modest recipe that works well with α-[^{35}S]thio-dATP labeling is presented here.[24]

Solutions

The normal deoxynucleotide triphosphates are best available as 100 mM solutions from Pharmacia-PL Biochemicals. These should be stored at −80°. They contain 50 mM Tris–HCl (pH 7.0) and their concentration has been confirmed by A_{260}. Stocks diluted to 10 mM and then 0.5 mM are used in the recipe below. All are stored at −20°.

ddNTP 2× stock solutions are prepared in small aliquots by diluting 10 mM stock solutions in water to these final concentrations: ddGTP, 0.2 mM; ddGTP/10, 0.02 mM; ddATP, 0.04 mM; ddTTP, 0.8 mM; ddCTP, 0.12 mM.

dNTP mixes are prepared from dNTP 0.5 mM stock and buffer (10 mM Tris–HCl at pH 8.0, 0.1 mM Na$_2$EDTA) in small aliquots and stored at −20°. They are labeled G′, A′, T′, C′, and I′.

	Microliters to make				
Solution	G′	A′	T′	C′	I′
0.5 mM dGTP	5	37.5	50	50	0
dATP	0	0	0	0	0
0.5 mM dTTP	50	37.5	5	50	50
0.5 mM dCTP	50	37.5	50	5	50
0.5 mM dITP	0	0	0	0	7.5
Buffer	37.5	37.5	37.5	37.5	35
Total	142.5	150	142.5	142.5	142.5

[22] S. A. Williams, B. E. Slatko, L. S. Moran, and S. M. DeSimone, BioTechniques 4, 138 (1986).

[23] A. T. Bankier and B. G. Barrell, in "Techniques in Nucleic Acid Biochemistry" (R. A. Flavell, ed.), Vol. B5, Elsevier, Limerick, Ireland. 1983.

[24] B. Shelton-Inloes and J. E. Sadler, personal communication.

dNTP/ddNTP mixes are prepared by mixing equal volumes of dNTP mixes with ddNTP 2× stocks. (Use the ddG/10 mix with the dITP mix.) Each N'/ddN mix = 50 μl N' (dNTP mix) + 50 μl ddNTP mix.

The chase solution is 1 mM in each of the four normal dNTPs, prepared from dNTP stock solutions and water.

Procedure

Annealing. Mix in a 500-μl Eppendorf tube: 600 ng template DNA, 2 μl (2.5 ng) primer, 10× 1.4 μl annealing buffer, and water to make 13.3 μl total. (10× annealing buffer is 100 mM Tris–HCl at pH 8.0, 50 mM MgCl$_2$.) Vortex and centrifuge briefly. Incubate in a 50–60° oven in a sand bath for 30–60 min. Chill on ice.

Prereaction. Add to each annealing reaction: 2.4 μl (about 20 μCi) α-[^{35}S]thio-dATP (Amersham; 650 Ci/mmol, 250 μCi/.03 ml) and 1.25 μl Klenow fragment (2 units/μl), for 17 μl total. (The actual amount of Klenow fragment to use should be determined empirically for each batch of enzyme, as described below.) Mix and set on ice.

Reactions. Label five 1.5-ml Eppendorf tubes per template with G, A, T, C, or I, and with the experimental template identification. To each tube add 3 μl of the prereaction mix. Then add 2 μl of the appropriate N'/ddN mix, a different one to each tube. Mix gently and minimally and incubate at 37° (air or sand).

After 12–20 min add 1 μl of chase solution to each tube, mix, and continue incubating at 37° for 15–20 min. For increased reliability, a small amount of chase solution should first be mixed with fresh enzyme, so that 1.5 μl of added chase solution contains 0.2–0.5 units of fresh Klenow fragment. Add 5 μl of stop/dye mix to each tube, vortex, and centrifuge briefly. [Stop/dye mix ("blue formamide"): 30 mg xylene cyanole, 30 mg bromphenol blue, 10 mM EDTA (0.2 ml of 0.5 M), and 9.8 ml formamide.]

The reactions are now ready for gel electrophoresis. They may be stored at −20° for a day or two without noticeable deterioration, but they should be freshly heated to 90° for 3 min immediately prior to each loading.

We wish next to call attention to a few minor points, in addition to those discussed previously[11,22] that affect the quality of dideoxy sequencing data.

Template Preparation

A common source of unsatisfactory dideoxy gel patterns may be traced to DNA template that seems to contain some inhibitors of the polymerase reaction. We suggest that the procedure of template extraction[11] should be modified as follows to make the preparation of high-quality template even more reproducible. (1) Take care not to overload the phenol extraction. A final yield of 16–20 μg of DNA from 250 μl of aqueous phase is optimal. More is not better, so the amount of PEG pellet actually extracted should be adjusted for the various growth rates of filamentous transducing phages and for packaging efficiencies of pseudophages. (2) At the first phenol extraction, remove only 90% of the aqueous phase to a new tube, leaving behind the interface material.

Use and Batches of Klenow Enzyme

The most common reason for poor dideoxy data is the particular lot of Klenow enzyme (large fragment of DNA polymerase I from E. coli[25,26]) used. For many commercial lots of this enzyme, there is only a narrow range of concentrations that give useful results. For some lots of enzyme, there is no concentration that will not give rise to false bands in the C channel, and the only known answer to this problem is to change to a different preparation. Only rare, excellent lots of enzyme have a range of useful concentrations that extends up to 30-fold above the minimal amount needed for complete reactions.

Determine the correct amount of Klenow enzyme to use for each new batch by setting up sequencing reactions with primer–template that is known to be well behaved and therefore has a known sequence, and then run several pairs of C and T channels (C is the first channel that will develop problems if the Klenow enzyme is unsatisfactory). Using concentration steps of a factor of 2 or 3, vary the amount of enzyme used in the reaction. Too little enzyme will result in bands at wrong positions due to incomplete extension. Too much enzyme (sometimes at a level only 3-fold above the minimal amount for complete reactions) causes a smooth background smear.

Remember to treat the Klenow fragment as an enzyme. Never allow the stock to sit in an ice bucket; dispense it only as needed from −20° storage. If dilutions must be made, make them immediately before use. Carry out all mixing gently so as to avoid undue air–solution interfaces (bubbles).

Gel Plate Preparation

Dry the gel plates thoroughly after polymerization is complete and they have been rinsed off on the outside.[22] Also be sure the gel device is

[25] H. Klenow, K. Overgaard-Hansen, and S. A. Patkar, Eur. J. Biochem. **22,** 371 (1971).
[26] P. Setlow, D. Brutlag, and A. Kornberg, J. Biol. Chem. **247,** 224 (1972).

dry on any vertical support surfaces that may contact the gel. Drops of water or buffer that remain on the glass surface during the run cause cool spots as they evaporate. These cool areas cause nonuniform mobility across the depth of the gel, and this causes the bands to blur.

A corollary to this advice is not to allow drafts to impinge on the gel plate surface during the run. The gel should be protected from air currents. One device geometry that accomplishes this with a safety/connection cover is available commercially.[7]

Sample Loading

Immediately before loading samples, use a Pasteur pipet to force upper buffer into the wells to rinse out urea. When loading the sample in blue formamide, do not allow it to fall down through buffer. Rather, put the pipet tip (we recommend a pulled melting point capillary) very near the bottom of the well and then introduce an approximate 1.4-mm depth of sample and withdraw the pipet to leave a sharp upper surface. Apply electromotive force immediately, without allowing the samples to stand in the wells. If you load slowly, load run each priming or two into the gel for 5 min before loading the next sct(s).

The Use of dITP to Resolve Compressions

"Compressions" are regions in a sequencing gel (whether dideoxy or chemically derived) that are unreadable or are read incorrectly, and are due to the fact that the commonly used conditions (7–8 M urea, warm gels) are not completely denaturing. Some small regions of high G and GC composition fold back on themselves, forming structures that are stabilized by intrastrand base pairs. Sequencing gels are fully and adequately denaturing for AT base pairs, which have two hydrogen bonds, but they are not adequately denaturing for GC base pairs, which have three hydrogen bonds. One proposed solution to this problem[27,28] is to include commercially available (Pharmacia-PL Biochemicals) dITP in the place of dGTP in all of the sequencing reactions. This is only partially, yet very usefully, successful. The following section describes how to use and interpret dITP for one extra sequencing reaction and channel, and discusses why it cannot be used in all four sequencing channels.

A single dITP reaction and channel is prepared by making two modifications to the standard G reaction.[11] First substitute dITP for dGTP. Second, although ddGTP is still used to cause the terminations, it com-

[27] D. R. Mills and F. R. Kramer, *Proc. Natl. Acad. Sci. U.S.A.* **76**, 2232 (1979).
[28] J. Gough and N. E. Murray, *J. Mol. Biol.* **166**, 1 (1983).

petes better with dITP than with dGTP, so that while the normal G reaction may contain a dGTP/ddGTP ratio of 17, one must use a 10-fold lower ratio of 1.7 for dITP/ddGTP.

dITP may not be used for all four channels due to the occurrence of two unexpected problems. The first problem only affects the C channel, since dITP works well (except for the second problem) in the A, T, and I reactions. The problem in the C channel is that weak C bands disappear completely. Apparently, the relative ability of Klenow fragment to accept ddC in place of dC is reduced at least 10-fold in the presence of dITP. This problem is not alleviated by lowering the concentration of dITP, nor by the presence of $\overline{Mn}^{2+}$. I have no explanation for this apparent allosteric effect of dITP.

The second problem with dITP is that in regions of high template structure (due to dyad symmetry in the sequence), Klenow fragment is halted or greatly slowed. This causes all channels containing dITP to have a dark band at several positions near the halt point, and no or only faint data bands above that point. My explanation for this is that dGTP may have a heretofore unnoticed template melting effect by forming base pairs with Cs in the template, and that the only two hydrogen bonds provided by dITP are not sufficient. This problem is very reproducible, and is nearly diagnostic for template stem structure of a certain stability.

dITP Data Interpretation. I recommend that six reactions–channels be employed for dideoxy sequencing, and loaded in the order ICATGC. The extra C channel is recommended because C is the most problematic reaction (the channel in which false bands first appear when the Klenow fragment is poor), and a difference in relative intensity of suspicious bands in the two C reactions is often informative. In addition, this recommended channel order is such that each band is never more than two channels away from the next nucleotide's band.

The absolute position of the bands in the I channel may not be accurate, since the entire nucleotide chain contains I in place of each G, and this causes a mobility shift of up to two positions. However, the number and spacing of I bands are absolutely accurate, much more accurate than the normal G channel in regions of compression. The interpretation procedure is to scan for differences in band number or band spacing between the G channel and the I channel. Usually, they are identical. Every few hundred nucleotides, however, they are different, and this difference both signals a compression problem and provides the data necessary to a correct reading of the sequence.

To read the data in a region of compression, first write down the I band positions as, for instance, I_I__II (i.e., I, space, I, space, space, I, I). (This sort of space count is *not* accurate for reaction–channels with the normal deoxynucleotides.) Then, fill the spaces with nucleotides from the nearest available bands from the other three channels, using them up as you read up the gel in 5′ to 3′ order. This method is very reliable, and can nearly always prevent the need to sequence the other strand. If problem two arises (a block in the I channel), then this useful type of data must unfortunately be forgone for the stretch of DNA above the block.

The substitution of dc^{7}GTP (7-deaza-dGTP) for dGTP has been suggested recently as an aid to resolving those compressions that are due to intrastrand G–G interactions, which are troublesome Hoogsteen base pairs.[29] It was however reported that compressions due to GC base pairs still remain. I am not yet experienced with the use of this compound, and thus cannot say whether it has either of the problems dITP has.

Single-Strand Vectors for Large Fragments

All the directed dideoxy sequencing strategies benefit from having large regions of substrate DNA to operate on at one time. The most widely used M13 vectors, the M13mp series,[30] are very prone to deletion of passenger DNA fragments that exceed a few kilobases in length. The mWB series of vectors[11] is at least 10 times less prone to this problem, and has been used routinely to carry 4–14 kb of DNA. It is not known why the mWB series are less prone to deletion, but it may have something to do with the fact that they carry their foreign DNA at the (probably therefore inactivated) *ori*− site (origin of minus strand replication) which has been shown to be dispensable.[31] The lack of a wild-type minus strand origin causes the mWB series to grow slowly and to a yield of 3–5 times less than the M13mp series, but this is actually a relative convenience. To partially avoid deletions, it is recommended that M13mp series constructs not be grown more than 6 hr. To gain a sufficient yield, it is recommended that the mWB series not be grown less than 16 hr.

Pseudophage Vectors

Pseudophage are plasmids that are packageable by M13 helper in trans by virtue of carrying the M13 origin of single-stranded replication and single-strand packaging[32,33] Several of these vectors are now available,

[29] S. Mizusawa, S. Nishimura, and F. Seela, *Nucleic Acids Res.* **14,** 1319 (1986). dc^{7}GTP is available from American Bionetics, Emeryville, CA.
[30] J. Messing, this series, Vol. 101, p. 20.
[31] M. H. Kim, J. C. Hines, and D. S. Ray, *Proc. Natl. Acad. Sci. U.S.A.* **78,** 6784 (1981).
[32] N. D. Zinder and J. D. Boeke, *Gene* **19,** 1 (1982).
[33] A. Levinson, D. Silver, and B. Seed, *J. Mol. Appl. Genet.* **2,** 507 (1984).

but it seems that not just any stretch of DNA containing the M13 intercistronic (origin) region is equally efficient at being packaged. One proposed optimal origin region is available in the form of cassettes (convenient DNA fragments with various linkers).[33] Many pseudophage plasmid vectors are available commercially which carry not only the M13 single-strand origin, but also promoters for T7, T3, and/or SP6 RNA polymerase.

In most cases pseudophage will not suffer the deletion problem of purely M13 vectors, since most of the time they are propagated simply as plasmids. If they are converted to single strands only for sequencing, then even a significant amount of deletion formation at this stage would not even be noticed on a sequencing gel (unless a common deletion boundary occurs in the sequenced region, an unlikely event). Not all M13 or f1 or fd phage are ideal as a helper phage for the purpose of packaging pseudophage DNA. Currently, the most efficient helpers seem to be M13K07[34] and R408.[34a] These helper phages actually package themselves less efficiently than a pseudophage plasmid in the same cell.

Infection of Female (F⁻) E. coli

We have recently discovered[35] that M13-transducing phage and M13-packaged pseudophage infect F⁻ cells at the useful frequency of 10^{-5} relative to F⁺ cells. This efficiency of infection is increased some 50-fold in the presence of 10% PEG, 0.5 M NaCl in a concentrated adsorption step. (Variations in adsorption time and temperature have no effect.) This route of transfection is a convenience, since one need not arrange that a recipient cell be male or competent in order to introduce a plasmid that is available as pseudophage.

Managing Sequence Data

As each gel is read, the new data of 50–400 nucleotides must be incorporated into the master sequence for the project. There are basically four computer-assisted steps to this assimilation of each new gel.

1. Check the gel for accurate reading. At least two people should read each gel, and one of them should read it twice. Each reading takes 3–5 min for 300 nucleotides, using a semiautomated gel reader.[7] Then all of

---

[34] J. Vieira and J. Messing, this series, Vol. 153, p. 3.
[34a] M. Russel, S. Kidd, and M. R. Kelley, *Gene* **45,** 333 (1986).
[35] W. M. Barnes and J. Welsch, in preparation.

the readings of each gel should be compared two by two, and disagreements should be resolved to the satisfaction of each party by reference to the original autoradiograph.

2. Check to be sure this gel is not merely read through into the M13 vector sequences. In the case where linker has been inserted to close a deletion, the data should be scanned for this linker, whose likelihood of actually occurring in a new sequence is low. This step may be put off until after the next two steps, as a final check.

3. Compare the new gel to the master sequence using a program such as COMPARE.[7,8] The advantage of the directed strategies is that only one strand of the data need be checked. For the random method, each strand must be scanned.

4. Integrate the new gel into the master sequence in two ways, depending on whether it is found to have any homology with the master sequence.

If an homology is found, any disagreements between this gel and the master sequence should be resolved in favor of one or the other by reference to the original films. Then, the new block of gel data is inserted into the master sequence at its proper location. The COMPARE program output indicates the exact boundaries of overlaps (redundantly sequenced DNA). These overlaps may be handled in several ways. For the data base system of CompuGene, all gel data is kept in clear, readable, original form in the master sequence file, but overlaps between adjacent gels are enclosed by parentheses inserted by the scientist. Note that no gel data is actually deleted for overlaps; the data within parentheses may still be examined and forms a permanent record of the raw data. The CompuGene software to print out and otherwise analyze the finished DNA sequence ignores the material in parentheses, whether DNA or comments and notes.

Another advantage to the directed (nested deletion) strategies becomes apparent if the COMPARE program discovers no homology between the new gel and the master sequence: the new gel can still be integrated into the master sequence using position information from the size of the deletion, which is inversely correlated with the size of the final construct. Filler data characters, such as dashes(—), are used on each side of a block of gel data that is not contiguous with other sequence data. Integration of the gel into the master sequence, insertion of parentheses and insertion of dashes may all be performed using one of the many available editor (word processing) programs.

Any data management system must carry out these data assimilation steps, but some of the available software carries out these steps more automatically. The method described above is one of the least automatic; the computer software finds the data and highlights the homologies and differences, but the scientist must do the integration with a general editor program. Other software systems[3] automate these steps to a greater degree, and show up to a dozen overlapping gels on the screen at one time, having put them in their relative location (and determined which strand they are, for the random methods) automatically.

[58] Direct Sequencing of Denatured Plasmid DNA

By Robert C. Mierendorf and Diana Pfeffer

A number of authors have reported methods for direct sequencing of double-stranded plasmid DNA using DNA polymerase I (Klenow fragment) and chain-terminating dideoxynucleotides.[1-8] In practice, recombinant plasmid DNA is prepared using standard methods and converted to a single-stranded form prior to sequencing. The denatured form is stable enough to allow the annealing of appropriate primers, and sequencing is carried out as first described by Sanger et al.[9] for single-stranded phage templates.

The direct method offers several advantages over other commonly used approaches. Sequencing by the chemical degradation of labeled DNA molecules[10] and by using single-stranded phage M13 templates for dideoxy chain termination reactions[11] both involve further isolation and/ or subcloning of DNA fragments beyond the initial cloning step. The direct method avoids these manipulations, and thus saves time and is especially convenient when analyzing large numbers of constructions. In

[1] M. Smith, D. W. Leung, S. Gilliam, C. R. Astell, D. L. Montgomery, and B. D. Hall, *Cell (Cambridge, Mass.)* **16**, 753 (1979).
[2] R. B. Wallace, M. J. Johnson, S. V. Suggs, K. Miyoshi, R. Bhatt, and K. Itakura, *Gene* **16**, 21 (1981).
[3] J. Viera and J. Messing, *Gene* **19**, 259 (1982).
[4] L.-H. C. Guo, R. C. A. Yang, and R. Wu, *Nucleic Acids Res.* **11**, 5521 (1983).
[5] E. Y. Chen and P. H. Seeburg, *DNA* **4**, 165 (1985).
[6] R. G. Korneluk, F. Quan, and R. A. Gravel, *Gene* **40**, 317 (1985).
[7] M. Haltiner, T. Kempe, and R. Tjian, *Nucleic Acids Res.* **13**, 1015 (1985).
[8] M. Hattori and Y. Sakaki, *Anal. Biochem.* **152**, 232 (1986).
[9] F. Sanger, S. Nicklen, and A. R. Coulson, *Proc. Natl. Acad. Sci. U.S.A.* **74**, 5463 (1977).
[10] A. M. Maxam and W. Gilbert, *Proc. Natl. Acad. Sci. U.S.A.* **74**, 560 (1977).
[11] J. Messing, this series, Vol. 101, p. 20.

METHODS IN ENZYMOLOGY, VOL. 152

addition, a longer sequence can be obtained from one plasmid preparation compared with one preparation of single-stranded phage DNA, since both strands can be sequenced from opposite directions using appropriate vectors and primers.

Reported procedures for sequencing double-stranded plasmids differ mainly in the method of template preparation; some recommend a linearization step followed by heat denaturation, whereas others favor denaturation of supercoiled DNA with alkali and subsequent neutralization. Our experience has been that either of these approaches can be used with success, but that both require good quality starting material. For simplicity, we routinely use the alkali denaturation method because it avoids the linearization step, which involves digestion with a restriction enzyme followed by extraction with phenol and ethanol precipitation.

We have found the twin promoter pGEM™ plasmid vectors (Promega), a representative of which is shown in Fig. 1, to be convenient for cloning and sequencing. These plasmids contain promoters for both bacteriophage SP6 and T7 RNA polymerases in opposite orientations flanking a region of multiple cloning sites and are commonly used for transcription of cloned sequences *in vitro*. The two phage promoters also serve as convenient, specific priming sites for DNA sequencing reactions, since they flank any DNA inserted between them and share little sequence homology. The SP6 promoter primer we have used is a 19-mer with the sequence 5'-GATTTAGGTGACACTATAG-3'. The T7 promoter primer

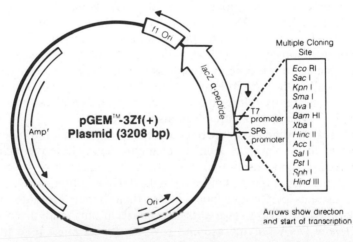

FIG. 1. Map of pGEM™-3Zf(+) plasmid (Promega), a representative vector that is convenient for direct sequencing of cloned DNA fragments. This vector allows color screening of recombinants and, if desired, production of single-stranded plasmid DNA *via* the phage f1 origin of replication.

is a 20-mer with the sequence 5'-TAATACGACTCACTATAGGG-3'. Primers are also commercially available for direct sequencing of pBR322 and pUC recombinant plasmids and inserts cloned in bacteriophage λgt11.

[*Editors' Note.* Plasmids containing promoters for SP6, T7, and T3 RNA polymerases in various combinations are now commercially available. Their respective polymerases can also be purchased.]

Use of Reverse Transcriptase vs Klenow DNA Polymerase

Whereas Klenow DNA polymerase has been most commonly used for dideoxy sequencing reactions, it is well known that this enzyme can have difficulty in reading accurately through certain DNA sequences, for example those that are high in dG : dC content.[9,12] In many cases where ambiguities cannot be resolved or corrections made by sequencing the opposite DNA strand with Klenow DNA polymerase, it is possible to use avian myeloblastosis virus (AMV) reverse transcriptase, with appropriate changes in the deoxy/dideoxynucleotide ratios, to determine the sequence accurately.

In our experience using denatured plasmid templates, the majority of sequences normally encountered in both prokaryotic and eukaryotic DNA are read equally well by either enzyme. However, with certain template regions one enzyme or the other has produced clearer information. (1) For example, regions rich in dA : dT residues are read more accurately by Klenow DNA polymerase. (2) Regions with homopolymeric tracts of dG or dC residues (such as those present in cDNA recombinant constructions) are read more accurately with reverse transcriptase. (3) When there are multiple dC residues in a row, reverse transcriptase reads them all with even intensity. With Klenow DNA polymerase the first dC is consistently fainter than the following dCs and can sometimes be missing in normal exposures (see also refs. 9,12).

Procedural modifications such as increasing the reaction temperature and amount of enzyme, as well as changing deoxy/dideoxynucleotide ratios, have not significantly improved the performance of either enzyme on difficult regions of templates we have used. [In addition, AMV reverse transcriptase has performed much better than cloned Moloney murine leukemia virus (MMLV) reverse transcriptase in our hands.] However, since Klenow DNA polymerase and reverse transcriptase tend to have difficulties on different types of sequences, they can be used in parallel to obtain more accurate and complete information for a given template. This

---

[12] A. J. H. Smith, this series, Vol. 65, p. 560.

is especially useful when inserts are too large to allow complementary sequence to be determined from the opposite strand using the opposite promoter primer. For these reasons we describe sequencing reaction conditions for both Klenow DNA polymerase and reverse transcriptase.

Method[13]

Plasmid Preparation

A critical element in obtaining good sequence is the purity of the plasmid template. The following alkaline extraction procedure, modified from that described by Birnboim and Doly[14] and Maniatis et al.,[15] is a rapid, reproducible method that yields suitable plasmid DNA from small overnight cultures.

1. Grow cells overnight in 5 ml of LB broth containing 100 μg/ml ampicillin at 37° with vigorous shaking.

2. Transfer 1.5 ml of the culture to a 1.5-ml microcentrifuge tube and centrifuge at 10,000 g for 1 min.

3. Carefully remove the supernatant, leaving the pellet as dry as possible.

4. Resuspend the cells by vortex mixing in 100 μl of an ice-cold solution containing 50 mM glucose, 10 mM EDTA, and 25 mM Tris–HCl (pH 8.0).

5. Incubate for 5 min at room temperature.

6. Add 20 μl of a freshly prepared solution containing 0.2 N NaOH, 1% sodium dodecyl sulfate (SDS), and mix by inversion, without vortexing. Incubate for 5 min at 0°.

7. Add 150 μl of ice-cold potassium acetate (pH 4.8). (The solution is 3 M with respect to potassium and 5 M with respect to acetate, prepared by adding 11.5 ml glacial acetic acid and 28.5 ml water to 60 ml of 5 M potassium acetate.) Mix by inversion for 10 sec, then incubate at 0° for 5 min.

8. Centrifuge at 10,000 g for 5 min.

9. Transfer the supernatant to a fresh 1.5-ml tube, avoiding the precipitate.

[13] All necessary reagents for sequencing pGEM plasmids, as well as pUC, M13, and pBR322 vectors, with both Klenow DNA polymerase and reverse transcriptase are commercially available in kit form (GemSeq K/RT™ and K/RT Universal sequencing systems; Promega, Madison, WI).

[14] C. Birnboim and J. Doly, *Nucleic Acids Res.* **7**, 1513 (1979).

[15] T. Maniatis, E. F. Fritsch, and J. Sambrook, "Molecular Cloning: A Laboratory Manual." Cold Spring Harbor Lab., Cold Spring Harbor, New York, 1982.

10. Centrifuge the supernatant again at 10,000 g for 5 min. Transfer the supernatant to another fresh tube.

11. Add RNase A to a final concentration of 20 μg/ml, and incubate at 37° for 20 min.

12. Add an equal volume of phenol–chloroform (1 : 1, saturated with 50 mM Tris–HCl at pH 8.0, 100 mM NaCl, 1 mM EDTA), mix by vortexing 30 sec, and centrifuge at 10,000 g for 30 sec. Transfer the aqueous phase to a fresh 1.5-ml tube.

13. Add 2.5 volumes of ethanol, mix, and incubate at −70° for 5 min.

14. Centrifuge at 10,000 g for 5 min. Remove the supernatant and rinse the pellet by adding 1 ml of prechilled 70% ethanol, mixing briefly, and centrifuging for 1 min. Carefully remove the supernatant and dry the pellet under vacuum.

15. Dissolve the pellet in 16 μl deionized water. Add 4 μl 4 M NaCl, mix, and then add 20 μl 13% polyethylene glycol (MW 8000). Mix well and incubate at 0° for 20 min.

16. Centrifuge at 10,000 g for 10 min. Remove supernatant and rinse and dry the pellet as in step 14.

17. Dissolve the DNA in 20 μl of deionized water. A yield of 1–3 μg should be expected.

Alkali Denaturation and Primer Annealing

1. Add 2 μl of 2 N NaOH, 2 mM EDTA to 20 μl of water containing 1–2 μg plasmid DNA in a siliconized, screw-capped microcentrifuge tube.

2. Incubate for 5 min at room temperature.

3. Neutralize by adding 3 μl sodium acetate (pH 5.2), and then add 7 μl deionized water. Mix, add 75 μl ethanol, mix again, and chill for 5 min at −70°.

4. Centrifuge at 10,000 g for 5 min, remove the supernatant, and rinse the pellet with 200 μl prechilled 70% ethanol. Dry the final pellet under vacuum.

5. To the dry pellet, add 6 μl deionized water, 1 μl of 10× buffer (100 mM Tris–HCl at pH 7.5, 500 mM NaCl for Klenow DNA polymerase; 340 mM Tris–HCl at pH 8.3, 500 mM NaCl, 60 mM MgCl$_2$, 50 mM dithiothreitol for reverse transcriptase), and 3 μl of the appropriate promoter primer (10 ng/μl). Anneal at 37° for 15–20 min.

Sequencing Reactions

1. Prepare four 1.5-ml screw-capped microcentrifuge tubes labeled C, A, T, and G for each set of sequencing reactions. Add 3 μl of the appropriate nucleotide mix to each tube.

Klenow DNA polymerase mixes

ddCTP mix: 66 μM ddCTP, 1.66 μM dCTP, 33 μM dTTP, 33 μM dGTP

ddATP mix: 300 μM ddATP, 33 μM dCTP, 33 μM dTTP, 33 μM dGTP

ddTTP mix: 117 μM ddTTP, 33 μM dCTP, 1.66 μM dTTP, 33 μM dGTP

ddGTP mix: 66 μM ddGTP, 33 μM dCTP, 33 μM dTTP, 1.66 μM dGTP

(all mixes contain 10 mM Tris–HCl at pH 7.5, 50 mM NaCl, 10 mM MgCl$_2$, 1 mM dithiothreitol)

Reverse transcriptase mixes

ddCTP mix: 100 μM ddCTP, 250 μM dCTP, 250 μM dTTP, 250 μM dGTP

ddATP mix: 3.6 μM ddATP, 250 μM dCTP, 250 μM dTTP, 250 μM dGTP

ddTTP mix: 200 μM ddTTP, 250 μM dCTP, 250 μM dTTP, 250 μM dGTP

ddGTP mix: 50 μM ddGTP, 250 μM dCTP, 250 μM dTTP, 250 μM dGTP

(all mixes contain 34 mM Tris–HCl at pH 8.3, 50 mM NaCl, 6 mM MgCl$_2$, 5 mM dithiothreitol)

2. Add 5 units of Klenow DNA polymerase or 5 units of reverse transcriptase to the annealing mixture (step 5, previous section).

3. Add 4 μl [α-^{32}P]dATP (10 mCi/ml; >400 Ci/mmol) or 5 μl [α-^{35}S]dATP (10 mCi/ml; >500 Ci/mmol) to the annealing mixture. Pipet up and down a few times to mix.

4. Add 3 μl of the primer–template–enzyme–label mixture to each of the nucleotide mixes and incubate at 37° for 15 min (20 min if using [α-^{35}S]dATP). Incubate at 42° for reverse transcriptase.

5. Add 1 μl of chase solution to each tube and incubate at 37° for another 15 min (42° for reverse transcriptase). (Chase solution: 2 mM each of dCTP, dATP, dTTP, and dGTP in the same buffer as the nucleotide mixes.)

6. Stop the reactions by adding 5 μl of 90% formamide containing 20 mM EDTA, 0.3% bromphenol blue, and 0.3% xylene cyanole. Mix well.

7. Heat the reactions at 70° for 3 min before loading the sequencing gel. Load 2.5 μl of each reaction. If multiple loading is desired to increase the amount of sequence read, the samples should be heated at 70° for 3 min immediately before each load to prevent possible renaturation.

The nucleotide mixes in step 1 have been optimized for using labeled dATP of a specific activity of 400–500 Ci/mmol, which results in a final dATP concentration of about 2 μM. If higher specific activities are used, the mixes should be supplemented with unlabeled dATP to achieve proper nucleotide ratios. Using a standard 6 or 8% sequencing gel with two loads, it is normal to read 200–250 bases for ^{32}P and 300–350 bases for ^{35}S from each end of the insert with this method. When the size of the insert is too large (>500 bp) to allow the complete sequence to be determined from both ends using both promoter primers, a series of progressive unidirectional deletion subclones can usually be constructed (depending on available restriction sites) using exonuclease III by the method described by Henikoff[16] or as described in this volume [57].

In summary, sequencing denatured plasmid DNA with either Klenow DNA polymerase or reverse transcriptase yields results comparable to those obtained from single-stranded M13 DNA with the corresponding enzymes. The method allows sequencing of both strands of a single plasmid isolate when a variety of vectors are used for recombinant constructions. The use of both sequencing enzymes provides an additional (or alternative) way to confirm, clarify, or extend a given sequence without additional template preparation.

[*Editors' Note*. Stephen A. Saxe (Laboratory of Cellular and Developmental Biology, National Institutes of Health) has modified the protocol for dideoxy sequencing of DNA templates to optimize the reading of G stretches. Others have previously suggested that reactions with Klenow fragment may be incubated at 50° (rather than 37°) but ambiguities are often still observed; the use of reverse transcriptase at 42° instead of Klenow fragment is not optimal for reading long AT stretches. Saxe suggests using 5 units of Klenow enzyme (rather than 1 unit) for *each* reaction with incubations at 37°. Excess Klenow fragment clearly promotes the reading of both GC and AT stretches.]

---

[16] S. Henikoff, *Gene* **28,** 351 (1984).

[59] Sequencing of RNA Transcripts Synthesized *in Vitro* from Plasmids Containing Bacteriophage Promoters

By ROBERT C. MIERENDORF and DIANA PFEFFER

The availability of plasmid vectors that contain promoters for bacteriophage RNA polymerases has made possible the efficient synthesis *in vitro* of homogenous RNA copies of virtually any cloned DNA sequence.[1,2] The RNA polymerases encoded by the bacteriophage SP6[3] and T7[4] recognize unique promoter sequences and are able to transcribe adjoining sequences with high specificity and fidelity. Defined RNA transcripts synthesized by this method are widely used as highly sensitive probes[2] and as substrates for the study of RNA processing[5] and translation.[6] We describe here a method of sequencing DNA cloned in these vectors based on the use of RNA intermediates generated *in vitro*.

Principle

The use of reverse transcriptase with appropriate oligodeoxynucleotide primers and chain-terminating dideoxynucleotides has been valuable for obtaining sequence information directly from RNA templates.[7–10] Historically, this method has been used for sequencing viral RNA, rRNA, and well-characterized mRNA, for which suitable oligodeoxynucleotide primers can be synthesized. The convenience of preparing pure RNA *in vitro* from plasmid constructions allows this method to be used with a wide variety of cloned DNA without prior sequence information. In the

[1] M. R. Green, T. Maniatis, and D. A. Melton, *Cell (Cambridge, Mass.)* **32**, 681 (1983).
[2] D. A. Melton, P. A. Krieg, M. R. Rebagliati, T. Maniatis, K. Zinn, and M. R. Green, *Nucleic Acids Res.* **12**, 7035 (1984).
[3] E. T. Butler and M. J. Chamberlin, *J. Biol. Chem.* **257**, 5772 (1982).
[4] P. Davanloo, A. H. Rosenberg, J. J. Dunn, and F. W. Studier, *Proc. Natl. Acad. Sci. U.S.A.* **81**, 2035 (1984).
[5] A. R. Krainer, T. Maniatis, B. Ruskin, and M. R. Green, *Cell (Cambridge, Mass.)* **36**, 993 (1984).
[6] P. A. Krieg and D. A. Melton, *Nucleic Acids Res.* **12**, 7057 (1984).
[7] P. H. Hamlyn, G. G. Brownlee, C. C. Cheng, M. J. Gait, and C. Milstein, *Cell (Cambridge, Mass.)* **15**, 1067 (1978).
[8] D. Zimmern and P. Kaesberg, *Proc. Natl. Acad. Sci. U.S.A.* **75**, 4257 (1978).
[9] A. C. Palmenberg, E. M. Kirby, M. R. Janda, N. L. Drake, G. M. Duke, K. F. Potratz, and M. S. Collett, *Nucleic Acids Res.* **12**, 2969 (1984).
[10] D. J. Lane, B. Pace, G. J. Olsen, D. A. Stahl, M. L. Sogin, and N. R. Pace, *Proc. Natl. Acad. Sci. U.S.A.* **82**, 6955 (1985).

METHODS IN ENZYMOLOGY, VOL. 152

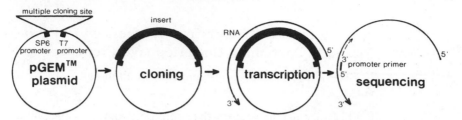

FIG. 1. Diagram of the steps involved in sequencing cloned DNA via RNA intermediates generated *in vitro* by bacteriophage RNA polymerases. In the example shown, T7 RNA polymerase is used for transcription of supercoiled recombinant plasmid DNA. The RNA transcript is then annealed with the oligonucleotide SP6 promoter primer and sequenced using dideoxynucleotide chain terminators and reverse transcriptase. To sequence beginning at the T7 promoter, SP6 RNA polymerase would be used and the transcript annealed with the T7 promoter primer.

pGEM™ plasmid vectors (Promega), promoters for both SP6 and T7 RNA polymerases are present in opposite orientations flanking a region of multiple cloning sites (see Fig. 1). This construction allows the synthesis of RNA corresponding to either strand of a given DNA insert using a single plasmid and either SP6 or T7 RNA polymerase.

As depicted in Fig. 1, the method we have used for sequencing is first to clone a DNA fragment into a pGEM plasmid, use the resulting recombinant to prepare an RNA copy that extends through the insert and the opposite promoter region, and then use a DNA primer homologous to the opposite promoter with reverse transcriptase and appropriate dideoxy/deoxynucleotide mixes in a chain termination sequencing reaction. The promoter primer thus directs the synthesis of chain terminated DNA through the inserted sequences beginning at the end opposite from the start of transcription.

Preparation of RNA Transcripts

Suitable transcripts can be prepared from supercoiled plasmids purified by standard methods. A modified alkaline lysis procedure (see previous chapter [58]) is satisfactory for preparing plasmid DNA from small overnight cultures. The transcription and sequencing reactions should be carried out under RNase-free conditions.

1. A 20-μl transcription reaction yields enough template for four sets of sequencing reactions (approximately 4 μg of RNA). In a sterile 1.5-ml microcentrifuge tube, combine the reagents at room temperature in the following order: 9 μl sterile deionized water, 4 μl 5× ribonucleotides (2.5

mM each of ATP, CTP, GTP, and UTP), 4 μl 5× transcription buffer (200 mM Tris–HCl at pH 7.5, 50 mM NaCl, 30 mM MgCl$_2$, 10 mM spermidine), 2 μl 100 mM dithiothreitol, and 0.5 μl (approximately 0.5 μg) supercoiled pGEM recombinant plasmid DNA.

2. Mix and warm the tube to 37°. Add 0.5 μl (5–10 units) of either SP6 or T7 RNA polymerase and incubate at 37° for 30 min. Transcripts can either be used directly in the annealing reactions or stored at −70°. If the transcripts are to be stored, the reaction should be extracted with one volume of phenol–chloroform (1 : 1, saturated with 10 mM Tris–HCl at pH 8.0, 100 mM NaCl, 1 mM EDTA), followed by the addition of 0.1 volume of 3 M sodium acetate (pH 5.2), and precipitation with 2.5 volumes of ethanol. After centrifugation, the pellet is dried under vacuum and resuspended in deionized water before being stored at −70°.

Annealing Template and Primer

1. In a screw-capped microcentrifuge tube, combine 1 μl deionized water, 1 μl 10× buffer (340 mM Tris–HCl at pH 8.3, 500 mM NaCl, 60 mM MgCl$_2$, 50 mM dithiothreitol), 3 μl of the appropriate promoter primer (10 ng/μl), and 5 μl (1 μg) of RNA transcript. If SP6 RNA polymerase was used for transcription, the T7 promoter primer is used for sequencing, and *vice versa*.

2. Heat the annealing mixture to 65° for 3 min. Allow the mixture to cool slowly to 42°. In a 500-ml reservoir, this usually takes about 40 min.

Sequencing Reactions

1. While the annealing mixtures are cooling, prepare four screw-capped microcentrifuge tubes labeled C, A, T, and G for each set of reactions. Add 2.5 μl of the appropriate nucleotide mix to each tube.

ddCTP mix: 100 μM ddCTP, 250 μM dCTP, 250 μM dTTP, 250 μM dGTP

ddATP mix: 3.6 μM ddATP, 250 μM dCTP, 250 μM dTTP, 250 μM dGTP

ddTTP mix: 200 μM ddTTP, 250 μM dCTP, 250 μM dTTP, 250 μM dGTP

ddGTP mix: 50 μM ddGTP, 250 μM dCTP, 250 μM dTTP, 250 μM dGTP

(all mixes contain 34 mM Tris–HCl at pH 8.3, 50 mM NaCl, 6 mM MgCl$_2$, 5 mM dithiothreitol)

2. Add 5 units avian myeloblastosis virus (AMV) reverse transcriptase to the annealing reaction.

3. Add 2.5 μl [α-^{32}P]dATP (10 mCi/ml: >400 Ci/mmol) or 5 μl [α-^{35}S]dATP (10 mCi/ml: >500 Ci/mmol) to the annealing reaction. Pipet up and down a few times to mix.

4. Add 3 μl of the template–primer–enzyme–label mixture to each of the nucleotide mixes and incubate at 42° for 15 min (20 min if using [α-^{35}S]dATP).

5. Add 1 μl of chase solution to each tube and incubate at 42° for another 15 min. (Chase solution: 2 mM each of dCTP, dATP, dTTP, and dGTP in the same buffer as the nucleotide mixes.)

6. Stop the reactions with 5 μl of 90% formamide containing 20 mM EDTA, 0.3% bromphenol blue, and 0.3% xylene cyanole. Mix well.

7. Heat the reactions at 70° for 10 min before loading the sequencing gel. Load 2.5 μl of each reaction. If multiple loading is desired to increase the amount of sequence read, the samples should be heated at 70° for 10 min immediately before each load to prevent possible renaturation.

Comments

This method has proved particularly useful in sequencing regions having high GC content, and thus provides an additional rapid approach to supplement the direct sequencing of denatured plasmid DNA with either Klenow DNA polymerase or reverse transcriptase (see this volume [58]). An additional advantage of the method is that the sequence can usually be clearly read within a few bases of the priming site. Limitations appear to be the increased frequency of "strong stops," or nucleotides which cause the appearance of bands in all four lanes of the gel, and difficulty reading through AU-rich regions of the template RNA.

Although this method was developed for sequencing RNA transcripts produced *in vitro* from plasmid templates, the same reaction conditions can be used for direct sequencing of naturally occurring RNA. For this application, specific oligodeoxynucleotide primers corresponding to regions of known sequence are annealed to the RNA sample, and sequencing is carried out as described above. In addition, the method can be directly applied to sequencing inserts in λ phage vectors containing SP6 and T7 promoters, such as those recently developed by Palazzolo and Meyerowitz.[11]

---

[11] M. J. Palazzolo and E. M. Meyerowitz, *Gene,* in press.

[60] Hybrid Selection of mRNA and Hybrid Arrest of Translation

By ROSEMARY JAGUS

Any recombinant DNA sequence is able to form stable hybrids with its corresponding mRNA. This property can be utilized for the unambiguous identification of a cloned gene and in the screening of large numbers of recombinant DNA molecules. Recombinant DNA can be used to select a specific mRNA from the total mRNA population for translation in a cell-free protein synthesizing system. Alternatively, identification of a gene can be based on the inhibition of translation of a specific polypeptide by hybridization of the cloned DNA to its corresponding mRNA. In addition, hybrid arrest of translation can be used to determine what sequences are essential for translation.

Precautions to Prevent RNA Degradation

Reagents. Diethyl pyrocarbonate (DEPC), as supplied, is stored at 4°, in the dark. Chelex is prepared as follows. Stir 15 g Chelex (Bio-Rad) in 150 ml 0.1 N NH$_4$OH for ~1 hr. Wash 6–10 times with 200 ml DEPC-treated water by suspension and filtration. Resuspend in 400 ml 0.1 N HCl for 10 min. Filter. Wash 6–10 times with DEPC-treated water. Wash 5 times with 50 mM potassium phosphate buffer (pH 7). Wash 3 times with DEPC-treated water. Check pH of final filtrate. Store damp resin in sterile bottle at room temperature.

Procedure. The availability of an active cell-free translation system is an essential requirement for these procedures, but the success of the procedures also depends on the adequate protection of mRNA from degradation. Procedures for the isolation of mRNA in an undegraded form are described in other chapters [20, 21], but a few general guidelines for the preparation of glassware and solutions are given here. All glassware is baked at 400° overnight. All solutions are made with MilliQ water and treated with prepared Chelex, 0.5 ml/100 ml (v/v), to remove metal ions, before filtration through Millipore filters (0.66 μm). Solutions are then made 0.1% with respect to diethyl pyrocarbonate, followed by autoclaving, or by incubation at 60° overnight for solutions labile to autoclaving. These precautions should minimize degradation of mRNA during isolation. Placental ribonuclease inhibitor (30 units/ml) can be used to prevent mRNA degradation during translation (see this volume [27]).

Hybrid Selection of mRNA

The mRNA encoded by a specific gene may be affinity purified by hybridization to its corresponding DNA immobilized on a solid support. Traditionally, hybrid selection of mRNA has been performed using DNA immobilized on nitrocellulose filters or to DNA covalently linked to activated cellulose, such as diazobenzyloxymethyl paper (DBM paper).[1]

DBM paper requires preparation and/or activation prior to use and has a lower capacity to bind DNA than does nitrocellulose. However, since DNA is covalently bound to this support, mRNAs prepared by this method contain no contaminating DNA and the filters can be used innumerable times. In contrast, nitrocellulose requires no preparation and has a higher binding capacity than DBM. However, the DNA is not irreversibly bound and gradually washes off. This means that the filters can be used only a limited number of times, and also that DNA may be present as a contaminant of the hybrid-selected mRNA preparation and inhibit its translation in cell-free systems.

In recent years, the availability of nylon membranes as matrices for binding nucleic acids has assumed importance in hybridization analyses.[2] These membranes have great mechanical strength, have very high capacities for nucleic acids, are able to bind smaller oligonucleotides, and have strong retention properties. They have, therefore, all the advantages of chemically activated cellulose papers without any apparent disadvantages. A method is given here, therefore, for the use of Gene Screen Plus (New England Nuclear), instead of nitrocellulose, for hybrid selection of mRNA. The method should be applicable to the use of other nylon membranes commercially available.

Immobilization of DNA on Gene Screen Plus

It is essential to linearize circular DNA before applying it to Gene Screen Plus. This is most easily done by digestion with an appropriate restriction enzyme. A method is given here for the direct application of DNA to Gene Screen Plus, along with a method for rapid transfer of DNA fragments from agarose gels.

Solutions. The following solutions should be treated as described earlier in this section to minimize degradation of mRNA: 1 M ammonium acetate (pH 5.5); 0.4 N NaOH; 0.25 N HCl; 2× SSC (60 mM sodium citrate at pH 7.5, 0.3 M NaCl).

Procedures. DNA is applied or transferred in 0.4 N NaOH, a step

[1] J. S. Miller, B. M. Paterson, R. P. Ricciardi, L. Cohen, and B. E. Roberts, this series, Vol. 101, p. 650.
[2] L. C. Reed and D. A. Mann, *Nucleic Acids Res.* **13**, 7207 (1985).

which promotes the covalent fixation of DNA to nylon membranes.[2] Radiolabeled DNA transferred to nylon membrane in alkali cannot be stripped from the membranes,[2] giving filters that can be used innumerable times for selection of specific mRNA sequences.

Direct Application to Gene Screen Plus. Soak Gene Screen Plus in water for 5 min, then in 0.4 N NaOH for 30 min. Denature restricted DNA in 0.4 N NaOH at room temperature for 10 min, then place in ice until ready to apply. DNA solution should be applied by spotting 10 μl at a time, drying between applications. Wash filter twice in 1 M ammonium acetate (pH 5.5), followed by two rinses in 2× SSC. Blot filter dry, air dry overnight. Baking is not necessary when using Gene Screen Plus.

Transfer from Agarose Gels. Restricted DNA is fractionated by agarose gel electrophoresis with 0.1% ethidium bromide in gel and running buffer. The position of each fragment is determined by use of a low-energy ultraviolet hand monitor held over the outside edges of the gel only, since even mild exposure to UV radiation is sufficient to render a significant proportion of the DNA molecules incapable of forming hybrids.[2] The positions of the band or bands of interest should be indicated by cutting notches and the gel should afterward be cut down to leave the required segments only.

DNA is transferred by the rapid procedure of Reed and Mann.[2] Soak gel in 2 volumes 0.25 N HCl with gentle agitation until bromphenol blue marker changes color (10–20 min). Rinse gel briefly in water. Rinse Gene Screen Plus in water for 5 min. Transfer DNA to membrane in 0.4 N NaOH (2 hr to overnight). Rinse membrane twice in 2× SSC. The strips of Gene Screen Plus thus generated may be cut into conveniently sized squares for use.

The upper amount of DNA that can be applied by these methods has not been determined, but is *at least* 100 μg/cm^2 for DNA applied directly. For restricted DNA transferred from agarose gels, gels (20 × 25 cm) may be loaded with 1–2 mg restricted DNA in a single horizontal well. The initial amount of DNA bound to the filter and that remaining subsequent to each selection may be determined by including trace amounts of ^{32}P-radiolabeled DNA along with the restricted DNA. Filters can be monitored by measuring Cerenkov radiation.

Immobilization of DNA on DBM Paper

Procedures for the preparation of DBM paper have been described in detail previously.[3,4] Immediately prior to the application of DNA (within

[3] J. C. Alwine, D. J. Kemp, and G. R. Stark, *Proc. Natl. Acad. Sci. U.S.A.* **74**, 5350 (1977).
[4] G. M. Wahl, M. Stern, and G. R. Stark, *Proc. Natl. Acad. Sci. U.S.A.* **76**, 3683 (1979).

5–10 min), activate paper by washing twice at 4° in water, followed by two washes with 20 mM NaOAc (pH 4). As with the use of Gene Screen Plus, DNA should be linearized before application.

Solutions. Sodium acetate (2 M, pH 5.5); 0.4 N NaOH; 1 N HCl; 25 mM sodium phosphate buffer (pH 6).

Procedure. Dissolve linearized DNA in 25 mM sodium phosphate buffer (pH 6). Heat at 80° for 1–2 min. Cool, add 4 volumes dimethyl sulfoxide. Apply to activated DBM paper and allow to air dry overnight. Wash 4 times with water. Wash 4 times with 0.4 N NaOH for 10 min at 37°. Wash 4 times with 2 M sodium acetate (pH 5.5). Wash 4 times with water and 4 times with hybridization buffer (see following section). In contrast to Gene Screen Plus, only 12–25 μg DNA/cm^2 should be applied.

Prehybridization and Hybridization

Solutions. Prehybridization buffer: 50% formamide (deionized), 0.75 M NaCl, 0.1 M PIPES–NaOH (pH 6.4), 8 mM EDTA, 1% glycine (if using DBM paper), 0.5% nonfat dried milk (if using Gene Screen Plus), 100 μg/ml calf liver tRNA, 0.5% sodium dodecyl sulfate (SDS). Hybridization buffer for high abundance class mRNAs: 65% formamide (deionized), 0.4 M NaCl, 0.01 M PIPES–NaOH (pH 6.4), 8 mM EDTA, 0.5% SDS. Hybridization buffer for low abundance mRNA: 50% formamide (deionized), 0.4 M NaCl, 0.01 M PIPES–NaOH (pH 6.4), 8 mM EDTA, 0.5% SDS. Wash buffer: 1× SSC, 2 mM EDTA, 0.5% SDS.

Procedure. One of the common problems with hybrid selection is degradation of RNA by contaminants in the formamide. Formamide should be deionized before use by stirring 10 ml with 1 g AG-501 SA (Bio-Rad) for 1 hr at room temperature. Filter the deionized formamide, snap-freeze single-use aliquots, and store at −70°. Once thawed for use, unused formamide should be discarded. Batches of deionized formamide should be tested before use, by assessment of its ability to decrease the translatability of mRNA.

Prehybridization of filters should be carried out in the appropriate prehybridization buffer at 37° overnight. Prior to hybridization, discard prehybridization buffer, wash 10 times with wash buffer at 60°. Rinse briefly in the hybridization buffer chosen.

For best results, hybridization should be carried out using poly(A)$^+$ RNA (this volume [25]), or if the mRNA is not polyadenylated, with a fraction enriched for the mRNA of interest. Optimal hybridization conditions should be chosen in consideration of the sequence (if known) and abundance of mRNA. This may have to be determined empirically. The variable parameters for the hybridization reaction are the concentrations of formamide and input RNA, and the temperature and duration of hybridization. Hybridization buffers for high and low abundance class mRNAs are given. For high abundance class mRNA, such as those coding for globin, actin, tubulin, hybridization should be carried out with 65% formamide, hybridization buffer for 2–4 hr at 48°. RNA concentration should be between 20 and 100 μg/ml for poly(A)+ RNA or 0.3–0.5 mg/ml for total RNA. For low abundance class mRNA, hybridization should be carried out with 50% formamide, hybridization buffer for 6–18 hr at 37°. RNA concentration should be at least 100 μg/ml for poly(A)+ RNA, or 5–10 mg/ml for total RNA.

The hybridization time should be optimized for each mRNA selected. Hybridization times should be as short as possible to minimize RNA degradation.

Rinsing/Elution

Solutions. 1× SSC: 15 mM sodium citrate at pH 7.5, 0.15 M NaCl. Buffer A: 5 mM Tris–HCl at pH 7.5, 1 mM EDTA. Solution B: 5 mM KCl, 2 mM EDTA, calf liver tRNA 10 μg/ml (Boehringer-Mannheim).

Procedure. In a 1.5-ml microfuge tube, wash filters 10 times with 1× SSC containing 0.5% SDS at 50°, followed by 5 washes with 1× SSC at 50°. Wash once with buffer A at 50° for 40 sec. This can be done by shaking or vortexing, allowing the filter to settle, and removing the solution. Elute by boiling 1 min in solution B followed by snap-freezing in dry ice–ethanol. Thaw, remove eluate, and repeat. Pool eluates, add two volumes ethanol and 0.1 volume 2 M potassium acetate, and store at −20° overnight. Recover eluted mRNA by centrifugation at 10,000 g for 30 min. Wash pellet twice with 70% ethanol. Lyophilize final pellet. Reconstitute in desired volume of water.

Hybrid Arrest of Translation

When sequences within mRNA essential for translation are hybridized to their complementary DNA sequences, translation of the encoded polypeptide is blocked.[5,6] This provides an alternative method for the identification of recombinant DNA molecules containing part or all of a sequence coding for a specific polypeptide. In addition, hybrid arrest can be used to identify the sequences within the mRNA that are required for translation.

[5] B. M. Paterson, B. E. Roberts, and E. L. Kuff, *Proc. Natl. Acad. Sci. U.S.A.* **74,** 4370 (1977).
[6] N. D. Hastie and W. A. Held, *Proc. Natl. Acad. Sci. U.S.A.* **75,** 1217 (1978).

Solutions. Hybridization buffer: 80% formamide (deionized), 0.01 M PIPES–NaOH at pH 6.4, 0.4 M NaCl, 8 mM EDTA.

Procedure. Linearize plasmid DNA by digestion with appropriate restriction enzymes. Extract DNA sequentially with phenol and chloroform–isoamyl alcohol (24:1) to remove contaminating proteins and salts (see this volume [4]). Each DNA preparation to be used for hybrid arrest should be titrated to determine the DNA concentrations that result in arrest of translation without causing inhibition of the cell-free translation system. The RNA used should be poly(A)$^+$ or otherwise enriched for the mRNA of interest to allow good rates of incorporation by cell-free translation system. The RNA concentration used should be at least twice that needed to give sufficient incorporation of [^{35}S]methionine for a clear signal on fluorographs, but below the concentration required for saturation, since this is a subtractive assay.

Ethanol-precipitate DNA and RNA together along with 10 μg calf liver tRNA. Precipitate at $-70°$ overnight. Collect precipitate by centrifugation at 10,000 g for 30 min. Wash pellet with 70% ethanol and dry under vacuum. Dissolve precipitate in 2–3 μl water, boil for 30 sec, and snap-freeze in dry ice–ethanol. Thaw and centrifuge for 10 sec to ensure that all material is in the bottom of tube and add 22–23 μl hybridization buffer and incubate at 37° for 6–18 hr.

Terminate the hybridization reaction by adding 200 μl ice-cold water. Transfer one half of the reaction volume to another tube and dissociate the hybrids by boiling for 60 sec, followed by snap-freezing in dry ice–ethanol. Thaw. Add 2 M sodium acetate (pH 5.5) to both reactions (hybridized and heat dissociated) to give a final sodium acetate concentration of 0.2 M. Precipitate with 2.5 volumes ethanol and dry final pellet under vacuum. Determine translation products of each by adding translation components directly to dried pellets. Analyze translation products directly or after immunoprecipitation as described in this volume [31].

[61] Northern and Southern Blots

By Geoffrey M. Wahl, Judy L. Meinkoth, and Alan R. Kimmel

The ability to fractionate nucleic acids and to determine which of them has sequences complementary to an array of DNA or RNA molecules is one of the most powerful tools of molecular biology. The Southern blot,[1]

---

[1] E. M. Southern, *J. Mol. Biol.* **98,** 503 (1975).

METHODS IN ENZYMOLOGY, VOL. 152

named for its inventor, is a method for transferring size-fractionated DNA from a gel matrix to a solid support followed by hybridization to a labeled probe. The identical process for RNA playfully became known as the Northern blot.[2,3] The use of dried agarose gels as the immobilized phase facetiously became known as the Unblot.[4-6] As a reflection of the versatility of the technique, there are now dot blots (this volume [62]), spot blots, slot blots, fast blots (this volume [62]), Western blots (proteins transferred to filters and detected with immunological probes) and no doubt more to come. All are invaluable tools for investigating and analyzing mRNAs, clones, genes, fragments, flanking sequences, repetitive elements, and the like. For example, a Southern blot of a restricted genomic clone must be almost identical (except for the end fragments linked to the vector) with a blot of similarly restricted genomic DNA, hybridized to the same probe. A cloned cDNA must be capable of hybridizing to the mRNA from which it supposedly was derived. Blots, then, are often key elements in establishing the identity of nucleic acids of interest.

Southern's technique for "blotting" electrophoretically fractionated DNA from an agarose gel to nitrocellulose by passive diffusion established that transfers were faithful replicas of high-resolution gel patterns. Passive transfer techniques consist of the following steps: electrophoresis, denaturation (often preceded by fragmentation to ensure transfer of all DNA, independently of initial size), neutralization, transfer, fixation onto nitrocellulose, and detection of specific immobilized sequences by hybridization.[7] Similar procedures are used for blotting RNA (Northern blots) with minor modifications.

The mechanism of binding of nucleic acids to nitrocellulose is unknown but it is believed to be noncovalent. Although other solid supports made of nylon (a variety are commercially available) are gaining in popularity, nitrocellulose still remains the solid support of choice for both RNA and DNA. It does, however, have two disadvantages: (1) unlike nylon, it is extremely friable, particularly when dry; (2) it binds fragments smaller than 200–300 bases poorly. Practically, one avoids the first problem by keeping filter membranes moist at all stages with the exception of the fixation step; drying is essential for permanent binding to nitrocellu-

[2] J. C. Alwine, D. J. Kemp, and G. R. Stark, *Proc. Natl. Acad. Sci. U.S.A.* **74,** 5350 (1977).
[3] J. C. Alwine, D. J. Kemp, B. A. Parker, J. Reiser, J. Renart, G. R Stark, and G. M. Wahl, this series, Vol. 68, p. 220.
[4] M. Purrello and I. Daluzs, *Anal. Biochem.* **128,** 393 (1983).
[5] T. M. Shinnick, E. Lund, O. Smithies, and F. R. Blattner, *Nucleic Acids Res.* **2,** 1911 (1975).
[6] S. G. S. Tsao, C. F. Brunk, and R. E. Pearlman, *Anal. Biochem.* **131,** 365 (1983).
[7] J. Meinkoth and G. Wahl, *Anal. Biochem.* **138,** 267 (1984).

lose. To solve the second problem, one uses diazotized cellulose,[8,9] a solid support with reactive groups that make possible covalent binding of nucleic acids, or one employs nylon filters which require ultraviolet (UV) irradiation to cross-link either DNA or RNA to the solid support.[10]

Preparation of Blots

DNA Transfers (Southern Blots) to Nitrocellulose (Steps 1–6) or Diazophenylthioether (DPT) Paper (Step 7)

1. Fractionate DNA electrophoretically in an agarose gel as described in this volume [8]. If the purpose of the blot is to probe mammalian genomic DNA for single-copy genes, each lane must contain 10–40 μg of restricted DNA. Best results are obtained when these gels are run very slowly, namely 0.7 V/cm overnight to allow the inevitable overload of salts to diffuse. In theory, one could remove salt by precipitating genomic digests before electrophoresis, but, because of losses and because redissolving large fragments >10 kb may require long periods, most investigators omit such a step.

If cloned DNA is to be studied by Southern blotting, nanogram quantities of DNA are sufficient (5–10 ng of the smallest fragments are required). These values are chosen to be compatible with detection in ethidium bromide-stained gels.

2. Briefly float nitrocellulose (cut to the approximate size of gel) on water to wet thoroughly; then soak in blotting buffer (20× SSC, 20× SSPE, or 1 M ammonium acetate) while performing the remaining steps (5 min minimum) (see below for definition of SSC and SSPE buffers). A discussion of the use of nylon filters follows the Northern blot protocol.

3. If fragments >8 kb are to be transferred, soak gel in 0.25 M HCl for 7.5–10 min or expose to UV light to fragment the DNA (agitate gently to ensure even exposure of gel to HCl). Caution must be used with the UV light approach since the efficiency of nicking can decrease over time due to solarization of the filter. One can monitor transfer efficiency by staining the gel following transfer (see step 6). Step 3 is omitted when DNA is <8 kb in size. This step can also be omitted if DNA is nicked during isolation (e.g., yeast DNA is often nicked during isolation).

4. Rinse gel with water to remove excess acid; then denature DNA by soaking gel in 1.5 M NaCl, 0.5 M NaOH (2 × 15 min).

[8] J. Reiser, T. Renart, and G. R. Stark, *Biochem. Biophys. Res. Commun.* **85,** 1104 (1978).
[9] E. T. Stellwag and A. E. Dahlberg, *Nucleic Acids Res.* **8,** 299 (1980).
[10] G. Church and W. Gilbert, *Proc. Natl. Acad. Sci. U.S.A.* **81,** 1991 (1984).

5. Neutralize by soaking in 1.0–3.0 M NaCl, 0.5 M Tris–HCl (pH 7.4) (2 × 15 min).

6. Arrange three or four sheets of filter paper (Whatman 3 MM or Schleicher and Schuell No. 470) or an even-surfaced sponge cut to about 2–3 in. larger than the gel in all directions in a dish. Saturate the paper or the sponge with blotting buffer and place the gel on top. The use of large volumes of blotting buffer is unnecessary. Cover all exposed paper surfaces with Parafilm or plastic wrap. Be sure to fit the Parafilm or plastic wrap snugly against the gel. Place the nitrocellulose on top of the gel. Remove bubbles. (A pipet or a glass stirring rod used as a rolling pin does the job.) Add one or two sheets of filter paper saturated with blotting buffer followed by a 1-in. (2.5 cm) stack of paper towels and a light weight to keep all layers compressed. The Parafilm or plastic wrap will prevent the transfer buffer from reaching the paper towels except by passing through the gel and nitrocellulose membrane. Transfer is usually complete after 1–2 hr but is often allowed to proceed overnight. (The efficiency can be checked by restaining the gel, after transfer, in 100 ml water, 1 μg/ml ethidium bromide.) A good rule of thumb is that transfer can be stopped when the gel thickness decreases to approximately 1 mm since the gel concentration at this point is sufficiently high to prevent further transfer. Place blot between sheets of paper towels or thick blotting paper and bake in a 65–80° vacuum oven until completely dry (minimum time, 15 min; longer than 24 hr results in brittle nitrocellulose). Blots can be stored dry indefinitely in plastic bags (but they can chip and break if handled roughly) or in prehybridization buffer.

7. For transfer to DPT paper,[11] fragment the DNA (step 3) as for transfer to nitrocellulose. Rinse gel with water and then soak gel in 1 M sodium acetate (pH 4.0) (1 × 30 min). Rinse gel in water, and then soak in 20 mM sodium acetate (1 × 30 min). Transfer time and apparatus are the same as in step 6 except that blotting buffer is 20 mM sodium acetate (pH 4.0). Baking is not required.

The Southern blots are now ready for use.

RNA Transfers (Northern Blots) to Nitrocellulose

1. Fractionate RNA in any of the denaturing gel systems specified in this volume [8]. For best results do not stain the gel.[12] Marker lanes may be cut from the gel and stained separately using the instructions in chapter [8] or denatured end-labeled DNA may be used as size standards. If the

[11] B. Seed, *Nucleic Acids Res.* **10,** 1799 (1982).
[12] P. Thomas, *Proc. Natl. Acad. Sci. U.S.A.* **77,** 5201 (1980).

purpose of the blot is to identify rare mRNAs (>0.01%), use 1–10 μg of poly(A)$^+$ RNA or the equivalent whole-cell RNA (polyadenylated RNA is usually 1–5% of the total).

2. Transfer the gel directly to nitrocellulose[11] (see Southern blots, step 6). Transfer is usually carried out for at least 4 hr. Use 10–20× SSC or 10–20× SSPE for RNA transfers. Fragmentation of RNA is not required; RNA molecules at least 10 kb in length transfer quantitatively.[7] Dry the blot as described in step 6. The Northern blots are now ready for hybridization.

Nylon Filters for Nucleic Acid Transfer

Recently a wide variety of nylon-based filters have been produced which, like nitrocellulose, have strong affinity for nucleic acids. They have been used with considerable success for DNA and RNA blot hybridization. The primary advantage of nylon filters compared with nitrocellulose filters is their structural stability. Whereas dry nitrocellulose filters are very brittle, it is nearly impossible to tear nylon filters.

The techniques for blotting of nucleic acids onto nylon are similar to those for nitrocellulose. However, since nylons differ, the reader should adhere to the manufacturer's recommendations. Finally, it should be noted that nucleic acids have been successfully eluted from gels and transferred to filter matrices using commercially available electrophoretic transfer devices. The manufacturer's directions should be consulted for details of the procedure.

Screening Blots

The prehybridization and hybridization of Southern and Northern blots, the choice and labeling of probes, and the washing of blots should be carried out using the conditions specified for screening libraries in this volume [45]. Instructions for detecting signals, namely, handling filters and autoradiographing them, are found in this volume (see [45] and [7], respectively). After the experiment has been completed, the bound radioactivity can be eluted and the filter reutilized (see this volume [45]). There are, however, techniques applied to blots that are not normally used for analyzing libraries. These are discussed below.

Sandwich Hybridization

The prehybridized blot is hybridized sequentially with two probes, the first unlabeled and the second labeled. In the first stage, the target sequence in the blot is hybridized with unlabeled plasmid sequences or single-stranded phage DNA which is attached to sequences complementary to the target DNA of interest. In the second stage, labeled sequences complementary to the vector (phage or plasmid) are hybridized with the vector tails which protrude from the hybridized target sequences. The latter approach has the advantage that it requires the synthesis of only one labeled probe, namely the vector, for detecting all target sequences. A potential advantage is that the signals produced from short and long inserts should be the same since it is the labeled vector sequences which produce the signal. This type of approach should prove to be beneficial for gene quantitation experiments since signal strength variations due to differences in probe sizes are eliminated. In practice, difficulties are encountered with poor signal-to-noise ratios (often high backgrounds with modest signal). One reason may be that the probe sizes required to form adequate "sandwiches" are also large enough to cause high backgrounds (see this volume [9]).

RNA Probes

The identification of several bacteriophage-encoded RNA polymerases and the sequencing of their promoters has spawned a new technology for producing RNA probes. These phage polymerases display undetectable levels of background RNA synthesis on either intact or nicked DNA molecules lacking the phage promoters. Cloning vectors are now available in which the promoters for a single polymerase, or for two different polymerases, lie adjacent to a multiple restriction site linker. Transcription with any of the available polymerases enables one to produce large quantities of high specific activity RNA probes which correspond to either the coding or noncoding strands. Since the polymerases can initiate more than once from each template, the quantity of RNA produced is in large part determined by the concentration of the rate-limiting nucleotide. Hence, large quantities of low specific activity or unlabeled RNA useful for *in vitro* translation (when the RNA is synthesized with cap nucleotide as in this volume [30]) or microinjection can be produced if sufficient quantities of ribonucleotide triphosphate precursors are employed. RNA with a specific activity useful for nucleic acid hybridization (i.e., greater than 2×10^8 cpm/μg) can be obtained when high specific activity labeled precursor is used.

The size of the transcript is critically dependent on the concentration of nucleotide precursors.[13] For example, when UTP or ATP are used as

[13] D. A. Melton, P. A. Kreig, M. R. Rebagliati, T. Maniatis, K. Zinn, and M. R. Green, *Nucleic Acids Res.* **12**, 7035 (1984).

the labeled precursors at a concentration of 2 μM, only 5–10% of the transcripts were observed to be full length (i.e., 1850 bases).[13] By contrast, 90% of the transcripts were full length (1850 bases) when the triphosphates were present at greater than 50 μM.[13] We have observed the synthesis of transcripts larger than 10 kb when the labeled nucleotide is present at 500 μM (see this volume [35]). The effects of using CTP and GTP at low concentrations are less pronounced than those of UTP and ATP (e.g., 50 and 30% of transcripts were 1850 bases when GTP or CTP, respectively, were used at 2 μM). It is possible, therefore, to produce large RNA molecules *in vitro* with any of the bacteriophage RNA polymerases currently available if sufficient quantities of the ribonucleotide precursors are included in the reaction.

The next section will focus on the preparation of high specific activity RNA probes to be used for hybridization with DNA or RNA blots. Other uses for RNA probes such as *in situ* hybridization to tissue sections [68] and bacterial colony screening [35, 45] are dealt with elsewhere in this volume.

Probe Synthesis

Transcription reactions are set up with all components except the polymerase and RNase inhibitor (RNasin) at *room temperature* and *in the order indicated* to prevent precipitation of DNA by the spermidine present in the transcription buffer. Add the following components for a 20-μl reaction containing approximately 1 μg of DNA template:

4 μl of 5× transcription buffer (5× buffer contains 200 mM Tris–HCl at pH 7.5, 0–50 mM NaCl, 30 mM MgCl$_2$, 10 mM spermidine, 500 μg/ml bovine serum albumin)

200 μl 100 mM dithiothreitol

1 μl RNasin (30 units per μl)

4 μl of a mixture of the cold rNTPs (excluding the one to be used for labeling) at a concentration of 2.5 mM each

2 μl of 120 μM cold nucleotide corresponding to the labeled nucleotide (to give a final concentration of ~15 μM; less cold nucleotide may be added if higher specific activity probe is required)

1 μl DNA (~1 μg)

5 μl labeled nucleotide (50 μCi, ~600 Ci/mmol)

1 μl polymerase (either SP6, T7, or T3, 1–5 units)

We have typically used UTP as the labeled nucleotide since it is more stable than GTP. In order to ensure the synthesis of full-length transcripts, it should be used at a concentration exceeding 12 μM. The specific activity of the probe can be increased by using greater quantities of undiluted labeled nucleotide. However, the concentration of the labeled nucleotide should be at least 12 μM or else a significant reduction in the average length of the probe and the number of picomoles of label incorporated can occur.[13] Reactions are incubated at 37–40° for 1–2 hr. After 1 hr at least 50% of the label should be incorporated into trichloroacetic acid-precipitable material (methods can be found in this volume [6]). RNA probes can be used for at least 1 week if stored in the presence of carrier RNA either as an ethanol precipitate or in 50% ethanol at 4°. However, less stringent hybridization conditions may have to be employed for probes used after 1 week due to radioautolysis of short RNA molecules.[14]

We have not found it necessary to remove the DNA template with RNase-free DNAse since the template should remain double stranded and should not interfere with the hybridization reaction (i.e., no denaturation step is required for RNA probes since they are single stranded). Consequently, it is only necessary to phenol extract the reaction to stop it, and then to ethanol precipitate the transcripts (this volume [4, 5]). Generally, probes prepared by ethanol precipitation contain unincorporated nucleotide triphosphate, but we have not observed this to add to hybridization background. However, due to the presence of unincorporated label, an accurate estimate of the amount of radioactivity incorporated can only be obtained by trichloroacetic acid precipitation of a small amount of the reaction. Alternatively, three ethanol precipitations using 2 M ammonium acetate have been reported to remove unincorporated label.[14]

Use of RNA Probes for Southern Blots

Several protocols have been described for using RNA probes on DNA blots. Due to a lack of consensus about which method provides the highest sensitivity and least noise, each of the protocols will be presented.

Procedure of Church and Gilbert.[15] DNA is transferred electrophoretically to nylon membranes and the fragments are bound covalently using ultraviolet irradiation (0.16 kJ/m^2).

Prehybridization: Incubation is performed at 65° for 5 min in 1% crystalline-grade bovine serum albumin, 1 mM EDTA, 0.5 M sodium phosphate at pH 7.2, 7% SDS (1 M sodium phosphate at pH 7.2 is made by dissolving 134 g of Na$_2$HPO$_4$·7H$_2$O and 4 ml of 85% H$_3$PO$_4$ per liter). Thirty milliliters of solution is needed for a 30 × 40 cm blot.

Hybridization: The prehybridization solution is removed and replaced with a solution of identical composition containing ~2 × 10^7 cpm/ml of probe. Hybridization is carried out in a volume of 5 ml for a 30 × 40 cm blot at 65° for 8–24 hr.

Washing: All washes are performed at 65° in preheated buffers except the last one in which ribonuclease (RNase) is employed, which is at 37°.

---

[14] K. Zinn, D. DiMaio, and T. Maniatis, *Cell (Cambridge, Mass.)* **34**, 865 (1984).
[15] G. M. Church and W. Gilbert, *Proc. Natl. Acad. Sci. U.S.A.* **81**, 1991 (1984).

Wash twice for 5 min each in 1 liter of 0.5% fraction V-grade bovine serum albumin, 1 mM Na$_2$EDTA, 40 mM sodium phosphate (pH 7.2), 5% sodium dodecyl sulfate (SDS). Wash eight more times (1 liter each) in 1 mM Na$_2$EDTA, 40 mM sodium phosphate (pH 7.2), 5% SDS for 5 min each. Remove the SDS by three additional washes in 100 mM sodium phosphate at pH 7.2 and then treat with 30 ml of pancreatic RNase A (10 μg/ml) in 0.3 M NaCl, 10 mM Tris–HCl (pH 7.5), 1 mM Na$_2$EDTA for 15 min. This extensive, very stringent washing protocol was designed to produce the extremely low background required for genomic sequencing studies. It may be possible to reduce the number of washes and washing volumes significantly and still obtain acceptable background for standard Southern blots where more target is available for hybridization.

Probe removal: Church and Gilbert[15] state that RNA probes can be removed by a 15-min, 65° incubation in 2 mM Tris–HCl (pH 8.2), 1 mM EDTA, 0.1% SDS. Longer treatments at higher temperatures may be required to remove the probe when greater quantities of immobilized target sequences are hybridized.

Procedure of Amasino.[16] This protocol employes 10% PEG (M_r 6–7.5 × 10^3) to accelerate the hybridization rate of RNA probes. It is contended that PEG (M_r 6–7.5 × 10^3) gives a far greater enhancement of hybridization signal than does dextran sulfate (M_r 5 × 10^5) with RNA probes, while both of these agents show equivalent enhancements with nick-translated DNA probes. The reason for the very substantial (>10×) differential effect of these polymers on RNA hybridization is not known.

Prehybridization and hybridization: Both are performed at 42° in the same solution without changing it. Blots are incubated in the following mixture for 5–10 min prior to addition of probe: 0.25 M sodium phosphate (pH 7.2), 0.25 M NaCl, 7% SDS, 1 mM EDTA, 50% formamide, 10% PEG (M_r 6–7.5 × 10^3). Hybridization is performed with 1–5 × 10^5 cpm/ml probe for 4–24 hr depending upon the application.

Washing: Blots are washed once for 5 min at 25° in 2× SSC (1× SSC: 0.15 M NaCl, 0.015 M sodium citrate at pH 70, 0.2% SDS); twice in 0.25 M sodium phosphate (pH 7.2), 0.2% SDS, 1 mM EDTA for 20–60 min at 65°; and twice in 0.05 M sodium phosphate (pH 7.2), 0.2% SDS, 1 mM EDTA for 20 min at 65°.

"Standard" Conditions. Prehybridization and hybridization: Both are performed in the same solution, and the blot is incubated in this solution for 5 min or until addition of probe. The hybridization solution contains 50% formamide, 5× SSPE (1×: 0.18 M NaCl, 0.01 M sodium phosphate at pH 7.7, 1 mM EDTA), 0.1% SDS, 2× Denhardt's reagent [1×: 0.02%

[16] R. Amasino, *Anal. Biochem.* **152,** 304 (1986).

bovine serum albumin, 0.02% poly(vinylpyrolidone), 0.02% Ficoll], 100 μg/ml denatured carrier DNA (e.g., salmon sperm), 200 μg/ml carrier RNA (e.g., yeast tRNA). After the brief prehybridization step, 1×10^6 to 1×10^7 cpm/ml of probe is added, and hybridization is carried out for 12–24 hr at 42°.

Washing: Immerse filters in 0.1× SSPE, 0.1% SDS preheated to 50°. Wash twice in this buffer for 15 min each. Monitor the background and wash at 60° in the same buffer for a more stringent treatment. Treatment with RNase as described in the method of Church and Gilbert can also reduce background.

RNA Probes for RNA Blots

RNA–RNA hybrids are substantially more stable than DNA–RNA or DNA–DNA hybrids. Consequently, stringent hybridization and washing conditions must be employed when RNA probes are used to analyze Northern, slot, or dot blots. Some probes may share sufficient homology with ribosomal RNAs over small regions that they will hybridize with moderate stability to these very abundant RNAs. In a slot or dot blot experiment, this spurious hybridization to ribosomal RNAs can obscure hybridization to an RNA of interest. Importantly, DNA probes spanning the same region show no such hybridization.[17] Therefore, extreme caution must be used when employing RNA probes to detect RNA molecules with analytical methods which do not employ size fractionation. The following protocol was abstracted from Zinn et al.,[14] with minor modifications.

Prehybridization and hybridization: The prehybridization is performed for at least 10 min at 60° in 50% formamide 5× SSPE, 5× Denhardt's reagent, 0.2% SDS, 200 μg/ml denatured carried DNA, 200 μg/ml yeast RNA. For hybridization, the prehybridization mix is diluted with 0.2 volume of 50% dextran sulfate, 5×10^6 to 1×10^7 cpm/ml of probe is added, and incubation continued for 10–12 hr at 60°. Alternatively, one may choose not to add dextran sulfate, or to use the PEG procedure (described above for DNA blots) under the more stringent conditions required for RNA–RNA hybridization reactions.

Washing: Place the filters directly into 0.1× SSPE, 0.1% SDS preheated to 60°. Wash twice in this buffer for 30 min or longer for each wash.

---

[17] G. M. Wahl, S. Albanil, K. Ignacio, and D. Richman, in "Medical Virology IV" (L. M. de la Maza and E. M. Peterson, eds.), p. 31. Lawrence Erlbaum Associates, London, 1985.

[62] Fast Blots: Immobilization of DNA and RNA from Cells

By CARL COSTANZI *and* DAVID GILLESPIE

Dot blotting can be a valuable technique in molecular cloning for rapid screening of transformants, transfectants, cell cultures, blood cells, and even solid tissue samples. This chapter is meant to provide the reader with practical, easy to follow directions for immobilizing nucleic acids directly from cells onto nitrocellulose membranes. We present two of the most widely used procedures for DNA immobilization and two for RNA immobilization along with a few variations that have appeared in the literature. The variety is not meant to confuse, only to provide freedom of choice. Each method outlined has been shown to be at least relatively quantitative with respect to appropriate controls and dilutions. The recipe of choice may differ with specific needs (for theory, see refs. 1 and 2 and this volume [43]).

Materials

Filtration is carried out under house vacuum on a minifiltration apparatus (either Minifold or Slot Blotter, Schleicher and Schuell; Hybridot, BRL) through prewet BA85 nitrocellulose (Schleicher and Schuell) backed with prewet blotter paper or paper towel. The final volume of sample applied to each well should not exceed 400 μl.

All required chemicals and reagents are readily available. Some less common materials and their suppliers are, respectively, the detergents Nonidet P-40 (NP-40), *N*-laurylsarcosine–sodium salt (SD sarc), Brij-35, and sodium deoxycholate (NaDOC), Sigma; sodium iodide, anhydrous, crystalline, Sigma; vanadyl–ribonucleoside complexes, either Bethesda Research Laboratories, New England BioLabs, or this volume [21].

DNA Immobilization

DNA must first be made single stranded before it will bind to nitrocellulose. The reannealing or "snap-back" character of DNA can make quantitative binding a problem. This is in fact a serious problem when denaturing covalently closed circular DNAs and can be alleviated with a prior linearization step. Another problem that must be addressed is the

[1] J. Meinkoth and G. Wahl, *Anal. Biochem.* **138,** 267 (1984).
[2] D. Gillespie and J. Bresser, *BioTechniques* **2,** 184 (1983).

coimmobilization of proteins and RNA which could cause inefficient DNA immobilization and serious background problems when hybridizations are carried out.

Recipe 1: Immobilization of Extracted DNA

Kafatos et al.[3] removed unwanted protein by proteinase K digestion followed by phenol–chloroform extraction. The DNA was denatured and the RNA was concomitantly degraded by a 1-hr incubation at 60–70° in 0.3–0.4 M NaOH. The pH and ionic strength were then adjusted with ammonium acetate and the resulting solution filtered through nitrocellulose. To bind the DNA irreversibly to the nitrocellulose, in this procedure, the filter must be baked in a vacuum oven at 80° for 1–2 hr. Use the following as a step-by-step recipe.

1. *Starting with cell pellet.* Resuspend cells to about 5×10^6 cells/ml of 10 mM Tris–HCl (pH 7.4).

2. Add 1.0 volume of fresh proteinase K stock solution [0.4 M Tris–HCl at pH 8.0, 100 mM EDTA, 1% sodium dodecyl sulfate (SDS), 200 μg/ml proteinase K], and incubate for 1 hr at 50–60°.

3. Extract with 1.0 volume of buffered phenol–chloroform (1 : 1 v/v, buffered with 10 mM Tris–HCl at pH 8.0, 1 mM EDTA, 100 mM NaCl).

4. Extract aqueous supernatant with 1.0 volume of chloroform–isoamyl alcohol (24 : 1 v/v).

5. *Starting with purified DNA.* To aqueous supernatant add 0.1 volume of 3 M NaOH to a final concentration of 0.3–0.4 M NaOH, incubate for 1 hr at 60–70°, then cool to room temperature (20–25°), and proceed.

6. Add 1.0 volume of 2 M ammonium acetate (pH 7.0) to a final concentration of 1 M.

7. Serially dilute into 1 M ammonium acetate (generally, at least four 10-fold dilutions are used to determine a range from which smaller dilutions can be made for precise quantitations).

8. Filter through nitrocellulose that has been completely prewet in water and then soaked in 1 M ammonium acetate. (See description of nitrocellulose under materials.)

9. Air dry the filter and bake it for 1–2 hr at 80° *in vacuo.*

Recipe 2: Direct DNA Immobilization in NaI

Bresser et al.[4] degraded proteins with a proteinase K treatment similar to that used above; however, the product was not subsequently extracted.

---

[3] F. C. Kafatos, C. W. Jones, and A. Efstratiadis, *Nucleic Acids Res.* **7,** 1541 (1979).
[4] J. Bresser, J. Doering, and D. Gillespie, *DNA* **2,** 243 (1983).

Binding to nitrocellulose of peptides produced by protease K was inhibited under the ionic conditions used to bind DNA; namely a solution of concentrated NaI. What has been referred to as "supersaturated NaI" and "saturated NaI" are called here 10 M and 5 M NaI, respectively. In practice, NaI concentrations above 80% of saturation at 25° maximally provide nucleic acid immobilization on nitrocellulose (J. Bresser, unpublished results). DNA was denatured by heating to 90–100° in concentrated NaI which simultaneously degraded any RNA. Reannealing was prevented by filtering the resulting solution while still hot (have the boiling water bath next to the blotting apparatus). This procedure bound the nucleic acid "irreversibly" so that no baking was required. The recipe for direct DNA immobilization in NaI is as follows.

1a. *Starting with bacterial cell pellet.* Resuspend cells to about 10^6–10^7 cells/ml of 10 mM Tris–HCl (pH 7.4) with 2 mg/ml lysozyme (fresh), incubate 10 min at room temperature (20–25°), and then proceed to step 2.

1b. *Starting with eukaryotic cell pellet.* Resuspend cells to about 10^6–10^7 cells/ml of 10 mM Tris–HCl (pH 7.4).

2. Add 1.0 volume of fresh proteinase K stock solution (see recipe 1) and incubate for 20 min at 37°.

3. Freeze at −70° in dry ice–ethanol bath and thaw in 37° water bath; repeat freeze–thaw cycle two more times and proceed.

4. *Starting from purified DNA.* Add 1.0 volume of hot 10 M NaI (75 g NaI brought up to 50 ml with water in a 50-ml screw-capped tube and dissolved in a boiling water bath with intermittent but vigorous agitation).

5. Serially dilute (see recipe 1) into 5 M NaI (made by 1 : 1 dilution of 10 M NaI into water) and incubate for 10 min at 90–100°.

6. Filter while hot onto nitrocellulose that has been completely prewet in water and then soaked in 6× SSC. (See Materials for description of nitrocellulose).

7. Wash filter with three changes of 70% ethanol at room temperature for 5 min each change.

8. Wash filter for 10 min at room temperature in fresh acetic anhydride solution (0.25 ml of acetic anhydride into 100 ml 0.1 M triethanolamine in water, prepared just before use.) This wash is recommended for crude DNA and is optional when purified DNA is used.

9. Air dry.

Porteus[5] described a modification of the NaI technique yielding increased hybridization signals, at least with large numbers of cells, by including SDS, sonication, and phenol extraction in the procedure.

[5] D. J. Porteus, *Somatic Cell Mol. Genet.* **11,** 445 (1985).

RNA Immobilization

RNA immobilization can be confounded by RNase contamination. Treatment of all pipet tips, tubes, and solutions with diethyl pyrocarbonate followed by autoclaving is recommended to reduce ubiquitous RNase contamination. Vanadyl–ribonucleoside complexes should be added whenever possible as an RNase inhibitor (see this volume [2]). In our opinion, cycloheximide should be used during cell harvesting and cycloheximide plus vanadyl–ribonucleoside complexes during cell lysis. DNA and protein coimmobilization must also be avoided as before.

Recipe 3: Cytodot Technique

The White and Bancroft "cytodot" technique[6] minimizes the DNA coimmobilization problem by separating the nucleus and cytoplasm and spotting the cytoplasmic fraction. Protein coimmobilization can be reduced by phenol–chloroform extractions.[1] RNA is denatured in 7.4% formaldehyde and immobilized in 1–3 M NaCl as follows.

1. *Starting with cell pellet.* Resuspend cells to about 10^7–10^8 cells/ml of ice-cold 10 mM Tris–HCl (pH 7.0), 1 mM EDTA, 10 mM vanadyl-ribonucleoside complexes.

2. Add 0.1 volume of 5% Nonidet P-40 and incubate 5 min on ice; repeat this step once.

3. Centrifuge at 15,000 g for 2.5 min to pellet nuclei.

4. Extract supernatant with 1.0 volume of buffered phenol–chloroform (1 : 1, v/v, buffered with 10 mM Tris–HCl at pH 7.0, 1 mM EDTA, 100 mM NaCl).

[*Editors' Note.* See this volume [4] for the use of a buffer at pH 5 to extract RNA preferentially; contaminating DNA is avoided.]

5. Extract aqueous supernatant with 1.0 volume of chloroform–isoamyl alcohol (24 : 1 v/v) and proceed.

6. *Starting with purified RNA.* Transfer cytoplasmic fraction or aqueous supernatant from chloroform extraction to tube with 0.3 volume of 20× SSC (6× final concentration) and 0.2 volume 37% formaldehyde (7.4% final concentration), and incubate for 15 min at 60° (can be stored at −70°).

7. Serially dilute (see recipe 1) into 15× SSC.

8. Filter through nitrocellulose that has been completely prewet in water then soaked in 6× SSC. (See Materials for description of nitrocellulose.)

9. Air dry filter and bake it for 1–2 hr at 80° *in vacuo.*

[6] B. A. White and F. C. Bancroft, *J. Biol. Chem.* **257**, 8569 (1982).

Steps 4 and 5 were introduced by Meinkoth and Wahl.[1] Other variations on the White and Bancroft procedure include that of Catanzaro et al.[7] who started with solid tissue which was homogenized in 10 volumes of 4 M guanidine thiocyanate, 5 mM sodium citrate (pH 7.4), 100 mM 2-mercaptoethanol, 0.5% SD sarc. The resulting homogenate was put into the above recipe at step 6. Cheley and Anderson[8] resuspended the cell pellet at 10^6 cells/ml of 7.6 M guanidine hydrochloride, 0.1 M potassium acetate (pH 5.0), homogenized the slurry with five passes through a 21-gauge needle, precipitated the nucleic acids in 0.6 volumes of 95% ethanol at −20° for 12 hr, pelleted at 5000 g for 20 min, and dissolved the pellet in 7.4% formaldehyde, 10× SSC as in step 6. And Thomas[9] used 1 M glyoxal, 50% (v/v) dimethyl sulfoxide, 10 mM sodium phosphate (pH 7.0) for 1 hr at 50° in place of 7.4% formaldehyde to denature the RNA.

Recipe 4: Quick Blot Procedure

Bresser et al.[4] again used concentrated NaI along with the detergents Brij-35 and sodium deoxycholate (NaDOC) to inhibit protein coimmobilization. The detergents as well as the absence of a denaturation step minimize the DNA coimmobilization. This "quick-blot" procedure is not efficient for ribosomal and transfer RNA. Again, no baking step is required in the following recipe.

1. *Starting with cell pellet.* Resuspend cells to 10^6–10^7 cells/ml of Hanks' balanced salt solution (Gibco) with 50 μg/ml cycloheximide and 10 mM vanadyl–ribonucleoside complexes.

2. Add 1.0 volume of fresh proteinase K stock solution (see recipe 1) and incubate for 30 min at 37°.

3. Add 0.05 volumes of 10% Brij 35 (0.5% final concentration) and mix.

4. Add 0.05 volume of 10% NaDOC (0.5% final concentration), mix, and proceed.

5. *Starting with purified mRNA.* Add 1.0 volume of hot 10 M NaI (fresh, see recipe 2).

6. Serially dilute (see recipe 1) into 5 M NaI containing 0.5% Brij 35.

7. Filter through nitrocellulose that has been completely prewet in water and then soaked in 6× SSC. (See Materials for description of nitrocellulose.)

8. Wash filter with three changes of 70% ethanol at room temperature for 5 min each change.

[7] D. F. Catanzaro, N. Mesterovic, and B. J. Morris, *Endocrinology (Baltimore)* **117,** 872 (1985).

[8] S. Cheley and R. Anderson, *Anal. Biochem.* **137,** 15 (1984).

[9] P. S. Thomas, *Proc. Natl. Acad. Sci. U.S.A.* **77,** 5201 (1980).

9. Wash filter for 10 min at room temperature in fresh acetic anhydride solution (0.25 ml of acetic anhydride into 100 ml of 0.1 M triethanolamine.) This wash is optional for purified mRNA.

10. Air dry.

Several investigators add Brij and DOC prior to protease K digestion. Aitken et al.[10] stored samples at −70° after step 4 and before step 5. They also reported a need to soak the nitrocellulose membrane for an extended period (24 hr) in 6× SSC. Kothary et al.[11] used 0.5% SDS during deproteinization with protease K. Klug et al.[12] denatured purified RNA for 5 min at 65° in 50% formamide, 5% formaldehyde, and MOPS buffer (20 mM MOPS, 5 mM sodium acetate, 1 mM EDTA at pH 7.0) prior to the addition of 10 M NaI (step 5).

Conclusion

We have presented the four most widely used procedures for immobilizing nucleic acids onto nitrocellulose. Each has its own merits and demerits but all have been shown to give at least relatively quantitative results. The judicious use of control samples, control probes, and serial dilutions are critical for accurate quantitations. Using the standard hybridization conditions one should expect to detect from a spot of 10^5 cells unique sequence genes using recipe 1 or 2 and mRNA with an abundance of 0.01 to 0.10% of total poly(A)$^+$ RNA using recipe 3 or 4. RNA blots are the most difficult to master. Catanzaro et al.[7] and Junker and Pederson[13] compared the cytodot and quick blot techniques for RNA immobilization and found them to be equivalent. Efficiency of RNA immobilization varies widely from one investigator to the next, but results have been internally comparable. RNase-free conditions must be maintained. Labeled RNA can be used to monitor immobilization efficiency. Hybridization conditions for dot blot filters are identical to conditions for Southern and Northern blots (see this volume [43, 61]. Hybrid detection is done by autoradiography (see this volume [7]) or the dots can easily be cut and counted by liquid scintillation.

Acknowledgments

Support from grants GM27264 and CA29545-05 is acknowledged. We would also like to thank V. Brown for her help in preparing this chapter.

[10] S. C. Aitken, M. E. Lippman, A. Kasid, and D. R. Schoenberg, *Cancer Res.* **45**, 2608 (1985).
[11] R. K. Kothary, E. A. Burgess, and E. P. M. Candido, *Biochim. Biophys. Acta* **783**, 137 (1984).
[12] G. Klug, N. Kaufmann, and G. Drews, *Proc. Natl. Acad. Sci. U.S.A.* **82**, 6485 (1985).
[13] S. Junker and S. Pederson, *Exp. Cell Res.* **158**, 349 (1985).

[63] Identification of Genomic Sequences Corresponding to cDNA Clones

By Nikolaus A. Spoerel and Fotis C. Kafatos

Hybridization Stringency and the Effect of Probe Length

The general methods applicable to the isolation of genomic sequences from phage λ or cosmid libraries have been described in this volume [15–17, and 18]. Here we want to discuss strategies for the investigation of genes that occur in several identical or nonidentical copies per genome, or that share a common conserved domain with other genes. The methods discussed are applicable both to the identification of the genes in Southern blots and to their isolation from libraries. Furthermore, the methods are well suited for the analysis of homologous genes in different species.

A high proportion of genes in eukaryotes are known to be members of multigene families. To name just a few, well-characterized families include globins,[1,2] immunoglobulins,[3,4] histones,[5,6] chorion proteins,[7] actins, and tubulins.[8,9] In addition, recent research has demonstrated that genes which overall are distantly related or even unrelated may share a common domain, detectable usually only on the basis of protein sequence homology but in some cases also by cross-hybridization. A prominent example are the genes encoding the homeo-box domain.[10,11]

Carefully controlled hybridization conditions and well-tailored probes are powerful tools in the isolation and analysis of genes which share a common domain or are members of multigene families. This chapter consists of a short review of recommended strategies and relevant parameters, which have been discussed in more detail earlier.[12,13] Using three

[1] F. S. Collins and S. M. Weissman, *Prog. Nucleic Acid Res. Mol. Biol.* **31,** 315 (1984).
[2] N. Proudfoot, *Nature (London)* **321,** 730 (1986).
[3] L. Hood, M. Kronenberg, and T. Hunkapiller, *Cell (Cambridge, Mass.)* **40,** 225 (1985).
[4] T. Hunkapiller and L. Hood, *Nature (London)* **323,** 15 (1986).
[5] C. C. Hentschel and M. L. Birnstiel, *Cell (Cambridge, Mass.)* **25,** 301 (1981).
[6] R. W. Old and H. R. Woodland, *Cell (Cambridge, Mass.)* **38,** 624 (1984).
[7] M. R. Goldsmith and F. C. Kafatos, *Annu. Rev. Genet.* **18,** 443 (1984).
[8] R. A. Firtel, *Cell (Cambridge, Mass.)* **24,** 6 (1981).
[9] M. E. Buckingham, *Essay Biochem.* **20,** 77 (1985).
[10] W. McGinnis, M. S. Levine, E. Hafen, A. Kuroiwa, and W. J. Gehring, *Nature (London)* **308,** 428 (1984).
[11] M. P. Scott and A. J. Weiner, *Proc. Natl. Acad. Sci. U.S.A.* **81,** 4115 (1984).
[12] G. A. Beltz, K. A. Jacobs, T. H. Eickbush, P. T. Cherbas, and F. C. Kafatos, this series, Vol. 100, p. 266.

examples from our analysis of the silkmoth chorion locus, we demonstrate how powerful carefully tailored short single-stranded probes can be in the analysis of closely related gene copies.

The course of any hybridization reaction is determined by the concentration of the reacting species and by the second-order rate constant k, which is a function of the sequence complexity. The stability of the resulting duplex is measured by the melting temperature T_m. The effects of various parameters and reaction conditions (e.g., base composition, salt concentration, denaturants) have been investigated. They generally affect T_m and k in the same direction (see this volume [43]). Reaction conditions have therefore been summarized in terms of the "criterion," which is equal to T_m minus T, where T is the temperature of the reaction. In general, k reaches a maximum at a criterion of 20–25° below T_m and becomes severely depressed at a criterion of <5°.[14] Although frequently it is preferable to determine T_m empirically, especially for short probes (see legend to Fig. 1), a good starting formula for calculating the expected T_m is

$$T_m = 81.5 + 0.41(G + C) + 16.6 \log[Na^+]$$
$$- 0.63(\% \text{ formamide}) - \frac{300 + 2000[Na^+]}{d}$$

where $G + C$ is the percentage of guanine and cytosine; $[Na^+]$, molarity of Na^+ or equivalent monovalent cation; and d, length of the hybridized duplex in nucleotides.[15]

Because of the cooperative nature of base pair formation, as well as the nonuniform base composition of real sequences, denaturation of DNA fragments usually does not proceed in a single step, but involves successive melting of specific regions or "melting domains" within the fragment. Melting domains, which can be visualized by electron microscopy,[16] range in length from less than 100 bp to about 2 kb. A DNA fragment of around 400 bp usually has more than one melting domain. At appropriately stringent criteria, certain domains will be double stranded if the strands are perfectly matched, but will be melted if the strands have mismatches.

Hybridization of two single DNA chains starts with nucleation of a short double-helical stretch and proceeds by a zipper-like process.[17] The former process is slow so that the number of possible nucleation sites at a

[13] J. M. Meinkoth and G. Wahl, *Anal. Biochem.* **138**, 267 (1984).
[14] T. I. Bonner, D. J. Brenner, B. R. Neufeld, and R. J. Britten, *J. Mol. Biol.* **81**, 123 (1973).
[15] E. T. Bolton and B. J. McCarthy, *Proc. Natl. Acad. Sci. U.S.A.* **48**, 1390 (1962).
[16] R. B. Inman, *J. Mol. Biol.* **18**, 464 (1966).
[17] J. G. Wetmur and N. Davidson, *J. Mol. Biol.* **31**, 349 (1968).

given hybridization temperature determines the rate of the reaction. The number of possible nucleation sites is directly related to the number and length of melting domains which can remain double stranded at a given hybridization temperature. At most criteria, the number of possible nucleation sites and therefore the kinetics of the reaction are affected by the degree of mismatching.

In general, discrimination between hybrids with different degrees of homology can be achieved either kinetically or by the degree of hybridization at the completion of the reaction. Whenever feasible, it is preferable to use an excess of filter-bound DNA, so as to take advantage of the kinetic differences and enhance discrimination through competition for a limited amount of probe; in that situation discrimination is effective both during the course of the reaction and at saturation.

Frequently however (e.g., for unique or low-copy genes in genomic Southern blots [61]), the probe is in excess. In that situation, discrimination between well-matched and mismatched sequences is not as effective with long probes as with short probes. Long probes typically have many melting domains. If in the mismatched hybrids even only one of these domains is stable at the criterion used, at the completion of the reaction the long probe will be bound to both well-matched and mismatched strands, and discrimination will not be achieved. To achieve effective discrimination, it would be necessary to exploit the kinetic difference; i.e., hybridization should ideally be done for very short times. In practice, however, short hybridization times may not generate a sufficient autoradiographic signal.

With a short DNA probe encompassing only one melting domain, the situation is entirely different. If a high enough criterion is chosen, tailored to the possible mismatching within the probe sequence, discrimination is achieved both during and after completion of the reaction. Therefore, small probes, ideally single stranded (to avoid probe exhaustion by reannealing), can achieve discrimination among related sequences if hybridizations are performed for extended times at several different criteria.

Detection and Recovery of the Members of a Multigene Family

If a probe such as a cDNA already exists, it can be used in genomic Southern experiments to test whether the pertinent gene is a member of a gene family. Digestion of genomic DNA with a rare-cutting restriction enzyme and hybridization with a long probe under appropriately permissive conditions are usually satisfactory for initial experiments, although a more sophisticated design is necessary for a reasonable estimate of gene numbers (see below). When the existence of a gene family is known or suspected, as complete a collection of the family members as possible can be recovered by screening the appropriate library at a very permissive criterion. That collection (a sublibrary) can then be screened repeatedly under more stringent, carefully controlled conditions, to recover members with various degrees of homology to the probe. Under conditions of probe excess, as in Northern blots, genomic Southern blots, and the screening of libraries, discrimination between strongly and weakly homologous sequences generally declines (and the amount of cross-hybridization increases) as the hybridization time proceeds. However, in this case recovery of even distantly related sequences is in fact desirable, and therefore long hybridization times and long probes can be used. Although double-stranded probes can reanneal to themselves under these conditions, single-stranded probes (RNA or DNA) cannot self-anneal and consequently are exceptionally well suited to recovering even distantly homologous family members.

Ideally, a criterion of 40–50° below T_m would be desirable, when very distant homologs are to be recovered. To avoid unnecessary background, the criterion should not be relaxed more than necessary for the particular family, and a criterion of 30–35° below T_m is usually appropriate. Screening under more relaxed conditions is considerably facilitated when very short probes encompassing not more than one melting domain (usually less than 200 bp) are used (see also below).

Sublibraries of bacterial plasmid-containing cultures or of bacteriophage λ lysates can be conveniently arranged in microtiter plates. Rescreening is done using replicas of these microtiter plates, transferred to nitrocellulose filters with a replica plater.

Discrimination between Members of a Multigene Family

Considering the theoretical implications discussed above, two different strategies can be used to discriminate between individual members of sublibraries by plaque or colony hybridization under stringent conditions.

Long Probes. Long probes potentially contain a high number of melting domains. As discussed above, it is necessary to depend on kinetics to achieve discrimination. Therefore, a stringent hybridization and a stringent wash are not entirely equivalent. We find that a stringent hybridization is more discriminatory than an equally stringent wash. However, as conditions become more stringent, the well-matched domain of highest stability tends to dominate the kinetics of the reaction. Effectively, different degrees of homology to inherently less stable melting domains (e.g., A+T rich) become immaterial, with consequent loss of information. Stringent hybridizations with long probes can achieve finely graded discrimination between related sequences only on the basis of their most stable domains, and even then success depends critically on exact reproducibility in the concentration of target DNA, which should ideally be in excess.

Colony or plaque lifts with their inherent irreproducibility can only be used in rescreening to focus attention on a limited number of clones, which can then be subjected to more refined analysis by dot blot hybridization (this volume [62]). In dot blots, purified DNA is immobilized on filters in precisely equal amounts and in dots of uniform diameter, resulting in far greater reproducibility. In addition, conditions of filter-bound DNA excess that favor discrimination of related sequences can be easily achieved. For maximum discrimination, temperatures as high as 5° below the T_m of the relevant cross-hybrids can be used without difficulty.

An additional limitation of long probes is that discrimination is difficult between fragments of overall low homology and fragments that are highly homologous in a single short domain. In conclusion, long probes are only useful for initial characterization of a sublibrary, especially if no sequence information is available. More detailed comparisons should be done using carefully tailored short probes.

Short Probes (20–200 bp). Short probes offer major advantages, as already discussed. Furthermore, with short probes the analysis of a sublibrary can be focused on specific regions that might differ in functional importance. If probes derived from cDNA clones are to be used, care should be taken that the region probed is not interrupted by introns in the corresponding genomic sequence. The probes should ideally be single stranded. They can be DNA probes, either derived from subclones in M13 vectors or end-labeled oligonucleotides. Alternatively, RNA probes synthesized from bacteriophage promoters can be used.[18] In this case, hybridization reactions usually depend on a single melting domain, and thus it is reasonably easy to determine the desired conditions of stringency, for either detecting all homologs or for discriminating even among extremely similar sequences. Usually, a similar level of discrimination can be achieved by varying the temperature of hybridization or the temperature of the wash.

Under optimal conditions, a single mismatch can be detected by hybridization with a small probe (less than 50 bp). This is demonstrated in Fig. 1, which shows an example of discrimination by Southern analysis, from our work on the chorion gene families of silkmoths, named A, B, HcA, and HcB.[7] In this experiment, filter-bound DNAs were derived from six overlapping genomic clones from a region of the giant chorion

[18] G. M. Wahl, J. L. Meinkoth, and A. R. Kimmel, this volume [61].

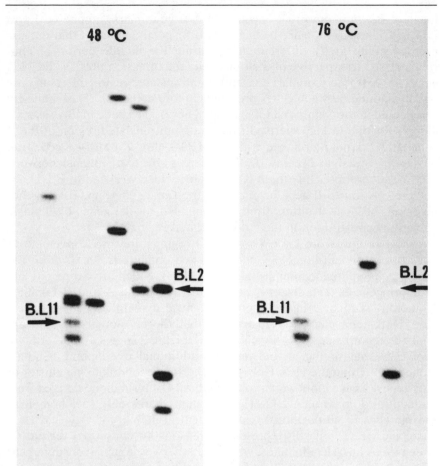

FIG. 1. Discrimination between related genes within a multigene family based on single nucleotide mismatches. Aliquots of approximately 200 ng of DNA from genomic clones B1 through B6 containing 14 A/B chorion gene pairs[19,20] were digested with suitable combinations of restriction enzymes, separated on an agarose gel, and transferred to zetaprobe membranes. The probe was derived from the 5' untranslated region of B.L11 (from −15 to +31) and subcloned into M13mp19. Labeled single-stranded probe was isolated on denaturing polyacrylamide gels and hybridized to the filter at 48° (30° below T_m) in 0.6 M Na⁺. The left panel shows this filter after washing at the same temperature in 0.3 M Na⁺, whereas the right panel shows the same filter after a 15-min wash at 76° (2° below T_m). This wash is able to discriminate on the basis of single nucleotide mismatches, e.g., against gene B.L2 (arrow) which differs from B.L11 by a single nucleotide change in the region probed. The melting T_m of the probe was determined by hybridization to multiple dot blots carrying immobilized B.L11 DNA at 45° in 0.6 M Na⁺ and washes at successively higher temperatures in 0.3 M Na⁺.

locus of *Bombyx mori* that encodes 14 A/B gene pairs, numbered L1 to L14.[19,20] These gene pairs belong to two respective classes that are expressed at distinctly different times during the middle period of choriogenesis.[21] Irrespective of their differential expression profiles, all 14 B genes are extremely similar; e.g., the 5′ untranslated region differs in only one nucleotide between genes prototypical for each class. This similarity is reflected in the Southern blot presented here. The probe in this case is a 48-bp subclone in M13, derived from the 5′ untranslated region of the B gene B.L12. Under a relaxed criterion of 30° below T_m, all 14 B genes are detected. A wash at 2° below T_m, however, is able to distinguish between the two subgroups. Only three B genes most closely related to B.L11 are detected. A fourth B gene belonging to the same subgroup, B.L2, is not detected, because it differs from the other B.L11-like genes by a single nucleotide substitution in the 5′ untranslated region.

Short probes from known conserved regions are also convenient to detect very distant homologs under relaxed conditions. In *B. mori,* the "high-cysteine" gene families (HcA and HcB), which are expressed late in choriogenesis, are distantly related to the middle A and B families. Homology is highest in the interior ("central domain") of these genes. HcA/HcB gene pairs are clustered within the chorion locus, and are flanked on both sides by A/B gene pairs.[19] Figure 2 shows a Southern blot similar to that of Fig. 1, but includes additional overlapping genomic clones containing the HcA/HcB genes and the second flanking cluster of A/B genes. These blots were probed with a 141-bp subclone derived from the central domain of A.L12. This fragment shares only 65% homology (mismatches in 49 positions) with the central domain of a typical HcA gene, HcA.12,[22] but still represents a reasonably conserved region between the A and HcA families, which are otherwise extensively diverged. Figure 2 demonstrates that cross-hybridization is easily detectable at a criterion of 30° below T_m. It is important to recognize that if the proteins encoded by two genes share an extensively conserved domain (20 or more amino acids), the genes can cross-hybridize with a probe directed toward that domain even if the silent sites are totally diverged.

Estimating Gene Copy Number

Finally, short probes are extremely useful for the quantification of gene copies in an organism. First, introns and restriction sites are not

[19] T. H. Eickbush and F. C. Kafatos, *Cell (Cambridge, Mass.)* **29,** 633 (1982).
[20] N. Spoerel, T. H. Eickbush, and F. C. Kafatos, in preparation.
[21] N. Spoerel, H. T. Nguyen, and F. C. Kafatos, *J. Mol. Biol.* **190,** 23 (1986).
[22] K. Iatrou, S. G. Tsitilou, and F. C. Kafatos, *Proc. Natl. Acad. Sci. U.S.A.* **81,** 4452 (1984).

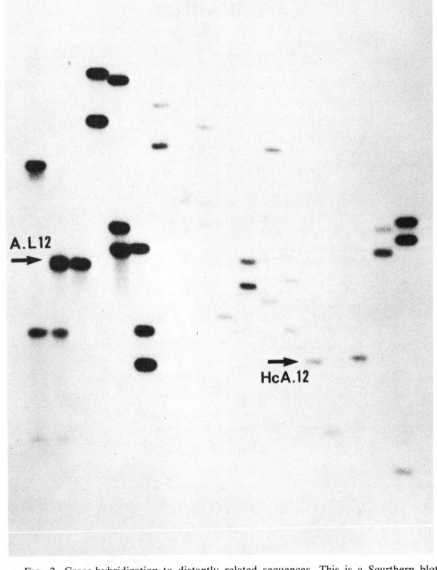

FIG. 2. Cross-hybridization to distantly related sequences. This is a Southern blot similar to the one shown in Fig. 1, but containing DNA from genomic clones B1 through B18,[19] which span the entire walk of the chorion locus. This blot was hybridized with an M13-derived probe specific for the central region of A.L12[20] (arrow) at 30° below T_m (65°) in 0.6 M Na⁺ and washed at the same temperature in 0.3 M Na⁺. The high-cysteine A genes, which are distantly related to A genes and clustered in the middle of the chorion locus, can be easily detected, including HcA.12 (35% sequence divergence).

A/B.L11 A/B.L12

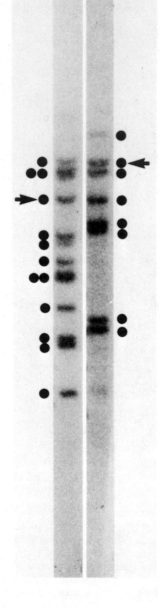

likely to fall within the small region to be probed, and thus usually only a single band is detected per gene copy. Second, under sufficiently permissive conditions with an excess of single-stranded probe and relatively long hybridization times, the intensities of self-hybrids and cross-hybrids are almost indistinguishable. The experiment shown in Fig. 3 demonstrates these aspects. It shows a Southern blot of genomic DNA from *B. mori,* hybridized with two different probes encompassing the 5′ flanking regions of A/B.L11 (231 bp) and A/B.L12 (247 bp).[21] At the criterion chosen (30° below T_m), the number of bands and therefore the number of (nonidentical) genes in each class can be quite easily quantified. In this case the gene counting has been confirmed by recovering from genomic libraries all 8 gene pairs homologous to A/B.L12 (up to 6% sequence divergence in the 5′ flanking region) and 9 of the 13 gene pairs homologous to A/B.L11 (up to 10% sequence divergence in the 5′ flanking region). One additional gene pair not detectable in Fig. 3 differs from A/B.L11 by 25% sequence divergence.[23]

Acknowledgments

We thank H. T. Nguyen for permission to quote unpublished results, and E. Fenerjian for secretarial assistance. This work was supported by grants from NIH and NSF to F.C.K. N.S. was a fellow of the Charles A. King Trust.

[23] H. T. Nguyen, N. Spoerel, and F. C. Kafatos, in preparation.

FIG. 3. Gene counting of two classes of A/B chorion gene pairs represented by A/B.L11 and A/B.L12. Genomic DNA from *B. mori* strain 703 was digested with *Hind*III and *Pst*I, fractionated on 0.7% agarose, transferred to zetaprobe membranes, and hybridized with single-stranded probes specific for the 5′ flanking regions of A/B.L11 or A/B.L12 (arrows).[21] Hybridization was at 30° below T_m in 0.6 M Na⁺, washes at the same temperature in 0.3 M Na⁺. Dots indicate the number of bands (and genes) detected with each probe. Most of these genes were cloned and sequenced (see text) and display sequence divergence of up to 10% (A/B.L11) or 5% (A/B.L12), respectively.

[64] Isolation of Full-Length Genes: Walking the Chromosome

By Nikolaus A. Spoerel and Fotis C. Kafatos

Genomic walking was pioneered in *Drosophila melanogaster* by W. Bender, P. Spierer, and D. S. Hogness.[1] Walking is the systematic isolation of DNA progressively farther away from a starting point. Genomic libraries are screened with a fragment of unique sequence from the end of a starting recombinant clone, leading to the isolation of recombinants containing, in addition, the adjacent sequences (a "walking step"). The walk proceeds as a linear sequence of such "steps," each using as probe a DNA fragment as far away as possible from the probe used in the previous step.

On a small scale, this technique can be used to expand a clone containing a fragment of a gene to a region containing the entire gene. On a larger scale, it is possible to clone arrays or clusters of genes of interest. Finally, it is possible to undertake a long walk, beginning from the nearest cloned locus and proceeding to a desired, unrelated gene, tens or hundreds of kilobases away. Such a strategy has been used extensively in *Drosophila*.

On the outset, some information about the starting clone used for the walk and about the gene to be recovered should be available. Most importantly, is it a single copy gene or a member of a multigene family? Are mutants or nearby genetic markers available? How large is the distance between starting clone and target? Walking will necessarily extend in both directions from the starting point, unless a particular direction can be distinguished, e.g., by using genetically marked chromosomal rearrangements.

In general, walking is considerably easier in *Drosophila* then in most other organisms. A major advantage is the existence of polytene chromosomes, which permit *in situ* hybridization[2,3] and convenient monitoring of the progress of the walk. A second advantage is the wealth of genetic and cytogenetic information which can be brought to bear. For example, even if no cloned probe is available near the target site, one can usually "jump" into the vicinity of it taking advantage of translocations, inversions, or deficiencies with one end near the target site and the other near a

[1] W. Bender, P. Spierer, and D. S. Hogness, *J. Mol. Biol.* **168,** 17 (1983).
[2] M. L. Pardue and J. G. Gall, *Methods Cell Biol.* **10,** 1 (1975).
[3] M. Levine, A. Garen, and B. Miller, *Proc. Natl. Acad. Sci. U.S.A* **80,** 2107 (1983).

cloned sequence. Finally, the sequence organization of the *Drosophila* genome has few interspersed repetitive elements which can lead the walk astray. Repetitive transposable elements are encountered, but they tend to differ in location in different strains, and thus can be jumped over merely by temporarily switching to a library prepared from another strain.

In organisms with extensively interspersed repetitive DNA, special precautions must be taken (see below). Chromosomal breakpoints and *in situ* hybridizations are usable as coarser tools in a few other species of interest, such as human and mouse. However, the most useful genetic markers in those species are DNA fragments defining restriction fragment length polymorphisms (RFLPs)[4] in the vicinity of the target gene.

Genomic Libraries

Suitable libraries have been prepared both from phage λ vectors allowing large insert sizes (20 kb) and from cosmids (this volume [17,18]). The preparation and characterization of libraries have been described.[5,6] Successful walking depends ultimately on the quality of the library used. If there is any doubt, it may be advisable to construct new libraries prior to undertaking a walk, in order to use the most suitable and advanced vectors, to judge the completeness of the library accurately, and to avoid possible rearrangements of the inserts. This is especially important if cosmid libraries are used. Completeness of the library is obviously essential, if the walk is not to be subject to the risk of ending prematurely. Cloning artifacts such as rearrangements can also block progress of the walk, although they may be circumvented by the use of multiple isolates, in conjunction with genomic Southern blots (this volume [61]) that establish the correct organization of the DNA.

Individual cosmid walking steps can cover much longer distances along the chromosomes (up to about 40 kb) than are possible with λ (about 20 kb). Cosmid libraries are, however, generally more difficult to handle than λ libraries at all steps of cloning and analysis. The library is difficult to store and screen, and the large inserts present more problems for restriction mapping and alignment. In addition, colony hybridization is less sensitive than plaque hybridization, and due to the low copy number of cosmids in the host cell, probes labeled with high specific activity are needed.

Some sequences present in eukaryotic DNA, especially those able to

---

[4] P. Little, *Nature* (*London*) **321**, 558 (1986).

[5] A.-M. Frischauf, this volume [17].

[6] A. G. DiLella and S. L. C. Woo, this volume [18].

form extensive secondary structure, cannot be cloned in standard rec^+ hosts, but require the use of hosts mutant in the *recB, recC,* and *sbcB* genes.[7] Specialized libraries constructed in such a host strain may be useful, if conventional cloning techniques prove to be unsuccessful (see this volume [14]).

Correct Walking Procedures

Restriction fragments can be easily isolated in low melting temperature agarose and labeled to high specific activity using random oligonucleotides as primers[8,9] (see this volume [8,10]). Unless the probe contains repeated sequences (in which case it is essentially unusable), a number of strong signals should be detected upon screening a good library, proportional to the number of genomes represented in the library. Since the individual clones represent random genomic fragments, they form an overlapping set. The clones comprising a set can be aligned by simple restriction enzyme mapping. Southern blot transfer and hybridization with a few appropriate restriction fragments can confirm that the individual clones are indeed homologous to each other. Finally, a restriction fragment at the end of the extended walk can be chosen for the next step.

Restriction mapping of phage λ clones by standard techniques can be rather slow and tedious. Hybridization of partial digests with highly labeled oligonucleotides complementary to the single-stranded right or left *cos* ends (*"cos* mapping")[10] can overcome these problems. The principle of this method is to end label indirectly either the left or the right arm of a recombinant λ clone using oligonucleotides complementary to the cos^L and cos^R ends. Partial restriction digests of the λ clone are hybridized in solution to the labeled oligonucleotide probes and separated in 0.5% agarose gels. The gels are dried on DEAE paper supports and the bands are visualized by autoradiography. Recently, the construction of a cosmid vector has been reported[11] that can be efficiently packaged *in vivo.* Such "phosmids" have single-stranded *cos* ends and can therefore be rapidly mapped using the "*cos* mapping" approach. Finally, rapid verification of homology between individual clones can be achieved using "restriction site footprinting."[12]

[7] A. R. Wyman and K. F. Wertman, this volume [14].

[8] A. P. Feinberg and B. Vogelstein, *Anal. Biochem.* **132,** 6 (1983).

[9] A. P. Feinberg and B. Vogelstein, *Anal. Biochem.* **137,** 266 (1984).

[10] H. R. Rackwitz, G. Zehetner, A. M. Frischauf, and H. Lehrach, *Gene* **30,** 195 (1984).

[11] P. F. R. Little and S. H. Cross, *Proc. Natl. Acad. Sci. U.S.A.* **82,** 3159 (1985).

[12] A. Coulson, J. Sulston, S. Brenner, and J. Karn, *Proc. Natl. Acad. Sci. U.S.A.* **83,** 7821 (1986).

Polymorphisms as well as repeated sequences can lead the walk astray. As already discussed, independent procedures can be used to confirm the correct progress of the walk (*in situ* hybridization, comparison with maps derived from genomic Southern blots), or to put the walk back on course (switching libraries and strains, "jumping"). Ultimately, the success of the walk rests on the ability to identify the target gene (see below).

Repeated Sequences

If a probe contains a repeated sequence, screening of the library will lead to too many positive signals and failure of that walking step. Such failure can be prevented by prior screening of the potential probes for the absence of repeated sequences. Screening procedures have to take two classes of repetitive sequences into account.

Highly repetitive sequences can be short (<1 kb) and extremely abundant (even greater than 10^4-fold). An easy diagnostic test for the presence of highly repetitive sequences is to use labeled total genomic DNA as probe for Southern transfers of restricted λ or cosmid clones ("Reverse" genomic Southern blots). Restriction fragments of unique DNA generally fail to give detectable signals by this test in eukaryotes, and can be identified in this way. Fragments that contain repeated sequences give detectable hybridization signals, corresponding in intensity to the size and frequency of the repeat.

Moderately repetitive elements, such as multigene families or transposable elements, may not be frequent enough to be detectable in reverse Southern blots. Such repeats can be identified by direct Southern blots. When a fragment containing such a repeat is hybridized to a Southern transfer of restricted genomic DNA, a multitude of bands are detected. Frequently some bands are very intense, generated by restriction fragments internal to the repetitive element, and the majority of bands are much weaker and correspond to junction fragments extending into adjacent chromosomal DNA.

Inserted transposons or the subsequent deletion variants are often responsible for cases of heterogeneity in the library. If more than one rearrangement is present in a certain region of the chromosome, it may be impossible to align individual clones unambiguously. If this occurs, it may be possible to jump over the scrambled region using a cosmid clone containing a large insert. Except in special cases (e.g., human libraries), use of an inbred or homozygous line of organism for the construction of the library is strongly recommended, to reduce the amount of heterogeneity encountered during the walk. As in the case of *Drosophila,* multiple libraries from distinct homozygous strains might be used to bypass such regions.

Specialized Techniques

Even if useful clones to start a walk may be separated by more than 1000 kb from the target gene, the new technologies of chromosome jumping or hopping may improve the speed of walking sufficiently to tackle such a distance.[13,14] To construct jumping libraries, chromosomal DNA is very lightly digested with rare-cutting restriction enzymes (e.g., *Bam*HI) or by complete digestion with *Not*I which has an 8-bp recognition site. Extremely long fragments (200–1000 kb) are isolated using pulsed-field gradient gel electrophoresis,[15] and circularized (at concentrations <0.5 μg/ml) in the presence of a selectable marker (e.g., a suppressor tRNA gene). Upon ligation, this marker is flanked directly by DNA sequences separated by 200–1000 kb in the genome. These sequences can be recovered in a "jumping library" using the suppressor tRNA gene as a selectable marker: after a second digest with a rare-cutter restriction enzyme, fragments carrying the tRNA gene are ligated into a suitable λ vector carrying several amber mutations. The entire process is repeated using different enzymes and a second library is prepared containing overlapping large fragments. By jumping from one library to the other and back, the fragments can be linked in an orderly walk.

If chromosomal deletions of the target gene are homozygous or hemizygous viable, clones in its vicinity may also be isolated directly by substraction hybridization. Such a procedure has been used successfully in the isolation of the locus for Duchenne muscular dystrophy.[16] Sequences present in the DNA of a normal individual, but not in an individual carrying a deletion of the Duchenne muscular dystrophy (DMD) locus on the X chromosome, were isolated by differential reassociation. Exploiting a phenol-accelerated competitive reassociation reaction,[17] *Mbo*I-cleaved human DNA was hybridized to a 200-fold excess of sheared DNA from a male patient carrying a visible deletion of the DMD region. Sequences deleted in this patient could be recovered from the reassociation reaction

[13] F. S. Collins and S. M. Weissman, *Proc. Natl. Acad. Sci. U.S.A.* **81,** 6812 (1984).
[14] A. Poustka, T. M. Pohl, D. P. Barlow, A.-M. Frischauf, and H. Lehrach, *Nature (London)* **325,** 353 (1987).
[15] D. C. Schwartz and C. R. Cantor, *Cell (Cambridge, Mass.)* **37,** 67 (1984).
[16] L. M. Kunkel, A. P. Monaco, W. Middlesworth, H. D. Ochs, and S. A. Latt, *Proc. Natl. Acad. Sci. U.S.A.* **82,** 4778 (1985).
[17] D. E. Kohne, S. A. Levinson, and M. J. Byers, *Biochemistry* **16,** 5329 (1977).

by the cloning of those reassociated DNA molecules maintaining *Mbo*I termini.

An alternative strategy to systematic walking is to isolate a sublibrary of genomic clones containing the genes of interest. This approach has been extremely useful when groups of genes are clustered in the genome, as for the genes of histocompatibility antigens in mouse and human or the chorion genes in silkmoths.[18] Individual members of the sublibrary are then placed into an ordered set of overlapping clones by restriction mapping and alignment of the individual restriction maps.

On a global scale, systematic analysis of restriction maps can be used to assemble a walk covering an entire genome. This approach is currently being used to assemble a complete walk through the genome of the nematode *Caenorhabditis elegans* (8×10^7 bp),[12] and similar projects are in the planning stage for a few other organisms, including the human genome (3×10^9 bp). Individual clones are labeled at sites of a restriction enzyme specific for a 6-bp sequence ("rare-cutter"), and the labeled DNAs are then restricted with an enzyme that cuts frequently (4-bp recognition sequence). The resulting labeled fragments are accurately sized on a sequencing gel. This information (the restriction site footprint) is stored in a computer and is then used to assemble individual clones using computer alignment methods.

Identifying the Target Gene

The ultimate success of a walk depends on the identification of the target gene. This can be achieved most directly by expression of the gene from cloned sequences in transformants (this volume [71]) or in some cases in *Escherichia coli* (this volume [69]). Genetic markers, such as closely linked restriction site polymorphisms or overlapping deletions that inactivate the target gene, have been used extensively. Finally, the characterization of transcripts in the region covered by the walk can lead to identification of the target gene.

Acknowledgments

We thank E. Fenerjian for secretarial assistance. This work was supported by grants from NIH and NSF to F.C.K. N.S. was a fellow of the Charles A. King Trust.

[18] G. A. Beltz, K. A. Jacobs, T. H. Eickbush, P. T. Cherbas, and F. C. Kafatos, this series, Vol. 100, p. 266.

[65] Cosmid Vectors for Genomic Walking and Rapid Restriction Mapping

By GLEN A. EVANS and GEOFFREY M. WAHL

The structural and functional analysis of complex genomes often requires cloning and determining a physical map for large regions of chromosomal DNA. Cosmid vectors are valuable tools for this analysis because of the ability to clone in a single vector fragments of genomic DNA ranging in size from 35 to 45 kb. The isolation of sequentially overlapping cosmid clones, using strategies known as "walking," have enabled physical linkage and extensive characterization of genes of the mouse major histocompatibility complex,[1,2] the mouse *t* locus,[3] and regions of gene amplification in drug-resistant mammalian cells.[4] Similar genomic "walking" on a larger scale may soon allow the isolation and structural analysis of genes associated with human heritable diseases such as Huntington's disease[5] and cystic fibrosis.[6]

"Walking" using traditional methods (see this volume [64]) is a time-consuming procedure. We have designed new cosmid vectors to expedite many of the steps involved in genomic walking, particularly the preparation of end-specific hybridization probes and the determination of restriction endonuclease maps of the cloned DNA.[7] This series of vectors designated pWE (for "walking easily"), is based on the pJB8 cosmid vector[8] containing the ColE1 origin of replication with the addition of bacteriophage promoter sequences on either side of the cloning site. The most useful pWE vectors, pWE15 and pWE16 (Fig. 1), contain bacteriophage T3 and T7 promoters flanking a unique *Bam*HI cloning site. By using the cosmid DNA containing a genomic insert as a template for either T3 or T7 polymerase, directional "walking" probes can be synthesized and used to

[1] M. Steinmetz, D. Stephan, and K. F. Lindhal, *Cell (Cambridge, Mass.)* **44,** 895 (1986).
[2] R. A. Flavell, H. Allen, L. C. Burkle, D. H. Sherman, G. L. Waneck, and G. Widera, *Science* **233,** 437 (1986).
[3] H. S. Fox, G. R. Martin, M. R. Lyon, B. Herrman, A. M. Frischauf, H. Lehrach, and L. Silver, *Cell (Cambridge, Mass.)* **40,** 63 (1985).
[4] M. Montoya-Zavala and J. J. Hamlin, *Mol. Cell. Biol.* **5,** 619 (1985).
[5] J. F. Gusella *et al., Nature (London)* **306,** 234 (1983).
[6] B. J. Wainwright, P. J. Scambler, J. Schmidtke, E. A. Watson, H. Y. Law, M. Farrall, H. D. J. Cooke, H. Eiberg, and R. Williamson, *Nature (London)* **318,** 384 (1985).
[7] G. M. Wahl, K. A. Lewis, J. C. Ruiz, B. E. Rothenberg, J. Zhao, and G. A. Evans, *Proc. Natl. Acad. Sci. U.S.A.* in press, 1987.
[8] D. Ish-Horowicz and J. F. Burke, *Nucleic Acids Res* **9,** 2989 (1981).

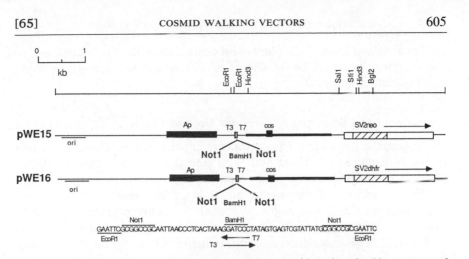

FIG. 1. The structural map of pWE15 and pWE16 cosmids and nucleotide sequence of the cloning regions. Cosmid pWE15 contains the *SV2-neo* gene and pWE16 contains the *SV2-dhfr* selectable and amplifiable gene. Ap denotes the β-lactamase; T3 and T7, the bacteriophage promoters for the synthesis of end-specific probes; *cos,* the cohesive termini from phage Charon 4A found in the cosmid pJB8; and *ori,* the plasmid origin of replication. Hatched boxes represent tn5 kanamycin/neomycin phosphotransferase (pWE15) or mouse dihydrofolate reductase cDNA sequence (pWE16), and open boxes present SV40 sequences. The heavy lines represent portions of the cosmid contributed by bacteriophage λ sequences. The T3 and T7 promoters and *Bam*HI cloning site were generated using a synthetic oligonucleotide inserted in the *Eco*RI sites of pWE2.[7] The sequence of the oligonucleotide is shown below the restriction map of the cosmid.

screen genomic cosmid libraries. Like other cosmids,[9] these vectors also contain the *SV2-neo*[10] or *SV2-dhfr*[11] genes which allow the expression, amplification, and rescue of cosmid clones in eukaryotic cells after DNA-mediated gene transfer. In addition, octameric recognition sequences for the restriction enzyme *Not*I have been placed near the *Bam*HI cloning site. Since *Not*I is infrequently represented in the mammalian genome, this enzyme will allow most genomic DNA inserts to be removed as single large restriction fragments. This feature is useful for the *in vitro* reconstruction of large genes (such as those extending over regions of DNA greater than the cloning capacity of cosmids), or the removal of vector sequences which might affect the expression of genes after microinjection into mouse embryos for the creation of transgenic mouse strains.[12] The presence of bacteriophage promoters flanking the insert DNA also allows

[9] Y. F. Lau and Y. W. Kan, *Proc. Natl. Acad. Sci. U.S.A.* **80,** 5225 (1983).

[10] P. J. Southern and P. Berg, *J. Mol. Appl. Genet.* **1,** 327 (1982).

[11] S. Subramani, R. Mulligan, and P. Berg, *Mol. Cell. Biol.* **1,** 854 (1981).

[12] R. L. Brinster, H. Y. Chen, M. E. Trumbauer, M. K. Yagle, and R. D. Palmiter, *Proc. Natl. Acad. Sci. U.S.A.* **82,** 4438 (1985).

the restriction map of the recombinant cosmid to be easily determined. In this chapter, we describe the use of pWE cosmids for genomic walking and restriction mapping.

Preparation and Screening of Genomic Cosmid Libraries

Genomic libraries in pWE vectors are constructed precisely according to the procedures of Dillela and Woo (see this volume [18]). Bacteriophage λ packaging extracts may contain significant amounts of EcoK restriction activity. To avoid the possibility that mammalian sequences containing an EcoK site might be absent or underrepresented in the library, we prepare genomic libraries using in vitro packaging extracts which lack EcoK restriction activity. (Gigapak Gold, Stratagene Cloning Systems, San Diego, CA).

Preparation of Cosmid DNA from Minilysates

Cosmid clones isolated from genomic libraries constructed in pWE vectors give a high yield of DNA when grown in bacterial strain DH5. For most purposes, including the synthesis of RNA probes for restriction analysis and genomic walking and traditional analysis of restriction fragments in agarose gels, the quantity of DNA prepared from 1.5- to 2-ml minicultures is sufficient. Though cosmid DNA can be prepared from larger cultures and purified by CsCl banding techniques (see this volume [13]), this is usually not necessary for molecular mapping and expression studies. To prepare cosmid DNA for the synthesis of end-specific probes:

1. Inoculate a 2.5-ml culture with a single colony from a fresh agar plate. LB broth [13] containing 20 μg/ml ampicillin (pWE15 or pWE16), or 25 μg/ml kanamycin sulfate (pWE15) should be used (the SV2-neo gene of pWE15 allows selection and plating of cosmid libraries using kanamycin which is more stable than ampicillin). Incubate the culture at 37° with vigorous shaking for no longer than 6–8 hr. Longer incubation periods give consistently lower yields of cosmid DNA.

2. Collect the bacterial cells from 1.5 ml of culture by centrifugation for 2 min in a microfuge. Remove the supernatant by aspiration and resuspend the cells in 300 μl of STET buffer (8% sucrose, 5% Triton X-100, 50 mM EDTA, 50 mM Tris–HCl at pH 8.0, prepared in diethyl pyrocarbonate-treated water as in chapter [2]). Save the remainder of the culture at 4° as a culture stock.

3. Add 25 μl of fresh lysozyme solution (10 mg/ml in STET buffer) to the resuspended cells and vortex vigorously. Place the microfuge tube in

boiling water for 2 min to lyse the bacteria. Allow the solution to cool for 2–5 min and collect the precipitate by centrifugation in a microfuge for 10 min. Remove the gelatinous pellet with a sterile toothpick and discard it.

4. Add 325 μl of 2-propanol to the cleared lysate, mix by vortexing, and allow the nucleic acid precipitate to form at room temperature for 5 min. Collect the precipitated nucleic acid by centrifugation in the microfuge for 10 min. Remove the alcohol by aspiration, air dry the pellet, and resuspend the DNA in 25 μl of sterile diethyl pyrocarbonate (DEPC)-treated water. An amount of 2 to 4 μl of this DNA solution is usually sufficient for restriction endonuclease analysis.

5. If the DNA preparation is to be used as a template for bacteriophage polymerases, it must be treated to remove contaminating ribonucleases. Extract the DNA once with phenol–chloroform–isoamyl alcohol (25 : 24 : 1, saturated with 50 mM Tris–HCl at pH 8.0), and once with chloroform. Adjust the aqueous phase to 0.4 M sodium acetate (pH 5.5) and precipitate the DNA with ethanol. Dissolve the precipitated DNA in DEPC-treated sterile water at a concentration of 1 mg/ml and store at −20°.

Synthesis of RNA Walking Probes

To decrease the probability that repetitive sequences might be present in the walking probe, the cosmid DNA is digested with a restriction enzyme that does not cut within the promoter sequences but generates relatively small restriction fragments. HaeIII and RsaI are useful restriction enzymes for this purpose and usually generate templates of 50–1000 nucleotides in length.

1. Digest the cosmid DNA with HaeIII or RsaI according to the specifications recommended by the manufacturers.

2. RNA probes are synthesized from this template DNA as follows. Mix 4 μl 5× reaction buffer (180 mM Tris–HCl at pH 7.5, 30 mM MgCl$_2$, 10 mM spermidine, prepared in DEPC-treated water); 1 μl RNasin (27 units/μl); 1 μl 0.2 M dithiothreitol; 2.5 μl of 2.5 mM GTP, 2.5 mM ATP, 2.5 mM CTP; 1.2 μl of 100 μM UTP; and 100 μCi [α-^{32}P]UTP (800 Ci/mmol). Add 1–2 μg of digested template DNA, 10–20 units of T7 or T3 RNA polymerase and DEPC-treated water to bring the final volume to 20 μl. Incubate at 37° for 1–2 hr and terminate the reaction by phenol–chloroform–isoamyl alcohol extraction, chloroform extraction, and ethanol precipitation. An amount of 1–2 μg of template cosmid DNA typically yields 1–10 × 10^7 cpm of hybridization probe. The integrity of the probe and efficiency of synthesis should be determined by analyzing an aliquot

of the probe on a 5% polyacrylamide–6 M urea sequencing gel [56,57]. It is not usually necessary to remove the template DNA before using the probe for library screening.

3. Genomic cosmid libraries are plated from nonsimplified stocks stored at $-70°$ at a density of 350 colonies/cm^2 on 24 × 24 cm square culture dishes containing LB agar with 25 μg/ml kanamycin sulfate (pWE15) or 25 μg/ml ampicillin (pWE16). Procedures for screening cosmid libraries and the preparation of replica filters for hybridization have been described in this volume [18,44,45].

4. Prehybridize in 0.25 M sodium phosphate buffer (pH 7.2), 0.25 M NaCl, 1% sodium dodecyl sulfate (SDS), 50% formamide, 2× Denhardt's solution (see this volume [45]), 25 μg/ml denatured salmon sperm DNA, and 200 μg/ml yeast tRNA for 10 min at 42°. Following prehybridization, 1–5 × 10^7 cpm of end-specific RNA probe is added to the hybridization solution and incubation continued for 12 hr at 42°.

5. After hybridization, the filters are washed in 0.1× SSPE (1× SSPE: 0.18 M NaCl, 10 mM NaH$_2$PO$_4$ at pH 7.4, 1 mM EDTA), 0.1% SDS at 65° degrees with three or four changes of washing solution over 2–4 hr. The filters are then dried and exposed to X-ray film. Positive clones are visible after 6–8 hr of exposure.

6. After isolation and purification of positive clones, the library filters may be reused for hybridization in the second or subsequent "steps" without further washing or removing the hybridized probe. Repetitive hybridizations with sequential walking probes will show new positive colonies representing each step of the walk.

Restriction Mapping

In addition to the synthesis of walking probes, pWE vectors allow restriction maps of the genomic DNA cloned in the cosmid to be determined by transcription from the bacteriophage promoters. Two methods have been derived for mapping insert DNA which are well suited to determining regions of overlap during genomic walking.

Method I. Analysis of RNA Transcripts from Partial Digestion Products of Recombinant Cosmid DNA

The restriction map of a cosmid clone can be determined by first creating a series of partial digestion products with one or several restriction endonucleases and then using this partially digested DNA as a template for RNA synthesis. The resulting RNA transcripts are analyzed on a denaturing agarose gel and the length of the transcript corresponds to the distance from the bacteriophage promoter to the appropriate restriction site. Since transcription from these cosmid clones may generate RNA molecules up to 20 kb, the use of T3 and T7 polymerase in separate reactions allows the map to be determined from both ends of the insert. Because *Eco*RI restriction sites are present near the riboprobe promoter in each vector, this is a convenient restriction enzyme for mapping.

1. Cosmid DNA is prepared from 1.5 ml of culture as described above. Partial restriction enzyme digestion products may be generated by a number of methods including varying the amount of restriction endonuclease and the incubation time.

2. RNA synthesis for restriction mapping is carried out by mixing at room temperature 4 μl of 5× reaction buffer, 1 μl RNasin, 1 μl of 0.2 M dithiothreitol, 1 μl of 0.4 M GTP, CTP, and ATP, 2 μl of 10 mM UTP, 10 μCi of [α-^{32}P]UTP (800 Ci/mmol), and 5 μl of yeast tRNA (10 mg/ml). Add 1–2 μg template DNA and DEPC-treated water to bring the final reaction volume to 20 μl. Add 10–20 units T7 or T3 polymerase and incubate 1 hr at 37°.

3. Add 1 μl of 10 mM UTP, 1 μl of 0.4 M GTP, CTP, and ATP, and 10–20 units of T3 or T7 polymerase and incubate an additional 30 min at 37°. Terminate the reaction by extracting once with phenol–chloroform–isoamyl alcohol, once with chloroform, and precipitating with ethanol.

4. Analyze the transcription products in a formaldehyde–agarose gel (this volume [8]), dry the gel, and visualize the transcripts by autoradiography (this volume [7]).

Method II. Restriction Mapping by Blot Hybridization with Time-Limited RNA Transcripts

In this method, cosmid DNA is digested to completion with a restriction endonuclease, and the DNA fragments separated in a 1% agarose gel and transferred to nitrocellulose according to the Southern procedure (this volume [61]). Radiolabeled RNA transcripts are then synthesized from undigested cosmid DNA and reactions terminated after varying lengths of time. These time-limited probes are hybridized to the Southern filters and the order of restriction fragments from the RNA promoter is defined by the time of appearance and intensity of hybridization to a particular fragment. This method is somewhat less precise than the previous restriction mapping method, and interpretation may be complicated by the presence of repetitive sequences within the cloned genomic DNA. However, it is very useful for simultaneously mapping several overlapping cosmid clones, in that the regions of overlap are quickly identified. Again, as transcription from T3 or T7 promoters may not generate transcripts much longer than 20 kb, mapping should be carried out from each end of the insert.

1. Prepare cosmid DNA from 1.5-ml cultures and digest with appropriate restriction endonucleases using conditions recommended by the manufacturer. Analyze the digested DNA in a 0.8% agarose gel and transfer to a nitrocellulose filter as described (this volume [61]).

2. Prepare time-limited RNA probes using the reaction conditions described for Method I. Remove aliquots at 2, 5, 10 and 20 min and terminate the reaction by phenol–chloroform–isoamyl alcohol extraction.

3. For each time point, hybridize 10^5–10^6 cpm of probe to a replica filter for 2–4 hr at 42° using the hybridization conditions described above. After washing the drying the filter, the order of appearance of hybridizing bands reflects the order of restriction fragments from the end of the insert.

Alternate Strategies for Restriction Mapping pWE Cosmids

In addition to determining restriction maps using the RNA polymerase promoters inherent in the vector, alternative strategies have been devised which avoid transcription-related problems and artifacts. Recombinant plasmids using pWE15 or pWE16 vectors (which do not contain an internal NotI site) may be digested to completion with NotI, and then used to generate a series of partial digestion fragments with another restriction enzyme. These partial digestion products are then separated on an agarose gel and blotted to a nitrocellulose filter. The filter is hybridized with small walking-type RNA probes synthesized from the T3 or T7 promoters as described above, or with ^{32}P-labeled oligonucleotides specific for the T3 or T7 promoter sequences. (These oligonucleotides are commercially available as T3 or T7 sequencing primers.) This procedure will selectively label the partial fragments terminating with T3 or T7 promoter and allow the localization of restriction sites relative to each end of the insert.

Acknowledgments

The authors wish to thank K. Lewis, B. Rothenberg, J. Zhao, J. Ruiz, A. Albi, and C. Landel for their contributions to the development of these vectors and Stratagene Cloning Systems (San Diego, California) for supplying reagents and assistance. This work was supported by NIH Grants HD18012, GM33868 (GAE) and GM22754 (GMW), and funds from the G. Harold and Leila Y. Mathers Charitable Foundation. GAE is a Pew Scholar in the Biomedical Sciences.

[66] Mapping of Gene Transcripts by Nuclease Protection Assays and cDNA Primer Extension

By Frank J. Calzone, Roy J. Britten, and Eric H. Davidson

Background Information

Introduction

An important problem often faced in the molecular characterization of genes is the precise mapping of those genomic sequences transcribed into RNA. This requires identification of the genomic site initiating gene transcription, the location of genomic sequences removed from the primary gene transcript during RNA processing, and knowledge of the sequences terminating the processed gene transcript. The objective of the protocols described here is the generation of transcription maps utilizing relatively uncharacterized gene fragments. The basic approach is hybridization of a single-stranded DNA probe with cellular RNA, followed by treatment with a single-strand-specific nuclease that does not attack DNA–RNA hybrids, in order to destroy any unreacted probe sequences. Thus the probe sequences included in the hybrid duplexes are protected from nuclease digestion. The sizes of the protected probe fragments determined by gel electrophoresis correspond to the lengths of the hybridized sequence elements. Methodology for nuclease mapping of RNA transcripts with genomic DNA probes was first introduced by Berk and Sharp.[1,2] We have adapted this technology for the purpose of obtaining transcription maps of new gene isolates that include 20–30 kb of genomic DNA, using probes as large as 2.5 kb. The information provided here will also be useful for less demanding mapping experiments. The use of RNA rather than DNA probes to map gene transcripts is presented in this volume [67].

Nuclease Selection

Three different nucleases suitable for mapping RNA transcripts by nuclease protection assays are currently available from commercial sources. The most frequently selected and least expensive is nuclease S_1[3,4] isolated from *Aspergillus oryzae*. Commercial preparations of this

[1] A. J. Berk and P. A. Sharp, *Cell (Cambridge, Mass.)* **12**, 721 (1977).
[2] A. J. Berk and P. A. Sharp, *Proc. Natl. Acad. Sci. U.S.A.* **75**, 1274 (1978).
[3] T. Ando, *Biochim. Biophys. Acta* **114**, 158 (1966).
[4] V. M. Vogt, *Eur. J. Biochem.* **33**, 192 (1973).

METHODS IN ENZYMOLOGY, VOL. 152

endonuclease display high specificity for single-stranded DNA or RNA under conditions of high ionic strength and low temperature, though they require calibration. Digestion of a hybrid molecule consisting of genomic DNA and mature mRNA such as that represented in Fig. 1a with nuclease S_1 would result in two genomic probe fragments that in denaturing gels have the lengths of the two exons. Nuclease S_1 and mung bean nuclease, an endonuclease similar to nuclease S_1,[5] can be used interchangeably in RNA transcript mapping assays, and therefore for simplicity mung bean nuclease is not here considered further.

A third nuclease which has special applications in RNA transcript mapping experiments is *Escherichia coli* exonuclease VII.[6,7] Exonuclease VII hydrolyzes single-stranded DNA from both the 5′ and 3′ directions. Unlike nuclease S_1, exonuclease VII requires free 5′ and 3′ termini to initiate digestion. Thus exonuclease VII can be used to map hybrids susceptible to nuclease S_1 cleavage due to a relatively high percentage of mismatched bases or unstable AT-rich sequences. Exonuclease VII protection assays are also useful in mapping short introns when performed in conjunction with nuclease S_1 experiments. For example the intron loop in the hybrid shown in Fig. 1a would not be hydrolyzed by exonuclease VII. After exonuclease VII digestion a probe fragment would be detected in a denaturing gel the length of which is equal to the intron plus exons 1 and 2. If the lengths of exons 1 and 2 are known from nuclease S_1 protection assays, the length of the intron may be determined by subtraction.

Single-Stranded DNA Probes

In mapping gene transcripts there are several clear advantages to using single-stranded probes for nuclease protection assays. A factor of major importance is that the hybridization reactions can be performed under well-characterized conditions, maximizing hybrid stability and the rate of duplex formation. In contrast, the overriding concern in hybridizations with double-stranded DNA probes is the suppression of probe renaturation, and this frequently results in suboptimal conditions for RNA–DNA duplex formation. Thus, hybridizations with double-stranded DNA probes are usually performed in 80% formamide at a relatively narrow range of temperatures slightly above the melting temperature (T_m) of duplex DNA.[8] Under these conditions the parameters affecting hybrid stability are not simply predictable, and preliminary experimentation is

[5] D. Kowalski, W. D. Kroeker, and M. Laskowski, Sr., *Biochemistry* **15**, 4457 (1976).
[6] J. W. Chase and C. C. Richardson, *J. Biol. Chem.* **249**, 4545 (1974).
[7] J. W. Chase and C. C. Richardson, *J. Biol. Chem.* **249**, 4553 (1974).
[8] J. Casey and N. Davidson, *Nucleic Acids Res.* **5**, 1539 (1977).

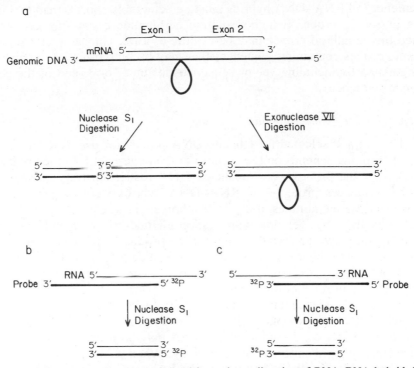

FIG. 1. Diagrams of products formed in nuclease digestion of RNA–DNA hybrid duplexes. (a) Comparison of nuclease S_1 and exonuclease VII activities. The drawing shows a hybrid duplex formed between a spliced RNA and an overlapping genomic DNA probe containing exons 1 and 2 separated by an intron. Nuclease S_1 digestion generates two probe fragments corresponding to the lengths of exons 1 and 2; exonuclease VII digestion does not degrade the intron loop in the genomic probe, resulting in a single probe fragment the length of which is the sum of the lengths of exons 1 and 2, and the intron. (b) Mapping the 5' terminus of RNA. To map the 5' terminus of an RNA directly, the probe is 5'-end labeled with ^{32}P as shown. The length of the probe fragment generated by nuclease S_1 (or exonuclease VII) treatment maps the distance between the 5' end of the probe and the 5' RNA terminus (see text). (c) Mapping the 3' terminus of RNA. The 3' terminus of an RNA can be mapped directly with 3'-end-labeled probes as shown. The strategy is analogous to the 5'-end-mapping experiments described in (b).

required to determine the best temperature of incubation.[8] In addition the rate of hybridization in 80% formamide is significantly lower (~12-fold).[8] Such complications can lead to delays while optimal conditions are worked out for each uncharacterized probe sequence.

Single-stranded probes are frequently contaminated with a small amount of complementary-strand DNA which may protect the full-length probe sequence from nuclease attack. Therefore, the length of the probe sequences in RNA–DNA hybrids must be detectably shorter than that of the untreated probe. If high-resolution acrylamide denaturing gels are used this length difference may be as little as 10%. Uncharacterized genomic probes containing sequences that may be fully transcribed can be lengthened by including vector sequences at the 5' or 3' end of the genomic sequence.

Strategies for Probe Labeling

The method selected for labeling DNA probes for use in RNA transcript mapping depends on the purpose of the experiment. In the original protocol[1,2] the probes were unlabeled DNA restriction fragments. The probe sequences protected in RNA–DNA hybrids were transferred to nitrocellulose membranes after electrophoresis in agarose gels, and detected by the DNA gel blot hybridization method. More recently,[9] most investigators have preferred to analyze the products of nuclease protection assays with high-resolution urea–acrylamide gels. For this purpose the DNA probes are labeled directly, thus avoiding transfer of probe fragments to nitrocellulose membranes. As shown in Fig. 1b, probes labeled at the 5' end (see this volume [10]) with ^{32}P can be used to locate the 5' end of RNA transcripts or 5' intron–exon junctions. Conversely, as indicated in Fig. 1c, 3'-end-labeled probes may be used to map 3' transcript termini or 3' intron–exon junctions. When a Maxam–Gilbert sequencing ladder[10] with the end-labeled probe is used for DNA size markers, the probe fragments generated in nuclease protection assays may be mapped with great accuracy. To prepare end-labeled probes for nuclease protection assays the sequences of the genomic fragment must be highly characterized since the end-labeled nucleotides must be included in the hybrid duplexes if they are to be protected at least partially from hydrolysis. This condition is not typically met when mapping the transcripts of new gene isolates.

Unlike end-labeled probes, the sequences in continuously labeled probes need not be well characterized before nuclease protection assays. Any segment of a continuously labeled probe hybridized to RNA will remain labeled after the nuclease treatment, including those which do not contain 5' or 3' probe termini. However, when the full length of the probe sequence is labeled, nuclease protection assays will only determine the overall length of the transcribed regions in each gene fragment. Additional information is usually necessary to locate the precise coordinates of the probe sequence contained in a RNA–DNA hybrid. For this purpose, we present a relatively simple method for using restriction enzymes to locate

[9] R. F. Weaver and C. Weissman, *Nucleic Acids Res.* **7**, 1175 (1979).
[10] A. M. Maxam and W. Gilbert, *Proc. Natl. Acad. Sci. U.S.A.* **74**, 560 (1977).

the transcribed sequences in uncharacterized single-stranded DNA probes. A different method for mapping the 5' end of gene transcripts with uncharacterized single-stranded DNA probes has been described by Hu and Davidson.[11]

Preliminary Characterization of the Gene

The design and interpretation of nuclease protection experiments with previously uncharacterized gene isolates require prior knowledge of the cell type or tissue distribution of the gene transcripts, the approximate transcript size, the orientation (5' → 3') of transcription, the approximate transcript prevalence, and the haploid gene number. Of primary importance is the identification of a source for the quantity of RNA necessary for transcript mapping experiments. Although not strictly quantitative, preliminary characterizations of gene transcripts by the RNA gel blot hybridization method (see this volume [61]) are usually sufficient for the initiation of nuclease protection experiments. A rough estimate of the transcript prevalence may be obtained using as standards, probes for other gene transcripts that have been titered by more quantitative methods. Any transcript that can be observed in RNA gel blot hybridizations should be easily detectable in the more sensitive nuclease protection assay. To assign the orientation of gene transcription, RNA gel blot hybridizations should be performed with probes that are specific for each strand of a gene fragment (see below, Preparation of Continuously Labeled Probe, or this volume [61]). Probes containing multicopy sequences may cross-react with divergent gene transcripts, and thus it is essential that the haploid copy number of all gene probes be determined. The haploid copy number of a probe sequence is usually inferred from a simple characterization of genomic restriction fragments[12] detected by the DNA gel blot hybridization method (see this volume [63]).

Materials and Methods

The solutions are listed in order of their appearance in the text. Hin buffer (10×) is 0.5 M NaCl, 0.1 M MgCl$_2$, 70 mM Tris–HCl (pH 7.4), 30 mM dithiothreitol (DTT). Sequencing primer (5 ng/μl M13 primer 15-mer) was obtained from New England BioLabs; however, other sequencing primers are acceptable. Labeling dNTP for the probe synthesis reaction contained 5 mM dCTP, 5 mM dGTP, 5 mM dTTP. Chase dNTP contained 10 mM dATP, 5 mM dCTP, 5 mM dGTP, 5 mM dTTP. [α-^{32}P]dATP, 10

[11] C. M. Hu and N. Davidson, *Gene* **42**, 21 (1986).
[12] J. J. Lee, R. J. Shott, S. J. Rose, III, T. L. Thomas, R. J. Britten, and E. H. Davidson, *J. Mol. Biol.* **172**, 149 (1984).

mCi/ml, ~400 Ci/mmol aqueous was obtained from Amersham or Dupont/ New England Nuclear). Denaturation buffer is 4.0 M NaOH, 0.2 M EDTA. Agarose gel sample buffer is 15% (w/v) Ficoll (type 400), 0.25% (w/v) bromphenol blue, 0.25% xylene cyanole. TAE gel buffer is 40 mM Tris–acetate, 1 mM EDTA (pH 7.8). DE81 elution buffer is 0.2 M NaOH, 1.5 M NaCl, 5 μg/ml denatured calf thymus DNA (highly purified). Sodium acetate (3 M, pH 5.5) was used to lower the pH of probe preparations. For hybridizations, the highest quality formamide commercially available was obtained and deionized with AG501-X8 mixed-bed resin (Bio-Rad). Hybridization buffer (5×) is 2 M NaCl, 0.2 M PIPES (pH 6.5), 5 mM EDTA. S_1 buffer is 0.25 M NaCl, 30 mM potassium acetate (pH 4.5), 1 mM ZnSO$_4$, 5% (v/v) glycerol. Native calf thymus DNA (1 mg/ml) was obtained from Sigma, purified extensively by proteinase K treatment and phenol extraction, and was sheared to 1–5 kb. To prepare denatured calf thymus DNA, native calf thymus DNA was boiled for 5–10 min. Yeast RNA (10 mg/ml) was also obtained from Sigma and highly purified. Exonuclease VII buffer is 30 mM KCl, 10 mM Tris–HCl (pH 7.4), 10 mM EDTA, 10 mM 2-mercaptoethanol. Nuclease stop buffer A is 4 M ammonium acetate, 50 mM EDTA. Nuclease stop buffer B is 20% (w/v) sodium dodecyl sulfate (SDS), 5 mg/ml yeast RNA. Sevag is chloroform and isoamyl alcohol (24 : 1). Phenol–Sevag is phenol equilibrated with 50 mM Tris (pH 7.4) and Sevag (1 : 1). Sequencing gel loading buffer is 90% (v/v) formamide, 10 mM NaOH, 10 mM EDTA, 0.1% (w/v) bromphenol blue, 0.1% (w/v) xylene cyanole. DNA isolation buffer is 0.1 M NaCl, 50 mM Tris–HCl (pH 7.4), 1 mM EDTA, 0.5% SDS. RTase (reverse transcriptase) buffer (10×) is 0.5 M Tris–HCl (pH 8.5 at 23°), 0.1 M MgCl$_2$, 0.4 M KCl, 10 mM dithiothreitol. Actinomycin D (4 mg/ml in 80% ethanol) was obtained from Boehringer-Mannheim. cDNA dNTP is 5 mM dATP, 5 mM dCTP, 5 mM dGTP, 5 mM dTTP.

Recommended enzyme sources include large fragment DNA polymerase I: Boehringer-Mannheim, IBI, or Bethesda Research Laboratories (BRL); nuclease S_1: Boehringer-Mannheim, BRL; exonuclease VII: BRL; avian myeloblastosis virus (AMV) reverse transcriptase: Life Sciences, Boehringer-Mannheim, BRL; RNasin: Promega Biotec, Boehringer-Mannheim; and proteinase K: Merck, Boehringer-Mannheim.

Preparation of Continuously Labeled Probes

Synthesis Reactions

The preparation of single-stranded DNA probes is greatly simplified by the use of M13 phage single-stranded DNA templates. A number of useful M13 cloning vectors have been developed by Messing and co-workers[13,14] (as of this writing MP-20/21 are the latest), and are commercially available from several sources. We recommend purification of M13 phage by isopynic centrifugation in CsCl (1.3 g/ml) before deproteinization to minimize contamination of probes with degraded template fragments. The purified M13 DNA template is annealed to sequencing primer in an Eppendorf tube containing 4 μl of 10× Hin buffer, 1 μl (1 μg) of M13 template, 2 μl (10 ng) of sequencing primer, and 11 μl of water. Renaturation is initiated by heating at 60° for 2 min followed by incubation at 37° for 10 min. To synthesize a labeled complementary strand of the annealed template, 5 μl (50 μCi) of [^{32}P]dATP, 2 μl of labeling dNTP, and 3 μl (15 units) of large fragment DNA polymerase I are added to the mixture. The concentration of Hin buffer in the final reaction is 1×. After incubation at 37° for 25 min, 2 μl of chase dNTP is added to the mixture followed by incubation for an additional 30 min. To minimize the amount of uncopied M13 template, which can be relatively difficult to separate from the probe at the gel purification step, a ≥5-fold molar excess of sequencing primer and unlabeled nucleotides are included in the synthesis reaction. The amount of [^{32}P]dATP in the synthesis reaction is slightly more than twice the amount needed to copy a genomic insert of 500 nucleotides (probe composition assuming an average of 25% AMP). In most cases 50–80% of the labeled nucleotide is incorporated into DNA.[15]

The nascent probe is separated from the vector strand by digestion with a restriction enzyme that cuts the subclone at a single site 3′ to the labeled strand of the genomic insert. The polylinker of MP-20/21 contains several restriction sites that may be used for this purpose. The restriction enzyme selected to release the labeled genomic strand will depend on the restriction sites used to insert the genomic fragment in the vector. For example, a HindIII–EcoRI gene fragment cloned in MP-21 could be released by adding 40 μl of 1× Hin buffer containing 20 units of HindIII to the synthesis reaction followed by incubation at 37° for 90 min. Alternatively, the NarI site in MP-20/21 may be used to add a 230-nucleotide 3′

---

[13] J. Messing, this series, Vol. 101, p. 20.

[14] J. Norrander, T. Kempe, and J. Messing, Gene 26, 101 (1983).

[15] A useful formula to calculate the amount of labeled nucleotide A (in microcuries) needed to copy fully a genomic insert of length L_i (in nucleotides) is

$$A = 2 \frac{L_i}{L} \frac{MCS}{330}$$

Here L is the total length of the M13 template in nucleotides, M is the mass (μg) of template, C is the fraction of the probe that is the labeled nucleotide, S is the label specific activity (μCi/μM), and 330 is the molecular mass of the average nucleotide (μg/μM).

trailer of vector sequences to the probe if there are no *Nar*I sites in the genomic insert. After restriction enzyme digestion, denaturation of the synthetic duplex releases three single-strand DNA fragments. The M13 template becomes a linear single-stranded DNA with a length equal to the vector plus the genomic insert, and is unlabeled. The synthetic copy is a single-stranded DNA molecule approaching the length of the M13 vector sequences in size. The labeled probe is the length of the genomic insert plus any short regions of vector sequence added to the 5' or 3' end. The relatively small size of the probe allows it to be easily separated from the M13 template and synthetic vector fragment. Contamination of labeled probes at the gel purification step with complementary template strands is the principle source of hybridization background in nuclease protection assays. Duplexes formed between the labeled probe sequence and degraded template produce a smear of fragments in high-resolution gels that interfere in the detection of RNA–DNA hybrids. Therefore all possible precautions should be taken to avoid degradation of the M13 template during probe synthesis and gel purification.

Gel Purification

The probe preparation is denatured by the addition of 6 μl of denaturation buffer. After incubation at room temperature for 2 or 3 min, 20 μl of agarose gel sample buffer is added and the sample is loaded on a 1–2% (depending on the probe length) native agarose gel in TAE gel buffer. To avoid renaturation of the probe and M13 template, the gel is run at relatively high voltage for 15–20 min before setting to a convenient voltage. After electrophoresis, the probe is located by autoradiography (at the workbench) with X-ray film double wrapped with aluminum foil. The film is pressed firmly against the gel, which is covered with plastic wrap, using a glass plate. Exposure time is usually 5–10 min. The majority of label should be located in the probe band. Vector sequences near the top of the gel should contain little isotope. An uncopied circular template strand (~8200 nucleotides) migrates as a ~2-kb dsDNA fragment. A block of agarose directly ahead of the probe band is cut out of the agarose gel with a razor blade and a piece of DE81 filter paper (Whatman) is placed against the probe band. The agarose block is replaced and electrophoresis is continued until the probe has been completely transferred to the filter (check elution by autoradiography as described above). The filter is then removed to a 1.5-ml Eppendorf tube and eluted with 0.4 ml of DE81 elution buffer. Neutral high-salt buffers will not elute fragments <500 nucleotides efficiently. The filter is shredded with an 18-gauge needle and vigorously vortexed. After 10 min a vent is made in the top of the tube with a needle, and then a small incision is cut into the bottom of the tube with the tip of an 18-gauge needle. The tube is placed on top of a 2063 Falcon tube and the buffer is collected by centrifugation at ~1000 g. Elution is repeated with an additional 0.4 ml of elution buffer and is monitored with a hand-held counter. The probe is neutralized with 0.2 ml of 3 M sodium acetate (pH 5.5) and stored frozen. Further purification steps such as phenol extraction or ethanol precipitation are usually unnecessary. The concentration of complementary strand contamination should be no greater than 10% of the probe and is usually much less.

Probe Specific Activity and Stability

A disadvantage of labeling single-stranded DNA probes internally with ^{32}P is that isotope decay results in strand scission, though probe degradation can be caused by other factors as well, including effects of radiation from other molecules and degradation caused by trace nuclease contamination and chemical hydrolysis. With time the probe becomes increasingly fragmented, reducing the fraction of probe molecules that form a discrete band in high-resolution acrylamide gels. The period of time an internally labeled probe is usable for nuclease protection assays depends on its specific activity and the length of sequence protected in hybrid, as well as on its stability. As an example consider a probe (25% dAMP) synthesized with [^{32}P]dATP (410 Ci/mmol). The specific activity of such a probe would be about 6.6×10^8 dpm/μg. Our experience has been that at this specific activity probes generating a 500-nucleotide fragment in nuclease protection assays are usable for about 5 days following gel purification of the probe. To detect larger fragments, probe life can be increased by reducing the specific activity of the labeled nucleotide in the synthesis reaction. Unlike the case when end-labeled probes are utilized, the molar sensitivity (minimum number of gene transcripts detectable) of continuously labeled probes increases proportionately with the length of the probe sequence protected in hybrid. For transcripts protecting more than ~100 nucleotides the molar sensitivity of continuously labeled probes at 6.6×10^8 dpm/μg exceeds the sensitivity of end-labeled probes.

Nuclease Protection Mapping Procedures

Preliminary Considerations

The amount of total RNA required for a nuclease protection assay is determined by the specific activity of the probe and the concentration of the complementary transcript in total RNA. At 6.6×10^8 dpm/μg, $\geq 10^{-2}$ pg of ^{32}P-labeled DNA probe is theoretically detectable by conventional autoradiography methods, using high-resolution acrylamide gels. In practice we have found that the maximum sensitivity of nuclease protection assays is reduced to $\sim 10^{-1}$ pg of complementary RNA transcript by probe degradation and hybridization background (see Gel Purification). However, subpicogram sensitivity is not generally necessary in mapping gene transcripts. Where possible the concentration of specific poly(A) transcripts the prevalence of which is $<10^{-6}$ of the mass of total RNA should be enriched by oligo(dT)-cellulose chromatography (20- to 50-fold enrichment). The maximum amount of bulk RNA [or poly(A) RNA] that should be included in a nuclease protection assay is about 20 μg. Greater quantities of RNA require excessive amounts of nuclease S_1 to digest completely, and in exonuclease VII experiments an RNase A step may be required to prevent overloading high-resolution acrylamide gels with bulk RNA.

Selection of Hybridization Conditions

The factors determining hybrid stability and the kinetics of nucleic acid reassociation are reviewed in detail in this volume [43], and elsewhere.[16,17] A commonly used solvent for hybridization reactions contains 50% formamide, 0.4 M NaCl, 1 mM EDTA, and 40 mM PIPES (pH 6.5).[8] To estimate the T_m of hybrids accurately in this solvent it would be necessary to know the G + C composition of the duplex,[18] the hybrid length,[19] and the percentage and location of any base pair mismatches.[20] For RNA–DNA hybrids of average G + C composition the rate of annealing in 50% formamide solutions is optimal at 10–15° below the T_m.[8] The effect of formamide is to reduce hybrid T_m,[21] thus avoiding probe and RNA degradation that occurs at high temperatures of incubation.

In practice the sequence characteristics of hybrids that will form in transcript mapping experiments with newly isolated gene fragments is usually unknown a priori. In this situation an initial estimate of hybrid T_m

[16] R. J. Britten, D. E. Graham, and B. R. Neufeld, this series, Vol. 29, p. 363.

[17] R. J. Britten and E. H. Davidson, in "Nucleic Acid Hybridization: A Practical Approach" (B. D. Hames and S. J. Higgins, eds.), p. 3. IRL Press, Washington, D.C., 1985.

[18] In aqueous solutions containing 0.4 M NaCl, T_m = 0.41 (% G + C) + 74.9°. This relationship holds for long DNA duplexes of 30–75% G + C. RNA–DNA hybrids of average (G + C) composition that are perfectly matched are 5–10° more stable than the corresponding DNA hybrid. For effect of formamide on T_m see note 21 and refs. 8, 16.

[19] The effect of hybrid length can be estimated by $D = 500/L$. Here D (in degrees Centigrade) is the reduction in T_m and L is the length in nucleotides of the base-paired duplex.

[20] Various data indicate that 1% mismatch reduces T_m by 1° (see ref. 16).

[21] For DNA, the reduction in T_m is linearly dependent on the formamide concentration; however, the reduction in T_m is greater for dA–dT duplexes (0.75° per 1% formamide) than for dG–dC duplexes (0.5° per 1% formamide). The relationship between T_m depression and formamide concentration is not linear for RNA–DNA hybrids.[8]

is made using the average genomic DNA G + C composition, which for many species can be obtained from literature data (in most animal DNAs the G + C content is about 40%). For the initial experiment the hybrid length is assumed to be relatively short (50 bp). The effect of base mismatch on T_m is not a concern when working with inbred lines or RNA and gene fragments derived from the same individual's genome, and can be ignored in the preliminary estimate of hybrid T_m. As an example, using the relationships outlined in footnotes 18–22 a reasonable temperature of incubation for initial hybridization reactions with uncharacterized gene fragments of the sea urchin *Strongylocentrotus purpuratus* (39% G + C) is ~43°.[22]

The concentration of single-stranded DNA probe in the hybridization reaction is in excess over the concentration of complementary RNA transcript. The kinetics of such reactions have been well characterized.[23] Under standard conditions the hybridization rate is dependent solely on the concentration of probe, and the rate constant k ($M^{-1}sec^{-1}$) of the reaction may be estimated for each probe using published standards.[23,24] Thus for each hybridization it is possible to select the probe concentration and time of incubation necessary to recover quantitatively an RNA transcript in a hybrid. For the case of a completely hybridized 500-nucleotide single-stranded probe of unique sequence, k is ~3.7×10^3 in 0.4 M NaCl and 50% formamide.[25] The reaction is essentially complete (i.e., 95%)

---

[22] The T_m of the average *S. purpuratus* 50-bp genomic DNA fragment in 50% formamide and 0.4 M NaCl is ~50° [T_m = 74.9 + 0.41 (39% G + C) − 0.62 (50% formamide) − 500/50]. In 50% formamide the T_m of RNA–DNA hybrids is 5–10° greater than the T_m of DNA, and the optimal temperature of incubation is 10–15° below T_m for RNA–DNA hybrids. Therefore ~43° (50° + 6° − 12.5°) is a reasonable first estimate of the temperature of incubation for uncharacterized probes.

[23] $S = e^{-kC_0t}$. Usually described as pseudo-first-order kinetics, where S is the fraction of complementary RNA that remains unpaired, C_0 is the probe concentration in nucleotides (moles per liter), and t is the time in seconds. A useful calculation of C_0t is $C_0t = Pt/83$, where P is the probe concentration in micrograms per milliliter, t is the time in hours, and 83 is a conversion factor. At 50% formamide the reaction is retarded about 2× compared to standard conditions (i.e., 0.18 M Na$^+$, aqueous medium).

[24] G. A. Galau, R. J. Britten, and E. H. Davidson, *Proc. Natl. Acad. Sci. U.S.A.* **74,** 1020 (1976).

[25] Under standard conditions[24] k = (5400/L_i) × 170 M^{-1} sec^{-1} ≈ 1840 M^{-1} sec^{-1}, where L_i is the complexity (length of unique sequence) of the probe (i.e., here 500 nucleotides), 5400 nucleotides is the length of the ϕX174 standard, and 170 M^{-1} sec^{-1} is the rate constant for the reaction of a ϕX174 DNA tracer with excess RNA.[24] The fragment sizes for which this measurement was obtained are ~400–500 nucleotides. For a completely complementary probe the rate will change as the square root of the probe length. For our case there is no length correction. In aqueous medium 0.4 M NaCl yields an acceleration factor of about 4 compared with standard conditions,[16] while 50% formamide retards the reaction ~2-fold. Thus in 0.4 M NaCl, 50% formamide medium the reaction rate is accelerated ~2-fold relative to standard conditions, and k is ~3.7×10^3 for the 500-nucleotide probes.

when incubated to a probe DNA C_0t of 8.1×10^{-4} (ln $0.05/-3.7 \times 10^3$). For a commonly used probe concentration such as 30 ng/ml, half of the reaction would occur in about a half-hour and 95% reaction in about $2\frac{1}{4}$ hr. The rate of hybridization is of course retarded at suboptimal temperatures, or at temperatures $<10°$ below hybrid T_m. This effect may be severe for hybrids of extremely high or low G + C composition. An additional factor that is unknown in advance of the result of the experiment and that could result in a several-fold retardation of the reaction rate is the length of complementary sequence that can form duplex, if this is only a minor fraction of the length of the probe.[26] Thus as a precaution, hybridizations with uncharacterized probes are carried to 5–10 times the DNA C_0t ideally required for 95% completion. This can usually be accomplished conveniently by an overnight incubation. Large excess of probe should be avoided in hybridization reactions with single-stranded DNA probes due to hybridization background in probe preparations (see Gel Purification). We recommend a moderate excess of probe over complementary RNA (~5-fold) in hybridization reactions for transcript mapping experiments.

Hybridization Protocol

Coprecipitate RNA (≥ 1 pg complementary transcript) and probe (≥ 5 pg) in 70% ethanol (0.1 M NaCl) in a plastic microcentrifuge tube treated with dimethydichlorosilane. Amounts many times larger than these minimal values may be used, depending on RNA prevalence, reaction volume, probe specific activity, etc. A control reaction containing an equal amount of an RNA that will not react with the probe should be included in each set of hybridization reactions. A suitable control RNA for reactions with most probes is a highly purified preparation of low molecular weight yeast RNA that can be obtained commercially. The precipitate is collected by centrifugation, and salt is removed by addition of excess 70% ethanol and recentrifugation. The nucleic acid pellet is dried in a Savant Speed Vac centrifuge lyophilizer. This pellet is resuspended in 62.5% formamide. After the pellet is dissolved, 5× hybridization buffer is added to a final concentration of 1×. To avoid evaporation the reaction is taken up in a dimethydichlorosilane-treated glass capillary pipet and both ends are flame sealed. Hybridization is initiated by incubation at 85° for 3 min. Heating also assures that the RNA and probe are fully resuspended. The sealed capillary tubes are then incubated in a water bath at the appropriate temperature.

[26] G. A. Galau, M. J. Smith, R. J. Britten, and E. H. Davidson, *Proc. Natl. Acad. Sci. U.S.A.* **74,** 2306 (1977).

Nuclease S₁ Digestions

Until recently the suitability of commercial preparations of nuclease S_1 for transcript mapping experiments was not tested by manufacturers. It was not unusual to obtain nuclease S_1 contaminated with significant levels of nuclease activity that destroys both DNA and DNA–RNA duplexes. Overdigestion of hybrids with such preparations of nuclease S_1 is a relatively common reason for failure in obtaining probe protection. We recommend calibration of each preparation of nuclease S_1 under the exact conditions selected for protection assays. The amount of enzyme used (unless very highly purified) should be just sufficient to digest 95% of an internally labeled and denatured DNA, as assayed by precipitation in 4% (w/v) trichloroacetic acid (TCA), without solubilizing a significant amount of labeled DNA duplex (see Synthesis Reactions; TCA precipitation is described in this volume [6]). RNA–DNA duplexes formed between gene probes and RNAs isolated from natural sources may contain mismatched bases, as a result of DNA sequence polymorphisms. Cleavage of mismatched bases by nuclease S_1 can be suppressed by lowering the incubation temperature for digestion to 13° or increasing salt concentration to reduce transient local denaturation of duplex, and by avoiding over digestion of hybrids.

Hybridization reactions (usually 5 to 10 μl) are expelled into 20 μl of water in a 1.5-ml Eppendorf tube, mixed, and placed in ice. A 300-μl aliquot of ice-cold nuclease S_1 buffer containing freshly added denatured calf thymus DNA (25 μg/ml), native calf thymus DNA (25 μg/ml), and the appropriate amount of nuclease S_1 is added to each tube. The reactions are incubated at 13° for 1.5 hr.

Nuclease S_1 reactions are terminated by the addition of 40 μl of nuclease stop buffer A and 4 μl of nuclease stop buffer B, followed immediately by extraction with an equal volume of phenol–Sevag. The protected fragments are precipitated by addition of 700 μl of absolute ethanol. The precipitate is collected by centrifugation, and the pelleted nucleic acid is washed once with 70% ethanol, and dried. The dry pellet is resuspended in ~10 μl of sequencing gel loading buffer. The sample is heated to 90° for 3–4 min just before loading on a 8 M urea (4–6%) acrylamide gel (19 : 1, acrylamide : bisacrylamide). Usually one-half of the sample is held in reserve for additional gel runs. After electrophoresis the gel is dried or covered with plastic wrap and the protected fragments are detected by autoradiography at −70° with Kodak XR-5 X-ray film and one Dupont Cronex intensifier (this volume [7]).

An example of a nuclease S_1 mapping experiment with a genomic DNA probe for exon 1 and part of intron 1 of the *S. purpuratus* actin gene

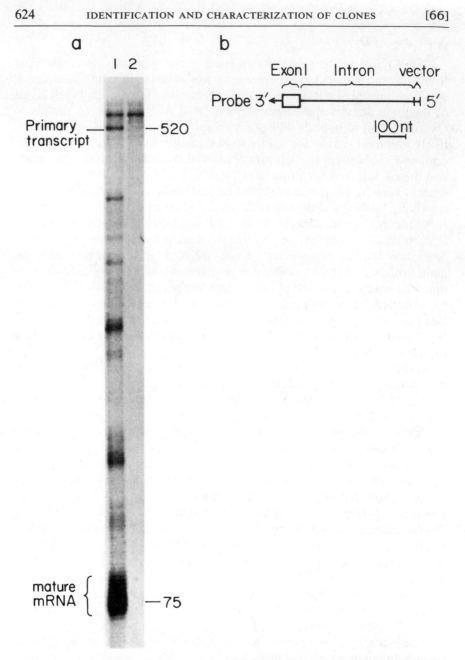

FIG. 2. Nuclease S_1 protection assay. A continuously labeled, single-stranded DNA probe (5×10^8 cpm/μg) was reacted with 10 μg of total RNA from sea urchin pluteus-stage embryos (lane 1) or yeast RNA (lane 2). After treatment with nuclease S_1 the protected fragments were separated by electrophoresis in a 8 M urea–6% acrylamide gel. The probe

CyIIIa[27] is shown in Fig. 2.[28,29] Hybrids with exon 1 of the mature mRNA are represented by the set of probe fragments that are ~75 nucleotides in length. The heterogeneity of fragment length is due to an activity in nuclease S_1 preparations that attacks RNA–DNA hybrid termini. As shown by this example, transcripts cannot be mapped to the precision of a single nucleotide by this method. Hybrids that formed with unspliced *CyIIIa* primary actin gene transcripts are represented by the relatively faint band at ~520 nucleotides.

Exonuclease VII Digestions

Unlike nuclease S_1, overdigestion with commercial preparations of exonuclease VII does not decrease hybrid yield. Thus preliminary characterization of the enzyme is not necessary. We have found that excess exonuclease VII is required to digest long (>500 nucleotides) 5' single-stranded DNA tails. The enzyme does not appear to solubilize single-stranded RNA to an appreciable extent.

Hybridization reactions are expelled into 20 μl of water as described in the nuclease S_1 protocol. A 130-μl aliquot of ice-cold exonuclease VII buffer containing 120 units/ml exonuclease VII is added to each hybridization reaction and the reactions are incubated at 37° for 1–2 hr. The amount of enzyme needed is at least 20-fold greater than would be required using the unit definition of the manufacturer. Exonuclease VII reactions are terminated and prepared for electrophoresis in the manner described for nuclease S_1 assays except that the volumes of all buffers and organic reagents added to the reaction are reduced by 50%, and yeast RNA carrier is not added if the amount of total RNA in the hybridization reaction is greater than 10 μg. The migration of protected fragments may appear to be disturbed by overloading the gel with RNA, in which case the sample may be treated with RNase A (DNase-free). RNase A reactions should be performed in a minimal volume, and should contain ≥0.3 M Na$^+$ to pro-

---

[27] R. Akhurst, F. J. Calzone, J. J. Lee, R. J. Britten, and E. H. Davidson, *J. Mol. Biol.* **193** (in press).

[28] J. J. Lee, F. J. Calzone, R. J. Britten, and E. H. Davidson, *J. Mol. Biol.* **188**, 173 (1986).

[29] J. J. Lee, F. J. Calzone, R. J. Britten, and E. H. Davidson, in preparation.

---

was a 540-nucleotide gene fragment of *S. purpuratus* actin gene *CyIIIa*[27] diagrammed in (b). The ~75-nucleotide fragments are derived from hybrids with exon 1 of the mature mRNA.[28] The 520-nucleotide fragment represents the primary gene transcript. Intermediate-size fragments are the result of processing intermediates or sequence polymorphism. The prevalence of the *CyIIIa* actin mRNA is 0.1–0.2% of total messenger RNA.[28] Nuclear transcripts represent ~1% of the total actin gene transcripts.[29]

tect hybrids. Excessive amounts (>2 μg) of RNase A should be removed by phenol–Sevag extraction before electrophoresis.

Restriction Enzyme Mapping of Protected Probe Fragments

For restriction enzyme mapping experiments, probe fragments obtained in nuclease protection assays are reannealed to complementary M13 templates. To prepare protected probe sequences for hybridization with M13 template, the nuclease S_1 digestion reaction is terminated and extracted with phenol–Sevag as described above, and ethanol precipitated. The sample is resuspended in 200 μl of DNA isolation buffer containing 60 μg/ml denatured calf thymus DNA and 400 μg/ml proteinase K. After incubation for 1 hr at 37° the sample is extracted twice with phenol–Sevag and once with Sevag. The probe fragments are then precipitated in 70% ethanol (0.1 M NaCl), collected by centrifugation, washed once with excess 70% ethanol, and dried. Exonuclease VII is more easily extracted. For restriction mapping assays the exonuclease VII reaction is terminated by addition of proteinase K to a final concentration of 200 μg/ml, SDS to 1% w/v, and yeast RNA to 60 μg/ml. After incubation for 1 hr at 37° the probe fragments are purified by phenol–Sevag extraction and ethanol precipitation.

The purified hybrids from nuclease S_1 and exonuclease VII reactions are resuspended in 50 μl of water, mixed with 6 μg ($\geq$1000-fold molar excess) of complementary single-stranded M13 template (vector plus genomic insert), and precipitated in 70% ethanol (0.1 M NaCl). The precipitate is resuspended in 16 μl of 62.5% formamide followed by addition of 4 μl of 5× hybridization buffer. The sample is sealed in a glass capillary pipet and hybridization is initiated by heating to 85° for 4 min. After incubation at 42° for 1.5 hr the hybridization reaction is expelled into 80 μl of water and the formamide and salts are removed by ethanol precipitation. The mixture is resuspended in 10 mM Tris–HCl (pH 7.4), 0.1 mM EDTA. Each restriction enzyme digest receives 2 μl of the reaction mixture and 3 μl of a solution containing a restriction enzyme in the appropriate buffer. After incubations for 1–2 hr (use a small tube to avoid excessive evaporation) three volumes of DNA isolation buffer are added followed by one extraction with phenol–Sevag, and ethanol precipitation. The precipitate is dissolved in 5 μl of sequencing gel loading buffer. The probe fragments are separated by electrophoresis in high-resolution urea–acrylamide gels and detected as described in nuclease S_1 digestions. The restriction enzymes selected for mapping studies should generate two to four probe fragments. When the restriction map of a probe sequence is unavailable, a suitable restriction enzyme can be identified in a prelimi-

nary set of digestions with enzymes expected to cut genomic DNA relatively frequently (e.g., *Dde*I, *Ava*II, *Rsa*I, *Sau*3AI). Each test digestion
should contain about 1×10^4 dpm of probe reannealed to a large excess of
complementary M13 template as described above. To map the restriction
sites in the probe and probe fragments obtained in nuclease protection
experiments, the 5′ sequences of the probe are preferentially labeled by
increasing the amount of M13 template in the probe synthesis reaction to
a point at which the complementary transcript of the genomic insert is not
fully labeled. The 5′ probe fragments will then contain more label per unit
length than 3′ fragments of comparable base composition. As shown in
Fig. 3, differential labeling of probe restriction fragments by this protocol
is easily detected by visual inspection of gel autoradiograms. The restriction fragments can be ordered (5′ → 3′) solely on the basis of label intensity. As shown in Fig. 3, frequently the data obtained for a single restriction enzyme are sufficient to map the probe fragments that bear sequences
complementary to RNA.

Interpretation of Nuclease Protection Data

The initial nuclease protection assays should confirm the orientation
of transcription determined by RNA gel blot hybridization, and should
provide a more quantitative estimate of transcript prevalence which will
be useful for subsequent experiments. The experiment shown in Fig. 3
demonstrates that RNA transcripts can be mapped with high precision
using relatively large (2.5-kb) genomic probes that have not been extensively characterized by restriction enzyme mapping. The possibility of
employing large genomic probes makes feasible the application of the
technique to the task of mapping gene transcripts of large (20–30 kb)
genomic regions. The interpretation of nuclease protection data is relatively straightforward for single-copy gene fragments. As shown in Fig. 2
the high specific activity of continuously labeled genomic probes allows
detection of primary as well as processed gene transcripts. Typically the
concentration of primary gene transcript in total cellular RNA is much
lower than the concentration of mature mRNA. Thus probes containing
intron sequences are identified in nuclease protection experiments by
bands in the gel autoradiogram that are relatively minor when compared
to the intensity of probe fragments containing exon sequences. Primary
transcripts should protect the full length of genomic probes containing
only internal gene sequences, while gene fragments which contain the
start of transcription or the 3′ end of processed transcripts plus overlapping nontranscribed DNA will be partially protected. The size of the
prevalent gene transcripts predicted from nuclease protection data should

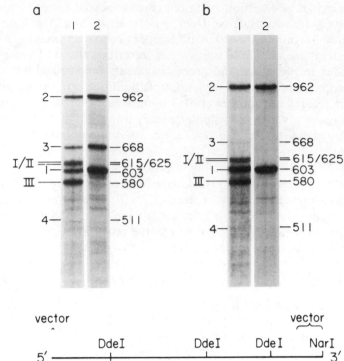

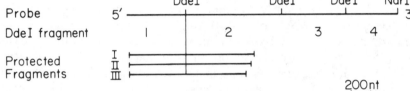

FIG. 3. Example of restriction mapping differentially labeled genomic probe fragments. The results of an exonuclease VII experiment to map the genomic sequences in a 3.7-kb maternal repetitive sequence transcript of *S. purpuratus* are shown to illustrate the restriction mapping technique. The genomic probe contained sequences complementary to the first 2060 nucleotides of the 3.7-kb transcript and 420 nucleotides of 5' flanking sequences. In reactions with egg poly(A) three large probe fragments labeled I–III in (c) were detected after exonuclease treatment (data not shown). For restriction mapping the protected fragments were reannealed with a complementary M13 template and digested with *Dde*I before gel electrophoresis. To determine the arrangement of *Dde*I sites in the probe shown in (c) the 5' sequences of the probe were selectively labeled by limiting the amount of [^{32}P]dATP in the probe synthesis reactions. For the experiment shown in (a) the amount of label in the probe synthesis reaction was sufficient to label 50–70% of the genomic insert. The order of the *Dde*I fragments, labeled 1–4 in (a) and (c), was determined from the labeling intensity of each fragment in the yeast RNA control (lane 2). *Dde*I fragment 1 contained the highest intensity of label and was therefore assigned the 5' position. Labeling was progressively reduced in *Dde*I fragments 2 and 3, and absent from *Dde*I fragment 4. The *Dde*I fragments detected in lane 2 were the result of M13 template present in the probe preparation which

approximate ($\pm 10\%$) the transcripts detected in RNA gel blot hybridization analyses. A potential complication may arise in mapping transcripts with some probes due to the presence of short repetitive sequences that are abundant in eukaryotic genomes.[30] A given short repeat sequence may be represented in many different genes, usually in introns or nontranslated regions of exons. The concentration of repeat sequence transcripts in total cellular RNA is frequently comparable to prevalent mRNA, and thus in some cases a short repeat sequence transcript may be confused with a short exon of a bona fide mRNA. Therefore, probes containing short repeat sequences should be identified by genomic DNA and RNA gel blot hybridization analyses.

Mapping of 5' Termini by cDNA Primer Extension

Preliminary Comments

To map the nucleotide initiating gene transcription with greater precision, a probe is hybridized to a sequence near the 5' end of the gene transcript and elongated with reverse transcriptase as shown in Fig. 4. For high-resolution mapping this primer should be located within ~200 nucleotides of the 5' terminus of the transcript. If the gene sequence near the start of transcription is known, it is most convenient to synthesize a 20- to 30-nucleotide primer chemically. The primer is labeled at the 5' end by phosphorylation with polynucleotide kinase and [γ-^{32}P]ATP, and the

[30] F. J. Calzone, H. T. Jacobs, C. N. Flytzanıs, J. W. Posakony, and E. H. Davidson, *Biol. Fert.* **3**, 347 (1985).
[31] F. J. Calzone, J. J. Lee, N. Le, R. J. Britten, and E. H. Davidson, in preparation.

protects the full length of a fraction of the probe molecules from exonuclease VII. In (b) the amount of label in the probe synthesis reaction was further reduced to cover 30–50% of the genomic insert. In this case only *Dde*I fragments 1 and 2 in lane 2 are labeled. Lane 1 in (a) and (b) shows the fragments obtained in reactions with egg poly(A) RNA. The fragments labeled I–III map the 5' termini of the different RNA–DNA hybrids. The fragments mapping the 3' hybrid terminus comigrate with *Dde*I fragment 1 (see c). A key result leading to the map of the probe fragments shown in (c) is that fragments I–III were detected when the sequences of *Dde*I fragments 3 and 4 were unlabeled. A repetitive element in the 5' end of the 3.7-kb transcript reacts with complementary copies of the same element present in other poly(A) RNAs. The RNA–RNA duplexes prevent hybridization of the genomic probe with the 5' sequences of the 3.7-kb transcript. Heterogeneity in the length of the RNA–RNA duplexes is responsible for the differences in the lengths of fragments I–III. It should be noted that if hybridization background in the probe preparation is less than 10%, a restriction digest of untreated probe reannealed to M13 template should be included in each mapping experiment.

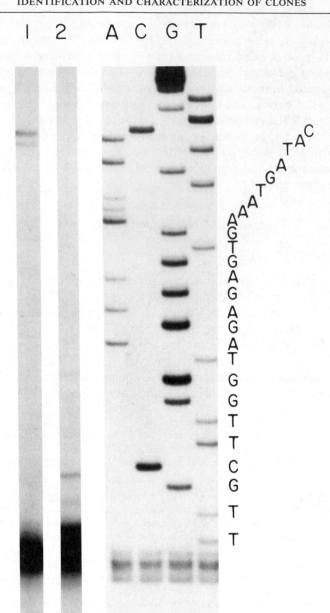

FIG. 4. Primer extension reaction. The results of an experiment to map the 5′ start of the *CyIIIa* actin gene transcript described in Fig. 2 is shown. The primer extension reaction (lane 1) contained 2 μg of total cellular poly(A) RNA (3–6 fmol *CyIIIa* actin mRNA) and 30

labeled primer is usually purified by electrophoresis in a high-percentage acrylamide denaturing gel. The theoretical maximum specific activity of a primer labeled at the 5' end with [γ-^{32}P]ATP (7000 Ci/mmol) is 1.5×10^7 dpm/pmol primer, although specific activities of ~8×10^6 dpm/pmol primer are more typical for chemically synthesized primers. Most commercial preparations of polynucleotide kinase contain unlabeled ATP as a stabilizing agent, and thus large excesses of this enzyme in labeling reactions can significantly lower the primer specific activity. When DNA sequence information is unavailable a continuously labeled primer can be prepared using a genomic restriction fragment identified as above by nuclease protection procedures. For high-resolution mapping such a primer should be $\leq$100 nucleotides in length. Single-stranded primers are preferred, but they are not an absolute necessity.

The criteria for selecting the amount of RNA and primer for extension reactions are the same as described earlier. We recommend a moderate excess of primer (5- to 10-fold) in the hybridization reactions. As with nuclease protection procedures the lower limit of RNA detectable in primer extension reactions is not solely determined by the probe specific activity. The other major factors determining sensitivity are the efficiency of hybridization of primer and RNA and the efficiency of primer elongation. We have found that at least 0.1 fmol of RNA (for example, 66 pg of a 2-kb transcript) is detectible using a 5'-end-labeled primer at 8×10^6 dpm/pmol. If possible the concentration of polyadenylated transcripts should be increased by oligo(dT)-cellulose chromatography. In general primer extension reactions carried out with poly(A) RNA display lower background transcription, and less incomplete chain elongation, which is commonly observed when using total cellular RNA. An optimal temperature of incubation is necessary to assure maximal efficiency in the hybridization of the primer to RNA. Unfortunately the empirically derived formulas for the estimation of RNA–DNA hybrid T_m are not accurate for very short DNA primers. Therefore, the hybridization temperature must be determined by preliminary experimentation. The commonly used temperatures range from 40 to 60° in 0.1–0.4 M monovalent cation.

Hybridization and Primer Extension Reactions

Mix $\geq$0.5 fmol of 20- to 30-nucleotide end-labeled primer and $\geq$0.1 fmol complementary RNA transcript [up to 10 μg of bulk poly(A) RNA] in

---

fmol of a 21-nucleotide primer labeled at the 5' terminus with ^{32}P (1200 dpm/fmol primer). Hybridization was at 40° for 6 hr. The control reaction (lane 2) contained yeast total cellular RNA. For the dideoxynucleotide sequencing reactions 5×10^6 dpm of 21-nucleotide primer was annealed with 8 μg of complementary M13 template. From Akhurst et al.[27]

a plastic microcentrifuge tube. The final volume is 4 μl. Add 1 μl of 5×
hybridization buffer. Seal the reaction in a silanized glass capillary pipet
and heat to 70° for 3 min. Incubate the reaction mixture at 40–60° for ≤6
hr. Expel the reaction into a microcentrifuge tube containing 5 μl of 10×
RTase buffer, 0.5 μl of actinomycin D, 5 μl of cDNA dNTP, 1 μl of
RNasin, and 32.5 μl of water. Add 30 units (1 μl) of AMV reverse
transcriptase and incubate the reaction at 37° for 1 hr. Terminate the
reaction by adding 5 μl of nuclease stop buffer A and 2 μl of nuclease stop
buffer B. Extract the sample with phenol–Sevag, ethanol precipitate, and
resuspend the extended primer in 10 μl of sequencing gel loading buffer.
A parallel reaction containing nonreacting RNA is included as a control.
The samples are denatured at 85° for 5 min, and analyzed by electrophoresis in 8 M urea–acrylamide DNA sequencing gels. A set of dideoxynucleotide DNA sequencing reactions using the labeled primer and a complementary M13 template is used for size markers (see Fig. 4).

Primer extension assays should map the 5′ terminus of a gene transcript to the accuracy of a single nucleotide. However, it is not unusual to
observe reverse transcripts that are shorter than full length. It has been
suggested that methylation of the ultimate or pentultimate nucleotides of
the gene transcript may interfere with reverse transcription, generating
cDNA transcript one or two nucleotides short. Heterologous cDNA transcripts due to fold-back synthesis or to cross-reaction of the primer with
other transcripts can be identified by failure of such cDNAs to comigrate
with the dideoxy sequencing ladder due to differences in base composition. If there are no introns between the location of the primer in the gene
sequence and the start of transcription, then the length of the primer
extension product should confirm the initiation site located by nuclease
protection using labeled gene probes as above.

Acknowledgment

Research from this laboratory was supported by NIH Grant HD-05753. F.J.C. was
supported by a Senior Fellowship by the American Cancer Society, California Division.

[67] A Molecular Titration Assay to Measure Transcript Prevalence Levels

By James J. Lee and Nancy A. Costlow

Background Information

Introduction

The molecular characterization of specific genes often includes experiments that assay for the presence and accumulation of particular transcripts. The methods in general use to assay for these transcripts involve the transfer of cellular RNA to a support matrix followed by hybridization with a specific probe.[1,2] The presence of transcripts is then detected by autoradiography. Existing protocols also attempt to use these blot methods to measure transcript prevalence.[3,4] The quantitative aspects of these protocols, however, are limited by the inability to control for such major factors as the efficiencies of transfer and binding of RNA to the support matrix as well as the uncertainties associated with the hybridization kinetics of filter-bound nucleic acids.[5] The molecular titration assay described here is an extension of techniques first used to study changes in the prevalence level of transcripts from different repeat families in the sea urchin *Strongylocentrotus purpuratus*.[6,7] This assay avoids the use of a solid support matrix and thus precludes many of the problems associated with the blot methods. The objective of these protocols is to provide a sensitive assay system with which to measure the absolute prevalence level of particular transcripts. By utilizing single-stranded RNA probes in solution hybridization reactions, the methodology presented here also provides a level of sensitivity comparable to other nuclease protection assays.[8]

[1] J. G. Williams and P. J. Mason, in "Nucleic Acid Hybridization: A Practical Approach" (B. D. Hames and S. J. Higgins, eds.), p. 139. IRL Press, Oxford, 1985.
[2] J. Meinkoth and G. Wahl, this volume [61].
[3] J. G. Williams and M. M. Lloyd, *J. Mol. Biol.* **129**, 19 (1979).
[4] M. Dworkin and I. B. Dawid, *Dev. Biol.* **76**, 435 (1980).
[5] J. Meinkoth and G. Wahl, *Anal. Biochem.* **138**, 267 (1984).
[6] R. H. Scheller, F. D. Costantini, M. R. Kozlowski, R. J. Britten, and E. H. Davidson, *Cell* **15**, 189 (1978).
[7] J. J. Lee, F. J. Calzone, R. J. Britten, R. C. Angerer, and E. H. Davidson, *J. Mol. Biol.* **188**, 173 (1986).
[8] F. J. Calzone, R. J. Britten, and E. H. Davidson, this volume [66].

Basic Approach

The assay described here is a molecular titration of cellular RNA with excess single-stranded probe. The procedure consists of a series of hybridization reactions with equal amounts of probe but increasing amounts of input cellular RNA. Because the probe is present in excess, a plot of radioactive probe driven into hybrid (as assayed by nuclease sensitivity) versus the input of cellular RNA is linear, and the amount of probe hybridized simply reflects the amount of cellular RNA used. Given an experimentally determined value for the slope (m), the number of transcripts in an RNA population complementary to a specific probe can be determined.

An attractive feature of these reactions is their independence of the kinetics of hybridization. This independence is the result of two factors. First, the hybridizations are performed such that the probe is present in at least 10-fold molar excess over corresponding transcripts in an RNA population. The form of the pseudo-first-order kinetic reaction can be described as follows:

$$D/D_0 = e^{-kR_0 t} \tag{1}$$

In this equation D/D_0 is the fraction of transcripts remaining unreacted at time t, R_0 the starting (and final) probe concentration, and k the observed rate constant for duplex formation. An obvious consequence of Eq. (1) is that the concentration of transcripts in an RNA population has no affect on the rate at which the reaction proceeds. Thus the rate of hybridization is the same whether the transcripts of interest are rare or prevalent in the RNA sample. Second, the hybridization reactions are also routinely carried out to completion (>95%). As a result, the titration method is not subject to the uncertainties associated with the kinetics of reaction termination.

Single-Stranded Probes

A variety of labeling procedures exist that generate single-stranded probes which can be used in the titration assay.[9,10] Constraints on the probes include that they be single stranded and of known specific activity. The quantitative aspects of the titration assay also require that the probes be well characterized. Physical parameters, such as the probe's specificity, its position within the transcription unit (i.e., its relationship to intron–exon borders), and the length of probe sequences which will hybrid-

[9] N.-t. Hu and J. Messing, *Gene* **17**, 271 (1982).
[10] M. R. Green, T. Maniatis, and D. A. Melton, *Cell* **32**, 681 (1983).

ize to transcript, *must* be known. Specific sequence information is also helpful since it allows a more precise determination of the probe-specific activity. These criteria can be met with both DNA and RNA probes.

Titration assays using single-stranded DNA probes require the use of nucleases capable of distinguishing between unhybridized probe (i.e., single-stranded DNA) and probe driven into RNA–DNA heteroduplexes. Two commercially available enzymes that can be used are nuclease S_1 and mung bean nuclease. These nucleases digest both single-stranded DNA and RNA as well as duplexes containing mismatched bases. Variability in the activity of both nucleases and their sensitivity to slight changes in reaction conditions limit the use of single-stranded DNA probes in titration assays. In addition, the background levels of radioactive DNA that is acid precipitable and resistant to nuclease S_1 or mung bean nuclease are as high as 5% of the input tracer.[8] These technical drawbacks, together with the difficulties of generating single-stranded DNA probes free of contaminating complementary sequences,[8] restrict the accuracy and/or sensitivity of titration assays using these probes. In contrast, single-stranded RNA probes, and the use of ribonuclease to remove unhybridized probe from the assay reactions, have eliminated many of the problems associated with DNA probes. For example, single-stranded RNA probes can be synthesized and isolated free of any contaminating complementary sequences. When used in hybridization reactions with cellular RNA, the high stability of the resulting RNA–RNA duplexes and their near-absolute resistance to ribonuclease digestion render these probes less sensitive to particular assay conditions. We have found that the background level of acid-precipitable, RNase-resistant radioactive RNA is reproducibly less than 0.5% of the input tracer, which, combined with the high specific activity commercially available for ribonucleotide precursors, means that extremely low transcript levels can be measured.

Preliminary Characterization of the RNA Sample

Methods designed to detect the presence of transcripts rely upon the ability of a labeled probe to hybridize with specific transcripts in an RNA sample. A preliminary concern is the availability of a sufficient mass of RNA to assay for the transcripts of interest. This is especially important in cases where the transcripts under study are present at extremely low levels. RNA gel blot hybridization data provide a convenient starting point for choosing a source of RNA and for determining the approximate prevalence and the molecular weight of particular transcripts. The high sensitivity of the titration assay permits the determination of the prevalence of any transcript detectable by the gel blot methods. In fact, as shown in a later section of this chapter, the titration assay easily detects and quantifies transcripts which go undetected in the qualitative procedures.

Unlike the gel blot methods, whose sensitivity depends heavily on the integrity of the transcripts in the RNA sample, molecular titrations can be performed with RNAs that are somewhat degraded. This results from the fact that the RNA is not selected by size during the titration. The extent of degradation tolerated by the assay is determined by the ability of the resulting RNA fragments to hybridize with the probe. It has been our experience that RNA samples degraded to an average fragment length range of 500–1000 nucleotides are still useful substrates for the titration assay.

Materials

The solutions are listed roughly in the order of their appearance in the text. Plasticware is made RNase free by washing with a 0.02% diethyl pyrocarbonate (DEP, Sigma) solution in water followed by autoclaving. The RNase-free plasticware is then treated with dichlorodimethylsilane (Sigma) and again autoclaved. Stock solutions are made RNase free by treatment with 0.02% DEP followed by autoclaving. Note that solutions which contain primary amines (e.g., Tris and EDTA) cannot be DEP treated. Proteinase K buffer, 10×, is 0.1 M Tris–HCl (pH 7.4), 1 M NaCl, 0.01 M EDTA, and 1% sodium dodecyl sulfate (SDS). Sevag is chloroform and isoamyl alcohol (24 : 1). Phenol–Sevag is double-distilled phenol equilibrated with 50 mM Tris–HCl (pH 7.4) and Sevag (1 : 1). RNA polymerase buffer 10×, is 400 mM Tris–HCl (pH 7.4), 60 mM MgCl$_2$, 100 mM dithiothreitol, 40 mM spermidine. The 10× NTP mix used in the probe synthesis reaction contains 5 mM each of ATP, CTP, and GTP. [α-^{32}P]UTP (10 mCi/ml, ~400 Ci/mmol aqueous) is from Amersham or Dupont/New England Nuclear. Yeast total RNA (Sigma) is resuspended in 1× proteinase K buffer, proteinase K is added to a final concentration of 100 μg/ml, and after a 1-hr incubation at 65°, the RNA is extracted twice with equal volumes of phenol–Sevag and once with an equal volume of Sevag. The RNA-containing aqueous phase is dialyzed against 10 volumes of 10 mM Tris–HCl (pH 7.4), 0.1 mM EDTA for 24 hr (changing buffer three times) at room temperature. This highly purified preparation of yeast total RNA is ethanol precipitated and finally resuspended in DEP-treated water at a concentration of 10 mg/ml. Sephadex G-50 (Sigma) is hydrated in 0.1% SDS, 0.02% DEP followed by autoclaving. The autoclaved Sephadex suspension is then made 10 mM Tris–HCl (pH 7.4), 0.1 mM EDTA. Hybridization mix, 5×, contains 2 M NaCl, 125 mM PIPES (pH 6.8), 5 mM EDTA. Analytical-grade formamide (Mallinckrodt), deionized with AG501-X8 mixed-bed resin (Bio-Rad), is stored at 4° and has a shelf-life of several months. Light mineral oil is from Mallinckrodt. SET, 1×, is 0.15 M NaCl, 0.03 M Tris–HCl (pH 8.0), 2 mM EDTA. Solid trichloroacetic acid (TCA, Mallinckrodt) is dissolved in water as a 100% (w/v) stock. Glass fiber filters (GF/C) are from Whatman.

Recommended enzyme sources are SP6 or T7 RNA polymerase: Promega Biotec or Boehringer-Mannheim; placental ribonuclease inhibitor (RNasin): Promega Biotec; proteinase K: Merck or Boehringer-Mannheim; RNase-free DNase I: Promega Biotec; ribonuclease A (sequencing quality): Cooper/Worthington or Boehringer-Mannheim; ribonuclease T_1 (sequencing quality): Cooper/Worthington or Boehringer-Mannheim.

Single-Stranded RNA Probe Preparation

Synthesis Reactions

Single-stranded RNA probes are prepared using plasmid vectors containing a polylinker bordered by a sequence-specific promoter of a bacteriophage RNA polymerase. A number of useful vectors are commercially available from several sources. We recommend using vectors that feature a polylinker flanked on each side by a different phage-specific promotor (e.g., the bluescribe vectors of Stratagene, San Diego, CA). These vectors allow the synthesis of RNA probes complementary to either strand of a cloned sequence from a single plasmid construct. We also recommend using CsCl-banded plasmid DNA for the probe synthesis reactions, although plasmid DNA isolated using rapid miniprep procedures as described in this volume [13] can be used. The probe synthesis reactions are carried out with linearized plasmid as template. The plasmid subclone is linearized by restriction enzyme digestion at a single site downstream of the insert. This site should be oriented such that the synthesized probe will be truncated by polymerase runoff. To synthesize a labeled probe that hybridizes to mRNA, the antisense strand of the cloned insert must be used as the template in the polymerase reaction. Keep in mind that like all other known RNA polymerases, the bacteriophage enzymes transcribe in a 5′ → 3′ direction.

Restriction endonuclease digestions of plasmid templates are terminated by incubation at 68° for 3 min. The linearized plasmid is then diluted with 5–10 volumes of water, 10× proteinase K buffer is added to 1×, and proteinase K is added to a final concentration of 100 µg/ml. After a 1-hr incubation at 65° the plasmid DNA is extracted twice with equal volumes of phenol–Sevag followed by a single extraction with an equal volume of Sevag. The linearized plasmid is precipitated at −20° with the addition of

2 volumes of ethanol. Proteinase K treatment of the digested plasmid is critical, because nearly all commercial preparations of restriction enzymes are contaminated with nonspecific ribonucleases. The plasmid DNA is resuspended in DEP-treated water at a concentration of 1 μg/ml. Probe synthesis reactions are generally carried out in 20-μl reaction volumes using 1 μg of plasmid DNA. A typical reaction is performed in an Eppendorf tube and contains 2 μl of 10× RNA polymerase buffer, 1 μl (1 μg) of plasmid template, 1 μl (30 U) of placental ribonuclease inhibitor, 2 μl of 10× NTP mix, 10 μl of [α-^{32}P]UTP (100 μCi), and 4 μl of water. The synthesis reaction is initiated with the addition of 1 μl (15 U) of SP6 or T7 RNA polymerase and incubated for 1 hr at 37°. Typically, >50% of the input [α-^{32}P]UTP is incorporated into trichloroacetic acid (TCA)-insoluble material (TCA precipitations are described elsewhere in this volume[11]). The use of labeled UTP as the only source of this nucleotide limits the mass yield of RNA probe. Under these conditions, transcription of a 400-nucleotide insert (assuming an average composition of 25% UMP residues) will produce approximately 150 ng of RNA probe with a specific activity of 6.6×10^5 dpm/ng. The synthesis reaction is terminated with a 30-sec incubation at 68°. To the reaction mixture 10 μl of yeast total RNA (100 μg), 3 μl 10× RNA polymerase buffer, 1 μl placental ribonuclease inhibitor (30 U), and water are added to a final volume of 50 μl. The plasmid template is removed with the addition of RNase-free DNase I (see this volume [21]) to a final concentration of 20 μg/ml. This reaction is incubated at 37° for 15 min. The DNase I digestion is stopped by diluting the reaction mix to 100 μl with DEP-treated water, followed immediately by extraction with an equal volume of phenol–Sevag. The phenol–Sevag-extracted probe is then extracted with an equal volume of Sevag and the resulting aqueous phase is loaded onto a 1-ml Sephadex G-50 spin column.

Column Purification

Spin columns are prepared in 1-ml disposable syringes by loading a Sephadex G-50 suspension into the syringe by gravity. A pad of dichloro-dimethylsilane-treated glass wool at the base of the syringe prevents Sephadex from escaping. The fully loaded column is allowed to run dry, at which time the syringe is suspended in a 15-ml plastic tube and spun in a clinical centrifuge at top speed for 45 sec. The centrifugal force eliminates all remaining buffer in the column and compresses the Sephadex to a volume of ~0.85 ml. The extracted RNA probe is applied directly to this

[11] S. L. Berger, this volume [6].

prespun column. An Eppendorf tube, whose cap has been removed, is put into a 15-ml tube and the column is suspended in this tube. This assembly is spun in the clinical centrifuge at top speed for 45 sec. The labeled RNA probe is recovered, free of unincorporated nucleotide triphosphates, in a volume equal to that applied to the column. The RNA probe is precipitated from 0.2 M sodium acetate at $-20°$ with the addition of 2 volumes of ethanol. The resulting pellet is washed with 70% ethanol and resuspended in DEP-treated water.

Probe Length and Stability

Using the conditions described here we have routinely synthesized intact labeled transcripts $\leq$2000 nucleotides in size. Before using a probe preparation, however, the size of the transcripts should be checked. This is an important step because some eukaryotic sequences act as efficient stop signals for the phage RNA polymerases. Probe preparations in which a significant fraction of the probe molecules are not full length should be avoided. These probes underrepresent sequences actually present in the insert, and as a result it is too difficult to control for many of the hybridization parameters. The subsequent fragmentation of an initially full-length probe (e.g., autoradiolysis with time) is of little consequence, since all sequences of the insert remain equally represented and since the assay does not size select for nuclease-protected probe molecules forming a discrete band on a gel. We have found that full-length probe preparations which are 1–2 weeks old can still be used in a titration assay.

Quantitation of Transcript Prevalence in an RNA Population

Preliminary Considerations

The sensitivity of the titration assay is dependent on the specific activity of the probe and the signal-to-noise of the assay. It is our experience that the ribonuclease background (i.e., noise) is generally $\leq$0.5% of the input tracer and this background is independent of the amount of probe used in the hybridization. Care must be taken to use an amount of probe which is in excess (>10-fold) of the transcripts of interest yet not so excessively high as to increase the noise to an undesirably high level. Besides reducing the background, the sensitivity of the assay can also be maximized by increasing the hybridization signal. The easiest way to produce an increase is to use probes of higher specific activity. The unfortunate drawback of synthesizing probes of these higher specific activities, however, is that many more dpms of probe are necessary to achieve probe excess and reaction completion. Thus there is little advantage to synthesizing probes of increased specific activity. Alternatively, the hybridization signal can be effectively increased by increasing the amount of cellular RNA used in the hybridization. The greater hybridization signal is obtained without a corresponding increase in the background noise. The only constraints on increasing the signal in this manner are the availability of cellular RNA and the achievable concentration of cellular RNA in the hybridization reaction. For example, sea urchin total embryonic RNA can be isolated in mg quantities and be resuspended to a maximum concentration of 5 mg/ml. Higher concentrations of RNA are possible if the RNA is sheared before resuspension. Cellular RNA sheared to a fragment length of 400–500 nucleotides is still a good substrate for the titration assay and can be resuspended to a concentration of up to 20 mg/ml.

Estimation of Transcript Prevalence

An estimate of the number of transcripts present must be made in order to determine the amount of probe needed to achieve an effective excess. The preliminary guess of transcript prevalence can usually be made from RNA gel blot or dot blot data. In its simplest form the guess is made by a comparison of input total RNA and autoradiographic exposure times with similar data using transcripts whose prevalence is independently known. The importance of being in effective tracer excess can be seen in Fig. 1. This figure represents an idealized form of the titration of excess single-stranded tracer with increasing amounts of input cellular RNA. The plot is divided into two major regions (I and II). The division occurs at the point at which the slope of the curve is discontinuous. This inflection point represents the mass of cellular RNA which contains enough transcripts to eliminate the effective tracer excess in the hybridization reaction. Thus in region I (effective tracer excess) the amount of probe driven into duplex is proportional to the input mass of cellular RNA, whereas after the inflection point, region II, the system saturates and the amount of probe driven into hybrid can no longer increase. Without an estimate of transcript prevalence, a set of test hybridizations with a given amount of probe is necessary to determine the point of inflection. Although inconvenient, this is the most accurate method of ensuring that the titration is performed under the condition of effective tracer excess.

Selection of Hybridization Conditions

The factors determining hybrid stability and the kinetics of nucleic acid reassociation are reviewed in detail in this volume,[8] and else-

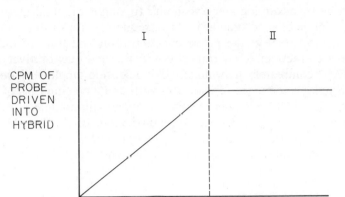

FIG. 1. Tracer hybridization as a function of cellular RNA input. The curve presented is an idealized form of probe hybridization (ordinate) versus increasing mass input of cellular RNA (abscissa). The plot is divided into two distinct regions (I and II) by the vertical dashed line. Region I is linear with a positive slope of m, while region II is also linear but has a slope of zero.

where[12,13] (see also this volume [43]). The solvent used for the hybridization reactions of the titration assay contains 50% formamide, 0.4 M NaCl, 1 mM EDTA, and 25 mM PIPES (pH 6.8). The effect of formamide is to reduce the melting temperature of the hybrids (T_m), thus avoiding probe and RNA degradation that occurs at high temperatures of incubation. To estimate accurately the T_m of a hybrid in this solution it is necessary to use well-characterized probes (e.g., GC composition and hybrid length). Unfortunately, studies of RNA–RNA hybridization are not nearly as complete as either DNA–DNA or DNA–RNA reactions.[14] The absence of data about the T_m of RNA–RNA duplexes makes it difficult to determine the appropriate temperature of hybridization. For hybrids of average GC composition it is our experience that under the above conditions the optimal temperature of hybridization using RNA probes 500–1000 nucleotides in length is 50°. This temperature may be changed, however, to meet the specific characteristics of any given probe.

The rate of hybridization reactions depends solely on the concentration of probe. The rate constant, $k(M^{-1}\ sec^{-1})$, may be estimated for

[12] R J. Britten, D. E. Graham, and B. R. Neufeld, this series, Vol. 29E, p. 393.
[13] R. J. Britten and E. H. Davidson, in "Nucleic Acid Hybridization: A Practical Approach" (B. D. Hames and S. J. Higgins, eds.), p. 3. IRL Press, Oxford, 1985.
[14] J. Casey and N. Davidson, *Nucleic Acids Res.* **5,** 1539 (1977).

each probe by assuming kinetics similar to single-stranded DNA probes and by using published standards.[15] Consequently, for each hybridization it is possible to select the probe concentration and time of incubation necessary to recover RNA transcripts in hybrids quantitatively. As an example, a completely hybridized 500-nucleotide single-stranded DNA probe of unique sequence reassociates with a rate constant (k) of ~3.7 × 10^3 in 0.4 M NaCl.[8,12,13] This reaction is essentially complete (i.e., 95%) when incubated to C_0t of 8.1 × 10^{-4} (ln 0.05/3.7 × 10^3) with respect to the probe. Thus for a commonly used probe concentration such as 4 ng/ml, half reaction would occur in about 4 hr, and 95% reaction in about 18 hr.

Hybridization Protocol

The titration assay consists of a series of hybridizations between a constant amount of RNA probe and increasing amounts of input cellular RNA. We have found that five or six data points are required to produce a line from which the slope can be accurately determined. The hybridization reactions are performed in dichlorodimethylsilane-treated 1.5-ml plastic microcentrifuge tubes. Reactions typically contain 0.05 to 0.20 ng of tracer and various amounts (0 to 100 μg) of input cellular RNA. These two components, plus added carrier RNA to maintain a constant amount of RNA in each set of reactions, are precipitated from 0.2 M ammonium acetate at −20°. Augmenting the total RNA content of each reaction to an equal level serves the following functions. Since each tube in a series has the same input mass of RNA, recovery and resuspension of the RNA from tube to tube are uniform. In addition, by normalizing the input mass of RNA, each reaction is identical with respect to the mass of RNA digestable by ribonuclease. The final RNA content of a series is usually set at 50 or 100 μg. The resulting pellets are freed from residual salt by drying under vacuum in a Savant Speed Vac concentrator. The dried pellets are taken up in 10 μl of 100% deionized formamide and 6 μl of water. Once dissolved, 4 μl of 5× hybridization mix is added, and the reactions are covered with 50 μl of mineral oil. Hybridization is initiated by incubation at 85° for 5 min. Heating also ensures that the RNA and probe are fully resuspended. The microcentrifuge tubes are then incubated in a water bath at the appropriate temperature.

[15] G. A. Galau, R. J. Britten, and E. H. Davidson, *Proc. Natl. Acad. Sci. U.S.A.* **74**, 1020 (1976).

Ribonuclease Digestion and Recovery of Hybrids

Commercial sources of highly purified (i.e., contaminant-free) ribonucleases have been available for some time.[16] These enzymes are specific for single-stranded RNA and function in the absence of both ATP and divalent cations. Two of these enzymes, ribonuclease A and ribonuclease T_1, which cleave 3' to pyrimidines and G residues, respectively, have been shown to be active and retain their specificity for single-stranded RNA in little more than Tris-buffered water containing an adequate concentration of monovalent cations. The specificity of these enzymes for single-stranded RNA and their high activity in simple buffers make them ideal tools for separating hybridized probe (double-stranded RNA) from unreacted probe (single-stranded RNA) in a titration assay. The use of only one of these enzymes, however, is insufficient to digest internally labeled RNA completely, as assayed by precipitation in 5% (w/v) TCA. Consequently, both enzymes must be used to digest unhybridized probe in a titration assay. The amount of enzyme to use is determined by separately calibrating each enzyme as follows. Increasing amounts of ribonuclease A or T_1 are added to samples containing only yeast total RNA under conditions that mimic those of the assay. TCA precipitations measure the maximum percentage digestion of the input tracer in each case. The amount of enzyme at these maxima are used to remove unhybridized probe in the titration assay. For example, in reactions which contain a probe specific for an actin 3' untranslated region from the sea urchin and 50 μg of yeast total RNA, we have found that 65 μg of ribonuclease A is needed to achieve maximal digestion (~80%) of the input tracer. Under the same conditions, maximal digestion (>95%) of the tracer occurs with 150 units of ribonuclease T_1. Using these predetermined amounts of each enzyme in an actual titration assay results in ≥99.5% digestion of the actin gene-specific probe.

The hybridization reactions are quenched on ice until treatment with nuclease. Unhybridized tracer is removed from each reaction with the addition of a 300-μl aliquot of 2.5× SET containing appropriate amounts of ribonuclease A and ribonuclease T_1. These reactions are incubated at 37° for 1 hr. Afterward, 300 μg of yeast total RNA is added as carrier and each entire reaction is immediately precipitated by the addition of an equal volume of ice-cold 10% (w/v) TCA. The precipitates are kept on ice for 10 min and are collected on GF/C filters. Their radioactivity is determined by scintillation counting. Since the labeling isotope is ^{32}P, dried

[16] H. Donis Keller, A. M. Maxam, and W. Gilbert, *Nucleic Acids Res.* **4**, 2527 (1977).

GF/C filters can be counted in a variety of scintillation cocktails (e.g., Aquasol New England Nuclear, Dupont) with maximal counting efficiency.

Unit Conversion and Interpretation of Titration Data

A plot of raw titration data (cpm of RNase-resistant probe versus mass unit of cellular RNA) should yield a line whose slope (m) is calculated by the least squares methods. Since optimal results from the titration assay are based in part on the initial guess of transcript prevalence, a linear plot may not always be obtained. If, for example, the transcript prevalence is much higher than the initial guess the condition of tracer excess may not exist. Under these conditions a plot of the assay data may prematurely level off (see Fig. 1), or in the extreme case never show any increase in hybridization signal throughout the range of input mass of cellular RNA used. Alternatively, if the transcript prevalence is much lower than the initial guess, the hybridization background may be too high when compared to the actual signal. The resulting scatter will make the titration data unintelligible. Either of the above conditions can be corrected simply by adjusting the amount of probe and/or the range of cellular RNA used in the titration assay.

The background hybridization, which should be subtracted from each of the raw data points of the assay, is represented by the y intercept of this plot. The data points, after subtracting this value, are then replotted to determine the slope (m). We have found that this treatment of background noise is more representative of the true background of the assay than using controls that contain only carrier RNA. This reflects the fact that all of the data points of the assay are used to produce the extrapolated background whereas carrier control values are generally obtained from only one or two samples.

The slope of the linear portion of a titration curve has the units cpm of ribonuclease-resistant probe per mass unit of cellular RNA. Data reduction, using appropriate reduction factors, transform this slope to an absolute prevalence level of the transcripts of interest. These reduction factors are listed in Table I and summarized in the following relationship:

$$T = \frac{n}{\alpha\beta\delta} m\Sigma \tag{2}$$

where T is the number of transcripts per cell, m is the experimentally determined slope of the titration curve, α is the ^{32}P fractional counting efficiency, β is the probe specific activity (dpm/ng), δ is the molecular weight of the hybridizable portion of the probe (ng/mmol), n is Avoga-

TABLE I

UNIT CONVERSIONS FOR REDUCTION OF TITRATION DATA

Input units	Conversion factor	Resulting units
Cpm of RNase-resistant probe	^{32}P counting efficiency	Dpm of RNase-resistant probe
Dpm of RNase-resistant probe	Specific activity of probe^{-1} (ng/dpm)	Nanograms of probe
Nanograms of probe	Probe molecular weight^{-1} (nmol/ng)	Nanomoles of probe
Nanomoles of probe	Avogadro's constant (6.023 × 10^{14} molecules/nanomole)	Number of transcripts
Number of transcripts	Transcript molecular weight (pg/transcript)	Picograms of transcripts
Mass unit of cellular RNA	Mass of RNA per cell	Cell

dro's constant, and Σ is the mass of RNA per cell. If needed, this equation can be expanded to transform the prevalence measurement into the mass of transcripts present. This transformation is accomplished by multiplying the value of T by the molecular weight of the transcripts of interest.

The unit conversion represented by Σ requires specific knowledge of the RNA content of the cell, cell fraction, or embryo under study. These data exist for many cell types and are compiled in Table II. If the RNA content per cell is unknown, this measurement can be made as follows. The RNA content of a cell is experimentally determined as the amount of RNA recovered from a known number of cells, normalizing for the efficiency of RNA isolation. The number of nuclei (i.e., the number of cells) contributing to an RNA preparation is determined by measuring the mass of DNA present. This determination is made by DAPi fluorescence measurements (DAPi, 4,6-diamidino-2-phenylindole[17]) of the cell, cell fraction, or embryo lysate. DAPi is a fluorescent dye that changes its spectral properties upon binding DNA. This change in fluorescence occurs in a wide range of buffers, including homogenization buffers containing low concentrations of ionic detergents (e.g., SDS). The amount of DNA (i.e., the number of nuclei) in an unknown sample is determined from the ratio of the changes in fluorescence between the unknown and a known DNA standard. The efficiency of RNA isolation is determined by *in vivo* labeling total RNA of growing cells or embryos with a [^{3}H]ribonucleotide precursor (e.g., [^{3}H]uridine). The fraction of [^{3}H]RNA (in dpm) recovered during the RNA isolation procedure is the RNA isolation efficiency. The number of cells contributing to a preparation of RNA is the product of the

[17] C. F. Brunk, K. C. Jones, and T. W. James, *Anal. Biochem.* **92**, 497 (1979).

TABLE II
TOTAL RNA CONTENT IN VARIOUS CELLS

Organism and source of total RNA	Total RNA/cell (ng)	Reference
Strongylocentrotus purpuratus, egg or embryo	2.8	[a]
Xenopus laevis, egg or embryo	4800	[b]
Drosophila melanogaster, egg or embryo	190	[c]
Mouse, egg	0.46	[d,e]
NIH 3T3 cells	2.5×10^{-2}	[f]
Mouse L cells	2.6×10^{-2}	[g]
Rat, liver	2.1×10^{-2}	[h]
Rat, kidney	1.0×10^{-2}	[h]
Rat, spleen	3.2×10^{-3}	[h]
Rat, thymus	1.7×10^{-3}	[h]
Tetrahymena thermophila	0.3	[i]
Saccharomyces cerevisiae	3.2×10^{-3}	[j]
Escherichia coli	1×10^{-4}	[k]

[a] A. S. Goustin and F. H. Wilt, *Dev. Biol.* **82,** 32 (1981).

[b] M. A. Taylor and L. D. Smith, *Dev. Biol.* **110,** 230 (1985).

[c] K. V. Anderson and J. A. Lengyel, *Dev. Biol.* **70,** 217 (1979).

[d] G. Kaplan, S. L. Abreu, and R. Bachvarova, *J. Exp. Zool.* **220,** 361 (1982).

[e] K. B. Clegg and L. Pikó, *Dev. Biol.* **95,** 331 (1983a).

[f] S. Tavtigian and B. Wold (unpublished observations).

[g] B. P. Brandhorst and E. H. McConkey, *J. Mol. Biol.* **85,** 451 (1974).

[h] H. Von Euler and L. Hahn, *Arch. Biochem.* **17,** 285 (1948).

[i] F. J. Calzone, R. C. Angerer, and M. A. Gorovsky, *J. Biol. Chem.* **258,** 6887 (1983).

[j] A. F. Croes, *Planta* **76,** 209 (1967a).

[k] J. D. Watson, *in* "Molecular Biology of the Gene," 3rd ed. W. A. Benjamin, Menlo Park, CA, 1976.

RNA isolation efficiency and the DAPi fluorescence measurement of the total number of nuclei.

The top section of Fig. 2b provides an example of the data obtained using the titration assay. This figure contains five separate titration assays performed with the actin *CyIIIa* gene-specific probe from the sea urchin *Strongylocentrotus purpuratus*. The plots represent the titration of *CyIIIa* gene transcripts in total RNA from *S. purpuratus* embryos of five different developmental stages. The changes in *CyIIIa* transcript prevalence

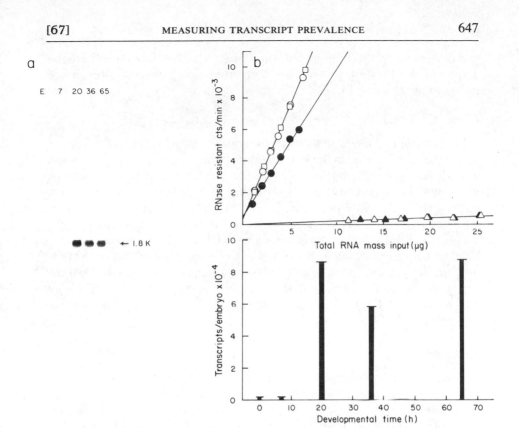

FIG. 2. Developmental patterns of actin *CyIIIa* transcript accumulation in *S. purpuratus* embryos. (a) Embryo total RNAs (10 μg per lane) were fractionated electrophoretically in a denaturing formaldehyde–agarose gel and blotted to nitrocellulose. The RNAs loaded in each lane were as follows: (left to right) total RNA from unfertilized eggs (E), from embryos 7, 20, 36, and 65 hr postfertilization. A ^{32}P-labeled *CyIIIa* RNA probe was prepared (see text; probe length 131 nucleotides, specific activity 8×10^5 dpm/ng) and hybridized with the blotted embryo RNAs for 20 hr at 50°, 0.40 M Na$^+$ in 50% formamide. (b) *CyIIIa* RNA probe [same probe as used in (a)] was used in tracer excess titrations of *S. purpuratus* unfertilized egg and embryo RNAs. Measurements were carried out on total RNAs prepared from (△) unfertilized eggs, and from embryos (▲) 7, (○) 20, (●) 36, and (□) 65 hr postfertilization. The upper portion of this panel represents probe hybridization (ordinate) as a function of increasing quantity of egg or embryo RNA (abscissa). Hybridizations were carried out at 50°, 0.40 M Na$^+$ in 50% formamide as described in the text. The quantity of RNA–RNA hybrid was assayed in each sample by digestion with RNase A and RNase T$_1$ in 0.375 M Na$^+$ followed by precipitation in 5% trichloroacetic acid. The data are presented as RNase-resistant counts per minute versus total RNA input (in micrograms). The slopes of the lines from these data were used to calculate the embryonic prevalence of the *CyIIIa* actin mRNAs. The results of these calculations are displayed in the histogram shown in the lower portion of this panel.

are illustrated in the bottom portion of Fig. 2b. The raw data (i.e., the slopes of the titration plots) were converted into transcript prevalence measurements using Eq. (2).

Limits of Detection

A determining factor in the sensitivity of the assay is the lower limit of probe that can be used in the hybridization. Since the probe concentration is inversely proportional to the hybridization time, the least amount of probe which can be conveniently used in a 20-μl hybridization reaction is 0.03 ng (RNA C_0t 95% occurring in ~48 hr). This amount of probe (specific activity 6.6×10^5 dpm/ng) would produce a titration assay background (0.5% of input tracer) of 100 dpm [$(0.03)(6.6 \times 10^5)(0.005)$]. At this level of background, the minimal amount of radioactivity that could reliably be detected is 200 dpm or 0.1–0.3 pg of complementary RNA transcript. Even at this lower limit, it has been our experience that the assay can reproducibly quantify transcripts to yield prevalence data for extremely rare RNA species.

Discussion

The molecular titration assay provides a system that allows the detection and quantitation of low prevalence transcripts with the speed and ease of qualitative blot methods. Figure 2a shows an RNA gel blot of total RNAs from different stages of *S. purpuratus* development. The RNA gel blot (10 μg of total RNA per lane) is hybridized with labeled actin *CyIIIa* RNA probe. Figure 2b shows titration assays in which these same probe and embryonic total RNAs were used. It is clear that whereas the blot analysis fails to detect *CyIIIa* transcripts in total RNA from either egg or 7 hr postfertilization embryo, the titration assay easily detects these transcripts. The titration assay also eliminates the cumbersome steps of gel electrophoresis and blot transfer. The synthesis of a probe, and the titration of transcripts from several RNA samples, can be accomplished in as little as 2–3 days. The cellular RNA under study is not size selected and the sensitivity and accuracy of the assay are not dependent on the integrity of the cellular RNA sample. This method, therefore, provides a rapid and convenient assay by which even extremely rare RNA species can be detected and quantified.

Acknowledgments

Research from this laboratory was supported by NIH Grant HD-05753 awarded to E. H. Davidson. J.J.L. was supported by an NIH training grant (GM07616). N.A.C. was supported by an American Cancer Society fellowship.

[68] Demonstration of Tissue-Specific Gene Expression by *in Situ* Hybridization

By Lynne M. Angerer, Kathleen H. Cox, and
Robert C. Angerer

In situ hybridization has emerged as a valuable tool for the identification of individual cells expressing specific genes. Recently, methods have become sufficiently sensitive to detect mRNAs present at the level of only a few molecules per cell. When mRNAs are expressed in only a small fraction of the cells in a mixed population, *in situ* hybridization may be the most sensitive nucleic acid hybridization technique available. Our laboratory has shown that antisense RNA probes offer a unique combination of advantages for detection of individual mRNAs by *in situ* hybridization.[1] Most importantly, antisense probes provide a large increase in sensitivity due to the absence of competing probe self-reassociation. Single-stranded RNA probes can be labeled to high specific activity, and a relatively large fraction of the labeled precursors can usually be converted to RNA product. Sense transcripts can be used to measure nonspecific background. Such backgrounds can be reduced by posthybridization RNase digestion under conditions where probe–target hybrids are resistant. The high stability of RNA–RNA duplexes allows use of higher posthybridization wash temperatures to achieve a given fidelity of base pairing (stringency), which also reduces backgrounds. RNA transcripts can be synthesized from truncated templates such that they are essentially devoid of vector sequences. This sequence purity maximizes the signal to noise ratio since lower probe concentrations are required to saturate target RNAs.

Tissue Preparation

Most of the diversity of opinion and practice of *in situ* hybridization revolves around procedures for fixing tissue, and for prehybridization treatments designed to increase access of probe to target mRNAs. *In situ* hybridization is usually applicable to tissue fixed under conditions that provide good histological detail. Fixed tissues are stable for long periods, allowing both future comparisons and retrospective studies. Because the magnitude of the signal depends on the fraction of target RNA retained in tissues throughout the procedure, it is generally observed that cross-

---

[1] K. H. Cox, D. V. DeLeon, L. M. Angerer, and R. C. Angerer, *Dev. Biol.* **101**, 485 (1984).

linking fixatives such as formaldehyde and glutaraldehyde are superior. Probe penetration and target accessibility are facilitated by use of proteases and probes of short fragment length.[2-4] The biochemical composition of tissues is highly variable, and there is no reason to expect that one procedure will be best for all. However, the procedure given here has been successfully employed, with minor modifications, in a variety of systems including sea urchin and *Drosophila* embryos, mammalian culture cells, unicellular prokaryotes, and plant tissues. In adapting *in situ* methods to new systems, it is very important to recognize that individual steps comprising any overall protocol are interdependent. Individual steps that work very well in the context of one complete method are disastrous in different combinations. Thus, tissues fixed less extensively (e.g., using lower concentrations, shorter times, or formaldehyde instead of glutaraldehyde) may require less protease digestion or none, and considerable loss of target RNAs may occur after protease pretreatment of tissues fixed with non-cross-linking fixatives such as ethanol–acetic acid.

As described below, we routinely use 1% glutaraldehyde to fix sea urchin embryos and tissues. Most target mRNA is retained and accessible to hybridization probes after appropriate protease treatments.[1] We and others have also used formaldehyde,[4-8] but we have not directly compared maximum signals obtainable with these two fixatives.

Fixation with Glutaraldehyde

Suspend cells or small tissue fragments in at least 5 volumes of ice-cold 1% glutaraldehyde, 50 mM phosphate buffer (pH 7.5), and sufficient NaCl to control the osmolality of the fixative, avoiding excessive shrinking or swelling. The osmolality of 1% glutaraldehyde is equivalent to 0.375% NaCl.[9] Change fixative and incubate on ice for a total of 1 hr. Wash in at least 10 volumes of phosphate–NaCl buffer at 0° for 30 min. Repeat this wash once. Single cells or large tissue blocks may require adjusting times or fixative concentrations proportionately.

[2] M. Brahic and A. T. Haase, *Proc. Natl. Acad. Sci. U.S.A.* **75**, 6125 (1978).

[3] L. M. Angerer and R. C. Angerer, *Nucleic Acids Res.* **9**, 2819 (1981).

[4] D. J. Brigati, D. Myerson, J. J. Leary, B. Spalholz, S. Z. Travis, C. K. Y. Fong, G. D. Hsiung, and D. C. Ward, *Virology* **126**, 32 (1984).

[5] R. C. Angerer, M. Stoler, and L. M. Angerer, *in* "In Situ Hybridization: Applications in Neurobiology" (K. Valentino, J. Eberwine, and J. Barchas, eds.). Oxford Univ. Press, London and New York, 1987.

[6] E. Hafen, M. Levine, R. L. Garber, and W. J. Gehring, *EMBO J.* **2**, 617 (1983).

[7] M. Jamrich, K. A. Mahon, E. R. Gavis, and J. G. Gall, *EMBO J.* **3**, 1939 (1984).

[8] M. E. Akam, *EMBO J.* **2**, 2075 (1983).

[9] G. Millonig and V. Marinozzi, *Adv. Opt. Electron Microsc.* **2**, 251 (1968).

Embedding in Paraffin and Sectioning

Wash one time with isotonic NaCl to prevent precipitation of phosphate in ethanol solutions. Resuspend tissue in the same solution and add 99% ethanol with stirring to a final concentration of 50% at room temperature. After 15 min, resuspend tissue in fresh 50% ethanol containing enough salt to prevent swelling. Incubate 15 min. Suspend in 70% ethanol at room temperature. Decant and add fresh 70% ethanol after 15 min. These initial dehydration steps were developed for a marine invertebrate embryo and may not be optimal for all tissues. When in doubt, carry out these steps in higher ethanol and/or salt concentrations that avoid excessive swelling.

Dehydrate the tissue through ethanol solutions (85, 95, 100, 100%; 100% is 99% stored over molecular sieve) at room temperature, each for 30 min, using at least 10 tissue volumes. Allow tissue to settle between changes and aspirate off the supernatants. Continue with two 30-min changes of xylene, followed by 45 min in paraffin. (Use paraffin mixture in Appendix.) Infiltrate with paraffin, using three changes for 15 min each at 58°. Use recently melted paraffin that has been degassed either in a vacuum oven or with a standard pump equipped with cold trap. Prolonged heating of paraffin or heating at temperatures above 60° damages its sectioning properties.

Place small pieces of tissue in molds containing melted paraffin. For suspensions of cells or embryos, transfer a drop of a concentrated suspension quickly to an embedding mold using a prewarmed wide-bore Pasteur pipet. Allow the drop to solidify and then heat at 58° just until the surface begins to glisten. Immediately add liquid paraffin just to cover the drop. Hold at room temperature until a skim forms and then fill the boat with liquid paraffin and allow the block to solidify at room temperature. Small quantities of tissue can be embedded in plastic electron microscopy molds (BEEM capsules) or microcentrifuge tubes. Processing of cell suspensions or small amounts of tissue may be facilitated by first embedding them in low melting temperature agarose (0.8% final concentration). Cut the agarose blocks into small pieces before fixation and embedding. Tissue can be stored in paraffin at 4° and subsequently remelted. This allows mixing of different samples in the same block for the best comparison of relative signals. Tissue is stable at least several years in paraffin, without loss of hybridizability of target RNAs or deterioration of morphology. Fixed tissue may be stored in 70% ethanol tightly capped at 4° for several months, although it does not appear to be as stable as in paraffin.

Slide Preparation. Clean slides in Chromerge at least overnight. Insert them vertically in a test tube rack (40 slides/rack) and rinse at least 15 min each in running tap and running distilled water. Finally rinse slides individually by holding them with forceps under a forceful stream of distilled water. Air dry in a dust-free place. Coating the slides with poly(L-lysine) by a modification of the procedure of McClay[10] provides best retention of tissue: Prepare 500 μg/ml poly(L-lysine) stock in 100 mM Tris–HCl (pH 8.0). Store frozen indefinitely. Dilute 1 : 10 with water, distribute into slide mailers, and incubate slides for 10 min at room temperature. At least four sets of slides can be coated from each volume of polylysine. Air dry in a dust-free place. Slides will have spots of polylysine on them. Coated slides keep at least several weeks at room temperature.

Sectioning. This is easier to learn by observation, and a few hours in an histology laboratory is probably a good investment.

Mount the paraffin block by pressing the back of it on a heated chuck or holder. Trim the tissue block with a sharp razor blade into a truncated pyramid with a trapezoidal face, so that the leading edge of each section is slightly wider than the trailing edge of the previous section, which it displaces from the knife blade. Make the block face as small as possible without losing tissue to minimize the area to be covered by probe and to facilitate sectioning.

Using a standard rotary microtome, section at 5 μm nominal thickness. The most frequent sectioning problems are dull blades, poor-quality paraffin, and static electricity, which causes sections to jump back to the block instead of ribboning from the knife. Cooling the block and knife by spraying with Cryoquick sometimes helps. Ribbons can be picked up via a drop of water on the end of a spatula and spread on the surface of 45° degassed distilled water to the original block face size. (Water must be degassed to avoid formation of bubbles between sections and slide.) Mount sections by inserting a coated slide into the bath near the sections at a 45° angle to the surface of the water, touching an edge of the ribbon to the slide and slowly withdrawing the slide from the bath. Keep the sections 1–5 cm from the left edge of the slide so that they are sure to be covered by autoradiographic emulsion but are not too close to the left end where the emulsion is thicker and drying artifacts may occur.

Place slides on a slide warmer at 40° for at least 2 hr. Shorter times may lead to more loss of sections. Store slides for up to a week in a clean, dry slide box at room temperature.

Sections as thin as about 0.5 μm can be cut with glass knives on an ultramicrotome. Adjacent thin sections are particularly useful for compar-

---

[10] D. R. McClay, G. M. Wessel, and R. B. Marchase, *Proc. Natl. Acad. Sci. U.S.A.* **78,** 4975 (1981).

ing, in the same cells, distributions of different mRNAs using two different probes or for confirming the reproducibility of patterns obtained with a single probe, especially if signals are low. Thin sections also allow more experiments from limited quantities of tissue.

Prehybridization Treatments

Deparaffinization and Hydration of the Sections. Immerse labeled slides in glass slide carriers in fresh xylene, two changes for 10 min each. Agitate gently either on a shaking platform, or with a magnetic stirring bar under the slide rack. Pass the slides sequentially for about 15 sec each through 99, 99, 95, 85, 70, 50, and 30% ethanol, water, water.

Proteinase K Digestion. Place slides in 100 mM Tris–HCl, 50 mM EDTA (pH 8.0, 37°), 1 μg proteinase K/ml for 30 min. A 10 mg/ml enzyme stock made in the same buffer is stored frozen in small aliquots, which are used once and discarded. Although 1 μg/ml is often close to the optimal concentration for glutaraldehyde-fixed material, different extents of digestion may be required for different tissues and/or fixation conditions. It is useful to try several concentrations to determine empirically the best compromise between tissue morphology and hybridization signal.

Acetic Anhydride. Acetylation of amino groups in tissues reduces electrostatic binding of probe.[11] In the sequence of steps given here, this treatment also serves to block binding of probe to polylysine and to inhibit proteinase K activity. Wash slides briefly in distilled water and then in freshly prepared 0.1 M triethanolamine–HCl (pH 8.0), at room temperature. Add undiluted acetic anhydride to an empty dry staining dish to give a final concentration of 0.25% (v/v) when sufficient triethanolamine buffer is added to cover the slides. Rapidly dip the slides up and down several times to mix the acetic anhydride. Incubate at room temperature for 10 min. Wash briefly in 2× SSC and dehydrate sections by passing slides through the graded ethanol series up to 99% ethanol. Place slides vertically in a test tube rack on a paper towel to dry. Cover to keep dust free.

Denaturation for Detection of DNA. DNA is denatured after proteinase K and acetic anhydride treatments by immersing the dehydrated slides in 95% formamide, 0.1× SSC for 15 min at 65°. The slides are then transferred to ice-cold 0.1× SSC for 5 min, dehydrated through ethanol solutions, and subsequently handled as for mRNA hybridization. We

---

[11] W. Hayashi, I. C. Gillam, A. D. Delaney, and G. M. Tener, *J. Histochem. Cytochem.* **36,** 677 (1978).

have used this procedure only with formaldehyde-fixed tissue and do not yet know what fraction of DNA sequence is available for hybridization.

Hybridization

Probe Preparation

Procedures for *in vitro* transcription of RNAs from vectors containing SP6 promoters and purification of transcripts have been discussed elsewhere in more detail[12-14] (see, also, this volume [30,59,61,65]). Shorter probes give higher *in situ* hybridization signals.[2,3,8] Fragments of about 150 nucleotides hybridize efficiently without large decreases in thermal stability.

Dissolve the probe in 50 μl water and add an equal volume of 0.2 M carbonate buffer, pH 10.2 (80 mM NaHCO$_3$/120 mM Na$_2$CO$_3$), which is made fresh or stored frozen in small aliquots. Incubate at 60° for time t given by the following relationship[1]:

$$t = (L_0 - L_f)/kL_0L_f$$

where t is the time in minutes; L_0, initial length in kilobases; and L_f, desired length in kilobases. The rate k is approximately 0.11 strand scissions/kb per minute.

Neutralize by adding 3 μl 3 M sodium acetate (pH 6.0) and 5 μl 10% (v/v) glacial acetic acid. Precipitate with ethanol as described in this volume [5]. Dissolve in a small volume of water and store frozen. ^{35}S-labeled probes should be dissolved in 10 mM Tris–HCl, 1 mM EDTA (pH 8.0), containing 20 mM dithiothreitol. Determine the size of hydrolyzed RNAs in denaturing formaldehyde or methylmercuric hydroxide 2–2.5% agarose gels (this volume [8]).

Setting up the Hybridization

Preparation of the Hybridization Mix. Final concentrations are 50% formamide, 0.3 M NaCl, 20 mM Tris–HCl (pH 8.0), 1 mM EDTA, 1× Denhardt's solution [0.02% each bovine serum albumin, Ficoll, polyvinylpyrollidone], 500 μg/ml yeast tRNA, 10% dextran sulfate. Add 500 μg/ml poly(A) for cDNA probes that contain poly(U) tracts. Dextran

[12] D. Melton, P. Krieg, M. Rebagliati, T. Maniatis, K. Zinn, and M. R. Green, *Nucleic Acids Res.* **12,** 7035 (1984).

[13] P. A. Krieg and D. A. Melton, *in* "Promega Notes," No. 15377. Published by Promega, Madison, WI, 1985.

[14] R. C. Angerer, K. H. Cox, and L. M. Angerer, *Genet. Eng.* **7,** 43 (1985).

sulfate increases hybridization signals 5-fold[5] independent of hybridization time. Include 100 mM dithiothreitol for ^{35}S-labeled probes. First combine nucleic acids in water and heat at 80° for 2 min. Then add the remaining components combined in a mix. Count a small aliquot by aqueous liquid scintillation counting (dextran sulfate quenches ^{3}H severely on filters) to ensure that probe concentrations are correct before proceeding.

Probe Concentration and Hybridization Time. The kinetics of *in situ* hybridization have been discussed.[1,14] Using a fixed hybridization time (16 hr), signals increase linearly with probe concentration until saturation is achieved. Maximum signal-to-noise ratios are obtained at probe concentrations just sufficient to saturate target RNAs; higher concentrations only increase backgrounds. In our system saturation is achieved at 0.2–0.3 μg/ml probe per kilobase of probe complexity. Higher complexity probes give higher signals at saturation, but also proportionately higher backgrounds, since higher RNA concentrations are required for saturation. Thus, longer probes require shorter exposures to achieve the same signal, but do not result in higher signal-to-noise ratios.

Setting up the Slides. To pretreated slides which are completely dry, spot the hybridization mix (4 μl/cm^2) at the center of the area containing sections. Using a diamond pencil, cut siliconized, baked (150°, 2 hr) coverslips to maximize section coverage and minimize probe amounts. Using fine-tipped forceps, lay the coverslip over the drop of mix and slowly decrease the angle between the coverslip and slide so that bubbles will be forced out. Coverslips and slides must be very clean to minimize bubbles. Do not press down on the coverslip because this results in a thinner layer of probe solution and decreased signals. To seal, immediately immerse the slides in mineral oil in a covered plastic box. Incubate at 45° which is approximately 25° below the T_m of hybrids formed *in situ* with 150-nucleotide probe fragments of typical GC content.[1] These conditions allow the maximal hybridization rate, and stringency can be further controlled during posthybridization wash steps.

Washes

Remove mineral oil by draining slides, positioned lengthwise, on a paper towel. Place in Coplin jars or glass slide racks and wash with three changes of chloroform. Surface tension keeps the coverslips on, provided the slides are not too vigorously agitated. Do not manually remove coverslips since the shear force generated may damage tissue. To remove coverslips, drain excess chloroform *briefly* and immerse slides in 4× SSC at room temperature, using three or four changes for a total of about 15 min to loosen coverslips. The coverslips should come off by the second wash, so that at least the last wash is effective in removing the bulk of unhybridized probe.

Nonspecifically bound probe is hydrolyzed with 20 μg/ml RNase A in 0.5 M NaCl. 10 mM Tris–HCl (pH 8.0), 1 mM EDTA, for 30 min at 37°. This is done in five-slot slide mailers to conserve enzyme. RNase is stored frozen as a 10 mg/ml stock, and may be refrozen. In cases of low signals, backgrounds may be further reduced by inclusion of 1 unit/ml RNase T_1.

The final washes are (1) RNase A buffer without enzyme (about 10–15 ml/slide), 30 min, 37°, done in a staining dish or Coplin jar; (2) 2× SSC, 30 min, room temperature, with up to 40 slides inserted vertically in a test tube rack that is immersed in 4 liters of 2× SSC in a plastic rodent cage on a platform shaker or magnetic stirrer; (3) 0.1× SSC, 500 ml/20 slides, 50°, 15 min, and (4) 4 liters of 0.1× SSC, 30 min, at room temperature. The 50° wash achieves approximately the same stringency as the hybridization conditions. Even higher temperatures (60–65°) can be used to improve background without significant loss of signal.

Dehydrate the sections by immersing the slides in glass carriers for about 10 sec each in 30, 50, 70, 85, 95% ethanol, each containing 300 mM ammonium acetate as a precaution against denaturation of *in situ* hybrids. Pass the slides through two changes of 99% ethanol and air dry.

Modifications in Washing Protocol for ^{35}S*-Labeled Probes.* Wash in 4× SSC as described above with addition of 10 mM dithiothreitol. Dehydrate the sections through graded ethanol concentrations containing 300 mM ammonium acetate. Place slides in Coplin jars containing hybridization buffer (50% formamide, 0.3 M NaCl, 20 mM Tris–HCl, 1 mM EDTA at pH 8.0, plus 10 mM dithiothreitol) and incubate at a temperature in the range of 50–65° for 10 min. The higher temperature is only about 5° below the *in situ* T_m values and has been shown also to reduce significantly ^{3}H and ^{35}S backgrounds without loss of signal.[15] Transfer to cold 2× SSC and proceed as described above, beginning with the RNase digestion step. Dithiothreitol is not included in RNase and subsequent washes.

Autoradiography

Darkroom Specifications

All steps are most easily carried out using the "Duplex Super Safelight" equipped with FDY filters in the top slots and FDW filters in the

[15] L. Angerer, D. DeLeon, K. Cox, R. Maxson, L. Kedes, J. Kaumeyer, E. Weinberg, and R. Angerer, *Dev. Biol.* **112,** 157 (1985).

bottom slots. Emulsion grains do not increase after 3 hr of exposure of dipped slides. Alternatively, work in absolute darkness or use a safelight equipped with a Wratten red filter 1.

Handling Emulsion

Stock Emulsion. Melt Kodak NTB-2 in a 45° water bath and dilute with an equal volume of 600 mM ammonium acetate, prewarmed to 45°. Distribute in 10-ml aliquots in plastic scintillation vials. Wrap with aluminum foil and store in a lighttight container at 4°.

Dipping Slides. Melt 10 ml of emulsion in 45° water bath (30–45 min). Pour the emulsion into a dipping chamber (slide mailer or preferably the chamber sold by Electron Microscopy Sciences). About 35 slides can be coated over at least half their length with this volume in the EMS dipping chamber. Allow emulsion to sit for 10–15 min to let bubbles rise and then remove surface bubbles by dipping several blank slides. Dip slides by immersing slowly and smoothly twice, allowing about 2 sec for each dip. Blot the bottom edges on a paper towel for 1–2 sec to remove excess emulsion and place the slide vertically to dry in a test tube rack. Thirty minutes after the last slide is dipped, transfer the rack to a covered chamber containing dampened paper towels for at least 3 hr to remove latent grains from the emulsion. We use a plastic rodent cage bottom, covered with a large sheet of Parafilm. Avoid plastic wraps that generate sparks. Remove and allow the slides to dry slowly and completely (about 45 min). Do not attempt to dry slides too rapidly (under vacuum, for example), because this leads to stretching artifacts in the form of grains along tissue edges.

Autoradiographic Exposure. Transfer slides to black plastic slide boxes, handling them only by their ends or edges, because touching or scratching the emulsion produces grains. The box should contain a packet of Drierite and is then sealed in bags which also contain desiccant. Slides are exposed at 4° in a larger lighttight box.

Developing. Allow the black box to come to about 15° while still sealed (about 15 min at room temperature). This avoids condensation on the slides which will reduce grain density. Remove slides to a glass slide carrier and immediately develop for 2.5 min in D19 developer. Longer developing times and higher temperatures apparently preferentially produce grains in the emulsion background. Stop in 2% acetic acid, for 15–30 sec, and fix in Kodak Fixer (not Rapid Fix) for 5 min. The developer, stop, and fix must all be at the same temperature, 15°, or the emulsion may crack and flake. Rinse in distilled water 15 min at 15° and in cold running tap water for 30 min.

Staining and Mounting Coverslips

Consult Rogers[16] for discussion of staining in conjunction with autoradiography. Many standard staining protocols (e.g., Giemsa, methylene blue, methyl green-pyronin) are applicable. Procedures which use prolonged acid destaining can bleach grains[16] and we have confirmed this observation. In many cases phase-contrast microscopy affords good morphological detail, and darkfield is good for visualizing grains, and thus elaborate staining is not required.

To mount coverslips, dip slides in several changes of fresh xylene and drain excess xylene by blotting the edge of the slide on a paper towel. To the moist slide add one to four drops of Permount (depending on coverslip size), apply the coverslip, press gently to force out bubbles, and wipe excess Permount from the edges. Allow Permount to harden by placing the slide on a 40° slide warmer for several hours. Clean slides with xylene to remove excess Permount, and with Formula 409 (or equivalent) to remove emulsion from the back side of the slide.

Choice of Radioisotope

Except for very abundant targets, we synthesize probes using [³H]UTP and [³H]CTP at 25–60 Ci/mmol, yielding probe specific activities of about $1-2 \times 10^8$ dpm/μg. ^{3}H provides the best resolution of any isotope, but the lowest autoradiographic efficiency (approximately 0.02 grains/disintegration in 5-μm sections.[17,18] Other advantages are stability and long half-life.

SP6 polymerase incorporates ^{35}S-substituted XTPs at a 2- to 3-fold reduced rate, so they should be diluted with unlabeled thiolated nucleotide. Using one ^{35}S-substituted nucleoside triphosphate, it is possible to obtain specific activities 10-fold higher than with ^{3}H precursors, and the autoradiographic efficiency is about 5 times higher. These combined factors indicate that signals can be obtained much more quickly, but again the sensitivity is limited by background. The resolution is adequate for most applications, except possibly those requiring discrimination between nucleus and cytoplasm in small cells.

Note that it is unwise to hybridize very high specific activity ^{35}S-labeled probes at concentrations far below those required to saturate

[16] A. W. Rogers, "Techniques in Autoradiography." Elsevier/North-Holland Biomedical Press, Amsterdam, 1979.

[17] G. L. Ada, J. J. Humphrey, B. A. Askonas, H. O. McDevitt, and G. V. Nossal, *Exp. Cell Res.* **41,** 557 (1966).

[18] S. R. Pelc and M. G. E. Welton, *Nature (London)* **216,** 925 (1967).

target RNAs. In these cases better resolution and similar signal levels can be achieved at lower cost by using saturating concentrations of lower specific activity ^{3}H-labeled probes.

^{32}P-labeled probes of very high specific activity can be obtained, and such probes have been used quite successfully in conjunction with X-ray film autoradiography to identify organs, or regions of organs, containing individual mRNAs. However, the resolution is usually insufficient for finer-scale localization of labeled cells. Direct comparison in our laboratory indicates that the autoradiographic efficiency (grains per disintegration) using our standard procedure is not much higher than obtained with ^{3}H.

Controls

Different experimental systems afford a variety of potential control experiments. Here we list several of the most basic and generally applicable.

Heterologous probes. We strongly recommend the use of heterologous probes of the same specific activity, concentration, and fragment length to control for variations in background among cell types or tissues.

mRNA accessibility in different cells. Accessibility is a difficult question to address experimentally. In some cases, hybridization patterns are essentially all or none, and specificity is inferred by demonstrating that different probes hybridize to different subsets of cells. Use of a general probe, such as poly(U), expected to hybridize to all cells at similar levels, can control for large differences in hybridization efficiency among cell types.[19] In other cases, differences of severalfold in signals may be observed, and the question of relative hybridization efficiency becomes more significant. One approach is to show that the ratio of grain densities over different regions is the same for different fixation conditions, and/or extents of proteinase digestion, or reaches a constant value as proteinase concentration is increased.

Comparisons to external data. In some cases probes may be available for RNAs whose concentrations in different stages of development or physiological conditions have been measured by other techniques. Immunohistochemical data on localization of the protein product may also be suitable for comparison.

Demonstration that signals derive from hybridization to RNA or DNA. Prehybridization treatments with RNase or DNase have frequently been used as controls to identify the target nucleic acids as DNA or

[19] D. A. Lynn, L. M. Angerer, A. M. Bruskin, W. H. Klein, and R. C. Angerer, *Proc. Natl. Acad. Sci. U.S.A.* **80**, 2656 (1983).

RNA.[20,21] However, it is difficult to remove cellular RNA or DNA completely from sections, and it is necessary to demonstrate that reduction of signals by RNase pretreatment is not due to degradation of probe by residual RNase. Because transcription of cellular genes is usually asymmetric, sense-strand probes are specific for DNA and provide a better way of discriminating between RNA and DNA targets. Using glutaraldehyde-fixed tissue we have never detected signals from DNA in the absence of intentional denaturation even, for example, with sense probes complementary to 3600 kb of histone repeat DNA.[22]

Sensitivity

The sensitivity of *in situ* hybridization depends directly on the signal-to-noise ratio that can be achieved. The important determining factors are RNA retention and accessibility, hybridization and hybrid detection efficiencies, and suppression of nonspecific sticking of probe. Using radioactively labeled probes and autoradiographic detection of signals, it is possible to measure directly the signal achieved for an mRNA target of known concentration.[1] Similar calculations show that at saturation ^{3}H- and ^{35}S-labeled probes complementary to 500 nucleotides of an mRNA (at maximum specific activities of 2.5×10^8 and 2.5×10^9 dpm/μg, respectively) would yield about 0.036 and 1.8 grains/100 μm^2 per day for an mRNA density of only 1 copy/cell (a fixed cell is about 25 μm^2 in our system). A grain density of 10 grains/100 μm^2 (i.e., about 5 days with ^{35}S-labeled probes) is readily detectable in the absence of nonspecific binding. This emphasizes the fact that the current limitation on sensitivity is not signal but the level of nonspecific binding of probe.

Those seeking more information should consult Angerer *et al.*[5]

Appendix: Sources of Materials

Here is a list of the more unusual equipment and chemicals used in our protocol.

Equipment

Heat block, VWR 13259-005 and 13259-130
Paraffin oven, Boekel 131700
Super Duplex Safelight, VWR TM72882-10 with FDY top and FDW
   bottom filters

[20] D. G. Capco and W. R. Jeffery, *Dev. Biol.* **89,** 1 (1982).
[21] D. L. Venezky, L. M. Angerer, and R. C. Angerer, *Cell (Cambridge, Mass.)* **24,** 385 (1981).
[22] D. V. DeLeon, K. H. Cox, L. M. Angerer, and R. C. Angerer, *Dev. Biol.* **100,** 197 (1983).

Enzymes and Chemicals

Proteinase K, EM Laboratories 24568-10

Glutaraldehyde, Sigma G5882 (vial used once and discarded)

Poly(L-lysine),[23] Sigma P1399

Paraffin: 1 part Paraplast (VWR 15159-409) to 1 part Tissue Prep 2 (Fisher)

NTB-2 emulsion, Eastman Kodak 1654433

All other chemicals are reagent grade. All solutions are prepared with water that is charcoal filtered, deionized, stirred with 0.1% diethyl pyrocarbonate (DEP) overnight, and autoclaved. Stock solutions of chemicals and Millipore filtered and DEP treated (unless reactive with DEP).

Accessories

Embedding molds, VWR 15160-157

BEEM capsules, Polysciences 0224

Slide mailers, Thomas 6707-M25

Test tube rack, VWR 60939-005

Staining dish and slide carrier, Wheaton; American Scientific Products 900300

Coplin jar, Wheaton; VWR 2546-000

Dipping chamber, Electron Microscopy Sciences 07051

Acknowledgments

We wish to express our appreciation to Donna Venezky DeLeon for excellent technical assistance in the development of these protocols. Modifications for the use of ^{35}S-labeled probes were aided by a generous gift of nucleotides from New England Nuclear. Research in the laboratory of R.C.A. and L.M.A. is supported by NIH GM 25553. R.C.A. is the recipient of a Research Career Development Award from the National Institutes of Health. K. H. Cox was supported in part by an Institutional Grant from the American Cancer Society.

[23] C. N. Berger [*The Embo J.* **5**, 85 (1986)] describes the use of 3-aminopropyltriethoxysilane as an alternative.

[69] Expression, Identification, and Characterization of Recombinant Gene Products in *Escherichia coli*

By ALLAN R. SHATZMAN *and* MARTIN ROSENBERG

There are numerous gene products of biological interest which cannot be obtained from natural sources in quantities sufficient for detailed biochemical and physical analysis. One solution to this problem has been the development of recombinant vector systems which are designed to achieve efficient expression of cloned genes in *Escherichia coli*.

Escherichia coli has been and continues to be the "workhorse" of the gene expression field. No other system has been developed which can express such a large number of known gene products at levels sufficient for detailed biochemical analysis or product development, as well as express undefined coding sequences (open reading frames) in amounts sufficient to determine the identity of the gene product and to characterize its function.

In general, the rationale used in the design of these *E. coli* expression systems involves the insertion of the coding region of interest into a multicopy vector system (usually a plasmid) such that the region is efficiently transcribed and translated. Since most genes do not naturally contain the proper signals for ribosome recognition and translation initiation in *E. coli,* special procedures must be devised to supply this information. This is done either by fusing the gene directly to a bacterial ribosome binding site to synthesize an authentic gene product or, as is more commonly done, to the N-terminal coding region of a bacterial gene to synthesize a fusion protein.

Once the gene is cloned into such an expression vector the gene product may be produced in *E. coli* in either a constitutive or inducible manner, and the synthesis monitored in several ways including direct visualization following resolution by gel electrophoresis, antibody detection, or functional assay. The expressed protein can be purified, using any of the above assay procedures, in quantities enabling biochemical studies to be performed that were previously impossible due to limitations in protein availability.

This chapter describes (1) how a gene may be adapted and inserted so that it is appropriately placed under the control of regulatory elements which permit its expression in *E. coli*; (2) methods of inducing expression; and (3) techniques by which the identity and characteristics of a cloned gene product may be proven.

Adapting a Gene for Expression in *E. coli*

Expression of a heterologous gene or gene fragment in *E. coli* requires that the coding sequence be placed under the transcriptional and translational control of regulatory elements recognized by the bacterial cell. The pAS vectors (Fig. 1) were designed specifically to direct gene expression by providing regulatory signals derived from the phage λ.[1] Phage elements were chosen because of their high efficiency and ability to be regulated. The system provides a promoter which can be fully controlled, an antitermination mechanism to help ensure efficient transcription across any

[1] A. Shatzman and M. Rosenberg, *Ann. N.Y. Acad. Sci.* **478,** 233 (1986).

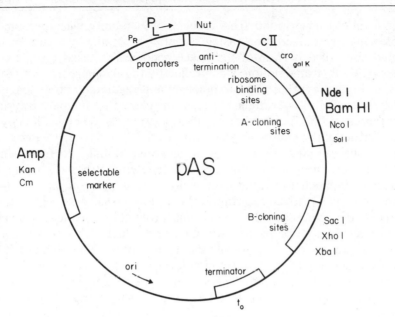

FIG. 1. Schematic diagram of the pAS plasmid vector (1) for the overproduction of heterologous gene products. The phage λ P_L and/or P_R promoters are provided to direct efficient transcription. N utilization site(s) and a transcription terminator (t_o) are provided to maximize transcription and plasmid stability. The λ *cII*, *cro*, or *E. coli galK* gene ribosome binding sites are provided to direct efficient translation initiation. Unique *Nde*I, *Bam*HI, *Nco*I, and *Sal*I restriction sites have been engineered adjacent to the ribosome binding site (A-cloning sites) to permit easy insertion of genes adjacent to the translational regulatory information. Several other unique restriction sites (B-cloning sites) are present to further facilitate gene insertion and manipulation. Ampicillin (Amp), kanamycin (Kan), and chloramphenicol (Cm) resistance markers have been inserted in multiple positions and orientations. The vector contains a ColE1 origin of replication. The diversity of lettering sizes used to represent the functional elements is a relative measure of the degree of use and data which have been generated with each of these elements (i.e., P_L is present and has been used far more often than P_R).

gene insert, high vector stability, several choices of antibiotic selection, and relatively easy insertion of the gene of interest adjacent to efficient translation regulatory information.

Efficient Regulated Transcription

The phage P_L promoter is used in vectors because of its high efficiency.[2] Plasmids carrying P_L are often unstable, presumably due to the

[2] K. McKenney, H. Shimatake, D. Court, U. Schmeissner, C. Brady, and M. Rosenberg, *in* "Gene Amplification and Analysis" (J. G. Chirikjian and T. S. Papas, eds.), Vol. 2, p. 383. Elsevier/North-Holland, New York, 1981.

high level of transcription. This problem of instability was overcome by repressing P_L transcription, using bacterial hosts that contain an integrated copy of a portion of the phage genome (bacterial lysogens). In these cells, P_L transcription is controlled by the phage repressor protein (cI), a product synthesized continuously and regulated autogenously by the lysogen.[3] Moreover, by using a lysogen carrying a temperature-sensitive repressor ($cI857$),[4] P_L-directed transcription can be controlled by the temperature of the culture and hence transcription can be activated at any time. Cells carrying the vector can be grown initially to high density without expression of the cloned gene at low temperature (32°), and subsequently induced to synthesize the product at high temperature (42°). Alternatively, a lysogen carrying a wild-type repressor gene (cI^+) can also be induced to synthesize the desired gene product by chemical means (see section on Expression of the Cloned Gene Product). This ability to control gene expression, coupled with the rapidity of the induction procedure and the efficiency of P_L, ensures high-level expression of the product in a relatively short time period. These features are particularly useful for the expression of gene products that may be lethal and/or rapidly turned over in bacteria.

In addition to providing a strong, regulatable promoter, the system also ensures that P_L-directed transcription efficiently traverses any gene insert. This is accomplished by providing both the phage λ antitermination function, N, and sites on the P_L transcription unit necessary for N utilization (Nut site). N expression from the host lysogen reduces transcriptional polarity, thereby helping to ensure that transcription traverses the entire P_L transcription unit[5] even in the presence of most transcription termination signals.

Efficient Translation

In addition to efficient transcription, the coding information must also be translated efficiently. This is accomplished either by placing the inserted information immediately adjacent to a translation initiation signal recognized in *E. coli*[6] or by fusing it in-frame to a gene or gene fragment which expresses well in *E. coli*.

[3] M. Ptashne, K. Backman, M. Humayun, A. Jeffrey, R. Maurer, B. Meyer, and R. Sauer, *Science* **194**, 156 (1976).
[4] R. Sussman and F. Jacob, *C. R. Hebd. Seances Acad. Sci.* **254**, 1517 (1962).
[5] M. Rosenberg, D. Court, D. Wulff, H. Shimatake, and C. Brady, *in* "The Operon" (J. Miller, ed.), p. 345. Cold Spring Harbor Lab., Cold Spring Harbor, New York, 1978.
[6] L. Gold, D. Pribnow, T. Schneider, S. Shinedling, B. Singer, and G. Stormo, *Annu. Rev. Microbiol.* **35**, 365 (1981).

The translation initiation signal we have utilized is that of the phage λ gene, *c*II. The entire coding region of this gene was removed, leaving only its initiator f-met codon and upstream translational regulatory sequences.[6] Additionally, these vectors have been engineered to provide *Nde*I, *Bam*HI, *Nco*I, *Asp*718, *Sal*I, *Sma*I, *Xma*I, *Nru*I, *Hpa*I, and/or *Eco*RV restriction sites adjacent to the translation initiation signal (see Table I). Thus, the system allows essentially any gene to be adapted for insertion into the vector so as to produce an authentic (or nearly authentic) gene product. Those genes which contain restriction sites compatible with the sites on the vector may be inserted directly into the vector. As most genes do not contain appropriately positioned restriction sites, it is often necessary to adapt existing restriction cloning sites within the gene for fusion to the translation initiation signals provided on the vectors.

Treatment of the *Bam*HI, *Sal*I, or *Asp*718 cloning sites on the vector with a single-strand-specific nuclease permits precise fusion of any coding region directly to the translation initiation codon (ATG).[7] This procedure results in the expression of gene products which contain only an initiating methionine fused to the protein encoded by the insert. Alternatively, the single-stranded overhanging ends produced at these restriction sites can also be filled in with DNA polymerase to result in a 4-bp insertion into the coding region. This information must be considered when fusing DNA inserts to the resulting blunt-end cloning site since it will result in additional amino acids being incorporated into the N-terminus of the protein (see Table I). The *Nco*I site on the vector allows direct fusion of coding information to the ATG codon. This is achieved by using a DNA polymerase fill-in reaction, at the site. We find that this is a more efficient way to achieve a blunt end immediately adjacent to the initiating ATG than is the single-stranded nuclease method described above. Use of the *Nde*I site positioned just upstream of the ATG requires that the gene insert provide its own initiating ATG codon. The *Nde*I site has proved most useful when the gene insert is adapted into the vector by rebuilding a portion of its 5'-terminal region using synthetic oligonucleotides. It is important to note that although the many restriction sites engineered into the vector provide cloning flexibility, some have the disadvantage of adding up to six codons to the N-terminus of the gene insert (see Table I).

Translational Fusions

Often, the efficient synthesis of a chimeric protein (by gene fusion) meets the needs of the experimenter. For example, this approach has

---

[7] M. Rosenberg, Y. S. Ho, and A. Shatzman, this series, Vol. 101, p. 123.

TABLE I
CLONING SITES FOR INSERTION OF CODING SEQUENCES ADJACENT TO
TRANSLATION INITIATION SIGNALS

Gene insertion site[a]	Cloning notes
*Bam*Hi 1. 5′ CAT**ATG**gatcc 3′ 3′ GTATACCTAGg 5′	All permit fusion directly to initiation codon following single-strand-specific nuclease treatment, or ligation to gene restricted with restriction enzyme, producing complimentary ends. Fill-in adds one codon plus one base pair
*Asp*718 2. 5′ CAT**ATG**gtacc 3′ 3′ GTATACCATGg 5′	
*Sal*1 3. 5′ CAT**ATG**tcgac 3′ 3′ GTATACAGCTg 5′	
*Nco*1 4. 5′ Cc**atgg** 3′ 3′ GGTACc 5′	Fill-in allows direct fusion to initiation codon
*Nde*1 5. 5′ CA**tatg**gatcc 3′ 3′ GTATacctagg 5′	Especially useful for cloning and expressing genes in which the 5′ end is rebuilt with synthetic oligonucleotides
*Sma*1 6. 5′ CAT**ATG**GATCCCggg 3′ 3′ GTATACCTAGGGccc 5′	Allows fusion of gene to translational start site in alternate reading frames but adds two to three amino acids to gene product
*Xma*1 5′ CAT**ATG**GATCccggg 3′ 3′ GTATACCTAGGGCCc 5′	
*Nru*1 7. 5′ CAT**ATG**TTCGcgaagttaacggatatcg 3′ 3′ GTAtACAAGCgcttcaattgcctatagc 5′	Allows blunt-end ligation to translation initiation signals in all three reading frames but adds up to six amino acids to the gene product
*Hpa*1 5′ CAT**ATG**TTCGCGAAGTTaacggatatcg 3′ 3′ GTATACAAGCGCttCAAttgcctatagc 5′	
*Eco*RV 5′ CAT**ATG**TTCGCGAAGTTAACGGATatcg 3′ GTATACAAGCGCTTCAATTGCCTAtagc	

[a] Each number listed (1–7) indicates an individual member of the pAS vector system designed specifically for cloning genes at these restriction sites which are unique in that vector. All insertion sites are adjacent to the *c*II ribosome binding site (rbs) except for site 3 (*Sal*1) which is adjacent to the *cro* rbs. The ATG initiation codon is in bold print. The recognition sequence of the restriction enzymes is overscored and underscored. Lowercase letters indicate the bases removed following restriction by these enzymes.

been used in the development of antisera to specific proteins which have diagnostic potential. Gene fusion technology has also been used for clone selection when an antisera to the gene product is already available. In addition, if the function of the gene is preserved in the chimera, detection and characterization of the gene product are also achievable. Fusing a heterologous gene to a well-characterized bacterial gene such as *lacZ* (for β-galactosidase) is a commonly used approach.[8–10] *lacZ* fusions generally result in the synthesis of very large fusion proteins, composed primarily of the β-galactosidase protein. These large protein fusions often have limited utility; however such fusions have been used successfully to identify and define a variety of different gene products.[8–10]

In an effort to minimize the problems associated with such fusions, we have constructed vectors in which translational fusions are made to the amino terminal region of either the protein encoded by the nonstructural influenza virus gene, *NS1*, or the protein encoded by the *E. coli* galactokinase (*galK*) gene. These gene products were chosen because they accumulated to high levels and were stable when expressed in *E. coli* using the pAS system.[7,11] Since gene expression can often be enhanced by gene fusion technology, it was reasoned that fusion of heterologous genes to the amino-terminal end of the *NS1* or *galK* gene products might result in stable high-level expression. This indeed was the case. For example, the human α_1-antitrypsin gene, which normally expresses very poorly in bacteria, accumulates to at least 20% of total cellular protein when produced as a fusion with the first 80 amino acids of *NS1*. Similar results have been obtained with other gene products fused to *galK*. Insertions into these fusion vectors can be made in all three reading frames using *Nco*I, *Sma*I, and *Xma*I restriction sites which are provided at the point of insertion. Moreover, high-titer antisera have been developed which recognize the amino-terminal portions of *NS1* and *galK*, thus permitting immunological detection of the resulting chimeric gene products (J. Young and C. Debouck, personal communication).

Expression of the Cloned Gene Product

Expression of the cloned gene product requires that the expression vector be transformed into a host lysogen and induced at an appropriate time in the growth cycle. The pAS expression vector system has been

[8] M. Casadaban, A. Martinez-Arias, S. Shapira, and J. Chou, this series, Vol. 100, p. 293.
[9] M. Rose and D. Botstein, this series, Vol. 101, p. 167.
[10] L. Guarente, this series, Vol. 101, p. 181.
[11] J. Young, V. Desselberger, P. Palese, B. Ferguson, A. Shatzman, and M. Rosenberg, *Proc. Natl. Acad. Sci. U.S.A.* **80**, 6105 (1983).

used in combination with a variety of *E. coli* K12 lysogenic hosts. In each case, the host lysogen carries a replication defective phage and, thus, induction of the expression system does not lead to phage development. Gene expression from various pAS derivatives has been examined in these K12 backgrounds and the results indicate that the host background can play an important role in the final yield of accumulated gene product. For the most part, we do not understand the reasons for the rather dramatic differences seen in product yield from different host strains. However, product stability often appears to be a determining factor. Recently, host lysogenic strains have been developed which are defective in certain proteolytic functions (i.e., Lon⁻ and Htpr⁻). These strains are obtained as temperature-sensitive conditional lethals and, thus, are compatible with our temperature induction system. These specialized host strains have had a dramatic impact on the expression of several (but not all) gene products in *E. coli*.[12,13]

In addition to the genotype of the host strain, the mode of induction can also affect the overall accumulation of gene product.[14] For example, a temperature induction in which cells are grown at 32° to an A_{650} of 0.6 and then rapidly shifted to 42° is known to result in a generalized heat shock response in *E. coli*. At least some of the proteins induced by this response are proteases.[15] In contrast, chemical induction with a DNA-damaging agent such as nalidixic acid leads to an SOS response by the host.[16] In this system, replication-defective *E. coli* lysogens (cI^+) containing the desired plasmid are grown at 37° to an A_{650} of 0.4, at which point naladixic acid is added to a final concentration of 60 μg/ml (from a 60 mg/ml stock in 1 N NaOH). Cells are then grown for 5 or more hours and then harvested. This mode of induction leads to a cellular state different from that of the temperature shift and these variations can lead to significant differences in gene product accumulation.

Identification of the Cloned Gene Product

Following the induction of cultures carrying the desired expression vector, cells may be analyzed in a variety of ways to detect the presence of the cloned gene product. Most typically, when the gene is expressed efficiently, the presence of the novel gene product can be determined

[12] R. Watt, A. Shatzman, and M. Rosenberg, *Mol. Cell. Biol.* **5**, 448 (1985).
[13] Y. S. Ho, D. Wulff, and M. Rosenberg, *in* "Regulation of Gene Expression—25 Years On" (I. R. Booth and C. F. Higgins, eds.), p. 79. Cambridge Univ. Press, London and New York, 1986 (in press).
[14] J. Mott, R. Grant, Y. S. Ho, and T. Platt, *Proc. Natl. Acad. Sci. U.S.A.* **82**, 88 (1985).
[15] T. Baker, A. Grossman, and C. Gross, *Proc. Natl. Acad. Sci. U.S.A.* **81**, 6779 (1984).
[16] J. Little and D. Mount, *Cell (Cambridge, Mass.)* **29**, 11 (1982).

directly by observing a new, inducible, prominent protein band not present from control cultures in an acrylamide gel system. For example, a 1-ml aliquot of the cell pellet is resuspended and boiled in 50 μl of a gel loading buffer (50 mM Tris–HCl at pH 6.8, 10% glycerol, 5% 2-mercaptoethanol, 2% sodium dodecyl sulfate, and 0.05% bromphenol blue). This material is then resolved by standard SDS–polyacrylamide gel electrophoretic analysis (see this volume [31]). The gel can then be stained with a variety of standard protein dyes (e.g., Coomassie brilliant blue) and analyzed for the appearance of a new band which migrates at the position predicted from the amino acid sequence deduced from the DNA sequence. Care must be taken to compare the staining pattern with that of the controls containing an induced culture which carries the parent expression vector. An uninduced culture also provides an additional and valuable control. This ensures that the new protein band derives from the gene insert and does not originate from cellular or other vector sequences.

Functional Identification

The functional identity of any gene insert can be determined and/or confirmed in several ways including (1) direct detection of a novel function or activity imparted to the bacterial host or by genetic complementation of the appropriate mutant host; (2) assay of whole-cell extracts for the activity of the cloned gene product; and (3) assay after partial or complete purification of the cloned gene product. There are abundant examples of how these three modes of functional identification have been used to demonstrate expression.[17–24]

Immunologic Identification

The purification of sufficient amounts of a gene product allows one to generate antigen specific mono- or polyclonal antisera easily. Such antisera can be used to diagnose and characterize the synthesis, structure,

[17] M. Karin, R. Najarian, A. Haflinger, P. Valenzuela, J. Welch, and F. Fogel, *Proc. Natl. Acad. Sci. U.S.A.* **81,** 337 (1984).
[18] T. Butt, E. Sternberg, J. Gorman, P. Clark, D. Hamer, M. Rosenberg, and S. Crooke, *Proc. Natl. Acad. Sci. U.S.A.* **81,** 3332 (1984).
[19] J. Wang, C. Queen, and D. Baltimore, *J. Biol. Chem.* **257,** 13181 (1982).
[20] B. Ferguson, M. Pritchard, J. Field, D. Rieman, R. Greig, G. Poste, and M. Rosenberg, *J. Biol. Chem.* **260,** 3652 (1985).
[21] D. Slamon, J. deKemion, I. Verma, and M. Cline, *Science* **224,** 256 (1984).
[22] E. Scolnick, A. Papageorge, and T. Shih, *Proc. Natl. Acad. Sci. U.S.A.* **76,** 5355 (1979).
[23] R. Sweet, S. Yokoyama, T. Kamata, J. Feramisco, M. Rosenberg, and M. Gross, *Nature (London)* **311,** 273 (1984).
[24] B. Krippl, B. Ferguson, N. Jones, M. Rosenberg, and H. Westphal, *Proc. Natl. Acad. Sci. U.S.A.* **82,** 7480 (1985).

and function of the protein both *in vivo* and *in vitro*. One approach is to use bacterial expression to produce the desired heterologous gene product in its entirety, as a fusion, or as a fragment of the gene. The protein may then be purified and used to produce high-titer mono- or polyclonal antisera. These antisera have been used to (1) map natural expression of the gene product with respect to cell type, subcellular distribution, and temporal regulation; (2) determine relative levels of expression in various cell types; (3) study protein processing and stability; (4) map immuno-dominant domains; (5) purify by immunoaffinity both the native and modi-fied forms of the protein; and (6) provide an *in vivo* diagnostic reagent for examining tissue distribution and expression of the gene product. These techniques have been applied to the study of the human adenovirus *Ela* gene product.[25–34]

In addition to the characterization of known genes, the same immuno-logic techniques can be used to identify, detect, and characterize "un-known genes" or so-called open reading frames (orfs). In this case, any DNA fragment is cloned into the expression vector so as to permit its potential translation in all three possible reading frames. Following ex-pression of the orf, the protein/peptide product can be used to obtain antisera which serve as a probe to examine (1) whether the DNA fragment encodes an expressed gene; (2) in what cell types and at what stages in cell development the gene is expressed; and (3) the various levels at which the gene product is synthesized in various cell types. These analyses have

[25] B. Ferguson, N. Jones, and M. Rosenberg, *Science* **224,** 1343 (1984).

[26] B. Ferguson, B. Krippl, N. Jones, J. Richter, H. Westphal, and M. Rosenberg, *Cancer Cells* **3,** 265 (1985).

[27] B. Ferguson, B. Krippl, O. Andrisani, N. Jones, H. Westphal, and M. Rosenberg, *Mol. Cell. Biol.* **5,** 2653 (1985).

[28] J. Richter, P. Young, N. Jones, B. Krippl, M. Rosenberg, and B. Ferguson, *Proc. Natl. Acad. Sci. U.S.A.* **82,** 8434 (1985).

[29] S. Yee, D. Rowe, M. Tremblay, M. McDermott, and P. Branton, *J. Virol.* **46,** 1003 (1983).

[30] L. Ratner, W. Haseltine, R. Patarca, K. Livak, B. Starcich, S. Josephs, E. Doran, J. Rafalski, E. Whitehorn, K. Baumeister, L. Ivanoff, S. Petteway, M. Pearson, J. Lauten-berger, T. Papas, J. Ghrayeb, N. Chang, R. Gallo, and F. Wong-Stall, *Nature (London)* **313,** 277 (1985).

[31] S. Wain-Hobson, P. Sonigo, O. Danos, S. Cole, and M. Alizon, *Cell (Cambridge, Mass.)* **40,** 9 (1985).

[32] R. Sanchez-Pescador, M. Power, P. Barr, K. Steimer, M. Stempien, S. Brown-Shimer, W. Gee, A. Renard, A. Randolph, J. Levy, D. Dina, and P. Luciw, *Science* **227,** 484 (1985).

[33] J. Allan, J. Colgan, T. Lee, M. McLane, P. Janki, J. Groopman, and M. Essex, *Science* **230,** 810 (1985).

[34] N. Kan, G. Franchini, F. Wang-Staal, G. DuBois, W. Robey, J. Lautenberger, and T. Papas, *Science* **231,** 1553 (1986).

helped to identify and characterize a variety of genes and their products. For example, several open reading frames were known to occur within the HTLV III viral genome.[30-32] The putative coding sequences for these orfs were cloned and expressed in *E. coli*[33] and the resulting gene products used to generate antisera. Both the bacterially synthesized proteins and the antisera generated from them have been used to identify and characterize these novel gene products from the HTLV III virus.[34] This approach was also used to study and characterize the immunodominant domain of the circumsporozoite protein (CSP) of *Plasmodium falciparum*.[35-37]

"Expression Libraries"

The ability to detect expression of any heterologous gene product synthesized in *E. coli* by the use of immunological screening techniques can be potentially applied to the isolation of specific genes from gene libraries. For example, the phage expression system (λgt11)[38] (also this volume [40]) has been utilized for cloning a number of genes, as well as for isolating DNA fragments encoding proteins of previously unknown structure and/or function.

The pAS vector system can also be used to screen libraries by expression. The expression library is constructed by inserting restriction (e.g., *Sau*3A) fragments from a genomic or cDNA library directly adjacent to the translational regulatory signals (e.g., at the *Bam*HI site) on the pAS expression vector. Alternatively, the DNA fragments can be inserted as fusions downstream of a gene product such as *NS1* or *galK* (see above). Individual clones can be induced to produce high levels of the protein encoded by the inserted fragment simply by incubating the colony at elevated temperature. Following this induction, the colonies are transferred to nitrocellulose and probed with gene-specific antisera (this volume [50]). It should be pointed out that following the induction step, genetic complementation of a mutant host and/or specific protein assay may also be used to isolate a clone from the library. Although this plasmid system is less efficient than λgt11 in terms of the number of clones ob

[35] J. Young, W. Hockmeyer, M. Gross, W. Ballou, R. Wirtz, J. Trosper, R. Beaudoin, M. Hollingdale, L. Miller, C. Diggs, and M. Rosenberg, *Science* **228**, 958 (1985).

[36] F. Zavala, A. Cochrane, E. Nardin, R. Nussenzweig, and V. Nussenzweig, *J. Exp. Med.* **157**, 1947 (1983).

[37] W. Ballou, J. Rothbard, R. Wirtz, D. Gordon, J. Williams, R. Gore, I. Schneider, M. Hollingdale, R. Beaudoin, W. Maloy, and W. Hockmeyer, *Science* **228**, 996 (1985).

[38] R. Young and R. Davies, *Proc. Natl. Acad. Sci. U.S.A.* **80**, 1194 (1983).

tained from a given amount of DNA, it is far more efficient with respect to the level of expression that can be obtained for each individual clone. This makes the vector system more sensitive and less dependent on having high-titer antisera.

Many of the DNA inserts do not contain an open reading frame or are not in the proper reading frame after insertion into the vector. In order to reduce the total number of clones screened, the expression vector can be adapted to select directly for open reading frames. For example, the pASK fusion vector system[1] was constructed by inserting the *E. coli galK* gene into the pAS vector immediately adjacent to the *c*II ribosome binding site. Vectors were constructed such that a unique *Bam*HI site occurs at the junction in each of the three reading frames (pASK 1–3). DNA fragments inserted between the *c*II ribosome binding site (rbs) and the *galK* gene potentially form a translational fusion in which *galK* coding information is fused downstream of the inserted coding sequence. Cells which contain pASK1 express *galK* at high levels and give rise to red colonies on indicator plates. Cells containing the pASK2 or pASK3 vectors do not express *galK* since the gene is out-of-frame with the *c*II rbs and thus appear as white colonies on the same indicator plates. DNA fragments which contain open reading frames (orfs) can be inserted at the *Bam*HI site in these vectors and shift the *galK* reading frame. Depending on the vector used and the nature of the frameshift, *galK*$^-$ (white) or *galK*$^+$ (red) colonies can be readily selected on indicator plates in the appropriate *galE*$^+$*T*$^+$*K*$^-$ host.[2] These clones can then be probed for expression using antisera as described above. The selection described here for clones bearing DNA inserts is analogous to that used in the λgt11 system. However, the λgt11 vector provides only a single site of insertion in one reading frame whereas the pASK system provides vectors for insertion in any of the three potential reading frames.

[*Editors' Note*. A variety of other vectors can be used to express heterologous proteins in bacterial systems. Examples include commercially available (Pharmacia) plasmids with the *tac* promoter. *tac* combines properties of the *trp* and *lac* promoters for inducible high expression *in vivo*.[39] Using another plasmid system, fusion proteins can be synthesized containing a collagen peptide sequence at the junction of the insert and the vector.[40] This site can be cleaved *in vitro* with collagenase to promote purification of the peptide sequence of interest. Other fusion systems result in the secretion of heterologous proteins into the periplasmic space of

[39] H. de Boer, L. J. Comstock, and M. Vasser, *Proc. Natl. Acad. Sci. U.S.A.* **80**, 21 (1983).
[40] J. Germino and D. Bastia, *Proc. Natl. Acad. Sci. U.S.A.* **81**, 4692 (1984).

E. coli, again facilitating their purification[41,42] (see also, this series, Vols. 101, 153, 154, and 155).]

Acknowledgments

We thank T. Berka, J. Young, C. Debouck, M. Gross, H. Johansen, J. Sutiphong, J. Auerbach, and J. Culp for supplying unpublished information to this chapter. We also thank M. McCullough for editing the manuscript.

[41] K. Talmadge and W. Gilbert, *Proc. Natl. Acad. Sci. U.S.A.* **79**, 1830 (1982).
[42] L. R. Liss, B. L. Johnson, and D. B. Oliver, *J. Bacteriol.* **164**, 925 (1985).

[70] Heterologous Gene Expression in Yeast

By GRANT A. BITTER

There are a number of techniques used to analyze a cloned DNA segment including restriction enzyme analysis, hybridization with oligonucleotide and mRNA sources, and DNA sequence determinations. These and other methodologies are described elsewhere in this volume. Although the DNA sequence of a clone is predictive of the amino acid sequence of the protein, a rigorous proof of identity of a cloned gene will include a demonstration that it encodes a protein possessing the appropriate biological activities. This will require incorporation of the cloned gene into a vector which programs expression in a given cell type. However, it is possible that synthesis of the correct protein is not accompanied by the expected biological activity due to incorrect processing or conformation of the heterologous protein. Consideration must be given, therefore, to the choice of an appropriate host cell type for expression.

Heterologous gene expression studies may utilize either prokaryotic (e.g., *Escherichia coli, Bacillus subtilis*) or eukaryotic (e.g., yeast or various mammalian cell) hosts. The expression of heterologous genes in *E. coli* and mammalian cell systems are reviewed elsewhere in this volume. This chapter will describe utilization of bakers' yeast, *Saccharomyces cerevisiae,* as a host for heterologous gene expression. Methods employed for efficient heterologous gene expression in yeast have been reviewed recently.[1] The reader is referred to this source for a detailed

[1] G. A. Bitter, K. M. Egan, R. A. Koski, M. O. Jones, S. G. Elliott, and J. C. Giffin, this series, Vol. 153 [33], in press.

description of this technology. This chapter will focus on the characteristics of heterologous proteins expressed in *S. cerevisiae* in relation to the properties of the same proteins produced in *E. coli* and mammalian cells.

Direct Expression

It is likely that proteins produced by direct expression (proteins lacking signal peptides and which therefore accumulate within the cytoplasm) have similar properties whether produced in prokaryotic or eukaryotic cells. Protein synthesis universally initiates with an AUG-methionine codon (except in certain bacterial mRNAs which utilize GUG-valine initiation codons). Cytoplasmic methionine aminopeptidases exist which remove the NH_2-terminal methionine in certain sequence contexts, and the specificity of these enzymes appear similar in both bacterial and eukaryotic cells.[2,3] Since the removal of NH_2-terminal methionine from cytoplasmic proteins has not been exhaustively investigated, however, it is possible that the specificity and efficiency of methionine removal may vary depending on the host cell. In addition, when proteins are produced at very high levels, the methionine may not be removed from true substrates with 100% efficiency, resulting in a heterogeneous population of protein products. Thus, for proteins which do not contain NH_2-terminal methionine in their fully processed native form, production in a heterologous system may yield an analog. There are many examples in which production of such analog (containing NH_2-terminal methionine) results in molecules with full biological activity (e.g., IFN-γ; Fieschko *et al.*[4]). However, there may be subtle differences between these analogs and the natural material, such as increased immunogenicity. The issue of NH_2-terminal specificity is thus complex for both prokaryotic and eukaryotic direct expression systems. Although it probably has minimal effects on biological activity of a heterologous protein, and thus is probably not critical in proving identity of a cloned gene, NH_2-terminal specificity should be carefully considered in choosing a host for production purposes.

A high level of expression of proteins in *E. coli* generally results in the compartmentalization of these products in inclusion bodies. These particulate "organelles" may be isolated and thus bring about a considerable

---

[2] F. Sherman and J. W. Stewart, *in* "The Molecular Biology of the Yeast Saccharomyces: Metabolism and Gene Expression" (J. N. Strathern, E. W. Jones, and J. R. Broach, eds.), p. 3010. Cold Spring Harbor Lab., Cold Spring Harbor, New York, 1982.

[3] F. Sherman, J. W. Stewart, and T. Susumu, *Bioessays* **3**, 27 (1986).

[4] J. C. Fieschko, K. M. Egan, T. Ritch, R. A. Koski, M. O. Jones, and G. A. Bitter, *Biotechnol. Bioeng.* (in press).

purification of the product. However, the desired protein is generally inactive and must subsequently be extracted with detergents and/or other solubilizing agents. In some instances, it is not possible to regain the biological activity of the protein (although the gene expressed does, in fact, encode the correct protein sequence). Thus, expression of heterologous genes in *E. coli* may yield false-negative results when used to prove identity of a cloned gene. The formation of insoluble aggregates analogous to *E. coli* inclusion bodies has not been observed in direct expression studies in eukaryotic cells. When human immune interferon (IFN-γ) was expressed as 5% of the total cell protein in yeast, more than 50% of the product was recovered in the soluble fraction after a 10,000 *g* centrifugation and, furthermore, the protein possessed full biological activity.[4] The formation of insoluble aggregates of IFN-γ in yeast has not been investigated at higher expression levels, however.

The data available to date thus indicate that direct expression of proteins in yeast has no apparent advantages over expression in *E. coli* with regard to NH_2-terminal amino acid specificity. Limited evidence exists indicating that efficient expression in yeast is not accompanied by the formation of insoluble, inactive aggregates. There is, however, one class of proteins that appears to be particularly suited to expression in *S. cerevisiae*. Hydrophobic membrane proteins (e.g., viral glycoproteins) are apparently quite toxic to *E. coli* and expressed only at very low levels, if at all. For example, expression of the gene encoding the major coat protein of hepatitis B virus (HBsAg) in *E. coli* resulted in no detectable antigen in cell lysates although antibodies to HBsAg were induced by injection of rabbits with the crude lysate.[5] In contrast, expression of HBsAg in either yeast[6] or mammalian[7] cells resulted in much higher expression levels. In both cell types, moreover, the HBsAg protein was incorporated into cellular membranes and could be isolated as lipoprotein particles with properties very similar to the 22-nm particles present in human serum. Several other hydrophobic membrane proteins have also been expressed in *S. cerevisiae*. The α subunit of the *Torpedo californica* acetylcholine receptor, upon expression in yeast, was inserted into the plasma membrane.[8] The major surface antigen of the sporozoite stage of

[5] P. MacKay, M. Pasek, M. Magazin, R. T. Kovacic, B. Allet, S. Staho, W. Gilbert, H. Schaller, S. A. Bruce, and K. Murray, *Proc. Natl. Acad. Sci. U.S.A.* **78,** 4510 (1979).
[6] P. Valenzuela, A. Medina, W. J. Rutter, G. Ammerer, and B. D. Hall, *Nature (London)* **298,** 347 (1982).
[7] A. Moriarty, B. H. Hoyer, J. W. Shih, J. L. Gerin, and D. H. Hamer, *Proc. Natl. Acad. Sci. U.S.A.* **78,** 2606 (1981).
[8] N. Fujita, N. Nelson, T. D. Fox, T. Claudio, J. Lindstrom, H. Reizman, and G. P. Hess, *Science* **231,** 1284 (1986).

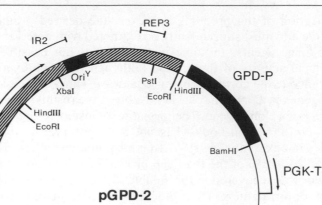

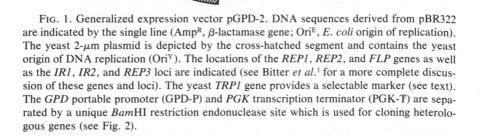

Fig. 1. Generalized expression vector pGPD-2. DNA sequences derived from pBR322 are indicated by the single line (Amp^R, β-lactamase gene; Ori^E, *E. coli* origin of replication). The yeast 2-μm plasmid is depicted by the cross-hatched segment and contains the yeast origin of DNA replication (Ori^Y). The locations of the *REP1*, *REP2*, and *FLP* genes as well as the *IR1*, *IR2*, and *REP3* loci are indicated (see Bitter *et al.*[1] for a more complete discussion of these genes and loci). The yeast *TRP1* gene provides a selectable marker (see text). The *GPD* portable promoter (GPD-P) and *PGK* transcription terminator (PGK-T) are separated by a unique *Bam*HI restriction endonuclease site which is used for cloning heterologous genes (see Fig. 2).

Plasmodium knowlesi was also successfully expressed in yeast.[9] In this case, the antigen was largely associated with the 25,000 *g* and 150,000 *g* particulate fractions of cell lysates and could be solubilized with Triton X-100, indicating an association with cell membranes. These results indicate that yeast is an appropriate host cell type for confirming the identity of putative clones encoding viral coat proteins or cell membrane and/or surface proteins.

Figure 1 depicts a restriction endonuclease map of a generalized yeast expression vector, pGPD-2. This vector[10] has been used for expression of

---

[9] S. Sharma and N. G. Godson, *Science* **228**, 879 (1985).
[10] G. A. Bitter and K. M. Egan, *Gene* **32**, 263 (1984).

a large number of heterologous genes in *S. cerevisiae* and has the following properties. It includes the plasmid pBR322 with an intact origin of replication and β-lactamase gene. Recombinant DNA manipulations may thus be performed in, and plasmids amplified and purified from, *E. coli*. The vector also includes the entire yeast 2-μm plasmid cloned at the *Eco*RI site in the large unique region. In this construction, the 2-μm plasmid provides an intact replication and amplification system such that it can be stably maintained at high copy numbers in yeast (see Bitter *et al.*[1] for a more complete discussion of plasmid stability and copy number). pGPD-2 also contains the yeast *TRP1* gene which allows selection of cells containing the plasmid when the host contains a chromosomal *trp1* mutation. Transformed yeast cells are selected by their ability to grow in medium lacking tryptophan. The "expression cassette" of pGPD-2 consists of a transcription promoter and termination element separated by a unique *Bam*HI restriction endonuclease recognition site. The *GPD* portable promoter is cloned as a *Hin*dIII–*Bam*HI fragment and results in efficient transcription of sequences downstream of the *Bam*HI site. Vector pGPD-2 uses a DNA sequence from the 3' end of the yeast *PGK* gene to effect termination of transcripts initiated from the *GPD* promoter. Heterologous gene expression is obtained in yeast by cloning the gene into the unique *Bam*HI site of pGPD-2 according to the strategies outlined below. Several other laboratories[11–14] have developed yeast expression vectors with different functional elements but general characteristics similar to pGPD-2.

Certain genes in *S. cerevisiae* contain introns, and the splicing of such mRNAs has been studied. However, the splicing of mRNAs from higher organisms produced in yeast is inefficient and/or aberrant.[15] Therefore, at this time, heterologous gene expression in yeast, as in *E. coli*, is limited to genomic clones lacking introns, cDNAs, or chemically synthesized genes. Genes containing introns can be expressed, and the transcript correctly processed, in mammalian cells (see this volume [71]).

Expression vector pGPD-2, and most of the yeast expression vectors referenced above, include the transcription start site within the promoter element but lack translation start signals. Therefore the requirements for

[11] R. A. Hitzman, F. E. Hagie, H. L. Levine, D. V. Goeddel, G. Amerer, and B. D. Hall, *Nature (London)* **293,** 717 (1981).
[12] R. Derynck, A. Singh, and D. V. Goeddel, *Nucleic Acids Res.* **11,** 1819 (1983).
[13] C. G. Goff, D. T. Moir, T. Kohno, T. Gravius, R. A. Smith, E. Yamasaki, and A. Taunton-Rigby, *Gene* **27,** 35 (1984).
[14] A. Miyanohara, A. Toh-E, C. Nozaki, F. Hamada, N. Ontomo, and K. Matsubara, *Proc. Natl. Acad. Sci. U.S.A.* **80,** 1 (1983).
[15] J. D. Beggs, J. VandenBerg, A. Van Ooyen, and C. Weissman, *Nature (London)* **283,** 835 (1980).

heterologous protein expression programmed by pGPD-2 are that the heterologous gene be cloned in the correct orientation and include the appropriate sequences for translation initiation. There are no consensus sequences in yeast mRNA untranslated leader regions analogous to bacterial ribosome binding sites. In eukaryotes, translation initiates almost exclusively at the first AUG on the mRNA. Therefore a functional translation initiation site in yeast refers to a DNA sequence in which the ATG-methionine codon of the heterologous gene represents the first AUG triplet in the mRNA. Although there are no required specific sequences, the sequence context of the untranslated leader is likely to affect the efficiency of translation. Thus optimal translation initiation efficiency in yeast appears to be correlated with A-rich and G-deficient untranslated leaders.[16] An A residue at position −3 is found in most yeast mRNAs. Other sequences within the untranslated leader region will probably support translation initiation, but with variable efficiencies. Similarly, yeast mRNA untranslated leaders are generally 30–50 nucleotides in length although much longer leaders are still functional.[10]

The manipulations involved in cloning heterologous genes in pGPD-2 are depicted in Fig. 2. The heterologous gene must be cloned in the correct orientation such that the DNA sequence proximal to the *GPD* promoter is

$$\overset{*}{n}(n)_x nnnATGnnn\text{---}$$

The star represents the transcription start site for the portable promoter. The absolute requirement for translation initiation is that the ATG initiation codon for the heterologous gene be the first such triplet downstream from the transcription start site. Cloned genes from certain genomic clones may be isolated as restriction endonuclease fragments which can be expressed without further modifications. In other instances, however, there may not be an appropriate restriction site 5′ to the desired ATG translation start site. In these cases, a synthetic linker may be utilized to fuse a restriction site 3′ to the ATG initiation codon to the expression vector. The synthetic linker should encode the same amino acids as the native gene between the ATG and restriction site utilized for cloning. However, the synthetic linker may utilize a different DNA sequence both within the coding region and the untranslated leader. Utilization of a synthetic linker presents an opportunity to optimize translation initiation. Evidence has been obtained that utilization of codons thought to be opti-

[16] G. Amerer, R. Hitzeman, F. Hagie, A. Barta, and B. D. Hall, in "Proceedings of the Third Cleveland Symposium on Macromolecules: Recombinant DNA" (A. G. Walton, ed.), p. 185. Elsevier, Amsterdam, 1981.

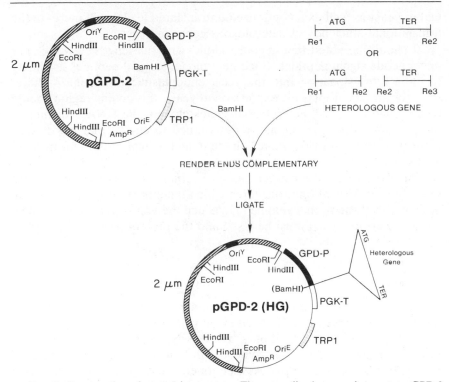

FIG. 2. Construction of expression vectors. The generalized expression vector pGPD-2 (Fig. 1) is digested with *Bam*HI for cloning the heterologous gene which includes an appropriate translation initiation sequence (see text) as well as a translation termination codon (TER). *RE1*, *RE2*, and *RE3* refer to three different, but nonspecified, restriction sites. Chemically synthesized heterologous genes may incorporate *Bam*HI cohesive termini for facile cloning. Certain cloned genes may be isolated as fragments for direct cloning. In other cases, the 5' end of the heterologous gene may need to be replaced with a chemically synthesized segment (see text). For many heterologous genes, the restriction enzyme-generated 5' and 3' termini will not be cohesive with the *Bam*HI-generated termini of pGPD-2. In such instances, the ends of the vector and heterologous gene must be rendered cohesive or blunt. This may be accomplished by incorporation of *Bam*HI linkers on the heterologous gene, utilization of synthetic linkers, partial end-fill of the vector and heterologous gene, or rendering both the vector and heterologous gene blunt (see Bitter *et al.*[1] for a more complete discussion of these methods). The cloning procedure utilized determines whether the *Bam*HI site is regenerated in the resultant expression vector [designated pGPD-2(HG) in this figure]. In all cases, the DNA sequence at the *GPD* promoter/heterologous gene fusion should be examined (see text).

mal in yeast as well as an optimized untranslated leader will increase the translational efficiency of heterologous genes in yeast.[10]

It should be noted that genomic clones and full-length cDNA clones may encode signal peptides. Such heterologous signal peptides are likely to initiate translocation into the yeast endoplasmic reticulum.[17] Therefore, for direct cytoplasmic expression in yeast, this coding region should be eliminated. For all heterologous genes cloned, the DNA sequence at the 5' end of the gene should be determined to ensure that the desired ATG translation start is, in fact, the first such triplet in the untranslated leader.

Once the desired construct has been obtained, it is introduced into yeast by transformation procedures which have been described in detail elsewhere[18,19] and in this volume [53]. Since the expressed protein will be intracellular, the yeast must be lysed and the protein product assayed as described previously.[1]

Secretion

The advantages of producing a protein in a secretion system include the following. Specific NH_2 termini may be produced which lack the methionine from the translation start. This NH_2-terminal specificity is a result of cleavage of the heterologous protein from the hybrid-secreted protein precursor (*vide infra*). As mentioned above, proteins produced by direct expression in *E. coli* often result in insoluble aggregates which must be subjected to a renaturation scheme in order to obtain the correct disulfide structure. Evidence exists indicating that proteins produced in a yeast secretion system will assume the correct disulfide structure.[20] This is presumably due to the presence of protein disulfide-isomerase (EC 5.3.4.1) in the endoplasmic reticulum of eukaryotic secretory cells.[21] This enzyme is believed to catalyze the accurate and efficient formation of disulfide bonds in secreted eukaryotic proteins. Finally, many eukaryotic secretory proteins contain carbohydrate additions. N-linked carbohydrate is covalently attached to Asn residues within the sequence context Asn-X-Ser or Asn-X-Thr where X is any amino acid except proline. O-linked carbohydrate may also be attached to serine or threonine residues.

[17] G. A. Bitter, in "Microbiology 1986" (D. Schlessinger, ed.), p. 330. Am. Soc. Microbiol., Washington, D.C., 1986.
[18] A. Hinnen, J. B. Hicks, and G. R. Fink, *Proc. Natl. Acad. Sci. U.S.A.* **75,** 1929 (1978).
[19] H. Ito, Y. Fukuda, and A. Kimara, *J. Bacteriol.* **153,** 163 (1983).
[20] K. M. Zsebo, H. S. Lu, J. C. Fieschko, L. Goldstein, J. Davis, K. Duker, S. V. Suggs, P. H. Lai, and G. A. Bitter, *J. Biol. Chem.* **261,** 5858 (1986).
[21] R. B. Freedman, *Trends Biochem. Sci.* **106,** 438 (1984).

However, the recognition sequence(s) for this latter modification has not been well defined. The presence of carbohydrate additions is required for the solubility and/or enzymatic activity of many proteins. In addition, this modification is required for the *in vivo* biological activity of certain glycoproteins and, furthermore, the specific structure of the carbohydrate addition may be critical.

The *S. cerevisiae* prepro-α-factor leader region (encoded by the *MFα1* gene) has been used to direct protein secretion from yeast.[20,22–27] The leader region consists of a hydrophobic NH_2-terminal 20–22 amino acid segment with properties similar to other characterized signal peptides. There is a 61–63 amino acid prosegment of unknown function which contains three sites of N-linked glycosylation. The prosegment is flanked by a spacer peptide which includes recognition sequences for two yeast proteases. The enzyme encoded by the yeast *KEX2* gene cleaves the precursor on the carboxyl side of Lys-Arg. The Glu-Ala dipeptides at the NH_2 terminus of the excised protein are removed by dipeptidyl aminopeptidase A which is encoded by the yeast *STE13* gene.

Two general vectors, pαC2 and pαC3, have been developed[20,23] for constructing prepro-α-factor/foreign gene fusions. These vectors, in addition to the prepro-α-factor leader coding region, contain the native *MFα1* promoter and transcription terminator elements. Thus, once the gene fusion is constructed, it will program synthesis of the hybrid protein in yeast. The gene fusion must first be transferred (as a *Bam*HI fragment) to a vector capable of selection and replication in yeast. Other promoter elements may be utilized to drive expression of the gene fusion according to strategies outlined previously.[1]

Both cloning vectors pαC2 and pαC3 contain a *Hin*dIII recognition sequence for construction of the fusion of the prepro-α-factor leader and foreign gene coding regions. The heterologous gene may thus be cloned as a *Hin*dIII fragment. Alternatively, the orientation in the ligation may be directed by cloning the heterologous gene as a *Hin*dIII–*Sal*I fragment. (It has been demonstrated[20,23] that the DNA sequences between the *Hin*dIII site and *Sal*I site of pαC2 or pαC3 may be deleted while retaining *MFα1*

[22] A. J. Brake, J. P. Merryweather, D. G. Coit, U. A. Heberlein, F. R. Masiarz, G. T. Mullenbach, M. S. Urdea, P. Valenzuela, and P. J. Barr, *Proc. Natl. Acad. Sci. U.S.A.* **81,** 4642 (1984).

[23] G. A. Bitter, K. K. Chen, A. R. Banks, and P.-H. Lai, *Proc. Natl. Acad. Sci. U.S.A.* **81,** 5330 (1984).

[24] A. Singh, J. M. Lugovoy, W. J. Kohr, and L. J. Perry, *Nucleic Acids Res.* **12,** 8927 (1984).

[25] A. Miyajima, M. W. Bond, K. Otsu, K. Arai, and N. Arai, *Gene* **37,** 155 (1985).

[26] G. P. Vlasuk, G. H. Bencen, R. M. Scarborough, P. K. Tsai, J. L. Whang, T. Maack, M. J. F. Camargo, S. W. Kirsher, and J. A. Abraham, *J. Biol. Chem.* **261,** 4789 (1986).

[27] R. Green, M. D. Schaber, D. Shields, and R. Kramer, *J. Biol. Chem.* **261,** 7558 (1986).

sequences necessary for transcription termination and polyadenylation). The fusion sequences encoded by the hybrid genes are depicted in Fig. 3. Assembly of the gene fusions involves restricting pαC2 or pαC3 with HindIII or HindIII and SalI and performing a three-piece ligation as depicted in Fig. 4. The vector will encode a hybrid protein with signals for secretion and processing. Therefore, the cloning manipulations must be very precise and involve the use of synthetic linkers (as depicted in Fig. 4) to create the gene fusion. An oligonucleotide complementary to DNA upstream from the processing sites can be used as a primer to confirm the sequence of the gene fusions.[20] Once the DNA sequence of the gene fusion is confirmed, it may be transferred to a yeast–E. coli shuttle vector (such as pYE; Bitter et al.[1]) as a BamHI fragment and introduced into yeast by transformation.

The amino acid sequence at the fusion point of hybrid proteins constructed as described in the legend to Fig. 4 is also depicted in Fig. 3. Hybrid proteins derived from pαC2 contain a spacer peptide with recognition sites for both the KEX2 and STE13 gene-encoded proteases. It has been observed that when such hybrid proteins are produced at high levels (e.g., from multicopy vectors), the dipeptidyl aminopeptidase A activity

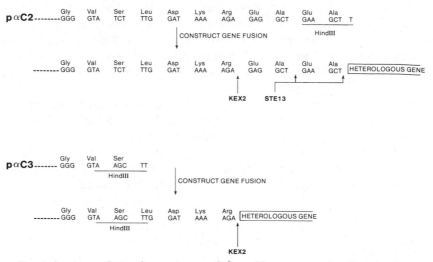

FIG. 3. Sequence of secretion vector gene fusions. The sequence of pαC2 and pαC3 up to the HindIII site used for construction of the gene fusions is depicted. The gene fusions are constructed as described in the legend to Fig. 4 and result in the hybrid protein sequences depicted. When pαC2 is used for secretion vector construction, the hybrid protein contains recognition sites for both the KEX2 and STE13 gene-encoded proteases. Plasmid pαC3 is used for construction of gene fusions which include only the KEX2 gene-encoded protease recognition site.

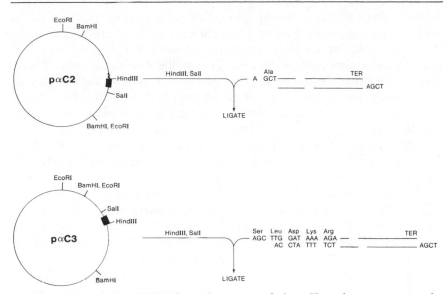

FIG. 4. Construction of *MFα1*/heterologous gene fusions. Heterologous genes may be cloned as either a *Hin*dIII or *Hin*dIII–*Sal*I fragment. The orientation of the ligation may be directed by cloning as a *Hin*dIII–*Sal*I fragment and this approach is depicted in the figure. The orientation of the *Bam*HI fragment is different in pαC2 and pαC3. The enclosed box represents the last α-factor peptide coding region of *MFα1*. The squiggle in pαC2 represents the spacer peptide. The required DNA sequence at the 5′ end of the heterologous gene is depicted for constructions in either pαC2 or pαC3. For pαC2, a *Hin*dIII cohesive end is utilized, but regeneration of the *Hin*dIII site will occur only when a T occupies the first base of the first codon of the heterologous gene. The construction in αC3 regenerates the *Hin*dIII cloning site. The desired coding region for the heterologous gene begins in-phase immediately after the alanine codon in pαC2 or the arginine codon in pαC3. The segment cloned must include the termination codon (TER) for the heterologous genes. The heterologous genes may be chemically synthesized to yield the precise gene fusion. When previously cloned genes are employed, it will be necessary to use a synthetic linker to create the gene fusion between the *Hin*dIII site of pαC2 or pαC3 and an internal restriction site of the heterologous gene. Similarly, the *Hin*dIII or *Sal*I cohesive terminus at the 3′ end of the heterologous gene may be supplied by a synthetic linker.

within the cell becomes rate limiting, and foreign proteins containing NH$_2$-terminal Glu-Ala dipeptide extensions are produced.[22,23] In using plasmid pαC3 for construction of the hybrid gene, the hybrid protein processing is independent of the *STE13* gene product, and the desired NH$_2$ terminus is produced following cleavage by the *KEX2* encoded enzyme.[20] It should be noted that for some hybrid precursors, incomplete cleavage by the *KEX2* encoded protease has been observed.[20]

The secreted protein product may be assayed after removal of cells by centrifugation or filtration. Since yeast culture medium is acidic, it may be necessary to buffer the medium for certain protein products. It should be noted that different proteins are secreted with different efficiencies using the prepro-α-factor leader.[20] Thus some proteins are efficiently secreted into the culture medium while others accumulate intracellularly. Therefore, if the assay for the protein product is of low sensitivity it may be necessary to concentrate the culture medium first.

It has been demonstrated that mammalian glycoproteins secreted using the prepro-α-factor leader accumulate N-linked carbohydrate.[28] This modification may be required for the solubility and/or enzymatic activity of certain proteins. The core structure of yeast oligosaccharides is identical to that present on mammalian glycoproteins.[29] Most yeast strains incorporate outer chains of mannose residues. These structures are larger and of composition different from the oligosaccharides of mammalian glycoproteins. The formal possibility exists, therefore, that certain glycoproteins produced in a yeast secretion system will be biologically inactive. In cases in which no biological activity is detected upon expression of a heterologous gene in a yeast secretion system, the biological activity of the same protein produced in a mammalian secretion system should be examined.

[28] S. G. Elliott *et al.,* in preparation.
[29] C. E. Ballou, *in* "The Molecular Biology of the Yeast Saccharomyces: Metabolism and Gene Expression" (J. N. Strathern, E. W. Jones, and J. R. Broach, eds.), p. 335. Cold Spring Harbor Lab., Cold Spring Harbor, New York, 1982.

[71] Use of Eukaryotic Expression Technology in the Functional Analysis of Cloned Genes

By BRYAN R. CULLEN

The purpose of this chapter is to describe ways in which eukaryotic expression technology can be used to identify and to analyze the function of cloned eukaryotic genes. As in the preceding two chapters, the assumption is made that the clone of interest has been sequenced and an open reading frame has been identified. Although expression of genomic sequences will be briefly discussed, in general it is assumed that the sequence of interest is a cDNA.

Eukaryotic expression technology offers a number of advantages which in essence devolve from the fact that the introduced gene is expressed in a host cell which closely approximates its natural milieu. Proteins encoded by the exogenously introduced genes are correctly folded by the host cell translational machinery and form appropriate disulfide bonds. Further, posttranslational processing of the protein by the host cell (glycosylation, phosphorylation, etc.) will normally closely approximate that observed in nature. The host cell can also recognize localization signals within the expressed protein and will correctly transport it to the nucleus, cell surface, extracellular medium, etc. All of these attributes can be difficult or impossible to reproduce in a heterologous prokaryotic host such as *Escherichia coli* and greatly facilitate the production and analysis of a functional eukaryotic gene product. Eukaryotic expression technology is made even more attractive by the ease and rapidity with which some introduced heterologous gene products can be synthesized in microgram or even milligram quantities.

This chapter is divided into three sections. The first section describes several possible strategies for maximizing heterologous gene expression in the cells of higher eukaryotes. The second section deals with potential assays for gene expression based on function, and the third section describes some immunological approaches. Overall, the focus is on the use of techniques which yield information not obtainable from heterologous gene expression in bacteria or yeast.

Expression Strategies

Mammalian cells and the cells of other higher eukaryotes are able to take up and express exogenously added DNA. Although the mechanism of DNA-mediated gene transfer remains uncertain, several properties of this process are now evident. Only a proportion of the cells exposed to the transfection solution are "competent" to take up DNA. Cells that are competent assimilate relatively high levels and this DNA then becomes linked into very large concatemers which have been termed *transgenomes*.[1] Transgenomes are expressed by the transfected cell; however, they lack functional centromeres and are lost from the cell population with first-order kinetics. Expression of transfected DNA therefore demonstrates a sharp peak 48–72 hr posttransfection which is termed *transient expression*. In a small fraction of the cells the transgenome becomes randomly linked to the cellular genome and is then stably maintained. If an excess of a nonselected gene of interest was mixed with a dominant selectable gene prior to transfection, these genes normally become associated during the transfection process, so that the nonselected gene is

---

[1] G. Scangos and F. H. Ruddle, *Gene* **14**, 1 (1981).

present and expressed in the majority of the selected colonies.[2] This process, which I term *simple cotransfection,* therefore results in cells which demonstrate stable or constitutive expression of the gene of interest.

Transient Expression

The level of transient expression in a transfected culture is determined in large part by the number of cells which take up the transfected gene and by the number of copies of that gene then expressed by the competent cell. The level of transient expression therefore varies dramatically with the cell line and transfection technique used. A cell line which gives very high levels of transient gene expression is COS.[3] COS cells are African green monkey kidney cells which have been transformed by an origin-defective mutant of simian virus 40 (SV40). These cells express high levels of SV40 T antigen and will therefore replicate to very high copy number any introduced plasmid containing an SV40 origin of replication.[4] This high copy number, combined with the high competence of these cells for DNA-mediated gene transfer, results in a high level of heterologous gene expression at 48–72 hr posttransfection (Table I). As will be discussed below, this level of gene expression is frequently sufficient for the functional and immunological analysis of the transfected gene product and has in fact been used for the cloning of eukaryotic genes by functional expression.[5,6] In addition, transient expression may be very useful for the synthesis of gene products which demonstrate toxicity at high levels of constitutive expression.

Constitutive Gene Expression

Cotransfection of cells with a plasmid containing a selectable marker and an excess (generally 10- to 20-fold) of a nonselected gene of interest permits the isolation, at ~2 weeks posttransfection, of colonies which constitutively express the latter gene. In general, these colonies only contain one or a few copies of the nonselected gene. Further, transfected

[2] M. Wigler, R. Sweet, G. Sim, B. Wold, A. Pellicer, E. Lacy, T. Maniatis, S. Silverstein, and R. Axel, *Cell (Cambridge, Mass.)* **16,** 777 (1979).
[3] Y. Gluzman, *Cell (Cambridge, Mass.)* **23,** 175 (1981).
[4] P. Mellon, V. Parker, Y. Gluzman, and T. Maniatis, *Cell (Cambridge, Mass.)* **27,** 279 (1981).
[5] G. G. Wong *et al., Science* **228,** 810 (1985).
[6] T. Yokota, N. Arai, F. Lee, D. Rennick, T. Mosmann, and K. Arai, *Proc. Natl. Acad. Sci. U.S.A.* **82,** 68 (1985).

TABLE I
EXPRESSION LEVELS OBTAINED USING DIFFERENT EUKARYOTIC
EXPRESSION STRATEGIES[a]

Cell line	Expression mode	Expression strategy	IL-2 production	
			Units	μg
COS	Transient	DEAE-dextran transfection	20,480	~1
CHO	Transient	CaPO$_4$ transfection	768	~0.04
CHO	Constitutive	Simple cotransfection	1,580	~0.08
CHO	Constitutive	dhfr-linked coamplification	401,000	~20
C127	Constitutive	BPV-based vector	19,320	~1

[a] The expression strategies used are described in the text. All use the expression vector pBC12BI except the last, which uses a BPV-based expression vector.[41] IL-2 production is given in IL-2 units (and micrograms) of protein secreted per 10[6] cells over a 48-hr period from 24 to 72 hr posttransfection (transient) or postseeding (constitutive). These values have been corrected for the number of cells transfected (transient) or present in the culture at the time of IL-2 harvest (constitutive). IL-2 has a specific activity of ~2 × 10[4] units/μg.

DNA which is stably maintained by cells is frequently expressed inefficiently and may indeed become methylated and transcriptionally inactive.[7] For these reasons, simple cotransfection generally results in relatively low levels of constitutive gene expression (Table I). The average level of expression of a coselected gene may be somewhat improved by tighter linkage to the selectable marker by, for example, cloning both genes into a single vector. However, unless the choice of the cell line to be used for expression is constrained by a requirement for a particular mutation or differentiation specific attribute (see below), simple cotransfection offers few advantages relative to the high-level, transient gene expression obtainable using COS cells at only 48 hr posttransfection.

Eukaryotic ribosomes appear able, at low efficiency, to reinitiate translation and can therefore express the second gene of a dicistronic mRNA, albeit at a low level.[8] A good way to improve cotransfection efficiency may therefore be to insert the selectable marker gene into the 3′ noncoding region of the nonselected gene. The selectable marker should be inserted such that the initiation codon of the selectable (3′) gene is immediately downstream of the termination codon of the nonselected (5′) gene of interest. Intervening cryptic AUGs should be avoided. Selection for a resistant phenotype, if possible at all, then requires the efficient

[7] L. Hwang and E. Gilboa, *J. Virol.* **50**, 417 (1984).
[8] D. S. Peabody and P. Berg, *Mol. Cell. Biol.* **6**, 2695 (1986).

expression of the dicistronic mRNA and, hence, the gene of interest located 5' to the selectable marker on the same mRNA. The *neo* gene, derived from the prokaryotic transposon Tn5, confers resistance to the aminoglycoside antibiotic G418 and is a dominant selectable marker in all mammalian cells.[9] Selection for G418 resistance, when the *neo* gene is located in the 3' position of a dicistronic mRNA, can result in levels of 5', nonselected gene expression $\geq$10-fold higher than simple cotransfection. Alternative methods for achieving high level constitutive gene expression are discussed below.

Eukaryotic Expression Vectors

A eukaryotic expression vector should be designed for efficient use in the widest possible range of cell lines and should be as versatile and compact as possible. Desirable features include (1) a prokaryotic origin of replication and a selectable marker functional in *E. coli*; (2) an SV40 origin of replication and a plasmid backbone deleted of "poison" sequences[10]; (3) a powerful transcription control region active in a wide range of eukaryotic cells; (4) a genomic (viral or cellular) polyadenylation signal and site[11]; and (5) an intronic sequence.[12]

Several useful eukaryotic expression vectors based on SV40-derived sequences have been described, of which the most widely used is perhaps the pSV2 vector.[9,11,13–15] For illustrative purposes, I will focus on two general-purpose expression vectors termed pBC12BI and p91023(B). pBC12BI, the structure of which is detailed in Fig. 1,[16–17a] is a good example of a compact and efficient eukaryotic expression vector. This vector is useful for heterologous gene expression in a wide range of cell types including COS cells (Table I). The larger expression vector p91023(B)[5] also contains all the essential features of a general-purpose expression vector but in addition contributes the tripartite leader of the adenovirus late mRNAs to the 5' end of transcribed, heterologous genes.

[9] P. J. Southern and P. Berg, *J. Mol. Appl. Genet.* **1,** 327 (1982).
[10] M. Lusky and M. Botchan, *Nature (London)* **293,** 79 (1981).
[11] R. J. Kaufman and P. A. Sharp, *Mol. Cell. Biol.* **2,** 1304 (1982).
[12] D. J. Hamer and P. Leder, *Cell (Cambridge, Mass.)* **18,** 1299 (1979).
[13] S. Subramani, R. Mulligan, and P. Berg, *Mol. Cell. Biol.* **1,** 854 (1981).
[14] R. C. Mulligan and P. Berg, *Science* **209,** 1422 (1985).
[15] R. S. McIvor, J. M. Goddard, C. C. Simonsen, and D. W. Martin, *Mol. Cell. Biol.* **5,** 1349 (1985).
[16] J. Hanahan, *J. Mol. Biol.* **166,** 577 (1983).
[16a] B. R. Cullen, K. Raymond, and G. Ju, *J. Virol.* **53,** 515 (1985).
[17] P. T. Lomedico, *Proc. Natl. Acad. Sci. U.S.A.* **79,** 5798 (1982).
[17a] Y. Morinaga, T. Franceschini, S. Inouye, and M. Inouye, *Bio Technology* **2,** 636 (1984).

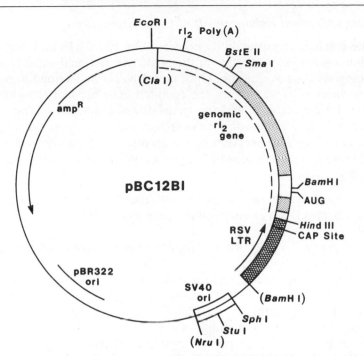

FIG. 1. Structure of expression vector pBC12BI. The pBC12BI vector is ~4.1 kb in size and contains a bacterial origin of replication and a β-lactamase (amp^R) gene derived from pXF3,[16] a pBR322 derivative which lacks sequences inhibitory to DNA replication in COS cells.[10] pBC12BI also contains an SV40 origin of replication (SV40 ori) and the powerful Rous sarcoma virus (RSV) long terminal repeat (LTR) transcription control region.[16a] In addition, the vector contains a complete copy of the genomic rat preproinsulin II (rI_2) gene,[17] which contributes an intron and an efficient polyadenylation signal–site. Heterologous cDNA sequences containing a translation initiation codon are normally inserted between the unique HindIII and BamHI sites in an anticlockwise orientation. Similarly, heterologous genomic sequences are normally inserted between HindIII and SmaI. If the investigator wishes to use the vector initiation codon, the cDNA should be inserted in frame into the unique vector BamHI site located immediately 3' to the AUG (variants of pBC12BI offering all three frames at this site exist). This will result in an initial translation product containing an additional five or six N-terminal amino acids derived from the insulin signal peptide. If desired, this may be corrected by the use of oligonucleotide-directed deletion mutagenesis.[17a] (The vector pBC12BI is available from the author upon request.) Cross-hatched areas, RSV LTR; stippled, intronic sequences. Restriction sites in parentheses were deleted during vector construction.

This nontranslated leader, in conjunction with the adenovirus VA RNAs encoded elsewhere on p91023(B), results in the enhanced translation of mRNAs transcribed from this vector.[18]

[18] R. J. Kaufman, *Proc. Natl. Acad. Sci. U.S.A.* **82**, 689 (1985).

Posttranscriptional Enhancement of Gene Expression

The enhanced translatability conferred by p91023(B) on inserted heterologous genes is an example of the posttranscriptional enhancement of heterologous gene expression. Once an mRNA is transcribed from a vector, the level of expression of the encoded gene product is significantly affected by the stability and the translational efficiency of the mRNA. Relatively little is known about the factors governing posttranscriptional regulation of gene expression in eukaryotic cells, and the rules I will propose are therefore somewhat empirical. With this in mind I suggest the following caveats.

1. Homopolymer tails inserted into the 5' noncoding region of an mRNA during cDNA cloning reduce gene expression from 2- to over 10-fold.[19]

2. "Cryptic" AUG initiation codons in the leader of an mRNA may reduce gene expression.[8]

3. 5' and 3' noncoding regions may contain sequences which specifically destabilize mRNAs.[20,21] In particular, several growth factor and oncogene mRNAs contain a 3' AU-rich sequence which may play a role in the posttranscriptional regulation of these genes.[22]

4. An mRNA may be poorly translated if the 5' noncoding region contains excessive secondary structure[23] or if the initiation codon diverges from the consensus sequence GXXAUGG.[24]

The general picture which emerges is that the cloned cDNA should be trimmed of as much mRNA noncoding sequence as possible prior to insertion into the expression vector. An even better approach may be to replace entirely, for example, the 5' noncoding sequence and AUG of the gene of interest with the 5' noncoding sequence and AUG of an mRNA which is very efficiently expressed. For example, pBC12B1 contains a unique *Bam*HI site located immediately 3' to the initiation codon of the efficiently expressed rat preproinsulin II gene (Fig. 1). Use of the rat preproinsulin 5' noncoding region and initiation codon to initiate translation of the human interleukin-2 (IL-2) gene results in an ~10-fold enhancement in the expression of secreted, correctly processed IL-2 relative to a similar pBC12BI-based construction containing the natural IL-2 AUG and leader region. The concept that noncoding sequences derived

[19] C. C. Simonsen *et al., ICN-UCLA Symp. Mol. Cell. Biol.* **25,** 1 (1981).

[20] P. H. Rabbitts, A. Forster, M. A. Stinson, and T. H. Rabbitts, *EMBO J.* **4,** 3727 (1985).

[21] T. Morris, F. Marashi, L. Weber, E. Hickey, D. Greenspan, J. Bonner, J. Stein, and G. Stein, *Proc. Natl. Acad. Sci. U.S.A.* **83,** 981 (1986).

[22] G. Shaw and R. Kamen, *Cell* **46,** 659 (1986).

[23] J. Pelletier and N. Sonenberg, *Cell (Cambridge, Mass.)* **40,** 515 (1985).

[24] M. Kozak, *Cell (Cambridge, Mass.)* **44,** 283 (1986).

from stable, efficiently expressed mRNAs can enhance the level of expression of a gene of interest may have general applicability.

Expression of Genomic Eukaryotic Sequences

A unique advantage of eukaryotic expression systems is the ability to express cloned sequences which retain intronic regions. Genomic sequences introduced into higher cells by DNA-mediated gene transfection are normally transcribed, spliced, and expressed appropriately.[12,25] The expression of cloned genomic copies of genes is therefore very similar to expression of cDNA copies, but is made technically difficult by the frequently very large size of genomic genes. If the exonic structure of the gene is known, the coding exons can be inserted into expression vectors such as pBC12BI (Fig. 1). Alternatively, the genomic clone may contain all the sequences required for expression within the transfected cell. However, it should be noted that enhancers and promoters may be highly tissue specific[26] and may therefore be poorly utilized in a transfected, heterologous cell.

Methods for Introducing DNA into Cells

A number of methods exist for the introduction of DNA into eukaryotic cells. No single procedure is ideal in all cases, the method of choice being determined by the target cell. In this section is a detailed protocol for an efficient DEAE-dextran-mediated transfection procedure for COS cells. In addition, a simple and efficient calcium phosphate coprecipitation technique[27,28] is described for introduction of DNA into several fibroblastic and epithelial cell lines. Many other variations on the DEAE-dextran and calcium phosphate procedure have been published which may offer advantages in certain instances.[15,29,30]

Two other useful procedures not described in detail here are protoplast fusion[31] and electroporation.[32] In protoplast fusion, E. coli cells harboring the expression vector are first treated with chloramphenicol to amplify plasmid copy number, and are then converted to protoplasts by

[25] P. W. Gray and D. V. Goeddell, *Proc. Natl. Acad. Sci. U.S.A.* **80**, 5842 (1983).

[26] M. D. Walker, T. Edlund, A. M. Boulet, and W. J. Rutter, *Nature (London)* **306**, 557 (1983).

[27] F. L. Graham and A. J. van der Eb, *Virology* **52**, 456 (1973).

[28] A. D. Miller, T. Curran, and I. M. Verma, *Cell (Cambridge, Mass.)* **36**, 51 (1984).

[29] C. Queen and D. Baltimore, *Cell (Cambridge, Mass.)* **33**, 741 (1983).

[30] C. M. Gorman, L. F. Moffat, and B. H. Howard, *Mol. Cell. Biol.* **2**, 1044 (1982).

[31] R. M. Sandri-Goldin, A. L. Goldin, M. Levine, and J. C. Glorioso, *Mol. Cell. Biol.* **1**, 743 (1981).

[32] H. Potter, L. Weir, and P. Leder, *Proc. Natl. Acad. Sci. U.S.A.* **81**, 7161 (1984).

treatment with lysozyme. These protoplasts are then directly fused to the recipient cells using polyethylene glycol. Electroporation, in contrast, uses short bursts of very high-voltage electricity to introduce exogenous DNA directly into the recipient cell line. Both approaches can be effective at introducing DNA into a wide range of cells, including cells refractory to transfection such as primary or lymphoid cells. Protoplast fusion is, however, somewhat time-consuming, and electroporation requires the use of special electrical equipment. These procedures should therefore be considered only after the simpler transfection techniques described below fail.

DEAE-Dextran-Mediated Transfection of COS Cells

Sterile Solutions. DNA dissolved in 100 mM NaCl, 10 mM Tris–HCl (pH 7.5) (~500 ng/35-mm plate); phosphate-buffered saline (PBS; Ca^{2+}, Mg^{2+}-free; GIBCO); DEAE-dextran (Pharmacia, MW ~500,000), 10 mg/ml in PBS (autoclaved); 40 mM chloroquine (Sigma) in PBS, prepared fresh.

1. Seed 35-mm tissue culture plates with 3×10^5 COS cells per plate. By the next day the plate should be just subconfluent.

2. Prepare transfection cocktail in a sterile Eppendorf tube. Add supercoiled plasmid DNA to PBS (190 μl DNA volume), vortex, and then add 10 μl of DEAE-dextran and mix.

3. Aspirate tissue culture medium from culture dish and rinse COS cells with 2 ml of PBS warmed to 37°. Aspirate PBS.

4. Add transfection cocktail and distribute evenly by tilting plate. Incubate at 37° for 30 min with occasional gentle shaking to prevent drying.

5. Add 2 ml of tissue culture medium supplemented with 80 μM chloroquine and incubate at 37° for 2.5 hr.

6. Aspirate supernatant medium and replace with 1 ml of tissue culture medium containing 10% DMSO for 2.5 min. Aspirate and add 2 ml of fresh medium.

7. Use cells or medium for analysis 48–72 hr posttransfection.

Comments. This is a very robust, simple procedure which is specifically tailored for transient expression in COS cells and which introduces DNA into 5–10% of the transfected cells. For different culture sizes simply scale up or down, e.g., use 2 ml of transfection cocktail for a 100-mm culture dish. Toxicity during the transfection process is not expected, but EDTA present in the DNA solution may result in cell detachment, and chloroquine at 80 μM can be toxic on subconfluent cultures. Maximum levels of gene expression are obtained using ~500 ng of plasmid DNA per

35-mm plate, and 50 ng per plate yields ~25% of maximal response. Although plasmid DNA purified by banding in CsCl gradients is preferred, DNA prepared by any technique, including alkaline minilysate[33] (see also this volume [13]), may be used as this transfection procedure is not markedly affected by contaminants such as tRNA. This technique may therefore also be used to screen single or pooled clones for functional expression.

Calcium Phosphate Transfection

Sterile Solutions. 2.0 M CaCl$_2 \cdot$ 2H$_2$O (Mallinckrodt); 2.0 M NaCl; 1.0 M Na(PO$_4$), pH 7.0; 0.5 M HEPES (pH 7.1 ± 0.05) (Sigma); NTE (150 mM NaCl, 10 mM Tris–HCl at pH 7.4, 1 mM EDTA).

1. For stable transformation experiments, the plasmid DNAs to be transfected should be linearized at a site distal to the gene to be expressed; *Pvu*I in the *amp*^{R} gene is frequently convenient. The DNA should then be extracted with aqueous phenol–chloroform and ethanol precipitated without carrier (see this volume [4, 5]). Resuspend DNA in NTE at ~100 µg/ml.

2. Seed 60-mm tissue culture plates with ~3 × 10^5 cells per plate in 4 ml of tissue culture medium. The culture should be less than or about one-quarter confluent when transfected the following day.

3. Prepare fresh transfection buffer from stocks [1 ml 0.5 M HEPES, 8.1 ml water, 0.9 ml 2 M NaCl, 20 µl 1 M Na(PO$_4$)] and mix thoroughly.

4. Add NTE (175 µl final DNA volume) to a sterile Eppendorf tube. Add 5 µg of sterile, high molecular weight carrier DNA. Add 5 µg of total linearized plasmid DNA and vortex (volume at this point is 175 µl). Add 25 µl of 2 M CaCl$_2$ and mix thoroughly.

5. Add 200 µl of transfection buffer dropwise and mix thoroughly but gently. Incubate at room temperature for 20–30 min.

6. Add mixture drop by drop directly into medium in the 60-mm culture plate and mix by swirling. Incubate at 37° for 16 hr.

7. Aspirate medium and replace with fresh growth medium.

8. At 48 hr posttransfection the cells or medium may be harvested for analysis or the cells may be placed under selective conditions. In the latter case, detach the cells by treatment with trypsin-EDTA, prepare a single cell suspension, and then reseed onto two 100-mm dishes in the presence of the selective agent.

9. If the dominant selectable marker used is the *neo* gene, cells are selected for resistance to the antibiotic G418[9] (Geneticin-Gibco). As the level of antibiotic appropriate for selection varies somewhat from cell line

---

[33] H. C. Birnboim and J. Doly, *Nucleic Acids Res.* **7**, 1513 (1979).

to cell line, the investigator should determine the minimum level of G418 sufficient to kill untransfected cells in 7–10 days. This level is normally between 0.2 to 1.0 mg/ml. Resistant colonies should be detectable ~14 days posttransfection.

10. Selection for expression of some marker genes, such as dihydrofolate reductase (dhFr) or thymidine kinase, requires the use of a mutant cell line. The most commonly used dhFr⁻ cell line is of CHO origin[34] and requires medium supplemented with hypoxanthine (10^{-4} M) and thymidine (10^{-5} M) for growth. Selection simply involves plating the cells into medium lacking these supplements.[13] Selection may occur more rapidly if dialyzed fetal calf serum (GIBCO) is used in media preparation. Colonies should be readily visible ~10 days posttransfection. Subsequently, the level of dhFr expression may be further amplified by exposure of the pooled dhFr⁺ cells to 10^{-7} M methotrexate (see below) followed by stepwise selection in increasingly higher levels of methotrexate.

Comments. This simple and effective calcium phosphate procedure works well with a variety of mammalian cell lines including HeLa, Ltk⁻, CHO, CV1, NIH, 3T3, 208F, and C127 and should yield a transformation frequency of ~10^{-4}. For cotransfection experiments use a 10:1 or 20:1 molar ratio of the nonselected to the selectable gene. Although primarily intended for constitutive expression studies, this procedure also yields moderate levels of transient expression in many cell lines (Table I). Calcium phosphate transfection is very sensitive to deviations from the ideal and to impurities. In particular the following should be noted. (1) The HEPES buffer must be pH 7.1 ± 0.05.[27] (2) Plasmid DNA should be purified by ispycnic centrifugation over *two* cesium chloride gradients and should be free of contaminants such as tRNA. (3) Carrier DNA, which facilitates establishment of stable transformation,[35] must also be of high purity. Best results are obtained with high-molecular-weight mammalian carrier DNA prepared by the investigator as described elsewhere in this volume [15, 18] and purified as described above for plasmid DNA. Commercial DNA preparations should be avoided.

Troubleshooting Heterologous Gene Expression in Eukaryotes

In order to be confident that problems with low gene expression levels are not due to difficulties with transfection, it is useful to test the system

[34] G. Urlaub and L. A. Chasin, *Proc. Natl. Acad. Sci. U.S.A.* **77**, 4216 (1980).
[35] M. Wigler, S. Silverstein, L. Lee, A. Pellicer, T. Cheng, and R. Axel, *Cell (Cambridge, Mass.)* **11**, 223 (1977).

with a readily assayable indicator gene. The most widely used indicator gene encodes the enzyme chloramphenicol acetyltransferase (CAT).[30] This gene is of prokaryotic origin and has no functional equivalent in eukaryotic cells. Intracellular CAT levels can be readily and sensitively quantitated using commercially available reagents and this gene is therefore extremely useful in comparisons of transfection efficiency. A minor problem is that CAT mRNA is unstable in eukaryotic cells so that it is difficult to extrapolate from a particular CAT activity to an expected level of synthesis of a second transfected gene. For this reason, I have instead used a second indicator gene, the human IL-2 gene.[36] This small (~15 kDa) lymphokine is normally expressed only by activated human T cells and is a good model gene for eukarotic expression. Constructions which express the IL-2 gene in transfected cells result in the synthesis and secretion of properly processed, secreted IL-2 into the supernatant medium. This medium can then be assayed for IL-2 activity using a standard, quantitative bioassay accurate to ~2 units (i.e., 100 pg) per milliliter.[36] Results obtained using this indicator gene are given in Table I and may be used by the investigator to estimate very roughly the potential level of expression of other heterologous genes cloned into pBC12BI or other, similar expression vectors.

If the expression level obtained after transfection of an indicator gene is as expected, levels of expression of the heterologous gene much lower than expected probably reflect posttranscriptional events in the transfected cells. Initially, the investigator should compare the steady-state level of the mRNA of the transfected heterologous gene with, for example, the steady-state level of the mRNA encoded by the rat preproinsulin II gene expressed in a parallel transfection using the parental pBC12BI vector. This insulin mRNA is efficiently expressed in transfected cells and will share a common 3' terminus with mRNAs encoding inserted, heterologous genes (Fig. 1). Both mRNAs can therefore be quantified using a single preproinsulin gene specific probe.[17] If the level of mRNA encoded by the heterologous gene is comparable to the insulin mRNA level, low expression may reflect poor translation. Both low mRNA stability and poor translatability may be addressed as described above. Low expression may also reflect an intrinsic property of the cloned gene product, such as toxicity or protein instability, and the investigator should then consider the use of an alternative eukaryotic expression strategy or expression in a heterologous host such as *E. coli* or yeast.

[36] R. J. Robb, this series, Vol. 116, p. 493.

Alternative Expression Strategies

To this point I have discussed expression strategies which depend on DNA-mediated gene transfer. A number of other approaches have been developed for both transient and constitutive gene expression. Several of these approaches require the use of specialized cell lines, viruses, or vectors and the reader is directed to the literature for a more detailed description of the techniques involved.

Lytic Virus Vectors. In several lytic DNA viruses, regions of the genome can be replaced with heterologous sequences without significant effect on the ability of the virus to replicate in tissue culture. Alternatively, these regions may express a function which can be complemented in trans using a helper virus. Examples include vaccinia virus,[37] adenovirus,[38] and the insect baculoviruses.[39] Due to the large size of these viruses, the gene to be expressed is normally introduced by targeted homologous recombination and is then selected based on a change in viral phenotype. Vaccinia-based vectors appear to be most useful as vectors for the presentation of foreign antigens to the immune systems of various animal or human hosts.[37] Adeno- and baculovirus based vectors are able to infect cultured cells efficiently and can express high levels of inserted heterologous genes.[38,39] Although the lytic replication cycle of these viruses results in only transient expression, these vectors nevertheless promise great utility in the production of proteins which are normally intracellular or which are toxic when overexpressed.

Episomal Viral Vectors. A number of the DNA tumor viruses are able to enter a semilatent state in which the genome is stably maintained within the nucleus of the infected cell with little or no cytopathic effect. Bovine papillomavirus (BPV) does not demonstrate lytic growth on murine cells in culture but instead enters a high copy number, generally episomal state which is marked, in cell lines such as NIH 3T3 and C127, by the loss of contact inhibition of growth.[40] Vectors have been constructed which contain all or part of the BPV genome attached to an expression vector similar to those described above.[40,41] Calcium phosphate-mediated transfection of C127 or NIH 3T3 cells with a BPV-based expression vector results in colonies of transformed cells. These colonies, which can be

[37] G. L. Smith, M. Mackett, and B. Moss, *Nature (London)* **302,** 490 (1983).
[38] M. Yamada, J. A. Lewis, and T. Grodzicker, *Proc. Natl. Acad. Sci. U.S.A.* **82,** 3567 (1985).
[39] G. E. Smith, M. D. Summers, and M. J. Fraser, *Mol. Cell. Biol.* **3,** 2156 (1983).
[40] N. Sarver, P. Gruss, M. Law, G. Khoury, and P. M. Howley, *Mol. Cell. Biol.* **1,** 486 (1981).
[41] G. N. Pavlakis and D. H. Hamer, *Proc. Natl. Acad. Sci. U.S.A.* **80,** 397 (1983).

selected on the basis of their growth phenotype, will maintain the heterologous expression vector as part of a BPV-based episome. Copy number is generally ≥50 per cell and high levels of constitutive expression can be obtained. An example using a derivative of the BPV-based vector pBPVMT1[41] and the human IL-2 gene is given in Table I. Recently, a second class of expression vectors based on the Epstein–Barr virus genome have been described.[42] These vectors are able to maintain an episomal state in human lymphoid cells and are of even greater potential utility.

Retroviral Vectors. Retroviruses of murine origin have several features which make them attractive as vectors. These include a very wide host range, efficient infection of cells, integration into the host cell genome in a defined fashion, lack of cytopathic effect, and a large genome capacity of which only a small part is required in cis. Several vectors have been constructed which can efficiently transfer a selectable marker, such as *neo*, as well as a second nonselected gene into heterologous cells.[7,43,44] The low copy number of the integrated provirus in the infected cell results, however, in only moderate levels of gene expression and retroviral vectors are therefore most useful for the highly efficient introduction of genes into cells which are refractory to transfection, such as primary lymphoid cells, and into stem cells as a possible approach to the correction of an inborn genetic deficiency.[43,44]

Gene-Linked Coamplification. Gene amplification is a frequent mechanism by which cells increase the expression of a gene product to a level necessary for survival. In cultured cells, this overexpression can sometimes be selected for by the use of inhibitors of an essential enzymatic function. One system involves the amplification of the gene for the enzyme dihydrofolate reductase (dhFr) in response to increasing concentrations of the drug methotrexate. Cells harboring over 1000 copies of the *dhFr* gene can be selected by sequential increases in the concentration of methotrexate to high levels.[45]

CHO cells which are defective for *dhFr* expression can be cotransfected with the *dhFr* gene and with a second nonselected gene as described above.[46,47] Because these genes become linked during the trans-

[42] B. Sugden, K. Marsh, and J. Yates, *Mol. Cell. Biol.* **5,** 410 (1985).

[43] A. D. Miller, E. S. Ong, M. G. Rosenfeld, I. Verma, and R. M. Evans, *Science* **225,** 993 (1984).

[44] R. Mann, R. C. Mulligan, and D. Baltimore, *Cell (Cambridge, Mass.)* **33,** 153 (1983).

[45] R. T. Schimke, *Cell (Cambridge, Mass.)* **37,** 705 (1984).

[46] R. J. Kaufman and P. A. Sharp, *J. Mol. Biol.* **159,** 601 (1982).

[47] R. J. Kaufman, L. C. Wasley, A. J. Spiliotes, S. D. Gossels, S. A. Latt, G. R. Larsen, and R. M. Kay, *Mol. Cell. Biol.* **5,** 1750 (1985).

fection process, the nonselected gene will be coamplified with the *dhFr* gene during selection for growth in increasingly high concentrations of methotrexate. This process can result in CHO cells containing ≥100 actively transcribed copies of the coamplified linked gene, and coamplification, using either dhFr or other, potentially dominant markers such as adenosine deaminase, appears to be the most powerful current technique for achieving very high levels of constitutive gene expression in mammalian cells[15,46–48] (Table I).

Assays for Functional Gene Expression

Secreted Proteins

Mammalian cells are able to recognize signals contained within the sequence and, perhaps, structure of proteins which determine the subcellular localization of that protein. Thus, fibroblastic cells such as COS and CHO are fully capable of secreting functional, correctly processed, differentiation-specific proteins such as IL-2, growth hormone, and tissue plasminogen activator.[41,47] These proteins are secreted via the constitutive pathway and the researcher may simply assay the supernatant medium for the presence of the biological activity which characterizes the protein of interest.

A problem arises in the case of hormones such as insulin, β-endorphin, and cholecystokinin which are secreted via the regulated pathway[49] and which undergo secondary, differentiation-specific proteolytic processing in the producing cells. Because fibroblastic cells such as COS do not express the regulated pathway, these cells will express and secrete the gene as a prohormone. Thus, COS cells transfected with the rat insulin gene secrete high levels of proinsulin but no detectable insulin.[17,50] If the prohormone can be detected and analyzed this may not be a problem. Alternatively, the researcher may wish to transfect a cell line such as AtT-20, a mouse anterior pituitary line which expresses the regulated secretion pathway and which is able to correctly process and secrete proteins such as insulin.[51]

[48] R. J. Kaufman, P. Murtha, D. E. Ingolia, C. Yeung, and R. E. Kellems, *Proc. Natl. Acad. Sci. U.S.A.* **83**, 3136 (1986).

[49] R. B. Kelly, *Science* **230**, 25 (1985).

[50] P. Gruss and G. Khoury, *Proc. Natl. Acad. Sci. U.S.A.* **78**, 133 (1981).

[51] H. H. Moore, M. D. Walker, F. Lee, and R. B. Kelly, *Cell (Cambridge, Mass.)* **35**, 531 (1983).

Cell Surface Glycoproteins

As is the case with signals which specify secretion, signals which specify cell surface expression also appear to be generally recognized by eukaryotic cells. A number of cell surface proteins have now been expressed including human T-cell antigens such at T8 and T4,[52] cell surface receptors for IL-2 and nerve growth factor,[53,54] and several viral cell surface glycoproteins.[55] Indeed, expression of heterologous genes on the cell surface of transfected mammalian cells is so efficient that it has been used as a tool in the expression cloning of several genes encoding cell surface glycoproteins.[52,54]

Detection of the cell surface expression of a gene of interest can be by immunological means (see below) or by function. Thus, constitutive cell surface expression of the influenza hemagglutinin gene confers the ability to bind erythrocytes and to demonstrate cell fusion at low pH.[55] Expression of the gene for the IL-2 receptor in mouse L cells confers the ability to bind IL-2 at low affinity but not high affinity.[53,56] Expression of the same IL-2 receptor gene in a lymphoid cell line confers the ability to bind IL-2 at both high and low affinity.[56] This phenomenon is believed to be due to a requirement for expression of a second lymphoid cell-specific protein for demonstration of high-affinity binding. This is an example of the general observation that expression of a particular phenotype, such as the ability to bind a ligand, may require the expression of more than a single gene product by the cell.

Intracellular Proteins

If the gene of interest expresses a dominant selectable marker or can complement a mutation in a cell, it may be possible to demonstrate expression of the gene by transfecting the appropriate cell line using the calcium phosphate procedure. A number of genes isolated from prokaryotes can confer a dominant selectable drug resistance on eukaryotic cells. These genes include the widely used *neo* gene described above.[9] Also useful as selectable markers are the *hgr* gene, which confers resis-

[52] D. R. Littman, Y. Thomas, P. J. Maddon, L. Chess, and R. Axel, *Cell* (*Cambridge, Mass.*) **40**, 237 (1985).
[53] W. J. Leonard *et al.*, *Nature* (*London*) **311**, 626 (1984).
[54] M. V. Chao, M. A. Bothwell, A. H. Ross, H. Koprowski, A. A. Lanahan, C. R. Buck, and A. Sehgal, *Science* **232**, 518 (1986).
[55] J. Sambrook, L. Rogers, J. White, and M. J. Gething, *EMBO J.* **4**, 91 (1985).
[56] M. Hatakeyama, S. Minamoto, T. Uchiyama, R. R. Hardy, G. Yamada, and T. Taniguchi, *Nature* (*London*) **318**, 467 (1985).

tance to hygromycin B,[42] and the gene for XGPRT, which confers resistance to mycophenolic acid.[14] All of these genes may be used in the simple cotransfection protocol described above.

A second approach to the demonstration of gene function is the genetic complementation of an auxotrophic mutation. Simple examples include thymidine kinase (TK) in TK$^-$ mouse L cells,[35] hypoxanthine–guanine phosphoribosyltransferase (HGPRT) in HGPRT$^-$ cells such as those of Lesch–Nyhan patients,[14,57] and dihydrofolate reductase in dhFr$^-$ CHO cells.[46] Use of a cell line lacking an enzyme activity of interest may also facilitate the biochemical detection of the expression of the gene encoding that enzyme. A trivial example of this is the prokaryotic CAT enzyme described above,[30] which has no eukaryotic counterpart. A second example is the enzyme galactokinase (galK) which is normally expressed in most cells but for which a mutant galK$^-$ cell line exists. Transfection of these cells with expression vectors containing the galK gene results in the transient expression of galK activity which can be readily detected biochemically.[16a] The most difficult cases involve attempts to demonstrate the functional expression of a gene in a cell which itself demonstrates the enzyme activity of interest. If the protein encodes a biological activity which can be detected in situ after, for example, starch gel electrophoresis it may be possible to chose to transfect the cloned gene into cells which express an enzyme variant with a significantly different electrophoretic mobility. Examples of this approach include the demonstration of the expression of the human purine nucleoside phosphorylase gene and murine adenosine deaminase gene after transfection of CHO cells.[15,48]

A special case of a dominant gene expression assay involves the use of viral or cellular genes whose expression results in the acquisition of the transformed phenotype, so-called oncogenes. Most commonly, the demonstration of the transforming potential of a cloned gene involves calcium phosphate-mediated transfection of an indicator cell line, such a NIH 3T3, which demonstrates contact inhibition of growth and which therefore forms very even cell monolayers in a confluent culture. Upon expression of some oncogenes, such as the v-ras genes, these cells acquire the ability to continue proliferation after reaching confluency and generate "foci" of overgrown, piled-up cells within the background monolayer. Variations of this assay have been extremely useful in the expression cloning and analysis of a number of cellular and viral oncogenes.[58]

[57] E. H. Szybalska and W. Szybalski, *Proc. Natl. Acad. Sci. U.S.A.* **48**, 2026 (1962).
[58] G. M. Cooper, *Science* **218**, 801 (1982).

Viral Complementation

As is the case with somatic cells, it is also possible to identify the function of a particular viral gene product based on its ability to complement a viral mutation when expressed in trans. This complementation will then lead to the production of infectious virus and, frequently, to the production of wild-type virus due to the frequency of homologous recombination or marker rescue in viral infections.[37-39] Examples include complementation of the envelope (*env*) gene deficiency of the replication-defective Bryan strain of RSV by transfection of an *env* gene expression plasmid[16a] and complementation of a replication-defective *tat⁻* genomic clone of human immunodeficiency virus (HIV) after transfection into a cell line which constitutively expresses the *tat* gene product.[59]

Immunological Assays of Gene Expression

The previous section reviewed approaches useful in the demonstration of gene expression by protein function. An alternative approach relies on the use of immunological techniques to characterize the expressed gene product. As is the case with expression in *E. coli* and yeast, Western blotting may be used to demonstrate the existence of a protein of the expected size which reacts with a specific antiserum and which is present only in the transfected culture. However, proteins of higher cells produced in their natural milieu express additional information contained within the primary sequence of the protein which determines both the posttranslational processing and subcellular localization of the protein within the transfected cell. The cells of higher eukaryotes appear able to follow instructions for subcellular localization correctly; however, posttranslational processing (N- and O-linked glycosylation, phosphorylation, sulfation, ADP-ribosylation, etc.) does appear to vary somewhat from cell line to cell line. Both approaches yield functional information not available from heterologous systems. Subcellular localization is most easily determined by immunofluorescence while posttranslational processing may be examined by the use of immunoprecipitation in the presence and absence of tunicamycin, an inhibitor of N-linked glycosylation.

The protocols described here for these two techniques assume transient expression in COS cells, but require little or no modification for other transient or constitutive expression strategies. In addition, the use of rabbit polyclonal antiserum is assumed. Readers are directed to more

[59] A. I. Dayton, J. G. Sodroski, C. A. Rosen, W. C. Goh, and W. A. Haseltine, *Cell (Cambridge, Mass.)* **44,** 941 (1986).

specialized texts for modifications required when using, for example, mouse monoclonal antibodies.

Indirect Immunofluorescence

1. Seed COS cells into a 35-mm tissue culture plate containing a sterile coverslip or into tissue culture chamber slides (Lab-Tek, Miles Scientific) so that they are approximately half confluent next day. On the next day, transfect as described in Methods for Introducing DNA into Cells. Include a negative control obtained by transfection with the parental vector.

2. At 48 hr posttransfection, aspirate medium and wash monolayer 3× in cold PBS. Fix cells to glass slide by immersion in 95% ethanol–5% acetic acid at −20° for 10 min. Wash 3× in cold PBS.

3. Incubate the cell monolayer with normal goat serum at 4° for 16 hr to block nonspecific binding.

4. Wash 3× with cold PBS. Add primary antibody diluted in either BSA buffer [1% bovine serum albumin, 0.3% gelatin, 25 mM Na(PO$_4$) at pH 7.5, 0.15 M NaCl] for cell surface fluorescence or Tween buffer [0.5 M NaCl, 1.0% bovine serum albumin, 5 mM Na(PO$_4$) at pH 6.5, 0.5% Tween 20] for cytoplasmic or nuclear fluorescence. Incubate at 4° for ≥1 hr.

5. Wash 3× with cold Triton buffer (0.3% Triton in PBS) and add fluor-conjugated goat anti-rabbit IgG diluted in Triton buffer. Incubate for ≥0.5 hr at 4°.

6. Wash extensively with cold PBS and mount using 50% glycerol in PBS. Examine under ultraviolet light.

Comments. This protocol is of general utility for cell surface, cytoplasmic, or nuclear proteins and is both simple and robust. Nevertheless, the procedure does require the empirical determination of the appropriate primary and secondary antibody dilutions to use in order to achieve the best possible signal-to-noise ratio. An example in which the nuclear localization of the c-*myc* oncogene protein product has been demonstrated can be found in Butnick *et al.*[60]

If the transformed cell expresses an introduced cell surface glycoprotein, immunofluorescence can be performed on viable cells without a fixation step. These cells may then be analyzed and sorted based on the level of the gene product they express on the cell surface by use of a fluorescence-activated cell sorter (FACS). This is a very useful way to isolate populations of cells which express high levels of a cell surface

---

[60] N. Z. Butnick, C. Miyamoto, R. Chizzonite, B. R. Cullen, G. Ju, and A. M. Skalka, *Mol. Cell. Biol.* **5**, 3009 (1985).

protein for subsequent use in analytical experiments.[61] Cloned genes encoding cell surface proteins could also be used in the FACS-based selection of cell populations expressing high levels of a cotransfected, linked gene encoding an intracellular or secreted protein.

Immunoprecipitation

1. Set up 35-mm plates of COS cells and transfect as described in Methods for Introducing DNA into Cells.

2. At 48 hr posttranscription, aspirate the tissue culture medium and replace with 1 ml of serum-free (FCS⁻), methionine-free (met⁻) medium (GIBCO). If desired, tunicamycin (Boehringer-Mannheim), 5 μg/ml, may be added. Preincubate for 1 hr at 37°.

3. Aspirate medium and replace with 1 ml of FCS⁻, met⁻ medium containing 200 μCi of [³⁵S]methionine (Amersham, ≥800 Ci/mmol). Add tunicamycin at 5 μg/ml if desired. Incubate cells at 37° for 15 min.

4. Remove medium and replace with FCS⁻, met⁺ medium for chase.

5. To harvest cells, aspirate medium and add 1 ml of room temperature radioimmune precipitation assay (RIPA) buffer [0.1% sodium dodecyl sulfate (SDS), 1% Triton X-100, 1% sodium deoxycholate, 0.15 M NaCl, 0.01 M Tris–HCl at pH 7.4, 1 mM EDTA, 0.25 mM phenylmethylsulfonyl fluoride (prepared as a 250× stock in acetone)]. Incubate with gently swirling for 5 min. Remove residual monolayer by scraping and transfer suspension to a 1.5-ml Eppendorf tube. Centrifuge in a microfuge for 15 min at full speed to clarify lysate.

6. Transfer cleared lysates into fresh Eppendorf tubes. If desired, quantify incorporated radioactivity by trichloroacetic acid (TCA) precipitation. To do this, add 10 μl of the sample to 200 μl of PBS in an Eppendorf tube. Add 2 μl of a 5% BSA solution and vortex. Add 100 μl of a 50% TCA solution and vortex. Collect the precipitated protein on a GF/A filter (Whatman) moistened with 5% TCA. Rinse the sample tube with 5% TCA and add to filter. Wash filter extensively with 5% TCA followed by absolute ethanol. Dry filter under heat lamp and count incorporated radioactivity in a scintillation apparatus.

7. Remove aliquots of each sample (by volume or incorporated [³⁵S]methionine) and incubate 2–16 hr with rabbit polyclonal antibody at 4° in a siliconized Eppendorf tube.

8. Add 200 μl of a 2% suspension of *Staphylococcus aureus* protein A (Staph A-Pansorbin, Calbiochem, La Jolla, CA) and incubate at 4° for 20 min.

[61] P. Kavathas and L. A. Herzenberg, *Proc. Natl. Acad. Sci. U.S.A.* **80,** 524 (1983).

9. Pellet protein A complex by centrifuging 15 sec in a microfuge. Remove and discard supernatant solution and resuspend pellet in 0.5 ml RIPA buffer by vortexing. Repeat two times.

10. Add 20 μl of 2× dissociation buffer (0.125 M Tris–HCl at pH 6.8, 4% SDS, 20% glycerol, 10% 2-mercaptoethanol, 0.2% bromphenol blue) to final protein A pellet and freeze at −20°.

11. Dissociate immune complexes by heating at 100° for 2 min. Pellet protein A by centrifugation and resolve labeled material by electrophoresis on a discontinuous SDS–polyacrylamide gel (this volume [31]). Prepare gel for autoradiography using a fluorographic agent of choice and expose to flashed X-ray film at −70° using a intensifying screen.

Comments. This procedure, which is based on that of Curran and Teich,[62] is only one example of many possible methods for the collection of immune complexes and the reader is referred to more specialized texts for alternate procedures. Again, the ideal level of rabbit antibody to use to obtain the best signal-to-noise ratio must be determined empirically. Although the demonstration of the correct subcellular location of a gene product and the confirmation of a particular pattern of posttranslational processing do not in themselves provide definitive evidence for the identity of a given gene product, these data may nevertheless facilitate analysis of a particular recombinant clone.

Acknowledgments

I wish to thank G. Ju, R. Chizzonite, and T. Curran for helpful discussions and advice on optimization of the protocols given. I also thank J. Farruggia and D. Thiele for their secretarial help.

[62] T. Curran and N. Teich, *Virology* **116,** 221 (1982).

[72] Identification of Regulatory Elements of Cloned Genes with Functional Assays

By NADIA ROSENTHAL

The recent development of sophisticated cloning procedures to isolate selected eukaryotic genes has resulted in an expanding repertoire of sequenced genomic segments, each of which potentially includes nucleotide signals controlling gene expression. From comparative sequence analysis it is evident that many genes share common consensus sequences present upstream of the transcription start site. These include the TATAA element, an AT-rich sequence which appears to set the 5' initiating nucleotide for transcription 25–30 bases downstream,[1-3] a GC-rich oligomer associated with certain viral and cellular genes,[4,5] and another element with a consensus sequence CCAAT approximately 80 bases upstream of many transcription start sites.[6] From numerous studies it appears that the position of these promoter elements relative to the transcription start site of a gene is inflexible.

Recently, a novel class of transcriptional regulatory elements, called *enhancers*, has been characterized in both viral and cellular genomes. These elements differ in several ways from the promoter-associated sequences described above. First, the position and orientation of enhancers relative to the transcription start site appear to be flexible; enhancer sequences can dramatically increase transcription of a linked gene from a position either upstream or downstream of a transcription unit, in an orientation-independent manner.[7,8] Enhancers have been identified in the gene introns[9-11] or in regions several thousands of bases away from a transcription unit.[12] Second, unlike promoter elements, enhancers do not share extensive sequence homology and therefore cannot be easily identified on the basis of DNA sequence data alone.

In practice, enhancers fall into two categories: those that are generally active in a wide variety of differentiated and undifferentiated cells, and those whose optimal activity is restricted to a particular tissue type. Ex-

[1] J. Cordon, B. Wasylyk, A. Buchwalder, P. Sassone-Corsi, C. Kedinger, and P. Chambon, *Science* 209, 1406 (1980).
[2] P. K. Gosh, P. Lebowitz, R. J. Frisque, and Y. Gluzman, *Proc. Natl. Acad. Sci. U.S.A.* 78, 100 (1981).
[3] Y. Gluzman, J. F. Sambrook, and R. J. Frisque, *Proc. Natl. Acad. Sci. U.S.A.* 77, 3898 (1980).
[4] S. L. McKnight and R. Kingsbury, *Science* 217, 316 (1982).
[5] R. D. Everett, D. Baty, and P. Chambon, *Nucleic Acids Res.* 11, 2447 (1983).
[6] A. Efstratiadis, J. W. Posakony, T. Maniatis, H. M. Lawn, C. O'Connell, R. A. Spritz, J. K. DeRiel, B. G. Forget, S. M. Weissman, J. I. Slightom, A. E. Blechl, O. Smithies, F. E. Baralle, C. C. Shoulders, and N. J. Proudfoot, *Cell (Cambridge, Mass.)* 21, 653 (1980).
[7] P. Moreau, R. Hen, D. Wasylyk, R. Everett, M. Gaub, and P. Chambon, *Nucleic Acids Res.* 9, 6047 (1981).
[8] M. Fromm and P. Berg, *Mol. Cell. Biol.* 3, 991 (1983).
[9] S. Gillies, S. Morrison, V. Oi, and S. Tonegawa, *Cell (Cambridge, Mass.)* 33, 717 (1983).
[10] J. Banerji, S. Rusconi, and W. Shuffner, *Cell (Cambridge, Mass.)* 27, 299 (1981).
[11] C. Queen and D. Baltimore, *Cell (Cambridge, Mass.)* 33, 741 (1983).
[12] M. Theisen, A. Stief, and A. Sippel, *EMBO J.* 5, 719 (1986).

amples of tissue-specific enhancers have been found from both viral and cellular sources, and the list is constantly expanding. In addition, enhancers in viruses such as simian virus 40 (SV40),[13,14] polyoma,[15,16] murine sarcoma virus (MSV),[17,18] or Rous sarcoma virus (RSV),[19,20] which have been shown to be active in a number of cell types, display varying degrees of species specificity in line with the host range of the parent virus.

Several recent reports indicate the presence of negative regulatory elements associated with both viral and cellular genes, which repress transcription in a position- and orientation-independent fashion.[21–23] In some cases, the action of these negative enhancers (silencers, dehancers) may be analogous to operator–repressor function in prokaryotes, or they may serve as protection from random activation by adjoining gene control sequences in large genomes.

In order to assign roles to these regulatory elements, functional assay systems have been developed in which cloned sequences can be reintroduced into eukaryotic cells in culture and tested for transcriptional activity. The popularity of cell culture systems for the analysis of cloned gene expression derives principally from the relative simplicity of the experimental procedures involved and from the reproducibility of the assays. Also, although promoter function has been successfully reconstituted *in vitro*, regulatory elements such as enhancers were first characterized in tissue culture, and subsequent attempts to reconstitute enhancers function *in vitro* have been only partially successful. Introduction of cloned genes into animal systems such as in germ-line integration of transgenic mice is an option which maximizes the possibility of correct expression of the exogenous gene, since the introduced sequences are exposed to a complete spectrum of cell-specific signals during embryonic develop-

[13] C. Benoist and P. Chambon, *Nature (London)* **290,** 304 (1981).
[14] P. Gruss, R. Dhar, and G. Khoury, *Proc. Natl. Acad. Sci. U.S.A.* **78,** 943 (1981).
[15] M. Katinka, M. Yaniv, M. Vasseur, and D. Blangy, *Cell (Cambridge, Mass.)* **20,** 393 (1980).
[16] C. Tyndall, G. LaMantia, C. M. Thacker, J. Favaloro, and R. Kamen, *Nucleic Acids Res.* **9,** 6231 (1981).
[17] L. A. Laimins, G. Khoury, C. Gorman, B. Howard, and P. Gruss, *Proc. Natl. Acad. Sci. U.S.A.* **79,** 6453 (1982).
[18] M. Kriegler and M. Botchan, *Mol. Cell. Biol.* **3,** 325 (1983).
[19] L. A. Laimins, P. Tsichlis, and G. Khoury, *Nucleic Acids Res.* **12,** 6427 (1984).
[20] B. R. Cullen, K. Raymond, and G. Ju, *J. Virol.* **53,** 515 (1985).
[21] L. A. Laimins, M. Holmgren-Konig, and G. Khoury, *Proc. Natl. Acad. Sci. U.S.A.* **83,** 3151 (1986).
[22] U. Nir, M. Walker, and W. Rutter, *Proc. Natl. Acad. Sci. U.S.A.* **83,** 3180 (1986).
[23] E. Remmers, J. Q. Yang, and K. Marcu *EMBO J.* **5,** 899 (1986).

ment. Although the technique is complex and the setup time-consuming and expensive compared to cell culture assays, generation of transgenic mice may be the only alternative when culture lines of a particular cell lineage are not available. With an appropriate cell line in hand, however, it is possible to demonstrate relatively quickly with an expression assay that a specific set of nucleotides functions as a transcriptional regulatory element. Through *in vitro* modification of the sequence, the role of particular nucleotides in the transcriptional activation of a linked gene can then be characterized in cell culture. While novel expression vectors are constantly being developed, several standard approaches currently in use are presented below in the second section. The third section describes procedures for the introduction of expression vectors into culture cells. The fourth section includes protocols for two reporter gene assays. In the final section, additional applications of expression assays are briefly discussed.

Detection of Regulatory Elements in Expression Assays

Marked Genes

In theory, if a cloned gene is small enough (less than 10 kb), one can introduce a genomic fragment containing coding and flanking sequences in a suitable vector into culture cells and assay for transcriptional activity by mRNA analysis. With subsequent deletions, regulatory sequences can be pinpointed. In some cases, this straightforward approach is not possible. For example, if the host cell and species from which the cloned gene was isolated are identical or closely related, transcripts from the introduced gene cannot be distinguished from the corresponding endogenous gene product. It therefore becomes necessary to mark the cloned gene in order to assay its transcription. This can be done by inserting a synthetic nucleotide linker into an exon in the genomic clone at a unique restriction enzyme recognition site (see Fig. 1a). Transcripts for the marked gene can then be scored by S_1 analysis (this volume [66]), using an end-labeled cDNA probe that spans the modified site. With standard urea–polyacrylamide gel analysis, full-length probe protection by the unmodified exon sequence in the endogenous message can be distinguished from partial, shorter protection by transcripts of the transfected gene bearing the linker gene sequence.[24]

Alternatively, a fragment of foreign DNA, such as a prokaryotic plas-

[24] C. Hunt, J. Ro, D. Dobson, H. Y. Min, and S. Spiegelman, *Proc. Natl. Acad. Sci. U.S.A.* **83,** 3786 (1986).

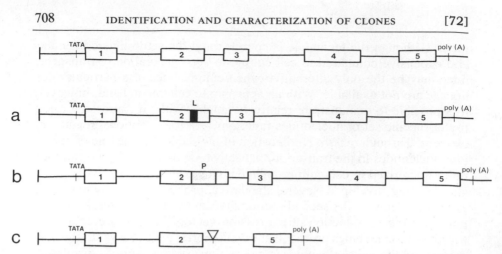

FIG. 1. Strategies for detection of regulatory elements in cloned genes. Marked genes: (a) a synthetic oligonucleotide linker (L) inserted into a unique restriction site in an exon sequence; (b) a fragment of prokaryotic DNA (P) inserted at the same position; (c) deletion of sequences including exons 3 and 4 to form a minigene.

mid or viral sequence, is inserted into the coding sequence of the transfected gene[25] (Fig. 1b). Northern blot analysis with probes homologous to the foreign DNA allow analysis of transcripts from introduced gene without interfering signals from endogenous gene activity (this volume [61]).

Minigenes

The advantage of testing intact or slightly modified genomic segments for expression is that potential regulatory sequences internal to the gene transcript are not eliminated from the constructions. If a gene is large, this approach becomes cumbersome due to difficulties in cloning and propagating large fragments. In this case, deletion of an internal portion of the coding sequences is an alternative approach that often results in a functional minigene transcription unit. The choice of sequences to be eliminated varies but usually includes deletion of internal exons, with the resection occurring between two introns, leaving transcription initiation and termination signals intact[26] (see Fig. 1c). Truncating a gene in this way not only results in a smaller, more manageable fragment but also allows discrimination of endogenous and introduced gene transcription products on the basis of messenger RNA size differences in Northern blot analysis.

[25] M. Garabedian, B. Sheperd, and P. Wensink, *Cell (Cambridge, Mass.)* **45,** 859 (1986).
[26] R. Scott, T. Vogt, M. Croke, and S. Tilghman, *Nature (London)* **310,** 562 (1984).

There are several potential problems with this approach. First, regulatory elements have been found in intragenic sequences,[9–11,27,28] and therefore it is possible that by deleting an internal portion of a gene, one may eliminate an important region of control. Second, since an internal deletion alters the transcript, this approach may lead to mRNA instability and thus result in an artifactual negative signal in a functional assay. The stability of an mRNA can be tested with an enhancer known to activate a large number of promoters in a wide variety of cell types, such as the SV40 enhancer.[7,8] If the minigene transcript is stable, but not activated, introduction of the enhancer into the minigene plasmid and transfection of the resulting construct into an appropriate cell type should activate the promoter and produce abundant transcripts. Conversely, inability to detect transcript accumulation even with an endogenous enhancer is an indication that the deletion in the minigene has led to messenger RNA instability, and a different deletion scheme should be tested.

Reporter Genes

This type of assay relies on the linkage of putative regulatory sequences to a reporter gene whose transcription is detected after transfection into cultured cells. The reporter function can be a eukaryotic gene, such as globin,[29,30] whose expression is normally limited to a restricted tissue, so that in any other cell type its transcription is not obscured by a background of endogeneous gene expression. The effect of successive delections and point mutations in the putative regulatory region under study can then be scored by measuring reporter mRNA abundance using Northern blot hybridization or S_1 nuclease analysis. As a control for transfection efficiency, a second gene whose transcription rate remains constant can be included in the same construction, or on a separate co-transfected plasmid.[27,28,30] With appropriate double probes, test gene mRNA levels can be normalized to those transcribed from the control gene. Alternatively, the protein product of an animal viral gene, such as SV40 T antigen, is used as a reporter function that is scored by either cellular transformation or immunofluorescence.[5] Finally, bacterial genes coding for easily assayed enzyme activities can be used as reporter genes.

[27] P. Charnay, R. Treisman, P. Mellon, M. Chao, R. Axel, and T. Maniatis, *Cell (Cambridge, Mass.)* **38,** 251 (1984).

[28] S. Wright, A. Rosenthal, R. Flavell, and F. Grosveld, *Cell (Cambridge, Mass.)* **38,** 265 (1984).

[29] P. Mellon, V. Parker, Y. Gluzman, and T. Maniatis, *Cell (Cambridge, Mass.)* **27,** 279 (1981).

[30] R. Treisman, M. Green, and T. Maniatis, *Proc. Natl. Acad. Sci. U.S.A.* **80,** 7428 (1983).

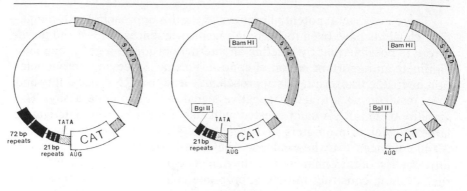

pSV2CAT   pA10CAT2   pA10CAT3M

FIG. 2. CAT expression vectors. All vectors contain the bacterial chloramphenicol ace-tyltransferase gene (CAT), followed by an SV40 t intron and poly(A) addition site. pSVC-2CAT[31] includes the SV40 72-bp repeated enhancer element, 21-bp repeated promoter element, and early gene TATA box, and serves as a positive control in CAT assays. pA10CAT2[17] has no enhancer sequences, but retains the SV40 early gene promoter. Putative enhancer sequences can be inserted at either BglII or BamHI sites. pA10CAT3M[31a] lacks any regulatory sequences, and is therefore a vector designed for testing promoter elements (in the BglII site). Additional enhancer elements can be inserted at the BamHI site. pSV-2CAT and pA10CAT2 are in a pBR322 amp[R] background. PA10CAT3M is in a PML amp[R] background.[31b]

A popular example is the bacterial gene coding for chloramphenicol ace-tyltransferase (CAT).[31] After transfection of the reporter gene construct, the kinetics of CAT activity in crude cell extracts are analyzed (a detailed protocol is given in the fourth section). A representative set of available CAT vectors is shown in Fig. 2.[31a,31b] In these vectors, putative enhancer elements can be tested in either position or orientation relative to the CAT gene (pA10CAT2), or enhancer–promoter regions under study can be fused to the CAT coding sequences at the 5′ end of the gene (pA-10CAT3M). The SV40 enhancer–promoter linked to the CAT coding se-quence (pSV2CAT) serves as a positive control.

As an alternative, the bacterial β-galactosidase gene has been success-fully used as a reporter function.[32–34] This assays allows quick and accu-rate quantitation of enzyme accumulation in transfected cells. A simple colorimetric assay outlined in the fourth section is performed on a crude

[31] C. M. Gorman, L. F. Moffat, and B. Howard, Mol. Cell. Biol. 2, 1044 (1982).
[31a] L. A. Laimins, P. Gruss, R. Pozzatti, and G. Khoury, J. Virol. 49, 183 (1984).
[31b] M. Lusky and M. Botchan, Nature (London) 293, 79 (1981).
[32] C. Hall, E. Jacob, G. Ringold, and F. Lee, J. Mol. Appl. Genet. 2, 101 (1983).
[33] P. Herbomel, B. Bourachot, and M. Yaniv, Cell (Cambridge, Mass.) 39, 653 (1984).
[34] T. Edlund, M. Walker, P. Barr, and W. Rutter, Science 230, 912 (1985).

cell extract prepared in the same way as for a CAT assay. A potential disadvantage of the system is the presence of an endogenous β-galactosidase isozyme in some culture cells so that, unlike the CAT assay, a negative control extract prepared from mock-transfected cells must always be included in the experimental design.

A recently developed series of vectors carrying the human growth hormone (hGH) gene as a reporter function[35] offers several advantages over either the CAT or the β-galactosidase systems. The major difference between the assays is that because the hGH gene product is secreted into the cell culture medium, reporter gene expression can be continuously monitored from the same population of cells. This feature of the hGH expression vectors makes kinetic analysis of gene function more convenient, as only a single dish of cells is needed to derive a time course of transcriptional activity, or to assay the effects of transcriptional modulators on the regulatory sequences under study. The radioimmunoassay for hGH performed on an aliquot of medium is quick and the components are commercially available. Further, the hGH transient expression system has been reported to be about 10-fold more sensitive than the CAT system in mouse L cells.[35] The potential disadvantages of the system may include variation of hGH secretion efficiencies in different cell types, or possible spurious biological effects of the hormone on cellular functions. Despite these possible drawbacks, the hGH expression vector series promises to be an important new addition to the currently available reporter gene systems.

The development of alternative reporter gene vectors whose expression can be assayed independently in the same cell facilitates the introduction of an invariant internal control in transient expression assays. The regulatory sequence under study is linked to the first reporter gene, while the second reporter gene is driven by a strong enhancer–promoter combination such as the RSV long terminal repeat (LTR)[36] or the SV40 early control region included in the pSV2 vector series.[31] Since both the CAT and β-galactosidase enzyme activities can be measured in the same extract, these two assays are routinely paired.[33,34] Alternatively, as a normalizing control, the hGH system[35] offers the possibility of monitoring parameters such as transfection efficiency and variability between cell cultures, before the cells are harvested.

The virtue of scoring protein rather than transcription products of reporter genes are that the assays are generally quicker and often more

[35] R. Selden, K. Burke Howie, M. E. Rowe, H. Goodman, and D. Moore, *Mol. Cell. Biol.* **6**, 3173 (1986).

[36] C. Gorman, R. Padmanabhan, and B. Howard, *Science* **221**, 551 (1983).

easily quantifiable. The disadvantage, as in the selectable marker assays, is that transcriptional activation by putative regulatory elements linked to the gene is measured indirectly by the presence of a protein product rather than by the accumulation of mRNA. While these two parameters are usually directly proportional, complications can arise, for example if the transcript starts at an unpredicted site and includes an upstream translation initiation codon not in frame with the reporter gene product. As with minigenes, it is possible to discriminate between the absence of transcription and a faulty, truncated, or unstable transcript by assaying reporter gene activity in the presence of a strong enhancer inserted into the construct at either the 5' or 3' end of the gene.

Enhancer Traps

The definitive properties of enhancer elements (relative position- and orientation-independent activation of transcription) implies that their location in a cellular gene locus is not predictable. This is in contrast to promoter elements, which by definition lie within a few hundred base pairs of the transcription start site. If the gene under study is large, identification of enhancer-like transcriptional regulatory elements by any of the above strategies can therefore be laborious. In order to simplify the search for enhancer sequences in a given segment DNA, it is sometimes quicker to test genomic subfragments in an enhancer trap assay. In theory, any reporter gene driven by a known invariant promoter can be a vector for enhancer trap assays, if a set of reporter gene constructs bearing fragments from the genomic locus are isolated, propagated, and tested individually.

Ideally, an enhancer trap that is self-propagating and selectable on the basis of reporter gene function would have broader applications. To date, papovaviral genomes with a deletion in the enhancer region have been used as enhancer traps, since in principle, selection on the basis of viral viability and isolation of plaques would allow isolation of activated genomic fragments inserted into the enhancerless viral vector.

This approach has not been particularly successful in isolating random cellular sequences with enhancer activity. In one example, integration and activation of an enhancerless polyoma viral vector allowed identification of adjacent DNA sequences that substituted for the viral enhancer.[37] In other studies, however, deletion of the SV40 enhancer from an otherwise intact viral genome, linkage of random cellular DNA fragments, and

---

[37] M. Fried, M. Griffiths, B. Davies, G. Bjursell, G. LaMantia, and L. Lania, *Proc. Natl. Acad. Sci. U.S.A.* **80**, 2117 (1983).

subsequent propagation in permissive cells resulted in viable viral revertants with no inserted cellular sequences. Instead, a revertant in the control experiment carried reiterations of sequences spanning both sides of
the deleted enhancer region.[38] Other experiments in which duplication of
a mutated enhancer restored its activity[39] puts into question whether the
actual nucleotide sequence or simply the reiterated structure of viral enhancers is the primary functional determinant in their transcription activation.

As a method for detection of enhancer activity in more limited DNA
sequences, however, enhancerless viral vectors appear to be particularly
well suited, and have been used to identify enhancer elements in several
animal viral genomes.[40]

Methods of Vector Introduction into Culture Cells

Transient Expression and Stable Transformation

Following introduction of expression vector DNA into cell cultures,
transient expression of the exogenous genes can be measured for up to 72
hr posttransfection. During this period the transfected plasmids are presumably converted to nonreplicating minichromosomes which are gradually degraded. The virtues of transient assays include the convenience of
the short time between transfection and harvest. This is ideal for the
design of expression studies such as deletion analysis of a regulatory
sequence, requiring sequential testing of many constructs.

If continuous monitoring of gene transcription is desired, however, to
measure induction or modulation of expression during a period longer
than the 72-hr time span of a transient assay, the construct must first be
integrated stably into the host cell genome. Among the earliest assays of
enhancer activity were tests in which a selectable gene was linked to a
putative enhancer sequence and the resulting construct was introduced
into culture cells under the appropriate selective pressure. An increased
number of resistent colonies relative to a negative control was an indirect
indication of transcriptional activation by the linked enhancer element.[41]
It is also possible to cointroduce a selectable gene driven by a separate
enhancer–promoter (such as the bacterial neomycin-resistance gene un-

[38] C. Swimmer and T. Shenk, *Proc. Natl. Acad. Sci. U.S.A.* **81,** 6652 (1984).
[39] W. Herr and Y. Gluzman, *Nature (London)* **313,** 711 (1985).
[40] M. Boshart, F. Weber, G. Jahn, K. Dorsch-Häsler, B. Fleckenstein, and W. Schaffner, *Cell (Cambridge, Mass.)* **41,** 521 (1985).
[41] M. R. Capecchi, *Cell (Cambridge, Mass.)* **22,** 479 (1980).

der the control of SV40 early regulatory sequences)[42] with any of the vector types described above, producing a drug-resistant cell population in which both the selected and nonselected genes are integrated within several weeks posttransfection. The selectable gene can be directly linked to the test construct or cotransfected on a separate vector by simply mixing the two plasmids. Integration of endogenous DNA is a rare event, but with an appropriate high ratio of selected to nonselected DNA (up to 10:1), both sequences are incorporated, often in concatemers, into the host genome.[43]

Stable transformation assays suffer from a number of potential complications. The most serious drawback is that the assay is based on long-term effects, which may involve uncontrolled parameters, such as the influence of variable integration sites of the selected gene in the host cell genome. For this reason, test gene expression is often measured in a mass culture of stable transformants. Individual clones, corresponding to separate integration events, can also be propagated, checked for number of integrated test gene copies, and tested for expression.

Transfection Procedures

For transient assays, direct DNA transfection in the presence of facilitating agents such as calcium phosphate[44] or DEAE-dextran[45] is most commonly used. These chemical methods do not work efficiently with all cell types and are most applicable to adherent cells. Furthermore, unlike calcium phosphate coprecipitation, the DEAE-dextran method is not suitable for long-term generation of stable cell lines, presumably because of its toxicity. Electroporation, the introduction of DNA molecules by a transient electric impulse passed through a cell culture,[46] appears to be ideal for both transient and stable assays on cells that are refractory to chemical transfection procedures, and for suspension cultures.[47] Finally, several viral vector systems including lytic SV40 vectors,[48] papilloma episomes,[49] and retroviral-based vectors[50] have the advantage of high-efficiency delivery via mass infection of culture cells. These vectors are

[42] P. Southern and P. Berg, *J. Mol. Appl. Genet.* **1**, 327 (1982).
[43] M. Wigler, R. Sweet, G. K. Sim, B. Wold, A. Pellicer, E. Lacy, T. Maniatis, S. Silverstein, and R. Axel, *Cell (Cambridge, Mass.)* **16**, 777 (1979).
[44] F. Graham and A. van der Eb, *Virology* **52**, 456 (1973).
[45] Z. Sompayral and K. Danna, *Proc. Natl. Acad. Sci. U.S.A.* **78**, 7575 (1981).
[46] H. Potter, L. Weir, and P. Leder, *Proc. Natl. Acad. Sci. U.S.A.* **81**, 7161 (1984).
[47] F. Toneguzzo, A. Hayday, and A. Keating, *Mol. Cell. Biol.* **6**, 703 (1986).
[48] R. Mulligan, B. Howard, and P. Berg, *Nature (London)* **277**, 105 (1979).
[49] P. Howley, N. Sarver, and M. F. Law, this series, Vol. 101, p. 387.
[50] R. Mann, R. Mulligan, and P. Berg, *Cell (Cambridge, Mass.)* **33**, 153 (1983).

complex because in order to maintain viability, viral control sequences must be retained in the vector, which can potentially interfere with the scoring of the activity of the regulatory element being tested. Further, in some cases viral infection may change the cellular environment in which the regulatory sequences are to be tested (for example, viral oncogene expression may transform the cell).

In any transfection procedure, efficiencies of plasmid delivery and expression can vary as a function of cell type density and condition. DNA purity can also affect transfection efficiency. For example, RNA contamination significantly reduces the transfection efficiency of the calcium phosphate coprecipitation technique. DNA configuration is another variable: supercoiled plasmid, as opposed to circular or linear forms, appears to be more efficiently expressed in both calcium phosphate and DEAE-dextran-mediated transfection. Since any of these variables can give rise to differences in expression levels equal to or greater than differences due to the linked regulatory elements being studied, a cotransfected control transcription unit such as a reporter gene driven by a strong enhancer is a valuable addition to the experimental design, and reduces the need for multiple experiments to obtain an accurate result.

Calcium Phosphate Coprecipitate Transfection Method

Included here is a model procedure for the transfection and assay of expression vectors using calcium phosphate coprecipitation on adherent cell culture monolayers. The protocols are modified versions of published techniques.[44]

Cells. Approximately 5×10^5 cells are seeded in a 100-mm culture dish 20 hr before addition of DNA. The cells are refed with fresh medium 4 hr before transfection.

DNA. The amount of test plasmid to be transfected is empirically determined and depends to some extent on the relative activity of the regulatory elements driving the reporter gene. For example, as little as 1 μg of a plasmid with an SV40 enhancer-driven CAT gene (pSV2CAT) has been used for consistently high expression.[31,51] If the strength of a putative regulatory sequence is unknown, it is advisable to start with a maximal amount of test plasmid (10–25 μg per 100-mm dish) and subsequently reduce that value if desired, always keeping the total amount of transfected DNA constant using a neutral "filler" plasmid. The filler DNA should be prepared in the same manner as the test DNA. This ensures uniformity in the quality of the calcium phosphate coprecipitate. For

[51] H. R. Schöler and P. Gruss, *Cell (Cambridge, Mass.).* **36,** 403 (1984).

optimal transfection efficiency, the DNA to be transfected should be protein and RNA free, and in supercoil form. Plasmid preparations of consistently high quality can be most reliably achieved by two successive CsCl gradient purifications.

Precipitate Preparation

1. Plasmids up to 25 μg total per 100-mm dish of cells are mixed with a fresh, sterile-filtered CaCl$_2$ solution to a final concentration of 124 mM in a total volume of 0.5 ml.

2. A 0.5-ml solution of sterile filtered 2× HBS (50 mM HEPES, 280 mM NaCl, 1.5 mM Na$_2$HPO$_4$ at pH 7.1) is placed in a separate tube. The pH of the HBS solution drops over time and should be adjusted before each use.

3. The DNA–calcium solution is added dropwise to the HBS solution, which is either continuously vortexed or gently bubbled with a stream of N$_2$ through a 1-ml sterile pipet. The solution should become slightly opaque.

4. After 30 min at room temperature the DNA–calcium phosphate precipitate is applied dropwise to medium covering the cells. If the same DNA precipitation is to be used for multiple dishes, a master mix of precipitate can be distributed into aliquots, since formation of the precipitate is not volume dependent. Depending on the cell type, the level of DNA uptake is increased by leaving the precipitate on the cultures for up to 24 hr.[52] Since the overall viability of the cell culture may be affected by longer incubation times, optimal conditions for DNA uptake must be determined for each cell type.

5. DNA uptake is significantly increased by a glycerol shock 4–12 hr posttransfection, although some cells do not tolerate this procedure. (An alternative shock procedure using dimethyl sulfoxide is described in Howley et al.[49]) The glycerol shock solution consists of 15% glycerol in 1× HBS and should be sterile filtered. To shock the cells, medium containing the DNA precipitate is removed, the cell monolayer is washed once with serum-free media, and then 3 ml glycerol shock solution is added directly to the monolayer. After exactly 3 min the glycerol is removed and replaced by 10 ml serum-free medium, followed by a second wash. The cells are then incubated for 24–48 hr in growth medium before harvesting for transient assays, or before applying selective medium for long-term transformation.

[52] D. Spandidos and N. Wilkie, *EMBO J.* **2**, 1193 (1983).

Reporter Gene Assays

As previously discussed, the detection of transient reporter gene activity 12–72 hr posttransfection allows rapid testing of putative regulatory sequences. In addition, the efficiency of a given transfection procedure in an individual cell type can be initially assessed using a reporter gene driven by an invariant strong enhancer. The following protocols for the detection of bacterial enzymes, chloramphemicol acetyltransferase (CAT) and β-galactosidase, are modified procedures from previously published reports.[31,33]

CAT Assay

Extract Preparation

1. To prepare the crude cell extract in which the CAT enzyme activity will be assayed, the cells are first washed 5 times with Ca^{2+}-, Mg^{2+}-free phosphate-buffered saline.

2. One milliliter of TEN solution (40 mM Tris–HCl at pH 7.5, 10 mM EDTA, 150 mM NaCl) is applied to the monolayer.

3. After 5 min the loosened cells are scraped from the dish and pelleted in an Eppendorf table-top centrifuge.

4. The supernatant is removed and replaced with 150 μl of 250 mM Tris–HCl at pH 8. The pellet is resuspended and subjected to three freeze–thaw cycles to lyse the cells.

5. A 10-min incubation at 60° at this point inactivates any endogenous acetylases that can counteract CAT activity, but does not inactivate the CAT enzyme itself.[53]

6. After the denatured cellular debris is removed by another centrifugation, the supernatant is transferred to a new tube and stored at −20°.

Enzyme Assay. The enzyme assay is performed using a portion of the extract (10–150 μl). For strong enhancer–promoter sets, such as the SV40 regulatory regions included in pSV2CAT, as little as 10 μl extract is sufficient to generate a consistent signal. If the activity of the regulatory region to be tested is unknown, a preliminary assay using 50 μl of the extract is advisable.

1. The following reagents are combined: cell extract (10–150 μl), 20 μl of freshly made acetyl-CoA at 3.5 mg/ml, 1–10 μl [^{14}C]chloramphenicol, and 250 mM Tris–HCl (pH 8) to a total reaction mixture of 180 μl. Since the addition of acetyl-CoA initiates the reaction, it should be added last.

[53] M. Mercola, J. Goverman, and K. Calame, *Science* **227,** 266 (1985).

2. The reaction mixture is incubated at 37°. Incubation times of 30–60 min are routine, but longer incubation can give higher signals particularly if an additional aliquot of acetyl CoA solution is added after the first hour. For definitive comparisons of CAT activity between different transfections, it is imperative that a time course study be done on each reaction, since it is the kinetics rather than the final percentage conversion that is the true measure of enzyme activity. This is easily done by removing aliquots from the reaction mixture at set time points and proceeding directly to the extraction procedure (step 3) to stop enzyme activity.

3. At the end of the reaction time, [^{14}C]chloramphenicol products are extracted with two successive 0.5-ml aliquots of ethyl acetate, which are pooled and evaporated under vacuum. The lower aqueous phase is discarded.

4. The reaction products are redissolved in 30 μl ethyl acetate immediately before application to a 25-mm silica TLC plate. A 10 μl aliquot spotted on the plate is usually sufficient.

5. Chromatography is carried out in chloroform–methanol (95 : 5) in a closed tank.

6. The plate is then dried and autoradiographed. Because chloramphenicol has two potential acetylation sites, two acetylated forms are produced by the enzyme reaction, running faster than the unacetylated form. A third higher spot corresponds to double acetylation and is seen only in conditions of high CAT activity. After autoradiography the spots can be cut out from the plate and counted to obtain exact conversion values.

β-Galactosidase Assay

Extract Preparation. Cells are harvested according to the protocol given for the CAT assay above, but step 5 is eliminated because the 60° incubation inactivates the enzyme. If this assay is to be used as a normalizing control, a portion (10–30%) can be reserved, and the rest of the extract can be frozen at −20° or tested immediately for CAT activity after treatment at 60°.

Enzyme Assay

1. The following stocks are prepared: 100× magnesium buffer (100 mM MgCl$_2$, 5 M 2-mercaptoethanol); 0.1 M sodium phosphate buffer (adjust 100 ml of 0.1 M Na$_2$HPO$_4$ to pH 7.3 at 37° using 0.1 M NaH$_2$PO$_4$); ONPG substrate solution [ONPG (*o*-nitrophenyl-β-D-galactopyranoside), 4 mg/ml in sodium phosphate buffer].

2. The following reagents are combined: 10–150 μl cell extract, 3 μl 100× magnesium buffer, 66 μl ONPG solution, and sodium phosphate buffer to a final volume of 300 μl.

3. The reaction mix is incubated at 37° for 30 min or until a yellow color is visible. The reaction is stopped by adding 0.5 ml of a 1 M Na$_2$CO$_3$ solution, and the color reaction is measured in a spectrophotometer at 410 nm.

A positive control for the enzyme assay can be run using commercially available β-galactosidase. The cell lysate in the above reaction is then substituted with 1 μl of a 100 μg/ml enzyme solution (in sodium phosphate buffer).

Additional Applications of Expression Assays

Once a promoter and/or enhancer element controlling transcription of a given gene is defined by functional assays in cell culture, the trans-acting factors required for this activity must be identified. Ideally an *in vitro* assay in which expression could be reconstituted would allow purification of transcription factors from nuclear extracts. Although several *in vitro* expression assays have been reported,[54,55] none of these studies has succeeded in mimicking the dramatic effects on transcription (up to several orders of magnitude) that enhancers exhibit *in vivo*. To date, various techniques for the detection of protein–DNA binding *in vitro* are available (see this volume [73]). These assays are powerful tools for the isolation of cellular factors which bind with high affinity to specific gene control sequences, yet they can only correlate binding with functional activity. Of obvious importance would be a functional assay in which putative trans-acting molecules, with specific affinity for known regulatory elements, could be tested.

In Vivo Competition

In this approach, the binding of regulatory sequence-specific factors is scored indirectly by an *in vivo* competition assay. The principle of this technique is to cointroduce in a transient assay a test construct, usually a reporter gene driven by the regulatory sequences under study, together with an excess of plasmids harboring copies of identical or different regu-

[54] P. Sassone-Corsi, J. Dougherty, B. Wasylyk, and P. Chambon, *Proc. Natl. Acad. Sci. U.S.A.* **81**, 308 (1984).
[55] H. Schöler, U. Schlokat, and P. Gruss, *in* "Eukaryotic Transcription," *Cold Spring Harbor Curr. Commun.*, p. 101 (1985).

latory elements not linked to any functional transcription unit. If DNA-binding factors necessary for transcriptional activation recognize both test and competitor elements, and if these factors are of limited amount in the nucleus, then the number of molecules available to activate transcription of the test gene should be effectively reduced by the presence of multiple competitor sequences, and test gene expression should decrease. This approach, first developed to study viral enhancer-binding factors,[51] has since been used to characterize the putative cellular factors binding to several additional viral and cellular regulatory elements.[53,56]

Although *in vivo* competition can identify specific sequences in regulatory elements recognized by trans-acting factors, it does not afford a way to identify or isolate the molecules themselves. As a functional assay, however, it can supplement *in vitro* DNA binding studies, and can be used to identify factors common to more than one regulatory element.

In Vivo Transaction of Transfected Genes

In a recent study,[57] a promising technique has been developed to transactivate *in vivo* a tissue-specific gene coding for the immunoglobulin heavy chain, stably integrated into a fibroblast genome. By mass microinjection of B-cell nuclear extracts via erythrocyte ghost-mediated fusion, the silent transfected gene was activated in the fibroblast background. This result suggests not only that positive regulatory transacting factors are involved in immunoglobulin gene expression, but that as a rule stably integrated transfected genes may be more susceptiable to activation by cellular factors supplied in trans than their endogenous counterparts. Whether this approach has general applications for the study of other tissue-specific genes remains to be seen.

Note Added in Proof: A promising new reporter gene encoding firefly luciferase has recently been characterized.[58] In this assay, emission of light in the presence of substrates ATP, O_2, and luciferin serves as a reporter function for linked transcriptional regulatory sequences.[59] Although luciferase activity in intact cells is lower than in cell lysates, improved methods of luciferin uptake may ultimately provide a simple sensitive assay for scoring reporter gene activity in individual cells.

[56] H. R. Schöler, A. Haslinger, A. Hegy, H. Holtgreve, and M. Karin, *Science* **232,** 76 (1986).

[57] H. Maeda, D. Kitamura, A. Kudo, K. Araki, and T. Watanabe, *Cell (Cambridge, Mass.)* **45,** 25 (1986).

[58] J. R. de Wet, K. V. Wood, M. DeLuca, D. R. Helsinki, and S. Subramani, *Molec. Cell Biol.* **7,** 725 (1987).

[59] D. W. Ow, K. V. Wood, M. DeLuca, J. R. de Wet, D. R. Helsinki, and S. H. Howell, *Science* **234,** 856 (1987).

[73] Interaction of Protein with DNA *in Vitro*

By LOTHAR HENNIGHAUSEN and HENRYK LUBON

One of the important means to control eukaryotic gene expression involves the binding of proteins to specific sites in the promoter and other regulatory regions of the gene.[1] This chapter is devoted to methodology for identifying DNA sequences that are bound *in vitro* by proteins in crude nuclear extracts. We present four methods of detection: nitrocellulose filter binding, mobility shift, exonuclease III protection, and DNase I protection. In addition to providing assay conditions, we discuss some applications of the different methods. Very often data from protein–DNA interaction studies permit ambiguous interpretations. We therefore discuss control experiments designed to help the researcher to evaluate data.

Radioactive Labeling and Isolation of DNA Fragments

In the nitrocellulose filter-binding assay protein–DNA interactions are studied using a mixture of DNA fragments derived from restriction digests of plasmid, cosmid, or phage DNA. The fragments are routinely labeled at both 5' ends with [γ-^{32}P]ATP using T4 polynucleotide kinase or at both 3' ends with [α-^{32}P]dATP and the Klenow fragment of DNA polymerase I, applying standard methods (see this volume [10]). The DNA fragments to be assayed by mobility shift are labeled only at one 3' or one 5' end. This can be done conveniently by digesting a fragment labeled at both ends with a second restriction enzyme, followed by a thorough deproteinization for 15 min at 37° with proteinase K at a concentration of 100 μg/ml. After stopping the reaction by heating it to 60° for 3 min, the fragments are recovered by phenol–chloroform extraction and ethanol precipitation (see this volume [4, 5]). They are separated by agarose gel electrophoresis and purified using electroelution or phenol extraction (see this volume [8]). DNA fragments labeled at only one end are versatile for the mobility shift assays because after recutting of the fragments with different restriction enzymes, binding sites can be mapped with high resolution without a further gel purification step. In general, DNA labeled by nick translation should not be used in the filter binding and mobility shift assay because some proteins tend to bind nonspecifically to nicks and thereby increase the nonspecific binding.

[1] W. S. Dynan and R. Tjian, *Nature (London)* **316**, 774 (1985).

Whereas DNA fragments to be assayed by DNase I can be labeled at either a 3' end or 5' end, exonuclease III protection requires a 5' end label. Since exonuclease III digests linear double-stranded DNA preferentially from 3' recessive ends, the strategy involved in preparing fragments is predetermined. If there are no convenient restriction endonuclease recognition sites in the vicinity of suspected protein-binding sites, we suggest subcloning fragments created by 4-bp cutters (e.g., HaeIII, AluI, RsaI) into a polylinker. This creates the opportunity for labeling restriction sites within the polylinker.

Preparation of Nuclei

The method described here works well with suspension cells (e.g., HeLa or lymphoblastoid cells) but can also be applied to tissue slices and attached cells (e.g., MCF-7). In the latter case, cells should be scraped off the plate using a rubber policeman. For a pilot experiment we normally start with 10^9 suspension cells. This will give a yield of 5–10 mg crude nuclear proteins.

Suspension cells are harvested by centrifugation for 10 min at 1800 rpm and 4° in a Sorvall HB-4 rotor. Pelleted cells are resuspended in 10 volumes of Dulbecco's phosphate-buffered saline (without calcium and magnesium) at 4° and centrifuged as above. Nuclei from pelleted cells are prepared as follows: cells are resuspended in 5 pellet volumes of 0.3 M sucrose in buffer A. Buffer A contains 10 mM HEPES–KOH at pH 7.9, 10 mM KCl, 1.5 mM MgCl$_2$, 0.1 mM EGTA, 0.5 mM dithiothreitol (DTT), 0.5 mM phenylmethylsulfonyl fluoride (PMSF), and 2 μg/ml each of antipain, leupeptin, and pepstatin A (Sigma). Cells are lysed by 8–12 strokes with a B pestle in a Dounce glass homogenizer (Kontes) and 1–2 strokes in the presence of 0.3–0.4% Nonidet P-40. Completion of lysis (100%) is monitored using a phase-contrast microscope. The homogenate is then centrifuged at 1200 g for 10 min and the pelleted nuclei are washed twice in 0.3 M sucrose in buffer A without Nonidet P-40.

Preparation of Crude Nuclear Extracts

We prepare the extracts using a modification of the procedure by Dignam et al.[2] Nuclei are resuspended with an all-glass Dounce homogenizer (10 strokes, B pestle) in 2.5 pelleted nuclei volumes of 400 mM NaCl, 10 mM HEPES–KOH at pH 7.9, 1.5 mM MgCl$_2$, 0.1 mM EGTA, 0.5 mM DTT, 5% glycerol, and 0.5 mM PMSF. The resuspended nuclei

---

[2] J. D. Dignam, R. M. Lebovitz, and R. G. Roeder, Nucleic Acids Res. **11**, 1475 (1983).

are stirred slowly for 30 min at 4° followed by centrifugation for 60 min at 100,000 g using a Type 70Ti rotor. After dialyzing the supernatant for 2–4 hr against 50 volumes of 20 mM HEPES–KOH at pH 7.9, 75 mM NaCl, 0.1 mM EDTA, 0.5 mM DTT, 20% glycerol, and 0.5 mM PMSF the extract is cleared by centrifugation at 25,000 g for 15 min, which removes precipitated material completely and partially removes lipid. The protein concentration in the supernatant will be between 5 and 10 mg/ml (determined by the method of Bradford[3] using the Bio-Rad kit). The supernatant (about 5–10 mg protein per 10^9 cells) should be frozen in small aliquots in liquid nitrogen and can be stored at $-70°$ for several weeks. Repeated freezing and thawing should be avoided, because in our experience some DNA-binding proteins will loose their activity upon refreezing. Other DNA-binding proteins (e.g., nuclear factor 1) are stable for several months and do not lose their activity on refreezing.

Filter Binding Assay

The nitrocellulose filter binding assay has been widely used to assay for protein–nucleic acid interactions.[4–9] Whereas in the presence of magnesium ion, double-stranded DNA fails to bind to nitrocellulose, most proteins and therefore also protein–DNA complexes are retained on nitrocellulose filters upon filtration. The filter binding assay allows the detection of a binding site in a mixture of several DNA restriction fragments[6–9] and the performance of accurate kinetic studies (e.g., determination of binding constant).[5] Furthermore the speed of the assay makes it a convenient tool for monitoring the purification of high-affinity DNA binding proteins.[10,11] The disadvantages of the assay is its inability to detect low-affinity DNA-binding proteins in crude nuclear extracts. In addition, no information about the number of proteins bound to a particular fragment can be obtained.

The described method successfully detected nuclear factor 1, a high-

[3] M. Bradford, *Anal. Biochem.* **72**, 248 (1976).

[4] M. Nierenberg and P. Leder, *Science* **145**, 1399 (1964).

[5] A. D. Riggs, H. Suzuki, and S. Bourgeois, *J. Mol. Biol.* **48**, 67 (1970).

[6] U. Siebenlist, L. Hennighausen, J. Battey, and P. Leder, *Cell (Cambridge, Mass.)* **37**, 381 (1984).

[7] U. Borgmeyer, J. Nowock, and A. E. Sippel, *Nucleic Acids Res.* **12**, 4295 (1984).

[8] L. Hennighausen, U. Siebenlist, D. Danner, P. Leder, P. Rosenfeld, D. Rawlins, and T. Kelly, *Nature (London)* **314**, 298 (1985).

[9] L. Hennighausen and B. Fleckenstein, *EMBO J.* **5**, 1367 (1986).

[10] P. J. Rosenfeld and T. J. Kelly, *J. Biol. Chem.* **261**, 1398 (1986).

[11] J. F. X. Diffley and B. Stillman, *Mol. Cell. Biol.* **6**, 1363 (1986).

affinity and sequence-specific DNA-binding protein, in crude nuclear extracts from a variety of eukaryotic cell types and tissues.[6–11]

Method

When searching for nuclear DNA-binding proteins it is convenient to start with plasmid, cosmid, or phage DNA that contains the recombinant fragments of interest. We normally start with a mixture of DNA fragments with an average size of 1 kb. Before applied in the assay, nitrocellulose filters (13 mm) should be boiled twice for 10 min in water, a step which reduces nonspecific binding of DNA. As controls, the input DNA (mixture of labeled fragments) and the DNA retained by the filter in the presence of bovine serum albumin should be separated electrophoretically in the same gel as the other samples. This allows a quantification of the amount of radioactivity incorporated into each fragment as well as estimation of the degree of nonspecific binding.

One microgram of crude nuclear proteins is preincubated for 15 min at room temperature with 1 μg of poly(dI-dC) (Pharmacia) as competitor DNA in 100 μl with 5 μg bovine serum albumin and a buffer composed of 10 mM HEPES–KOH at pH 8.0, 100 mM NaCl, 10 mM MgCl$_2$, 0.1 mM EDTA, and 2 mM DTT. After incubation in the presence of 1 ng of ^{32}P-labeled probe for another 30 min, 200 μl of the above buffer is added to the reaction mixture which is then filtered through a nitrocellulose filter using a standard filter suction device. After the filtration step, filters are washed two times with 1 ml of the same buffer. The flow rate in the initial filtration step and in the washing steps should not exceed 2 ml/min. It is important that the filters do not dry out in any step because this can result in tight binding of the DNA. DNA is extracted from the filters at 37° in Eppendorf tubes for 4–6 hr with 500 μl 0.25% sodium dodecyl sulfate (SDS) and proteinase K at 60 μg/ml. After removing the filter and adding 0.05 volume 5 M NaCl and 2 μg *Escherichia coli* tRNA, polynucleotides are ethanol precipitated, dried, and redissolved in 0.2% SDS, 10 mM Tris–HCl at pH 8.0, 1 mM EDTA, and 0.01% bromphenol blue. The DNA is then subjected to gel electrophoresis. Fragments with a size between 150 and 1000 bp are separated in a 1.5% agarose gel; smaller fragments are separated in a polyacrylamide gel (see this volume [8]). After drying the gel, filter-retained fragments are visualized by autoradiography (see this volume [7]).

Even if a DNA fragment contains a strong binding site (e.g., the nuclear factor 1 binding sites in the immediate early gene 1 of the human cytomegalovirus) only 5–20% of this fragment is retained on the filter.[9] Analyzing the flow-through after the filtration step for the absence of a restriction fragment would be less successful.

Specificity

Each DNA molecule is a nonspecific target for DNA-binding proteins. If the largest fragment (target) is preferentially retained on the filter it is necessary to exclude nonspecific binding due to the large size of the fragment. In this case the starting DNA can be recut with a different set of enzymes, which should result in specific filter retention (i.e., 5–10 times over background binding) of one of the smaller fragments. If a site of protein–DNA interaction has been detected on a specific DNA fragment the position of the binding element can be narrowed using different sets of restriction endonucleases[9] or using fragments shortened *in vitro*.[8] The sequence specificity of binding can be confirmed in a competition analysis.[6]

If a number of sites with different affinities for a binding protein have been identified, it is possible to determine their relative affinities in an internal competition assay.[9] For this purpose a limiting amount of nuclear proteins is incubated with a mixture of fragments containing the different binding sites. With increasing DNA concentrations the strong binding sites will compete preferentially for the binding proteins and the weak sites will not be retained on the filter any longer.

A negative result could reflect the absence of binding sites, the presence of very weak binding sites (for their identification, see next paragraph), an inactive nuclear extract, or inadequate conditions. An evaluation of the extract is subjective because different DNA binding proteins have different stabilities. It is, however, possible to set up working conditions by determining nuclear factor 1 (NF1) activity. Strong binding sites for NF1 have been described in a number of cellular and viral genes such as (1) the long terminal repeat (LTR) of the mouse mammary tumor virus, (2) the human c-*myc* gene, (3) the human IgM gene, (4) the adenovirus terminal repeat, (5) the immediate early gene promoter of the human and simian cytomegalovirus, and (6) the chicken lysozyme gene (see refs. 6–11). NF1 activity serves also as a positive control in the three other methods to be described.

Mobility Shift Assay

A recent alternative to the filter-binding assay is the mobility shift assay.[12–14] It is based on the retarded mobility of protein–DNA complexes during polyacrylamide gel electrophoresis. This method is advantageous because it is more sensitive than the filter binding assay by at least one

[12] M. Fried and D. Crothers, *Nucleic Acids Res.* **9,** 6505 (1981).
[13] M. Garner and A. Revzin, *Nucleic Acids Res.* **9,** 3407 (1981).
[14] F. Strauss and A. Varshavsky, *Cell (Cambridge, Mass.)* **37,** 889 (1984).

order of magnitude. Furthermore, the number of complexes observed (not all of them might reflect specific protein–DNA interaction) provides information about the bound proteins. In many cases satisfactory results can only be obtained using a single DNA fragment of 300 bp or less per assay, and thereby restricting the application to regions that have been functionally predefined (e.g., promoters, enhancers). At the other end of the scale the limiting factor is the size of the binding site. We successfully performed mobility shift assays with fragments as small as 25 bp. Due to the speed of the method it is feasible to assay a number of fragments in competition experiments. This facilitates detection of sequences recognized by proteins with similar binding requirements.

Method

Crude nuclear extracts contain many DNA-binding proteins. Upon gel electrophoresis, high molecular weight complexes formed between the DNA probe and specific as well as nonspecific binding proteins will remain near the origin. Preincubation of nuclear proteins with either heterogeneous (e.g., *E. coli*) or homogeneous [e.g., poly(dI-dC)] DNA prior to adding the probe reduces the buildup of nonspecific high molecular weight aggregates. Therefore specific (and also nonspecific) protein–DNA complexes of lower molecular weight enter the gel, but migrate more slowly than the naked DNA probe. A simple copolymer such as poly(dI-dC) elicits a more complex binding pattern than a heterogeneous competitor and therefore increases the sensitivity of the assay. Poly(dI-dC) can be purchased from various vendors, but it is necessary to keep in mind that the efficacy of competition for nonspecific binding can vary among different batches. Nuclear extracts from some tissues (e.g., mammary glands) are rich in nucleases and the binding reaction has to be performed in lower concentrations of Mg^{2+}.

In order to detect binding sites with unknown affinity we routinely incubate 1 μg of nuclear proteins with increasing amounts of poly(dI-dC) (none, 100, 200, and 500 ng, and 1, 2, 5, 10, and 20 μg) in 25 μl of buffer composed of 10 mM Tris–HCl (pH 7.5), 50–100 mM KCl, 5 mM $MgCl_2$, 1 mM DTT, 1 mM EDTA, 12.5% glycerol, and 0.1% Triton X-100 for 30 min at room temperature. After adding 1 ng of the labeled probe and incubating for an additional 20–40 min, the samples are separated in a low ionic strength 4% polyacrylamide gel (acrylamide–bisacrylamide, 30 : 1) containing 1 mM EDTA, 3.3 mM sodium acetate, and 6.7 mM Tris–HCl (pH 7.5). Prior to loading the sample, the gel (0.15 × 16 cm) should be run for 2 hr at 20 mA and the buffer should be recirculated between the compartments. In an initial experiment we carry out electrophoresis at 30 mA and room temperature until the bromphenol blue reaches the bottom of the gel. After soaking the gel for 20 min in 5% glycerol to retard cracking, it is transferred to Bio-Rad filter paper (catalog no. 165-0921), dried under vacuum in a conventional gel dryer at 80° for 1 hr, and subjected to autoradiography. The naked DNA probe and the protein–DNA complexes should be clearly visible after a few hours exposure.

The glycerol concentration in the assay can be increased to 25% without altering formation of the complex. This feature will gain importance in the exonuclease III protection assays described in the next section.

Specificity of Binding

If one or more potential protein–DNA complexes are visualized in the low ionic strength polyacrylamide gel, it is necessary to demonstrate specificity of binding. This can be done in a competition assay using a defined amount of poly(dI-dC) that elicits the complexes, and increasing concentrations of unlabeled cloned competitor DNA. Preincubation of the nuclear proteins with poly(dI-dC) and increasing amounts (1- to 100-fold molar excess) of unlabeled probe over its labeled homolog prior to adding the labeled probe should reduce formation of specific complexes by more than 90%. This rule of thumb is applicable only if the amount of a specific binding protein in the assay is the limiting component. Unrelated cloned fragments similar in size to the probe should compete less and only nonspecifically. By increasing the salt concentration or the poly(dI-dC) concentration it is possible to estimate the relative affinity (or abundance) of the binding proteins. Once sites of protein–DNA interaction have been defined, the mobility shift assay is very useful for identifying the binding factors in a number of cell types and tissues. Figure 1 shows a mobility shift assay with crude nuclear proteins from mammary epithelial cells and the promoter fragment of a milk protein gene.[15] In the absence of poly (dI-dC) as competitor the protein–DNA complexes do not enter the gel (lane 1). Nonspecific protein–DNA interactions are reduced with increasing concentrations of poly(dI-dC) and specific protein–DNA complexes, as well as noncomplexed probe, can enter the gel. At a thousandfold excess of poly(dI-dC) over probe, five protein–DNA complexes are visible. The uncomplexed probe has the highest mobility (lane 13). Specificity of binding has to be determined in a competition assay.

[15] S. M. Campbell, *Nucleic Acids Res.* **12**. 8685 (1984).

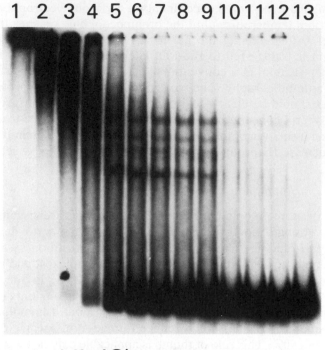

Poly (dI-dC) ⟶

FIG. 1. Electrophoretic mobility shift assay. The 112-bp *Xba*I–*Kpn*I (nucleotide −88 to +24) fragment overlapping the murine whey acidic protein gene promoter[15] was isolated and radioactively labeled. One microgram nuclear extract from mammary epithelial cells, 0.1 ng of the labeled fragment, and increasing amounts of poly(dI-dC) were incubated and separated in a native polyacrylamide gel as described in the text. Lane 1–12: binding occurred in the presence of 0-, 22-, 43-, 65-, 170-, 340-, 680-, 1360-, 2720-, 5470-, 10,880- and 21,760-fold weight excess of poly(dI-dC) relative to probe. Lane 13: naked DNA probe. At a 1000-fold excess of poly(dI-dC) relative to probe, five protein–DNA complexes that migrate more slowly than the probe are visible.

Exonuclease Protection

Exonuclease III (Exo III) from *E. coli* was discovered by Richardson and Kornberg.[16] In its first successful application for protein–DNA protection studies, SV40 T antigen binding sites were mapped.[17] The enzyme, a 3′ to 5′ exonuclease, ideally degrades a double-stranded DNA fragment

[16] C. C. Richardson and A. Kornberg, *J. Biol. Chem.* **239**, 242 (1964).
[17] D. Shalloway, T. Kleinberger, and D. M. Livingston, *Cell (Cambridge, Mass.)* **20**, 411 (1980).

processively from the 3' end to the 5' end until its passage is blocked by bound proteins.[17-19] Recent improvements increased the sensitivity of the assay and it is now possible to detect sequence-specific DNA binding activities in crude eukaryotic nuclear extracts.[20] In a typical assay using crude nuclear proteins, not all the specific binding sites are protected. Therefore not just the borders of the binding sites that are closest to the ends are detected, but also those located in the center of the fragments. The presence of endogenous nucleases and phosphatases as well as variable amounts of DNA-binding activities make it necessary to establish optimal assay conditions for each nuclear extract and each DNA fragment. For example, using the conditions described by Wu[20] it is possible to protect 50% of a heat-shock activator protein binding site (0.3 ng of a 250-bp DNA fragment) with 25 μg of nuclear extract from *Drosophila* cells. On the other hand, nuclear extract from mammary epithelial cells is rich in nucleases and only small amounts of protein can be incubated with the probe. Using 1 μg of crude nuclear proteins from mammary epithelial cells we protected about 1% of a strong protein-binding site in the promoter region of a milk protein gene.

Method

The DNA probe should be 300 bp or smaller, thereby permitting accurate mapping of protein binding sites from both ends. It is also crucial that the fragments are virtually free of nicks, because Exo III recognizes the free 3' ends in those structures and degrades the DNA starting from those internal recognition points as well as from the terminal 3' ends. The amount of competitor DNA needed to visualize specific protein DNA complexes should be determined in the mobility shift assay and can be transferred to the Exo III protection analysis.

Crude nuclear proteins (1 μg), the appropriate amount of poly(dI-dC), and 1 μg mixed oligodeoxyribonucleotides p(dN)$_5$ (Pharmacia) are incubated in a total volume of 50 μl for 15 min at room temperature in binding buffer (10 mM Tris–HCl at pH 7.5, 80 mM NaCl, 5 mM MgCl$_2$, 0.1 mM EDTA, 1 mM DTT, 5% glycerol, 0.1% Triton X-100). After further incubation for 15–20 min in the presence of about 1 ng of the labeled probe, Exo III is added and the reaction is allowed to continue for another 10 minutes at 30°. The reaction is terminated by the addition of an equal

[18] C. Vocke and D. Bastia, *Cell (Cambridge, Mass.)* **35**, 495 (1983).
[19] D. von der Ahe, S. Janich, C. Scheidereit, R. Renkawitz, G. Schutz, and M. Beato, *Nature (London)* **313**, 706 (1984).
[20] C. Wu, *Nature (London)* **317**, 84 (1985).
[21] H. Ohlsson and T. Edlund, *Cell (Cambridge, Mass.)* **45**, 35 (1986).

volume of 20 mM EDTA and 1% SDS. In parallel, we perform the assay in the absence of nuclear proteins. After purification by phenol extraction and ethanol precipitation (in the presence of 2 μg tRNA) the reaction products can be separated in polyacrylamide–8 M urea sequencing gels and visualized by autoradiography (see this volume [7,56]).

When we perform the experiment with a given fragment for the first time, we need to titrate the amount of Exo III. Low amounts of Exo III will not cut the DNA to completion, which results in the appearance of specific subfragments that can be mistaken for protein-induced stops. Specific protein-induced stops can only be mapped accurately if the naked DNA is digested extensively. As a guideline, we suggest using between 100 and 2000 units of enzyme in several parallel assays. It is possible to add those amounts of Exo III, because an increase of glycerol to 25% does not affect the binding properties.

Specificity

It can not necessarily be assumed that fragments generated by extensive digestion of the protein–DNA complexes with Exo III reflect the 3' boundaries of the binding sites. Even digestion of naked DNA with the enzyme can give rise to distinct fragments that are probably caused by sequence-specific stops of the enzyme.[17,19,22] As in the mobility shift assay, competition experiments with the unlabeled homologous fragment can distinguish between specific and nonspecific interactions.

If the notion is true that "exo stops" found on naked DNA are caused by secondary structure in the DNA, it can be expected that some of the pauses observed upon protein binding do not reflect the binding site itself, but a secondary structure induced in cis by bound proteins. A novel technique developed in our laboratory called "restriction competition" can be applied to distinguish between the two kinds of stops. The method is based on the possibility of cutting the DNA fragment with restriction endonucleases and thereby destroying protein binding sites. The unlabeled competitor fragment homologous to the probe is cut with different restriction endonucleases that map in the vicinity of protein-induced "exo stops." In a typical restriction competition assay we preincubate the nuclear proteins with poly(dI-dC) and increasing amounts (up to a 100-fold excess over probe) of either the intact homologous fragment or the restricted fragments. The uncut fragment competes for all the binding factors that induce specific stops and thereby causes an "exo stop" pattern comparable to naked DNA. The cut competitor fragments do not bind the

[22] W. Linxweiler and W. Horz, *Nucleic Acids Res.* **10**, 4845 (1982).

proteins whose binding sites were affected by cleaving the DNA. This will result in the disappearance of only certain fragments. Restriction competition therefore allows the correlation of specific "exo stops" with individual DNA-binding proteins.

Figure 2 shows an Exo III protection assay using crude nuclear extracts from different cell types and the promoter (upstream) region of a milk protein gene. "Exo stops" seen upon digestion of naked DNA are probably caused by DNA secondary structure. Additional, protein-induced "exo stops" can be detected upon incubation of the probe with crude nuclear extracts from mammary epithelial and HeLa cells (marked by arrows). The different extracts induce a very similar pattern of "exo stops." This indicates that the binding factors visualized in this experiment are common to the cell types assayed. It should be noted that these reactions are performed in DNA excess and that the vast majority of the radiolabel has run off the gel.

DNase I Protection

Accurate information about DNA sequences protected by DNA-binding proteins comes from DNase I protection analysis.[23] Although a number of reports show DNase I protection using either purified or enriched sequence-specific DNA-binding proteins, the method was extended to crude nuclear proteins only recently.[21,24,25] Naturally DNase I protection in crude nuclear extracts is not as clean as with purified proteins, but nevertheless it is the most direct way to show protein binding to specific DNA sequences.

Method

In the following we will present conditions that allow us to detect protection in a number of milk protein genes and viral genes. Depending on the affinity and abundance of the particular DNA binding proteins under investigation it will be necessary to adjust the NaCl and poly(dI-dC) concentrations. One microgram poly(dI-dC) and increasing amounts of crude nuclear protein (10, 20, 40, 80, 160 μg) are incubated in a total volume of 50 μl binding buffer (50 mM NaCl, 0.1 mM EDTA, 20 mM HEPES–KOH at pH 7.5, 0.5 mM DTT, 10% glycerol) for 15–20 min in ice. Upon adding 0.5–1 ng of the labeled probe, incubation continues for 10 min at room temperature. The $MgCl_2$ and $CaCl_2$ concentrations in the

[23] D. Galas and A. Schmitz, *Nucleic Acids Res.* **5**, 3157 (1978).
[24] F. K. Fujimura, *Nucleic Acids Res.* **14**, 2845 (1986).
[25] A. G. Wildeman et al., *Mol. Cell. Biol.* **6**, 2098 (1986).

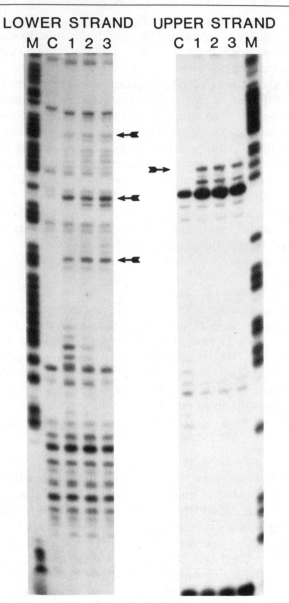

FIG. 2. Exonuclease III protection analysis. The 87-bp *Dra*I–*Xba*I (nucleotide −175 to −88) fragment overlapping the upstream regulatory region of the murine whey acidic protein gene[15,26] was 5' end labeled. Three hundred picograms of the labeled probe were incubated with 5 µg poly(dI-dC) and 1 µg crude nuclear extract from HeLa cells (lanes 1), MCF-7 cells (lanes 2), or lactating mammary glands (lanes 3). The samples were digested with 520 units exonuclease III and the reaction products were separated in a sequencing gel (this volume [56]). The "exo stops" seen on the probe in the absence of nuclear proteins (lane c) are

solution are adjusted to 5 and 1 mM, respectively, and 0.1–0.5 units pancreatic deoxyribonuclease I (DNase I) is added. The DNase I stock solution (2 μg/μl in 150 mM NaCl and 50% glycerol) is diluted with 25 mM NaCl, 10 mM HEPES–KOH at pH 7.5, and 0.5 mM DTT. After 30 sec incubation at room temperature, 100 μl stop buffer (0.375% SDS, 15 mM EDTA, 100 mM NaCl, 100 mM Tris–HCl at pH 7.6), sonicated *E. coli* DNA to 50 μg/ml, and proteinase K to 100 μg/ul are added to the reaction mixture. Incubation continues for 15 min at 37° and 2 min at 90° followed by a thorough phenol and chloroform extraction and ethanol precipitation. The reaction products are separated in polyacrylamide–8 M urea sequencing gels and subjected to autoradiography.

Specificity

The reaction conditions described above allowed us to detect strong protein binding sites in milk protein genes[26] and the promoter region of the human cytomegalovirus immediate early gene 1.[27] Figure 3 shows DNase protection of a protein binding site in the promoter (upstream) region of the rat α-lactalbumin gene. This site was detected in the presence of 25 μg nuclear extract made from mammary glands from pregnant animals. By increasing the amount of nuclear extract in the assay it is possible to detect additional weaker sites of protein–DNA interaction (e.g., TATA box), but due to the concomitant higher nonspecific binding, protection is not as clear. As in the previous methods, specificity of binding can be verified in competition assays using the homologous unlabeled fragment and an unrelated fragment. Increasing the poly(dI-dC) concentration (up to 5000-fold weight excess over labeled probe) in the presence of a fixed protein concentration can improve protection and provide information about the affinity (or abundance) of the binding protein. Nuclear extracts prepared from tissue contain endogenous nucleases which might cleave the probe. As a control we therefore routinely incubate the probe with the extract in the absence of DNase I and separate the reaction products in a denaturing gel.

[26] H. Lubon and L. Hennighausen, *Nucleic Acids Res.* **15,** 2103 (1987).
[27] P. Ghazal, H. Lubon, B. Fleckenstein, and L. Hennighausen, *Proc. Natl. Acad. Sci. U.S.A.,* in press.
[28] P. K. Qasba and S. K. Safaya, *Nature (London)* **308,** 377 (1984).

probably due to secondary structure within the DNA. The arrows mark protein-induced "exo stops." Assays with fractions enriched for the DNA-binding proteins will result in an enhancement of the protein-induced stops, but not the nonspecific ones. Lanes M: A/G Maxam–Gilbert sequencing reaction of the respective fragment (see this volume [56]); lower strand, the fragment was labeled at the *Xba*I site; upper strand, the fragment was labeled at the *Dra*I site.

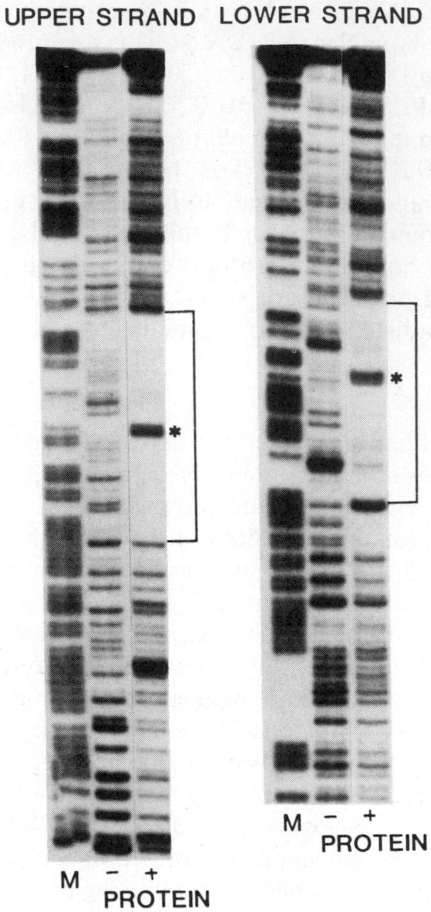

UPPER STRAND LOWER STRAND

FIG. 3. DNase I protection analysis. The 207-bp *Sau*3A-*Bam*HI fragment (nucleotides −202 to +5) overlapping the promoter of the rat α-lactalbumin gene[28] was labeled at either 5′ end. Three hundred picograms of the labeled probe was incubated with 1 μg poly(dI-dC) and 25 μg crude nuclear proteins prepared from mammary glands from late-pregnancy animals and subjected to DNase I digestion. The reaction conditions are described in the text. The left panel (the fragment is labeled at the *Sau*3A site) shows the protected region on the upper strand (that which is transcribed into mRNA) and the right panel (the fragment is labeled at the *Bam*HI site) shows protection of the lower strand. In lane M, the A/G sequence pattern (Maxam and Gilbert technique) of the respective fragments is shown. The protected sequence (−123 to −83) indicated by brackets harbors a sequence conserved among several milk protein genes.[26] Its position on each strand is measured relative to the location of the labeled terminus. Here the protected region begins 79 bp from the *Sau*3A end (upper strand) and 88 bp from the *Bam*HI end (lower strand). Asterisks indicate protein-induced DNase I hypersensitive sites.

[*Editors' Note*. In Chapter 73, methods were presented for determining sequence specific binding of proteins *in vitro* to cloned regulatory elements of DNA. Chapter 74 contains high resolution methods for mapping the location of proteins bound to DNA *in vivo*; the difference in sensitivity to DNase I of DNA in chromatin compared with naked DNA is used to predict specific protein–DNA interactions. A different version[1,2] of this technique involves digestion of chromatin with DNase I, isolation of DNA, and complete cleavage with a restriction endonuclease that cuts rarely, as in [74]. Then, the fragments are separated electrophoretically in gels [8], transferred to nitrocellulose [61] and hybridized to a ^{32}P-labeled probe [43, 45] that abuts the particular restriction site chosen as the point of reference. The visualization step is called indirect end labeling because the end is generated by a restriction enzyme and "labeling" of all subfragments sharing that end is achieved by hybridization. Improvements in Chapter 74 include solution hybridization and the use of short, highly radioactive, highly purified probes.]

[1] S. A. Nedospasov and G. P. Georgiev, *Biochem. Biophys. Res. Commun.* **92,** 532 (1980).
[2] C. Wu, *Nature (London)* **286,** 854 (1980).

[74] *In Vivo* Footprinting of Specific Protein–DNA Interactions

By P. David Jackson *and* Gary Felsenfeld

We describe in this chapter a method for mapping the pattern of cleavages within specific unique DNA sequences in a complex genome at near single base resolution. The method has been used to accurately map the sites of sequence-specific protein–nucleic acid interactions ("footprints"; Fig. 1A) upon regulatory sequences in various members of the α- and β-globin gene families within chicken cells and cell nuclei.[1-3] We examined the pattern of cleavages introduced into the genomic copies of β- and α-globin gene promoters by exogenously added nuclease, but any reagent (endogenous as well as exogenous) that cleaves DNA is a candidate for this analysis. The method can be applied to the study of any gene for which *in vivo* structural information is desirable. The method described here is an improved version of that previously published.[1]

The method (Fig. 1B) is related to that of Berk and Sharp for mapping mRNA cap sites[4] with modifications that increase sensitivity and specific-

[1] P. D. Jackson and G. Felsenfeld, *Proc. Natl. Acad. Sci. U.S.A.* **82,** 2296 (1985).
[2] B. Kemper, P. D. Jackson, and G. Felsenfeld, *Molec. Cell. Biol.* **7,** 2059 (1987).
[3] P. D. Jackson, B. E. Emerson, and G. Felsenfeld, unpublished observation.
[4] A. J. Berk and P. A. Sharp, *Cell (Cambridge, Mass.)* **12,** 721 (1977).

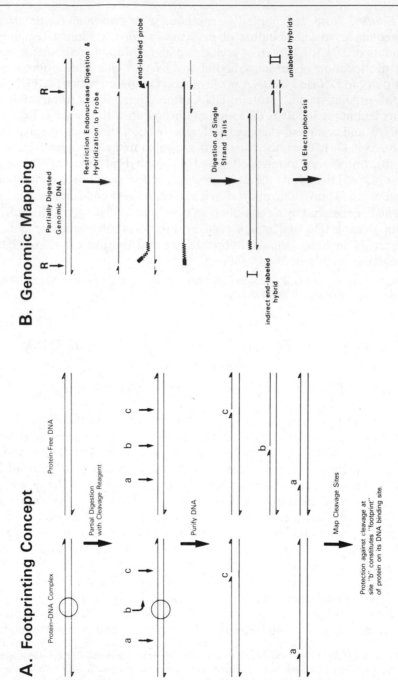

ity. It (along with other mapping methods[5,6]) also adapts the "indirect end-labeling" concept[7] to allow the high-resolution mapping of genomic cleavage sites with respect to an experimentally introduced reference point, the "indirect end label."[8]

Genomic DNA in cell nuclei or in whole cells is treated with nucleases or other DNA cutting agents to delineate the DNA contact points of site-specific DNA binding proteins and other chromatin structures that protect DNA from attack (Fig. 1A). The partially digested genomic DNA (Fig. 1B, top) is then purified, cut with a restriction endonuclease to reduce the size of the DNA fragment containing the target sequence, and hybridized to a highly purified 5'-end-labeled single-strand DNA probe complementary to the site of interest (Fig. 1B, step 1). The probe is prepared by primed synthesis on a single-strand template (Fig. 2). The probe–genomic DNA hybrids are heat treated at low ionic strength to eliminate nonspecific hybridization and are then digested with a single-strand-specific nu-

---

[5] C. Wu, *Nature (London)* **309**, 229 (1984).
[6] G. M. Church and W. Gilbert, *Proc. Natl. Acad. Sci. U.S.A.* **81**, 1991 (1984).
[7] C. Wu, *Nature (London)* **286**, 854 (1980).
[8] "Indirect end label" concept as used here is diagrammed in Fig. 1B. Briefly, cleavages in the target genomic strand are mapped with respect to the "end label" on the *probe* strand hybridizing to the target strand. As the end label reference point is not covalently attached to the mapped strand, the label is referred to as "indirect."

FIG. 1. (A) Footprint concept. Binding of proteins to DNA affects the accessibility of DNA sequences at the binding site to digestion with nucleases or other DNA cleavage reagents. Partial digestion of protein–DNA complexes and protein-free DNA reveals a site that is normally cleaved in protein-free DNA but is protected from cleavage by protein binding (site b) as well as sites that are unaffected by protein binding (sites a and c). The protection against cleavage at site b constitutes the footprint of the protein bound to its recognition site on the DNA molecule. (B) Genomic mapping. Information on the patterns of cleavage and protection along a genomic DNA sequence that can identify footprints of *in vivo* protein–DNA interactions is obtained by mapping the cleavage sites with respect to an experimentally introduced indirect end-label reference point. Genomic DNA from nucleic or from whole cells is partially digested with the DNA cleavage reagent of choice. Protein-free DNA from the same source is treated similarly. Both samples are subsequently purified and further digested (step 1) with restriction endonucleases (R in diagram) chosen to uncouple the sequence of interest from the extended adjacent lengths of genomic DNA that might otherwise interfere with hybridization. The restriction endonuclease-cleaved DNA is then hybridized (step 1) to a single-strand 5'-end-labeled DNA probe complementary to the sequence to be mapped. After hybridization, the hybridization mixture is heated to denature nonspecific hybridization products and the single-strand tails trimmed from the hybrids and unhybridized probe and genomic strands degraded by digestion with a single-strand-specific nuclease (step 2). The resultant mixture is resolved by electrophoresis in nondenaturing polyacrylamide gels (step 3) and the indirect end-labeled hybrid bands (molecule I) visualized by autoradiography.

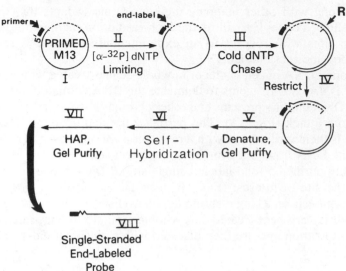

FIG. 2. Preparation of highly purified single-strand 5'-end-labeled DNA probe. Single-strand end-labeled probe is made by primed synthesis on a single-strand template. Single-strand template is primed (I) by hybridization of a short oligonucleotide (primer) to vector sequences (solid semicircle) directly adjacent to the template sequence for the genomic probe to be synthesized (dashed semicircle). 5'-End-label is introduced (II) by extending the primer using large-fragment DNA polymerase I and limiting $[\alpha\text{-}^{32}P]dNTP$ such that label is extended only into the first 10–30 nucleotides of the nascent probe sequence complementary to the genome. Probe synthesis is completed (III) by continuing the synthesis with a massive excess of unlabeled nucleotides on through the rest of the probe template sequence (dashed semicircle) and into the adjacent vector template sequences (solid semicircle). The probe–template complex is digested (IV) with a restriction endonuclease (R) (*Hin*dIII in the example in the text) chosen to maximize the size difference between the subsequently denatured probe and template strands, and the probe strand purified away from the template by denaturing gel electrophoresis (V). The probe is further purified to remove the few remaining template strands (and template fragments) by sequential self-hybridization (VI) and hydroxyapatite (HAP) chromatography (VII) which separates the pure single-strand probe from contaminating template–probe duplex molecules. Intact probe molecules (VIII) are separated from the radiolytically degraded probe molecules that accumulate during probe purification by a second round of denaturing gel electrophoresis (VII) to yield single-strand end-labeled probe (VIII) for genomic mapping and footprinting (Fig. 1, A and B).

clease (Fig. 1B, step 2). This enzymatic step degrades unhybridized probe molecules and trims single-strand tails from the hybrids leaving a population of radioactive hybrid molecules that extend from the probe 5' end label to the proximal cut in the complementary genomic strand, thereby allowing the mapping of these genomic cuts with respect to the end label reference point (Fig. 1B, molecule I). The population of labeled hybrid molecules thus generated are sized by electrophoresis on polyacrylamide gels under nondenaturing conditions (Fig. 1B, step 3) and the gels dried and autoradiographed to display the genomic cutting pattern comprising the footprint relic of *in vivo* DNA–protein interactions.

Cleavage and Subsequent Preparation of Genomic DNA

The choice of whether to perform the footprint analysis on whole cells or cell nucleic depends to some extent on the nature of the DNA cleavage reagent employed. Nuclei are permeable to most cleavage reagents whereas cells are more selective. Reagents such as dimethyl sulfate,[6] psoralen,[9] neocarcinostatin,[10–13] and related drugs can be used to cleave DNA in intact cells. We have used nucleases[1–3] as probes of chromatin structure in isolated nuclei and in concentrated lysed whole cells and we have used neocarcinostatin chromophore[14,15] as a structural probe in intact cells. Nuclease digestion of isolated nuclei is technically the easiest experiment, but it is important to appreciate the lability of some chromatin structures in the design and interpretation of experiments on isolated nuclei, since structural components can be lost during isolation procedures. The exact procedure to be used in cell or nuclear preparation will vary with the tissue or cell type to be analyzed but in general the samples should be handled gently and quickly, and kept cold and protease free to stabilize nuclear structures. Aliquots of the preparation are treated with a series of concentrations of the cleavage reagent designed to produce one or two cuts per 500 bases within the sequence of interest. For DNase I digestion of protein-free DNA or nuclease hypersensitive sites in isolated nuclei these concentrations are in the range 5–200 ng enzyme/mg DNA for 30 min at 20° in 50 mM NaCl, 20 mM HEPES (pH 7.5), 5 mM MgCl$_2$, and 1 mM CaCl$_2$ with protease inhibitors when necessary. Digestions are terminated by the addition of Na$_2$EDTA (pH 8.0), to 10 mM, NaCl to 0.5 M, and protease K to 100 μg/ml. After 30 min at 37° sodium dodecyl sulfate is added to 0.25% and protease K to 200 μg/ml and digestion allowed to proceed for 24 hr at 37°. Cutting frequency can be assayed by Southern blotting[16] to follow the appearance of cutting within a hypersen-

[9] M. M. Becker and J. C. Wang, *Nature* (*London*) **309**, 682 (1984).

[10] L. S. Kappen, I. H. Goldberg, and T. S. A. Samy, *Biochemistry* **18**, 5123 (1979).

[11] T. Hatayama, I. H. Goldberg, M. Takeshita, and A. P. Grollman, *Proc. Natl. Acad. Sci. U.S.A.* **75**, 3603 (1978).

[12] P. D. Jackson and G. Felsenfeld, unpublished observations.

[13] L. S. Kappen, M. A. Napier, I. H. Goldberg, and T. S. A. Samy, *Biochemistry* **19**, 4780 (1980).

[14] L. S. Kappen, M. A. Napier, and I. H. Goldberg, *Proc. Natl. Acad. Sci. U.S.A.* **77**, 1970 (1980).

[15] L. S. Kappen and I. H. Goldberg, *Biochemistry* **19**, 4786 (1980).

[16] E. M. Southern, this series, Vol. 68, p. 152.

sitive site[7,17] of interest or to follow the disappearance of a short (~500 bp) restriction fragment containing the site of interest. The indirect end-label Southern blotting method for low-resolution mapping of cutting within hypersensitive sites is well documented.[7,16,17] Briefly, purified genomic DNA from, e.g., nuclease-digested nuclei is digested with restriction endonucleases that will cut at least 500 and not more than 4000–6000 base pairs away from possible nuclease-hypersensitive sites near the gene of interest. The DNA is resolved by electrophoresis in agarose gels (see this volume [8]), transferred to a solid support [61], and hybridized to a radioactive probe [45].[7,16] The 300- to 600-bp probe is chosen such that it hybridizes specifically with the first 300–600 bases adjacent to the restriction endonuclease cut at one end of the genomic restriction fragment and such that it does not overlap the restriction site. The combination of restriction endonuclease digestion of genomic DNA and hybridization to short probes specific for one end of the genomic restriction fragment allows the mapping of genomic cleavage sites (e.g., hypersensitive sites) with respect to the apparent end label (indirect end label) provided by the short probe. Deproteinized, unrestricted genomic DNA is used to prepare controls displaying the cleavage pattern of the cleavage reagent on protein-free DNA and to make up a mixture of various restriction endonuclease-digested samples to act as size standards (Fig. 5, lanes 1–2 and lanes marked STD, respectively). We have found that genomic DNA serves quite well as substrate for the preparation of protein-free DNA controls, obviating the need to use cloned sequences for this purpose as was previously described.[1]

Genomic DNA is prepared for high-resolution mapping by deproteinization following by restriction endonuclease digestion and treatment with ribonuclease A to eliminate endogenous RNA that would otherwise compete for radioactive probe during hybridization. The deproteinized genomic DNA is treated with 1 μg of heated ribonuclease A (100° for 15 min, then cooled to room temperature) per 20 μg of genomic DNA in 10 mM Tris–HCl (pH 8.0), 1 mM Na$_2$EDTA (TE buffer) at 37° for 30 min and followed directly by restriction endonuclease digestion, phenol–chloroform extractions to remove protein (this volume [4]), alcohol precipitation (this volume [5]), and resuspension of the DNA to 2 g/liter in TE.

The restriction endonuclease digestion is chosen to isolate the genomic DNA sequence of interest on a small (300–1000 bp) fragment within the genomic digest. In the example of the mapping method presented here, digestion with *Pvu*II serves to center the ~200-bp 5' nucle-

[17] J. D. McGhee, W. I. Wood, M. Dolan, J. D. Engel, and G. Felsenfeld, *Cell* (*Cambridge, Mass.*) **27**, 45 (1981).

ase-hypersensitive region of the chicken erythrocyte β^A-globin gene on a 580-bp fragment. Having the target sequence on such a small genomic DNA fragment limits the ability of otherwise long genomic DNA to compete kinetically with single-strand probe for hybridization to the complementary genomic strand,[18] and similarly lowers the initiation rate of strand displacement reactions which would favor the longer, more stable genomic renaturation products over the shorter probe–genomic DNA hybrids at thermodynamic equilibrium. The loss of signal due to these phenomena is further limited by driving the hybridization reaction with a ~30-fold molar excess of probe over genomic complement. The preferential stability of genomic renaturation products and the resulting negative effects of strand displacement on signal intensity can be completely eliminated by choosing the 5′ end of the genomic restriction fragment to coincide with that of the probe.

Probe Preparation and Purification

Construction of Probe Template

5′-End-labeled single-strand probe is prepared by primed synthesis on a single-strand template followed by isolation and purification of the newly synthesized probe strand. The basic procedure is as previously described[1] (Fig. 2) with technical modifications that increase probe yield.

The template consists of ≥300 bp of the genomic sequence of interest cloned into a vector which yields filamentous phage particles containing single-strand DNA copies during infectious growth in bacterial cultures. We have used the M13 phage vectors developed by Messing[1,3,19] as well as pTZ plasmid vectors (P-L Biochemicals) bearing filamentous phage origins of replication[2,20–22] with equal success for the preparation of template. Recombinant phage are prepared from infected bacterial cultures[19–21] (see this volume [13]) and the single-strand DNA template purified from the phage by phenol–chloroform extraction and alcohol precipitation.

The choice of genomic fragment to be cloned and how to orient it within the vector sequence is made on the basis of two criteria: (1) each template will produce a probe that is strand specific (i.e., two templates will be required to examine completely the cutting pattern on both strands

[18] J. G. Wetmur and N. Davidson, *J. Mol. Biol.* **31**, 349 (1968).
[19] J. Messing, this series, Vol. 101, p. 20.
[20] L. Dente, G. Cesareni, and R. Cortese, *Nucleic Acids Res.* **11**, 1645 (1983).
[21] B. P. H. Peeters, J. G. G. Schoenmakers, and R. N. H. Konings, *Gene* **46**, 269 (1986).
[22] D. A. Mead and B. Kemper, *in* "Vectors: A Survey of Molecular Cloning Vectors and Their Uses" (D. T. Denhardt and R. L. Rodriguez, eds.). Butterworth, Boston, in press.

within a given genomic sequence), and (2) the mapping technique has best resolution 60–220 bases from the 5′-end-labeled region of the probe complementary to the genome (Fig. 3). As with any mapping technique, it is desirable to choose the end-labeled region such that it will avoid intense cutting sites in the genome. In the worst case, in which 100% of the genomic DNA is cut within the region complementary to the 5′-end-labeled stretch of the probe, it is very difficult to map the other genomic cut sites properly. Placement of the 5′-end label so that it is complementary to areas less heavily cut just adjacent to hypersensitive regions in chromatin eliminates this problem. The m19XB probe used here (Figs. 3–5) is about 300 bases long and extends at each end about 50 bases beyond the sequence of the 200-bp 5′-nuclease-hypersensitive region of the β^A-globin gene.

We have used synthetic primers complementary either to vector sequences directly adjacent to the cloned genomic sequences (e.g., 17-base "universal sequencing primer" used as primer for the m19XB probe), or to regions within the cloned genomic sequences. The amount of [α-^{32}P]dNTPs necessary to extend label into probe sequences complementary to genomic DNA is directly proportional to the distance between the 3′ end of the primer and the 5′ end of genomic complement in the probe sequence. It is a consideration of economy to minimize this distance when designing a cloning strategy. A related point is that shorter labeled stretches will suffer less radiolytic damage and thereby increase probe yield, as will become obvious in descriptions of probe purification below.

Probe Synthesis

Priming of Single-Strand Template (Fig. 2, step I). Approximately 0.75 μg of M13 phage DNA or 0.35 μg of single-strand pTZ DNA with a 300-base insert is required for each 250 μg of genomic DNA analyzed. This starting amount represents a 1000-fold molar excess of the cloned template sequence over its chicken genomic complement and compensates for radiolytic and other losses incurred during probe purification. Template DNA is primed by mixing template at a DNA concentration of 0.167 g/liter with a 2.5-fold molar excess of synthetic oligonucleotide primer in 1× SMT buffer (1× SMT: 42.0 mM NaCl, 5.8 mM MgCl$_2$, 5.8 mM Tris–HCl at pH 7.5). The mixture is heated to 55° for 5 min, incubated at 37° for 2 hr, and then frozen on dry ice and stored at −20° until use, when the primer is extended using the large fragment of DNA polymerase I.

5′-End Labeling of Probe Strand (Fig. 2, step II). Labeling of probe within regions complementary to the genomic sequence is confined to 10–

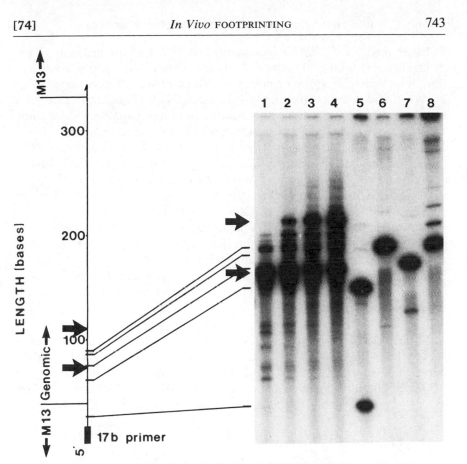

FIG. 3. Primed synthesis titrations with limiting [α-^{32}P]dNTPs. Pilot primed syntheses are performed as described on 0.75 μg of m19XB template using 3.52 pmol (lane 1, 6–8), 4.22 pmol (lane 2), 5.06 pmol (lane 3), and 6.08 pmol of each [α-^{32}P]dNTP. Lanes 1–4 are without a chase with excess unlabeled dNTPs. Lanes 5–8 are chased with unlabeled dNTPs and are then digested with different restriction endonucleases: *Sma*I (lane 5), *Ban*I (lane 6), *Apa*I (lane 7), and *Mnl*I (lane 8). The samples are denatured by heating in formamide and fractionated electrophoretically in a polyacrylamide gel of 12.5% acrylamide–0.625% bisacrylamide containing 45 m*M* Tris–borate (pH 8.3), 1.25 m*M* Na$_2$EDTA, 8.5 *M* urea. A map of the region of primed synthesis is diagrammed at the far left, with the position of kinetic stops in the incorporation of label marked with arrows. The restriction standards (lanes 5–8) are correlated with the map by connecting lines. The 3' end of the mapped probe strand is shown cleaved at the *Hind*III site as during probe purification described in text. The sizes of the major kinetic stops are at 73 bases (37 bases of label at 5' end of genomic sequence) and 110 bases. The restriction standards are at 25, 62, 74, 85, and 89 bases. The bands in lanes 6–8 illustrate radiolytic strand breakage of the probe, since a smear of breakage products appears 17 bases (the length of the unlabeled primer) below the major band. The intensity of the smear is proportional to the time between synthesis and electrophoresis, which for this pilot experiment was 12 hr.

40 bases near the 5' end by incubating first with limiting amounts of [α-^{32}P]dNTPs and taking advantage of inherent kinetic blocks to polymerization, which provide stopping points for all templates we have tested (Fig. 3). We found it necessary to adjust labeled dNTP concentrations over a range of ~2-fold for individual templates to obtain suitable end-labeling; the differences among templates are probably related to template secondary structures (e.g., self-complementary stem and loop structures).[1] The m19XB probe used in the analysis of genomic DNA shown in Figs. 4 and 5 is synthesized by adding primed M13–m19XB template (0.75 μg in 4.5 μl) to 16.7 μl of a mixture containing 3.33 pmol (10 μCi) each of α-^{32}P-labeled dATP, dCTP, dGTP, and dTTP (3000 Ci/mmol), 1.2× SMT, 0.15 mM 2-mercaptoethanol, and 1.2 units of large-fragment DNA polymerase I (any polymerase preparation suitable for dideoxynucleotide-terminated DNA sequencing is acceptable in this application).

Extension of Probe Strand with Unlabeled Nucleotides (Fig. 2, step III). After incubation at 20° for 1 hr, the probe strand is extended through the remaining genomic sequence and into adjacent M13 vector sequence with greatly reduced incorporation of radioactivity (Fig. 2, step III) by adding 3.75 μl of a mixture containing 2.0 nmol each of the four unlabeled dNTPs in 1× SMT, 0.15 mM 2-mercaptoethanol, and 1.25 units of DNA polymerase I large fragment. Incubation is continued at 20° for 20 min, at 37° for 5 min, and finally at 68° for 10 min to denature polymerase. The extent of probe 5'-end label is approximately as in Fig. 3, lane 1.

Probe Purification

Restriction Endonuclease Digestion of Probe–Template Complex (Fig. 2, step IV). The probe–template complex is digested with a restriction endonuclease chosen to maximize the size difference between the denatured single strands of the newly synthesized probe and its M13 template (Fig. 2, step IV). For the m19XB probe, 10 units of *Hin*dIII endonuclease, 2.0 μl of buffer (0.5 M Tris–HCl at pH 8.0, 50 mM MgCl$_2$, bovine serum albumin at 1 g/liter, and water to 40 μl are added to the above reaction mixture and incubated at 37° for 1 hr. *Hin*dIII cleaves once, yielding a linearized template strand of ~7500 bases and an end-labeled probe strand of ~350 bases including adjacent M13 sequences and primer (Fig. 3). The reaction is stopped by the addition of Na$_2$EDTA (pH 8.0) to 15 mM final concentration, and the DNA purified by extraction with Tris-neutralized phenol and chloroform. Traces of chloroform are removed by extraction with two volumes of n-butanol, the aqueous phase diluted with 2 ml of TE, and the sample concentrated and desalted by centrifugation in a

[74] *In Vivo* FOOTPRINTING 745

Centricon-30 disposable concentration device according to the manufacturer's directions (Amicon Corporation).

Denaturing Gel Purification of Probe Strand (Fig. 2, step V). The retained sample, in a volume of 50–100 μl, is denatured by dilution to 1 ml with deionized formamide (plus 5 μl 0.2 M NaEDTA at pH 8.0 and xylene cyanole FF dye to 0.01 g/l) and heated at 68° for 5 min prior to loading in a single slot on a 1.5-mM thick polyacrylamide gel (5% acrylamide, 0.25% bisacrylamide) containing 45 mM Tris–borate (pH 8.3), 1.25 mM Na$_2$EDTA, 8.5 M urea. The gel is prerun prior to sample loading to heat it to ~50°. Avoiding alcohol precipitation and keeping the sample dilute and warm prior to electrophoresis increase probe yield by decreasing rehybridization of the template and probe strands and by minimizing nonspecific losses on surfaces often incurred during precipitation. The sample is fractionated electrophoretically at 30–40 V/cm for 2 hr and the gel autoradiographed for about 15 sec to localize the probe. The probe band is excised and electroeluted from the gel slice for 1 hr at 125 V into a small volume of a high ionic strength solution (4.75 M NaCl, 25 mM HEPES at pH 7.5, 5 mM Na$_2$EDTA, sonicated calf thymus DNA at 50 μg/ml) in an electroelution apparatus (International Biotechnologics, Inc., catalog no. 46000) containing the Tris–borate–EDTA electrode buffer as above. The electroeluted sample is desalted and concentrated in a Centricon-30 by two or three cycles of dilution with TE followed by centrifugation as above. The probe is pure enough at this point to use for the study of restriction digests containing short unique chicken DNA sequences at 0.1× genomic abundance (Fig. 4, lanes 1–5). Contaminating template sequences remain, however, and the signal-to-noise ratio obtained using this probe is improved (Fig. 4, lanes 6–11) by continuing the probe purification described below. The analysis of mixtures of restriction endonuclease-digested genomic DNA (as in Figs. 4 and 5) provides an important test of the validity of the mapping information provided by the technique. Each of the restriction endonuclease cleavages introduced into the genomic DNA should be mapped uniquely and at the correct distance with respect to the probe end label for the mapping information to be valid. This point will be covered in more detail in the discussion of Figs. 4 and 5 contained in the section on genomic footprinting and validation of results at the end of this chapter.

Removal of Contaminating Template Strands by Sequential Self-Hybridization and Hydroxyapatite Chromatography (Fig. 2, steps VI and VII). The probe is purified further by self-hybridization and hydroxyapatite (HAP) chromatography to remove the small amount of contaminating template sequences. To the probe in 60 μl of TE is added 6.67 μl of 10×

HB (10× HB: 3.18 M NaCl, 200 mM HEPES at pH 7.5, 1.0 mM Na$_2$EDTA). This mixture is flame sealed in a glass capillary, heated to 107° for 5 min in a saturated NaCl bath, and transferred directly to a 68° water bath for incubation overnight. During this self-hybridization, probe strands in excess will hybridize with contaminating template strands. The pure single-strand probes may be separated from these duplexes by their differential affinity for HAP which binds the duplex molecules more tightly than the single strands in 150 mM NaPO$_4$ at pH 7.[23]

Hydroxyapatite (Bio-Rad) is prepared by suspension and settling in 50 mM NaPO$_4$ (pH 7.0). The HAP is washed once in 1 M NaPO$_4$ (pH 7.0), and then equilibrated in 50 mM NaPO$_4$ (pH 7.0). The self-hybridized probe is diluted into 0.2 ml of sonicated calf thymus DNA (0.2 g/liter) in 50 mM NaPO$_4$ (pH 7.0), and applied to 0.1 ml of the gently packed HAP (centrifuged 1 min at 250 g in swinging bucket rotor) in a tightly capped polypropylene centrifuge tube. The HAP is suspended in the DNA solution by gentle vortexing or inversion and intermittently resuspended during a 15-min incubation at 68°. The HAP is sedimented at 250 g for 1 min and washed six times with 2 ml of 50 mM NaPO$_4$ (pH 7.0), at 68°. Greater than 90% of the radioactivity should bind to the HAP, and ^{32}P in the wash buffer should reach a low constant value after three or four washes. The probe is eluted from the matrix with four 68° washes of 0.5 ml 150 mM NaPO$_4$ (pH 7.0). Combined washes containing ~90% of the eluted radioactivity (only a small percentage of the ^{32}P should remain associated with HAP at this point) are filtered by centrifugation through 0.45-μm pore size Centrex cellulose acetate filters (Schleicher & Schuell) to remove any stray HAP and the filtered probe desalted and concentrated by dilution with TE and centrifugation in a Centricon-30 device as before.

Final Removal of Radiolytic Degradation Products by Denaturing Gel Electrophoretic Purification of Intact Probe Strand (Fig. 2, steps VII and VIII). The probe in ~50 μl of TE is carefully concentrated to ~5 μl by repeated extraction with 2 volumes of n-butanol and the sample diluted to 100 μl with the same proportions of formamide, EDTA, and xylene cyanole FF used before. The sample is heated prior to electrophoresis in a polyacrylamide–urea–Tris–borate–EDTA gel as before with smaller well size to accommodate the smaller sample volume. The reason for performing this second gel purification of the probe is made clear upon autoradiography of the gel. It is apparent that during the time since electrophoresis ceased in the first gel (~20 hr) considerable radiolysis has occurred. Up to 50% of the probe will appear in a smear of radioactivity just below

[23] G. Bernardi, *Procedures Nucleic Acid Res.* **2**, 455 (1971).

the remaining tight band of intact probe. It is a conservative choice to excise only the intact probe band, but it thereby assures the integrity of the probe that now goes into the analysis of genomic DNA. The probe (Fig. 2, step VIII) is electroeluted, concentrated, and desalted as before and is now ready for hybridization to genomic DNA. The probe should be used immediately as storage only decreases the signal-to-noise ratio, eliminating the gains made by the second probe purification gel.

Analysis of Genomic DNA

Hybridization of Genomic DNA with Probe (Fig. 1B, step 1)

To make more efficient use of the effort expended in probe purification, we generally analyze 40–50 25-μg samples at a time with the probe preparation scaled accordingly. Hybridization of single-stranded 5'-end-labeled probe with genomic DNA is accomplished by mixing 0.1 ng ($\sim$3 $\times$ 10^5 dpm) of probe with 25 μg of genomic DNA (a 30-fold molar excess of probe over its chicken genomic complement) in 20 μl of 1$\times$ HB. The mixture is flame sealed in a glass capillary, heat denatured (5 min, 107°), and directly transferred to a 68° bath for a 12-hr hybridization. Genomic DNA is 60–70% renatured and hybridization to probe is complete in this time. The reaction is stopped by expelling the mixture into 14 μl of TE and precipitating the DNA with 40 μl of isopropanol at room temperature. The pellet is washed in 70% ethanol, dried, and dissolved in 15.6 μl of 15 mM NaCl, 1.5 mM HEPES (pH 7.5).

Heat Denaturation of Nonspecific Hybrids

Cross-hybridization of the probe to related sequences in the genome is eliminated by incubation in this low ionic strength solvent at 70° for 30 min. The temperature of incubation is probe dependent as is discussed below. This heat denaturation step is necessary to eliminate high backgrounds of nonspecific hybridization, the magnitude of which varies as a function of probe sequence. Hybrids of probe with the specific genomic sequence are not denatured under these conditions.

This step is analogous to the "high-stringency wash" step in Southern blotting[16] and the considerations governing the choice of conditions are the same for both. Higher G + C content (the probe used here is ~ 70% G + C) and greater homology to nonspecific genomic sequences dictate incubation at a higher temperature (e.g., 70° for the m19XB probe). The temperature of incubation is chosen empirically by observation of the influence of temperature upon the signal-to-noise ratio obtained in the analysis of standard genomic samples (e.g., a mixture of restriction endonuclease digests of genomic DNA as in Fig. 4). Since little is known in many cases about the extent of cross-reacting sequences within a genome, one can start such a survey of temperature effects near the melting temperature of the specific hybrid according to its content of G + C bases and its length, and work down in temperature until the background from nonspecific hybridization becomes unacceptable again. In practice, T (in degrees Centigrade) = 41 + 0.41 (% G + C bases in probe) is a reasonable upper limit, at which temperature both specific and nonspecific hybrids of less than about 50 base pairs are denatured.[24,25] Omission of the heat denaturing step is a potential source of artifact.

Digestion of Single-Strand DNA Regions

After, heating, the hybridized samples are digested with a calibrated activity of mung bean nuclease[26] to trim off single-strand regions from the hybrids, as well as unhybridized single-strand probe and residual denatured genomic DNA (Fig. 1B, step 2). The pattern of mung bean nuclease-resistant fragments is relatively insensitive to 2-fold changes in enzyme activity but new batches of enzyme should be calibrated in the assay (Fig. 4). Overdigestion leads to nibbling in at the ends of duplex molecules; underdigestion leads to increasing levels of contamination with nonhybridized probe. The amount of nuclease used should be 1.5- to 2-fold more than the minimum necessary to degrade unhybridized probe molecules. This low nuclease activity (6.5 units of nuclease/25 μg of genomic DNA for the enzyme preparations we have used) will assure the elimination of the unhybridized species and the effective trimming of single-strand regions from the hybrids and yet minimize the nibbling in at the ends of hybrid molecules that can decrease resolution (Fig. 4, lane 6 vs. lane 7). A mixture of restriction endonuclease digests of genomic DNA is assayed in parallel with the other genomic samples to provide size standards that control for the effects of mung bean nuclease on fragment size (Fig. 5, lanes marked STD).

A potential artifact can arise through the action of the single-strand-specific nuclease at sites where the probe is not complementary to the genomic sequence. Sequence mismatches in the hybrid might result from genomic sequence polymorphisms or inaccurate copying of template dur-

[24] J. Marmur and P. Doty, *J. Mol. Biol.* **5**, 109 (1962).

[25] C. R. Cantor and P. R. Schimmel, "Biophysical Chemistry," Part III, Chapter 22. Freeman, San Francisco, California, 1980.

[26] M. Laskowski, Sr., this series, Vol. 65, p. 263.

ing probe synthesis (see discussion of alternative probe preparation strategies below). The nuclease can attack these defects in the duplex if they are large enough, and yield fragments unrelated to the genomic cleavage pattern under investigation.

Digestion of single-strand regions is effected by adding 6.5 units of mung bean nuclease (P-L Biochemicals) in 10.4 μl of 125 mM sodium acetate (pH 5.3), 2.5 mM magnesium acetate, 2.5 mM L-serine, 2.5 mM 2-mercaptoethanol (added fresh), 0.0125% Triton X-100 (Sigma), 0.25 mM zinc acetate, and incubating at 40° for 60 min. The reaction is stopped by adding 3.5 μl of 2.5 M NaCl, 0.5 M Tris–HCl (pH 8.5), followed by precipitation with 36 μl 2-propanol. The pellet is washed with 70% ethanol, dried, and redissolved in 4–8 μl of 10 mM Na$_2$EDTA (pH 8.0), 6% Ficoll, 0.06% xylene cyanole FF to yield 25–12.5 μg genomic DNA equivalents of hybrid molecules per 4 μl to be loaded on the gel.

Resolution of the Trimed Radioactive Hybrids on Nondenaturing Polyacrylamide Gels (Fig. 1B, step 3)

The choice of resuspension volume depends on considerations of sensitivity and resolution. At 25 μg/4 μl, sensitivity is increased, but the higher load of DNA decreases the resolution, especially above 180 base pairs. Regardless of choice, the concentration of all experimental samples and the genomic restriction standards should be identical in order to equalize the effects of sample load on fragment migration during gel electrophoresis. Samples in 4 μl are loaded in 0.8 × 5 mm wells on a 0.8 × 330 × 380 *nondenaturing* gel (8% acrylamide–0.4% bisacrylamide containing electrophoresis buffer) and fractionated electrophoretically in a buffer containing 90 mM Tris–borate (pH 8.3), 2.5 mM Na$_2$EDTA at 18 V/cm for 7 hr. Glass plates used to form the gel slab should be pretreated with dichlorodimethylsilane and rinsed with ethanol and water prior to gel casting. Nondenaturing gels are used to ensure the integrity of labeled fragments that may have been internally nicked by radiolysis or mung bean nuclease.

After electrophoresis, the glass gel-casting plates are separated and a sheet of cellophane (Bio-Rad) that has been thoroughly wetted is closely apposed to the gel with care taken to eliminate air bubbles from the gel–cellophane interface. A sheet of heavy filter paper is laid over the cellophane for mechanical support and the gel *carefully* freed from the glass plate by turning the glass plate side of the gel sandwich up and prying the gel–cellophane laminate free from the glass, allowing it to fall gently on the supporting heavy filter paper. A sheet of Mylar (0.0005 in. thick, or

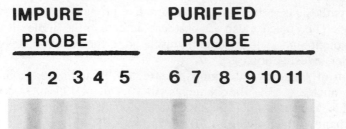

other transparent heat-stable film) is layered without air pockets over the upper naked gel surface, and the four-layered sandwich dried in a vacuum slab gel drying apparatus (e.g., Bio-Rad or Hoefer apparatus) at 80°. The transparent dried cellophane–gel–Mylar laminate (heavy filter paper removed) is sandwiched between two Dupont Cronex Lightening Plus intensifying screens and exposed to a sheet of flashed[27] (also this volume [7]) Kodak XAR-5 film for ~14 days at −80°. The use of a transparent gel backing (cellophane) and two intensifying screens effectively doubles the sensitivity of the method over that previously described.[1]

Potential Modifications to Mapping Methods

Further gains in sensitivity could be obtained if necessary by prior partial purification of the genomic sequences to be probed (e.g., by gel electrophoretic size fractionation of restriction endonuclease-digested genomic samples). This sequence purification has the potential to dramatically increase the signal-to-noise ratio by greatly decreasing the complexity of the analyzed genomic sample and by allowing a much greater specific sequence load on the final analytical gel.

The basic steps of this analysis (i.e., end-labeled single-strand probe synthesis and purification, hybridization, trimming of the hybrids with single-strand-specific nuclease and gel electrophoresis) might be accomplished using variations of the mapping technique described here. Some of these variations may represent improvements in ease of probe purification or increased resolution of the genomic analysis, but for the most part they remain speculative and untested.

---

[27] R. Lasky, this series, Vol. 65, p. 363.

FIG. 4. Mung bean nuclease digestion titrations of m19XB probe–chicken DNA restriction mix hybrids. Single-strand probe hybridized with a mixture of chicken DNA restriction endonuclease digests is processed as described in the text with the exception that different activities of mung bean nuclease are used to trim the hybrids. Two different purities of probe are used in the hybridization: impure probe (lanes 1–5) used directly after the first denaturing gel and highly purified probe (lanes 6–11) that has gone through the entire purification scheme. Mung bean nuclease concentrations used are 6.5 units (lanes 1, 6, and 11), 13 units (lanes 2 and 10), 26 units (lanes 3 and 9), 52 units (lanes 4 and 8), and 104 units (lanes 5 and 7). The position of the restriction endonuclease cuts in the genomic sequence are given at the right with respect to the start site of transcription. The actual lengths of the trimmed hybrid fragments extending from the probe 5' end at −4 to the genomic restriction positions marked at the right of the figure are 118, 138, 156, 196, 201, 224, and 299 base pairs from bottom to top radioactive fragment [e.g., (−122) − (−4) = 118 base pairs].

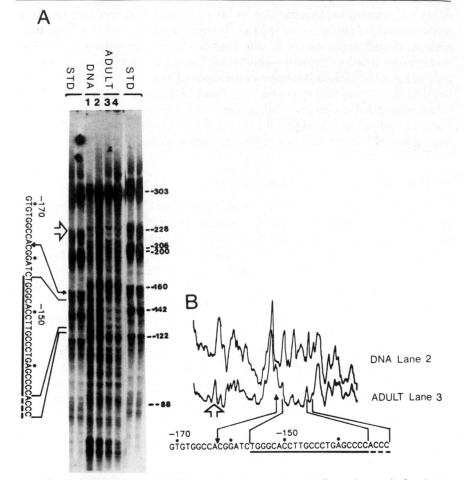

FIG. 5. Analysis of genomic DNase I cleavage patterns. (A) Genomic protein-free DNA samples (lanes 1 and 2) or purified nuclei from adult erythrocytes (lanes 3 and 4) were digested with DNase I. DNase I digestion increases by 2.5-fold going from right to left within each pair of samples. The genomic DNA in the standard lanes (STD) is as described in Fig. 3. Samples are analyzed with the m19XB probe as described in text. A region protected from DNase I cleavage in adult erythrocytes is mapped to the sequence on the left which, along with the standards, are numbered with respect to the initiation site for transcription on the chicken β^A-globin gene. Enhanced cleavage is indicated by the small arrow, and the protection adjacent to the 5' edge of the hypersensitive region by the large open arrow. (B) Densitometric scans of the indicated lanes from (A).

Probe purification could be effected by the incorporation of an affinity label (e.g., Hg-dUTP[28] or biotinylated dUTP) during probe synthesis followed by the separation of denatured probe and template strands by chromatography on matrices containing covalently bound sulfhydryl or streptavidin moieties. Alternatively, an RNA template for probe synthesis could be transcribed from a genomic clone in one of the vector systems utilizing phage RNA polymerase promoters.[21,22] Treatment with DNase I would yield pure RNA template, which could then serve as substrate for the synthesis and purification of probe using sequential primed synthesis with reverse transcriptase followed by alkaline hydrolysis of the RNA template. Direct end labeling of the RNA copy would provide even more facile access to a probe which, after genomic hybridization and nucleolytic trimming might provide a further advantage in the potential for elimination of gel overloading problems by DNase I treatment of hybridized, single-strand trimmed samples.

In practice, the incorporation of an affinity label is probably only to be preferred for the preparation of long (2-kb) probes for which gel purification is inefficient. The use of an RNA template is an intriguing possibility, but might suffer from the inherently lower fidelity of the RNA polymerase and reverse transcriptase reactions relative to those catalyzed by DNA polymerase I large fragment, which possesses a 3′–5′ exonucleolytic "editing" activity that limits misincorporation of bases.[29] Inaccurate transcription of the templates during probe synthesis would result in sequence mismatches in the probe–genomic DNA hybrid that could be recognized by mung bean nuclease or other single-strand-specific nucleases used to trim hybrids. In addition, altering the chemistry of the probe or using nucleases other than mung bean to trim the hybrids can introduce novel artifacts arising from the varying propensity of the nucleases for exo- or endonucleolytic attack upon duplex substrates. For example, the substitution of S_1 nuclease for mung bean nuclease to trim the hybrids has consistently led to poorer signal-to-noise ratios in our hands.

The problems to be overcome in the development of an alternative scheme for genomic mapping are obvious: signals introduced by errors in probe synthesis or by enzymatic degradation of the hybrid duplex must be small relative to the miniscule signals arising from the genomic cleavages to be mapped. Any alternative that can solve these problems can be used for high resolution genomic mapping when used in conjunction with a heat denaturation step to decrease signals due to cross-hybridization to related genomic sequences.

[28] P. D. Jackson and G. Felsenfeld, unpublished observations.
[29] D. Brutlag and A. Kornberg, *J. Biol. Chem.* **247**, 241 (1972).

Genomic Footprint and Validation of Mapping Results

The genomic restriction standards shown in Figs. 4 and 5A are mixtures of several different restriction endonuclease digests. Each digest should give a *single* band in the autoradiogram corresponding in length to the distance between the probe 5′-end label and the closest restriction endonuclease site in the complementary genomic strand. In examining these standard lanes, the validity of the mapping information gained from this technique becomes apparent. Approximately 95% of the radioactivity migrates in bands corresponding to the predicted lengths of the end-labeled hybrid molecules for each of the individual restriction digests in the standard. This demonstrates that the signals observed in an analysis of DNase I-cleaved genomic DNA (e.g., Fig. 5A, lanes 1–4) can be mapped with confidence onto the genomic DNA sequence with little interference from internal fragments (Fig. 1B, molecules labeled II) or from cross-hybridizing genomic sequences.

The probe used in this example of intranuclear genomic footprinting is complementary to the region extending from −6 to −303 bases from the site of transcription initiation in the chicken β^A-globin gene. The mapped genomic strand contains a string of 16 G residues extending from −181 to −196. A region protected from DNase I attack in adult erythrocyte nuclei can be seen (Fig. 5); it includes the sequence from −130 to −156. Enhanced cutting is observed at −162 on this sequence and depressed cutting at ~−230, the 5′ margin of the nuclease hypersensitive region. A densitometric scan (Fig. 5B) of lanes of comparably digested protein-free DNA (lane 2) and adult erythrocyte nuclei (lane 3) emphasizes the protection from DNase I attack afforded by proteins site-specifically bound to this sequence in adult erythrocyte nuclei. The existence of this site-specific protein DNA complex *in vivo* is supported by the observation of similar nuclease protection in this region shown by genomic mapping of the complementary genomic strand[1] and by *in vitro* nuclease protection experiments using partially purified nuclear factors reconstituted on cloned DNA.[30]

Genomic DNA marked with DNA cleavage reagents in either whole cells, cell nuclei, or protein-free DNA can be analyzed with this method to display high resolution maps of specific protein–DNA interactions. This provides valuable information about developmental changes in the structure of genetic regulatory elements *in vivo*. The structural information obtained about various functional states of a gene can be a powerful

[30] B. E. Emerson, C. D. Lewis, and G. Felsenfeld, *Cell (Cambridge, Mass.)* **41**, 21 (1985).

adjunct in the design and interpretation of both *in vitro* reconstitution and transcriptional studies, and *in vivo* transcriptional analyses.

Acknowledgments

We thank Byron Kemper for introducing us to the use of pTZ plasmids in this application, and Betty Canning for her superb help in preparing the manuscript.

Author Index

Numbers in parentheses are footnote reference numbers and indicate that an author's work is referred to although the name is not cited in the text.

A

Aaij, C., 70
Abelson, J., 433
Abraham, J. A., 681
Abreu, S. L., 646
Ada, G. L., 658
Adachi, Y., 489
Adamietz, P., 24, 216
Adams, A. E. M., 487, 488(27)
Adams, D. A., 118, 382, 384(16), 385(16)
Adams, D. S., 254
Adelman, J. P., 444, 446(9)
Adesnik, M., 254
Aggarwal, B. B., 444, 446(6)
Ahmed, A., 542
Ainley, W. M., 524
Aitken, S. C., 587
Akam, M. E., 650, 654(8)
Akhurst, R., 625, 631
Albanil, S., 581
Alberts, B. M., 154, 159(6)
Alexander, D. C., 373, 376(2), 384(2)
Alizon, M., 670, 671(31)
Allan, J., 670, 671(33)
Allen, H., 604
Allet, B., 675
Alwine, J. C., 569, 573
Amasino, R., 580
Ambrose, B. J. B., 382, 526, 527(18), 528, 529, 531, 532
Amerer, G., 675, 677, 678
Ammerer, A., 515
Amphlett, G. W., 330, 444
Amundsen, S. K., 178
Anderson, C. W., 267, 268(3), 296, 298, 299(2)
Anderson, K. V., 646
Anderson, R., 513, 586
Anderson, S., 110, 343, 443, 446(2), 447(2), 538, 539
Ando, T., 611

Andrews, A. T., 71
Andrisani, O., 670
Anfinsen, C. B., 21
Angerer, L. M., 404, 649, 650, 654, 655(1, 5, 14), 656, 659, 660
Angerer, R. C., 404, 633, 646, 649, 650, 654, 655(1, 5, 14), 656, 659, 660
Ansorge, W., 59, 526, 527
Arai, K., 681, 686, 720
Arai, N., 681, 686
Arber, W., 113
Armstrong, J., 290
Arraj, J. A., 133
Ashton, S. H., 262
Askonas, B. A., 658
Astell, C. R., 176, 556
Astrin, S. M., 470
Atkinson, T. C., 436
Auffray, C., 220
Aviv, H., 254, 257(10)
Axel, R., 284, 396, 516, 686, 694, 699, 700(35), 709, 714
Azuma, C., 381

B

Baas, F., 182
Bacchetti, S., 85
Bachmann, B. J., 133, 134, 180
Bachvarova, R., 646
Backman, K., 664
Baer, R., 539
Bahl, C. P., 343, 348(2)
Bailey, J. M., 80
Baker, T., 668
Balazs, I., 573
Baldari, C., 194
Baldwin, R. L., 177
Ballantine, M., 542
Ballard, A., 443, 446(4)
Ballou, C. E., 684

D

K

Kacicj, R., 105
Kaesberg, P., 563
Kafatos, F. C., 240, 325, 583, 588, 592(7), 593(19, 20), 594, 595(19, 20), 597, 603
Kaiser, A. D., 116
Kaji, A., 526
Kajihara, J., 434
Kalinina, N. F., 524(m, o), 525
Kallnins, A., 140
Kamata, T., 669
Kamen, R., 690, 706
Kan, N., 670, 671(34)
Kan, Y. W., 200, 211(5), 605
Kaplan, D., 110
Kaplan, G., 646
Kappen, L. S., 739, 740(15), 747(15)
Karamov, E. E., 117(31), 118
Karin, M., 220, 669, 720
Karn, J., 178, 184, 190, 600, 603(12)
Karyev, A. S., 524(t), 525
Kashdan, M. A., 55
Kasid, A., 587
Katinka, M., 706
Kato, M., 434
Kaufman, J. F., 249, 250(4), 251(4), 252(4), 253(4)
Kaufman, R. J., 330, 444, 688, 689, 697, 698, 700(46, 48)
Kaufmann, N., 587
Kaufmann, Y., 258
Kaumeyer, J., 656
Kavathas, P., 703
Kawashima, E. H., 402, 435, 437(10), 448
Kay, R. M., 697, 698(47)
Kaytes, P. S., 389
Kazazian, H. H., 448
Keating, A., 714
Kedes, L., 656
Kedinger, C., 705
Keem, K., 294, 410
Kellems, R. E., 698, 700(48)
Keller, W., 458
Kelley, D. E., 35, 36(2)
Kelley, S. V., 130, 137(1), 138(1)
Kelly, R. B., 698
Kelly, T. J., 723, 724(8), 725(8)
Kemp, D. J., 569, 573
Kempe, T., 556, 617

Kemper. B., 735, 739(2), 741(2), 755
Kenny, C., 138
Keyt, B., 445(11), 446
Khorana, H. G., 100, 108
Khoury, G., 696, 698, 706, 710
Kidd, V. J., 448
Kidwell, M. G., 397
Kielland-Brandt, M. C., 483
Kieny, M. P., 108, 547
Kikuchi, Y., 103
Kim, M. H., 553
Kimara, A., 680
Kimm, K., 514
Kimura, A., 482, 497(7)
Kinashi, T., 381
Kincaid, R., 460, 469(7)
King, P. V., 109
King, W., 470, 471(18), 472(18), 474, 478(18), 479(25)
Kingsbury, R., 284, 705
Kingston, I. B., 443, 446(2), 447(2)
Kirby, E. M., 563
Kirby, K. S., 22
Kirsher, S. W., 681
Kisiel, W., 359
Kitamura, D., 720
Kiwashima, E. H., 404
Kiyota, T., 434, 436(7)
Kleckner, N., 517
Klein, R. D., 116
Klein, W. H., 659
Kleinberger, T., 728, 729(17), 730(17)
Klenow, H., 550
Kleppe, K., 100
Klimasauskas, S., 130, 137(2)
Klofelt, C., 441
Klotz, L., 481
Klug, A., 110
Klug, G., 587
Knopf, J. L., 330, 444
Knudsen, P. J., 249, 250(4), 251(4), 252(4), 253(4)
Knudson, D. L., 402
Knutson, G. J., 330, 444
Kobayashi, I., 192
Kobilka, B. K., 446(13), 447
Kobori, J. A., 541
Kocher, D. C., 25
Koehler, K. A., 229
Kohij, V., 443, 446(4)

Subject Index

blunt ends, 183
with DNase I, 110
enzyme concentration dependence,
185, 186
with exonuclease III, 105, 106, 543,
545, 546
with exonuclease VII, 611–613, 625,
626
high-molecular-weight, 200, 201
with λ exonuclease, 106, 107
with mung bean exonuclease, 334,
335, 748, 749
nicks to gaps, 543, 544
with restriction endonucleases, 113–
114, 119, *see also* specific endonucleases
size fractionation, 183–189
with S₁ nuclease, 325, 326, 328, 329,
349, 352, 611–613, 623–625
for Southern blotting, 449, *see also*
Southern blot
time dependence, 186
with T4 polymerase, from 3′-termini,
107, 108, 330
vector arms, 195, 196
dot blot, 582–584, 592, *see also* Dot blot
double-stranded
absorbance, 50
cDNA cloning, 308–310
fluorescence determination, 50
gel electrophoresis, 62, 63
labeling of 3′ ends, 94, 104
nick translation, 91–94
resynthesis, 107, 108
from single-stranded RNA, 307–310
3′-end labeling with dideoxynucleoside
triphosphate, 104
enzymes for modifying and labeling, 94–
110
eukaryotic, palindromes, 176
expression in *E. coli*, 451, 452, 458–469,
661–673
extraction of replicative form, from
M13, 162
first strand, hairpin loop at 3′ end, 310
fluorescence determination, 50
fractionation
gel electrophoresis, 62
sample preparation, 68, 79
hydroxylapatite chromatography,

double- and single-strand separation, 430, 431, 745, 746
size markers, 69, 70
in sucrose gradients, 196, 493, 494
fragmentation, 183, 184
phosphatase treatment, 185–189
gel elution, 74, 75
genomic, *see also* Genomic cloning,
Genomic libraries
cleavage, 739–741
heat denaturation, 747, 748
high-resolution mapping, 740
hybridization, 747
isolation, 180–183
preparation, 739–741
from human blood, 182, 183, 204
from monolayer cell cultures, 182
from organs, 181
prevention of random association, 184,
185
purification
in CsCl gradients, 494
removal of RNA, 493–495
restriction mapping, 751–755
from yeast, 493, *see also* Yeast
high molecular weight, 180
preparation, 204, 205
hybridization, 579–581, *see also* Hybridization
immobilization, 582–584, 592
insert, preparation, subcloning, 520,
521
insert-to-vector ratio, 191
isolation, 721
low-melting temperature gel, 86
phenol extraction, 33, 35, 40, 41
labeling, 721, 722, *see also* DNA probe
λ phage
arms
preparation by enzymatic digestion,
195, 196
purification in sucrose gradients, 196
*Eco*RI sites, 17
isolation, 161
metabolism, 173, 174
purification
from λgt10, 362
from λgt11, 366
size, 199
libraries, *see* Libraries

O

P

Z

	U		C		A		G	
U	UUU	Phe	UCU	Ser	UAU	Tyr	UGU	Cys
	UUC	Phe	UCC	Ser	UAC	Tyr	UGC	Cys
	UUA	Leu	UCA	Ser	UAA	Ochre	UGA	(Umber)
	UUG	Leu	UCG	Ser	UAG	Amber	UGG	Trp
C	CUU	Leu	CCU	Pro	CAU	His	CGU	Arg
	CUC	Leu	CCC	Pro	CAC	His	CGC	Arg
	CUA	Leu	CCA	Pro	CAA	Gln	CGA	Arg
	CUG	Leu	CCG	Pro	CAG	Gln	CGG	Arg
A	AUU	Ile	ACU	Thr	AAU	Asn	AGU	Ser
	AUC	Ile	ACC	Thr	AAC	Asn	AGC	Ser
	AUA	Ile	ACA	Thr	AAA	Lys	AGA	Arg
	AUG	Met	ACG	Thr	AAG	Lys	AGG	Arg
G	GUU	Val	GCU	Ala	GAU	Asp	GGU	Gly
	GUC	Val	GCC	Ala	GAC	Asp	GGC	Gly
	GUA	Val	GCA	Ala	GAA	Glu	GGA	Gly
	GUG	Val	GCG	Ala	GAG	Glu	GGG	Gly